# *CHILTON'S* GUIDE TO
# BRAKES, STEERING
# and SUSPENSION

| | |
|---|---|
| **Vice President & General Manager** | John P. Kushnerick |
| **Executive Editor** | Kerry A. Freeman, S.A.E. |
| **Senior Editor** | Richard J. Rivele, S.A.E. |

**CHILTON BOOK COMPANY**
**Chilton Way, Radnor, PA 19089**

Manufactured in USA
©1985 Chilton Book Company
ISBN 0-8019-7644-8

34567890    43210987

# SAFETY NOTICE

Proper service and repair procedures are vital to the safe, reliable operation of all motor vehicles, as well as the personal safety of those performing repairs. This manual outlines procedures for servicing and repairing vehicles using safe effective methods. The procedures contain many NOTES, CAUTIONS and WARNINGS which should be followed along with standard safety procedures to eliminate the possibility of personal injury or improper service which could damage the vehicle or compromise its safety.

It is important to note that repair procedures and techniques, tools and parts for servicing motor vehicles, as well as the skill and experience of the individual performing the work vary widely. It is not possible to anticipate all of the conceivable ways or conditions under which vehicles may be serviced, or to provide cautions as to all of the possible hazards that may result. Standard and accepted safety precautions and equipment should be used when handling toxic or flammable fluids and safety goggles or other protection should be used during cutting, grinding, chiseling, prying, or any other process that can cause material removal or projectiles.

Some procedures require the use of tools specially designed for a specific purpose. Before substituting another tool or procedure, you must be completely satisfied that neither your personal safety, nor the performance of the vehicle will be endangered.

Part numbers listed in this reference are not recommendations by Chilton for any product by brand name. They are references that can be used with interchange manuals and aftermarket supplier catalogs to locate each brand supplier's discrete part number.

Although information in this manual is based on industry sources and is as complete as possible at the time of publication, the possibility exists that some car manufacturers made later changes which could not be included here. While striving for total accuracy, Chilton Book Company cannot assume responsibility for any errors, changes, or omissions that may occur in the compilation of this data.

# CONTENTS

## 1 Brakes    1

## 2 Exhaust    39

## 3 Suspension and Steering    47

## 4 Tune-Up    337

# 1 Brakes

## DRUM BRAKE SERVICE

Most drum brakes are self energizing, with automatic adjusters. Utilization of the frictional force to increase the pressure of shoes against the drum is called *self-energizing* action. Utilization of force in one shoe to apply the opposite shoe is called *servo* action.

### Brake Lining

Brake lining is made of asbestos impregnated with special compounds to bind the asbestos fibers together. Some linings are woven of asbestos threads and fine copper wire. With a few exceptions, most brake lining material is made from asbestos fibers ground up, pressed into shape and either riveted or bonded onto the brake shoe.

The primary shoe, sometimes called leading or forward brake shoe, is the shoe that faces toward the front of the car.

The secondary shoe, sometimes called the trailing or reverse brake shoe, faces the rear of the car.

### Backing Plate

Thorough brake work starts at the brake backing plate. Check the brake area for any indication of lubricant leakage. If the leakage is due to brake fluid, replace or rebuild the wheel cylinder. If the leakage consists of wheel bearing grease, replace the inner bearing seal. It may be necessary to replace the axle bearing or seal. To check the backing plate mounting, tap the plate clockwise and counterclockwise. If movement occurs in either direction, remove the backing plate and check for worn bolts or elongated bolt holes. Replace worn parts. A loose backing plate can usually be detected by listening for a "clicking" sound when applying the brakes while the car is moved forward and backward.

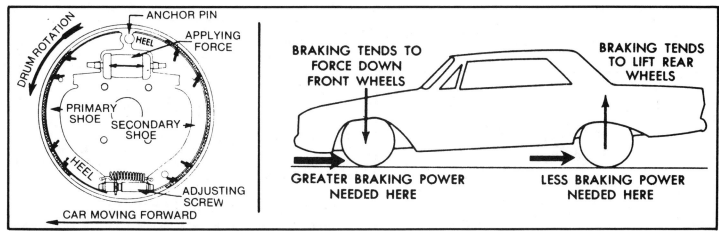

Action of the car and the car's brakes as the brakes are applied

1

# BRAKES

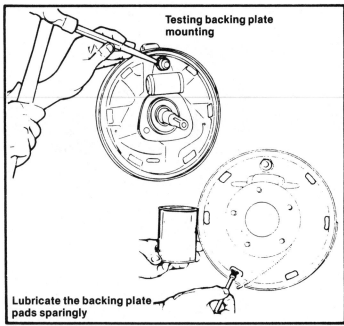

**Testing backing plate mounting**

**Lubricate the backing plate pads sparingly**

**Backing plate service**

## Wheel Cylinder

Wheel cylinders should be inspected for leakage. Carefully inspect the boots. If they are torn, cut, heat cracked or show evidence of leakage, the wheel cylinder should be replaced or overhauled. Don't gamble. If the cylinder doesn't look healthy, replace or rebuild.

### INSPECTION

1. Wash all parts in clean denatured alcohol. If alcohol is not available, use specified brake fluid. Dry with compressed air.
2. Replace scored pistons. Always replace the rubber cups and dust boots.
3. Inspect the cylinder bore for score marks or rust. If either condition is present, the cylinder bore must be honed. However, the cylinder should not be honed more than 0.003 inch beyond its original.
4. Check the bleeder hole to be sure that it is open.

### ASSEMBLY

1. Apply a coating of heavy-duty brake fluid to all internal parts.
2. Thread the bleeder screw into the cylinder and tighten securely.
3. Insert the return spring, cups, and pistons into their respective positions in the cylinder bore. Place a boot over each end of the cylinder.

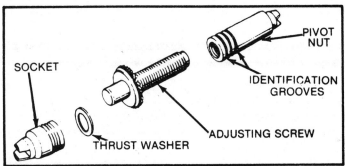

SOCKET

PIVOT NUT

IDENTIFICATION GROOVES

ADJUSTING SCREW

THRUST WASHER

**Typical adjusting screw assembly**

## Adjusting Screw Assembly

Disassemble the adjusting screw assembly. Using an electric wire brush, clean up the threads. Lubricate with brake fluid and reassemble the unit. Turn the threads all the way in by hand. If the threads bind at any point, replace the unit.

## Installing Brake Shoes

1. Preassemble the brake shoes adjusting screw assembly and spring, plus the parking lever assembly (rear brakes only).
2. Spread the assembly; place it on the backing plate. Make sure that the wheel cylinder sockets are in the proper position.
3. Install the retainer pin and spring on both shoes.
4. Install the shoe guide.
5. Install the adjusting cable.
6. Install the parking link and spring (rear only).
7. Install the primary retracting spring.
8. Install the secondary retracting spring.

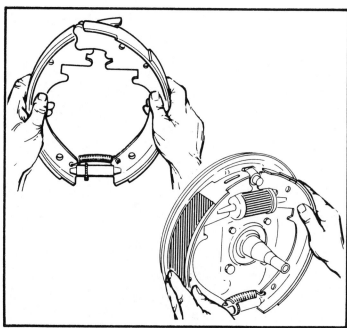

**Installing brake shoes**

## Brake Drums

### BRAKE DRUM TYPES

The *full-cast* drum has a cast iron web (back) of 3/16 to 1/4 inch thickness (passenger car sizes) whereas the *composite* drum has a steel web approximately 1/8 inch thick. These two types of drums, with few exceptions, are not interchangeable.

### BRAKE DRUM DEPTH

Place a straightedge across the drum diameter on the open side. The actual drum depth is the measurement at a right angle from the straightedge to that part of the web which mates against the hub mounting flange.

### ALUMINUM DRUMS

When replaced by other types, aluminum drums must be replaced in pairs.

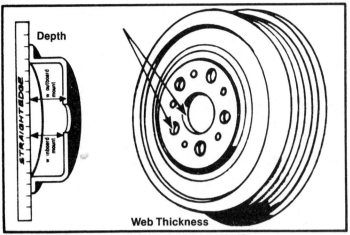

Brake drum measurement

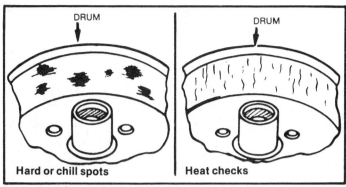

Brake drum inspection

## METALLIC BRAKES

Drums designed for use with standard brake linings should not be used with metallic brakes.

## BOLT CIRCLE

The circumference on which the centers of the wheel bolt holes are located around the drum-hub center is the bolt circle. It is shown as a double number (example: 6–5½). The first digit indicates the number of holes. The second number indicates the bolt circle diameter.

## REMOVING TIGHT DRUMS

Difficulty removing a brake drum can be caused by shoes which are expanded beyond the drum's inner ridge, or shoes which have cut into and ridged the drum. In either case, back off the adjuster to obtain sufficient clearance for removal.

## BRAKE DRUM INSPECTION

The condition of the brake drum surface is just as important as the surface to the brake lining. All drum surfaces should be clean, smooth, free from hard spots, heat checks, score marks and foreign matter imbedded in the drum surface. They should not be out of round, bellmouthed or barrel shaped. It is recommended that all drums be first checked with a drum micrometer to see if they are within oversize limits. If drum is within safe limits, even though the surface appears smooth, it should be turned not only to assure a true drum surface but also to remove any possible contamination in the surface from previous brake linings, road dusts, etc. Too much metal removed from a drum is unsafe and may result in:

1. Brake fade due to the thin drum being unable to absorb the heat generated.
2. Poor and erratic brake action due to distortion of drums.
3. Noise due to vibration caused by thin drums.

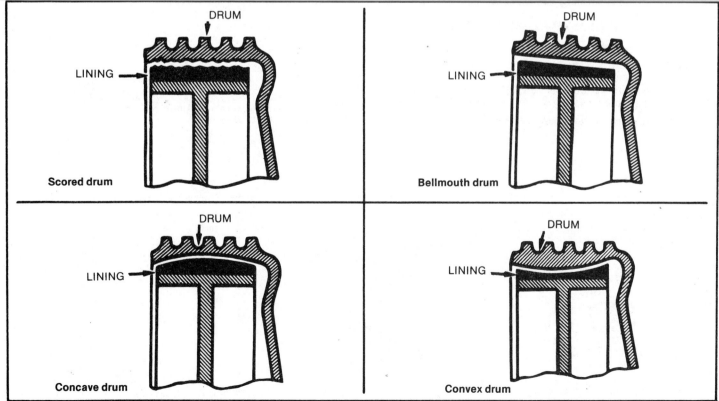

Checking brake drum wear

# BRANKES

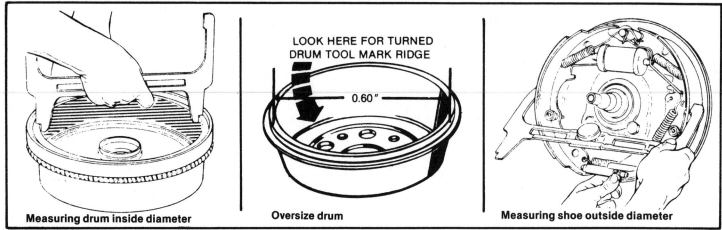

Measuring drum inside diameter

LOOK HERE FOR TURNED DRUM TOOL MARK RIDGE

0.60″

Oversize drum

Measuring shoe outside diameter

Checking brake drum-to-shoe clearance

4. A cracked or broken drum on a severe or very hard brake application.

Brake drum run-out should not exceed .005″. Drums turned to more than .060″ oversize are unsafe and should be replaced with new drums, except for some heavy ribbed drums which have an .080″ limit. It is recommended that the diameters of the left and right drums on any one axle be within .010″ of each other. In order to avoid erratic brake action when replacing drums, it is always good to replace the drums on both wheels at the same time.

If the drums are true, smooth up any slight scores by polishing with fine emery cloth. If deep scores or grooves are present which cannot be removed by this method, then the drum must be turned.

## Adjusting Drum Brakes
### PRELIMINARY ADJUSTMENT

1. Set a brake shoe adjustment gauge at .030 inch less than the brake drum diameter.
2. Center the gauge over the shoes at the greatest lining thickness and run out the adjuster until the new lining touches the gauge.
3. Install the drum.

**NOTE: Tight clearance can aggravate normal seating problems.**

Most service technicians prefer to set the initial brake adjustment with a gauge and then brake the vehicle backward and forward allowing the shoes and drums to seek the correct running clearance. Complete seating normally occurs within 1000 miles.

### ROUTINE ADJUSTMENT

1. Use a brake adjusting tool to expand the brake shoes against the drum. Raising the tool handle turns the star wheel adjuster in the proper direction to expand the shoes. Turn the adjuster until a heavy drag is felt while turning the wheel.
2. Depress the brake pedal hard several times and recheck wheel drag. Continue to depress brake pedal and recheck drag until a true heavy drag is obtained.
3. Turn the star wheel adjuster in the opposite direction until the wheel turns freely.
4. Drive the car, braking forward and backward, to allow the self-adjusters to obtain the best running clearance.

**NOTE: Exceptions to the preceding are General Motors' "H" body cars (Astre, Monza, Skyhawk, Starfire, Vega). These brakes are automatically adjusted when the parking brake is applied. After brake service, apply and release the parking brake until the brakes are correctly adjusted.**

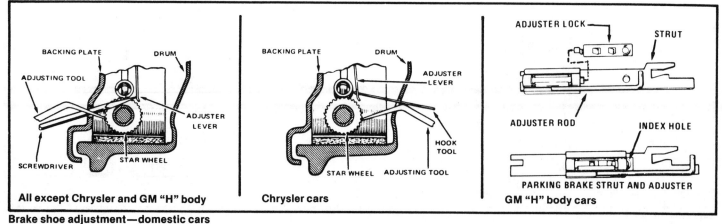

BACKING PLATE    DRUM

ADJUSTING TOOL

ADJUSTER LEVER

SCREWDRIVER    STAR WHEEL

All except Chrysler and GM "H" body

BACKING PLATE    DRUM

ADJUSTER LEVER

HOOK TOOL

STAR WHEEL    ADJUSTING TOOL

Chrysler cars

ADJUSTER LOCK    STRUT

ADJUSTER ROD    INDEX HOLE

PARKING BRAKE STRUT AND ADJUSTER

GM "H" body cars

Brake shoe adjustment—domestic cars

# BRAKE SYSTEM BLEEDING

If the master cylinder has been replaced it is more practical and safe to bleed most of the air out at the master cylinder. This can be done either on or off the car and prevents great masses of air from being passed through the system.

## Manual Bleeding

1. Fill the master cylinder with new fluid of the correct type.
2. On cars with power brakes pump the brake pedal several times to remove all vacuum from the power unit.
3. Pump the brake pedal to pressurize the system and, while holding the pedal down, release the hydraulic pressure at the wheel cylinder bleeder valve. The pedal must be held depressed until the bleeder valve is closed to prevent air from entering the system.
4. Repeat until a steady, clear (no air bubbles) flow of fluid is seen at the wheel cylinder.

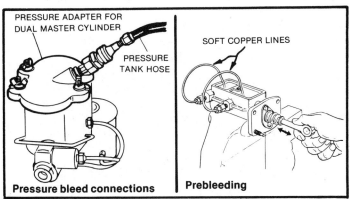

**Bleeding the master cylinder**

---CAUTION---

*The bleeder valve at the wheel cylinder must be closed at the end of each stroke, and before the brake pedal is released, to insure that no air can enter the system. It is also important that the brake pedal be returned to the full up position so the piston in the master cylinder moves back enough to clear the bypass outlets.*

## Pressure Bleeding

Pressure bleeding equipment should be of the diaphragm type, placing a diaphragm between the pressurized air supply and the brake fluid. This prevents moisture and other contaminants from entering the hydraulic system.

**NOTE: Front disc/rear drum equipped vehicles use a metering valve which closes off pressure to the front brakes under certain conditions. These systems contain manual release actuators which must be engaged to pressure bleed the front brakes.**

1. Connect the tank hydraulic hose and adapter to the master cylinder.
2. Close hydraulic valve on the bleeder equipment.
3. Apply air pressure to the bleeder equipment.

---CAUTION---

*Follow equipment manufacturer's recommendations for correct air pressure.*

4. Open the valve to bleed air out of the pressure hose to the master cylinder.

**NOTE: Never bleed this system using the secondary piston stopscrew on the bottom of many master cylinders.**

5. Open the hydraulic valve and bleed each wheel cylinder. Bleed rear brake system first when bleeding both front and rear systems.

# FLUSHING HYDRAULIC BRAKE SYSTEMS

Hydraulic brake systems must be totally flushed if the fluid becomes contaminated with water, dirt or other corrosive chemicals. To flush, simply bleed the entire system until *all* fluid has been replaced with the correct type of new fluid.

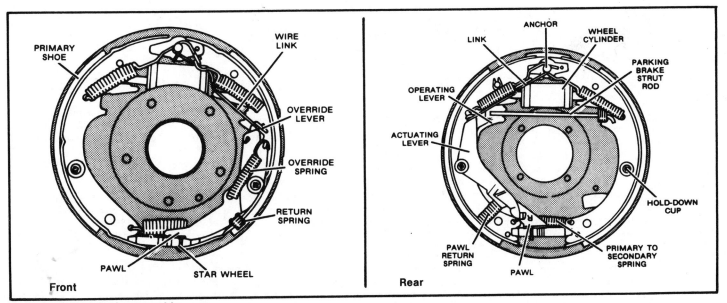

**Typical GM drum brake assembly**

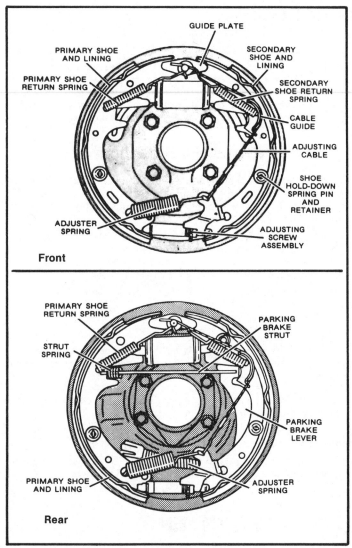

Typical AMC and Ford drum brake assembly

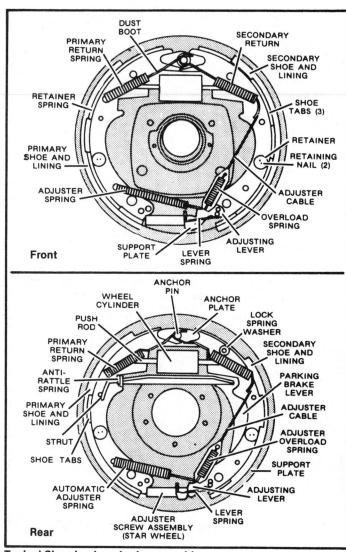

Typical Chrysler drum brake assembly

# DISC BRAKE SERVICE

Caliper disc brakes can be divided into three types: the four-piston fixed-caliper type, the single-piston floating-caliper type, and the single piston sliding-caliper type.

In the four piston type (two in each side of the caliper) braking effect is achieved by hydraulically pushing both shoes against the disc sides.

With the single piston floating-caliper type, the inboard shoe is pushed hydraulically into contact with the disc; the reaction force thus generated is used to pull the outboard shoe into frictional contact (made possible by letting the caliper move slightly along the axle centerline).

In the sliding caliper (single piston) type, the caliper assembly slides along the machined surfaces of the anchor plate. A steel key located between the machined surfaces of the caliper and the machined surfaces of the anchor plate is held in place with either a retaining screw or two cotter pins. The caliper is held in place against the anchor plate with one or two support springs.

All disc brake systems are inherently self-adjusting and have no provision for manual adjustment.

## Inspection

Disc pads (lining and shoe assemblies) should be replaced in axle sets (both wheels) when the lining on any pad is worn to 1/16 in. at any point. *If lining is allowed to wear past 1/16 in. minimum thickness severe damage to disc may result.*

**NOTE: State inspection specifications take precedence over these general recommendations.**

Note that disc pads in floating caliper type brakes may wear at any angle, and measurement should be made at the narrow end of the taper. Tapered linings should be replaced if the taper exceeds 1/8 in. from end to end (the difference between the thickest and thinnest points).

———CAUTION———

*To prevent costly paint damage, remove some brake fluid (don't re-use) from the reservoir and install the reservoir cover before replacing the disc pads. When replacing the pads, the piston is depressed and fluid is forced back through the lines to squirt out of the fluid reservoir.*

If the caliper is unbolted from the hub, do not let it dangle by the brake hose; it can be rested on a suspension member or wired onto the frame.

## Servicing the Caliper Assembly

1. Raise the vehicle on a hoist and remove the front wheels.
2. Working on one side at a time only, disconnect the hydraulic inlet line from the caliper and plug the end. Remove the caliper mounting bolts or pins, and shims if used, and slide the caliper off the disc.
3. Remove the disc pads from the caliper. If the old ones are to be reused, mark them so that they can be reinstalled in their original positions.
4. Open the caliper bleed screw and drain the fluid. Clean the outside of the caliper and mount it in a vise with padded jaws.

---CAUTION---

*When cleaning any brake components, use only brake fluid or denatured (Isopropyl) alcohol. Never use a mineral-based solvent, such as gasoline or paint thinner, since it will cause rubber parts to swell and quickly deteriorate.*

5. Remove the bridge bolts, separate the caliper halves, and remove the two O-ring seals from the transfer holes.
6. Pry the lip on each piston dust boot from its groove and remove the piston assemblies and springs from the bores. If necessary, air pressure may be used to force the pistons out of the bores, using care to prevent them from popping out of control.

7. Remove the boots and seals from the pistons and clean the pistons in brake fluid. Blow out the caliper passages with an air hose.
8. Inspect the cylinder bores for scoring, pitting, or corrosion. Corrosion is a pitted or rough condition not to be confused with staining. Light rough spots may be removed by rotating crocus cloth, using finger pressure, in the bores. Do not polish with an in-and-out motion or use any other abrasive.
9. If the pistons are pitted, scored, or worn, they must be replaced. A corroded or deeply scored caliper should also be replaced.
10. Check the clearance of the pistons in the bores using a feeler gauge. Clearance should be 0.002–0.006 in. If there is excessive clearance, the caliper must be replaced.
11. Replace all rubber parts and lubricate with brake fluid. Install the seals and boots in the grooves in each piston. The seal should be installed in the groove closest to the closed end of the piston with the seal lips facing the closed end. The lip on the boot should be facing the seal.
12. Lubricate the piston and bore with brake fluid. Position the piston return spring, large coil first, in the piston bore.
13. Install the piston in the bore, taking great care to avoid damaging the seal lip as it passes the edge of the cylinder bore.
14. Compress the lip on the dust boot into the groove in the caliper. Be sure the boot is fully seated in the groove, as poor sealing will allow contaminants to ruin the bore.
15. Position the O-rings in the cavities around the caliper transfer holes, and fit the caliper halves together. Install the bridge bolts (lubricated with brake fluid) and be sure to torque to specification.
16. Install the disc pads in the caliper and remount the caliper on the hub. Connect the brake line to the caliper and bleed the brakes. Replace the wheels. Recheck the brake fluid level, check the brake pedal travel, and road test the vehicle.

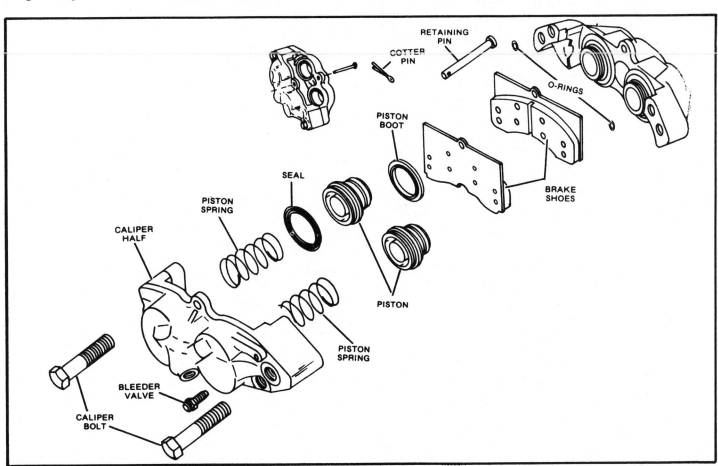

**Typical four piston caliper**

# BRAKES

## BRAKE SYSTEM TUNE-UP PROCEDURE

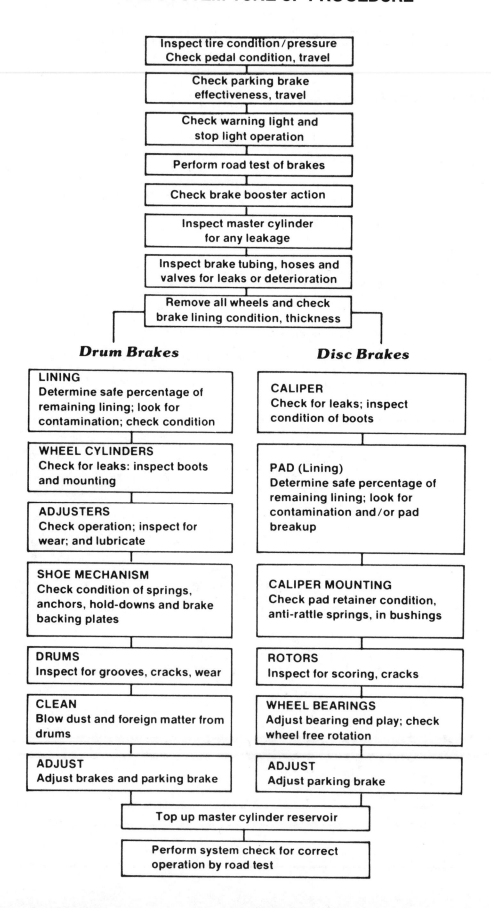

Inspect tire condition/pressure
Check pedal condition, travel

Check parking brake
effectiveness, travel

Check warning light and
stop light operation

Perform road test of brakes

Check brake booster action

Inspect master cylinder
for any leakage

Inspect brake tubing, hoses and
valves for leaks or deterioration

Remove all wheels and check
brake lining condition, thickness

### *Drum Brakes*

**LINING**
Determine safe percentage of
remaining lining; look for
contamination; check condition

**WHEEL CYLINDERS**
Check for leaks: inspect boots
and mounting

**ADJUSTERS**
Check operation; inspect for
wear; and lubricate

**SHOE MECHANISM**
Check condition of springs,
anchors, hold-downs and brake
backing plates

**DRUMS**
Inspect for grooves, cracks, wear

**CLEAN**
Blow dust and foreign matter from
drums

**ADJUST**
Adjust brakes and parking brake

### *Disc Brakes*

**CALIPER**
Check for leaks; inspect
condition of boots

**PAD (Lining)**
Determine safe percentage of
remaining lining; look for
contamination and/or pad
breakup

**CALIPER MOUNTING**
Check pad retainer condition,
anti-rattle springs, in bushings

**ROTORS**
Inspect for scoring, cracks

**WHEEL BEARINGS**
Adjust bearing end play; check
wheel free rotation

**ADJUST**
Adjust parking brake

Top up master cylinder reservoir

Perform system check for correct
operation by road test

# BRAKE PERFORMANCE DIAGNOSIS

| The Condition | The Possible Cause | The Corrective Action |
|---|---|---|
| **PEDAL GOES TO FLOOR** | (a) Fluid low in reservoir.<br>(b) Air in hydraulic brake system.<br>(c) Improperly adjusted brake.<br>(d) Leaking wheel cylinders.<br><br>(e) Loose or broken brake lines.<br><br>(f) Leaking or worn master cylinder.<br><br>(g) Excessively worn brake lining. | (a) Fill and bleed master cylinder.<br>(b) Fill and bleed hydraulic brake system.<br>(c) Repair or replace self-adjuster as required.<br>(d) Recondition or replace wheel cylinder and replace both brake shoes.<br>(e) Tighten all brake fittings or replace brake line.<br>(f) Recondition or replace master cylinder and bleed hydraulic system.<br>(g) Reline and adjust brakes. |
| **SPONGY BRAKE PEDAL** | (a) Air in hydraulic system.<br><br>(b) Improper brake fluid (low boiling point).<br>(c) Excessively worn or cracked brake drums.<br>(d) Broken pedal pivot bushing. | (a) Fill master cylinder and bleed hydraulic system.<br>(b) Drain, flush and refill with brake fluid.<br>(c) Replace all faulty brake drums.<br>(d) Replace nylon pivot bushing. |
| **BRAKES PULLING** | (a) Contaminated lining.<br>(b) Front end out of alignment.<br>(c) Incorrect brake adjustment.<br>(d) Unmatched brake lining.<br><br>(e) Brake drums out of round.<br>(f) Brake shoes distorted.<br>(g) Restricted brake hose or line.<br>(h) Broken rear spring. | (a) Replace contaminated brake lining.<br>(b) Align front end.<br>(c) Adjust brakes and check fluid.<br>(d) Match primary, secondary with same type of lining on all wheels.<br>(e) Grind or replace brake drums.<br>(f) Replace faulty brake shoes.<br>(g) Replace plugged hose or brake line.<br>(h) Replace broken spring. |
| **SQUEALING BRAKES** | (a) Glazed brake lining.<br>(b) Saturated brake lining.<br>(c) Weak or broken brake shoe retaining spring.<br>(d) Broken or weak brake shoe return spring.<br>(e) Incorrect brake lining.<br>(f) Distorted brake shoes.<br>(g) Bent support plate.<br>(h) Dust in brakes or scored brake drums. | (a) Cam grind or replace brake lining.<br>(b) Replace saturated lining.<br>(c) Replace retaining spring.<br><br>(d) Replace return spring.<br>(e) Install matched brake lining.<br>(f) Replace brake shoes.<br>(g) Replace support plate.<br>(h) Blow out brake assembly with compressed air and grind brake drums. |
| **CHIRPING BRAKES** | (a) Out of round drum or eccentric axle flange pilot. | (a) Repair as necessary, and lubricate support plate contact areas (6 places). |
| **DRAGGING BRAKES** | (a) Incorrect wheel or parking brake adjustment.<br>(b) Parking brakes engaged.<br>(c) Weak or broken brake shoe return spring.<br>(d) Brake pedal binding.<br><br>(e) Master cylinder cup sticking.<br>(f) Obstructed master cylinder relief port.<br>(g) Saturated brake lining.<br>(h) Bent or out of round brake drum. | (a) Adjust brake and check fluid.<br><br>(b) Release parking brakes.<br>(c) Replace brake shoe return spring.<br>(d) Free up and lubricate brake pedal and linkage.<br>(e) Recondition master cylinder.<br>(f) Use compressed air and blow out relief port.<br>(g) Replace brake lining.<br>(h) Grind or replace faulty brake drum. |
| **HARD PEDAL** | (a) Brake booster inoperative.<br>(b) Incorrect brake lining.<br>(c) Restricted brake line or hose.<br>(d) Frozen brake pedal linkage. | (a) Replace brake booster.<br>(b) Install matched brake lining.<br>(c) Clean out or replace brake line or hose.<br>(d) Free up and lubricate brake linkage. |
| **WHEEL LOCKS** | (a) Contaminated brake lining.<br>(b) Loose or torn brake lining.<br>(c) Wheel cylinder cups sticking.<br>(d) Incorrect wheel bearing adjustment. | (a) Reline both front or rear of all four brakes.<br>(b) Replace brake lining.<br>(c) Recondition or replace wheel cylinder.<br>(d) Clean, pack and adjust wheel bearings. |
| **BRAKES FADE (HIGH SPEED)** | (a) Incorrect brake lining.<br>(b) Overheated brake drums.<br>(c) Incorrect brake fluid (low boiling temperature).<br>(d) Saturated brake lining. | (e) Replace lining.<br>(b) Inspect for dragging brakes.<br>(c) Drain, flush, refill and bleed hydraulic brake system.<br>(d) Reline both front or rear or all four brakes. |
| **PEDAL PULSATES** | (a) Bent or out of round brake drum. | (a) Grind or replace brake drums. |
| **BRAKE CHATTER AND SHOE KNOCK** | (a) Out of round brake drum.<br>(b) Loose support plate.<br><br>(c) Bent support plate.<br>(d) Distorted brake shoes.<br>(e) Machine grooves in contact face of brake drum. (Shoe Knock).<br>(f) Contaminated brake lining. | (a) Grind or replace brake drums.<br>(b) Tighten support plate bolts to proper specifications.<br>(c) Replace support plate.<br>(d) Replace brake shoes.<br>(e) Grind or replace brake drum.<br><br>(f) Replace either front or rear or all four linings |
| **BRAKES DO NOT SELF ADJUST** | (a) Adjuster screw frozen in thread.<br>(b) Adjuster screw corroded at thrust washer.<br>(c) Adjuster lever does not engage star wheel.<br>(d) Adjuster installed on wrong wheel. | (a) Clean and free-up all thread areas.<br>(b) Clean threads and replace thrust washer if necessary.<br>(c) Repair, free up or replace adjusters as required.<br>(d) Install correct adjuster parts. |

# BRAKES

## BRAKE PERFORMANCE DIAGNOSIS

| The Condition | The Possible Cause | The Corrective Action |
|---|---|---|
| NOISE—Groan—Brake noise emanating when slowly releasing brakes (creep-groan). | (a) Not detrimental to function of disc brakes—no corrective action required. (Indicate to operator this noise may be eliminated by slightly increasing or decreasing brake pedal efforts.) | |
| RATTLE—Brake noise or rattle emanating at low speeds on rough roads, (front wheels only). | (a) Shoe anti-rattle spring missing or not properly positioned.<br>(b) Excessive clearance between shoe and caliper. | (a) Install new anti-rattle spring or position properly.<br>(b) Install new shoe and lining assemblies. |
| SCRAPING | (a) Mounting bolts too long.<br>(b) Loose wheel bearings. | (a) Install mounting bolts of correct length.<br>(b) Readjust wheel bearings to correct specifications. |
| FRONT BRAKES HEAT UP DURING DRIVING AND FAIL TO RELEASE | (a) Operator riding brake pedal.<br>(b) Stop light switch improperly adjusted.<br>(c) Sticking pedal linkage.<br>(d) Frozen or seized piston.<br>(e) Residual pressure valve in master cylinder.<br>(f) Power brake malfunction. | (a) Instruct owner how to drive with disc brakes.<br>(b) Adjust stop light to allow full return of pedal.<br>(c) Free up sticking pedal linkage.<br>(d) Disassemble caliper and free up piston.<br>(e) Remove valve.<br>(f) Replace. |
| LEAKY WHEEL CYLINDER | (a) Damaged or worn caliper piston seal.<br>(b) Scores or corrosion on surface of cylinder bore. | (a) Disassembly caliper and install new seat.<br>(b) Disassemble caliper and hone cylinder bore. Install new seal. |
| GRABBING OR UNEVEN BRAKE ACTION | (a) Causes listed under "Pull"<br>(b) Power brake malfunction. | (a) Corrections listed under 'Pull''.<br>(b) Replace. |
| BRAKE PEDAL CAN BE DEPRESSED WITHOUT BRAKING EFFECT | (a) Air in hydraulic system or improper bleeding procedure.<br>(b) Leak past primary cup in master cylinder.<br>(c) Leak in system.<br>(d) Rear brakes out of adjustment.<br>(e) Bleeder screw open. | (a) Bleed system.<br>(b) Recondition master cylinder.<br>(c) Check for leak and repair as required.<br>(d) Adjust rear brakes.<br>(e) Close bleeder screw and bleed entire system. |
| EXCESSIVE PEDAL TRAVEL | (a) Air, leak, or insufficient fluid in system or caliper.<br>(b) Warped or excessively tapered shoe and lining assembly.<br>(c) Excessive disc runout.<br>(d) Rear brake adjustment required.<br>(e) Loose wheel bearing adjustment.<br>(f) Damaged caliper piston seal.<br>(g) Improper brake fluid (boil).<br>(h) Power brake malfunction. | (a) Check system for leaks and bleed.<br>(b) Install new shoe and linings.<br>(c) Check disc for runout with dial indicator. Install new or refinished disc.<br>(d) Check and adjust rear brakes.<br>(e) Readjust wheel bearing to specified torque.<br>(f) Install new piston seal.<br>(g) Drain and install correct fluid.<br>(h) Replace. |
| BRAKE ROUGHNESS OR CHATTER (PEDAL PUMPING) | (a) Excessive thickness variation of braking disc.<br>(b) Excessive lateral runout of braking disc.<br>(c) Rear brake drums out-of-round.<br>(d) Excessive front bearing clearance. | (a) Check disc for thickness variation using a micrometer.<br>(b) Check disc for lateral runout with dial indicator. Install new or refinished disc.<br>(c) Reface rear drums and check for out-of-round.<br>(d) Readjust wheel bearings to specified torque. |
| EXCESSIVE PEDAL EFFORT | (a) Brake fluid, oil or grease on linings.<br>(b) Incorrect lining.<br>(c) Frozen or seized pistons.<br>(d) Power brake malfunction. | (a) Install new shoe linings as required.<br>(b) Remove lining and install correct lining.<br>(c) Disassemble caliper and free up pistons.<br>(d) Replace. |
| PULL | (a) Brake fluid, oil or grease on linings.<br>(b) Unmatched linings.<br>(c) Distorted brake shoes.<br>(d) Frozen or seized pistons.<br>(e) Incorrect tire pressure.<br>(f) Front end out of alignment.<br>(g) Broken rear spring.<br>(h) Rear brake pistons sticking.<br>(i) Restricted hose or line.<br>(j) Caliper not in proper alignment to braking disc. | (a) Install new shoe and linings.<br>(b) Install correct lining.<br>(c) Install new brake shoes.<br>(d) Disassemble caliper and free up pistons.<br>(e) Inflate tires to recommended pressures.<br>(f) Align front end and check.<br>(g) Install new rear spring.<br>(h) Free up rear brake pistons.<br>(i) Check hoses and lines and correct as necessary.<br>(j) Remove caliper and reinstall. Check alignment. |

# MANUFACTURER'S SPECIFIED BRAKE SPECIFICATIONS

## DOMESTIC PASSENGER CARS

**NOTICE:** Should state inspection regulations exceed manufacturer's specifications for lining or rotor/drum reserve, the state inspection regulation specification **must** be used.

| VEHICLE — YEAR, MAKE AND MODEL | BRAKE SHOE Min. Lining Thickness * | BRAKE DRUM DIAMETER Standard Size | BRAKE DRUM Machine To | BRAKE PAD Min. Lining Thickness | BRAKE ROTOR Min. Thickness Machine To | BRAKE ROTOR Discard At | Variation From Parallelism | Runout T.I.R. | DESIGN | CALIPER Mounting Bolts Torque (ft-lbs) | CALIPER Bridge, Pin or Key Bolts Torque (ft-lbs) | WHEEL Lugs or Nuts Torque (ft-lbs) | STEP 1 Tighten Spindle Nut (ft-lbs) | STEP 2 Back Off Retorque (in-lbs) | STEP 3 Lock, or Back Off and Lock |
|---|---|---|---|---|---|---|---|---|---|---|---|---|---|---|---|
| **AMERICAN MOTORS** | | | | | | | | | | | | | | | |
| 82-81 All exc. 6 cyl. Concord Wagon and Eagle | .030 | 9.000 | 9.060 | .030* | .815 | .810 | .0005 | .003 | 1 | 80 | 15 | 75 | 25 | 6 | Step 2 |
| 82-81 6 cyl. Concord Wagon and Eagle | .030 | 10.000 | 10.060 | .030* | .815 | .810 | .0005 | ■.003 | 1 | 80 | 15 | 75 | 25 | 6 | Step 2 |
| 80-78 Spirit exc. 4 cyl., Concord, Gremlin, exc. 4 cyl.; Pacer, Eagle, AMX | .030 | 10.000 | 10.060 | .062 | .815 | .810 | .0005 | ■.003 | 1 | 80 | 15 | 75 | 25 | 6 | Step 2 |
| 80-78 Spirit & Gremlin w/4 cyl. | .030 | 9.000 | 9.060 | .062 | .815 | .810 | .0005 | .003 | 1 | 80 | 15 | 75 | 25 | 6 | Step 2 |
| 78 Matador | .030 | 10.000 | 10.060 | .062 | — | 1.120 | .0005 | .003 | 1 | 80 | 15 | 75 | 25 | 6 | Step 2 |
| 77 Gremlin, Hornet, Pacer | .030 | 10.000 | 10.060 | .062 | .815 | .810 | .0005 | .003 | 1 | 80-90 | 15-18 | 75-90 | 25 | 6 | Step 2 |
| 77 Matador | .030 | 10.000 | 10.060 | .062 | — | 1.120 | .0005 | — | 1 | 80-90 | 15-18 | 75-90 | 22 | 6 | Step 2 |
| 76-75 Pacer w/drum brakes, Front | .030 | 10.000 | 10.060 | — | — | — | — | — | — | — | — | 60-90 | 22 | 6 | Step 2 |
| Rear | .030 | 10.000 | 10.060 | — | — | — | — | — | — | — | — | 75 | — | — | — |
| 76-75 Pacer w/disc brakes | .030 | 9.000 | 9.060 | .062 | 1.130 | 1.120 | .0005 | .003 | 1 | 80 | 15 | 60-90 | 22 | 6 | Step 2 |
| 76-75 Hornet & Gremlin w/6 cyl. | .030 | 9.000 | 9.060 | .062 | 1.130 | 1.120 | .0005 | .003 | 1 | 80 | 15 | 75 | 22 | 6 | Step 2 |
| 76-75 Hornet & Gremlin w/V8 and All Matador | .030 | 10.000 | 10.060 | .062 | 1.130 | 1.120 | .0005 | .003 | 1 | 80 | 15 | 75 | 22 | 6 | Step 2 |
| 74-72 Hornet, Gremlin & Javelin 6 cyl. w/drum brakes | .030 | 9.000 | 9.060 | — | — | — | — | — | — | — | — | 75 | 20 | 6 | Step 2 |
| 74-72 Ambassador, Matador, Hornet V8, Gremlin V8 & Javelin V8 w/drum brakes | .030 | 10.000 | 10.060 | — | — | — | — | — | — | — | — | 75 | 20 | 6▲ | Step 2 |
| 74-73 Hornet, Gremlin & Javelin w/6 cyl. w/disc brakes | .030 | 9.000 | 9.060 | .062 | — | .940 | .0005 | .005 | 2 | 80 | — | 75 | 20 | 6 | Step 2 |
| 74-72 Hornet V8, Gremlin V8, Javelin V8, All Matador & Ambassador w/disc brakes | .030 | 10.000 | 10.060 | .062 | — | .940 | .0005 | .005 | 2 | 80 | — | 75 | 20 | 6▲ | Step 2 |
| 72 Hornet, Gremlin & Javelin w/6 cyl. | .030 | 10.000 | 10.060 | .062 | — | .940 | .0005 | .005 | 2 | 80 | — | 75 | 20 | 12 | Step 2 |

■ Eagle - .004    ▲ 1972 - 12 in/lbs.    * .030" over rivet head; if bonded lining, use .062"

# BRAKES

## CHRYSLER CORP. — CHRYSLER — DODGE — PLYMOUTH

| YEAR, MAKE AND MODEL | BRAKE SHOE Min. Lining Thickness* | BRAKE DRUM Standard Size | BRAKE DRUM Machine To | BRAKE PAD Minimum Lining Thickness | BRAKE ROTOR Machine To | BRAKE ROTOR Discard At | Variation From Parallelism | Runout T.I.R. | DESIGN | CALIPER Mounting Bolts Torque (ft-lbs) | CALIPER Bridge, Pin or Key Bolts Torque (ft-lbs) | WHEEL Lugs or Nuts Torque (ft-lbs) | STEP 1 Tighten Spindle Nut (ft-lbs) | STEP 2 Back Off Retorque (in-lbs) | STEP 3 Lock, or Back Off and Lock |
|---|---|---|---|---|---|---|---|---|---|---|---|---|---|---|---|
| 82 Aries, Reliant, LeBaron Dodge 400 | .030 | 7.870 | 7.900 | .030 | .912 | .882 | .0005 | .004 | 6 | 70-100 | 18-22 | 85 | 20-25 | Handtight | Step 2 |
| 82 Cordoba, Diplomat, Gran Fury, Mirada, New Yorker, Imperial w/10" rear brake | .030 | 10.000 | 10.060 | .125 | .970 | .940 | .0005 | .004 | 5 | 95-125 | 15-20 | 85 | 20-25 | Handtight | Step 2 |
| w/11" rear brake | .030 | 11.000 | 11.060 | .125 | .970 | .940 | .0005 | .004 | 5 | 95-125 | 15-20 | 85 | 20-25 | Handtight | Step 2 |
| 82-78 Omni, Horizon | .030 | 7.870 | 7.900 | .030 | .461 | .431 | .0005 | .005 ■ | 8 | 70-100 | 25-40 | 85 | 20-25 | Handtight | Step 2 |
| 81 Aries, Reliant | .030 | 7.870 | 7.900 | .030 | .912 | .882 | .0005 | .004 | 6 | 70-100 | 18-22 | 85 | 20-25 | Handtight | Step 2 |
| 81-78 Aspen, Volare, LeBaron, Diplomat, St. Regis, Cordoba, Gran Fury, Magnum, Mirada, Newport, New Yorker, Imperial w/10" rear brakes | .030 | 10.000 | 10.060 | .030 | .970 | .940 | .0005 | .005 ■ | 5 | 95-125 | 15-20 | 85 | 20-25 | Handtight | Step 2 |
| w/11" rear brakes | .030 | 11.000 | 11.060 | .030 | .970 | .940 | .0005 | .005 ■ | 5 | 95-125 | 15-20 | 85 | 20-25 | Handtight | Step 2 |
| 78 Fury, Monaco w/10" rear brakes | .030 | 10.000 | 10.060 | .030 | .970 | .940 | .0005 | .004 | 5 | 95-125 | 25-40 | 85 | 20-25 | Handtight | Step 2 |
| w/11" rear brakes | .030 | 11.000 | 11.060 | .030 | .970 | .940 | .0005 | .004 | 5 | 95-125 | 25-40 | 85 | 20-25 | Handtight | Step 2 |
| 77-75 Chrysler, Gran Fury, Royal Monaco | .030 | 11.000 | 11.060 | .030 | 1.195 | 1.180 | .0005 | .004 | 5 | 95-125 | 25-35 | 85† | 20-25 | Handtight | Step 2 |
| 77-76 Volare & Aspen, exc. Wagon | .030 | 10.000 | 10.060 | .030 | .970 | .940 | .0005 | .004 | 5 | 95-125 | 15 | 85 | 20-25 | Handtight | Step 2 |
| 77-76 Volare & Aspen Wagon | .030 | 11.000 | 11.060 | .030 | .970 | .940 | .0005 | .004 | 5 | 95-125 | 15 | 85 | 20-25 | Handtight | Step 2 |
| 77-76 Fury, Coronet, Cordoba, Monaco, Charger SE | .030 | 11.000 | 11.060 | .030 | .970 | .940 | .0005 | .004 | ◆ | 95-125 | 25-40 | 85 | 20-25 | Handtight | Step 2 |
| 76-75 Valiant, Dart w/ft. drum: Front | .030 | 10.000 | 10.060 | — | — | — | — | — | — | — | — | 70 | 20-25 | Handtight | Step 2 |
| Rear | .030 | 9.000 | 9.060 | — | — | — | — | — | — | — | — | 70 | — | — | — |
| 76-75 Valiant, Dart w/ft. disc | .030 | 10.000 | 10.060 | .030 | .970 | .940 | .0005 | .004 | 5 | 95-125 | 15 | 70 | 20-25 | Handtight | Step 2 |
| 75 Imperial: Front | — | — | — | .030 | 1.195 | 1.180 | .0005 | .004 | 5 | 95-125 | 25-35 | 85 | — | Handtight | Step 2 |
| Rear | — | — | — | .030 | .970 | .940 | .0005 | .004 | 5 | 95-125 | 25-35 | 85 | — | — | — |
| 75 Fury, Coronet, Charger, Cordoba exc. Wagon | .030 | 10.000 | 10.060 | .030 | .970 | .940 | .0005 | .004 | 2 | 95-125 | 25-35 | 70 | 20-25 | Handtight | Step 2 |
| 75 Fury & Coronet Wagon | .030 | 11.000 | 11.060 | .030 | .970 | .940 | .0005 | .004 | 2 | 95-125 | 25-35 | 70 | 20-25 | Handtight | Step 2 |
| 74 Imperial Front | — | — | — | .030 | 1.195 | 1.180 | .0005 | .004 | 5 | 75-100 | 30-35 | 65 | 8 | Skip | 1 Slot |
| Rear | — | — | — | .030 | .970 | .940 | .0005 | .004 | 5 | 95-125 | 25-35 | 65 | — | — | — |
| 74-73 Chrysler, Dodge, Plymouth Full Size | .030 | 11.000 | 11.060 | .030 | 1.195 | 1.180 | .0005 | .004 | 5 | 75-100 | 25-35 | 65 | 8 | Skip | 1 Slot |
| 74-73 Charger, Coronet, Challenger, Belvedere, Satellite, Barracuda, Valiant & Dart w/ft. disc. | .030 | 10.000 | 10.060 | .030 | .970 | .940 | .0005 | .004 | ◆ | 75-100 | 25-35 | 65 | 8 | Skip | 1 Slot |

◆ 2 or 5    * .030" over rivet head; if bonded lining, use .062"    ■ 1982-80 - .004    † 1975 - 65 ft/lbs.

| Application | | | | | | | | | | | | | | | |
|---|---|---|---|---|---|---|---|---|---|---|---|---|---|---|---|
| **74-73 Valiant, Dart w/ft. drum:** | | | | | | | | | | | | | | | |
| Front | .030 | 10.000 | 10.060 | — | — | — | — | — | — | — | — | 65 | 8 | Skip | 1 Slot |
| Rear | .030 | 9.000 | 9.060 | — | — | — | — | — | — | — | — | 65 | — | — | — |
| **73-72 Imperial** | .030 | 11.000 | 11.060 | .030 | 1.195 | 1.180 | .0005 | .0025 | 5 | 75-100 | 30-35 | 65 | 8 | Skip | 1 Slot |
| **72 Chrysler, Monaco, Polara, Fury, GT, Suburban S.W. V.I.P.** | .030 | 11.000 | 11.060 | .030 | 1.195 | 1.180 | .0005 | .0025 | 5 | 75-100 | 30-35 | 65 | 8 | Skip | 1 Slot |
| **72 Charger, Coronet, Crestwood, SE, Super Bee, Belvedere, Satellite, GTX, Regent, Road Runner, Sebring, Barracuda, Challenger w/discs** | .030 | 10.000 | 10.060 | .030 | .970 | .940 | .0005 | .0025 | 5 | 75-100 | 30-35 | 65 | 8 | Skip | 1 Slot |
| **72 Charger, Coronet, Crestwood, SE, Super Bee, Belvedere, Satellite, GTX, Regent, Road Runner, Sebring, Barracuda, Challenger w/10″ drum** | .030 | 10.000 | 10.060 | — | — | — | — | — | — | — | — | 65 | 8 | Skip | 1 Slot |
| w/11″ drum | .030 | 11.000 | 11.060 | — | — | — | — | — | — | — | — | 65 | 8 | Skip | 1 Slot |
| **72 Dart, Demon, GT, GTS, Swinger, Valiant, Duster, Scamp, Signet, V100, V200** | | | | | | | | | | | | | | | |
| w/10″ drums | .030 | 10.000 | 10.060 | — | — | — | — | — | — | — | — | 55 | 6 | Skip | 1 Slot |
| w/9″ drums | .030 | 9.000 | 9.060 | — | — | — | — | — | — | — | — | 55 | 6 | Skip | 1 Slot |
| w/ft. disc | .030 | 10.000 | 10.060 | .030 | .790 | .780 | .0005 | .0025 | 9 | 50-80 | 70-80 | 55 | 6 | Skip | 1 Slot |
| **FORD MOTOR CO. — FORD — MERCURY — LINCOLN** | | | | | | | | | | | | | | | |
| **82 Lincoln Continental** Front | — | — | — | .125 | — | .972 | .0005 | .003 | 10 | 80-110 | 40-60 | 80-105 | 17-25 | 10-15 | Step 2 |
| Rear | — | — | — | .125 | — | .895 | .0005 | .004 | 11 | — | 29-37 | 80-105 | — | — | — |
| **82-81 Thunderbird, XR-7** | | | | | | | | | | | | | | | |
| w/9″ rear brake | .030 | 9.000 | 9.060 | .125 | — | .810 | .0005 | .003 | 10 | — | 30-40 | 80-105 | 17-25 | 10-15 | Step 2 |
| w/10″ rear brake | .030 | 10.000 | 10.060 | .125 | — | .810 | .0005 | .003 | 10 | — | 30-40 | 80-105 | 17-25 | 10-15 | Step 2 |
| **82-81 Cougar, Granada** | | | | | | | | | | | | | | | |
| w/9″ rear brake | .030 | 9.000 | 9.060 | .125 | — | .810 | .0005 | .003 | 10 | — | 30-40 | 80-105 | 17-25 | 10-15 | Step 2 |
| w/10″ rear brake | .030 | 10.000 | 10.060 | .125 | — | .810 | .0005 | .003 | 10 | — | 30-40 | 80-105 | 17-25 | 10-15 | Step 2 |
| **82-81 Escort, Lynx, EXP, LN7** | | | | | | | | | | | | | | | |
| w/7″ rear brake | .030 | 7.000 | 7.060 | .125 | — | .882 | .0005 | .003 | 44 | — | 18-25 | 80-105 | 17-25 | 10-15 | Step 2 |
| w/8″ rear brake | .030 | 8.000 | 8.060 | .125 | — | .882 | .0005 | .003 | 44 | — | 18-25 | 80-105 | 17-25 | 10-15 | Step 2 |
| **82-79 Mustang, Capri, Fairmont, Zephyr** w/9″ rear brake | .030 | 9.000 | 9.060 | .125 | — | .810 | .0005 | .003 | 10 | — | 30-40 | 80-105 | 17-25 | 10-15 | Step 2 |
| w/10″ rear brake | .030 | 10.000 | 10.060 | .125 | — | .810 | .0005 | .003 | 10 | — | 30-40 | 80-105 | 17-25 | 10-15 | Step 2 |
| **82-79 Lincoln Town Car, Mark VI, LTD, Marquis,** w/10″ rear brakes | .030 | 10.000 | 10.060 | .125 | — | .972 | .0005 | .003 | 10 | — | 40-60 | 80-105 | 17-25 | 10-15 | Step 2 |
| w/11″ rear brakes | .030 | 11.030 | 11.090 | .125 | — | .972 | .0005 | .003 | 10 | — | 40-60 | 80-105 | 17-25 | 10-15 | Step 2 |
| **80 Thunderbird, Cougar** | .030 | 9.000 | 9.060 | .125 | — | .810 | .0005 | .003 | 10 | — | 30-40 | 80-105 | 17-25 | 10-15 | Step 2 |
| **80-79 Granada, Monarch, Versailles w/o rear disc brakes** | .030 | 10.000 | 10.060 | .125 | — | .810 | .0005 | .003 | 1 | U105-L65 | 12-16 | 80-105 | 17-25 | 10-15 | Step 2 |
| w/rear disc brakes | — | — | — | .125 | — | .895 | .0005 | .004 | 11 | 90-120 | 12-16 | 80-105 | — | — | — |

◆ 2 or 5   * .030″ over rivet head; if bonded lining, use .062″   ■ 1982 - 80 - .004   † 1975 - 65 ft/lbs.

# BRAKES

| YEAR, MAKE AND MODEL | Brake Shoe Min. Lining Thickness | Brake Drum Dia. Standard Size | Brake Drum Dia. Machine To | Brake Pad Min. Lining Thickness | Rotor Min. Thickness Machine To | Rotor Discard At | Variation From Parallelism | Runout T.I.R. | Design | Caliper Mounting Bolts Torque (ft-lbs) | Caliper Bridge, Pin or Key Bolts Torque (ft-lbs) | Wheel Lugs or Nuts Torque (ft-lbs) | Step 1 Tighten Spindle Nut (ft-lbs) | Step 2 Back Off Retorque (in-lbs) | Step 3 Lock, or Back Off and Lock |
|---|---|---|---|---|---|---|---|---|---|---|---|---|---|---|---|
| 80-77 Pinto, Bobcat, Mustang II | .030 | 9.000 | 9.060 | .030 | — | .810 | .0005 | .003 | 1 | U105-L65 | 12-16 | 80-105 | 17-25 | 10-15 | Step 2 |
| 79 LTD II, Thunderbird, Cougar | .030 | 11.030 | 11.090 | .125 | — | 1.120 | .0005 | .003 | 1 | 90-120 | 12-16 | 80-105 | 17-25 | 10-15 | Step 2 |
| 79-77 Mark V w/o rear disc brakes | .030 | 11.030 | 11.090 | .125 | — | 1.120 | .00025 | .003 | 1 | 90-120 | 12-16 | 80-105 | 17-25 | 10-15 | Step 2 |
|    w/rear disc brakes | — | — | — | .125 | — | .895 | .0004 | .004 | 11 | 90-120 | 12-16 | 80-105 | — | — | — |
| 78 Ford, Lincoln, Mercury, Custom 500, LTD II, Cougar, Ranchero II, Squire, Thunderbird | .030 | 11.030 | 11.090 | .125 | — | 1.120 | .0005 ■ | .003 | 1 | 90-120 | 12-16 | 70-115 | 17-25 | 10-15 | Step 2 |
| 78 Fairmont, Zephyr exc. Wagon | .030 | 9.000 | 9.060 | .125 | — | .810 | .0005 | .003 | 10 | — | 30-40 | 70-115 | 17-25 | 10-15 | Step 2 |
|    Wagon | .030 | 10.000 | 10.060 | .125 | — | .810 | .0005 | .003 | 10 | — | 30-40 | 70-115 | 17-25 | 10-15 | Step 2 |
| 78-77 Maverick, Comet, Granada, Monarch, Versailles | .030 | 10.000 | 10.060 | .125 | — | .810 | .0005 | .003 | 1 | U105-L65 | 12-16 | 70-115 | 17-25 | 10-15 | Step 2 |
| 78-77 LTD II, Thunderbird, Cougar | .030 | 11.030 | 11.090 | .125 | — | 1.120 | .0005 | .003 | 1 | 90-120 | 12-16 | 70-115 | 17-25 | 10-15 | Step 2 |
| 77-73 Lincoln, Mark IV | | | | | | | | | | | | | | | |
|    w/o rear disc brakes | .030 | 11.030 | 11.090 | .125 | — | 1.120 | .0005 | .003 | 1 | 90-120 | 12-16 | 70-115 | 17-25 | 10-15 | Step 2 |
|    w/rear disc brakes | — | — | — | .125 | — | .895 | .0004 | .004 | 11 | 90-120 | 12-16 | 70-115 | — | — | — |
| 77-74 Ford, Mercury, Meteor | | | | | | | | | | | | | | | |
|    w/o rear disc brakes | .030 | 11.030 | 11.090 | .125 | — | 1.120 | .0005 | .003 | 1 | 90-120 | 12-16 | 70-115 | 17-25 | 10-15 | Step 2 |
|    w/rear disc brakes | — | — | — | .125 | — | .895 | .0005 | .004 | 11 | 90-120 | 12-16 | 70-115 | — | — | — |
| 76-75 Granada, Monarch, Maverick, Comet | | | | | | | | | | | | | | | |
|    w/o rear disc brakes | .030 | 10.000 | 10.060 | .125 | — | .810 | .0005 | .003 | 1 | U105-L65 | 12-16 | 70-115 | 17-25 | 10-15 | Step 2 |
|    w/rear disc brakes | — | — | — | .125 | — | .895 | .0005 | .003 | 11 | 90-120 | 12-16 | 70-115 | — | — | — |
| 76-75 Pinto, Bobcat, Mustang | .030 | 9.000 | 9.060 | .030 | — | .810 | .0005 | .003 | 1 | U105-L65 | 12-16 | 70-115 | 17-25 | 10-15 | Step 2 |
| 76-74 Torino, Montego, Cougar | .030 | 11.030 | 11.090 | .030 | — | 1.120 | .0005 | .003 | 1 | 90-120 | 12-16 | 70-115 | 17-25 | 10-15 | Step 2 |
|    w/10" rear brakes | .030 | 10.000 | 10.060 | .030 | — | 1.120 | .0005 | .003 | 1 | 90-120 | 12-16 | 70-115 | 17-25 | 10-15 | Step 2 |
| 76-73 Thunderbird | .030 | 11.030 | 11.090 | .125 | — | 1.120 | .0005 | .003 | 1 | U105-L65 | 12-16 | 70-115 | 17-25 | 10-15 | Step 2 |
| 74 Maverick, Comet | .030 | 10.000 | 10.060 | .125 | — | .810 | .0005 | .003 | 1 | U105-L65 | 12-16 | 70-115 | 17-25 | 10-15 | Step 2 |
| 74 Pinto, Mustang | .030 | 9.000 | 9.060 | .030 | — | .875 | .00025 | .002 | 1 | 90-120 | 12-16 | 70-115 | 17-25 | 10-15 | Step 2 |
| 73 Ford, Mercury, Meteor, Lincoln, Continental | .030 | 10.000 | 10.060 | .125 | — | 1.120 | .0005 | .003 | 1 | 90-120 | 12-16 | 70-115 | 17-25 | 10-15 | Step 2 |
| 73 Torino, Montego | .030 | 10.000 | 10.060 | .030 | — | 1.120 | .0005 | .003 | 1 | U105-L65 | 12-16 | 70-115 | 17-25 | 10-15 | Step 2 |
| 73-72 Pinto | .030 | 9.000 | 9.060 | .030 | .700 | .685 | .0007 | .003 | 21 | U125-L65 | 25-35 | 70-115 | 17-25 | 10-15 | Step 2 |
| 73-72 Cougar, Mustang | .030 | 10.000 | 10.060 | .030 | .890 | .875 | .0007 | .002 | 24 | U125-L65 | 25-35 | 70-115 | 17-25 | 10-15 | Step 2 |
| 73-72 Maverick, Comet w/9" drum | .030 | 9.000 | 9.060 | — | — | — | — | — | — | — | — | 70-115 | 17-25 | 10-15 | Step 2 |
|    w/10" drum | .030 | 10.000 | 10.060 | — | — | — | — | — | — | — | — | 70-115 | 17-25 | 10-15 | Step 2 |
| 72 Torino, Montego | .030 | 10.000 | 10.060 | .030 | — | 1.120 | .0007 | .003 | 1 | 90-120 | 25-35 | 70-115 | 17-25 | 10-15 | Step 2 |
| 72 Ford, Mercury, Meteor, Thunderbird, Lincoln, Mark IV | .030 | 11.030 | 11.090 | .030 | 1.135 | 1.120 | .0007 | .003 | 24 | U125-L105 | 25-35 | 70-115 | 17-25 | 10-15 | Step 2 |

■ Lincoln - .00025

14

## GENERAL MOTORS CORP — BUICK

| Model | | | | | | | | | | | | | WHEEL BEARING ADJUSTMENT | |
|---|---|---|---|---|---|---|---|---|---|---|---|---|---|---|
| 82 Century, Skyhawk | | | | | | | | | | | | | | |
|   w/vented disc | * | 7.880 | 7.899 | .030 | .830 | .815 | .0005 | .004 | 40 | — | 28 | 100 | NON | ADJUSTABLE |
|   w/solid disc | * | 7.880 | 7.899 | .030 | .444 | .429 | .0005 | .004 | 40 | — | 28 | 100 | NON | ADJUSTABLE |
| 82 Regal, LeSabre | * | 9.500 | 9.560 | .030 | .980 | .965 | .0005 | .004 | 4 | — | 35 | 80† | 12 Handtight | ½ Flat |
| 82-79 Riviera w/o rear disc brakes | * | 9.500 | 9.560 | .030 | .980 | .965 | .0005 | .004 | 4 | 32 | 35 | 100 | NON | ADJUSTABLE |
|   w/rear disc brakes | | — | — | .030 | .980 | .965 | .0005 | .004 | 41 | — | 30 | 100 | 12 Handtight | ½ Flat |
| 82-79 Electra, Estate Wagon | * | 11.000 | 11.060 | .030 | .980 | .965 | .0005 | .004 | 4 | — | 35 | 100 | NON | ADJUSTABLE |
| 82-80 Skylark | * | 7.880 | 7.899 | .030 | .830 | .815 | .0005 | .003 | 40 | — | 28 | 103 | 12 Handtight | ½ Flat |
| 81-79 Century, Regal, LeSabre | * | 9.500 | 9.560 | .030 | .980 | .965 | .0005 | .005 | 4 | — | 35 | 80† | 12 Handtight | ½ Flat |
| 80-76 Skyhawk | ** | 9.500 | 9.560 | .030 | .830 | .815 | .0005 | .004 | 15 | — | — | 80 | 12 Handtight | ½ Flat |
| 79 Skylark | ** | 9.500 | 9.560 | .030 | .980 | .965 | .0005 | .004 | 4 | — | 35 | 80 | 12 Handtight | ½ Flat |
| 78-77 Electra, Estate Wagon, Riviera | * | 11.000 | 11.060 | .030 | .980 | .965 | .0005 | .004 | 4 | — | 35 | 80■ | 12 Handtight | ½ Flat |
| 78-77 Century, LeSabre, Regal, Skylark | * | 9.500 | 9.560 | .030 | .980 | .965 | .0005 | .004 | 4 | — | 35 | 80■ | 12 Handtight | ½ Flat |
| 76-75 Riviera, Electra, Estate Wagon, LeSabre, Custom | * | 12.000 | 12.060 | .030 | 1.230 | 1.215 | .0005 | .005 | 4 | — | 30-40 | 90 | 19 132 | 1/16 Turn |
| 76-75 Apollo, Skylark | * | 9.500 | 9.560 | .030 | .980 | .965 | .0005 | .004 | 4 | — | 35 | 70 | 19 132 | 1/16 Turn |
| 75 Century, Regal | * | 9.500 | 9.560 | .030 | .980 | .965 | .0005 | .004 | 4 | — | 35 | 70 | 19 132 | 1/16 Turn |
| 75 Skyhawk | * | 9.000 | 9.060 | .030 | .455 | .440 | .0005 | .005 | 15 | — | 35 | 70 | 19 132 | 1/16 Turn |
| 74-72 Electra, Custom, LeSabre, Centurion, Riviera, Wildcat | * | 11.000 | 11.060 | .030 | 1.230 | 1.215 | .0005 | .005 | 4 | — | 30-40 | 75 | 12 Handtight | ½ Flat |
| 74-73 Estate Wagon | * | 12.000 | 12.060 | .030 | 1.230 | 1.215 | .0005 | .005 | 4 | — | 30-40 | 75 | 12 Handtight | ½ Flat |
| 74-73 Century, Regal | * | 9.500 | 9.560 | .030 | .980 | .965 | .0005 | .004 | 4 | — | 35 | 65-70 | 19 132 | 1/16 Turn |
| 72 GS, Gran Sport, Skylark, Sports Wagon | * | 9.500 | 9.560 | .030 | .980 | .965 | .0005 | .004 | 4 | — | 35 | 65-70■ | 19 132 | 1/16 Turn |

*.030" over rivet head, if bonded lining use .062"    ■ w/½" stud 100 ft/lbs.    † w/Aluminum whls. LeSabre 90 ft/lbs., Regal 100 ft/lbs.

## GENERAL MOTORS CORP. — CADILLAC

| Model | | | | | | | | | | | | | WHEEL BEARING ADJUSTMENT | |
|---|---|---|---|---|---|---|---|---|---|---|---|---|---|---|
| 82 Cimarron | .030 | 7.880 | 7.899 | .030 | .830 | .815 | .0005 | .004 | 40 | — | 28 | 100 | NON | ADJUSTABLE |
| 82 Fleetwood | .030 | 11.000 | 11.060 | .030 | .980 | .965 | .0005 | .004 | 4 | — | 28 | 100 | 12 Handtight | Step 2 |
| 82 Eldorado, Seville | | | | | | | | | | | | | | |
|   Front | — | — | — | .030 | .980 | .965 | .0005 | .004 | 4 | — | 28 | 100 | NON | ADJUSTABLE |
|   Rear | — | — | — | .030 | .980 | .965 | .0005 | .004 | 41 | 35 | 38 | 100 | | |
| 82-77 Fleetwood Limo, Commercial Chassis | * | 12.000 | 12.060 | .062 | 1.230 | 1.215 | .0005 | .004 | 4 | — | 30 | 100 | 12 Handtight | Step 2 |
| 81-79 Fleetwood, Brougham, DeVille, Seville, (RWD) | | | | | | | | | | | | | | |
|   w/o rear disc brakes | * | 11.000 | 11.060 | .062 | .980 | .965 | .0005 | .004 | 4 | — | 30 | 100 | 12 Handtight | Step 2 |
|   w/rear disc brakes (1979) | * | — | — | .062 | .980 | .965 | .0005 | .004 | 41 | 34 | 30 | 100 | | |

*.030" over rivet head, if bonded lining use .062"

# BRAKES

| Year, Make and Model | Brake Shoe Min Lining Thick. | Brake Drum Std. Size | Brake Drum Machine To | Brake Pad Min Lining Thick. | Rotor Machine To | Rotor Discard At | Variation From Parallelism | Runout T.I.R. | Design | Caliper Mounting Bolts Torque (ft-lbs) | Caliper Bridge, Pin or Key Bolts Torque (ft-lbs) | Wheel Lugs or Nuts Torque (ft-lbs) | Bearing Step 1 Tighten Spindle Nut (ft-lbs) | Bearing Step 2 Back Off Retorque (in-lbs) | Bearing Step 3 Lock, or Back Off and Lock |
|---|---|---|---|---|---|---|---|---|---|---|---|---|---|---|---|
| **81-79 Eldorado, Seville (FWD)** | | | | | | | | | | | | | | | |
| Front Disc | — | — | — | .062 | .980 | .965 | .0005 | .004 | 4 | — | 30 | 100 | NON ADJUSTABLE | | |
| Rear Disc | — | — | — | .062 | .980 | .965 | .0005 | .004 | 41 | 35 | 30 | 100 | NON ADJUSTABLE | | |
| **78-77 DeVille** | * | 11.000 | 11.060 | .062 | .980 | .965 | .0005 | .005 | 4 | — | 30 | 100 | 12 | Handtight | Step 2 |
| **78-77 Brougham, Seville:** Front | — | — | — | .062 | .980 | .965 | .0005 | .005 | 4 | — | 30 | 100 | 12 | Handtight | Step 2 |
| Rear | — | — | — | .062 | .910 | .905 | .0005 | .003 | 41 | 35 | 30 | 100 | 12 | Handtight | Step 2 |
| **78-77 Eldorado:** Front | — | — | — | .062 | — | 1.170 | .0005 | .008 | 4 | — | 30 | 130 | NON ADJUSTABLE | | |
| Rear | — | — | — | .062 | — | 1.170 | .0005 | .008 | 41 | 35 | 30 | 130 | NON ADJUSTABLE | | |
| **76 Seville** | * | 11.000 | 11.060 | .062 | .980 | .965 | .0005 | .005 | 4 | — | 30 | 80 | 15 | Handtight | Step 2 |
| **76-74 Calais, Brougham, DeVille, Fleetwood 75, Commercial Chassis** | * | 12.000 | 12.060 | .062 | 1.220 | 1.215 | .0005 | .005 | 4 | U60-L80 | 30-40 | 100 ▲ | 15 | Handtight | Step 2 |
| **76 Eldorado:** Front | — | — | — | .062 | 1.205 | 1.190 | .0005 | .008 | 4 | — | 30 | 130 | NON ADJUSTABLE | | |
| Rear | — | — | — | .062 | 1.205 | 1.190 | .0005 | .008 | 41 | 35 | 30 | 130 | NON ADJUSTABLE | | |
| **75-72 Eldorado** | * | 11.000 | 11.060 | .062 | 1.205 | 1.190 | .0005 | .008 | 4 | 35 | 30 | 130 | NON ADJUSTABLE | | |
| **73-72 Calais, Fleetwood 60, 75 DeVille, Commercial Chassis** | * | 12.000 | 12.060 | .062 | 1.220 | 1.215 | .0007 | .005 | 4 | U60-L80 | 30 | 130 | 15 | Handtight | Step 2 |

\* .030" over rivet head, if bonded lining use .062"  ▲ 1975-74 130 ft/lbs.

## GENERAL MOTORS CORP. — CHEVROLET

| Year, Make and Model | Brake Shoe Min Lining Thick. | Brake Drum Std. Size | Brake Drum Machine To | Brake Pad Min Lining Thick. | Rotor Machine To | Rotor Discard At | Variation From Parallelism | Runout T.I.R. | Design | Caliper Mounting Bolts Torque (ft-lbs) | Caliper Bridge, Pin or Key Bolts Torque (ft-lbs) | Wheel Lugs or Nuts Torque (ft-lbs) | Bearing Step 1 Tighten Spindle Nut (ft-lbs) | Bearing Step 2 Back Off Retorque (in-lbs) | Bearing Step 3 Lock, or Back Off and Lock |
|---|---|---|---|---|---|---|---|---|---|---|---|---|---|---|---|
| **82 Celebrity, Cabalier** | | | | | | | | | | | | | | | |
| w/vented disc | * | 7.880 | 7.899 | .030 | .830 | .815 | .0005 | .004 | 40 | — | 28 | 100 | NON ADJUSTABLE | | |
| w/solid disc | * | 7.880 | 7.899 | .030 | .444 | .429 | .0005 | .004 | 40 | — | 28 | 100 | NON ADJUSTABLE | | |
| **82 Camaro** w/rear drum brakes | * | 9.500 | 9.560 | .030 | .980 | .965 | .0005 | .004 | 4 | — | 21-35 | 80 † | 12 | Handtight | ½ Flat |
| w/rear disc brakes | — | — | — | .030 | .980 | .965 | .0005 | .004 | 41 | — | 30-45 | 80 † | 12 | Handtight | ½ Flat |
| **82 Malibu, Monte Carlo, El Camino** | * | 9.500 | 9.560 | .030 | .980 | .965 | .0005 | .004 | 4 | — | 35 | 80 † | 12 | Handtight | ½ Flat |
| **82-80 Citation** | * | 7.880 | 7.899 | .030 | .830 | .815 | .0005 | .003 | 40 | — | 28 | 103 | NON ADJUSTABLE | | |
| **82-79 Impala, Caprice** w/9½" rear brakes | * | 9.500 | 9.560 | .030 | .980 | .965 | .0005 | .004 | 4 | — | 35 | 80 | 12 | Handtight | ½ Flat |
| w/11" rear brakes | * | 11.000 | 11.060 | .030 | .980 | .965 | .0005 | .004 | 4 | — | 35 | 100 | 12 | Handtight | ½ Flat |
| **82-78 Chevette** | * | 7.874 | 7.899 | .030 | .390 | .374 | .0005 | .005 | 20 | 70 | 28 | 70 | 12 | Handtight | ½ Flat |
| **82-77 Corvette:** Front | — | — | — | .030* | 1.230 | 1.215 | .0005 | .005 | 13 | 70 | 130 | 70 † | 12 | Handtight | ½ Flat |
| Rear | — | — | — | .030* | 1.230 | 1.215 | .0005 | .005 | 13 | 70 | 60 | 70 † | NON ADJUSTABLE | | |
| **81-79 Malibu, Camaro, Nova, Monte Carlo, El Camino** | * | 9.500 | 9.560 | .030 | .980 | .965 | .0005 | .004 | 4 | — | 35 | 80 † | 12 | Handtight | ½ Flat |

\* .030" over rivet head, if bonded lining use .062"  † Aluminum whls; Corvette 80, Camaro 105, others 90.

| Year & Model | | | | | | | | | | | | | | | |
|---|---|---|---|---|---|---|---|---|---|---|---|---|---|---|---|
| 80-76 Monza | * | 9.500 | 9.560 | .030 | .830 | .815 | .0005 | .005 | 15 | — | — | 80† | 12 | Handtight | ½ Flat |
| 78 Caprice, Camaro, Impala, Nova exc. Wagon | * | 9.500 | 9.560 | .030 | .980 | .965 | .0005 | .004 | 4 | — | 35 | 80† | 12 | Handtight | ½ Flat |
| Wagon | * | 11.000 | 11.060 | .030 | .980 | .965 | .0005 | .004 | 4 | — | 35 | 100 | 12 | Handtight | ½ Flat |
| 78 Malibu, Monte Carlo, El Camino | * | 11.000 | 11.060 | .030 | .980 | .965 | .0005 | .004 | 4 | 70 | 35 | 80 | 12 | Handtight | ½ Flat |
| 77 Chevette | * | 7.880 | 7.899 | .030 | .456 | .441 | .0005 | .005 | 20 | — | 28 | 70 | 12 | Handtight | ½ Flat |
| 77 Impala, Caprice exc. Wagon | * | 9.500 | 9.560 | .030 | .980 | .965 | .0005 | .005 | 4 | — | 35 | 80 | 12 | Handtight | ½ Flat |
| Wagon | * | 11.000 | 11.060 | .030 | .980 | .965 | .0005 | .005 | 4 | — | 35 | 100 | 12 | Handtight | ½ Flat |
| 77 Chevelle, Monte Carlo, El Camino, Malibu | * | 11.000 | 11.060 | .030 | .980 | .965 | .0005 | .005 | 4 | — | 35 | 80 | 12 | Handtight | ½ Flat |
| 77 Camaro, Nova | * | 9.500 | 9.560 | .030* | .980 | .965 | .0005 | .005 | 4 | — | 35 | 80 | 12 | Handtight | ½ Flat |
| 77-76 Vega | * | 9.500 | 9.560 | .030 | .455 | .440 | .0005 | .005 | 15 | — | — | 80† | 12 | Handtight | ½ Flat |
| 76 Chevette | * | 7.870 | 7.899 | .030 | .448 | .433 | .0005 | .005 | 20 | 70 | 28 | 70 | 12 | Handtight | ½ Flat |
| 76-72 Corvette: Front | | — | — | .030* | 1.230 | 1.215 | .0005 | .005 | 13 | 75 | 130 | 75 | 12 | Skip | 1-1½ Slot |
| Rear | | — | — | .030* | 1.230 | 1.215 | .0005 | .005 | 13 | 75 | 50 | 75 | — | — | — |
| 76-72 Bel Air, Impala, Caprice | * | 11.000 | 11.060 | .030* | 1.230 | 1.215 | .0005 | .005 | 4 | — | 35 | 80 | 12 | Skip | 1-1½ Slot |
| 76-72 Nova | * | 9.500 | 9.560 | .030* | .980 | .965 | .0005 | .005 | 4 | — | 35 | 70 | 12 | Skip | 1-1½ Slot |
| 76-73 Chevelle, Camaro, Monte Carlo, El Camino | * | 9.500 | 9.560 | .030* | .980 | .965 | .0005 | .005 | 4 | — | 35 | 70 | 12 | Skip | 1-1½ Slot |
| 75 Monza | * | 9.000 | 9.060 | .030 | .455 | .440 | .0005 | .005 | 15 | — | — | 65 | 12 | Skip | 1 Flat |
| 75-72 Vega | * | 9.000 | 9.060 | .030 | .455 | .440 | .0005 | .005 | 15 | — | — | 65 | 12 | Skip | 1 Flat |
| 72 Chevelle, Camaro, El Camino | * | 9.500 | 9.560 | .030* | .980 | .965 | .0005 | .005 | 4 | — | 35 | 70 | 12 | Skip | 1-1½ Slot |

* .030″ over rivet, if bonded lining use .062″     † Aluminum whls; Corvette 80, others 90.

## GENERAL MOTORS CORP. — OLDSMOBILE

| Year & Model | | | | | | | | | | | | | | | |
|---|---|---|---|---|---|---|---|---|---|---|---|---|---|---|---|
| 82 Ciera, Firenza w/vented disc | * | 7.880 | 7.899 | .030 | .830 | .815 | .0005 | .004 | 40 | — | 28 | 100 | | NON ADJUSTABLE | |
| w/solid disc | * | 7.880 | 7.899 | .030 | .444 | .429 | .0005 | .004 | 40 | — | 28 | 100 | | NON ADJUSTABLE | |
| 82 Cutlass Supreme, 88 | * | 9.500 | 9.560 | .030 | .980 | .965 | .0005 | .004 | 4 | — | 35 | 100† | 12 | Handtight | ½ Flat |
| 82-80 Omega | * | 7.880 | 7.899 | .030 | .830 | .815 | .0005 | .003 | 40 | — | 28 | 103 | | NON ADJUSTABLE | |
| 82-79 Toronado w/o rear disc | * | 9.500 | 9.560 | .030 | .980 | .965 | .0005 | .004 | 4 | — | 35 | 100 | | NON ADJUSTABLE | |
| w/rear disc | | — | — | .030 | .980 | .965 | .0005 | .004 | 41 | 32 | 30 | 100 | — | — | — |
| 82-78 Custom Cruiser, 88 (w/403), 98 | * | 11.000 | 11.060 | .030 | .980 | .965 | .0005 | .004 | 4 | — | 35 | 100 | 12 | Handtight | ½ Flat |
| 81-78 Cutlass, 88 (w/o 403) | * | 9.500 | 9.560 | .030 | .980 | .965 | .0005 | .004 | 4 | — | 35 | 80 | 12 | Handtight | ½ Flat |
| 80-76 Starfire | * | 9.500 | 9.560 | .062 | .830 | .815 | .0005 | .005 | 15 | — | — | 80 | 12 | Handtight | ½ Flat |
| 79-78 Omega w/5 Speed | * | 11.000 | 11.060 | .030 | .980 | .965 | .0005 | .005 | 4 | — | 35 | 80 | 12 | Handtight | ½ Flat |
| 79-78 Omega w/o 5 Speed | * | 9.500 | 9.560 | .030 | .980 | .965 | .0005 | .005 | 4 | — | 35 | 80 | 12 | Handtight | ½ Flat |

* .030″ over rivet, if bonded lining use .062″     † 88 w/7/16″ stud; 80 ft/lbs.

# BRAKES

| YEAR, MAKE AND MODEL | BRAKE SHOE Minimum Lining Thickness | BRAKE DRUM DIAMETER Standard Size | BRAKE DRUM DIAMETER Machine To | BRAKE PAD Minimum Lining Thickness | BRAKE ROTOR MIN THICKNESS Machine To | BRAKE ROTOR MIN THICKNESS Discard At | BRAKE ROTOR Variation From Parallelism | BRAKE ROTOR Runout T.I.R. | DESIGN | CALIPER Mounting Bolts Torque (ft-lbs) | CALIPER Bridge, Pin or Key Bolts Torque (ft-lbs) | WHEEL Lugs or Nuts Torque (ft-lbs) | WHEEL BEARING STEP 1 Tighten Spindle Nut (ft-lbs) | WHEEL BEARING STEP 2 Back Off Retorque (in-lbs) | WHEEL BEARING STEP 3 Lock, or Back Off and Lock |
|---|---|---|---|---|---|---|---|---|---|---|---|---|---|---|---|
| 78-72 Toronado | * | | | .062 | 1.185 | 1.170 | .0005 | .002 | 4 | — | 35-40 | 130 | NON ADJUSTABLE | | |
| 77 Custom Cruiser, 88 (w/403), 98 | * | 11.000 | 11.060 | .030 | .980 | .965 | .0005 | .005 | 4 | — | 40 | 100 | 30 | Handtight | Step 2 |
| 77 Omega w/o 5 Speed | * | 9.500 | 9.560 | .030 | .980 | .965 | .0005 | .004 | 4 | — | 40 | 80 | 30 | Handtight | Step 2 |
| 77 88 (w/o 403) | * | 9.500 | 9.560 | .030 | .980 | .965 | .0005 | .004 | 4 | — | 40 | 80 | 30 | Handtight | Step 2 |
| 77-76 Omega w/5 Speed, Cutlass | * | 11.000 | 11.060 | .030 | .980 | .965 | .0005 | .004 | 4 | — | 40 | 80 | 30 | Handtight | Step 2 |
| 76 Omega w/5 Speed | * | 9.500 | 9.560 | .030 | .980 | .965 | .0005 | .004 | 4 | — | 40 | 80 | 30 | Handtight | Step 2 |
| 76 Cutlass | * | 11.000 | 11.060 | .030 | .980 | .965 | .0005 | .004 | 4 | — | 40 | 80 | 30 | Handtight | Step 2 |
| 76-72 88, 98 exc. Wagon & H.D. Pkg. | * | 11.000 | 11.060 | .062 | 1.230 | 1.215 | .0005 | .005 | 4 | — | 40 | 80 | 30 | Handtight | Step 2 |
| 76-72 88 Wagon & H.D. Pkg. | * | 12.000 | 12.060 | .062 | 1.230 | 1.215 | .0005 | .005 | 4 | — | 40 | 80 | 30 | Handtight | Step 2 |
| 75 Starfire | * | 9.000 | 9.060 | .062 | .455 | .440 | .0005 | .005 | 15 | — | 40 | 80 | 30 | Handtight | Step 2 |
| 75-72 Omega, Cutlass exc. Vista Cruiser | * | 9.500 | 9.560 | .062 | .980 | .965 | .0005 | .004 | 4 | — | 40 | 80 | 30 | Handtight | Step 2 |
| 75-72 Vista Cruiser | * | 11.000 | 11.060 | .062 | .980 | .965 | .0005 | .004 | 4 | — | 40 | 80 | 30 | Handtight | Step 2 |

* .030" over rivet head, if bonded lining use .062".

| GENERAL MOTORS CORP. — PONTIAC | BRAKE SHOE Minimum Lining Thickness | BRAKE DRUM DIAMETER Standard Size | BRAKE DRUM DIAMETER Machine To | BRAKE PAD Minimum Lining Thickness | BRAKE ROTOR MIN THICKNESS Machine To | BRAKE ROTOR MIN THICKNESS Discard At | BRAKE ROTOR Variation From Parallelism | BRAKE ROTOR Runout T.I.R. | DESIGN | CALIPER Mounting Bolts Torque (ft-lbs) | CALIPER Bridge, Pin or Key Bolts Torque (ft-lbs) | WHEEL Lugs or Nuts Torque (ft-lbs) | WHEEL BEARING STEP 1 Tighten Spindle Nut (ft-lbs) | WHEEL BEARING STEP 2 Back Off Retorque (in-lbs) | WHEEL BEARING STEP 3 Lock, or Back Off and Lock |
|---|---|---|---|---|---|---|---|---|---|---|---|---|---|---|---|
| 82 A6000, J2000 w/vented disc | * | 7.880 | 7.899 | .030 | .830 | .815 | .0005 | .004 | 40 | — | 28 | 100 | NON ADJUSTABLE | | |
| w/solid disc | * | 7.880 | 7.899 | .030 | .444 | .429 | .0005 | .004 | 40 | — | 28 | 100 | NON ADJUSTABLE | | |
| 82 Firebird w/rear drum brakes | * | 9.500 | 9.560 | .030 | .980 | .965 | .0005 | .004 | 41 | — | 21-35 | 80† | 12 | Handtight | ½ Flat |
| w/rear disc brakes | — | | | .030 | | | | | | 70 | 30-45 | 80† | | Handtight | ½ Flat |
| 82-81 T 1000 | * | 7.874 | 7.899 | .030 | .390 | .374 | .0005 | .005 | 20 | — | 28 | 70 | 12 | Handtight | ½ Flat |
| 82-80 Phoenix (F.W.D.) | — | 7.880 | 7.899 | .030 | .830 | .815 | .0005 | .003 | 40 | — | 35 | 103 | NON ADJUSTABLE | | |
| 82-77 Bonneville, Catalina, LeMans, Grand Prix, Grand Am, Safari w/9.5" rear brakes | * | 9.500 | 9.560 | .030 | .980 | .965 | .0005 | .004 | 4 | — | 35 | 80 | 12 | Handtight | ½ Flat |
| w/11" rear brakes | * | 11.000 | 11.060 | .030 | .980 | .965 | .0005 | .004 | 4 | — | 35 | 80▲ | 12 | Handtight | ½ Flat |

* .030" over rivet head, if bonded lining use .062".

† w/Aluminum whls. 105 ft/lbs.

▲ ½" stud 100 ft/lbs.   ½" stud 100 ft/lbs.

**Top specification table**

| Model | | | | | | | | | | | | | | |
|---|---|---|---|---|---|---|---|---|---|---|---|---|---|---|
| 81-77 Firebird, Ventura, Phoenix (R.W.D.) | * | 9.500 | 9.560 | .030 | .980 | .965 | .0005 | .004 | 4 | 35 | 80 | 12 | Handtight | ½ Flat |
| w/rear disc (1981-79) | — | — | — | .030 | .921 | .905 | .0005 | .004 | 41 | 30 | 80 | — | | ½ Flat |
| 80-76 Sunbird, Astre | * | 9.500 | 9.560 | .030 | .830 | .815 | .0005 | .004 | 15 | 35 | 80 | 12 | Handtight | ½ Flat |
| 76-72 Catalina, Bonneville, Gran Ville exc. Wagon | .125 | 11.000 | 11.060 | .125 | 1.230 | 1.215 | .0005 | .004 | 4 | 35 | 75 | 12 | Handtight | ½ Flat |
| Wagon, Grand Safari | .125 | 12.000 | 12.060 | .125 | 1.230 | 1.215 | .0005 | .004 | 4 | 35 | 75 | 12 | Handtight | ½ Flat |
| 76-74 Le Mans, Firebird, Grand Prix, Ventura exc. Wagon | .125 | 9.500 | 9.560 | .125 | .980 | .965 | .0005 | .004 | 4 | 35 | 70 | 12 | Handtight | ½ Flat |
| Wagon | .125 | 11.000 | 11.060 | .125 | .980 | .965 | .0005 | .004 | 4 | 35 | 70 | 12 | Handtight | ½ Flat |
| 73-72 Ventura | .125 | 9.500 | 9.560 | .125 | — | .980 | .0005 | .004 | 4 | 35 | 70 | Snug | Handtight | Step 2 |
| 73-72 Le Mans, Firebird, Grand Prix | .125 | 9.500 | 9.560 | .125 | .980 | .965 | .0005 | .004 | 4 | 35 | 70 | Snug | Handtight | Step 2 |

* .030" over rivet head, if bonded lining use .062"    ▲ ½" stud 100 ft/lbs.

# DOMESTIC LIGHT TRUCKS

**NOTICE:** Should state inspection regulations exceed manufacturer's specifications for lining or rotor/drum reserve, the state inspection regulation specification **must** be used.

| AMERICAN MOTORS — JEEP | | | | | | | | | | | | | | |
|---|---|---|---|---|---|---|---|---|---|---|---|---|---|---|
| 82 CJ, Scrambler | .030* | 10.000 | 10.060 | .062 | — | .815 | .001 | .005 | 1 | — | 30 | 75 | 50 | Skip | 1/6 Turn |
| 82 Cherokee, Wagoneer, J-10 | .030* | 11.000 | 11.060 | .062 | — | 1.215 | .001 | .005 | 4 | 35 | 30 | 75 | 50 | Skip | 1/6 Turn |
| 82 J-20 | .030* | 12.000 | 12.060 | .062 | — | 1.215 | .001 | .005 | 4 | 35 | 30 | ■75 | 50 | Skip | 1/6 Turn |
| 81-79 CJ | .030* | 10.000 | 10.060 | .062 | — | .815 | .001 | .005 | 1 | — | 15 | 75 | 50 | Skip | 1/3 Turn |
| 81-78 Cherokee, Wagoneer, J-10 | .030* | 11.000 | 11.060 | .062 | — | 1.215 | .001 | .005 | 4 | 35 | — | 75 | 50 | 420 | 1/3 Turn |
| 81-78 J-20 | .030* | 12.000 | 12.060 | .062 | — | 1.215 | .001 | .005 | 4 | 35 | 15 | ■75 | 50 | 420 | 1/3 Turn |
| 78-77 CJ | .030* | 11.000 | 11.060 | .062 | — | 1.120 | .001† | .005 | 1 | — | 15 | 75 | 50 | Skip | 1/4 Turn |
| 77-76 Cherokee, Wagoneer, J-10 | .030* | 12.000 | 12.060 | .062 | — | 1.215 | .0005 | .005 | 4 | — | 15 | 75 | 50 | Skip | 1/4 Turn |
| 76-72 J-20 | .030 | 11.000 | 11.060 | .062 | — | 1.215 | .0005 | .005 | 4 | — | — | ■75 | ◀ | Skip | 1/4 Turn |
| 76-72 CJ Front & Rear | .030 | 11.000 | 11.060 | — | — | — | — | — | — | — | — | 75 | 50 | Skip | 1/6 Turn |
| 75-74 Cherokee, Wagoneer, J-10 | .030 | 12.000 | 12.060 | .062 | 1.230 | 1.215 | .003 | .005 | 4 | — | 15 | 75 | ◀ | Skip | 1/4 Turn |
| 75-74 J-20 | .030 | 11.000 | 11.060 | .062 | 1.230 | 1.215 | .003 | .005 | 4 | — | 15 | ■75 | ◀ | Skip | 1/4 Turn |
| 73-72 all w/11" brakes | .030 | 11.000 | 11.060 | — | — | — | — | — | — | — | — | 75 | ◀ | Skip | 1/6 Turn |
| 73-72 all w/12" brakes | .030 | 12.000 | 12.060 | — | — | — | — | — | — | — | — | 75 | ◀ | Skip | 1/6 Turn |
| 73-72 8,000 GVW Camper | .030 | 12.125 | 12.185 | — | — | — | — | — | — | — | — | 75 | ◀ | Skip | 1/6 Turn |

* .030" over rivet head, if bonded lining use .062"    † 1977 .0005    ◀ while rotating whl., tighten nut until whl. binds.    ■ J20 (8400 GVW) 130 ft/lbs.

# BRAKES

## CHRYSLER CORP. — DODGE, PLYMOUTH

| YEAR, MAKE AND MODEL | Brake Shoe Min. Lining Thick.* | Drum Std. Size | Drum Machine To | Pad Min. Lining Thick. | Rotor Machine To | Rotor Discard At | Variation From Parallelism | Runout T.I.R. | Caliper Design | Caliper Mounting Bolts (ft-lbs) | Bridge Pin or Key Bolts (ft-lbs) | Wheel Lugs or Nuts (ft-lbs) | Step 1 Tighten Spindle Nut (ft-lbs) | Step 2 Back Off Retorque (in-lbs) | Step 3 Lock, or Back Off and Lock |
|---|---|---|---|---|---|---|---|---|---|---|---|---|---|---|---|
| 82 Rampage | .030 | 7.870 | 7.900 | .030 | .461 | .431 | .0005 | .004 | 8 | 70-100 | 25-40 | 85 | 20-25 | Handtight | Step 2 |
| 82-79 Ramcharger, Trail Duster 4 x 2 | .030 | 10.000 | 10.060 | .030 | 1.220 | 1.190 | .0005 | .004 | 5 | 110 | 17 | 105 | 8 | Handtight | Step 2 |
| 4 x 4 | .030 | 10.000 | 10.060 | .030 | 1.220 | 1.190 | .001 | .005 | 5 | 150 | 17 | 105 | 50 | 360-480 | 1/3 Turn |
| 78-74 Ramcharger, Trail Duster 4 x 2 | .030 | 11.000 | 11.060 | .030 | 1.220 | 1.190 | .0005 | .004 | 5 | 95 | 17 | 85-125 | 8 | Handtight | Step 2 |
| 4 x 4 | .030 | 11.000 | 11.060 | .030 | 1.220 | 1.190 | .0005 | .004 | 5 | 95 | 17 | 85-125 | 100 | Skip | 1 Slot |
| 82-75 D100/150 exc. w/9¼" R. axle | .030 | 10.000 | 10.060 | .030 | 1.220 | 1.190 | .0005 | .004 | 5 ◆ | 110 | 17 | 105 ● | 8 | Handtight | Step 2 |
| 76-75 D100 w/9¼" rear axle | .030 | 11.000 | 11.060 | .030 | 1.220 | 1.190 | .0005 | .004 | 5 | 100 | 17 | 105 | 8 | Handtight | Step 2 |
| 74-73 D100 exc. w/11" front brakes | .030 | 10.000 | 10.060 | .030 | 1.220 | 1.190 | .0005 | .004 | 5 | 100 | 17 | 65-85 | 20-25 | Handtight | Step 2 |
| 73-72 D100 w/11" front brakes | .030 | 11.000 | 11.060 | — | — | 1.180 | .0005 | .0025 | 2 | 75-100 | 30-35 | 65-85 | 20-25 | Handtight | Step 2 |
| 72 D100 | .030 | 10.000 | 10.060 | .030 | 1.220 | 1.190 | .001 | .005 | 5 | 110 | 17 | 70-90 | 20-25 | Handtight | Step 2 |
| 82-81 D200/250 w/3300 lb. F.A. | .030 | 12.000 | 12.060 | .030 | 1.220 | 1.190 | .0005 | .005 | 5 | 100 | 17 | 105 ● | 8 | Handtight | Step 2 |
| 82-79 D200/250 w/4000 lb. F.A., D300/350 | .030 | 12.000 | 12.060 | .030 | — | 1.180 | .001 | .005 | 2 | 75-100 | 30-35 | 105 | 8 | Handtight | Step 2 |
| 80-79 D200 w/3300 lb. F.S., GVW over 6200 lb. | .030 | 12.000 | 12.060 | .030 | 1.220 | 1.190 | .001 | .005 | 5 | 160 | 17 | 105 ● | 8 | Handtight | Step 2 |
| 80-79 D200 w/3300 lb. F.A., GVW 6200 lb. | .030 | 12.000 | 12.060 | .030 | 1.220 | 1.190 | .0005 | .004 | 5 | 100 | 17 | 105 | 8 | Handtight | Step 2 |
| 78-75 D200 w/6600 lb. GVW | .030 | 12.120 | 12.180 | .030 | 1.160 | 1.130 | .0005 | .004 | 5 | 100 | 17 | 105 ● | 8 | Handtight | Step 2 |
| 78-75 D200 over 6000 GVW, D300 | .030 | 12.120 | 12.180 | .030 | 1.160 | 1.130 | .0005 | .004 | 5 | 160 | 17 | 65-85 ● | 8 | Handtight | Step 2 |
| 74 D200 w/6000 lb. GVW | .030 | 12.000 | 12.060 | .030 | 1.160 | 1.130 | .0005 | .004 | 5 | 160 | 17 | 65-85 | 8 | Handtight | Step 2 |
| 74 D200 over 6000 GVW, D300 | .030 | 12.120 | 12.180 | .030 | 1.160 | 1.130 | .0005 | .004 | 5 | 100 | 17 | 70-90 | 8 | Handtight | Step 2 |
| 73 D200 w/6000 lb. GVW | .030 | 12.000 | 12.060 | .030 | — | 1.125 | .0005 | .004 | 5 | 100 | 17 | 70-90 | 8 | Handtight | Step 2 |
| 73 D200 over 6000 lb. GVW | .030 | 12.120 | 12.180 | .030 | — | 1.180 | .0005 | .004 | 5 | 100 | 17 | 70-90 | 20-25 | Handtight | Step 2 |
| 73 D200 w/12.57 rotor diameter | .030 | 12.000 | 12.060 | .030 | — | 1.125 | .0005 | .004 | 5 | — | — | 70-90 ● | 20-25 | Handtight | Step 2 |
| 73 D200 w/12.82 rotor diameter | .030 | 12.000 | 12.060 | .030 | — | — | .0005 | .004 | 5 | — | — | 70-90 | 20-25 | Handtight | Step 2 |
| 73-72 D300 | .030 | 12.000 | 12.060 | — | — | 1.125 | — | — | 2 | 75-100 | 30-35 | 79-90 | 20-25 | Handtight | Step 2 |
| 72 D200 w/6000 lb. GVW | .030 | 12.000 | 12.060 | .030 | — | 1.180 | .0005 | .0025 | 5 | 75-100 | 30-35 | 70-90 | 20-25 | Handtight | Step 2 |
| 72 D200 w/12.57 rotor diameter | .030 | 12.120 | 12.180 | .030 | — | 1.125 | .0005 | .0025 | 5 | 75-100 | 30-35 | 70-90 | 20-25 | Handtight | Step 2 |
| 72 D200 w/12.82 rotor diameter | .030 | 12.000 | 12.060 | .030 | 1.220 | 1.190 | .001 | .005 | 5 | — | — | 70-90 | 20-25 | Handtight | Step 2 |
| 82-79 W150 | .030 | 10.000 | 10.060 | .030 | 1.220 | 1.190 | .001 | .005 | 5 | 150 | 17 | 105 | 50 | 360-480 | 1/3 Turn |
| 78-77 W100 | .030 | 11.000 | 11.060 | .030 | 1.220 | 1.190 | .0005 | .004 | 5 | 100 | 17 | 105 | 50 | 360-480 | 1/3 Turn |
| 76-75 W100 | .030 | 11.000 | 11.060 | .030 | 1.220 | 1.190 | .0005 | .004 | 5 | 100 | 17 | 105 | 30-40 | Handtight | Step 2 |
| 74-72 W100 | .030 | 11.000 | 11.060 | — | — | 1.130 | .001 | — | 5 | — | — | 70-90 | 50 | Skip | 1/3 Turn |
| 82-79 W200/250 exc. w/Spicer 60 | .030 | 11.000 | 11.060 | .030 | 1.160 | 1.130 | .001 | .005 | 5 | 150 | 17 | 105 | 50 | 360-480 | 1/3 Turn |
| 82-79 W200/250 w/Spicer 60, W300/350 | .030 | 12.000 | 12.060 | .030 | 1.160 | 1.130 | .001 | .005 | 21 | 160 | 15 | 105 | 50 | 360-480 | 1/3 Turn |

● 5/8" whl. stud (82 - 75) 200 ft/lbs.; (74) 175 - 225 ft/lbs.; (73 - 72) 175 - 225 ft/lbs.; (75-72) F125 - 175 ft/lbs. w/dual rear whls. 325 ft/lbs. ■ 5/8" bolt 140-180 ft/lbs. ◆ (1980 - 75) 100 ft/lbs.

▲ w/dual rear whls. (78-76) 300-350 ft/lbs.; (75-72) F125-175 ft/lbs.: R300-350 ft/lbs. † 2 or 5.

| Model | 1 | 2 | 3 | 4 | 5 | 6 | 7 | 8 | 9 | 10 | 11 | 12 | 13 | 14 | 15 |
|---|---|---|---|---|---|---|---|---|---|---|---|---|---|---|---|
| 78-75 W200 exc. w/Spicer 60 | ⅓ Turn | 360-480 | 50 | 105 | 17 | 100 | 5 | .004 | .0005 | 1.130 | 1.160 | .030 | 12.060 | 12.000 | .030 |
| 78-75 W200 w/Spicer 60, W300 | ⅓ Turn | 360-480 | 50 | 105 | 15 | 160 | 21 | .004 | .0005 | 1.130 | 1.160 | .030 | 12.060 | 12.000 | .030 |
| 74-72 W200 exc. w/4500 lb. F.A. Front | ⅓ Turn | Skip | 50 | — | — | — | — | — | — | — | — | .030 | 12.180 | 12.120 | .030 |
| Rear | ⅓ Turn | Skip | 50 | — | — | — | — | — | — | — | — | .030 | 12.060 | 12.000 | .030 |
| 74-72 W200 w/4500 lb. F.A. W300 | ⅓ Turn | Skip | 50 | — | — | — | — | — | — | — | — | .030 | 12.060 | 12.000 | .030 |
| 82-76 B100/150, PB100/150 w/10" rear brakes | Step 2 | Handtight | 30-40 | 85-125 | 14-22 | 95-125 | 5 | .004 | .0005 | 1.190 | 1.220 | .030 | 10.060 | 10.000 | .030 |
| w/11" rear brakes | Step 2 | Handtight | 30-40 | 85-125 | 14-22 | 95-125 | 5 | .004 | .0005 | 1.190 | 1.220 | .030 | 11.060 | 11.000 | .030 |
| 75-73 B100, PB100 w/10" rear brakes | Step 2 | Handtight | 20-30 | 65-85 | 17 | 75-100 | 5 | .004 | .0005 | 1.190 | 1.220 | .030 | 10.060 | 10.000 | .030 |
| w/11" rear brakes | Step 2 | Handtight | 20-30 | 65-85 | 17 | 75-100 | 5 | .004 | .0005 | 1.190 | 1.220 | .030 | 11.060 | 11.000 | .030 |
| 72 B100 w/10" rear brakes | Step 2 | Handtight | 20-30 | 65-85 | 17 | 75-100 | † | .004 | .0005 | 1.190 | 1.220 | .030 | 10.060 | 10.000 | .030 |
| w/11" rear brakes | Step 2 | Handtight | 30-40 | 85-125 | 14-22 | 95-125 | † | .004 | .0005 | 1.190 | 1.220 | .030 | 11.060 | 11.000 | .030 |
| 82-78 B200/250, PB200/250 | Step 2 | Handtight | 20-30 | 65-85 | 17 | 75-100 | 5 | .004 | .0005 | 1.190 | 1.220 | .030 | 10.060 | 10.000 | .030 |
| 77-76 B200, PB200 | Step 2 | Handtight | 20-30 | 65-85 | 17 | 75-100 | 5 | .004 | .0005 | 1.190 | 1.220 | .030 | 11.060 | 11.000 | .030 |
| 75-73 B200, PB200 | Step 2 | Handtight | 20-30 | 65-85 | 17 | 75-100 | 5 | .004 | .0005 | 1.190 | 1.220 | .030 | 11.060 | 11.000 | .030 |
| 72 B200 w/10" brakes | Step 2 | Handtight | 20-30 | 65-85 | 17 | 75-100 | 5 | .004 | .0005 | 1.190 | 1.220 | .030 | 10.060 | 10.000 | .030 |
| w/11" rear brakes | Step 2 | Handtight | 20-30 | 65-85 | 17 | 75-100 | † | .004 | .0005 | 1.190 | 1.220 | .030 | 11.060 | 11.000 | .030 |
| 82-79 B300/350, CB300/350, PB300/350 w/3600 lb. F.A. | Step 2 | Handtight | 30-40 | 85-125 | 14-22 | 95-125 ■ | 5 | .004 | .0005 | 1.190 | 1.220 | .030 | 12.060 | 12.000 | .030 |
| w/4000 lb. F.A. | Step 2 | Handtight | 30-40 | 175-225 ▲ | 14-22 | 95-125 ■ | 5 | .004 | .0005 | 1.130 | 1.160 | .030 | 12.060 | 12.000 | .030 |
| 78-76 B300, CB300, PB300 | Step 2 | Handtight | 30-40 | 85-125 ▲ | 14-22 | 95-125 ■ | 5 | .004 | .0005 | 1.130 | 1.160 | .030 | 12.060 | 12.000 | .030 |
| 75-73 B300, CB300, PB300 w/10" brakes | Step 2 | Handtight | 20-30 | 65-85 ▲ | 17 | 140-180 | 5 | .004 | .0005 | 1.130 | 1.160 | .030 | 12.060 | 12.000 | .030 |
| 72 B300, CB300 w/11" brakes | Step 2 | Handtight | 20-30 | 65-85 ▲ | 17 | 140-180 | † | .004 | .0005 | 1.130 | 1.160 | .030 | 12.060 | 12.000 | .030 |

● ⅝" whl. stud (82 - 75) 200 ft/lbs.; 175 - 225 ft/lbs. (73 - 72) 125 - 175 ft/lbs. (75-72) F125-175 ft/lbs. R300-350 ft/lbs.  ▲ w/dual rear whls. (78-76) 300-350 ft/lbs. w/dual rear whls. 325 ft/lbs.  ■ 5/8" bolt 140-180 ft/lbs.  † 2 or 5

### FORD MOTOR CO.

| Model | 1 | 2 | 3 | 4 | 5 | 6 | 7 | 8 | 9 | 10 | 11 | 12 | 13 | 14 | 15 |
|---|---|---|---|---|---|---|---|---|---|---|---|---|---|---|---|
| 82-81 Bronco | 45° | Skip | 50 | 90 | 12-20 | 74-102 | — | .003 | .0007 | 1.120 | — | .030 | 11.091 | 11.031 | .030 |
| 80-76 Bronco | 90° | Skip | 50 | 90 | 12-20 | 50-60 | 1 | .003 | .0007 | 1.120 | — | .030 | 11.091 | 11.031 | .030 |
| 75-72 Bronco w/10" brakes | 90° | Skip | 50 | 90 | — | — | — | — | — | — | — | .030 | 10.060 | 10.000 | .030 |
| w/11" brakes | 90° | Skip | 50 | 90 | — | — | — | — | — | — | — | .030 | 11.060 | 11.000 | .030 |
| 82-80 F100 w/46-4900 GVW | ⅛ Turn | Skip | 22-25 | 90 | 12-20 | — | 1 | .003 | .0005 | .810 | — | .030 | 10.060 | 10.000 | .030 |
| 82-77 F100, F150, E100, E150 | ⅛ Turn | Skip | 22-25 | 90 | 12-20 | — | 1 | .003 | .0007 | 1.120 | — | .030 | 11.091 | 11.031 | .030 |
| 76-75 F100, F150, E100, E150 | 2 Slots | Skip | 17-25 | 90 | 12-20 | — | 1 | .003 | .0007 | 1.120 | — | .030 | 11.091 | 11.031 | .030 |
| 74 E100 | 2 Slots | Skip | 17-25 | 90 | — | — | — | — | — | — | — | .030 | 10.060 | 10.000 | .030 |
| 74-73 F100 | 2 Slots | Skip | 17-25 | 90 | 12-20 | — | 1 | .003 | .0003 | 1.120 | — | .030 | 11.091 | 11.031 | .030 |
| 72 E100 | ⅛ Turn | Skip | 22-25 | 90 | — | — | — | — | — | — | — | .030 | 11.091 | 11.031 | .030 |
| 82-77 F250 (6900 GVW std.) | — | — | — | — | — | — | — | — | — | — | — | — | 12.060 | 12.000 | — |
| 76-75 F250 (6900 GVW std.) | — | — | — | — | — | — | — | — | — | — | — | — | — | — | — |
| w/12" rear brakes | 2 Slots | Skip | 17-25 | 90 | 12-20 | — | 1 | .003 | .0007 | 1.120 | — | .030 | 12.060 | 12.000 | .030 |
| w/12⅛" rear brakes | 2 Slots | Skip | 17-25 | 90 | 12-20 | — | 1 | .003 | .0007 | 1.120 | — | .030 | 12.185 | 12.125 | .030 |

| VEHICLE | BRAKE SHOE | BRAKE DRUM | | BRAKE PAD | BRAKE ROTOR | | | | CALIPER | | | WHEEL | WHEEL BEARING SETTING | | |
|---|---|---|---|---|---|---|---|---|---|---|---|---|---|---|---|
| | | DIAMETER | | | MIN. THICKNESS | | | | | | | | STEP 1 | STEP 2 | STEP 3 |
| YEAR, MAKE AND MODEL | * Minimum Lining Thickness | Standard Size | Machine To | Minimum Lining Thickness | Machine To | Discard At | Variation From Paral-lelism | Runout T.I.R. | DESIGN | Mounting Bolts Torque (ft-lbs) | Bridge, Pin or Key Bolts Torque (ft-lbs) | Lugs or Nuts Torque (ft-lbs) | Tighten Spindle Nut (ft-lbs) | Back Off Retorque (in-lbs) | Lock, or Back Off and Lock |
| **74-73** F250 (6900 GVW std.) | | | | | | | | | | | | | | | |
| w/12″ rear brakes | .030 | 12.000 | 12.060 | .030 | — | 1.120 | .0007 | .003 | 1 | — | 12-20 | 90 | 17-25 | Skip | 2 Slots |
| w/12⅛″ rear brakes | .030 | 12.125 | 12.185 | .030 | — | 1.120 | .0007 | .003 | 1 | — | 12-20 | 90 | 17-25 | Skip | 2 Slots |
| **72** F250 (Std.) w/12″ brakes | .030 | 12.000 | 12.060 | — | — | — | — | — | — | — | — | 90 | 17-25 | Skip | 2 Slots |
| w/12⅛″ brakes | .030 | 12.125 | 12.185 | — | — | — | — | — | — | — | — | 90 | 17-25 | Skip | 2 Slots |
| **82-78** F250 (6900 GVW H.D.), F350, E250, E350 | .030 | 12.000 | 12.060 | .030 | — | 1.180 | .0007 | .003 | 19 | 74-102 | 12-20◆ | 90■ | 22-25 | Skip | ⅛ Turn |
| **77** F250 (6900 GVW H.D.), F350, E250, E350 | .030 | 12.000 | 12.060 | .030 | — | 1.214 | .0007 | .003 | 19 | 74-102 | 12-20◆ | 90■ | 22-25 | Skip | ⅛ Turn |
| **76** F250 (6900 H.D.), F350, E250, E350 | | | | | | | | | | | | | | | |
| w/12″ rear brakes | .030 | 12.000 | 12.060 | .030 | — | 1.180 | .0007 | .003 | 19 | 74-102 | 12-20◆ | 90■ | 17-25 | Skip | 2 Slots |
| w/12⅛″ rear brakes | .030 | 12.125 | 12.185 | .030 | — | 1.180 | .0007 | .003 | 19 | 74-102 | 12-20◆ | 90■ | 17-25 | Skip | 2 Slots |
| **75-72** F250 (6900 GVW H.D.), F350 | | | | | | | | | | | | | | | |
| w/12″ rear brakes | .030 | 12.000 | 12.060 | .030 | — | .940 | .001 | .003 | 23 | 55-72 | 17-23◆ | 90■ | 17-25 | Skip | 2 Slots |
| w/12⅛″ rear brakes | .030 | 12.125 | 12.185 | .030 | — | .940 | .001 | .003 | 23 | 55-72 | 17-23◆ | 90■ | 17-25 | Skip | 2 Slots |
| **75** E250, E350 | .030 | 12.000 | 12.060 | .030 | — | .940 | .001 | .003 | 23 | 55-72 | 17-23◆ | 90■ | 17-25 | Skip | 2 Slots |
| **74-72** E250 | .030 | 11.031 | 11.091 | — | — | — | — | — | — | — | — | 90 | 17-25 | Skip | 2 Slots |
| **74-72** E350 | .030 | 12.000 | 12.060 | — | — | — | — | — | — | — | — | 135■ | 17-25 | Skip | 2 Slots |
| **82-81** F100, F150 (4 x 4) | .030 | 11.031 | 11.091 | .030 | — | 1.120 | .0007 | .003 | 1 | 74-102 | 12-20 | 90 | 50 | Skip | 45° |
| **80-76** F100, F150 (4 x 4) | .030 | 11.031 | 11.091 | .030 | — | 1.120 | .0007 | .003 | 1 | 50-60 | 12-20 | 90 | 50 | Skip | 90° |
| **75-72** F100 (4 x 4) Front | .030 | 11.000 | 11.060 | — | — | — | — | — | — | — | — | 90 | 50 | Skip | 90° |
| Rear | .030 | 11.031 | 11.091 | — | — | — | — | — | — | — | — | 90 | 50 | Skip | 90° |
| **82-81** F250 (4 x 4) | .030 | 12.000 | 12.060 | .030 | — | 1.180 | .0007 | .003 | 19 | 74-102 | 12-20◆ | 90 | 50 | Skip | 90° |
| **82-81** F350 (4 x 4) | .030 | 12.000 | 12.060 | .030 | — | 1.180 | .0007 | .003 | 19 | 74-102 | 12-20◆ | 90■ | 50 | 372-468 | 135°/150° |
| **80-76** F250 (4 x 4) F350 (4 x 4) | .030 | 12.000 | 12.060 | .030 | — | 1.180 | .0007 | .003 | 19 | 74-102 | 12-20◆ | 90■ | 50 | Skip | 90° |
| **75-72** F250 (4 x 4) Front | .030 | 12.125 | 12.185 | — | — | — | — | — | — | — | — | 90 | 50 | Skip | 90° |
| Rear | .030 | 12.000 | 12.060 | — | — | — | — | — | — | — | — | 90 | 50 | Skip | 90° |
| **75-72** F350 (4 x 4) | .030 | 12.000 | 12.060 | .030 | — | .940 | .001 | .003 | 23 | 55-72 | 17-23◆ | 135■ | 50 | Skip | 90° |

◆ Caliper bridge bolt 155 - 185 ft/lbs. ■ 350 w/single rear whls. (82 - 78) 145 ft/lbs., (77 - 72) 135 ft/lbs., w/dual rear whls. (82 -78) 220 ft/lbs. (77 - 72) 210 ft/lbs.

| YEAR, MAKE AND MODEL | * Minimum Lining Thickness | Standard Size | Machine To | Minimum Lining Thickness | Machine To | Discard At | Variation From Paral-lelism | Runout T.I.R. | DESIGN | Mounting Bolts Torque (ft-lbs) | Bridge, Pin or Key Bolts Torque (ft-lbs) | Lugs or Nuts Torque (ft-lbs) | Tighten Spindle Nut (ft-lbs) | Back Off Retorque (in-lbs) | Lock, or Back Off and Lock |
|---|---|---|---|---|---|---|---|---|---|---|---|---|---|---|---|
| **GENERAL MOTORS CORP.** | | | | | | | | | | | | | | | |
| **82** S10/15 | .062 | 9.500 | 9.560 | .030 | .980 | .965 | .0005 | .004 | 4 | — | 21-35 | 90 | 12 | Handtight | ½ Flat |
| **82-76** Blazer, Jimmy (4 x 2) | .062 | 11.150 | 11.210 | .030 | 1.230 | 1.215 | .0005 | .004 | 4 | — | 35 | 75-100 | 12 | Handtight | ½ Flat |
| (4 x 4) | .062 | 11.150 | 11.210 | .030 | 1.230 | 1.215 | .0005 | .004 | 4 | — | 35 | 70-90 | 50 | 420 | ⅜ Turn |
| **75-74** Blazer, Jimmy (4 x 2) | .030 | 11.000 | 11.060 | .030 | 1.230 | 1.215 | .0005 | .005 | 4 | — | 35 | 75-100 | 12 | Handtight | ½ Flat |
| w/11″ rear brake (4 x 4) | .030 | 11.000 | 11.060 | .030 | 1.230 | 1.215 | .0005 | .005 | 4 | — | 35 | 70-90 | 50 | 420 | ¼ Turn |
| w/11⅛″ rear brake (4 x 4) | .030 | 11.150 | 11.210 | .030 | 1.230 | 1.215 | .0005 | .005 | 4 | — | 35 | 70-90 | 50 | 420 | ¼ Turn |
| **73-72** Blazer, Jimmy (4 x 2) | .030 | 11.000 | 11.060 | .030 | 1.230 | 1.215 | .0005 | .005 | 4 | — | 35 | 65-90 | Snug | Skip | ¼ Turn |
| (4 x 4) | .030 | 11.000 | 11.060 | .030 | 1.230 | 1.215 | .0005 | .005 | 4 | — | 35 | 55-75 | 50 | 420 | ¼ Turn |

| | | | | | | | | | | | | | | | |
|---|---|---|---|---|---|---|---|---|---|---|---|---|---|---|---|
| **82-76** Suburban, C,G, 10/15 | | | | | | | | | | | | | | | |
| w/1.0" rotor: | | | | | | | | | | | | | | | |
| w/11" rear brake | .062 | 11.000 | 11.060 | .030 | .980 | .965 | .0005 | .004 | 4 | — | 35 | 75-100 | 12 | Handtight | ½ Flat |
| w/11⅛" rear brake | .062 | 11.150 | 11.210 | .030 | .980 | .965 | .0005 | .004 | 4 | — | 35 | 75-100 | 12 | Handtight | ½ Flat |
| w/1.25" rotor: | | | | | | | | | | | | | | | |
| w/11" rear brakes | .062 | 11.000 | 11.060 | .030 | 1.230 | 1.215 | .0005 | .004 | 4 | — | 35 | 75-100 | 12 | Handtight | ½ Flat |
| w/11⅛" rear brakes | .062 | 11.150 | 11.210 | .030 | 1.230 | 1.215 | .0005 | .004 | 4 | — | 35 | 75-100 | 12 | Handtight | ½ Flat |
| **75-74** Suburban, C, G 10/15 | | | | | | | | | | | | | | | |
| w/11" rear brakes | .030 | 11.000 | 11.060 | .030 | 1.230 | 1.215 | .0005 | .005 | 4 | — | 35 | 75-100 | 12 | Handtight | ½ Flat |
| w/11⅛" rear brakes | .030 | 11.150 | 11.210 | .030 | 1.230 | 1.215 | .0005 | .005 | 4 | — | 35 | 75-100 | 12 | Handtight | ½ Flat |
| **73-72** Suburban, C, G10/15 | .030 | 11.000 | 11.060 | .030 | 1.230 | 1.215 | .0005 | .005 | 4 | — | 35 | 65-90 | Snug | Skip | ¼ Turn |
| **82-76** Suburban, C,G, 20/25, 30/35 under 8600 GVW | | | | | | | | | | | | | | | |
| w/11⅛" rear brakes | .062 | 11.150 | 11.210 | .030 | 1.230 | 1.215 | .0005 | .004 | 4 | — | 35 | 90-120 ▲ | 12 | Handtight | ½ Flat |
| w/13" rear brakes | .062 | 13.000 | 13.060 | .030 | 1.230 | 1.215 | .0005 | .004 | 4 | — | 35 | 90-120 ▲ | 12 | Handtight | ½ Flat |
| **82-76** C,G, 30/35 over 8600 GVW | .062 | 13.000 | 13.060 | .030 | 1.480 | 1.465 | .0005 | .004 | 1 | — | 15 | 90-120 ● | 12 | Handtight | ½ Flat |
| **75-74** Suburban, C, G 20/25, 30/35 under 8600 GVW | | | | | | | | | | | | | | | |
| w/11⅛" rear brakes | .030 | 11.150 | 11.210 | .030 | 1.230 | 1.215 | .0005 | .005 | 4 | — | 35 | 90-120 ▲ | 12 | Handtight | ½ Flat |
| w/13" rear brakes | .030 | 13.000 | 13.060 | .030 | 1.230 | 1.215 | .0005 | .005 | 4 | — | 35 | 90-120 ▲ | 12 | Handtight | ½ Flat |
| **75-74** C, G 30/35 over 8600 GVW | .030 | 13.000 | 13.060 | .030 | 1.480 | 1.465 | .0005 | .005 | 1 | — | 15 | 90-120 | 12 | Handtight | ½ Flat |
| **73-72** C 20/25 w/11⅛" rear brakes | .030 | 11.150 | 11.210 | .030 | 1.230 | 1.215 | .0005 | .005 | 4 | — | 35 | 90-120 | Snug | Skip | ¼ Turn |
| w/12" rear brakes | .030 | 12.000 | 12.060 | .030 | 1.230 | 1.215 | .0005 | .005 | 4 | — | 35 | 90-120 | Snug | Skip | ¼ Turn |
| w/13" rear brake | .030 | 13.000 | 13.060 | .030 | 1.230 | 1.215 | .0005 | .005 | 4 | — | 35 | 90-120 | Snug | Skip | ¼ Turn |
| **73-72** G 20/25 | .030 | 11.000 | 11.060 | .030 | 1.230 | 1.215 | .0005 | .005 | 4 | — | 35 | 75-100 | Snug | Skip | ¼ Turn |
| **73-72** C, G 30/35 w/13" rear brakes | .030 | 13.000 | 13.060 | .030 | 1.230 | 1.215 | .0005 | .005 | 4 | — | 35 | 90-120 | Snug | Skip | ¼ Turn |
| w/15" rear brakes | .030 | 15.000 | 15.060 | .030 | 1.230 | 1.215 | .0005 | .005 | 4 | — | 35 | 90-120 | Snug | Skip | ¼ Turn |
| **82-76** K 10/15, 20/25 | | | | | | | | | | | | | | | |
| w/11⅛" rear brakes | .062 | 11.150 | 11.210 | .030 | 1.230 | 1.215 | .0005 | .004 | 4 | — | 35 | 70-90 ■ | 50 | 420 | ⅜ Turn |
| w/13" rear brakes | .062 | 13.000 | 13.060 | .030 | 1.230 | 1.215 | .0005 | .004 | 4 | — | 35 | 70-90 ■ | 50 | 420 | ⅜ Turn |
| **75-74** K 10/15, 20/25 | | | | | | | | | | | | | | | |
| w/11" rear brakes | .030 | 11.000 | 11.060 | .030 | 1.230 | 1.215 | .0005 | .005 | 4 | — | 35 | 90-120 ▲ | 50 | 420 | ¼ Turn |
| w/11⅛" rear brakes | .030 | 11.150 | 11.210 | .030 | 1.230 | 1.215 | .0005 | .005 | 4 | — | 35 | 90-120 ▲ | 50 | 420 | ¼ Turn |
| w/13" rear brakes | .030 | 13.000 | 13.060 | .030 | 1.230 | 1.215 | .0005 | .005 | 4 | — | 35 | 90-120 ▲ | 50 | 420 | ¼ Turn |
| **73-72** K10/15, 20/25 | | | | | | | | | | | | | | | |
| w/11" rear brake | .030 | 11.000 | 11.060 | .030 | 1.230 | 1.215 | .0005 | .005 | 4 | — | 35 | 55-75 | 50 | 420 | ¼ Turn |
| w/11⅛" rear brake | .030 | 11.150 | 11.210 | .030 | 1.230 | 1.215 | .0005 | .005 | 4 | — | 35 | 55-75 | 50 | 420 | ¼ Turn |
| w/13" rear brake | .030 | 13.000 | 13.060 | .030 | 1.230 | 1.215 | .0005 | .005 | 4 | — | 35 | 55-75 | 50 | 420 | ¼ Turn |
| **82-77** K 30/35 | .062 | 13.000 | 13.060 | .030 | 1.480 | 1.465 | .0005 | .004 | 4 | — | 35 | 90-120 ● | 50 | 420 | ⅜ Turn |

▲ G 20/25, K 10/15 is 75-100 ft/lbs. ● w/8 bolt whl. 110-140 ft/lbs., w/10 bolt whl. 130-180 ft/lbs. ■ K20/25 is 90-120 ft/lbs.

| | | | | | | | | | | | | | | | |
|---|---|---|---|---|---|---|---|---|---|---|---|---|---|---|---|
| **INTERNATIONAL HARVESTER CORP.** | | | | | | | | | | | | | | | |
| **80-72** Scout II | .030 | 11.000 | 11.060 | .125 | — | 1.120 | .0005 | .005 | 1 | — | 12-18 | 70-90 | 30 | Skip | ¼ Turn |
| **75-74** Travel All (½ Ton) 100/150 | .030 | 11.000 | 11.060 | .125 | — | 1.120 | .0005 | .005 | 1 | — | 12-18 | 70-90 | To Drag | Skip | To Slot |
| **75-74** Travel All (¾ Ton) 200 | .030 | 12.000 | 12.060 | .125 | — | 1.120 | .0005 | .005 | 1 | — | 12-18 | 70-90 | To Drag | Skip | To Slot |

# IMPORTED PASSENGER CARS AND LIGHT TRUCKS

**NOTICE:** Should state inspection regulations exceed manufacturer's specifications for lining or rotor/drum reserve, the state inspection regulation specification **must** be used.

| YEAR, MAKE AND MODEL | BRAKE SHOE Min. Lining Thickness * | BRAKE DRUM DIAMETER Standard Size | BRAKE DRUM DIAMETER Machine To | BRAKE PAD Minimum Lining Thickness | BRAKE ROTOR MIN. THICKNESS Machine To | BRAKE ROTOR MIN. THICKNESS Discard At | BRAKE ROTOR Variation From Parallelism | BRAKE ROTOR Runout T.I.R. | DESIGN | CALIPER Mounting Bolts Torque (ft-lbs) | CALIPER Bridge, Pin or Key Bolts Torque (ft-lbs) | WHEEL Lugs or Nuts Torque (ft-lbs) | WHEEL BEARING STEP 1 Tighten Spindle Nut (ft-lbs) | WHEEL BEARING STEP 2 Back Off Retorque (in-lbs) | WHEEL BEARING STEP 3 Lock or Back Off and Lock |
|---|---|---|---|---|---|---|---|---|---|---|---|---|---|---|---|
| **AUDI** | | | | | | | | | | | | | | | |
| 82-80 4000 Series | .097 | — | 7.894 | .079 | .413 | .393 | — | .002 | 37 | 36 | 25 | 65 | ■ | — | — |
| 82-81 5000 Series exc. Turbo | .098 | 9.005 | 9.094 | .079 | — | .787 | .0008 | .002 | 37 | 83 | 18 | 80 | ■ | — | — |
| 80-78 5000 Series exc. Turbo | .098 | 9.005 | 9.094 | .078 | .807 | .807 | .0008 | .004 | 39 | 83 | — | 80 | ■ | — | — |
| 82-80 5000 Turbo Front | — | — | 9.094 | .079 | .807 | .787 | .0008 | .002 | 31 | 83 | 25 | 80 | ■ | — | — |
| Rear | — | — | — | .079 | .335 | .315 | — | .002 | 33 | 47 | 25 | 80 | — | — | — |
| 79-78 Fox | .097 | 7.87 | 7.90 | .078 | .413 | .393 | — | .002 | 37 | 43 | — | 65 | ■ | — | — |
| 77-73 Fox | .097 | 7.87 | 7.90 | .078 | .413 | .393 | — | .004 | 39 | 43 | — | 65 | ■ | — | — |
| **BMW** | | | | | | | | | | | | | | | |
| 82 528e Front | — | — | — | .138 | .803 | — | .0008 | .008 | 16 | 89 | 14-18 | 72-80 | 22-24† | 12● | Step 2 |
| Rear | — | — | — | .138 | .331 | — | .0008 | .008 | 17 | 43-48 | 14-18 | 72-80 | — | — | Step 2 |
| 81-79 528i Front | — | — | — | .080 | .846 | .827 | .0008 | .008 | 16 | — | — | 59-65 | 22-24† | 12● | Step 2 |
| Rear | — | — | — | .080 | .354 | .335 | .0008 | .008 | 17 | — | — | 59-65 | — | — | Step 2 |
| 82-79 633CSi Front | — | — | — | .080 | .846 | .827 | .0008 | .008 | 16 | — | — | 59-65 | 22-24† | 12● | Step 2 |
| Rear | — | — | — | .080 | .728 | .709 | .0008 | .008 | 17 | — | — | 59-65 | — | — | Step 2 |
| 82-78 733i Front | — | — | — | .080 | .846 | .827 | .0008 | .006 | 16 | 59-70 | — | 60-66 | 22-24† | 12● | Step 2 |
| Rear | — | — | — | .080 | .374 | .354 | .0008 | .006 | 17 | 44-49 | — | 60-66 | — | — | Step 2 |
| 82 320i w/ATE | .118 | 9.842 | 9.882 | .118 | .846 | .827 | .0008 | .008 | 17 | 58-69 | — | 59-65 | 22-24† | 12● | Step 2 |
| w/Girling | .118 | 9.842 | 9.882 | .118 | .480 | .461 | .0008 | .008 | 14 | 58-69 | — | 59-65 | 22-24† | 12● | Step 2 |
| 81-77 320i | .118 | 9.842 | 9.882 | .118 | .846 | .827 | .0008 | .008 | 17 | 58-69 | 33-40 | 59-65 | 22-24 | 12 | Step 2 |
| 78-77 530i Front | — | — | — | .080 | .846 | .827 | .0008 | .008 | 16 | 44-48 | 16-19 | 59-65 | 22-24 | 12 | Step 2 |
| Rear | — | — | — | .080 | .480 | .461 | .0008 | .008 | 17 | 58-69 | — | 59-65 | — | — | Step 2 |
| 79-77 630CSi Front | — | — | — | .080 | .846 | .827 | .0008 | .008 | 16 | 44-48 | — | 59-65 | 22-24 | 12 | Step 2 |
| Rear | — | — | — | .080 | .728 | .709 | .0008 | .008 | 17 | 58-69 | — | 59-65 | — | — | Step 2 |
| 76-72 2002 | .120 | 9.060 | 9.100 | .080 | .354 | .354 | .0008 | .008 | 16 | 58-69 | 16-19 | 59-65 | 22-24 | 12 | Step 2 |
| 74-72 2002tii | .120 | 9.060 | 9.100 | .080 | .459 | — | .0008 | .008 | 16 | 58-69 | 33-40 | 59-65 | 22-24 | 12 | Step 2 |
| **BUICK OPEL** | | | | | | | | | | | | | | | |
| 79-76 | .040 | 9.000 | 9.040 | .067 | — | .339 | .0006 | .006 | 14 | 36 | — | 50 | 21 | Handtight | Step 2 |
| 75 | .040 | 9.000 | 9.040 | .067 | .465 | .465 | .0004 | .004 | 17 | 72 | — | 65 | 18 | Skip | ¼ Turn |
| 74-72 | * | 9.060 | 9.090 | .067 | .404 | .394 | .0006 | .006 | 17 | 72 | — | 65 | 18 | Skip | ¼ Turn |

■ Seat brg. while turning wheel, back off nut until thrust washer can be moved slightly by screwdriver w/finger pressure. lock

† Keep nut stationary and tighten bearing cover two full turns.

● Loosen nut and retighten a maximum of 24 ft/lbs.

* .030" over rivet head; if bonded lining use .062".

## DODGE — COLT, CHALLENGER, D-50 PICKUP

| Model | | | | | | | | | | | | | | | | |
|---|---|---|---|---|---|---|---|---|---|---|---|---|---|---|---|---|
| 82-79 Colt (FWD) | .040 | 7.100 | 7.200 | .040 | | | | | .006 | 18 | 43-58 | — | 51-58 | 14 | 48 | Step 2 |
| 80-78 Colt exc. Wagon | .040 | 9.000 | 9.050 | .080 | | | | | .006 | 18 | 51-65 | — | ■ 51-58 | 14.5 | 43 | Step 2 |
| 80-78 Colt Wagon | — | — | — | — | | | | | — | — | — | — | — | — | — | — |
|   w/rear drum brakes | .040 | 9.000 | 9.050 | .040 | | | | | .006 | 34 | 29-36 | — | ■ 51-58 | 14.5 | 43 | — |
|   w/rear disc brakes | — | — | — | .040 | | | | | .006 | 27 | 51-65 ◄ | — | ■ 51-58 | 14.5 | 43 | Step 2 |
| 77-76 Colt | .040 | 9.000 | 9.050 | .080 | | | | | .006 | 18 | 51-65 ◄ | — | ■ 51-58 | 14.5 | 43 | Step 2 |
| 75-73 Colt | .040 | 9.000 | 9.050 | .080 | | | | | .006 | 18 | 51-65 ◄ | — | ■ 51-58 | 14.5 | 43 | Step 2 |
| 72 Colt | .040 | 9.000 | 9.050 | .080 | | | | | .006 | 14 | 30-36 | — | ■ 51-58 | 14.5 | 43 | Step 2 |
| 82-78 Challenger | | | | | | | | | | | | | | | | |
|   w/rear drum brakes | .040 | 9.000 | 9.050 | .040 | | | | | .006 | 34 | 51-65 | — | ■ 51-58 | 14.5 | 43 | Step 2 |
|   w/rear disc brakes | — | — | — | .040 | | | | | .006 | 27 | 29-36 | — | ■ 51-58 | 14.5 | — | — |
| 82-79 D-50 Pickup | .040 | 9.500 | 9.579 | .040 | | | | | .006 | 34 | 51-65 | — | 51-58 | 21.7 | 48 | Step 2 |

◄ Caliper to adapter shown, adapter to caliper 29-36 ft/lbs.  ■ w/aluminum wheels: - 58-72 ft/lbs.

## DATSUN

| Model | | | | | | | | | | | | | | | | |
|---|---|---|---|---|---|---|---|---|---|---|---|---|---|---|---|---|
| 82 Stanza | .059 | 8.000 | 8.050 | .080 | .630 | — | .630 | .0012 | .006 | 45 | 53-72 | 23-30 | 58-72 | 29-33 | Skip | ¼ Turn |
| 82-79 210 | .059 | 8.000 | 8.050 | .063 | .331 | — | .331 | .0012 | .005 | 38 | 53-72 | — | 58-72 | 22-25 | Skip | ¼ Turn |
| 78-74 B210 | .059 | 8.000 | 8.050 | .063 | .331 | — | .331 | .0012 | .005 | 38 | 53-72 | — | 58-72 | 18-22 | Skip | ¼ Turn |
| 73-72 1200 | .059 | 8.000 | 8.051 | .063 | .331 | — | .331 | .0015 | .005 | 38 | 33-41 | — | 60 | 22-25 | Skip | ¼ Turn |
| 82-79 310 | .059 | 8.000 | 8.050 | .079 | .339 | — | .339 | .0012 | .005 | 38 | 40-47 | 12-15 | 58-72 | 29-33 | Handtight | Step 2 |
| 78-76 F10 | .039 | 8.000 | 8.051 | .063 | .339 | — | .339 | — | .006 | 38 | 40-47 | — | 58-65 | 18-22 | Skip | ⅙ Turn |
| 81-78 510 | .059 | 9.000 | 9.055 | .080 | .331 | — | .331 | .0012 | .005 | 36 | 53-72 | 12-15 | 58-65 | 18-22 | Skip | ⅙ Turn |
| 73-72 510 | .059 | 9.000 | 9.055 | .040 | .331 | — | .331 | .0012 | .005 | 26 | 53-72 | — | 58-65 | 18-22 | Skip | ⅙ Turn |
| 76-75 610 | .059 | 9.000 | 9.055 | .063 | .331 | — | .331 | — | .005 | 38 | 53-72 | 12-15 | 58-65 | 18-22 | Skip | ⅙ Turn |
| 74-73 610 | .059 | 9.000 | 9.055 | .040 | .331 | — | .331 | .0012 | .005 | 26 | 53-72 | — | 58-65 | 18-22 | Skip | ⅙ Turn |
| 77-76 710 | .059 | 9.000 | 9.055 | .063 | .331 | — | .331 | .0012 | .006 | 36 | 53-72 | 12-15 | 58-65 | 22-25 | Skip | ⅙ Turn |
| 75-73 710 | .059 | 9.000 | 9.055 | .063 | .331 | .341 | .331 | .0012 | .006 | 30 | 53-72 | 16-23 | 58-72 | 18-22 | — | ⅙ Turn |
| 82 810 Maxima w/rear drum | .059 | 9.000 | 9.055 | .079 | .630 | — | .339 | .0028 | .006 | 46 | 53-72 | 12-15 | 58-72 | 18-22 | Skip | ⅙ Turn |
| 82 810 Maxima w/rear disc | .059 | — | — | .079 | .339 | — | .339 | .0028 | .006 | 30 | 28-38 | — | 58-72 | 18-22 | Skip | ⅙ Turn |
| 81 810 w/front disc | .059 | 9.000 | 9.055 | .080 | .413 | — | .413 | .0012 | .006 | 42 | 53-72 | 12-15 | 58-72 | 18-22 | Skip | ⅙ Turn |
| 81 810 w/rear disc | — | — | — | .079 | .413 | — | .413 | .0012 | .005 | 36 | 28-38 | 16-23 | 58-72 | 18-22 | Skip | ⅙ Turn |
| 80-77 810 | .059 | 9.000 | 9.055 | .079 | .339 | — | .339 | .0028 | .006 | 46 | 53-72 | 16-23 | 58-72 | 18-22 | Skip | — |
| 82 200SX front disc | — | — | — | .079 | .413 | — | .413 | .0028 | .005 | 36 | 28-38 | 12-15 | 58-72 | 18-22 | Skip | ⅙ Turn |
| 82 200SX rear disc | — | — | — | .063 | .339 | — | .339 | .0028 | .006 | 42 | 28-38 | — | 58-72 | — | — | ⅙ Turn |
| 81-80 200SX front disc | .059 | 9.000 | 9.055 | .060 | .331 | — | .331 | .0012 | .005 | 38 | 53-72 | — | 58-65 | 18-22 | Skip | ⅙ Turn |
| 81 200SX rear disc | .059 | 9.000 | 9.055 | .063 | .331 | — | .331 | — | .005 | 38 | 53-72 | — | 58-65 | 18-22 | Skip | ⅙ Turn |
| 82 280 ZX front disc | — | — | — | .080 | .709 | — | .709 | .0012 | .004 | 37 | 53-72 | 16-23 | 58-72 | 18-22 | Skip | — |
| 82 280 ZX rear disc | — | — | — | .080 | .339 | — | .339 | .0012 | .004 | 46 | 28-38 | 16-23 | 58-72 | 18-22 | Skip | ⅙ Turn |
| 81-79 280 ZX front disc | — | — | — | .080 | .709 | — | .709 | .0012 | .004 | 37 | 53-72 | 16-23 | 58-72 | 18-22 | Skip | — |
| 81-79 280 ZX rear disc | — | — | — | .080 | .339 | .423 | .339 | .0015 | .006 | 42 | 28-38 | — | 58-65 | 18-22 | Skip | ⅙ Turn |
| 78-72 280Z, 260Z, 240Z | .059 | 9.000 | 9.055 | .080 | .413 | — | .413 | .0012 | .004 | 14 | 53-72 | 12-15 | 58-65 | 18-22 | Skip | ⅙ Turn |
| 82 720 Pickup | .059 | 10.000 | 10.055 | .080 | .413 | — | .413 | .0012 | .006 | 36 | 53-72 | 12-15 | 87-108 | 25-29† | Skip | ⅙ Turn |
| 81 720 Pickup | .059 | 10.000 | 10.055 | .080 | .413 | — | .413 | .0028 | .006 | 36 | 53-72 | 12-15 | 87-108 | 25-29† | Skip | ⅙ Turn |

† 2WD only; 4WD 108-195 ft/lbs.

# BRAKES

| YEAR, MAKE AND MODEL | BRAKE SHOE Min. Lining Thickness * | BRAKE DRUM Standard Size | BRAKE DRUM Machine To | BRAKE PAD Min. Lining Thickness | ROTOR Min. Thick. Machine To | ROTOR Discard At | ROTOR Variation From Parallelism | ROTOR Runout T.I.R. | DESIGN | CALIPER Mounting Bolts (ft-lbs) | CALIPER Bridge, Pin or Key Bolts (ft-lbs) | WHEEL Lugs or Nuts Torque (ft-lbs) | WHEEL BEARING STEP 1 Tighten Spindle Nut (ft-lbs) | WHEEL BEARING STEP 2 Back Off Retorque (in-lbs) | WHEEL BEARING STEP 3 Lock, or Back Off and Lock |
|---|---|---|---|---|---|---|---|---|---|---|---|---|---|---|---|
| 80 720 Pickup | .059 | 10.000 | 10.055 | .080 | — | .413 | .0028 | .006 | 36 | 53-72 | 12-15 | 58-72 | 25-29† | Skip | 1/6 Turn |
| 79-78 620 Pickup | .059 | 10.000 | 10.055 | .080 | — | .413 | .0015 | .006 | 36 | 53-72 | 12-15 | 58-72 | 25-29† | Skip | 1/6 Turn |
| 77-72 620 Pickup | .059 | 10.000 | 10.055 | — | — | — | — | — | — | — | — | 58-65 | 22-25† | Skip | 1/6 Turn |

† 2WD only; 4WD 108-145 ft/lbs.

### FORD — CAPRI, COURIER, FIESTA

| YEAR, MAKE AND MODEL | BRAKE SHOE Min. Lining Thickness * | BRAKE DRUM Standard Size | BRAKE DRUM Machine To | BRAKE PAD Min. Lining Thickness | ROTOR Min. Thick. Machine To | ROTOR Discard At | ROTOR Variation From Parallelism | ROTOR Runout T.I.R. | DESIGN | CALIPER Mounting Bolts (ft-lbs) | CALIPER Bridge, Pin or Key Bolts (ft-lbs) | WHEEL Lugs or Nuts Torque (ft-lbs) | WHEEL BEARING STEP 1 (ft-lbs) | WHEEL BEARING STEP 2 (in-lbs) | WHEEL BEARING STEP 3 |
|---|---|---|---|---|---|---|---|---|---|---|---|---|---|---|---|
| 77-76 Capri | .030 | 9.000 | 9.050 | .100 | .460 | .450◆ | .0004 | .002 | 14 | 45-50 | — | 50-55 | 17-25 | ½ Turn | 12 in/lbs. |
| 74-72 Capri | .030 | 9.000 | 9.050 | .100 | — | .330 | — | .0035 | 14 | 40-50 | — | 50-55 | 17-25 | ½ Turn | 12 in/lbs. |
| 82-79 Courier | .039 | 10.236 | 10.244 | .276■ | .433 | — | .0005 | .004 | 35 | — | — | 58-65 | 17-25 | Skip | ¼ Turn |
| 78-77 Courier | .039 | 10.236 | 10.244 | .315■ | .433 | — | .0005 | .004 | 35 | — | — | 58-65 | 17-25 | Skip | ¼ Turn |
| 76-74 Courier | .039 | 10.236 | 10.244 | — | — | — | — | — | — | — | — | 58-65 | 17-25 | ½ Turn | 6-8ft/lbs. |
| 73-72 Courier | .039 | 10.236 | 10.244 | — | — | — | — | — | — | — | — | 58-65 | 17-25 | ½ Turn | 12 in/lbs. |
| 80-78 Fiesta | .060 | 7.000 | — | .060 | — | .340 | — | .006 | 22 | 38-45 | — | 63-85 | 15-18 | 180° | Handtight |

■ Measurement of shoe & lining   ◆ Machining not recommended   ▲ 74 is 85 ft/lbs.

### PLYMOUTH — ARROW, CHAMP, SAPPORO & ARROW PICKUP

| YEAR, MAKE AND MODEL | BRAKE SHOE Min. Lining Thickness * | BRAKE DRUM Standard Size | BRAKE DRUM Machine To | BRAKE PAD Min. Lining Thickness | ROTOR Min. Thick. Machine To | ROTOR Discard At | ROTOR Variation From Parallelism | ROTOR Runout T.I.R. | DESIGN | CALIPER Mounting Bolts (ft-lbs) | CALIPER Bridge, Pin or Key Bolts (ft-lbs) | WHEEL Lugs or Nuts Torque (ft-lbs) | WHEEL BEARING STEP 1 (ft-lbs) | WHEEL BEARING STEP 2 (in-lbs) | WHEEL BEARING STEP 3 |
|---|---|---|---|---|---|---|---|---|---|---|---|---|---|---|---|
| 80-76 Arrow w/rear drum brakes | .040 | 9.000 | 9.050 | .040 | — | .450 | — | .006 | 34 | 51-65 | — | 51-58■ | 14.5 | 43 | Step 2 |
| w/rear disc brakes | — | — | — | .040 | — | .330 | — | .006 | 27 | 29-36 | — | 51-58■ | — | — | — |
| 82-79 Arrow Pickup | .040 | 9.500 | 9.579 | .040 | — | .720 | — | .006 | 34 | 51-65 | — | 51-58 | 21.7 | 48 | Step 2 |
| 82-79 Champ (FWD) | .040 | 7.100 | 7.200 | .040 | — | .450 | — | .006 | 18 | 43-58 | — | 51-58■ | 14 | 48 | Step 2 |
| 82-78 Sapporo w/rear drum brakes | .040 | 9.000 | 9.050 | .040 | — | .430 | — | .006 | 34 | 51-65 | — | 51-58■ | 14.5 | 43 | Step 2 |
| w/rear disc brakes | — | — | — | .040 | — | .330 | — | .006 | 27 | 29-36 | — | 51-58 | — | — | — |

■ w/aluminum wheels; - 58-72 ft/lbs.

### FIAT

| YEAR, MAKE AND MODEL | BRAKE SHOE Min. Lining Thickness * | BRAKE DRUM Standard Size | BRAKE DRUM Machine To | BRAKE PAD Min. Lining Thickness | ROTOR Min. Thick. Machine To | ROTOR Discard At | ROTOR Variation From Parallelism | ROTOR Runout T.I.R. | DESIGN | CALIPER Mounting Bolts (ft-lbs) | CALIPER Bridge, Pin or Key Bolts (ft-lbs) | WHEEL Lugs or Nuts Torque (ft-lbs) | WHEEL BEARING STEP 1 (ft-lbs) | WHEEL BEARING STEP 2 (in-lbs) | WHEEL BEARING STEP 3 |
|---|---|---|---|---|---|---|---|---|---|---|---|---|---|---|---|
| 82-79 Strada | .060 | 7.293 | 7.336 | .060 | .368 | .350 | — | .006 | 33 | — | — | 65 | NON ADJUSTABLE | | |
| 82-74 X1/9 Front | — | — | — | .080 | .368 | .354 | .002 | .006 | 43 | 36 | — | 51 | NON ADJUSTABLE | | |
| Rear | — | — | — | .080 | .368 | .354 | .002 | .006 | 27 | 36 | — | 51 | 14.5 | 60 | 1/3 Turn |
| 82-75 2000 & 124 Front | — | — | — | .080 | .368 | .354 | — | .006 | 22 | 59 | — | 65 | — | — | — |
| Rear | — | — | — | .080 | .372 | .354 | — | .006 | 27 | 55 | — | 65 | 14.5 | 60 | 1/3 Turn |
| 74-72 124 Front | — | — | — | .080 | .372 | .354 | — | .006 | 22 | 36 | — | 51 | 14.5 | 60 | 1/3 Turn |
| Rear | — | — | — | .080 | — | .354 | — | .006 | 27 | 40 | — | 51 | 14.5 | 60 | 1/2 Flat |
| 81-79 Brava | .181 | 9.000 | 9.030 | .060 | .368 | .386 | .002 | .004 | 22 | 50 | — | 65 | 14.5 | 60 | 1/3 Turn |
| 78-75 131 Series | .181 | 9.000 | 9.030 | .060 | .368 | .354 | .002 | .006 | 22 | 50 | — | 65 | 14.5 | 60 | 1/3 Turn |
| 79-73 128 | .060 | 7.300 | 7.332 | .080 | .368 | .354 | .002 | .006 | 22 | 36 | — | 51 | NON ADJUSTABLE | | |

### HONDA

| Vehicle | | | | | | | | | | | | | | | |
|---|---|---|---|---|---|---|---|---|---|---|---|---|---|---|---|
| 82 Civic Hatchback w/1300 4 SPD | .079 | 7.090 | 7.130 | .063 | .350 | — | .0006 | .006 | 37 | 56 | 20 | 58 | 18 | 48 | Step 2 |
| exc. 1300 4 SPD | .079 | 7.090 | 7.130 | .063 | .390 | — | .0006 | .006 | 37 | 56 | 20 | 58 | 18 | 48 | Step 2 |
| 81-80 Civic CVCC Hatchback | .079 | 7.090 | 7.130 | .063 | .350 | .343 | .0006 | .006 | 37 | 56 | 20 | 58 | 18 | 48 | Step 2 |
| 79-75 Civic, CVCC | .079 | 7.080 | 7.130 | .063 | .354 | .343 | .0006 | .006 | 38 | 36-43 | — | 51-65 | NON ADJUSTABLE | | Step 2 |
| 74-73 Civic | .079 | 7.080 | 7.130 | .063 | .354 | — | .0006 | .006 | 38 | 36-43 | — | 51-65 | NON ADJUSTABLE | | Step 2 |
| 82-81 Civic Sedan | .079 | 7.090 | 7.130 | .063 | .390 | .437 | .0006 | .006 | 37 | 56 | 20 | 58 | 18 | 48 | Step 2 |
| 82-80 Civic Wagon | .079 | 7.870 | 7.910 | .063 | .390 | — | .0006 | .006 | 22 | 56 | — | 58 | 18 | 48 | Step 2 |
| 79-76 Civic Wagon | .079 | 7.870 | 7.910 | .063 | .449 | .437 | .0006 | .006 | 18 | 63 | 63 | 51-65 | NON ADJUSTABLE | | Step 2 |
| 82 Accord | .079 | 7.080 | 7.130 | .063 | .600 | — | .0006 | .006 | 37 | 56 | 20 | 80 | 18 | 48 | Step 2 |
| 81-79 Accord | .079 | 7.080 | 7.130 | .063 | .413 | — | .0006 | .006 | 22 | 56 | — | 58 | 18 | 36 | Step 2 |
| 78-76 Accord | .079 | 7.080 | 7.130 | .063 | .449 | .437 | .0006 | .006 | 22 | 58-66 | — | 50-65 | 18 | 36 | Step 2 |
| 82-79 Prelude | .079 | 7.080 | 7.130 | .063 | .413 | — | .0006 | .006 | 37 | 56 | 13 | 80 | 18 | 36 | Step 2 |

### ISUZU

| Vehicle | | | | | | | | | | | | | | | |
|---|---|---|---|---|---|---|---|---|---|---|---|---|---|---|---|
| 1982-81 I-Mark | .039 | 9.000 | 9.040 | .067 | — | .338 | — | .006 | 49 | 36 | — | 50† | 22 | Handtight | Step 2 |
| 1982-81 Pickup | .039 | 10.000 | 10.039 | .039 | .453 | .437 | .003 | .005 | 47 | 64 | 15 | 65 | 22 | Handtight | 1.8-2.6▲ |

† w/aluminum whls. 90 ft/lbs.   ▲ Radial pull in lbs. at lug nuts.

### LUV — CHEVROLET

| Vehicle | | | | | | | | | | | | | | | |
|---|---|---|---|---|---|---|---|---|---|---|---|---|---|---|---|
| 82-81 | .059 | 10.000 | 10.059 | .236 | .668 | .653 | .003 | .005 | 47 | 64 | 15 | 65 | 22■ | Handtight | Step 2 |
| 80-76 | .059 | 10.000 | 10.059 | .236 | .668 | .653 | .003 | .005 | 22 | 64 | — | 65 | 22■ | Handtight | Step 2 |
| 75-72 | .059 | 10.000 | 10.059 | — | — | — | — | — | — | — | — | 65 | 22 | Handtight | Step 2 |

■ Wheel bearing adjustment for 2 WD only

### MAZDA

| Vehicle | | | | | | | | | | | | | | | |
|---|---|---|---|---|---|---|---|---|---|---|---|---|---|---|---|
| 82-77 GLC | .040 | 7.874 | 7.914 | .040 | .472 | — | .003 | — | 35 | 33-40 | — | 65-80 | 14-18 | 1.0-1.3▲ | Step 2 |
| 82-79 626 | .040 | 9.000 | 9.040 | .040 | .472 | — | .004 | — | 37 | 33-40 | — | 65-80 | 14-18 | .8-1.9▲ | Step 2 |
| 82-80 RX7 | .040 | 7.874 | 7.914 | .040 | .670 | — | .004 | — | 32 | — | — | 65-80 | SEAT | 1.0-1.4▲ | — |
| w/rear disc | .040 | — | — | .040 | .354 | — | .004 | — | 32 | — | — | 65-80 | — | — | Step 2 |
| 79 RX7 | .040 | 7.874 | 7.914 | .040 | .670 | — | .004 | — | 35 | 33-40 | — | 65-80 | SEAT | 1.0-1.4▲ | Step 2 |
| 78-76 Cosmo Front | — | — | — | .276■ | .669 | — | .004 | — | 35 | 40-47 | — | 65-72 | SEAT | .9-2.2▲ | Step 2 |
| Rear | — | — | — | .276■ | .354 | — | .004 | — | 35 | 40-47 | — | 65-72 | SEAT | .9-2.2▲ | Step 2 |
| 77-74 808 | .040 | 7.874 | 7.914 | .256■ | .394 | — | .004 | — | 18 | — | — | 65-72 | SEAT | .9-2.2▲ | Step 2 |
| 73-72 808 | .040 | 7.874 | 7.914 | .256■ | .394 | — | .003 | — | 18 | — | — | 65 | SEAT | .9-2.2▲ | Step 2 |
| 78-74 RX4 | .040 | 9.000 | 9.040 | .276■ | .433 | — | .004 | — | 35 | 36-40 | — | 65-72 | SEAT | .9-2.2▲ | Step 2 |
| 77-75 RX3 | .040 | 7.874 | 7.914 | .256■ | .394 | — | .004 | — | 18 | — | — | 65-72 | SEAT | .9-2.2▲ | Step 2 |
| 73-72 RX3 | .040 | 7.874 | 7.914 | .256■ | .394 | — | .003 | — | 18 | — | — | 65 | SEAT | .9-2.2▲ | Step 2 |
| 73-72 RX2 | .040 | 7.874 | 7.914 | .276■ | .433 | — | .003 | — | 18 | — | — | 65 | SEAT | .9-2.2▲ | Step 2 |

■ Measurement of shoe & lining   ▲ Radial pull in lbs. at lug nuts.

27

# BRAKES

| Year, Make and Model | Brake Shoe Min. Lining Thickness * | Brake Drum Standard Size | Brake Drum Machine To | Brake Pad Min. Lining Thickness | Rotor Machine To | Rotor Discard At | Rotor Variation From Parallelism | Rotor Runout T.I.R. | Design | Caliper Mounting Bolts Torque (ft-lbs) | Caliper Bridge Pin or Key Bolts Torque (ft-lbs) | Wheel Lugs or Nuts Torque (ft-lbs) | WB Step 1 Tighten Spindle Nut (ft-lbs) | WB Step 2 Back Off Retorque (in-lbs) | WB Step 3 Lock or Back Off and Lock |
|---|---|---|---|---|---|---|---|---|---|---|---|---|---|---|---|
| 82 B2200 Pickup | .040 | 10.236 | 10.275 | .276 ■ | — | .748 | — | .004 | 35 | 40-47 | — | 58-65 | SEAT | 1.3-2.4▲ | Step 2 |
| 82-81 B2000 Pickup | .040 | 10.236 | 10.275 | .276 ■ | — | .433 | — | .004 | 35 | 40-47 | — | 58-65 | SEAT | 1.3-2.4▲ | Step 2 |
| 80-77 B1800/B2000 Pickup | .040 | 10.236 | 10.275 | .276 ■ | — | .433 | — | .004 | 35 | 40-47 | — | 58-65 | SEAT | 1.3-2.4▲ | Step 2 |
| 76-72 B1600 Pickup | .040 | 10.236 | 10.275 | — | — | — | — | — | — | — | — | 58-65 | SEAT | 1.3-2.4▲ | Step 2 |
| 77-74 Rotary Pickup | .040 | 10.236 | 10.275 | .276 ■ | — | .433 | — | .004 | 35 | — | — | 58-65 | SEAT | 1.3-2.4▲ | Step 2 |

■ Measurement of shoe & lining     Radial pull in lbs. at lug nuts.

## RENAULT

| Year, Make and Model | Brake Shoe Min. Lining Thickness | Brake Drum Standard Size | Brake Drum Machine To | Brake Pad Min. Lining Thickness | Rotor Machine To | Rotor Discard At | Rotor Variation From Parallelism | Rotor Runout T.I.R. | Design | Caliper Mounting Bolts Torque | Caliper Bridge Pin/Key Bolts Torque | Wheel Lugs/Nuts Torque | WB Step 1 | WB Step 2 | WB Step 3 |
|---|---|---|---|---|---|---|---|---|---|---|---|---|---|---|---|
| 82 Fuego | .020 | 8.996 | 9.035 | .276 | — | .354 | — | .003 | 29 | 74 | 42 | 59 | 22 | 1/6 Turn | ◄ |
| 82 18i | .020 | 8.996 | 9.035 | .276 | — | .354 | — | .003 | 29 | 74 | 42 | 59 | 22 | 1/6 Turn | ◄ |
| 81 18i | .020 | 8.996 | 9.035 | .276 | — | .354 | — | .003 | 29 | 74 | 42 | 59 | 22 | 1/6 Turn | ◄ |
| 80-74 Gordini | .203 ■ | 9.000 | 9.035 | .276 | — | .354 | .0004 | .008 | 43 | — | — | 45-60 | 20 | 1/4 Turn | ◄ |
| 82 LeCar | .203 ■ | 7.096 | 7.136 | .275 | — | .354 | — | .004 | 43 | 50 | — | 40-45 | 20 | 1/4 Turn | ◄◄ |
| 81-76 LeCar/5 Series | .203 ■ | 7.096 | 7.136 | .275 | — | .354 | — | .004 | 43 | 50 | — | 40-45 | 20 | 1/4 Turn | ◄◄ |

■ Measurement of shoe & lining   ▲ .001" - .002" end play   ▲ Radial pull in lbs. at lug nuts.

## SAAB

| Year, Make and Model | Brake Shoe Min. Lining Thickness | Brake Drum Standard Size | Brake Drum Machine To | Brake Pad Min. Lining Thickness | Rotor Machine To | Rotor Discard At | Rotor Variation From Parallelism | Rotor Runout T.I.R. | Design | Caliper Mounting Bolts Torque | Caliper Bridge Pin/Key Bolts Torque | Wheel Lugs/Nuts Torque | WB Step 1 | WB Step 2 | WB Step 3 |
|---|---|---|---|---|---|---|---|---|---|---|---|---|---|---|---|
| 82-79 900 Front | — | — | — | .040 | — | .461 | .0006 | .004 | 28 | — | — | 65-80 | NON | ADJUSTABLE | |
| Rear | — | — | — | .040 | — | .374 | .0006 | .004 | 17 | — | — | 65-80 | NON | ADJUSTABLE | |
| 78-75 99 Front | — | — | — | .080 | — | .461 | .0006 | .004 | 28 | — | — | 65-80 | NON | ADJUSTABLE | |
| 74-70 Front | — | — | — | .080 | — | .374 | .0006 | .004 | 17 | — | — | 65-80 | NON | ADJUSTABLE | |
| 78-70 99 Rear | — | — | — | .080 | — | .374 | .0006 | .004 | 17 | — | — | 65-80 | NON | ADJUSTABLE | |

■ Measurement of shoe & lining   ▲ Radial pull in lbs. at lug nuts.

## SUBARU

| Year, Make and Model | Brake Shoe Min. Lining Thickness | Brake Drum Standard Size | Brake Drum Machine To | Brake Pad Min. Lining Thickness | Rotor Machine To | Rotor Discard At | Rotor Variation From Parallelism | Rotor Runout T.I.R. | Design | Caliper Mounting Bolts Torque | Caliper Bridge Pin/Key Bolts Torque | Wheel Lugs/Nuts Torque | WB Step 1 | WB Step 2 | WB Step 3 |
|---|---|---|---|---|---|---|---|---|---|---|---|---|---|---|---|
| 82-80 All Models | .060 | 7.090 | 7.120 | .295 ■ | — | .394 | — | .004 | 48 | 36-51 | 33-54 | 58-72 | 36 | 1/8 Turn | 1.9-3.2▲ |
| 79-75 All Models | .060 | 7.090 | 7.120 | .060 | — | .330 | — | .006 | 27 | 36-51 | — | 58-72 | 36 | 1/8 Turn | 1.9-3.2▲ |
| 74-72 Sedan, Wagon Front | .060 | 9.000 | 9.040 | — | — | — | — | — | — | — | — | 40-54 | 36 | 1/8 Turn | 2.2-3.4▲ |
| Rear | .060 | 7.090 | 7.120 | .060 | — | — | — | — | — | — | — | 40-54 | 36 | 1/8 Turn | 2.2-3.4▲ |
| 74-72 Coupe | .060 | 7.090 | 7.120 | .060 | — | .330 | — | .006 | 27 | 36-51 | — | 40-54 | 36 | 1/8 Turn | 2.2-3.4▲ |

■ Measurement of shoe & lining   ▲ Radial pull in lbs. at lug nuts.

## TOYOTA

| Year, Make and Model | Brake Shoe Min. Lining Thickness | Brake Drum Standard Size | Brake Drum Machine To | Brake Pad Min. Lining Thickness | Rotor Machine To | Rotor Discard At | Rotor Variation From Parallelism | Rotor Runout T.I.R. | Design | Caliper Mounting Bolts Torque | Caliper Bridge Pin/Key Bolts Torque | Wheel Lugs/Nuts Torque | WB Step 1 | WB Step 2 | WB Step 3 |
|---|---|---|---|---|---|---|---|---|---|---|---|---|---|---|---|
| 82-80 Corolla | .040 | 9.000 | 9.040 | .040 | — | .453 | — | .006 | 32 | 40-54 | 58-68 | 66-86 | 21 | Handtight | .7-1.5 ■ |
| 79-75 Corolla 1600 | .040 | 9.000 | 9.040 | .040 | — | .350 | — | .006 | 22 | 40-54 | — | 65-86 | 21 | Handtight | .7-1.5 ■ |
| 74-72 Corolla 1600 | .040 | 9.000 | 9.040 | .040 | — | .350 | — | .006 | 22 | 48 | — | 70-75 | 21 | Handtight | .7-1.5 ■ |

■ Radial pull in lbs. at lug nuts.

| Vehicle | Lining | Drum Diam. | Drum Max | Disc Lining | Disc Min | | Runout | | | | | Wheel Nut | Torque | Bearing Adj. | End Play | ■ |
|---|---|---|---|---|---|---|---|---|---|---|---|---|---|---|---|---|
| 79-75 Corolla 1200 | .040 | 7.874 | 7.914 | .040 | .350 | — | — | .006 | 18 | 30-40 | — | 58-69 | 21 | Handtight | .7-1.5 | ■ |
| 74-72 Corolla 1200 | .040 | 7.874 | 7.914 | .040 | .350 | — | — | .006 | 22 | 40-54 | — | 65-86 | 21 | Handtight | .7-1.5 | ■ |
| 82-80 Tercel | .040 | 7.087 | — | .040 | .354 | .360 | — | .006 | 7 | 33-39 | 11-15 | 66-86 | 22 | Handtight | .8-1.9 | ■ |
| 82 Celica | .040 | 9.000 | 9.040 | .118 | .750 | — | — | .006 | 22 | 59-75 | 12-17 | 66-86 | 22 | Handtight | .8-1.9 | ■ |
| 81 Celica | .040 | 9.000 | 9.040 | .118 | .450 | — | — | .006 | 22 | 40-54 | — | 65-86 | 21 | Handtight | .7-1.5 | ■ |
| 80-77 Celica | .040 | 9.000 | 9.040 | .040 | .450 | — | — | .006 | 22 | 40-54 | — | 65-86 | 21 | Handtight | .7-1.5 | ■ |
| 76 Celica | .040 | 9.000 | 9.040 | .040 | .350 | — | — | .006 | 22 | 48 | — | 65-86 | 21 | Handtight | .7-1.5 | ■ |
| 75-72 Celica | .040 | 9.000 | 9.040 | .040 | .350 | — | — | .006 | 47 | 48 | — | 70-75 | 21 | Handtight | .7-1.5 | ■ |
| 82 Supra Front | — | — | — | .118 | .750 | — | — | .006 | 7 | 59-75 | 12-17 | 66-86 | 22 | Handtight | .8-1.9 | ■ |
| Rear | — | — | — | .118 | .669 | — | — | .006 | 22 | 30-39 | 12-17 | 66-86 | — | — | — | |
| 81-79 Supra Front | — | — | — | .118 | .450 | — | — | .006 | 27 | 40-54 | 11-15 | 65-86 | 21 | Handtight | .7-1.5 | ■ |
| Rear | .040 | 7.870 | 7.910 | .040 | .354 | — | — | .006 | 14 | 29-39 | — | 65-86 | 22 | Handtight | .8-1.9 | ■ |
| 82-81 Starlet | .040 | 9.000 | 9.040 | .040 | .354 | — | — | .006 | 14 | 67-87▲ | — | 65-86 | 21 | Handtight | .7-1.5 | ■ |
| 82-75 Corona | .040 | 9.000 | 9.040 | .040 | .450 | — | — | .006 | 22 | 67-87▲ | — | 65-86 | 21 | Handtight | .7-1.5 | ■ |
| 74 Corona deluxe | .040 | 9.000 | 9.050 | .040 | .450 | — | — | .006 | 22 | 53 | — | 65-86 | 21 | Handtight | .7-1.5 | ■ |
| 74 Std. Corona | .040 | 9.010 | 9.040 | .040 | .350 | — | — | .006 | 32 | 53 | — | 65-86 | 21 | Handtight | .7-1.5 | ■ |
| 73-72 Corona | .040 | 9.000 | 9.040 | .040 | .350 | — | — | .006 | 14 | 62-68 | — | 65-86 | 21 | Handtight | .8-2.2 | ■ |
| 82-81 Cressida | .040 | 9.000 | 9.040 | .040 | .669 | — | — | .006 | 14 | 67-87 | — | 65-86 | 21 | Handtight | .8-2.2 | ■ |
| 80-78 Cressida | .040 | 9.000 | 9.040 | .040 | .450 | — | — | .006 | 14 | 67-87 | — | 65-94 | 21 | Handtight | .7-1.5 | ■ |
| 76-73 Corona Mark II | .040 | 9.000 | 9.040 | .040 | .450 | — | — | .006 | 16 | 67-87 | — | 65-86 | 21 | Handtight | .7-1.5 | ■ |
| 73-72 Corona Mark II | .040 | 9.000 | 9.040 | .040 | .450 | — | — | .006 | — | 52-76 | — | 66-86 | 43 | 35-60 | 6.2-12.6 | ■ |
| 82-75 Land Cruiser | .060 | 11.610 | 11.650 | .060 | .748 | — | — | .005 | 14 | — | — | 65-87 | 43 | 35-60 | 3.9-5.0 | ■ |
| 74-72 Land Cruiser | .060 | 11.400 | 11.400 | — | — | — | — | .006 | — | 68-86 | — | 66-86 | 22 | Handtight | 1.3-3.8 | ■ |
| 82-75 Pickup 2 W.D. (4 x 2) | .040 | 10.000 | 10.060 | .040 | .453 | — | — | — | — | — | — | 65-87 | 22 | Handtight | 1.3-3.8 | ■ |
| 74-72 Pickup 2 W.D. (4 x 2) | .040 | 10.000 | 10.060 | — | — | — | — | — | — | — | — | — | — | — | — | |
| 72 Pickup 2 W.D. (4 x 2) | — | — | — | — | — | — | — | — | — | — | — | — | — | — | — | |
| w/9" Brakes | .040 | 9.000 | 9.060 | .040 | .453 | — | — | .006 | 16 | 55-75 | — | 65-87 | 22 | Handtight | 1.3-3.8 | ■ |
| 82-79 4 x 4 Pickup | .040 | 10.000 | 10.060 | .040 | .748 | — | — | .006 | 5 | 80-126 | 29-39 | 66-86 | 43 | 35-60 | 6.2-12.6 | ■ |
| 82-79 4 x 2 Cab & Chassis | .040 | 10.000 | 10.060 | .040 | — | — | — | — | — | — | — | 66-86 | 22 | Handtight | 1.3-3.8 | ■ |

**VOLKSWAGEN – F.W.D. VEHICLES EXCEPT VANAGON**

| Vehicle | Lining | Drum Diam. | Drum Max | Disc Lining | Disc Min | | Runout | | | | | Wheel Nut | Torque | Bearing | End Play | ■ |
|---|---|---|---|---|---|---|---|---|---|---|---|---|---|---|---|---|
| 82 Rabbit, Jetta, Scirocco | .098 | 7.086 | 7.105 | .250† | .413 | .393 | — | .002 | 8 | 50 | 30 | 65 | ■ | — | — | |
| 82 Rabbit Pickup | .098 | 7.894 | 7.894 | .250† | .413 | .393 | — | .002 | 8 | 50 | 30 | 65 | ■ | — | — | |
| 81-80 Rabbit, Jetta, Pickup w/Kelsey-Hayes Caliper | * | 7.086 | 7.105 | .080 | .413 | .393 | — | .004 | 8 | 36▲ | 30 | 65 | ■ | — | — | |
| 81-80 Rabbit, Jetta, Scirocco w/Girling Caliper | * | 7.086 | 7.105 | .080 | .413 | .393 | — | .004 | 38 | 36 | — | 65 | — | — | — | |
| 79-77 Rabbit w/drum brakes, Front | .040 | 9.059 | 9.079 | — | — | — | — | — | — | — | — | 65 | — | — | — | |
| Rear | * | 7.086 | 7.105 | — | — | — | — | — | — | — | — | 65 | — | — | — | |
| 79-75 Rabbit, Scirocco w/Girling Caliper | * | 7.086 | 7.105 | .080 | .452 | — | — | .004 | 38 | 36 | — | 65 | ■ | — | — | |
| 79-75 Rabbit, Scirocco w/Ate Caliper | * | 7.086 | 7.105 | .080 | .452 | .413 | .0008 | .004 | 39 | 43 | — | 65 | ■ | — | — | |
| 82 Quantum | .098 | 7.874 | 7.894 | .276† | .413 | .393 | — | .002 | 8 | 50 | 30 | 65 | ■ | — | — | |
| 81-73 Dasher | * | 7.850 | 7.900 | .080 | .413 | .393 | — | .004 | 37 | 43▲ | 25 | 65 | ■ | — | — | |
| 77 Dasher | * | 7.850 | 7.900 | .080 | .413 | .393 | .0008 | .004 | ▲ | 43 | — | 65 | ■ | — | — | |
| 76-74 Dasher | * | 7.850 | 7.900 | .080 | .433 | .413 | — | .004 | ▲ | 43 | — | 65 | ■ | — | — | |

■ Radial pull in lbs. at lug nuts.    ▲14" whl. shown, w/13" whl. 51-65 ft/lbs

■ Seat brg. while turning wheel, back off nut until thrust washer can be moved slightly by screwdriver w/finger pressure, lock.

■ Design 38 or 39    ▲ Measurement of shoe & lining    ▲ w/self locking bolt 50 ft/lbs

* .098 riveted: .059 bonded    † Measurement of shoe & lining

| VEHICLE — YEAR, MAKE AND MODEL | BRAKE SHOE Min. Lining Thickness * | BRAKE DRUM Standard Size | BRAKE DRUM Machine To | BRAKE PAD Minimum Lining Thickness | ROTOR Machine To | ROTOR Discard At | ROTOR Variation From Parallelism | ROTOR Runout T.I.R. | DESIGN | CALIPER Mounting Bolts Torque (ft-lbs) | CALIPER Bridge Pin or Key Bolts Torque (ft-lbs) | WHEEL Lugs or Nuts Torque (ft-lbs) | STEP 1 Tighten Spindle Nut (ft-lbs) | STEP 2 Back Off Retorque (in-lbs) | STEP 3 Lock. or Back Off and Lock |
|---|---|---|---|---|---|---|---|---|---|---|---|---|---|---|---|
| **VOLKSWAGEN — REAR DRIVE VEHICLES AND VANAGON** | | | | | | | | | | | | | | | |
| **TYPE I — Beetle, Super Beetle, Karman Ghia** | | | | | | | | | | | | | | | |
| 79-72 Super Beetle, Front | .100 | 9.768 | 9.803 | — | — | — | — | — | — | — | — | 87-94 | Snug-Up | Handtight | Step 2 |
| Rear | .100 | 9.055 | 9.094 | — | — | — | — | — | — | — | — | 87-94 | — | — | Step 2 |
| 78-72 Beetle, Front | .100 | 9.059 | 9.098 | — | — | — | — | — | — | — | — | 87-94 | Snug-Up | Handtight | Step 2 |
| Rear | .100 | 9.055 | 9.094 | — | — | — | — | — | — | — | — | 87-94 | — | — | Step 2 |
| 74-73 Karman Ghia | .100 | 9.055 | 9.094 | .080 | — | .335 | .0008 | .004 | 17 | 58-65 | — | 94 | Snug-Up | Handtight | Step 2 |
| 72 Karman Ghia | .100 | 9.055 | 9.094 | .080 | — | .335 | .0008 | .008 | 17 | 58-65 | — | 108 | Snug-Up | Handtight | Step 2 |
| **TYPE II — Van, Bus, Wagon, Vanagon, Transporter, Kombi** | | | | | | | | | | | | | | | |
| 82-80 Vanagon | .098 | 9.921 | 9.960 | .078 | .452 | .433 | — | .004 | ■ | 115 | — | 123 | Snug-Up | Handtight | Step 2 |
| 79-77 Van, Bus, Wagon, Transporter, Kombi | .098 | 9.921 | 9.960 | .080 | .492 | .453 | .0008 | .004 | ■ | 123 | — | 94 | Snug-Up | Handtight | Step 2 |
| 76-73 Van, Bus, Kombi, Wagon, Transporter | .100 | 9.921 | 9.960 | .080 | — | .472 | .0008 | .004 | ■ | 116 | — | 94 | 7 | Handtight | Step 2 |
| 72 Van, Bus, Kombi, Wagon, Transporter | .100 | 9.921 | 9.960 | .080 | .482 | .472 | .0008 | .004 | 17 | 72 | — | 94 | 7 | Handtight | Step 2 |
| **TYPE III — Square Back/Fastback** | | | | | | | | | | | | | | | |
| 73-72 All | .100 | 9.768 | 9.803 | .080 | — | .393 | .0008 | .008 | 17 | 56-65 | — | 94 | Snug-Up | Handtight | Step 2 |
| **TYPE IV — 411, 412 Sedan** | | | | | | | | | | | | | | | |
| 74-72 All | .100 | 9.768 | 9.803 | .080 | — | .393 | .0008 | .004 | 17 | 58-65 | — | 94 | Snug-Up | Handtight | Step 2 |

■ Design 14 or 17

| VEHICLE — YEAR, MAKE AND MODEL | BRAKE SHOE Min. Lining Thickness * | BRAKE DRUM Standard Size | BRAKE DRUM Machine To | BRAKE PAD Minimum Lining Thickness | ROTOR Machine To | ROTOR Discard At | ROTOR Variation From Parallelism | ROTOR Runout T.I.R. | DESIGN | CALIPER Mounting Bolts Torque | CALIPER Bridge Pin or Key Bolts Torque | WHEEL Lugs or Nuts Torque | STEP 1 Tighten Spindle Nut | STEP 2 Back Off Retorque | STEP 3 Lock. or Back Off and Lock |
|---|---|---|---|---|---|---|---|---|---|---|---|---|---|---|---|
| **VOLVO** | | | | | | | | | | | | | | | |
| 82-75 240 Series, Front w/ATE: vented | — | — | — | .062 | .897 | — | .0008 | .004 | 16 | 65-70 | — | 70-95 | 50 | Skip | 1/3 Turn to slot |
| Solid | — | — | — | .062 | .519 | — | .0008 | .004 | 16 | 65-70 | — | 70-95 | 50 | Skip | 1/3 Turn to slot |
| w/Girling: vented | — | — | — | .062 | .818 | — | .0008 | .004 | 16 | 65-70 | — | 70-95 | 50 | Skip | 1/3 Turn to slot |
| Solid | — | — | — | .062 | .519 | — | .0008 | .004 | 16 | 65-70 | — | 70-95 | 50 | Skip | 1/3 Turn to slot |
| Rear w/ATE | — | — | — | .062 | .331 | — | .0008 | .004 | 14 | 38-46 | — | 70-95 | — | — | — |
| w/Girling | — | — | — | .062 | .331 | — | .0008 | .004 | 17 | 38-46 | — | 70-95 | — | — | — |

| | | | | | | | | | | | | | | |
|---|---|---|---|---|---|---|---|---|---|---|---|---|---|---|
| 74-72 140 Series, Front Disc | — | — | — | .062 | .457 | — | .0012 | .004 | 16 | 65-70 | — | 70-100 | 50 | Skip | ⅓ Turn to slot |
| Rear w/Girling | — | — | — | .062 | .331 | — | .0012 | .006 | 14 | 45-50 | — | 70-100 | — | — | — |
| Rear w/Ate | — | — | — | .062 | .331 | — | .0012 | .006 | 17 | 45-50 | — | 70-100 | — | — | — |
| 82-76 260 Series, Front w/ vented disc | — | — | — | .062 | .818 | — | .0008 | .004 | 16 | 65-72 | — | 72-94 | 50 | Skip | ⅓ Turn to slot |
| w/solid disc | — | — | — | .062 | .519 | — | .0008 | .004 | 16 | 65-72 | — | 72-94 | 50 | Skip | ⅓ Turn to slot |
| Rear w/ATE | — | — | — | .062 | .331 | — | .0008 | .004 | 14 | 38-46 | — | 72-94 | — | — | — |
| w/Girling | — | — | — | .062 | .331 | — | .0008 | .004 | 17 | 38-46 | — | 72-94 | — | — | — |
| 75-72 160 Series, Front | — | — | — | .062 | .900 | — | .0012 | .004 | 16 | 65-70 | — | 70-95 | 50 | Skip | ⅓ Turn to slot |
| Rear w/Girling | — | — | — | .062 | .331 | — | .0012 | .006 | 14 | 38-47 | — | 70-95 | — | — | — |
| Rear w/ATE | — | — | — | .062 | .331 | — | .0012 | .006 | 17 | 38-47 | — | 70-95 | — | — | — |
| 73-72 1800 Sport Coupe, Front | — | — | — | .062 | — | .520 | .0012 | .006 | 16 | 65-70 | — | 70-100 | 50 | Skip | ⅓ Turn to slot |
| Rear | — | — | — | .062 | — | .331 | .0012 | .006 | 17 | 45-80 | — | 70-100 | — | — | — |

# DESIGN ILLUSTRATIONS

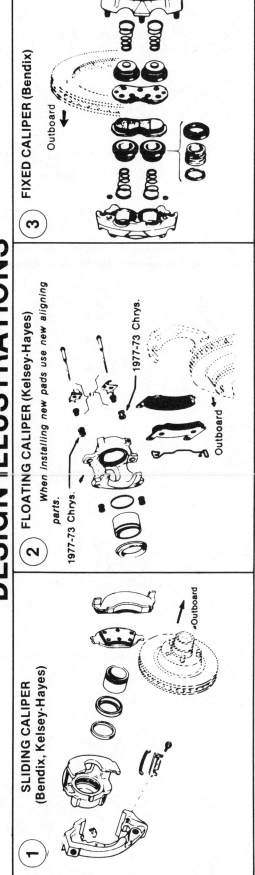

**① SLIDING CALIPER (Bendix, Kelsey-Hayes)**

Outboard

**② FLOATING CALIPER (Kelsey-Hayes)**
When installing new pads use new aligning parts.
1977-73 Chrys.
1977-73 Chrys.
Outboard

**③ FIXED CALIPER (Bendix)**
Outboard

31

# BRAKES

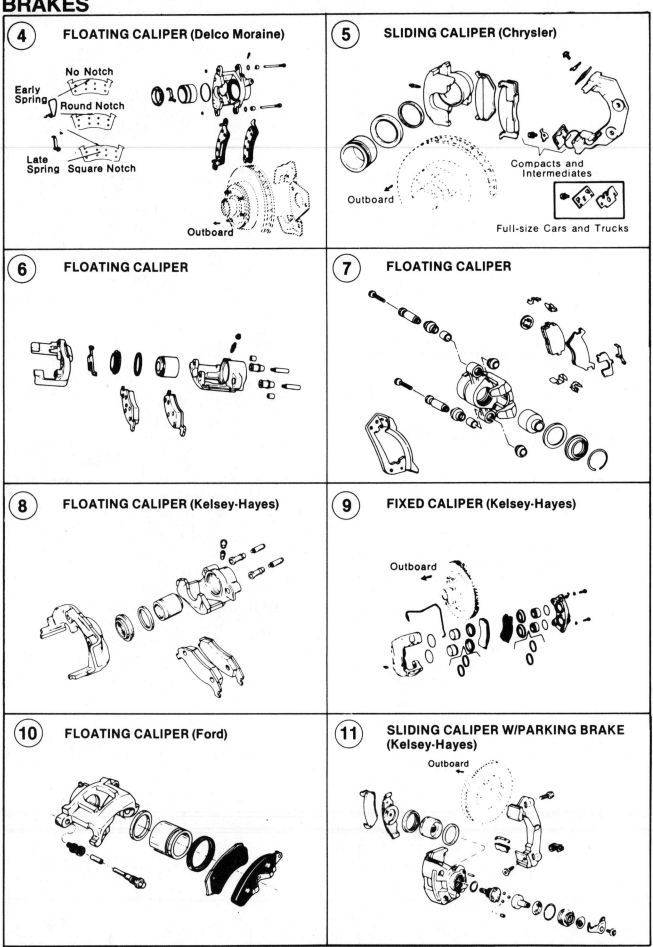

**4** FLOATING CALIPER (Delco Moraine)

No Notch

Early Spring

Round Notch

Late Spring    Square Notch

Outboard

**5** SLIDING CALIPER (Chrysler)

Outboard

Compacts and Intermediates

Full-size Cars and Trucks

**6** FLOATING CALIPER

**7** FLOATING CALIPER

**8** FLOATING CALIPER (Kelsey-Hayes)

**9** FIXED CALIPER (Kelsey-Hayes)

Outboard

**10** FLOATING CALIPER (Ford)

**11** SLIDING CALIPER W/PARKING BRAKE (Kelsey-Hayes)

Outboard

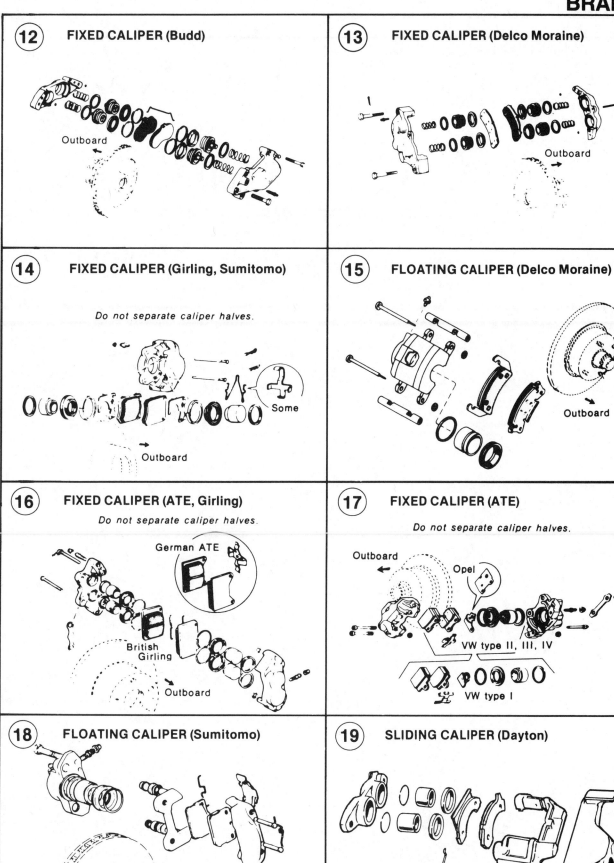

**(12) FIXED CALIPER (Budd)**

Outboard

**(13) FIXED CALIPER (Delco Moraine)**

Outboard

**(14) FIXED CALIPER (Girling, Sumitomo)**

*Do not separate caliper halves.*

Some

Outboard

**(15) FLOATING CALIPER (Delco Moraine)**

Outboard

**(16) FIXED CALIPER (ATE, Girling)**

*Do not separate caliper halves.*

German ATE

British Girling

Outboard

**(17) FIXED CALIPER (ATE)**

*Do not separate caliper halves.*

Outboard

Opel

VW type II, III, IV

VW type I

**(18) FLOATING CALIPER (Sumitomo)**

Outboard

**(19) SLIDING CALIPER (Dayton)**

# BRAKES

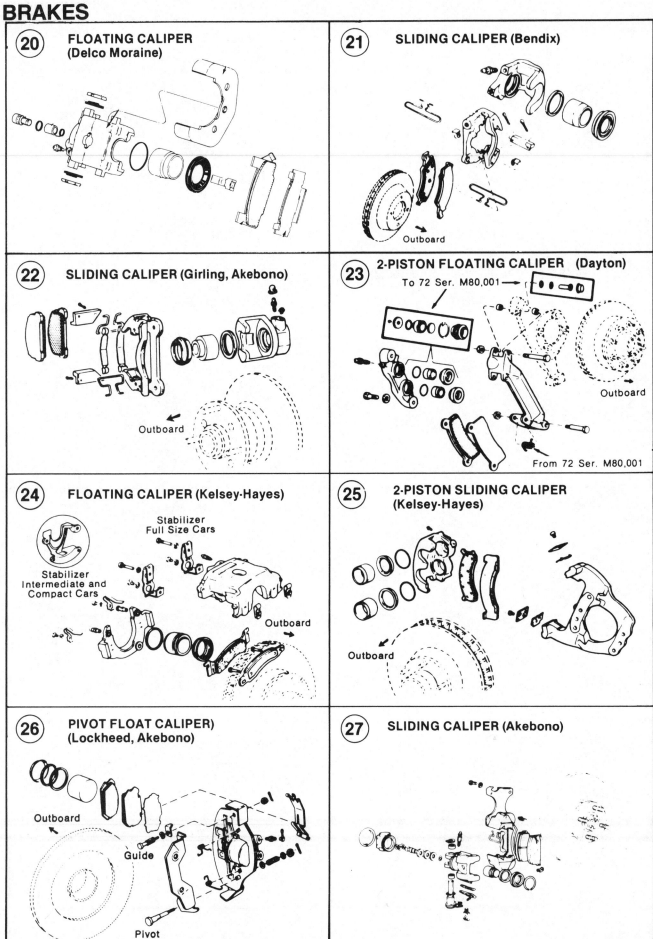

**20** FLOATING CALIPER (Delco Moraine)

**21** SLIDING CALIPER (Bendix)

Outboard

**22** SLIDING CALIPER (Girling, Akebono)

Outboard

**23** 2-PISTON FLOATING CALIPER (Dayton)

To 72 Ser. M80,001

Outboard

From 72 Ser. M80,001

**24** FLOATING CALIPER (Kelsey-Hayes)

Stabilizer Full Size Cars

Stabilizer Intermediate and Compact Cars

Outboard

**25** 2-PISTON SLIDING CALIPER (Kelsey-Hayes)

Outboard

**26** PIVOT FLOAT CALIPER) (Lockheed, Akebono)

Outboard

Guide

Pivot

**27** SLIDING CALIPER (Akebono)

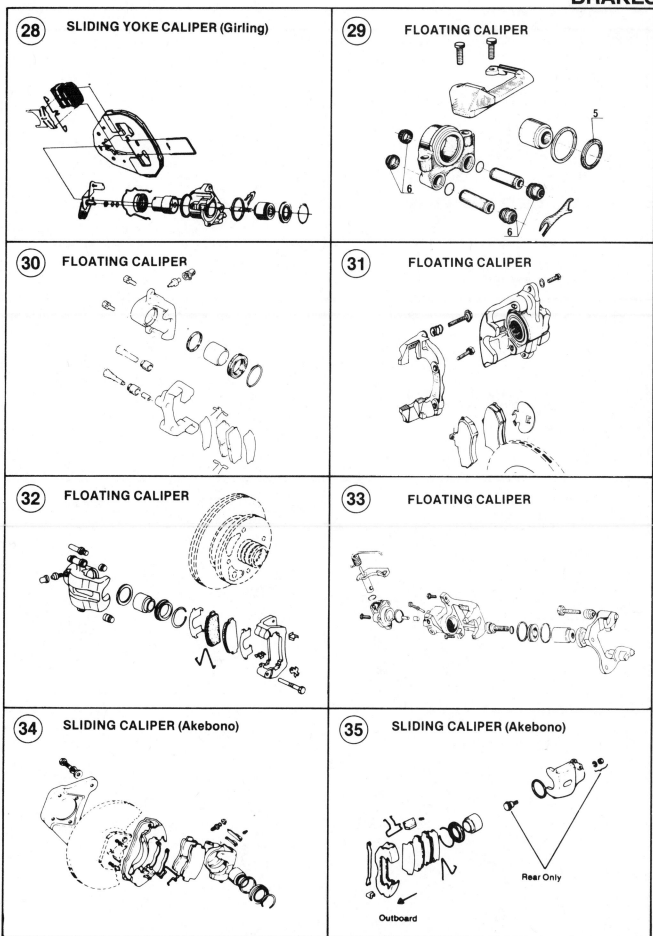

**28** SLIDING YOKE CALIPER (Girling)

**29** FLOATING CALIPER

**30** FLOATING CALIPER

**31** FLOATING CALIPER

**32** FLOATING CALIPER

**33** FLOATING CALIPER

**34** SLIDING CALIPER (Akebono)

**35** SLIDING CALIPER (Akebono)

Rear Only

Outboard

# BRANCHES

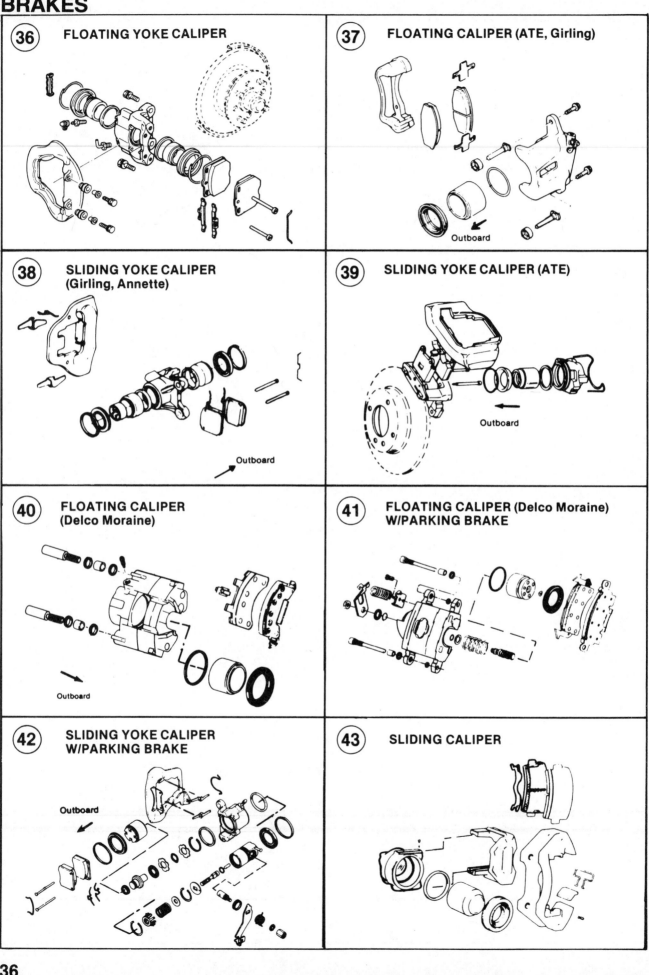

**36** FLOATING YOKE CALIPER

**37** FLOATING CALIPER (ATE, Girling)

Outboard

**38** SLIDING YOKE CALIPER (Girling, Annette)

Outboard

**39** SLIDING YOKE CALIPER (ATE)

Outboard

**40** FLOATING CALIPER (Delco Moraine)

Outboard

**41** FLOATING CALIPER (Delco Moraine) W/PARKING BRAKE

**42** SLIDING YOKE CALIPER W/PARKING BRAKE

Outboard

**43** SLIDING CALIPER

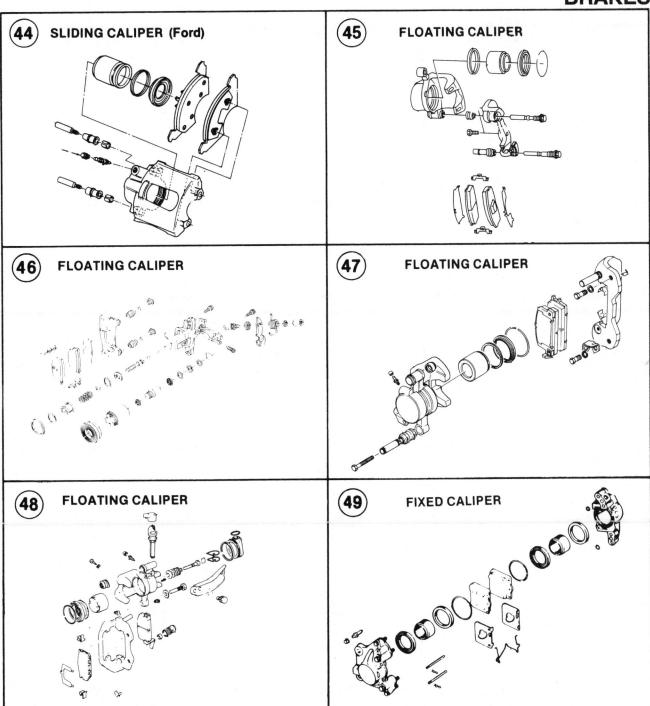

**44** SLIDING CALIPER (Ford)

**45** FLOATING CALIPER

**46** FLOATING CALIPER

**47** FLOATING CALIPER

**48** FLOATING CALIPER

**49** FIXED CALIPER

# 2 Exhaust

## EXHAUST SYSTEM

The exhaust system is composed of the exhaust manifold, the muffler, resonators or catalytic converter, the tubing connecting these parts, and the tubing (tailpipe) leading away from them. The physical integrity of the exhaust system is vital to the proper operation of modern emission-controlled engines as well as to the safety of the vehicle's occupants.

A complete inspection of the exhaust system should be made whenever the vehicle is raised on a lift to service other parts (brakes, steering, suspension and other chassis related items).

### Inspection

#### PHYSICAL COMPONENTS

1. Visually examine the mufflers, resonators, tailpipes, exhaust pipes, catalytic converters, heat shields and attaching hardware while the vehicle is on a lift. Damaged parts should be replaced.

2. Look for rusted or corroded surfaces that will damage components or cause exhaust leaks. Brackets, clamps and insulators should be replaced if they are damaged or badly corroded. Do not attempt to repair parts such as these. Slight cracking of rubber hangers is normal, but deep cracks or breaks are cause for replacement.

3. The exhaust system must be free of binding and vibrations. These are usually caused by loose, broken or misaligned clamps, brackets, insulators, shields, pipes or hangers. Check the individual components and adjust or replace individual items to maintain adequate clearance and/or alignment.

4. Check the inlet and outlet pipes, mufflers, resonators and catalytic converters for cracked joints or broken welds that could result in leaks. Any part should be replaced if there is reasonable certainty that its service life is limited.

MUFFLER    MAIN OXIDATION CATALYST

BALL-JOINT COUPLING    MINI OXIDATION CATALYST

Configuration of representative muffler, conventional catalytic converter, and mini converter

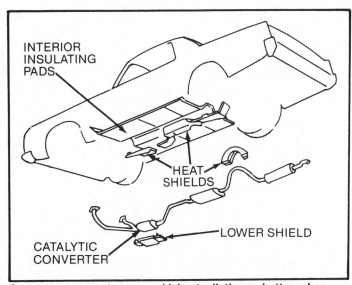

INTERIOR INSULATING PADS

HEAT SHIELDS

CATALYTIC CONVERTER    LOWER SHIELD

**Converter temperatures run high at all times, hotter when greater amounts of HC are present. Never undercoat a converter, and always replace heat shields if removed for service reasons, or if missing or damaged.**

5. On catalytic converter-equipped vehicles, check for missing or misaligned heat shields. Damaged shields should be replaced rather than repaired.

6. Do not overtighten clamps or hangars; it is very easy to crush the exhaust pipes.

## Exhaust System Leaks

The greatest danger from exhaust system leaks are exhaust gases entering the passenger compartment. This is because gasoline consists primarily of hydrogen (H) and carbon (C) which combine to form hydrocarbons (HC). Air consists primarily of nitrogen (N) and oxygen (O).

During normal combustion, HC combines with O to form carbon dioxide ($CO_2$) and water ($H_2O$). Unfortunately, combustion is never that perfect in a working engine, so exhaust gases form that contain (in addition to $CO_2$ and $H_2O$) carbon monoxide (CO), unburned hydrocarbons (HC) and a combination of nitrogen (N) and oxygen (O) in the form of oxides of nitrogen (NO). Since the amount of O present in the NO can vary to form versions of NO, the tailpipe version is identified as $NO_x$.

CO is the killer. It is odorless, and as little as 15 parts of CO per 10,000 parts O is enough to lead to blood poisoning and sickness.

Breathing exhaust gases in a closed garage for as little as three minutes can cause death. *Never take chances with exhaust fumes.*

To protect the mechanic and vehicle's occupants from the entrance and accumulation of carbon monoxide (CO) in the vehicle's passenger compartment, the following precautions should be observed.

1. Never run the engine in a closed garage or a confined, unventilated area longer than necessary to move the vehicle in or out.

2. If you must occupy a vehicle with the engine running for an extended period of time, adjust the heating or ventilation controls to bring outside air into the vehicle. Put the blower fan on high speed setting and set the heater or A/C controls in any position but OFF or MAX COOL.

3. Do not drive with the trunk lid or hatch lid open. Doing so will create a vacuum that will suck exhaust gases into the vehicle.

4. The tailgate window on station wagons should be closed if the engine is to be run for an extended period of time.

5. Check that the exhaust pipes are securely tightened at all connections. With the car running, you should be able to detect the "puffing" around a leaking area. Also, white deposits around a joint indicate an exhaust leak.

6. Inspect the openings in the body (especially around tailgate windows) for damage to gaskets and seals. Accident damage and deck lid lock or latch failure can also cause an exhaust leak into the passenger compartment.

## TOOLS

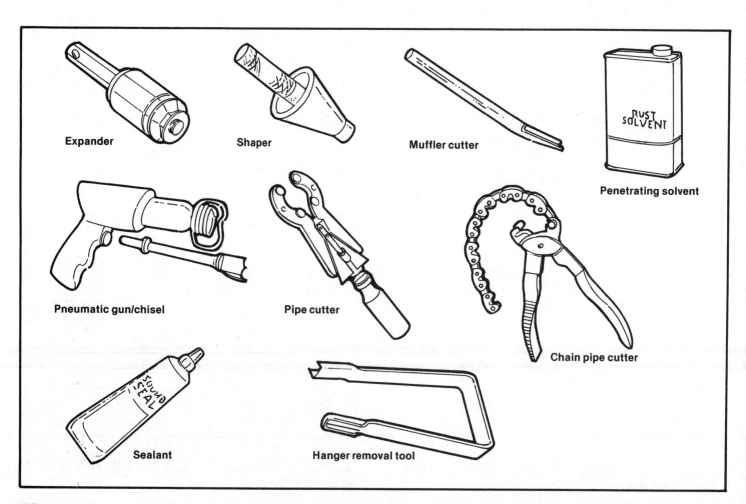

Expander

Shaper

Muffler cutter

Penetrating solvent

Pneumatic gun/chisel

Pipe cutter

Chain pipe cutter

Sealant

Hanger removal tool

## PROBLEM DIAGNOSIS

| Condition | Possible Cause | Correction |
|---|---|---|
| Leaking exhaust gases | Leaks at pipe joints | Tighten U-bolt nuts at leaking joints |
| | Damaged or improperly installed seals or packing | Replace seals or packing as necessary |
| | Loose exhaust pipe heat tube extension connections | Replace seals or packing as required. Tighten stud nuts or bolts to specifications |
| | Burned or rusted out exhaust pipe heat tube extensions | Replace heat tube extensions as required |
| Exhaust noises | Leaks at manifold or pipe connections | Tighten clamps at leaking connections. Replace gasket or packing as required |
| | Burned or blown out muffler | Replace muffler assembly |
| | Burned or rusted out exhaust pipe | Replace exhaust pipe |
| | Exhaust pipe leaking at manifold flange or pipe joints | Tighten attaching bolts/nuts |
| | Exhaust manifold cracked or broken | Replace manifold |
| | Leak between manifold and cylinder head | Tighten manifold to cylinder head nuts/bolts |
| | Missing heat control valve spring | Replace spring |
| Loss of engine power and/or internal rattles in muffler | Dislodged turning tubes and or baffles in muffler | Replace muffler |
| Engine hard to warm up or will not return to idle | Manifold heat control valve stuck open | Free manifold heat control valve with solvent |
| | Blocked crossover passage in intake manifold | Remove restriction or replace manifold |

## Restricted Exhaust Check

Restricted exhaust systems will result in loss of power and performance and popping back through the carburetor. The common symptoms of exhaust restriction are a severe—and often sudden—loss of power, a loud whistling or whooshing noise under the car, and a vacuum-controlled transmission that won't shift properly. Check the fuel and ignition systems first, and visually inspect the exhaust system for restrictions. If none are visible, perform the following test.

1. Attach a vacuum gauge and tachometer to the engine. At idle, the engine should read a steady vacuum of about 16–21 in./Hg.

2. Rapidly increase the engine speed to 2000 rpm. The vacuum should drop momentarily but quickly return to normal. If the vacuum settles below 16 in./Hg, there is a restriction in the exhaust system.

3. Disconnect the front exhaust pipe at the manifold and repeat the test. If the vacuum stays below 16 in./Hg, the restriction is in the exhaust manifold. If not, the restriction is in the pipes or muffler.

4. Check the manifold heat control valve before going any further. Tap the shaft with a hammer and lubricate the shaft with a

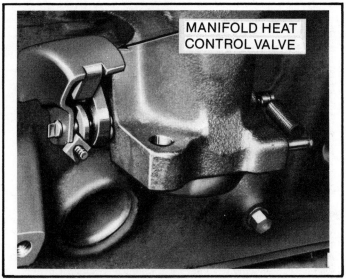

MANIFOLD HEAT CONTROL VALVE

**Exhaust manifold heat control valve**

penetrating oil. If the valve is free and the vacuum still does not return to normal, the restriction is in the exhaust manifold.

5. Restrictions in the manifold will normally show up early in the life of the vehicle, because they are due to casting flashes (residue from manufacturing). Casting flashes at the outer ends of the ports can usually be chipped away. However, if they cannot be removed, or if they are inaccessible, the manifold should be replaced.

6. If the vacuum settles around 16–21 in./Hg, the restriction is somewhere in the pipes, muffler, resonator or catalytic converter.

7. Reconnect the front pipe at the exhaust manifold and disconnect the muffler. Repeat the test. If the vacuum drops below 16 in./Hg, the catalytic converter is plugged, and it should be replaced or serviced.

8. If the vacuum is normal, the restriction could be in the tailpipe, exhaust pipe or muffler. The pipes can be probed for restrictions. Muffler restrictions could be caused by a loose baffle which sometimes can be heard by tapping or shaking the muffler.

## Exhaust System Service Tips

1. Always wear safety glasses.
2. Spray all nuts and bolts on existing clamps, hangers and brackets with penetrating oil. Allow a few minutes for the solvent to work before attempting to loosen. Always replace with new clamps, hangers and brackets.

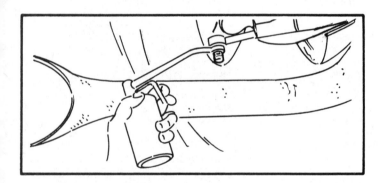

3. Remove clamps, hangers and brackets with an impact wrench or a ratchet socket wrench. Never loosen the muffler until the tailpipe has been disconnected.

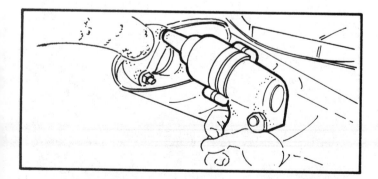

4. When replacing only the muffler on original equipment welded assemblies, use a cutter to cut the pipe as close as possible to muffler head. When servicing a welded connection, it should be cut and the new connection clamped when installing replacement parts. Also, coat slip joints with exhaust system sealer before assembling except at the catalytic converter.

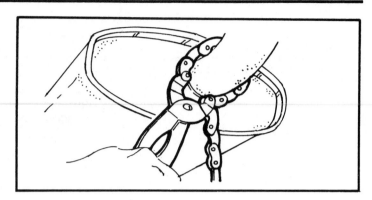

5. On any "muffler only" replacement, always disconnect the tailpipe from the muffler before removing the muffler.

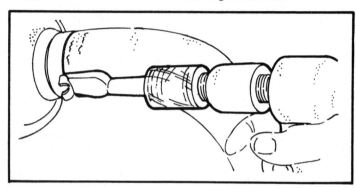

## Catalytic Converters

There are two major types of converters. The first is a pellet type that contains a shell that is filled with alumina beads that have been coated with platinum or palladium. The second is a "honeycomb" type design called a monolith, on which platinum or palladium is coated. The core construction provides a huge surface area of catalyst that promotes the desired chemical reaction. Neither type significantly restricts the passage of exhaust gases, unless the use of leaded fuel "plugs up" the converter, or the converter is physically damaged.

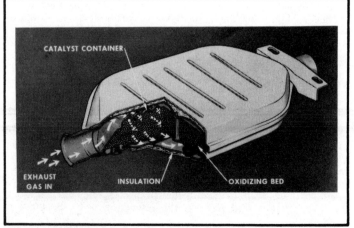

This is a pellet-type converter that is serviceable by replacement of the coated beads.

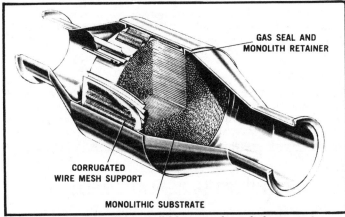

This is one version of a monolithic type of catalytic converter for dual exhaust systems. The honeycomb substrate presents the largest surface of catalyst to the passing exhaust gases.

## SERVICE AND DIAGNOSIS TIPS

There isn't much service associated with catalytic converters. The pellet type used on many cars can be recharged with new coated beads if necessary. Also, the lower covers can be replaced if damaged. In the case of monolithic (honeycomb) designs, the only service is to install a new unit.

There are no special problems regarding installation. In fact, most units are replaced in the same way as mufflers.

**NOTE: Some catalytic converters are equipped with fittings for air injection lines or oxygen sensor probes. If a converter containing such fittings is replaced, the new unit must match the old one, and all hoses or probes transferred from the old converter to the new one.**

### Diagnosis

Mechanics not used to working on catalytic converter-equipped cars should be aware of certain exceptions to standard service procedures, which may damage a catalytic converter.

1. When testing an engine component, do not crank the engine for more than 15 seconds at a time. Allow the engine to rest for one minute between tests.

2. Do not push-start cars equipped with a manual transmission and catalytic converter. If the engine fails to start, raw fuel will be pulled through the engine into the exhaust system. When the engine does start, the possible backfire could damage the expensive converter.

3. Avoid the use of gasoline additives, fuel system cleaners or tune-up solvents, unless specifically approved by the vehicle manufacturer. Many cleaners contain non-petroleum based solvents that will damage the catalyst. Combustion chamber cleaners may also loosen carbon deposits that could plug the converter.

4. Correct any dieseling problems as soon as possible.

5. When performing a power balance test, do not short out one cylinder for more than a few seconds at a time.

6. Do not turn off the ignition switch when the car is in motion.

Converters normally operate in the range of 1000–1500°F. Subject a converter to temperatures hotter than 2000°F for a long time and it will look like this one. Extended heavy-throttle driving with a missing cylinder did this one in.

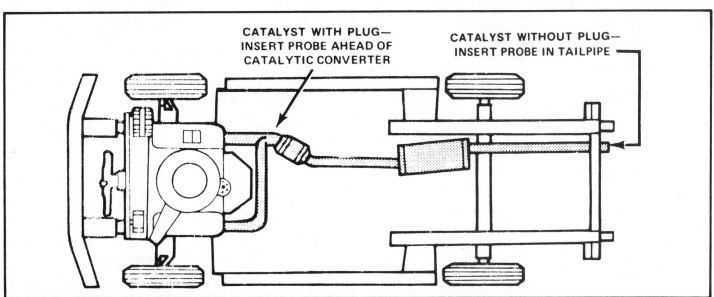

The most common infrared reading location is the tailpipe. A few converter-equipped cars have an analyzer probe tap ahead of the converter to allow exhaust gas readings before catalyst cleaning.

# EXHAUST

## Heat Build-Up

Catalytic converters run so hot that heat shields are normally required to protect vehicle components and prevent brush fires. At normal working temperatures, they can melt asphalt when the car is left idling at one spot for a long period of time. Production-installed catalytic converters may operate at temperatures approaching 2000°F. (Standard exhaust pipes run around 1000°F.) This is why it is necessary to observe certain precautions (see the preceding) when working on or around catalytic converter-equipped vehicles. Anything that contributes to more raw gasoline in the converter (misfiring, coasting, testing, etc.) can cause converter temperatures to approach dangerous levels. Overheating is the most common cause of catalytic converter restriction.

## Leaded Gas

Any lead in the exhaust gases—the result of burning leaded fuel—will poison the catalyst, making it less effective. And overrich mixtures (in a flooded engine, for example) can cause the catalytic converter to work itself to death in short order.

Lead deposits kill a converter two ways: chemically and mechanically. Depending upon the type of converter and the amount of lead in the fuel, you can ruin a converter with as little as one tankful of leaded gas. The lead deposits permanently coat the converter's expensive catalytic metals. Once these metals are lead-coated, they cannot oxidize or reduce the exhaust gas flowing around them.

A restricted exhaust was diagnosed on this car. The converter honeycomb passages were clogged with lead deposits.

A clogged converter will severely restrict the exhaust system. In extreme cases, the back-pressure will increase to the point that the engine will not run. Because of their honeycomb structure, monolithic converters are more susceptible to lead-deposit clogging than pellet converters. However, if enough lead runs through a pellet converter, the lead deposits will eventually fuse the pellets together.

## Objectionable Odors

Different gasolines have different amounts of sulfur in them. When the engine is too rich or too lean, the catalytic converter announces the presence of the sulfur with that rotten-egg odor. During some driving situations, a normal and temporary enrichment will cause a temporary rotten-egg odor. During deceleration, for example, the engine pulls fuel out of the idle circuit, and the smell of sulfur may be present until the engine burns off the extra fuel. Electronically controlled fuel systems usually lean themselves out during deceleration to stop decel emissions and odors. Power-circuit enrichment can also cause a momentary sulfur odor after the driver floors the pedal for passing gear.

## Physical Damage

The greatest peril to a catalytic converter is probably from external abuse rather than internal overheating, or even lead poisoning. Careless jacking or positioning of pads on lifts might bash in the converter's metal shell. The same kind of abuse is possible when driving the low-slung platinum filled container over in-shop obstructions.

## CONVERTER REPLACEMENT

Converters are engineered to last for at least 50,000 miles without requiring service. The bead-type converters (General Motors, for example) feature a drain plug so that worn out beads can be shaken out and new beads put in. See Converter Service following. Catalytic converters should be replaced with another of the same type and design.

Removing a converter permanently can upset the E.G.R. calibration and 1) aggravate a detonation problem or 2) *create* a detonation problem. Most cars on the road use back-pressure-transducer type E.G.R. It's a fact that all E.G.R. systems—and back-pressure types in particular—need a certain amount of exhaust back-pressure to work properly. Although it may not affect horsepower, removing a converter often reduces back-pressure just enough so that the E.G.R. won't open as much as it should. When the cooling effect of E.G.R. is thus reduced, the engine starts pinging. The cure for such detonation, of course, is to retard the timing, but the amount of timing retardation required may make the car run worse than ever. In fact, it may be impossible to eliminate the detonation entirely.

If you remove a bad converter from a car and have to order a new one, get proof that you ordered a new one. Silly as it might sound, your best defense against an emission control tampering charge is solid proof that you are replacing the bad part instead of permanently removing it.

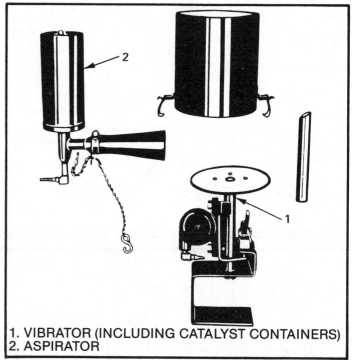

1. VIBRATOR (INCLUDING CATALYST CONTAINERS)
2. ASPIRATOR

Special tools for replacing the pellets in a catalytic converter

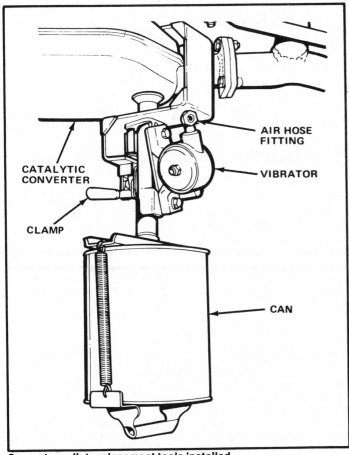

**Converter pellet replacement tools installed**

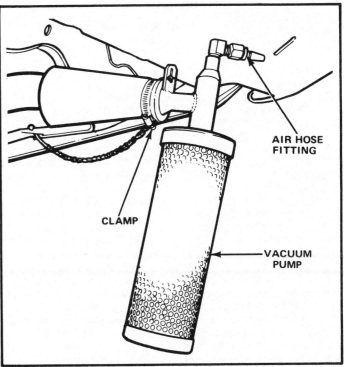

**Vacuum pump attached to exhaust pipe**

## CONVERTER SERVICE

If necessary, pellet-type catalytic converters can be serviced to avoid replacement of the entire converter unit. The converter can be drained of the old beads and be recharged with new coated pellets.

1. Raise and support the car.
2. Place the vacuum pump hose on the exhaust pipe and tighten the clamp.
3. Connect shop air of at least 80 psi to the vacuum pump fitting.
4. Remove the plug from the bottom of the converter.
5. Lock the vibrator in place on the catalytic converter.

6. Replace the shop air hose from the vacuum pump and connect it to the vibrator fitting.
7. Catalyst beads will drain into the can for approximately 10 minutes. If no beads fall from the converter, the unit should be replaced.
8. When the converter is empty, disconnect the shop air hose, remove the can and discard the beads.
9. Install a refill can on the vibrator and connect the vibrator to the air hose.
10. Beads will be drawn up into the converter and packed into place.

**NOTE: If beads come out of the tailpipe, the converter is defective and should be replaced.**

11. When the converter is full, disconnect the air hose and the vibrator from the converter.
12. Coat the threads of a plug with nickel based anti-seize compound and install the plug.
13. Remove the vacuum pump and lower the car.

# Preface to Suspension and Steering

Integrity and honesty are the two most important qualities in a wheel alignment mechanic. Followed by that, he must have knowledge, ability, and proper equipment. If the first two qualities are not fully realized, the next three cannot work to his advantage on any long term basis.

The belief that only a gifted few can really understand the "secrets" of front end geometry is false.

The purpose of this section is to present the principles of suspension and steering systems. A shortage of qualified alignment mechanics exists; a mechanic willing to learn all the aspects of suspension and steering systems, and work with accurate, modern equipment, can nearly guarantee himself a secure future.

Start by learning what the alignment angles are and what they do. The purposes and goals of an alignment are important. Finding the actual alignment adjustments is relatively easy. The Chilton service manuals show how to make the adjustments.

Probably the most common reason the mechanic may not check all the angles is that he lacks a machine that can measure and register the steering axis inclination, toe-out on turns, thrust line, or check an automobile for "square". Without these measurements there is no reasonably quick way of finding bent parts or to do a total alignment. A modern alignment machine not only finds and immediately reports all the angles, but will also respond immediately as the adjustments are made. If the mechanic has erred in any of the adjustments, the machine will detect and display the error before the adjustments are secured. This saves time.

If you have not yet purchased alignment equipment, or are considering updating existing equipment, you should become familiar with every aspect of the complete alignment service before deciding which equipment to buy. The machine should be capable of measuring all the angles described in this section.

Some consider the front suspension and steering systems to be one of the weaker parts of the automobile. Even under normal conditions, the suspension and steering systems must be able to withstand brutal forces. The suspension must be able to stay intact while the wheels are locked by the brakes, hitting holes and bumps, and at the same time support tremendous bouncing and shifting weight. After all these severe shocks and stresses, the suspension must return to the exact position from which it started. While an automobile is being driven, its suspension is constantly under variable stresses, strains and shocks. It must be able to withstand this for many thousands of miles.

Automobile suspension and steering systems are more advanced today. As it was with tune-up mechanics, who have not kept pace with new technology, the uninformed front end mechanics will, one by one, fall by the wayside, victims of keen, advanced, knowledgable competition. They will be replaced by mechanics who have learned not just about caster, camber, and toe, but about all the angles involved in the complete alignment procedure—mechanics that have the knowledge and the equipment to solve problems—front and rear.

Changes in the geometric angles or alignment may be caused by normal wear—a little here and a little there—but not enough at any one point to require parts replacement. Springs or torsion bars may sag a little.

One of the most damaging things to the quality the engineers designed into the suspension and steering systems is the mechanic that does not restore the proper angles during the alignment procedure. If you are to survive in the current marketplace, "Set the toe and let it go" must be a thing of the past.

Likewise, don't try to compensate for an unknown problem by changing a suspension angle. Do not compensate with caster for a power steering pull. Do not compensate with caster for a conical tire. Temporary corrections lead to later problems. Learn how to isolate and correct the actual problem.

Steering axis inclination has been a controversial angle in the suspension system, and is explained in this section. You will understand why toe out on turns must be checked with every alignment. You will find why the least thought of and discussed reason for an alignment is the most important…safety.

# 3 Suspension and Steering

## SUSPENSION AND STEERING

### Theory and Principles

#### EVOLUTION OF AUTOMOBILE SUSPENSION SYSTEMS

Decades ago, the front suspension of a carriage consisted of a single solid axle that pivoted in the center and was attached to, and turned by, a horse, which supplied the motive and directional power. The evolution of the horseless carriage brought higher speeds and new problems: a need for driver control, better braking, and directional stability. The single pivot front axle was difficult to control. It prohibited the use of front brakes, as the slightest imbalance of side-to-side braking effort would cause the vehicle to veer from one side to the other.

The single pivot axle was soon replaced by the Ackerman axle, which provided a fixed position axle that could not be turned for steering. Steering was through two pivot points, one on each end of the axle, and connected to it with vertical kingpins. Ackerman axles gave way to a later design known as the Elliot axle which provided a tilted kingpin with a double yoke on the spindle and a short axle to steer the wheel. As speeds increased and front brakes were needed, the Reverse Elliot axle with kingpin inclination (KPI) was developed, giving greater stability and permitting the introduc-

FRONT VIEW

BODY

PIVOT

BOTTOM VIEW

PIVOT

WHOLE AXLE TURNED

**Single pivot front axle**

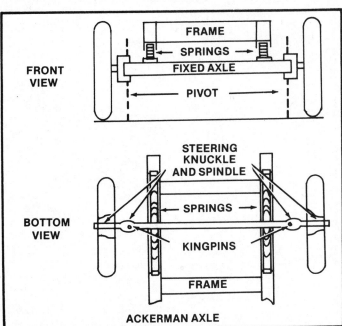

FRAME

SPRINGS

FRONT VIEW

FIXED AXLE

PIVOT

STEERING KNUCKLE AND SPINDLE

BOTTOM VIEW

SPRINGS

KINGPINS

FRAME

ACKERMAN AXLE

**The Ackerman axle design provided a fixed position axle that could not be turned for steering. It provided two pivot points, one on each end of the axle, with vertical kingpins**

# SUSPENSION AND STEERING

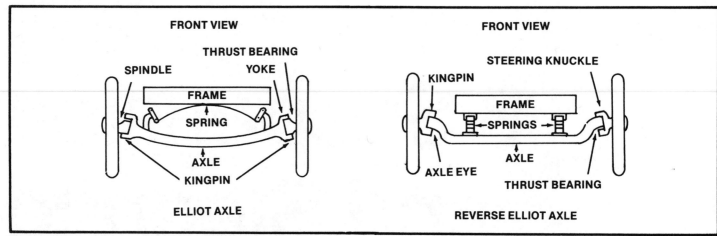

FRONT VIEW

SPINDLE
THRUST BEARING
YOKE
FRAME
SPRING
AXLE
KINGPIN

ELLIOT AXLE

FRONT VIEW

STEERING KNUCKLE
KINGPIN
FRAME
SPRINGS
AXLE
AXLE EYE
THRUST BEARING

REVERSE ELLIOT AXLE

**The straight (or "Reverse Elliot") front axle improved stability with a four point support instead of three, allowed a shorter turning radius and permitted the automobile to be set closer to the road**

tion of front brakes. The Reverse Elliot, or basic straight front axle, improved stability with a four point support instead of three, allowed a shorter turning radius, and permitted the automobile to be set closer to the road.

For many years, every automobile used a straight front axle. However, ride control engineers soon learned that a bump hit on one side transferred shock to the other side and jolted the whole automobile. They also learned that the greater amount of unsprung weight or weight not supported by springs (wheels, tires, brakes, axles, suspension pieces, etc.), the harsher the ride. Straight axles had to be heavy to be strong, and contributed significantly to unsprung weight. Other drawbacks were evident. If in the process of turning one wheel gripped the road and gave good sta-

bility while the other wheel hit a rough portion of the road, the shock was transferred to the stable wheel causing it to lose some of its grip. Because of the harsh ride and poor handling characteristics of the straight axle, new designs incorporating independent suspension were developed.

## INDEPENDENT SUSPENSIONS

There are several types of independent suspensions, but all share a common feature: Each front wheel is supported independently of the other, so that movement of one wheel, up or down, will not necessarily cause movement of the other. Independent suspensions

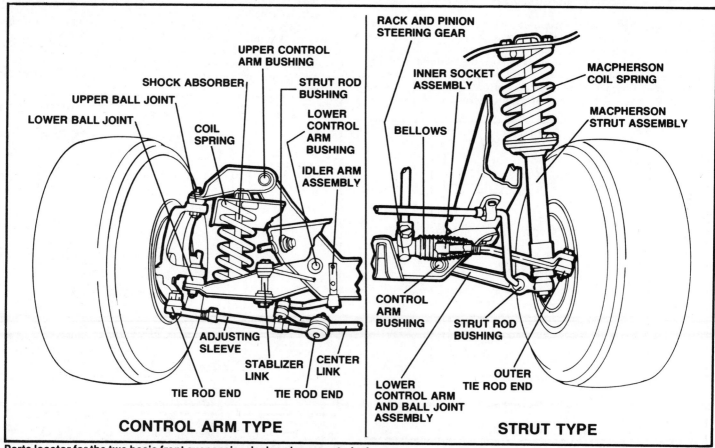

UPPER CONTROL ARM BUSHING
SHOCK ABSORBER
STRUT ROD BUSHING
UPPER BALL JOINT
LOWER BALL JOINT
COIL SPRING
LOWER CONTROL ARM BUSHING
IDLER ARM ASSEMBLY
ADJUSTING SLEEVE
STABLIZER LINK
CENTER LINK
TIE ROD END
TIE ROD END

**CONTROL ARM TYPE**

RACK AND PINION STEERING GEAR
INNER SOCKET ASSEMBLY
MACPHERSON COIL SPRING
MACPHERSON STRUT ASSEMBLY
BELLOWS
CONTROL ARM BUSHING
STRUT ROD BUSHING
LOWER CONTROL ARM AND BALL JOINT ASSEMBLY
OUTER TIE ROD END

**STRUT TYPE**

**Parts locator for the two basic front suspension designs in use on today's cars**

are much lighter than solid axles, reducing unsprung weight and greatly improving riding and handling characteristics.

Early independent suspensions were not without problems, the greatest of which was poor and uneven tire wear. In most of these designs, the upper and lower control arms were of equal length and the wheels were kept vertical, resulting in side-to-side tire scuff. Equal length control arms also caused the wheels to toe inward while turning, creating directional instability. A short arm on top and a long arm on the bottom corrected this problem.

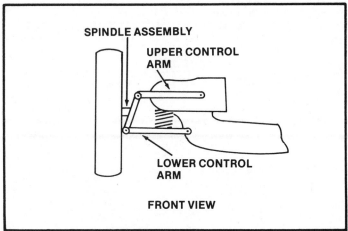

**Early independent suspension**

By the late 1960's, enginers had learned that by positioning the arms at slightly different angles, the wheel would move out away from the frame at the top as the spring was initially compressed, which improved handling and stability in a crosswind. By further changing the position of the arms, front end dipping could be somewhat relieved when the brakes were applied.

Although an automobile may be properly aligned to factory specifications, there are other factors involved that directly affect ride, handling, and tire wear. One of these factors is dampening. Softer springing to improve ride characteristics created a need for better dampening of road shock. The early shock dampeners (shock absorbers) were indirect acting. The later shock absorbers (called direct acting or airplane type) were mounted directly to the suspension and frame and controlled the amount of road shock transmitted to the vehicle. Shock absorbers that check compression only are known as single acting; shock absorbers that check rebound as well are known as double acting. The shock absorber action partially controls the spring action, thus protecting the spring and the suspension. This action also keeps wheel bounce to a minimum and gives the tire greater contact with the road.

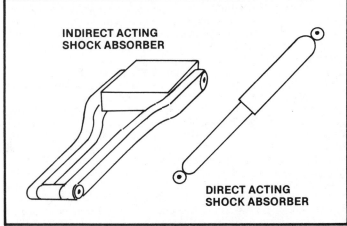

**Softer springing to improve riding created a need for better dampening of road shock**

Despite the use of shock absorbers, there were still some forces that could not be absorbed, such as large bumps and deep holes. These hazards were severe enough to damage the suspension components. The use of rubber bumpers attached to frame members and control arms reduced suspension damage by absorbing the harder shocks.

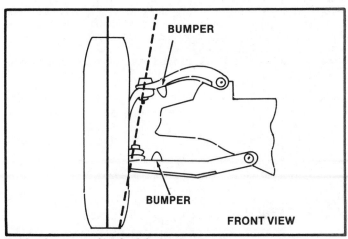

**Rubber bumpers absorbed the harder shocks**

Other problems encountered with independent suspensions involved excessive lean on turns, oscillation of the front of the car on a straight road, and side-to-side roll. Although the automobile had proper alignment in the static position, it showed signs of misalignment while in motion and exhibited excessive tire wear. Wheels had excessive tilt while cornering, which caused the outer edges of the tires to wear prematurely. To solve this problem, the stabilizer bar or "sway bar" was introduced. It was first installed in the rear and later added to the front. The stabilizer bar is a round metal bar mounted in rubber blocks and secured by brackets to the front portion of the frame. The ends of the bar are attached to the outer ends of the lower control arms. The twisting or torsional action of the bar permits the vehicle to corner at much higher speeds and remain reasonably flat.

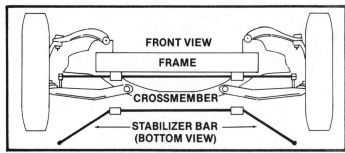

**The stabilizer bar reduced excessive lean on turns, oscillation of the front of the car on straight road, and side-to-side roll**

## A-ARM SUSPENSIONS

Early independent suspension systems used kingpins and bushings to connect the steering knuckle with the axle or control arms. The bushings above the kingpins were eccentric and also served to adjust the alignment of the suspension. However, failure to lubricate the suspension bushings meant that the alignment adjustments became frozen and expensive repairs followed.

The control arms were shaped as an "A" (hence the term A-arm). The wider the "A" was spread at its base, the greater the support and stability. In these early designs, the upper control arms also acted as indirect shock absorbers in many suspension systems. When balljoints replaced kingpins and bushings, the alignment adjustments were moved to other locations, usually where the A-arm mounted to the frame.

A further refinement was the addition of the brake or strut rod. The lower A-arm was redesigned as a single straight arm and the strut or brake rod was added to it. The strut rod connected the outer end of the lower arm and the front of the frame, creating a still wider "A" effect for greater stability. The strut rod also provided strength and prevented the single lower arm from buckling under or being torn off. This system permits the strut rod to be mounted in rubber where it connects to the front of the frame. When the wheel hits a rough portion of the road it can "give" slightly, reducing road shock transmitted to the frame and yielding a better ride.

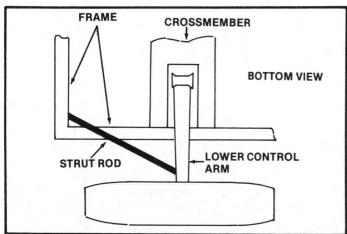

The strut rod provided greater stability and strength, and prevented damage to the single lower control arm

Conventional suspensions used a coil spring between the upper and lower control arms, to control suspension travel and reduce shock. Development of unitized and unibody construction permitted the use of a coil spring above the upper arm with both spring and shock absorber mounted on the upper arm instead of on the lower arm. This system is a short-and-long-arm design (the lower arm is longer than the upper control arm) but requires a very strong inner fender panel (spring tower) to absorb the spring and shock absorber forces. The advantage of the higher and wider mounting of the spring is greater resistance to the roll forces of the automobile. Coupled with a strut rod-type lower arm, this system provides excellent ride and handling characteristics.

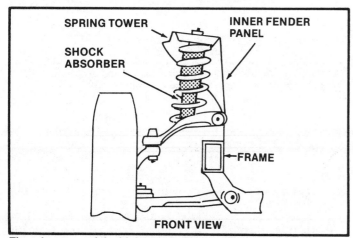

The advantage of the higher and wider mounting of the springs is greater resistance to the roll forces of the automobile. The darkened area shows typical shock absorber location

## MACPHERSON STRUTS

The single control arm suspension is better known as the "Mac-Pherson strut." This strut is a one-piece design incorporating a shock absorber and concentric coil spring, giving excellent anti-roll characteristics due to the high placement of the spring. Spindle load and road shocks are transferred directly to the spring without going through a control arm, producing a much smoother ride. In this type of suspension the stabilizer bar may sometimes serve as the strut rod. As with the coil spring on the upper arm, a strong all-welded inner fender is required.

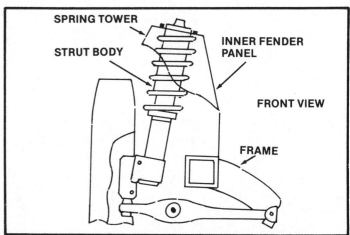

The MacPherson Strut provides for the spindle load and road shock to be transferred directly to the spring without going through a control arm. The shock absorber is located inside the strut body

Weight has become a very important factor in the design of modern automobiles. The strut suspension, combined with the unibody, permits the elimination of the heavy frame and the upper control arm. The unibody supports the upper end of the coil spring, eliminating the need for a frame horn or extension of the crossmember for coil spring support.

Space and accessibility to the serviceable areas of the automobile are also very important considerations in the manufacturing of smaller, lighter automobiles. When all available options are added (power steering, power brakes, air conditioning, and so on), available space becomes extremely limited. After an automobile has been sold and delivered to the customer, it must be readily serviceable. In the past, many small automobiles without strut-type suspensions posed problems in normal service procedures because of extremely cramped working conditions. The introduction of transversely mounted engines further complicated space problems. Because the strut suspension eliminates so many previously necessary parts, it occupies much less space while allowing the automobile to be "downsized" for even greater weight reduction, better economy, and greater ease of service.

The strut suspension, much simpler than previous mechanisms, generally is easier and less expensive to maintain because it contains fewer moving parts. There are many design variations in strut suspensions, but most have a coil spring at the top of the strut and a shock absorber within the strut body. Others have a coil spring supported by the control arm and still others have a shock absorber mounted independently of the strut body. Some types use a single control arm in combination with a longitudinal strut rod that attaches to the chassis, while others may use a wide double "A" type control arm with no strut rod. All modern strut suspensions eliminate the upper control arm, and all are mounted in rubber to minimize noise and road shocks, resulting in a smoother ride. Because the strut suspension reduces the number of pivot points, total free play in the moving parts has been reduced. Consequently, there is less loss of motion in the steering and suspension, resulting in more responsive handling characteristics. Another consequence, however, is greater sensitivity to any imbalanced or out-of-round conditions in the tires. Strut suspensions using a coil spring mounted at the top of the strut are often regarded as having superior ride and handling characteristics because of the wider and higher placement of the spring, but strut suspensions with the coil spring mounted at the control arm, independent of the strut, will ride and handle just as well under normal driving conditions.

## GEOMETRIC PRINCIPLES

Of the many different types of suspensions in use today, all have one thing in common: they all use the same principles of geometry. If the relevant geometric angles are thoroughly understood, the type of design is of little consequence to the mechanic. If the angles to be considered—caster, camber, toe, steering axis inclination, toe-out on turns, thrust line, and "squaring" of the automobile—are understood, making the adjustments is relatively easy.

When these angles are in harmony with one another, an automobile will follow the desired course without wander, scuffing, dragging, or slipping. The vehicle will be free-rolling. Steering will tend to return to the straight position and will be controllable with minimal effort.

We will examine each of these angles individually and see why they are used and how they are related. Changing one may have an effect on another. We will also examine other factors involved in proper total alignment. We will see how some of these factors may affect an automobile even after alignment adjustments have been completed.

## Suspension Designs

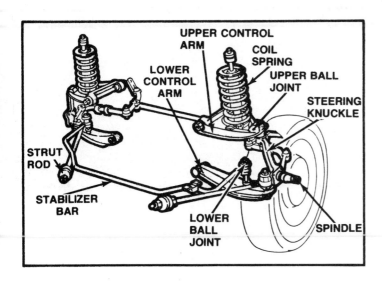

A coil spring is mounted on top of the control arm with the shock absorber in the center of the coil spring. Only the upper control arm is A-arm design. This design is used primarily on Ford cars

Typical unequal length A-arm suspension used on many American sedans. The shock absorber and coil spring are positioned between the upper and lower control arms. Note that the control arms (A-arms) are not the same length

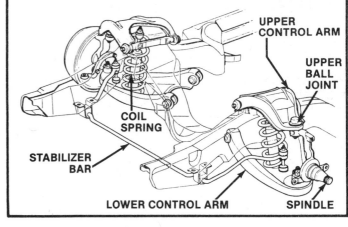

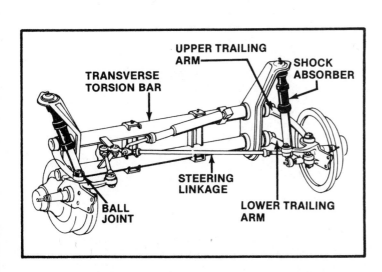

Transverse torsion bar front suspension used primarily by VW on rear engine cars

# SUSPENSION AND STEERING

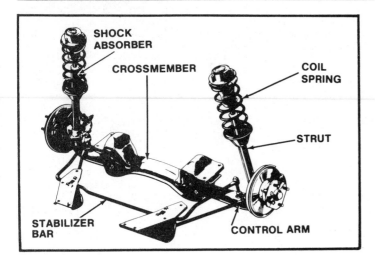

MacPherson strut type front suspension

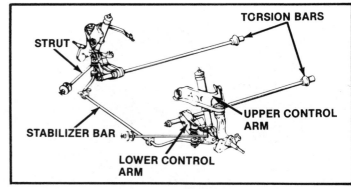

Unequal length A-arm suspension with torsion bars

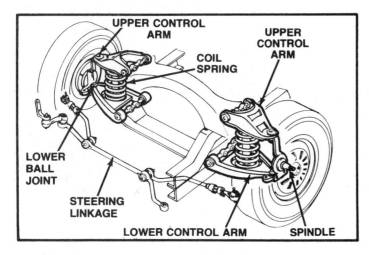

A truck's independent front suspension is very similar to the unequal length A-arm suspension used on passenger cars. It functions in the same manner, but the components are often beefier to handle the extra stress

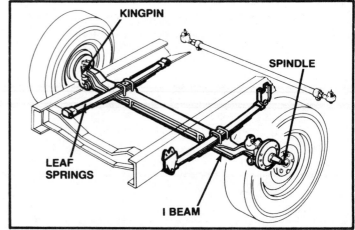

A straight I beam front suspension is uncomplicated and meant to handle heavy loads

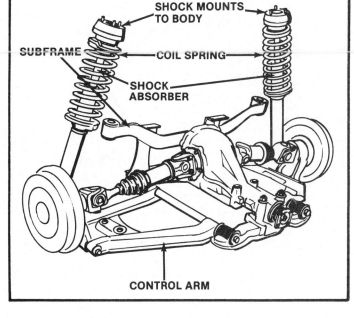

The twin I beam front suspension is used almost exclusively on Ford trucks. The coil spring is mounted between the frame and the I beam that carries each wheel. The I beam pivots at the other end, and a radius rod locates the fore-and-aft position of each beam

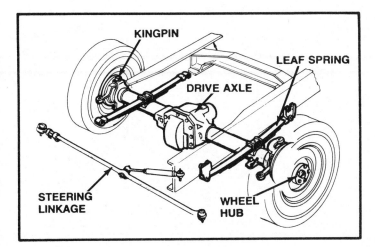

A four wheel drive front suspension is basically the same as an I beam suspension except that a front drive axle takes the place of an I beam

The rear strut suspension combines the coil spring and shock absorber in one unit, attached to the body and the wheel spindle. The lower control arm, strut and rear axle are usually mounted on some sort of subframe, which is attached to the car body

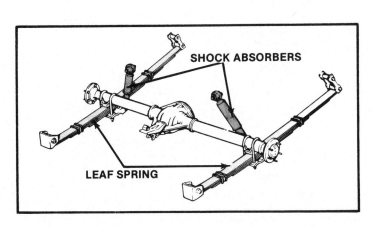

A basic, uncomplicated leaf spring rear suspension with shock absorbers to control vibration and vertical movement of the axle

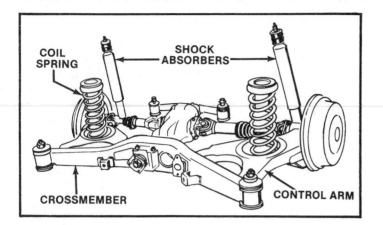

This type of independent rear suspension is used on many sports cars. Coil springs are used between the control arm and the vehicle body. The control arms pivot on a crossmember and are attached to the spindle at the other end. A shock absorber is attached to the spindle or the control arm

Swing arm rear suspensions are typical of rear-engined VWs. The rear wheels are independently sprung on diagonal arms or swing arms. The torsion bars are anchored at either end by splines which can also be used to adjust suspension height

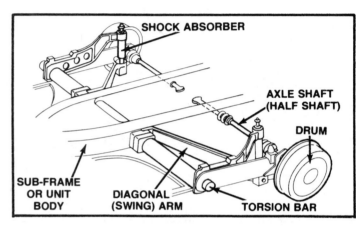

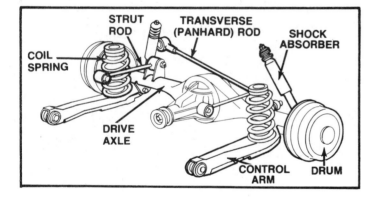

These non-independent rear suspensions differ from similar designs in that coil springs replace leaf springs and strut rods and control arms serve to position the rear axle

An independent (dead) rear axle is mostly used on front wheel drive cars. Trailing arms holding the wheel and shock absorber are attached to a rigid crossmember.

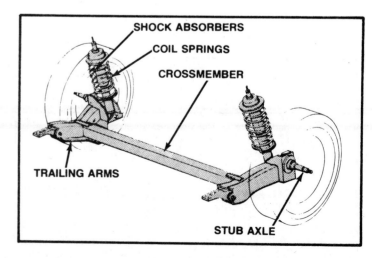

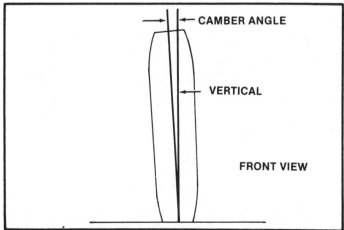

## Wheel Alignment
### PREALIGNMENT

#### Tire Check

Before an automobile is aligned, it should be given a thorough prealignment check. Proper tire pressure is very important, so pressure must be checked on each tire before an alignment is started. Incorrect tire pressure has a drastic effect on alignment

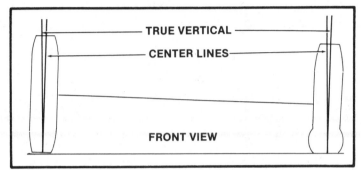

**Incorrect tire pressure can result in misalignment**

settings. For example, assume there are two underinflated tires on the left side of a car with normal pressure in the tires on the right side. The car would be lower in curb height on the left side, increasing the camber on that side and decreasing camber on the right side. No matter how painstakingly done, if the alignment were adjusted with the tires in this condition, camber would be altered when the tires were brought to proper pressure. Experiments have shown this can change camber as much as one-half degree, an excessive amount of change.

There are a number of other factors concerning tires that should be examined and noted while tire pressures are being checked. Look for uneven wear patterns caused by misalignment. Badly worn tires can also be indicative of other suspension problems, such as worn ball joints, bent suspension components, worn shocks or other components that can affect vehicle performance and/or alignment. See the section on Diagnosis and Troubleshooting.

Determine if tires are matched. If there are two different sizes, they should differ only front to rear. If there are two new tires and two good but older tires, the new tires should both be on the front or both be on the rear. Conventional tires should not be used with radial tires. A common misconception is that radial tires may be used with conventional tires by installing the first pair of radial tires in the rear. Mixing radial and bias ply tires is not a recommended practice. Different size footprints can create a safety hazard, especially when braking on wet, muddy, or gravelly surfaces. The best policy is to follow the tire or vehicle manufacturer's recommendations.

#### Suspension Check

In addition to checking tires, learn to do a thorough parts check. Loose wheel bearings, for example, not only distort the alignment readings, but are hazardous and can lead to expensive repairs. Loose steering parts greatly affect "running toe" as do other suspension parts. Loose balljoints or bushings not only change "running toe," but caster and camber as well. Bad stabilizer bar bushings or links can cause excessive vibration, especially on a rough corner. A bad strut rod bushing can cause an automobile to pull when the brakes are applied. The brakes may be fine, but they erroneously become suspect if the bad strut rod bushing remains undetected. The list of cause and effect is long. If loose parts are discovered during a prealignment check, alignment should not be adjusted until the necessary repairs are made. There is no way to set the static alignment correctly for the running alignment, given

the uncertain tolerances c⸱⸱ alignment on an automob⸱ parts will also compromis⸱ safety of the occupants.

Do not overlook any p⸱ the parts check. Jacking the heel of your hand wi⸱ to check that the idler a⸱ cylinder are bolted tightly ⸱⸱ sector shaft, because it can result in a ⸱⸱⸱ toe setting procedure, regardless of the method o⸱ ⸱⸱

#### Curb Height

After you have checked tires and parts, checking the curb height of the automobile is equally important. Even slight variations in curb height create problems, since suspension and steering parts are interrelated. If curb height is not correct, the entire system will be thrown out of alignment. Look for heavy objects in the trunk or passenger compartment if you suspect unequal weight distribution. The owner may have weighted the trunk with bags of sand for winter traction.

Check the springs. All four springs may have sagged, but not noticeably. Check the factory specifications for the proper curb height for the specific make and model automobile with which you are working. Physically measure this height; the eye just isn't accurate enough. If coil springs have sagged, they must be replaced in pairs. Torsion bars should be properly adjusted. The alignment procedure should be performed only after curb height has been corrected, because any change in curb height will change the alignment settings.

#### Shock Absorbers

Check the shock absorbers. A knee on the bumper will not yield a proper check. Check for leakage and drive the automobile to see how the shock absorbers react. At the same time, note any driving irregularities or peculiarities the automobile may have.

## Factors Affecting Wheel Alignment
### CONVENTIONAL SUSPENSIONS

#### Camber

Camber is the tilt of the wheel in or out at the top as viewed from the front and is measured in degrees from true vertical. When set at zero degrees of camber, the centerline of a wheel is at a 90° angle to the road surface. A wheel that is tilted out at the top is said to have positive camber. A wheel that is tilted in at the top is said to have

**Camber is the tilt of the wheel in or out at the top**

ero camber maximizes the area of the tire sur-
th the road surface.

l arms are of equal length, the wheel will remain
ause of the up and down motion, however, the arms
atively shorter and longer with respect to the horizontal
the road surface. This creates side to side scuffing and
ant tire wear. For example, as a wheel moves up or down it

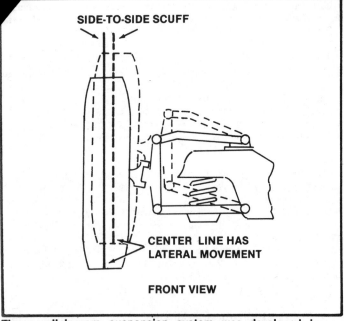

**The parallelogram suspension system was developed in Europe. The equal length upper and lower control arms caused side-to-side tire scuff**

will be brought closer to the chassis. Since it remains vertical, it will be forced to scuff at the road surface, and scuff again in the opposite direction when the wheel returns to the normal position.

By using a short upper control arm and a long lower control arm, the pivot point for the change in camber is located directly at the road surface. As the suspension moves up or down, the wheel stays in place at the road surface with respect to lateral travel, but moves in and out at the top. One might think that constantly changing camber would create excessive tire wear, but the movement is slight; indeed, locating the camber pivot point at the road surface actually reduces tire wear.

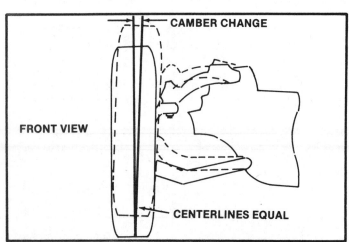

**As the suspension moves up and down the wheel stays in place at the road surface and moves in and out at the top**

Camber does not remain the same in a steered position as it is when the wheels are set for straight ahead. Many modern automobiles use high positive caster settings, which cause a change in camber while the wheels are steered. (Caster is the forward or rearward angle formed between true vertical and a line drawn through the steering pivot points to the road surface.) As an automobile is steered, positive caster creates positive camber on the inside wheel

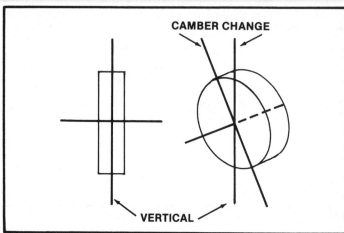

**Positive or negative caster causes a change in camber when the wheels are steered**

and negative camber on the outside wheel. Negative caster settings have the opposite effect. The sharper the automobile is steered, the greater the effect caster will have on camber. Consequently, this effect will be more pronounced at the lower speeds at which the sharpest turns are negotiated. In an automobile traveling at a high rate of speed around a slight curve, the effect of caster on camber will be quite small.

The lengths and positions of the upper and lower control arms cause camber to become negative as the spring is compressed. As the spring is expanded, camber changes to zero or slightly negative. Because of inertia, the body of an automobile will lean to the outside of a curve, thus compressing the outside spring. This is called weight transfer. Consequently, the forces that create side stress problems are also the forces that reasonably solve them. Changes in outside camber counteract side stress problems created in turns.

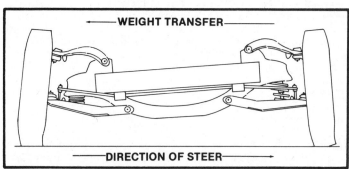

**In hard cornering, because of inertia, weight shifts to the outside wheel causing that wheel to have negative camber. The negative camber on the outside wheel partially counteracts the side stress forces**

Many alignment mechanics have not considered camber to be a directional control angle, but it is; a simple illustration shows why this is so. A cone shaped object will tend to roll in a circle, and so will a tire if forced to roll in a conical position. Thus an automobile with negative camber on the right side and positive camber on the left side will tend to pull to the left if this condition is severe enough, assuming the road is flat and all other alignment angles are set properly.

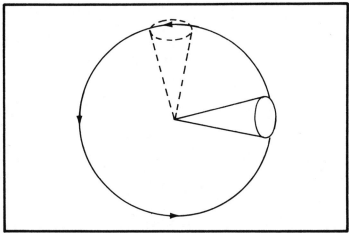

**A cone will roll in a circle**

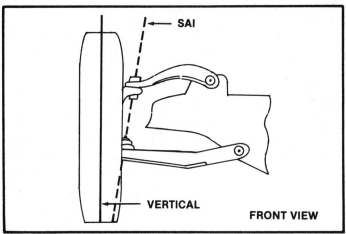

**Steering axis inclination**

Another simple example demonstrates how seriously camber affects directional control. The left sides of the tires in the illustration of misaligned camber are smaller in diameter and are also buckled, thus creating a greater rolling resistance on the left. The right sides of the tires are larger in diameter and are more free-rolling. This forced conical position of the tires creates a pulling condition that would become more severe if the brakes were applied. Thus, misaligned camber not only causes uneven tire wear—it can also be a serious safety hazard. Uneven tire wear, of course, is expensive, but an automobile that loses stability when the brakes are applied is dangerous.

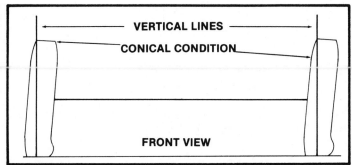

**A badly misaligned camber can create a conical condition**

Each vehicle manufacturer has established the proper static camber setting for maximum tire wear; this specification considers vehicle loads. Many specifications also call for less camber on the right side, to counter the effect of road crown. Due to the different tolerance specifications used in the manufacturing of different automobiles, camber specifications will vary from one automobile to another. There is no single camber specification that can be used successfully on all automobiles. For a proper camber adjustment, each wheel must be brought to the exact position of the thrust line before camber corrections or adjustments are made, and it probably will be necessary to reset the wheel to the thrust line position if a great amount of camber adjustment is made during the procedure. Altering camber specifications is not recommended.

## Steering Axis Inclination (SAI)

The next angle to be examined is steering axis inclination (SAI), also referred to as kingpin inclination (KPI) or balljoint inclination. Steering axis inclination is viewed from the front and is found by drawing a line (also called the pivot line) through the centers of the upper and lower balljoints and extending the line to the road surface. The degrees of difference from the true vertical to the pivot line is the SAI. Steering axis inclination is created by moving

the upper balljoint closer to the center of the automobile relative to the lower balljoint. (No consideration is given at this point to the position of this balljoint relative to the front or rear of the automobile; this is called caster, and is covered in the next section.)

When an automobile is steered, SAI produces an equal downward force on both front wheels. To see why this is so, look at the tip of the wheel spindle in the illustration. Because of the tilt of the steering knuckle and spindle, the spindle moves through an arc.

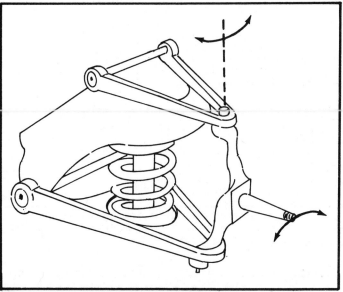

**When an automobile is steered in either direction, SAI produces an equal downward force on the wheels**

The highest point of the arc is when the spindle is centered—that is, when the wheels are straight ahead. If the spindle is turned from this position, it moves through its arc, resulting in a downward force on the wheel attached to it. Since the wheels cannot be forced into the road, the car is raised; it returns to its original lower position when the wheels are straightened out. The force of gravity causes the automobile to go straight ahead unless forced or steered in another direction. When the steering wheel is released after a turn, the weight of the car will make the automobile return to the straight ahead position.

If the inclination angle is less on one side than on the other, the automobile will pull to the lesser side. The automobile will steer easier to the side of least inclination, and if the condition is severe enough, the steering wheel may have to be helped to the straight ahead position.

On some earlier independent suspension systems, a large amount of SAI was used and hard steering resulted. Steering also returned to straight ahead quite readily. This was offset by the use of negative caster. Thus, it is possible to set an automobile with negative caster, and have it return to straight ahead without wandering. Ideally, a properly engineered SAI angle will give the wheels a tendency to return to the straight ahead position, giving stability and directional control, but without creating hard steering.

The introduction of SAI not only aided directional stability of the automobile—it also permitted the use of four wheel brakes. If the upper balljoint were directly over the lower balljoint, the pivot to the road surface would be vertical. If the automobile were in a

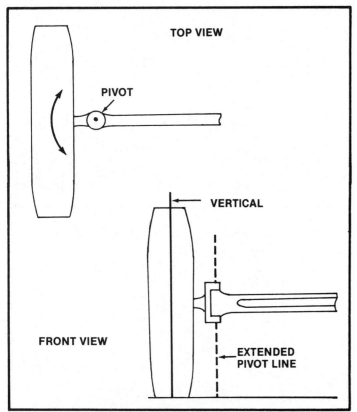

**Without SAI the wheel is forced to rotate around the pivot as it is steered. The extended pivot line does not point to the contact point of the wheel on the road surface**

static position and the wheels were steered, they could rotate around the pivot only if they were free to roll. If brakes were applied, the wheels could not roll, making it impossible for the wheels to rotate around the pivot. The same principle would apply if the automobile were in motion. The harder the braking action, the harder the automobile would steer. Further, irregular road conditions or unequal braking would cause serious pulling conditions.

Earlier engineers tilted the top of the kingpin inward so that the centerline of the pivot projected to the contact point of the wheel on the road surface. This new angle was termed "kingpin inclination," and it eliminated the need for the wheel to be free to roll as it was steered.

Another component of the SAI angle is scrub radius. Scrub radius is the distance between the pivot line and the centerline of the wheel at the tire's contact point with the road. Only one pivot point exists in the tire footprint. The rest of the tire footprint must rotate around this point as the wheels are steered. At first, engineers believed this point should be as close to the centerline of the tire as possible. Later, however, they discovered that a pivot point in the exact center of the tire caused a scrubbing action in opposite directions within the same tire footprint. That is, when the wheels

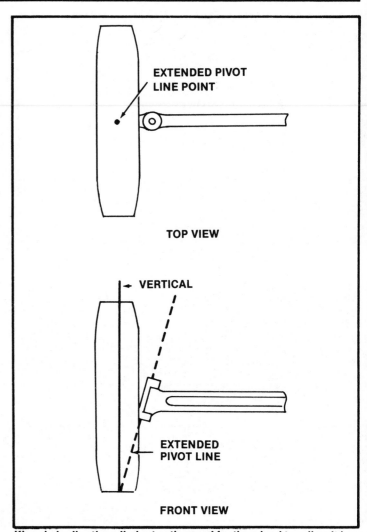

**Kingpin inclination eliminates the need for the wheel to roll as it is steered. The wheel can be steered in place**

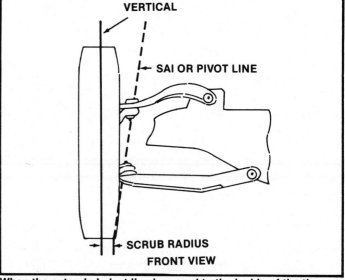

**When the extended pivot line is moved to the inside of the tire footprint, the distance from the center line is the scrub radius**

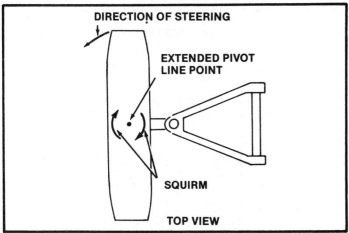

A pivot point in the exact center of the tire footprint causes scrubbing action in opposite directions when the wheels are steered. This is called squirm

were steered, tire surfaces were forced to scrub one direction on one side of the pivot and in the opposite direction on the other side. This scrubbing action is called squirm and it causes a slight instability in turns.

The remedy for squirm is to bring the pivot point to the inside edge of the tire, causing the scrubbing action to be in one direction only and relieving the squirm of the tire. The result is better stability while turning with no appreciable difference in the tire wear.

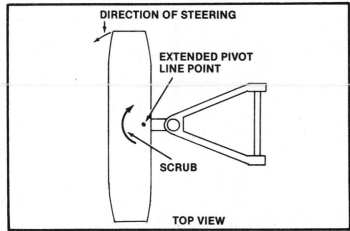

Scrubbing in one direction only relieves tire squirm

As we have seen, camber is the inward or outward tilt of the wheel at the top, and steering axis inclination is the inward tilt of the steering pivot line. The arithmetic sum of camber and SAI is called the included angle. Included angle, like caster and SAI, is referenced from the true vertical. It is a nonadjustable angle that cannot change unless the spindle assembly has been damaged. By determining the camber and SAI, the mechanic may then use the included angle to determine if suspension parts have been bent.

In the example shown here, the SAI is at 8° and camber at 0°, with an included angle of 8° (8° plus 0° equals 8°). If the camber angle were adjusted to +2° (positive), the SAI angle would become 6°. It makes no difference how this adjustment is made. The top balljoint must be moved out 2° or the bottom balljoint in 2° to change the camber to +2°. Since SAI as well as camber and the included angle are referenced from the true vertical, the SAI has been changed to 6°. The included angle did not change (6° plus 2° equals 8°). Camber was the intended adjustment, but in the process, the SAI angle also changed.

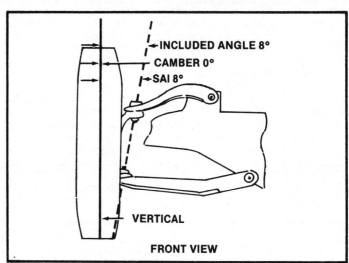

The arithmetic sum of camber and SAI equals the included angle

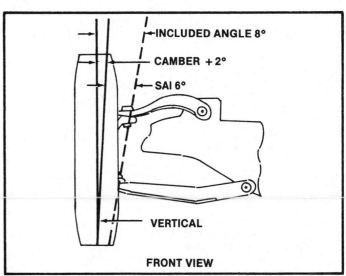

When the camber is adjusted, SAI also changes

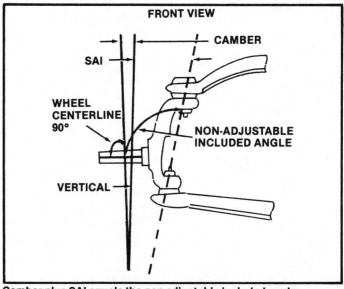

Camber plus SAI equals the non-adjustable included angle

# SUSPENSION AND STEERING
## THEORY AND PRINCIPLES

There has been some confusion as to whether the SAI angle is adjustable or not, as a result of the common misconception that the angle of the spindle assembly is the reference for the SAI. It is not. The part of the spindle that connects the upper and lower balljoints is referenced to the axle angle. This angle (the included angle) is manufactured into the spindle assembly and is not adjustable. If the included angle is referenced to the true vertical, the SAI and camber angles can be found. The line that is at a 90° angle to the axle is the centerline of the wheel. This line is measured against true vertical to determine the degree of camber. The line drawn through the portion of the spindle that connects the balljoints is referenced to the true vertical and is the SAI. Camber plus SAI equals the nonadjustable manufactured angle—the included angle. (When the camber is negative, it is subtracted from the SAI angle to find the included angle.)

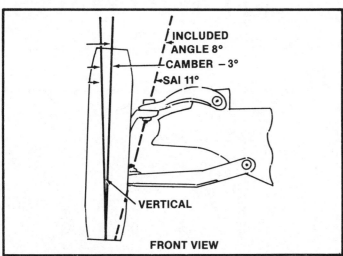

If the camber is negative, it is subtracted from the SAI angle to find the included angle

Knowing and understanding SAI, camber, and included angle can help the mechanic to quickly determine if there is a bent part and where it is located. For example, the factory specification requires an SAI of 6° and camber of +2°, the included angle will be 8°. If these specifications can be met through normal adjustment procedures, the spindle assembly, upper control arm and lower control arm are not bent. However, let's examine some cases where there are bent or damaged parts using the above specifications.

*Problem One:* Assume the SAI is 6° and the camber is 0°. SAI of 6° plus camber of 0° equal an included angle of 6°. Using the above specification of 8°, this is incorrect. If camber were adjusted out to +2°, the SAI would be reduced to 4°. SAI of 4° plus camber of +2° equal an included angle of 6°—still incorrect. If the included angle specification cannot be met, the spindle is bent.

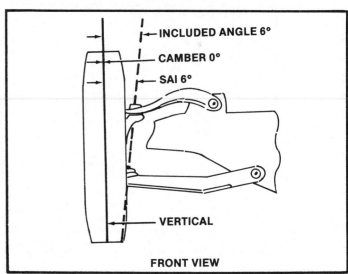

If the included angle specification cannot be met, the spindle is bent

*Problem Two:* Assume an SAI of 4° and a camber of +4°. The included angle is correct, indicating that the spindle is not bent, but SAI and camber are incorrect. Suppose that camber cannot be adjusted by more than 1° toward negative. The SAI would then be 5° and the camber +3°. Included angle remains correct, but SAI and camber are still incorrect. If the SAI is less than the specification and the included angle is correct, the lower control arm is bent.

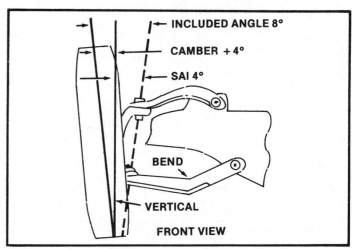

If, after maximum adjustment, camber is greater than the specification, SAI is less than the specification and the included angle is correct, the lower control arm is bent

## REFERENCE CHART

| Angles | | | Problem |
| SAI | Camber | Included angle | |
| --- | --- | --- | --- |
| Correct | Less than specification | Less than specification | Bent spindle |
| Less than specification | Greater than specification | Correct | Bent lower control arm |
| Greater than specification | Less than specification | Correct | Bent upper control arm |
| Less than specification | Greater than specification | Greater than specification | Bent lower control arm and spindle |

*Problem Three:* Suppose another automobile has an SAI of 8° and camber of 0° and camber can be adjusted to only + 1°. SAI would then be 7°. The included angle is correct, but SAI and camber are incorrect. If the SAI is greater than the specification and the included angle is correct, the upper control arm is bent.

If the included angle is incorrect, the spindle is bent. If the SAI is also incorrect, both the spindle and another part are bent. Parts usually do not stretch when bent, so something is probably shorter. An SAI greater than or less than the manufacturer's specification indicates where to look for damage. This does not take into account any bent frame rails or crossmembers.

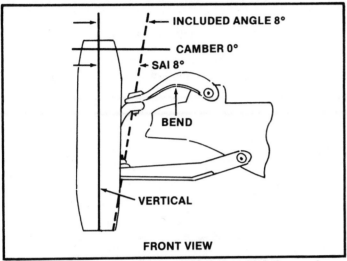

If, after a maximum adjustment, camber is less than the specification, SAI is greater than the specification, and the included angle is correct, the upper control arm is bent

## Caster

Caster is referenced from the side and is the angle formed by drawing a line through the upper and lower balljoints and extending this line to the road surface. Caster angle is measured in degrees from true vertical. If the upper balljoint is located more toward the rear of the automobile than the lower balljoint, caster is positive. If the upper balljoint is ahead of the lower balljoint, caster is negative.

Caster is not considered to be a wear angle, but is used to aid directional stability, determine the amount of effort required to turn the wheels from the straight ahead position, and assist in the returning action of the wheels to the straight ahead position.

To better understand the effect of caster angle on the steering of

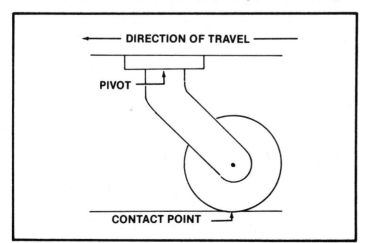

When permitted, the caster will always swing to the rear of the pivot

an automobile, consider the action of a caster wheel on a tool cabinet. If the caster wheel is permitted to pivot freely, the contact point is always behind the pivot point. By reversing the direction of the cabinet, the caster will swing around behind the pivot point.

In the accompanying illustration, line A has been drawn through the pivot to the floor. Another line, B, has been drawn through the axle of the wheel to the point where the pivot line touches the floor. Measuring in degrees from line B to the true vertical indicates the degree of positive caster.

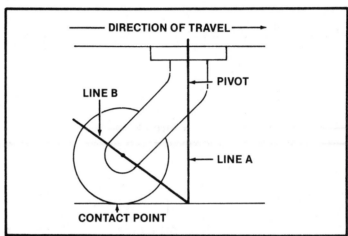

Measuring in degrees from line B to line A indicates the degree of caster

If the caster were locked so that it could not swing and the direction of movement of the cabinet were again reversed, the cabinet would not be directionally stable. Although the caster could not swing, it would want to do so. This illustrates negative caster and suggests the possible effect of negative caster when applied to an automobile.

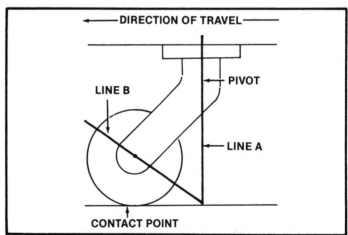

Locking the caster and changing the movement of the cabinet would create instability

If the caster fork were replaced by balljoints, line A would be drawn through the balljoints to find the angle of the pivot. In this case, the contact point is behind the pivot point showing positive caster. To determine the degree, measure from the true vertical.

If you could see down through the top of a tire, you would see the tire footprint on the road surface and also the part of the footprint to which the pivot line is pointing. In the case illustrated, the pivot point is exactly at the center of the tire footprint, indicating zero (0) degrees caster. This means the footprint of the tire on the road surface is divided evenly. The wheel would have no tendency

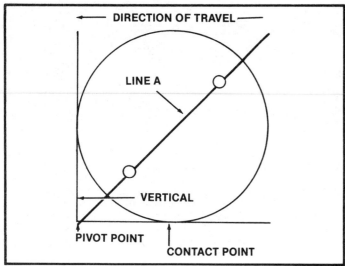

The contact point is behind the pivot point showing positive caster

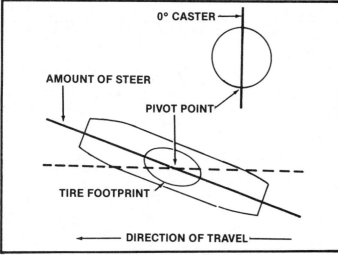

Zero caster contributes nothing to directional stability

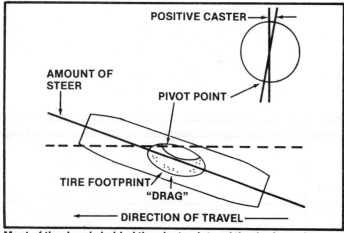

Most of the drag is behind the pivot point and the tire force that makes the wheel return to the straight ahead position

to turn in either direction, because zero caster contributes nothing to directional stability.

When caster is positive, the extended pivot point is ahead of most of the tire footprint. The portion of the tire footprint to the inside of the pivot point equals the portion to the outside. The remainder of the footprint (the dark area in the figure) must "drag" while being steered. Most of the drag or friction is behind the pivot point and is the tire force that makes the wheel return to the straight ahead position.

Negative caster is illustrated when the pivot point is to the rear of the footprint. Again, the portion of the inside of the footprint with respect to the pivot point is equal to the portion to the outside. In the case of negative caster, most of the drag is ahead of the pivot point. Thus, when the automobile is steered, friction tends to turn the wheel farther from the straight ahead position.

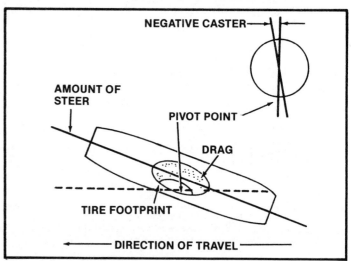

Most of the drag is ahead of the pivot point and will make the wheel tend to steer farther

If the tire forces were the only forces involved in caster and directional stability, any error in caster balance from side-to-side would be revealed while turning. However, other forces are involved. When the automobile is steered, positive caster forces the spindle, or axle, down on the inside wheel, which raises that side of the automobile. On the outside wheel the spindle is raised, which lowers that side of the automobile. When the steering wheel is

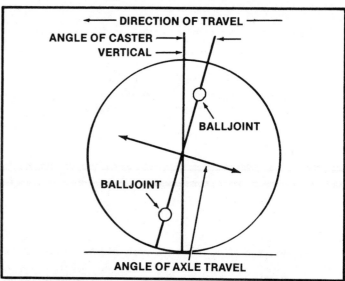

Gravity will tend to equalize the weight of the automobile forcing the front wheels to return to the straight ahead position

released, and the caster is equal on both sides, gravity will tend to equalize the weight of the automobile on the front wheels, forcing the wheels to return to the straight ahead position.

If an automobile is set with negative caster, the inside spindle will rise and permit that side of the automobile to lower. The outside spindle will lower and raise that side of the automobile. With a negative caster setting, the "drag" of the tire footprint is ahead of the pivot. If this condition were severe enough, the "drag" forces of the tire footprint could overcome the weight equalization and SAI forces, and the wheels would not tend to return to the straight ahead position. This would be an extreme condition but helps to illustrate the next point. If an automobile had positive caster on one side and negative caster on the other, a pulling condition would exist. If the automobile were steered to the side of positive caster, both sides of the automobile would raise, as shown earlier. Because of gravity, the automobile would seek its lowest point. Both sides of the automobile would be lowered if the wheels were steered to the side with negative caster. The "drag" force of the negative side footprint (which is mostly ahead of the pivot point) would be stronger than the drag force on the positive side footprint (which is mostly behind the pivot point). If the size of the drag in the negative side footprint were equal in area to the size of the drag in the footprint on the positive side, the amount of drag on the negative side being pushed would be greater. The uneven forces, coupled with the tendency of the automobile to seek the lowest possible point, would cause the automobile to continue to pull.

A pulling condition may exist even without positive caster on one side and negative caster on the other. Assuming all other angles are set to proper specifications and the road surface is flat, an automobile will pull to the side that is set with the least positive caster. With an imbalance in caster settings, pulling will be constant and not greatly affected by braking.

High pressure tires were used on very early automobiles, resulting in a small tire footprint. There was very little tire drag, and these automobiles were very unstable at higher speeds. Large amounts of positive caster increased stability, but also created hard steering and increased road shock at the steering wheel (road shock caused by positive caster results from the pivot line being pointed at irregularities of the road). A considerable amount of shock was transferred directly through the steering mechanism to the steering wheel. High pressure tires were replaced by "balloon" tires that had much larger footprints. The increase in drag permitted a reduction in positive caster settings, resulting in easier steering, equal or better stability, and much less road shock.

The development of radial tires has provided a much larger footprint, which increases steering drag and stability, but since more road shock is absorbed by the flexible sidewalls of the radial tire, caster can be increased to more positive settings. Hydraulic power steering systems cushion road shock and have overcome hard steering associated with high positive caster, but at the same time, hydraulic power steering systems resist sudden reversals of direction. This offsets the increased force with which the wheels return to the straight ahead position as a result of increased positive caster and explains why some automobile models have very high positive caster settings with power steering but less positive caster without power steering.

Caster angle is also influenced by the horizontal plane of the automobile. Rear springs that sag will increase the caster angle toward positive. Front springs that sag will decrease caster toward negative.

Altering the height of the rear of an automobile by extending rear shackle lengths or adding rear air shock absorbers pumped all the way up can change the intended front suspension angles to the point of making the automobile dangerous to drive. Raising the rear of the automobile not only upsets the front end geometry, it also increases rear end sway and weakens every aspect of safety designed into the automobile. Automobiles used in competitive racing that have had suspension and steering systems altered have been totally redesigned by experts. Complete systems have been changed in harmony and are geometrically designed to handle safely at high speeds under specific conditions. These automobiles would not drive or ride well at normal speeds under all conditions.

Caster specifications are not normally altered and sometimes are not adjustable. However, the results are predictable if the angles are altered even slightly. If the caster settings are equally decreased to the negative on an automobile without power steering, the automobile will tend to steer easier but will also tend to wander at high speeds, and consequently will not be as stable. The automobile will be more affected by crosswinds. An older person who drives only in town might find the automobile easier to steer. Nonetheless, the reduction in caster is not recommended. Increased positive caster can improve high speed stability but it also can create some adverse conditions. For example, road shock increases and a greater work load is placed on the power steering system, potentially shortening its life. It is for these reasons that changes in caster settings from the factory specifications are not recommended.

## Toe

Toe, viewed from the top, is the difference of the distance between the extreme fronts of the wheels and the distance between the extreme rears. If the distance is less between the wheels at the front than at the rear, the wheels are toed in. If the distance is greater, the wheels are toed out.

As far as tire wear is concerned, toe is the most critical of alignment adjustments. A wheel suffering from an incorrect toe setting will scuff along the road surface as it is forced to continue in the direction of travel. Any other adjustment made will affect toe, and thus toe is always the last alignment adjustment made.

In rear wheel drive automobiles, toe is set slightly inward while

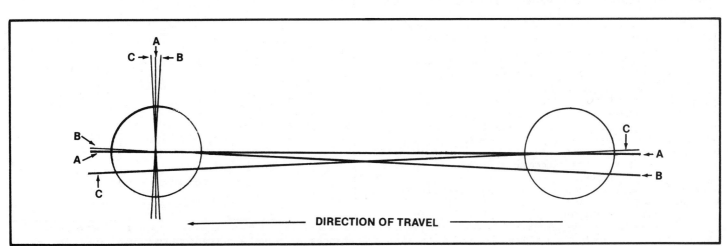

DIRECTION OF TRAVEL

**The caster angle is influenced by the horizontal plane of the automobile**

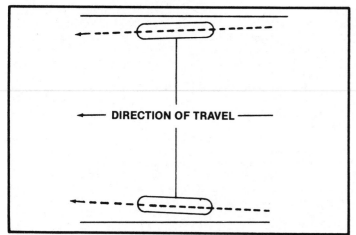

**Toe is the distance between the extreme fronts of the wheels compared to the distance between the extreme rears**

the car is in a stationary position, because toe changes as the automobile is driven. Rolling resistance of the tires, coupled with the rearward movement of the suspension, changes the setting of the steering linkage (and thus the toe) as the automobile moves forward. If the steering linkage is in the front of the suspension, it is expanded or pushed open. If the steering linkage is in the rear of the suspension, it is compressed. Either way, the toe opens when the automobile moves forward. The "running toe" (toe setting when the car is moving forward) will be zero when all the normal steering tolerances are taken up.

Front wheel drive automobiles have the same travel, but in the opposite way. Because the front wheels are pulling the automobile, they tend to pull inward toward each other. The specifications of these automobiles commonly call for toe-out in the static position.

The slightest amount of looseness in the steering linkage or suspension parts can create a very unfavorable effect on the "running toe." Looseness will cause running toe-out even if a proper static setting has been made. Although too much running toe-in or running toe-out will have a serious effect on tire wear, even a small amount of running toe-out can cause the car to wander and will seriously affect directional stability while braking.

Early model automobiles used a great amount of positive camber. For many years it was thought that positive camber would give the wheels a tendency to turn out due to the forced conical position of the tire, so additional toe-in was specified. However, when the "balloon" tire was introduced and higher road speeds were common, tires began to wear unevenly, showing greater wear on the outsides. Research and tests proved that there is no relationship between the angles of camber and toe. They are entirely independent.

In older cars equipped with a single beam axle, the steering linkage was a simple affair that attached one end of a tie-rod to the arm that emerged from the steering box (the pitman arm) and the other end to a steering arm at one wheel. The other wheel was connected by means of another tie-rod that ran from its steering arm to the other tie-rod. As the automobile was steered, the pitman arm pushed or pulled the tie-rod steering the one wheel. Simultaneously, the other tie-rod was pushed or pulled by the first so that both wheels were steered. The system included an adjustment to set toe. As the wheels were forced up and down by road irregularities, the entire system went up and down together. This was very simple and worked well, but when independent suspensions appeared, steering geometry changed.

Early independent suspension designs exhibited a tendency for the front wheels to close as the frame of the car dropped due to road irregularities. As the automobile moved upward from going over a rise, the wheels opened. Any additional weight in the automobile would lower the frame, lengthen the tie-rods, and spread the wheels at the rear.

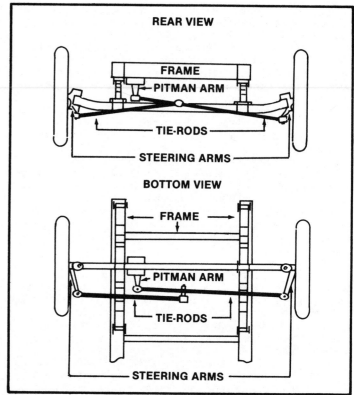

**Single beam axles had a very simple steering system**

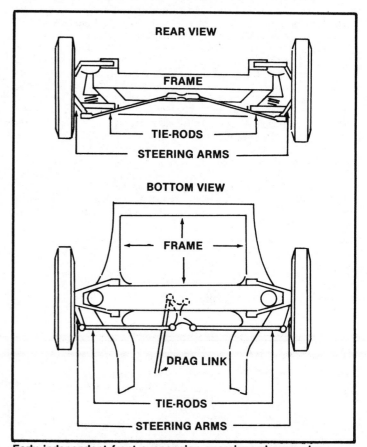

**Early independent front suspension experienced excessive changes in toe as the suspension was raised or lowered from load changes and road conditions**

In the modern automobile, tie-rods are nearly the same length as the lower control arm, and are positioned in the automobile at approximately the same angle as the lower control arms. If the frame is raised or lowered from any cause, the tie-rods assume the same position and angle as the lower control arms. Consequently, they do not change lengths (or the toe setting of the wheels). The inner ends of the tie-rods are attached to a center link (or, in a rack and pinion system, the steering rack), and only the center link (or rack) moves up and down with the frame.

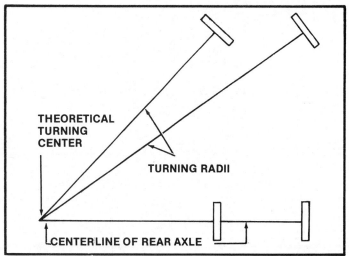

Toe out on turns prevents the tires from scuffing and adds stability when the automobile is steered

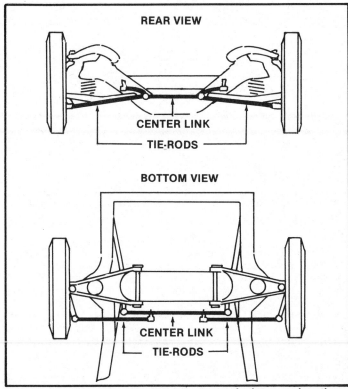

In modern automobiles, the tie rods are nearly the same length as the lower control arms, and are positioned at approximately the same angles as the lower control arms

Although there are many different types of steering systems in use today, they all conform to the same geometric principles. Some steering linkages are in front of the suspensions and some are behind. In most cases, the location of the steering linkage is determined by the available space in the design of the car. Regardless of the location of the steering system, it will be built in harmony with the suspension system.

Some suspension and steering systems are intentionally built with greater tolerances. Consequently, no single static toe setting can be used on every automobile. It is extremely important that static toe be set exactly to the manufacturer's specification for the specific automobile being aligned. It is also important that toe be set using the thrust line as the reference. Any deviation in the thrust line from the centerline will cause toe-out on turns and affect the running toe.

### Toe-Out On Turns

Zero toe is the ideal toe when driving straight ahead, but what happens to the toe when the automobile is steered? The wheels must toe-out as they are steered, and the farther the wheels are steered, the greater the toe-out must become. This change is called toe-out on turns. That is, when viewed from the top, the distance between the fronts of the two wheels is greater than the distance between the rears while the wheels are turned. The more sharply the wheels are turned, the greater the difference in the two dis-

tances. Toe-out is designed to prevent the tires from scuffing and to add stability in turns.

Although it might seem logical that zero toe would be the ideal setting under all conditions, a closer look at steering geometry demonstrates that this is not so. When an automobile is steered, both front wheels revolve around the same center. Regardless of the direction steered, the outside wheel steers in a larger arc than the inside wheel. The inside wheel is also ahead of the outside wheel and must steer more sharply to remain perpendicular to the radius. When the steering is returned to the straight ahead position, the wheels must return to the parallel position.

This is similar to the principle of a lever moving in a circle. In the illustration, the lever pivots at point A. If the other end moves from point B to point C, the lever has travelled half the distance along the horizontal line but not half the distance of the arc. If the lever is moved to point D, it has moved the other half of the horizontal line, but a much greater distance of the arc.

When we apply this principle to the automobile steering mechanism, we see (in the next figure) that both levers (steering arms) have travelled the same distance on the horizontal line, but because the levers are on different segments of their arc, the connecting lines (wheels) have turned at different angles. The centerline of the automobile and the mid-point of the rear axle determine the *theoretical* angle of the steering arms, which are not parallel to the

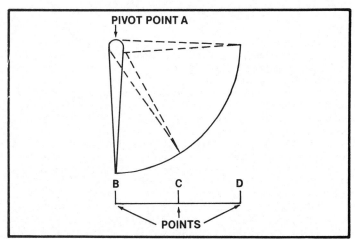

If the lever is moved to point C, the lever has travelled half the distance of the horizontal line, but not half the distance of the arc

# SUSPENSION AND STEERING

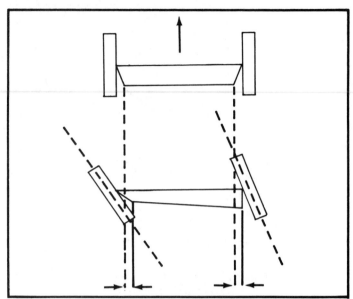

The levers (steering arms) are in different segments of their arcs demonstrating different turning angles

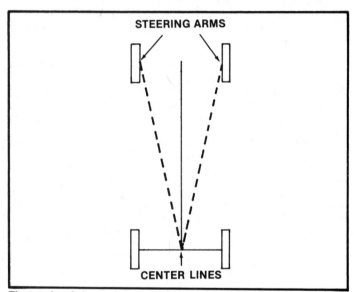

The angle of the steering arms is theoretically determined by the wheelbase of the automobile

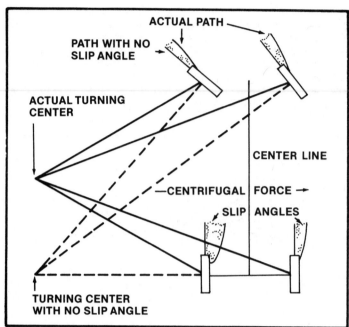

The actual turning center is considerably ahead of the theoretical turning center

front wheels. Each is angled inward, pointing directly to the center of the rear axle, and toe-out on turns becomes automatic as a result of this design feature.

Another consideration in determining the angle of the steering arms is the slip angle. As the speed of the automobile increases, centrifugal force acts upon the automobile and causes all the tires to slip somewhat. The tires are not actually slipping, but rather are twisting at the point of contact. Because of this twisting condition, the actual turning center is considerably ahead of the theoretical turning center, and the difference between centers is referred to as the slip angle.

If the front slip angle is greater than the rear slip angle, understeer exists. With an understeer condition, the steering wheel must be progressively turned inward to maintain the same turning radius. The greater the speed, the worse the condition. If the rear slip angle is greater than the front slip angle, oversteer exists. With an oversteer condition, the rear of the automobile will tend to slide to the outside of the turn at high speeds, and the steering wheel will

have to be turned to the outside of the turn to prevent the automobile from spinning around. Equal front and rear slip angles create neutral steer, which while desirable, is rarely attained. Varying speeds, varying amounts of fuel in the tank, and changes in number and location of passengers in the automobile all contribute to the changes in slip angle. An automobile decelerating into a turn may have understeer that changes to neutral steer as power is lightly applied through the turn. Steering characteristics may further change to oversteer as the automobile comes out of the turn with still more power applied. Modern automobiles with normal alignment settings, proper tire sizes, and proper tire pressures may display no perceptible symptoms of understeer or oversteer at normal road speeds. Nevertheless, varying degress of understeer and oversteer will occur, depending on variations in speed, weight distribution, and road conditions. An automobile with two underinflated front tires or two overinflated rear tires would produce an understeer condition, perhaps noticeable at normal road speeds. If the unequal tire inflation were reversed, the automobile would experience oversteer. Thus, proper tire size and inflation are critical to the handling ability originally engineered into the automobile.

Many other factors are involved in determining slip angles—engine size (power-to-weight ratio), tire size, tire pressures, suspension geometry, firmness of the suspension, stabilizer bars, weight distribution, weight transfer, center of gravity and the roll center are all considered in determining proper angles for the steering arms.

Different automobiles have different slip angles. Consequently, different automobiles with the same length wheelbase may have various toe-out-on-turns specifications. Use of the specification set forth by the manufacturer for the specific automobile being checked is important.

Toe-out on turns is a nonadjustable angle. The steering arms are installed with preset manufactured angles. If the toe-out on turns specification cannot be met, one or both steering arms is bent.

This is a critical wear angle which should be checked in every alignment procedure. An incorrect toe-out on turns will adversely affect tire wear and stability while steering.

## Thrustline and Centerline

### FRAME-TYPE AUTOMOBILES

Proper frame alignment is very important to proper alignment of all four wheels and while frame straightening is a special operation, the alignment mechanic should know how to check frame alignment. A misaligned frame can greatly affect the overall alignment, including the straightness of the steering wheel. The proper methods for correcting a bent frame are covered in Volume 3 of Chilton's *Mechanics' Handbook* series.

The centerline of the frame is the line from which all alignment references begin. The manufacturing of the frame requires that it be made within certain tolerances. The same applies to the installation of the front and rear suspensions. Since *tolerance* means "an acceptable margin of deviation from a given norm," all automobiles are not geometrically perfect or exactly alike. Some automobiles are manufactured and assembled on the outer limits of all tolerances and, while not geometrically perfect, are still within the general manufacturing tolerance of the manufacturer.

In most frame-type automobiles, adjustments provide compensation for the geometric imbalances. Most of these automobiles provide for front suspension adjustments, but few provide for rear suspension adjustments. Frame alignment is checked by referring to the frame centerline. However, if all the adjustments can be made to specification, and all other geometric angles are within tolerance, checking frame alignment is not necessary. If one or more angles cannot be brought to specification, a further check must be made to find the cause of the problem.

When an automobile is driven straight ahead, all four wheels should run parallel to the frame centerline. Front and rear axles should be at, or nearly at, 90° angles to the frame centerline.

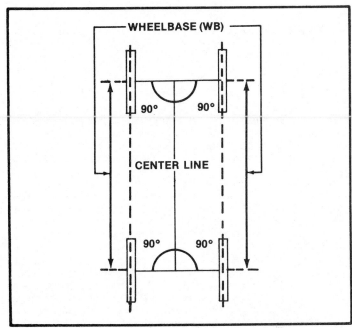

All four wheels should be parallel to the frame centerline

The wheelbase is the distance from the center of the front axle to the center of the rear axle. Any difference in side-to-side wheelbase measurement can affect alignment. Some common causes for deviations in wheelbase measurement are:

1. Front wheel set back
2. Rear axle shift
3. Front wheel set back and rear axle shift
4. A swayed frame
5. A diamond frame.

In a "set back" frame, one front wheel is set back closer to its corresponding rear wheel. Thus, the wheelbase on one side is shorter. If the condition is caused by a bent suspension and is severe enough, a proper alignment probably cannot be obtained. If

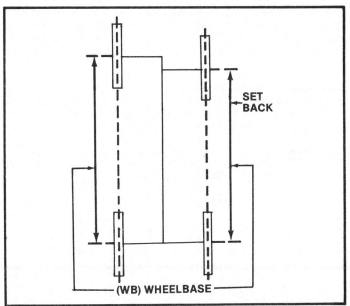

Set back damage. Although parallel to the frame centerline, the one wheel is set back

the set back is caused by a bent frame, a proper alignment is not possible. And if it is severe, the automobile likely will pull toward the shorter wheelbase.

"Rear axle shift" is a condition in which one side of the rear axle has shifted toward the front or the rear. This can be determined by measuring from a common point on the frame to the rear wheel on each side of the automobile and comparing the measurements. A modern alignment machine will find and measure rear axle shift in minutes. Regardless of the direction of the shift, or the cause, the

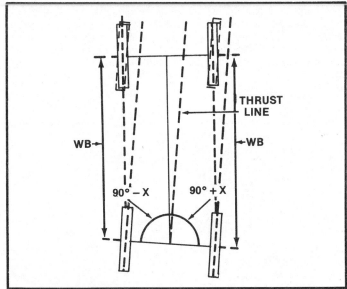

Axle shift. The left rear wheel has shifted toward the front of the automobile changing the thrust line which also changes the direction of travel

thrustline (see the next section) has been changed to a different angle relative to the frame centerline, changing the direction in which the automobile will tend to travel. The thrustline is the true direction of travel of the automobile. If the front wheels remain parallel to the frame centerline, but the thrustline has changed, the automobile will no longer tend to go straight. For the automobile to go straight ahead, the front wheels must be steered in the exact direction of the thrustline, resulting in "crabbing" under the conditions described here. Many automobiles have an acceptable amount of deviation in thrustline from its true position. This can be compensated for, and a straight steering wheel obtained, through normal alignment procedures.

A swayed frame occurs when the automobile has been hit at an angle in the front, resulting in a shorter opposite side wheelbase. The rear wheels will not track properly with respect to the front wheels. Such an automobile cannot be made to drive properly or safely until the frame alignment has been corrected.

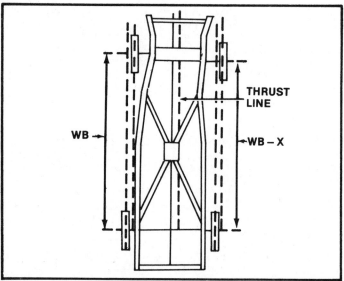

**Swayed frame. The automobile has been hit in the left front area changing the wheelbase and it no longer tracks properly**

A diamond frame means that one side frame rail has been driven back such that the frame is no longer rectangular. Any automobile with this problem must be sent to a frame or body shop before an alignment can be completed. Faulty driving and tracking symptoms caused by a diamond frame could be similar to those caused by rear axle shift, so identifying the difference between the two problems is important.

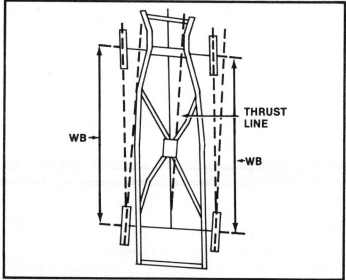

**Diamond frame. The automobile is no longer rectangular and the thrust line has been changed**

Rear axle offset does not affect side-to-side wheelbase measurement, but while the wheels remain parallel to each other and to the frame centerline, the rear wheels will not track properly as compared to the front wheels. However, the direction of the thrustline remains parallel to the centerline. The human eye cannot easily detect problems like this, but modern alignment machines are capable of distinguishing rear axle offset from rear axle shift.

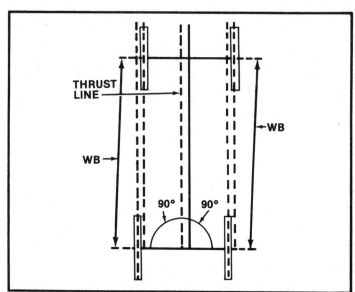

**You must be able to determine if rear axle offset is within tolerance**

## HOW TO DETECT FRAME PROBLEMS

Proper frame alignment is the foundation of properly functioning steering and suspension systems. Calibrated frame gauges quickly determine the extent and location of any bend in the frame. Even the slightest bend of any type will be detected. Frame gauges are of reasonable cost, can be installed in minutes, and are not difficult to learn to use and read.

The best alignment mechanics know all of the angles of suspension and steering systems and how to use and read frame gauges. They also have alignment machines that can read and measure all of the angles discussed in this chapter.

The alignment machine will indicate to the mechanic if the automobile is in "square." An automobile in "square" is actually rectangular in shape, but the term "square" indicates that: (1) the thrustline and centerline coincide, (2) all four wheels are parallel to each other and to the centerline, (3) front and rear axles will be at right angles to the centerline, and (4) front and rear axle midpoints will be at the vehicle centerline. With this knowledge and equipment, you can quickly and accurately analyze any alignment problem.

Assuming the front and rear suspension adjustments are correct, the next step is to adjust the steering. Whether the frame centerline, rear wheels, or the thrustline method is used as a reference in adjusting the toe, the steering box must be brought to "top dead center" with equal turns in either direction. After the steering wheel has been locked in this position, all toe adjustments are then made from this reference. (If the steering wheel cannot be centered with equal steering wheel turns in either direction, either the pitman arm is bent or the steering wheel has been removed and repositioned).

Obtaining a straight steering wheel is one of the more difficult tasks in the alignment procedure. Allowable tolerances of the manufacturing process have their greatest influence in the steering wheel adjustment. If the automobile is geometrically perfect, or nearly so, the centerline or rear wheels can be used as a reference to set toe and obtain a straight steering wheel. However, if the automobile is less than geometrically perfect, these two references will produce a crooked steering wheel. Since the thrustline is the true direction of travel of the automobile, toe is set with reference to the thrustline, which also assures a straight steering wheel.

## "UNITIZED" AND "UNIBODY" AUTOMOBILES

Because of the need to reduce weight and increase fuel efficiency, the heavier frame-type automobile is being replaced by the lighter, more fuel efficient "unitized body" and "unibody" chassis/body units. The body is bolted to the frame of a frame-type automobile.

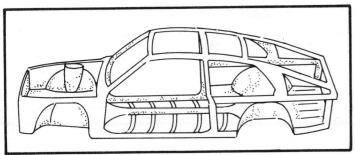

**Unitized body construction**

All alignment references begin at the frame centerline, with no references made to the body. In unitized body construction, a short box-frame extends toward the front and toward the rear to support the suspension, engine and driveline components. The body itself is the main structural member. Unibody construction is a single, completely spot-welded, fully integrated body frame assembly with inner fender panels braced within the structure to support the suspension and other components. These two types of construction have replaced the independent frame assembly. With unitized and unibody construction, all angles that are referenced to the true vertical pose no problems; however, these and all other angles must be referenced to the body centerline. This change of reference has created new challenges in the total alignment procedure.

If the equipment used to align these chassis/body units has not been specifically designed for this type of automobile construction, it is obsolete regardless of how new it is or how much it cost. Such equipment provides no means of finding the basic reference from which the alignment procedure must start. Frame gauges may be used to find the body centerline, but this information cannot be

fed to an outdated alignment machine. These automobiles have no benchmarks on the underside of the body as an accurate or quick reference from which misalignment might be physically measured. Therefore, the alignment machine must be able to find the body centerline and use it as a reference to bring the automobile to "square".

An automobile with a unitized body or unibody design and a solid axle rear suspension could have any of the rear alignment problems described in the previous section, regarding frame-type automobiles. These problems can only be found using an alignment machine with advanced capabilities.

A unitized or unibody automobile with independent rear suspension may also have some additional problems not previously discussed. For example, if one rear wheel were toed out, where would the thrustline be? The thrustline would be at the midpoint of the angle between the centerline of the properly aligned wheels and the driving direction of the misaligned wheel. The same theory would apply if one wheel were toed in.

Misaligned unibody toe in the rear wheels may cause the thrustline to deviate from the centerline. How is the variance measured? Deviation of the thrustline relative to the centerline, measured in degrees, equals one-half of the net difference in the toe of the two wheels. Three possibilities exist: (1) one rear wheel toed in or out with the other having zero toe, (2) both wheels toed in or out, and (3) one wheel toed in with the other toed out. The thrustline deviation can be stated in somewhat more specific terms. The toe of one rear wheel minus the toe of the other rear wheel equals the net toe of the rear wheel assembly. Dividing this net toe by two gives the average net toe of the rear wheel assembly and is the degree of deviation of the thrustline from the centerline. For example, if the left rear wheel is toed out by 3° and the right rear wheel is toed out by 1°, the net toe of the two rear wheels is 2° (3° − 1° = 2°) or an average of 1° toe out for the two wheels of the rear wheel assembly (2° ÷ 2° = 1°).

If the right rear wheel were toed in the same direction as the left rear wheel (that is, left rear wheel toed out at 3° and right rear wheel toed in at 1°), the two angles would be added together and then divided by two to find the thrustline deviation. In this case, the net toe would be 4° (3° + 1° = 4°) and the deviation of the thrustline would be 2° (4° ÷ 2° = 2°). In the latter example, the entire rear assembly would tend to steer to the left, and as with rear steering vehicles (lift trucks) the automobile would steer to the right. The driver would have to correct by steering the front wheels toward the left the same number of degrees as the thrustline deviation.

An automobile with faulty rear wheel alignment will tend to steer in a direction opposite to the direction of the net toe. For example, if the net toe is toward the left (left rear wheel toed out or right rear wheel toed in) the automobile will steer toward the right. If the automobile is steered in a direction opposite to the net toe, and the deviated thrustline, the automobile will tend to oversteer and the rear of the automobile may tend to spin outward. If the automobile is steered to the same direction of the net toe, and the deviated thrustline, the automobile will tend to understeer. The above conditions become more evident if the automobile is driven on wet, loose, or slippery surfaces.

If both rear wheels are toed in or toed out by exactly the same degree, thrustline deviation will be zero and the automobile will steer straight. However, the rear tires will be subjected to excessive scuffing wear, and the automobile will display instability in turns and on uneven road surfaces.

A rear wheel that is set back but is parallel to the body centerline may or may not materially affect the way the automobile drives, depending upon the severity of this condition. If the condition is pronounced, the automobile may tend to pull to the side of the shorter wheelbase. Nonetheless, the automobile is indeed out of square.

There are numerous ways in which the wheels of an independent rear suspension might be misaligned, and proper equipment is essential to the diagnostic process.

Many automobiles with unitized or unibody designs provide few

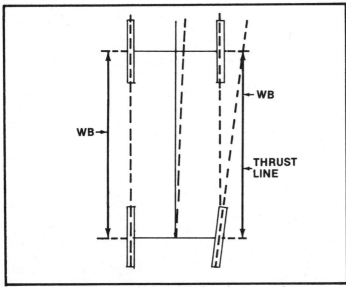

**The body centerline must be known before the toe can be properly set and the thrustline established**

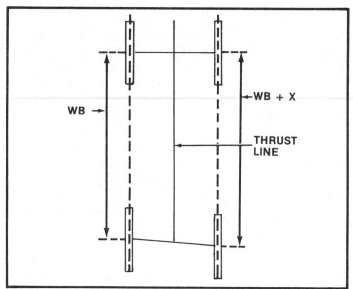

**The right rear wheel is set back but is still parallel to the body centerline**

suspension adjustments. In the past, these automobiles were difficult to align. Now, however, many automobile and parts manufacturers have made available offset bushing and various part kits to aid in performing normal alignment adjustments. But if the alignment procedure reveals that a unitized or unibody automobile has a bend in the structure, the automobile must be sent to a frame or body shop with proper equipment. Older frame-rack machines simply cannot detect body misalignments with any degree of accuracy because they were never designed to find the proper references on unibodies. Machines designed especially for unitized and unibody collision repair work, however, are available and can help restore the structure to a condition equal to the original specifications.

## MACPHERSON STRUT SUSPENSIONS

The basic geometric angles used in the strut suspension—caster, camber, steering axis inclination, and included angle—are the same

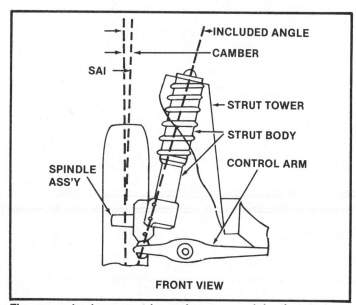

**The same basic geometric angles are used in the strut suspension**

as those for other types of suspensions. Under normal driving conditions, the wheels of the strut suspension will remain very close to the desired camber setting. When the coil spring is greatly compressed, however, camber will go slightly negative; when the spring is greatly expanded, camber will go slightly positive. These changes occur only under extreme conditions but actually improve the handling characteristics. Most strut suspensions respond in this manner.

Some strut suspensions provide no alignment adjustmens, while others provide for adjustment of camber only, or caster only. All have toe adjustments. If the misalignment is minor, and there are no caster or camber adjustments provided, an alignment kit (offset bushing kit) may be used.

Unibody construction with four wheel MacPherson strut suspension requires proper recognition and diagnosis of the alignment angles of the entire automobile. (See the MacPherson Strut Suspension Reference Chart.) Alignment equipment that cannot find the unibody centerline, front and rear, and that can align the wheels only to each other are obsolete for MacPherson strut work.

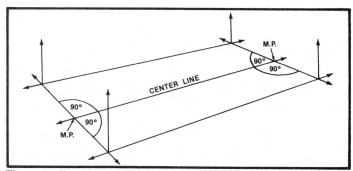

**The centerline of the unibody is the reference from which the entire alignment procedure begins**

Also, the rear wheels must be properly aligned (within the permissible tolerances as referenced to the centerline). The alignment of the rear wheels is extremely important because the remainder of the alignment procedure is referenced to the thrustline established by the alignment of the rear wheels. See the previous section on thrust and centerline.

No MacPherson strut alignment procedure can be done properly without finding the correct base reference prior to the alignment procedure. With unibodies, the correct base reference is the body centerline. Like any alignment procedure, proper MacPherson strut alignment is a three dimensional procedure: (1) side to side, (2) front to rear, and (3) vertical overview. All angles are interrelated and corrections must be made from the same base reference: the unibody centerline. Although caster, camber, SAI, and the included angle are referenced to true vertical, the four points representing the center of each wheel must be referenced to the body centerline (horizontal) as follows:

1. A line through the centers of the two front wheels and the line through the centers of the two rear wheels (as viewed from above) must be at right angles to the *centerline*.
2. The midpoints (M.P.) of the front and rear horizontal lines must lie on the centerline.

All must be within permissible tolerances.

Steering axis inclination (SAI) is the major directional control angle. Unibody automobiles with MacPherson strut suspension have increased SAI for greater stability, which also relieves torque steer in front wheel drive automobiles with MacPherson struts. Steering axis inclination must be measured in every alignment procedure. If the SAI is correct but the camber is incorrect, then the included angle will also be incorrect, indicating that the spindle and/or strut body is bent. The unibody structure probably has not been damaged, however. When the SAI specification is correct, the mechanic can be reasonably certain that the unibody strut tower is undamaged. If the SAI is less than the specification and the included angle is correct, either the control arm is bent or the strut tower is out at the top. A bent control arm can be found easily. However, a shifted strut tower can be nearly impossible to find without the proper equipment. Relying on hood and fender alignment to determine if the strut tower is bent is not accurate. Even if a bend in the unibody structure is detected, the bend cannot be accurately corrected without a machine designed with this capability.

If the SAI is less than specification and the camber is correct, but the included angle is incorrect, any or all of these three possibilities exist: (1) a bent spindle and/or a bent strut body, (2) a bent control arm, or (3) a strut tower out of location at the top.

If the SAI is greater than the manufacturer's specification, the wheel likely has been hit high from the side, pushing the top of the strut tower inward. The included angle would be used to determine if the strut body or spindle has been bent.

The included angle is used to determine if there is a bent component in the suspension. The SAI is used to determine if the unibody is bent or twisted. A bent control arm will change the SAI, but is usually readily detectable. If the SAI remains incorrect after the control arm has been replaced, the strut tower probably has shifted. Worn parts also affect the readings.

Caster or camber must never be corrected by bending any part of the strut assembly! Bending may crack or structurally weaken the strut and may cause it to break at any time, potentially resulting in a serious accident. Any bent strut assembly must be replaced. If the misalignment is somewhat more severe (that is, affecting more than simply caster or camber), the alignment machine will help in locating the bent part or parts. If the included angle specification cannot be met, a damaged spindle may be the culprit. However, in this case, the strut body may be bent as well. A bent strut body or bent spindle will change the included angle. In most cases, if the spindle is bent, the strut body will also be bent, and the entire strut unit must be replaced.

## REAR SUSPENSION ALIGNMENT

In the past, rear suspension alignment was generally ignored. Few domestic cars had any means of adjustment (independent rear suspension systems on mass-produced cars are a recent development).

However, any suspension system can be out of alignment, whether it has provisions for alignment adjustments or not. Both solid axle and independent suspensions can have worn or bent parts, resulting in misalignment.

Until recently, proper checking of rear alignment was a lengthy and difficult procedure. New equipment makes alignment work much easier and more accurate. Since the rear wheels are not steered, they are not subjected to the same stresses and forces as the front wheels, but the same alignment principles apply to front and rear suspensions. The main difference is that caster is not normally a consideration in rear wheel alignment.

Just as in the front suspension, toe in the rear suspension is viewed from the top. It is the distance between the extreme fronts and extreme rears of the wheels. Rear toe settings are referenced to the body or frame centerline. Independent rear suspensions have more wear points than solid axle rear suspensions, and because of this, loose or worn parts will affect the running toe of the rear wheels in independent systems.

Camber in the rear suspension is the same as camber in the front. It is viewed from the rear and is the inward or outward tilt of the top of the wheel as measured from the true vertical.

Misaligned running toe or camber will upset the handling qualities of the automobile, causing unpredictable and perhaps dangerous behavior under poor driving conditions. Rear wheels toed in or out tend to "steer" the rear of the car in the direction of misalignment. Incorrect camber can also cause some unwanted and dangerous steering, but more often results in abnormal tire wear and, because of the reduced tire contact patch, a tendency for the rear wheels to lock or hop under strong braking.

## MONO AXLE AND FOUR WHEEL DRIVE SUSPENSIONS

Mono axle front suspensions are used on many four wheel drive vehicles and heavy trucks. Four wheel drive vehicles equipped with mono front axles have greater road clearance because the spring is mounted above the axle. These units are subjected to many off-road uses and require added strength and higher frame and body placement. Because the mono axle is rigid, there is no camber change as the spring is compressed or expanded. The wheels remain in the pre-set camber position at all times. There are no caster/

## MACPHERSON STRUT SUSPENSION REFERENCE CHART

| Angles | | | Problem |
| --- | --- | --- | --- |
| SAI | Camber | Included Angle | |
| Correct | Less than specification | Less than specification | Bent spindle and/or bent strut body |
| Correct | Greater than specification | Greater than specification | Bent spindle and/or bent strut body |
| Less than specification | Greater than specification | Correct | Bent control arm or strut tower out at top |
| Greater than specification | Less than specification | Correct | Strut tower in at top |
| Greater than specification | Greater than specification | Greater than specification | Strut tower in at top and spindle and/or strut body bent |
| Less than specification | Greater than specification | Greater than specification | Bent control arm or strut tower out at top plus bent spindle and/or strut body |
| Less than specification | Less than specification | Less than specification | Strut tower out at top and spindle and/or strut body bent or bent control arm |

camber adjustments provided, although both should be checked; bent or worn parts will affect both.

Many four wheel drive vehicles are driven under adverse conditions and therefore require a periodic alignment and parts check. Alignment procedures should follow the manufacturer's recommendations. A serious alignment problem, such as a bent axle housing or spindle assembly, will necessitate replacement of the damaged parts. Inner and outer axle housing bearing seats are perfectly aligned so that a straight axle will turn freely. The slightest bend in the axle housing will misalign the bearings and force the axle to bend as it rotates. This places stress on the axle bearings which greatly shortens axle bearing life and requires greater power to turn the axle. The added friction creates heat which may cause multiple problems that could be quite costly. *A bent axle housing cannot be properly straightened; it must be replaced.* There is no way the bearing seats can be perfectly realigned by attempting to straighten the axle housing, even if the desired suspension alignment settings have been achieved. Any bent suspension or steering parts should be replaced. The accompanying table shows how to find bent parts.

to absorb bumps and road irregularities independently, while providing sturdy, simple construction. The outer ends of the I beams are attached to the spindle and to the radius arms. The inner ends are attached to a pivot bracket fastened to the frame near the opposite side of the vehicle. This type of construction provides good anti-roll characteristics. The use of a progressive coil spring (a spring that becomes increasingly stiff as it is compressed) provides good riding qualities on normal roads and sturdiness off the road. The radius arms permit the I beam to move up and down, stabilize any front to rear movement of the I beam, and help maintain the proper caster setting. The spindle is mounted to the I beam by a spindle bolt (kingpin). There are no ball joints.

The action of the I beam produces good handling characteristics. As the spring is compressed, the camber increasingly moves toward negative. The exact opposite occurs as the spring is expanded. The more the spring is expanded, the more positive the camber becomes. In cornering, body roll of the vehicle increases as speed and sharpness of the turn increase. Up to the critical point where the vehicle will either roll over or lose traction, these cornering forces compress the outside front spring (creating negative camber

## MONO AXLE REFERENCE CHART

| | Angles | | | Problem |
|---|---|---|---|---|
| SAI | Camber | | Included Angle | |
| Correct | Greater than specification | | Greater than specification | Bent spindle assembly |
| Greater than specification | Less than specification | | Correct | Bent axle housing |
| Less than specification | Greater than specification | | Correct | Bent axle housing |
| Less than specification | Greater than specification | | Greater than specification | Bent spindle assembly plus bent axle housing |

Some four wheel drive vehicles have independent front suspension. If the vehicle has upper and lower control arms, the reference chart under "Steering Axis Inclination" may be used to locate bent parts. If the suspension is twin I beam, the diagnostic portion of this section dealing with twin I beams will explain how to locate bent parts. Geometric principles apply equally to front wheel, four wheel, and rear wheel driven vehicles.

## Twin I Beam Suspension

The "Twin I Beam" front suspension was developed to combine independent front wheel action with the strength and dependability of the mono beam axle. Twin I beam axles allow each front wheel

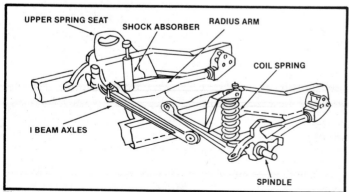

**The twin I beams are shown through the frame detail to show the mounting details**

in that wheel) and expand the inside spring (creating positive camber in the inside wheel). These changes tend to increase the cornering ability of the vehicle. Connection of the I beam to the opposite side of the vehicle provides changes in camber identical to those described with respect to roll forces, giving good resistance to crosswind forces.

Vehicle height is most important for the proper alignment of camber. Many twin I beam suspensions have been needlessly bent and re-aligned to achieve proper camber when the problem may have been caused by nothing more than a weak spring. As the spring weakens and sags it becomes more compressed, lowering vehicle height, which causes camber to go toward negative. In most cases, all that is required is that the weakened springs be replaced (they must be replaced in pairs). When the vehicle is restored to the proper height, the camber will usually be correct. Excessive vehicle height will cause positive camber. An incorrect set of springs will result in incorrect camber settings. In the case of *new* springs, however, vehicle height may be temporarily too high. A few hundred miles of driving will lower the height to normal. Camber should be correct after new springs have had a chance to settle in.

If the vehicle height is proper and a camber misalignment still exists, determine where the problem is before any corrective steps are taken. Specifically, check the SAI. If the SAI is correct and camber is not, the included angle is incorrect. As previously discussed, if the included angle is incorrect, the spindle assembly is bent. Bending an I beam to correct a camber problem makes no sense if the included angle is incorrect—bending does not correct the problem, but rather adds another. If the vehicle height and SAI are correct and the included angle is incorrect, the solution is to replace the spindle assembly. If the vehicle height is correct and the included angle is correct, but the SAI and camber are incorrect, the problem is a bent I beam. In the following reference table, vehicle height is assumed to be absolutely correct. *The relationships listed in the table will not be true if the vehicle height is incorrect.*

Twin I beam front suspensions have two unique design provisions, both of which affect service procedures. Some twin I beam suspensions have balljoints and provide for a camber adjustment.

## TWIN I BEAM REFERENCE CHART

| Angles | | | Problem |
|---|---|---|---|
| SAI | Camber | Included Angle | |
| Correct | Greater than specification | Greater than specification | Bent spindle assembly |
| Greater than specification | Less than specification | Correct | Bent I beam |
| Less than specification | Greater than specification | Correct | Bent I beam |
| Less than specification | Greater than specification | Greater than specification | Bent I beam plus bent spindle assembly |

The twin I beam suspension is a simple but very strong suspension. It is not easily bent out of alignment. Nonetheless, many "corrective bends" have been made because of improper vehicle height.

Occasionally, a bent I beam may be isolated as an alignment problem. At this point, and this point only, a corrective bend should be considered unless the I-Beam suspension is equipped with ball joints. The manufacturer does not recommend any type of corrective bending. However, when I beams have actually become bent by accident, corrective bends have been performed with satisfactory results. If the error of misalignment in the I beam is one degree or less, a *cold bend* may be performed, but only with the proper tools. *Under no circumstances whatsoever should heat be used to make this corrective bend!* If the error in misalignment caused by the I beam is more than one degree, replace the I beam.

If an error in caster is found, something is bent. Two possibilities exist: (1) a bent radius arm, or (2) a twisted I beam somewhere between where the radius arm is attached and the spindle assembly. In both cases, replace the damaged part. Do not attempt to straighten either part.

A particular problem often encountered with twin I beam suspension is tire cupping, which is sometimes called scalloping. Frequently, alignment is blamed for this condition. Altering a good alignment to a misalignment will not correct tire cupping problems. Three conditions usually cause tire cupping: (1) loose parts, (2) out-of-round or out-of-balance tires and wheels, and (3) bad or weak shock absorbers. These problems may occur in any combination. If there are no loose parts, the tires and wheels are in balance and are not out-of-round, the alignment is at specification, and the shock absorbers *appear* to be good, it is usually the shock absorbers that are at fault, regardless of their appearance. The I beam is nothing more than a very long control arm. The longer the control arm, the greater the leverage and the amount of unsprung weight. This increases the dampening required. If the shock absorber strength is less than it should be and a nearly undetectable out-of-balance condition exists, cupping will occur. Loose parts, even those within tolerance, together with a slightly weak shock absorber, can cause a cupping condition. These problems can be nearly impossible to solve unless the real culprit is identified—the weak shock absorber. The solution is to replace the shock absorbers with heavy duty shock absorbers. This will not greatly affect the riding qualities, and if everything else is correct, the cupping condition will be eliminated. Replacing standard size wheels and tires with larger ones increases the amount of unsprung weight and the need for dampening. The solution is to install a second pair of front shock absorbers giving a total of four front shock absorbers. Many twin I beam truck suspensions have provisions for the installation of the second pair.

*Under no circumstances should a "corrective bend" be made on this suspension.* It is of different construction (box type and not solid metal), and a bend of any kind will weaken the structure, creating a serious problem later on. If camber cannot be corrected through the normal adjustment procedure, follow the twin I beam diagnosis in this section. The problem can be isolated, and when the faulty part or parts are replaced, alignment will be proper.

Some twin I beam suspensions provide only one tie rod adjustment, making the correction of off-center steering wheel more difficult. If the error is not great, perhaps the manufacturing tolerances are on the outer limits but still within the specifications. If the error is great, the steeering wheel may have been removed and improperly replaced, or perhaps the pitman arm is bent or improperly installed, or a steering arm is seriously bent. The procedure to correct an off-center steering wheel on a twin I beam steering system is simple if you have the proper equipment and nearly impossible if you do not. With adequate alignment equipment, the cause for the off-center steering wheel can be located. The steering wheel will be dead center after proper alignment procedures have been completed with reference to the thrustline.

## SUMMARY

Problems can be isolated more quickly and more effectively if the automobile is given a test drive before alignment procedures are begun. Excessive tire wear may be seen at a glance, but most other problems must be diagnosed. A test route (including rough, smooth, crowned and flat road surfaces) can relate a great deal of information about the automobile in just a few minutes. Vibration in the automobile may suggest the need for alignment. However, a test drive may indicate that the wheels are out of balance or a universal joint has gone bad. Shock absorbers that are leaking obviously need to be replaced, but a test drive is an excellent way to determine how good or bad the shock absorbers are when there are no visible symptoms of wear. Every step of the prealignment check must be performed. A carefully developed routine will help assure that no part of the prealignment is overlooked.

The angles of the suspension are referenced to the true vertical and the vehicle's centerline. In the overall alignment procedure, angles must be adjusted individually as required. Proper balance among the geometric angles within the entire suspension system, front and rear, is important to optimum tire wear, good handling and safety. Equally important, alignment principles and practices should be thoroughly understood. This knowledge will prove to be invaluable when extraordinary problems are encountered.

In the total alignment procedure, the goal is to have all four wheels of the automobile adjusted to the position of a perfect rectangle, with all four wheels parallel to the longitudinal centerline of

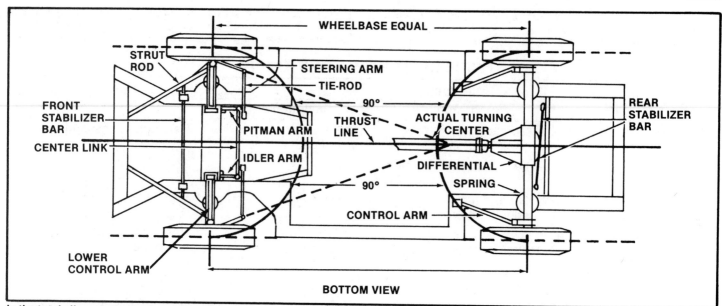

**In the total alignment procedure, all parts and all angles are considered**

the rectangle. The mid-points between the two front wheels and between the two rear wheels will lie on the centerline of the rectangle. (Many automobiles are designed and manufactured to have different tread or track widths). The thrustline will also lie on or coincide with the centerline if the alignment goal is achieved without error. Assuming all suspension angles (caster, camber, toe, SAI and included angle) are correct, an automobile, on a flat surface, will roll in a straight line with a straight steering wheel.

Because of manufacturing tolerances, this goal is rarely attained, but the questions affecting service of the vehicle are: Can the errors in geometry be located and measured? If these errors are found to be within all tolerances, can the thrustline and centerline be found and used as a reference? Can the automobile, with these errors, be made to roll in a straight line on a flat surface with a straight steering wheel? If the answers are yes, the vehicle can be successfully aligned.

## FRONT SUSPENSION AND STEERING LINKAGE
## TROUBLESHOOTING—REAR WHEEL DRIVE

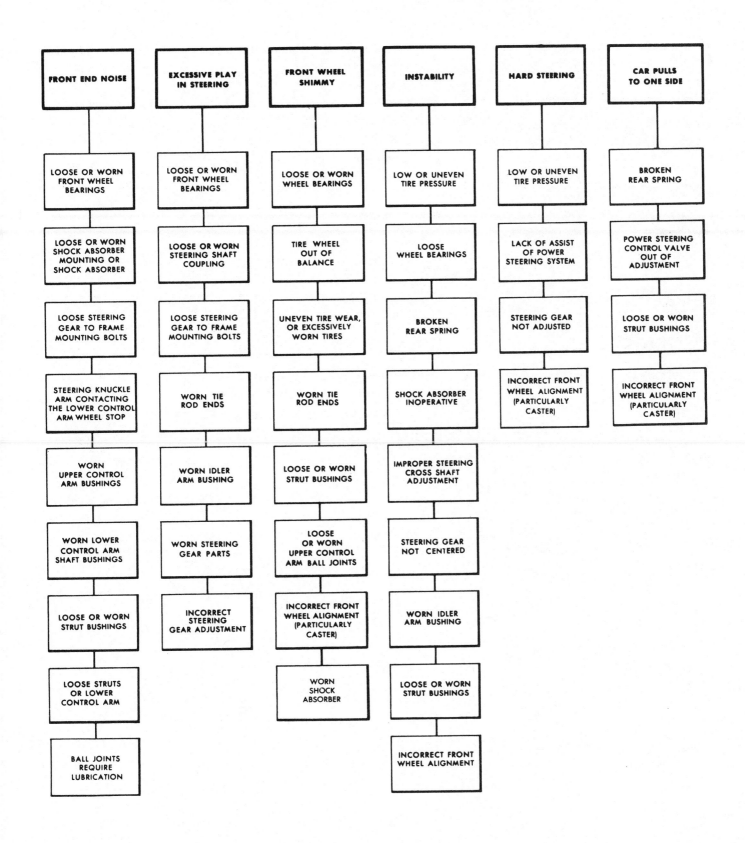

## FRONT SUSPENSION AND STEERING LINKAGE
## TROUBLESHOOTING—FRONT WHEEL DRIVE

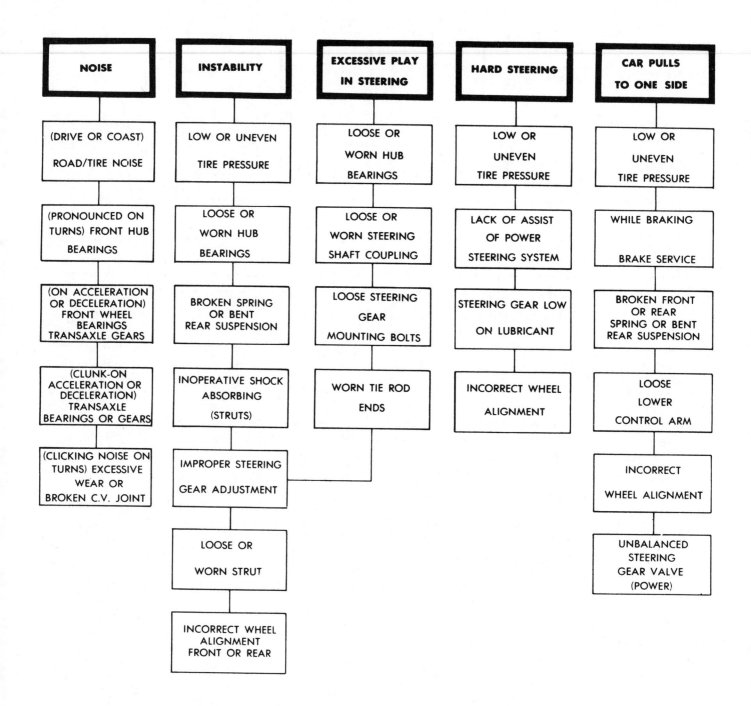

## TAPERED WHEEL BEARING TROUBLESHOOTING

CONSIDER THE FOLLOWING FACTORS WHEN DIAGNOSING BEARING CONDITION:

1. GENERAL CONDITION OF ALL PARTS DURING DISASSEMBLY AND INSPECTION.

2. CLASSIFY THE FAILURE WITH THE AID OF THE ILLUSTRATIONS.

3. DETERMINE THE CAUSE.

4. MAKE ALL REPAIRS FOLLOWING RECOMMENDED PROCEDURES.

**GOOD BEARING**

**BENT CAGE**

CAGE DAMAGE DUE TO IMPROPER HANDLING OR TOOL USAGE.

REPLACE BEARING.

**BENT CAGE**

CAGE DAMAGE DUE TO IMPROPER HANDLING OR TOOL USAGE.

REPLACE BEARING.

**GALLING**

METAL SMEARS ON ROLLER ENDS DUE TO OVERHEAT, LUBRICANT FAILURE OR OVERLOAD.

REPLACE BEARING — CHECK SEALS AND CHECK FOR PROPER LUBRICATION.

**ABRASIVE STEP WEAR**

PATTERN ON ROLLER ENDS CAUSED BY FINE ABRASIVES.

CLEAN ALL PARTS AND HOUSINGS, CHECK SEALS AND BEARINGS AND REPLACE IF LEAKING, ROUGH OR NOISY.

**ETCHING**

BEARING SURFACES APPEAR GRAY OR GRAYISH BLACK IN COLOR WITH RELATED ETCHING AWAY OF MATERIAL USUALLY AT ROLLER SPACING.

REPLACE BEARINGS — CHECK SEALS AND CHECK FOR PROPER LUBRICATION.

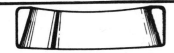

**MISALIGNMENT**

OUTER RACE MISALIGNMENT DUE TO FOREIGN OBJECT.

CLEAN RELATED PARTS AND REPLACE BEARING. MAKE SURE RACES ARE PROPERLY SEATED.

**INDENTATIONS**

SURFACE DEPRESSIONS ON RACE AND ROLLERS CAUSED BY HARD PARTICLES OF FOREIGN MATERIAL.

CLEAN ALL PARTS AND HOUSINGS, CHECK SEALS AND REPLACE BEARINGS IF ROUGH OR NOISY.

**FATIGUE SPALLING**

FLAKING OF SURFACE METAL RESULTING FROM FATIGUE.

REPLACE BEARING — CLEAN ALL RELATED PARTS.

## TAPERED WHEEL BEARING TROUBLESHOOTING

### BRINELLING

SURFACE INDENTATIONS IN RACEWAY CAUSED BY ROLLERS EITHER UNDER IMPACT LOADING OR VIBRATION WHILE THE BEARING IS NOT ROTATING.

REPLACE BEARING IF ROUGH OR NOISY.

### CAGE WEAR

WEAR AROUND OUTSIDE DIAMETER OF CAGE AND ROLLER POCKETS CAUSED BY ABRASIVE MATERIAL AND INEFFICIENT LUBRICATION. CHECK SEALS AND REPLACE BEARINGS.

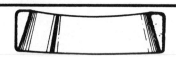

### ABRASIVE ROLLER WEAR

PATTERN ON RACES AND ROLLERS CAUSED BY FINE ABRASIVES.

CLEAN ALL PARTS AND HOUSINGS, CHECK SEALS AND BEARINGS AND REPLACE IF LEAKING, ROUGH OR NOISY.

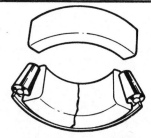

### CRACKED INNER RACE

RACE CRACKED DUE TO IMPROPER FIT, COCKING, OR POOR BEARING SEATS.

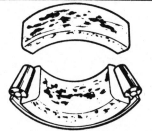

### SMEARS

SMEARING OF METAL DUE TO SLIPPAGE, SLIPPAGE CAN BE CAUSED BY POOR FITS, LUBRICATION, OVERHEATING, OVERLOADS OR HANDLING DAMAGE.

REPLACE BEARINGS, CLEAN RELATED PARTS AND CHECK FOR PROPER FIT AND LUBRICATION.

REPLACE SHAFT IF DAMAGED.

### FRETTAGE

CORROSION SET UP BY SMALL RELATIVE MOVEMENT OF PARTS WITH NO LUBRICATION.

REPLACE BEARING. CLEAN RELATED PARTS. CHECK SEALS AND CHECK FOR PROPER LUBRICATION.

### HEAT DISCOLORATION

HEAT DISCOLORATION CAN RANGE FROM FAINT YELLOW TO DARK BLUE RESULTING FROM OVERLOAD OR INCORRECT LUBRICANT.

EXCESSIVE HEAT CAN CAUSE SOFTENING OF RACES OR ROLLERS.

TO CHECK FOR LOSS OF TEMPER ON RACES OR ROLLERS A SIMPLE FILE TEST MAY BE MADE. A FILE DRAWN OVER A TEMPERED PART WILL GRAB AND CUT META, WHEREAS, A FILE DRAWN OVER A HARD PART WILL GLIDE READILY WITH NO METAL CUTTING.

REPLACE BEARINGS IF OVER HEATING DAMAGE IS INDICATED. CHECK SEALS AND OTHER PARTS.

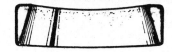

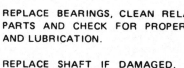

### STAIN DISCOLORATION

DISCOLORATION CAN RANGE FROM LIGHT BROWN TO BLACK CAUSED BY INCORRECT LUBRICANT OR MOISTURE.

RE-USE BEARINGS IF STAINS CAN BE REMOVED BY LIGHT POLISHING OR IF NO EVIDENCE OF OVERHEATING IS OBSERVED.

CHECK SEALS AND RELATED PARTS FOR DAMAGE.

## HOW TO READ TIRE WEAR

The way your tires wear is a good indicator of other parts of the suspension. Abnormal wear patterns are often caused by the need for simple tire maintenance, or for front end alignment.

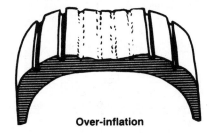

**Over-inflation**

Excessive wear at the center of the tread indicates that the air pressure in the tire is consistently too high. The tire is riding on the center of the tread and wearing it prematurely. Occasionally, this wear pattern can result from outrageously wide tires on narrow rims. The cure for this is to replace either the tires or the wheels.

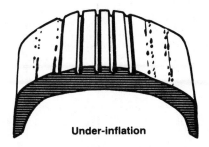

**Under-inflation**

This type of wear usually results from consistent under-inflation. When a tire is under-inflated, there is too much contact with the road by the outer treads, which wear prematurely. When this type of wear occurs, and the tire pressure is known to be consistently correct, a bent or worn steering component or the need for wheel alignment could be indicated.

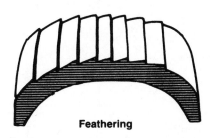

**Feathering**

Feathering is a condition when the edge of each tread rib develops a slightly rounded edge on one side and a sharp edge on the other. By running your hand over the tire, you can usually feel the sharper edges before you'll be able to see them. The most common causes of feathering are incorrect toe-in setting or deteriorated bushings in the front suspension.

**One side wear**

When an inner or outer rib wears faster than the rest of the tire, the need for wheel alignment is indicated. There is excessive camber in the front suspension, causing the wheel to lean too much putting excessive load on one side of the tire. Misalignment could also be due to sagging springs, worn ball joints, or worn control arm bushings. Be sure the vehicle is loaded the way it's normally driven when you have the wheels aligned.

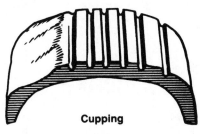

**Cupping**

Cups or scalloped dips appearing around the edge of the tread almost always indicate worn (sometimes bent) suspension parts. Adjustment of wheel alignment alone will seldom cure the problem. Any worn component that connects the wheel to the suspension can cause this type of wear. Occasionally, wheels that are out of balance will wear like this, but wheel imbalance usually shows up as bald spots between the outside edges and center of the tread.

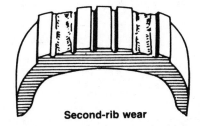

**Second-rib wear**

Second-rib wear is usually found only in radial tires, and appears where the steel belts end in relation to the tread. It can be kept to a minimum by paying careful attention to tire pressure and frequently rotating the tires. This is often considered normal wear but excessive amounts indicate that the tires are too wide for the wheels.

# MANUFACTURER'S SPECIFIED ALIGNMENT TOLERANCES
## DOMESTIC CARS

All vehicles should be set to preferred specifications when being aligned. The minimum and maximum setting specified are a guide to use when checking alignment. The manufacturers consider alignment within these tolerances acceptable for safe vehicle operation while still limiting abnormal tire wear.

| VEHICLE IDENTIFICATION<br>YEAR MODEL | ADJ. ILL. NO. | CASTER (Degrees) MIN. | PREF. | MAX. | CAMBER (Degrees) MIN. | PREF. | MAX. | TOE-IN (Inches) MIN. | PREF. | MAX. | TOE-IN (Millimeters) MIN. | PREF. | MAX. | TOE-OUT ON TURNS (Degrees) OUTSIDE WHEEL | INSIDE WHEEL | STRG. AXIS INCL. (DEG.) |
|---|---|---|---|---|---|---|---|---|---|---|---|---|---|---|---|---|
| RESET VEHICLE ALIGNMENT AS CLOSE TO PREFERRED SETTINGS AS POSSIBLE FOR BEST RESULTS. | | | | | | | | | | | | | | | | |
| **AMERICAN MOTORS** | | | | | | | | | | | | | | | | |
| (A) 1980 Eagle king pin inclination is 11½°. •Wheels at Full Turning Angle | | | | | | | | | | | | | | | | |
| 81-80 Spirit, Concord, AMX | (1)(2) | 0 | 1 | 2½ | | | | 1/16 | 1/8 | 3/16 | 1.6 | 3.2 | 4.8 | NA | 38• | 7 3/4 |
| Left Wheel . . . . . . . . | | | | | 1/8 | 3/8 | 3/4 | | | | | | | | | |
| Right Wheel . . . . . . . | | | | | − 1/8 | 1/8 | 1/2 | | | | | | | | | |
| 81-80 Eagle . . . . . . . . . (A) | (1)(2) | 2 | 2½ | 3 | − 1/8 | 3/8 | 5/8 | 1/16 (out) | 1/8 (out) | 3/16 (out) | 1.6 (out) | 3.2 (out) | 4.8 (out) | NA | 38• | 11 27/32 |
| 80-79 Pacer . . . . . . . . | (3) | 1 | 2 | 3½ | 0 | 1/4 | 3/4 | 1/16 | 1/8 | 3/16 | 1.6 | 3.2 | 4.8 | NA | 35• | 7 3/4 |
| 79 AMX, Concord, Spirit . . | (1)(2) | 0 | 1 | 2½ | 0 | 1/4 | 3/4 | 1/16 | 1/8 | 3/16 | 1.6 | 3.2 | 4.8 | NA | 38• | 7 3/4 |
| 78 Matador, AMX, Concord, Gremlin . . . . . . . | (1)(2) | 0 | 1 | 2 | | | | 1/16 | 1/8 | 3/16 | 1.6 | 3.2 | 4.8 | 22 | 25 | 7 3/4 |
| Left Wheel . . . . . . . . | | | | | 1/8 | 3/8 | 5/8 | | | | | | | | | |
| Right Wheel . . . . . . . | | | | | 0 | 1/8 | 1/2 | | | | | | | | | |
| 78 Pacer . . . . . . . . . . | (3) | 1 | 2 | 3 | | | | 1/16 | 1/8 | 3/16 | 1.6 | 3.2 | 4.8 | 22 | 25 | 7 3/4 |
| Left Wheel . . . . . . . . | | | | | 1/8 | 3/8 | 5/8 | | | | | | | | | |
| Right Wheel . . . . . . . | | | | | 0 | 1/8 | 1/2 | | | | | | | | | |
| 77-73 Gremlin, Hornet . . . . | (1)(2) | − 1/2 | 0 | 1/2 | | | | 1/16 | 1/8 | 3/16 | 1.6 | 3.2 | 4.8 | 22 | 25 | 7 3/4 |
| Left Wheel . . . . . . . . | | | | | 1/8 | 3/8 | 5/8 | | | | | | | | | |
| Right Wheel . . . . . . . | | | | | 0 | 1/8 | 1/2 | | | | | | | | | |
| 77-75 Pacer . . . . . . . . | (3) | 1/2 | 1 | 1½ | | | | 1/16 | 1/8 | 3/16 | 1.6 | 3.2 | 4.8 | 22 | 25 | 7 3/4 |
| Left Wheel . . . . . . . . | | | | | 1/8 | 3/8 | 5/8 | | | | | | | | | |
| Right Wheel . . . . . . . | | | | | 0 | 1/8 | 1/2 | | | | | | | | | |
| 77-73 Matador, Ambassador . . . . . . . . . | (1)(2) | 1/2 | 1 | 1½ | | | | 1/16 | 1/8 | 3/16 | 1.6 | 3.2 | 4.8 | 22 | 25 | 7 3/4 |
| Left Wheel . . . . . . . . | | | | | 1/8 | 3/8 | 5/8 | | | | | | | | | |
| Right Wheel . . . . . . . | | | | | 0 | 1/8 | 1/2 | | | | | | | | | |
| 72 Hornet, Javelin, AMX, Matador, Gremlin, Ambassador, Rebel . . . . . | (1)(2) | 1/2 | 1 | 1½ | | | | 1/16 | 1/8 | 3/16 | 1.6 | 3.2 | 4.8 | 22 | 25 | 7 3/4 |
| Left Wheel . . . . . . . . | | | | | 1/8 | 3/8 | 5/8 | | | | | | | | | |
| Right Wheel . . . . . . . | | | | | 0 | 1/8 | 1/2 | | | | | | | | | |
| **CHECKER MOTORS CORP.** | | | | | | | | | | | | | | | | |
| 81 All Models . . . . . . . . . | (24) | 1½ | 2 | 2½ | 1/4 | 1/2 | 3/4 | 1/16 | 3/32 | 1/8 | 1.6 | 2.4 | 3.2 | 17½ | 20 | 7 |
| 80-72 All Models . . . . . . . | (24) | 1½ | 2 | 2½ | 1/2 | 1 | 1½ | 1/16 | 3/32 | 1/8 | 1.6 | 2.4 | 3.2 | 17½ | 20 | 7 |
| **CHRYSLER CORP.** | | | | | | | | | | | | | | | | |
| (A) Check vehicle suspension height before performing alignment. | | | | | | | | | | | | | | | | |
| (B) Maximum left to right variation in caster not to exceed 1¼° when checking alignment. | | | | | | | | | | | | | | | | |
| (C) The engine must be running during toe adjustment of vehicles with power steering. | | | | | | | | | | | | | | | | |
| **CHRYSLER (A) (B) (C)** | | | | | | | | | | | | | | | | |
| 81-80 Newport, New Yorker | (7) | − 1/4 | 1 | 2¼ | − 1/4 | 1/2 | 1¼ | 0 | 1/8 | 5/16 | 0 | 3.2 | 8.0 | 18 | 20 | 8 |
| 79 Newport, New Yorker . . | (7) | − 1/2 | 3/4 | 2 | | | | 1/16 | 1/8 | 1/4 | 1.6 | 3.2 | 6.4 | 18 | 20 | 8 |
| Left Wheel . . . . . . . . | | | | | 0 | 1/2 | 1 | | | | | | | | | |
| Right Wheel . . . . . . . | | | | | − 1/4 | 1/4 | 3/4 | | | | | | | | | |
| 78-77 Newport, New Yorker | (7) | − 1/2 | 3/4 | 2 | | | | 1/16 | 1/8 | 1/4 | 1.6 | 3.2 | 6.4 | 18 5/16 | 20 | 9 |
| Left Wheel . . . . . . . . | | | | | 0 | 1/2 | 1 | | | | | | | | | |
| Right Wheel . . . . . . . | | | | | − 1/4 | 1/4 | 3/4 | | | | | | | | | |
| 76-74 Newport, New Yorker | (7) | − 1/2 | 3/4 | 1 3/4 | | | | 1/16 | 1/8 | 1/4 | 1.6 | 3.2 | 6.4 | 18 5/16 | 20 | 9 |
| Left Wheel . . . . . . . . | | | | | 0 | 1/2 | 1 | | | | | | | | | |
| Right Wheel . . . . . . . | | | | | − 1/4 | 1/4 | 3/4 | | | | | | | | | |
| 73-72 Newport, New Yorker | | | | | | | | | | | | | | | | |
| W/Power Steering . . . . . | (6) | 1/4 | 3/4 | 1¼ | | | | 3/32 | 1/8 | 5/32 | 2.4 | 3.2 | 4.0 | 18 3/4 | 20 | 7½ |
| Left Wheel . . . . . . . . | | | | | 1/4 | 1/2 | 3/4 | | | | | | | | | |
| Right Wheel . . . . . . . | | | | | 0 | 1/4 | 1/2 | | | | | | | | | |
| W/Manual Steering . . . . | (6) | − 1 | − 1/2 | 0 | | | | 3/32 | 1/8 | 5/32 | 2.4 | 3.2 | 4.0 | 18 3/4 | 20 | 7½ |
| Left Wheel . . . . . . . . | | | | | 1/4 | 1/2 | 3/4 | | | | | | | | | |
| Right Wheel . . . . . . . | | | | | 0 | 1/4 | 1/2 | | | | | | | | | |
| 81-80 Cordoba . . . . . . . . | (7) | 1 1/4 | 2½ | 3 3/4 | − 1/4 | 1/2 | 1¼ | 0 | 1/8 | 5/16 | 0 | 3.2 | 8.0 | 18 | 20 | 8 |
| 79-77 Cordoba . . . . . . . . | (7) | − 1/2 | 3/4 | 2 | | | | 1/16 | 1/8 | 1/4 | 1.6 | 3.2 | 6.4 | 18 | 20 | 8 |
| Left Wheel . . . . . . . . | | | | | 0 | 1/2 | 1 | | | | | | | | | |
| Right Wheel . . . . . . . | | | | | − 1/4 | 1/4 | 3/4 | | | | | | | | | |

| VEHICLE IDENTIFICATION (YEAR MODEL) | ADJ. ILL. NO. | CASTER (Degrees) MIN. | PREF. | MAX. | CAMBER (Degrees) MIN. | PREF. | MAX. | TOE-IN (Inches) MIN. | PREF. | MAX. | TOE-IN (Millimeters) MIN. | PREF. | MAX. | TOE-OUT ON TURNS (Degrees) OUTSIDE WHEEL | INSIDE WHEEL | STRG. AXIS INCL. (DEG.) |
|---|---|---|---|---|---|---|---|---|---|---|---|---|---|---|---|---|
| 76-75 Cordoba | (7) | − 1/2 | 3/4 | 1 3/4 | | | | 1/16 | 1/8 | 1/4 | 1.6 | 3.2 | 6.4 | 18 | 20 | 8 |
|   Left Wheel | | | | | 0 | 1/2 | 1 | | | | | | | | | |
|   Right Wheel | | | | | − 1/4 | 1/4 | 3/4 | | | | | | | | | |
| 81-80 LeBaron | (7) | 1 1/4 | 2 1/2 | 3 3/4 | − 1/4 | 1/2 | 1 1/4 | 0 | 1/8 | 5/16 | 0 | 3.2 | 8.0 | 18 | 20 | 8 |
| 79-77 LeBaron | (7) | 1 1/2 | 2 1/2 | 3 3/4 | | | | 1/16 | 1/8 | 1/4 | 1.6 | 3.2 | 6.4 | 18 | 20 | 8 |
|   Left Wheel | | | | | 0 | 1/2 | 1 | | | | | | | | | |
|   Right Wheel | | | | | − 1/4 | 1/4 | 3/4 | | | | | | | | | |
| 81 Imperial | (7) | 1 1/4 | 2 1/2 | 3 3/4 | − 1/4 | 1/2 | 1 1/4 | 0 | 1/8 | 5/16 | 0 | 3.2 | 8.0 | 18 | 20 | 8 |
| 75-74 Imperial | (7) | − 1/2 | 3/4 | 1 3/4 | | | | 1/16 | 1/8 | 1/4 | 1.6 | 3.2 | 6.4 | 18 1/4 | 20 | 9 |
|   Left Wheel | | | | | 0 | 1/2 | 1 | | | | | | | | | |
|   Right Wheel | | | | | − 1/4 | 1/4 | 3/4 | | | | | | | | | |
| 73-72 Imperial | | | | | | | | | | | | | | | | |
|   W/Power Steering | (5) | 1/4 | 3/4 | 1 1/4 | | | | 3/32 | 1/8 | 5/32 | 2.4 | 3.2 | 4.0 | 17 7/8 | 20 | 9 |
|   Left Wheel | | | | | 1/4 | 1/2 | 3/4 | | | | | | | | | |
|   Right Wheel | | | | | 0 | 1/4 | 1/2 | | | | | | | | | |
|   W/Manual Steering | (5) | − 1 | − 1/2 | 0 | | | | 3/32 | 1/8 | 5/32 | 2.4 | 3.2 | 4.0 | 17 7/8 | 20 | 9 |
|   Left Wheel | | | | | 1/4 | 1/2 | 3/4 | | | | | | | | | |
|   Right Wheel | | | | | 0 | 1/4 | 1/2 | | | | | | | | | |
| **DODGE DIV. (A) (B) (C)** | | | | | | | | | | | | | | | | |
| 81-80 St. Regis | (7) | − 1/4 | 1 | 2 1/4 | − 1/4 | 1/2 | 1 1/4 | 0 | 1/8 | 5/16 | 0 | 3.2 | 8.0 | 18 | 20 | 8 |
| 79 St. Regis | (7) | − 1/2 | 3/4 | 2 | | | | 1/16 | 1/8 | 1/4 | 1.6 | 3.2 | 6.4 | 18 | 20 | 8 |
|   Left Wheel | | | | | 0 | 1/2 | 1 | | | | | | | | | |
|   Right Wheel | | | | | − 1/4 | 1/4 | 3/4 | | | | | | | | | |
| 80 Aspen | | 1 1/4 | 2 1/2 | 3 3/4 | − 1/4 | 1/2 | 1 1/4 | 0 | 1/8 | 5/16 | 0 | 3.2 | 8.0 | 18 | 20 | 8 |
| 79-76 Aspen | (7) | 1 1/2 | 2 1/2 | 3 3/4 | | | | 1/16 | 1/8 | 1/4 | 1.6 | 3.2 | 6.4 | 18 | 20 | 8 |
|   Left Wheel | | | | | 0 | 1/2 | 1 | | | | | | | | | |
|   Right Wheel | | | | | − 1/4 | 1/4 | 3/4 | | | | | | | | | |
| 81-80 Mirada | (7) | 1 1/4 | 2 1/2 | 3 3/4 | − 1/4 | 1/2 | 1 1/4 | 0 | 1/8 | 5/16 | 0 | 3.2 | 8.0 | 18 | 20 | 8 |
| 81-80 Diplomat | | 1 1/4 | 2 1/2 | 3 3/4 | − 1/4 | 1/2 | 1 1/4 | 0 | 1/8 | 5/16 | 0 | 3.2 | 8.0 | 18 | 20 | 8 |
| 79-77 Diplomat | (7) | 1 1/2 | 2 1/2 | 3 3/4 | | | | 1/16 | 1/8 | 1/4 | 1.6 | 3.2 | 6.4 | 18 | 20 | 8 |
|   Left Wheel | | | | | 0 | 1/2 | 1 | | | | | | | | | |
|   Right Wheel | | | | | − 1/4 | 1/4 | 3/4 | | | | | | | | | |
| 79-77 Charger, Charger SE Magnum XE, Monaco, Royal Monaco | | | | | | | | | | | | | | | | |
|   W/Power Steering | (7) | − 1/2 | 3/4 | 2 | | | | 1/16 | 1/8 | 1/4 | 1.6 | 3.2 | 6.4 | 18 | 20 | 8 |
|   Left Wheel | | | | | 0 | 1/2 | 1 | | | | | | | | | |
|   Right Wheel | | | | | − 1/4 | 1/4 | 3/4 | | | | | | | | | |
|   W/Manual Steering | (7) | − 1 3/4 | − 1/2 | 3/4 | | | | 1/16 | 1/8 | 1/4 | 1.6 | 3.2 | 6.4 | 18 | 20 | 8 |
|   Left Wheel | | | | | 0 | 1/2 | 1 | | | | | | | | | |
|   Right Wheel | | | | | − 1/4 | 1/4 | 3/4 | | | | | | | | | |
| 76-74 Coronet, Charger | | | | | | | | | | | | | | | | |
|   W/Power Steering | (7) | − 1/2 | 3/4 | 1 3/4 | | | | 1/16 | 1/8 | 1/4 | 1.6 | 3.2 | 6.4 | 18 | 20 | 8 |
|   Left Wheel | | | | | 0 | 1/2 | 1 | | | | | | | | | |
|   Right Wheel | | | | | − 1/4 | 1/4 | 3/4 | | | | | | | | | |
|   W/Manual Steering | (7) | − 1 3/4 | − 1/2 | 1/2 | | | | 1/16 | 1/8 | 1/4 | 1.6 | 3.2 | 6.4 | 18 | 20 | 8 |
|   Left Wheel | | | | | 0 | 1/2 | 1 | | | | | | | | | |
|   Right Wheel | | | | | − 1/4 | 1/4 | 3/4 | | | | | | | | | |
| 73 Coronet, Charger | | | | | | | | | | | | | | | | |
|   W/Power Steering | (7) | 1/4 | 3/4 | 1 1/4 | | | | 3/32 | 1/8 | 5/32 | 2.4 | 3.2 | 4.0 | 18 1/2 | 20 | 8 |
|   Left Wheel | | | | | 1/4 | 1/2 | 3/4 | | | | | | | | | |
|   Right Wheel | | | | | 0 | 1/4 | 1/2 | | | | | | | | | |
|   W/Manual Steering | (7) | − 1 | − 1/2 | 0 | | | | 3/32 | 1/8 | 5/32 | 2.4 | 3.2 | 4.0 | 18 1/2 | 20 | 8 |
|   Left Wheel | | | | | 1/4 | 1/2 | 3/4 | | | | | | | | | |
|   Right Wheel | | | | | 0 | 1/4 | 1/2 | | | | | | | | | |
| 72 Coronet, Charger | | | | | | | | | | | | | | | | |
|   W/Power Steering | (6) | 1/4 | 3/4 | 1 1/4 | | | | 3/32 | 1/8 | 5/32 | 2.4 | 3.2 | 4.0 | 17 3/4 | 20 | 7 1/2 |
|   Left Wheel | | | | | 1/4 | 1/2 | 3/4 | | | | | | | | | |
|   Right Wheel | | | | | 0 | 1/4 | 1/2 | | | | | | | | | |
|   W/Manual Steering | (6) | − 1 | − 1/2 | 0 | | | | 3/32 | 1/8 | 5/32 | 2.4 | 3.2 | 4.0 | 17 3/4 | 20 | 7 1/2 |
|   Left Wheel | | | | | 1/4 | 1/2 | 3/4 | | | | | | | | | |
|   Right Wheel | | | | | 0 | 1/4 | 1/2 | | | | | | | | | |
| 81 Aries | | | | | | | | | | | | | | | | |
|   Front | (4) | | Fixed | | − 1/4 | 5/16 | 3/4 | 5/32 (out) | 1/16 (out) | 1/8 (in) | 4.0 (out) | 1.6 (out) | 3.2 (in) | NA | NA | 13 3/8 |
|   Rear | (48) | | Fixed | | − 1 | − 1/2 | 0 | 3/16 (out) | 0 | 3/16 (in) | 4.8 (out) | 0 | 4.8 (in) | | | |
| 81-78 OMNI, 024 | | | | | | | | | | | | | | | | |
|   Front | (4) | | Fixed | | − 1/4 | 5/16 | 3/4 | 5/32 (out) | 1/16 (out) | 1/8 (in) | 4.0 (out) | 1.6 (out) | 3.2 (in) | NA | NA | 13 3/8 |
|   Rear | (48) | | Fixed | | − 1/2 | − 1 | − 1 1/2 | 5/32 (out) | 3/32 (in) | 11/32 (in) | 4.0 (out) | 2.4 (in) | 8.7 (in) | | | |

| VEHICLE IDENTIFICATION YEAR MODEL | ADJ. ILL. NO. | CASTER (Degrees) MIN. | PREF. | MAX. | CAMBER (Degrees) MIN. | PREF. | MAX. | TOE-IN (Inches) MIN. | PREF. | MAX. | TOE-IN (Millimeters) MIN. | PREF. | MAX. | TOE-OUT ON TURNS (Degrees) OUTSIDE WHEEL | INSIDE WHEEL | STRG. AXIS INCL. (DEG.) |
|---|---|---|---|---|---|---|---|---|---|---|---|---|---|---|---|---|
| 76-74 Monaco, Royal Monaco .......... | (7) | − 1/2 | 3/4 | 1 3/4 | | | | 1/16 | 1/8 | 1/4 | 1.6 | 3.2 | 6.4 | 18 1/4 | 20 | 9 |
| Left Wheel ......... | (7) | | | | 0 | 1/2 | 1 | | | | | | | | | |
| Right Wheel ........ | (7) | | | | − 1/4 | 1/4 | 3/4 | | | | | | | | | |
| 73-72 Monaco, Polara..... | | | | | | | | | | | | | | | | |
| W/Power Steering ..... | (6) | 1/4 | 3/4 | 1 1/4 | | | | 3/32 | 1/8 | 5/32 | 2.4 | 3.2 | 4.0 | 18 3/4 | 20 | 7 1/2 |
| Left Wheel ......... | | | | | 1/4 | 1/2 | 3/4 | | | | | | | | | |
| Right Wheel ........ | | | | | 0 | 1/4 | 1/2 | | | | | | | | | |
| W/Manual Steering .... | (6) | − 1 | − 1/2 | 0 | | | | 3/32 | 1/8 | 5/32 | 2.4 | 3.2 | 4.0 | 18 3/4 | 20 | 7 1/2 |
| Left Wheel ......... | | | | | 1/4 | 1/2 | 3/4 | | | | | | | | | |
| Right Wheel ........ | | | | | 0 | 1/4 | 1/2 | | | | | | | | | |
| 76-74 Dart, Swinger, Challenger | | | | | | | | | | | | | | | | |
| W/Power Steering ..... | (6) | − 1/2 | 3/4 | 1 3/4 | | | | 1/16 | 1/8 | 1/4 | 1.6 | 3.2 | 6.4 | 18 1/2 | 20 | 7 1/2 |
| Left Wheel ......... | | | | | 0 | 1/2 | 1 | | | | | | | | | |
| Right Wheel ........ | | | | | − 1/4 | 1/4 | 3/4 | | | | | | | | | |
| W/Manual Steering .... | (6) | − 1 3/4 | − 1/2 | 1/2 | | | | 1/16 | 1/8 | 1/4 | 1.6 | 3.2 | 6.4 | 18 1/2 | 20 | 7 1/2 |
| Left Wheel ......... | | | | | 0 | 1/2 | 1 | | | | | | | | | |
| Right Wheel ........ | | | | | − 1/4 | 1/4 | 3/4 | | | | | | | | | |
| 73-72 Dart, Swinger, Demon, Challenger | | | | | | | | | | | | | | | | |
| W/Power Steering ..... | (6) | 1/4 | 3/4 | 1 1/4 | | | | 3/32 | 1/8 | 5/32 | 2.4 | 3.2 | 4.0 | 17 1/2 | 20 | 7 1/2 |
| Left Wheel ......... | | | | | 1/4 | 1/2 | 3/4 | | | | | | | | | |
| Right Wheel ........ | | | | | 0 | 1/4 | 1/2 | | | | | | | | | |
| W/Manual Steering .... | (6) | − 1 | − 1/2 | 0 | | | | 3/32 | 1/8 | 5/32 | 2.4 | 3.2 | 4.0 | 17 1/2 | 20 | 7 1/2 |
| Left Wheel ......... | | | | | 1/4 | 1/2 | 3/4 | | | | | | | | | |
| Right Wheel ........ | | | | | 0 | 1/4 | 1/2 | | | | | | | | | |
| **PLYMOUTH (A) (B) (C)** | | | | | | | | | | | | | | | | |
| 81-80 Gran Fury .......... | (7) | − 1/4 | 1 | 2 1/4 | − 1/4 | 1/2 | 1 1/4 | 0 | 1/8 | 5/16 | 0 | 3.2 | 8.0 | 18 | 20 | 8 |
| 77 Gran Fury .......... | (7) | − 1/2 | 3/4 | 2 | | | | 1/16 | 1/8 | 1/4 | 1.6 | 3.2 | 6.4 | 18 | 20 | 9 |
| Left Wheel ......... | | | | | 0 | 1/2 | 1 | | | | | | | | | |
| Right Wheel ........ | (7) | | | | − 1/4 | 1/4 | 3/4 | | | | | | | | | |
| 76-75 Gran Fury .......... | (7) | − 1/2 | 3/4 | 1 3/4 | | | | 1/16 | 1/8 | 1/4 | 1.6 | 3.2 | 6.4 | 18 5/16 | 20 | 9 |
| Left Wheel ......... | | | | | 0 | 1/2 | 1 | | | | | | | | | |
| Right Wheel ........ | | | | | − 1/4 | 1/4 | 3/4 | | | | | | | | | |
| 80 Volare ............. | | 1 1/4 | 2 1/2 | 3 3/4 | − 1/4 | 1/2 | 1 1/4 | 0 | 1/8 | 5/16 | 0 | 3.2 | 8.0 | 18 | 20 | 8 |
| 79-76 Volare ............. | (7) | 1 1/2 | 2 1/2 | 3 3/4 | | | | 1/16 | 1/8 | 1/4 | 1.6 | 3.2 | 6.4 | 18 | 20 | 8 |
| Left Wheel ......... | | | | | 0 | 1/2 | 1 | | | | | | | | | |
| Right Wheel ........ | | | | | − 1/4 | 1/4 | 3/4 | | | | | | | | | |
| 81 Reliant ............. | | | | | | | | | | | | | | | | |
| Front ............. | (4) | | Fixed | | − 1/4 | 5/16 | 3/4 | 5/32 (out) | 1/16 (out) | 1/8 (in) | 4.0 (out) | 1.6 (out) | 3.2 (in) | NA | NA | 13 3/8 |
| Rear ............. | (48) | | Fixed | | − 1 | − 1/2 | 0 | 3/16 (out) | 0 | 3/16 (in) | 4.8 (out) | 0 | 4.8 (in) | | | |
| 81-78 Horizon, TC3 ....... | | | | | | | | | | | | | | | | |
| Front Wheel ...... | (4) | | Fixed | | − 1/4 | 5/16 | 3/4 | 5/32 (out) | 1/16 (out) | 1/8 (in) | 4.0 (out) | 1.6 (out) | 3.2 (in) | NA | NA | 13 3/8 |
| Rear Wheel ....... | (48) | | Fixed | | − 1 1/2 | − 1 | 1/2 | 5/32 (out) | 3/32 (in) | 11/32 (in) | 4.0 (out) | 2.4 (in) | 8.7 (in) | | | |
| 78-77 Fury ............. | | | | | | | | | | | | | | | | |
| W/Power Steering ..... | (7) | − 1/2 | 3/4 | 2 | | | | 1/16 | 1/8 | 1/4 | 1.6 | 3.2 | 6.4 | 18 | 20 | 8 |
| Left Wheel ......... | | | | | 0 | 1/2 | 1 | | | | | | | | | |
| Right Wheel ........ | | | | | − 1/4 | 1/4 | 3/4 | | | | | | | | | |
| W/Manual Steering .... | (7) | − 1 3/4 | − 1/2 | 3/4 | | | | 1/16 | 1/8 | 1/4 | 1.6 | 3.2 | 6.4 | 18 | 20 | 8 |
| Left Wheel ......... | | | | | 0 | 1/2 | 1 | | | | | | | | | |
| Right Wheel ....... | | | | | − 1/4 | 1/4 | 3/4 | | | | | | | | | |
| 76-75 Fury ............. | | | | | | | | | | | | | | | | |
| W/Power Steering ..... | (7) | − 1/2 | 3/4 | 1 3/4 | | | | 1/16 | 1/8 | 1/4 | 1.6 | 3.2 | 6.4 | 18 | 20 | 8 |
| Left Wheel ......... | | | | | 0 | 1/2 | 1 | | | | | | | | | |
| Right Wheel ........ | | | | | − 1/4 | 1/4 | 3/4 | | | | | | | | | |
| W/Manual Steering .... | (7) | − 1 3/4 | − 1/2 | 1/2 | | | | 1/16 | 1/8 | 1/4 | 1.6 | 3.2 | 6.4 | 18 | 20 | 8 |
| Left Wheel ......... | | | | | 0 | 1/2 | 1 | | | | | | | | | |
| Right Wheel ........ | | | | | − 1/4 | 1/4 | 3/4 | | | | | | | | | |
| 76-74 Valiant, Duster ..... | | | | | | | | | | | | | | | | |
| W/Power Steering ..... | (6) | − 1/2 | 3/4 | 1 3/4 | | | | 1/16 | 1/8 | 1/4 | 1.6 | 3.2 | 6.4 | 18 1/2 | 20 | 7 1/2 |
| Left Wheel ......... | | | | | 0 | 1/2 | 1 | | | | | | | | | |
| Right Wheel ........ | | | | | − 1/4 | 1/4 | 3/4 | | | | | | | | | |
| W/Manual Steering .... | (6) | − 1 3/4 | − 1/2 | 1/2 | | | | 1/16 | 1/8 | 1/4 | 1.6 | 3.2 | 6.4 | 18 1/2 | 20 | 7 1/2 |
| Left Wheel ......... | | | | | 0 | 1/2 | 1 | | | | | | | | | |
| Right Wheel ........ | | | | | − 1/4 | 1/4 | 3/4 | | | | | | | | | |
| 74 Fury ............. | (7) | − 1/2 | 3/4 | 1 3/4 | | | | 1/16 | 1/8 | 1/4 | 1.6 | 3.2 | 6.4 | 18 5/16 | 20 | 9 |
| Left Wheel ......... | | | | | 0 | 1/2 | 1 | | | | | | | | | |
| Right Wheel ........ | | | | | − 1/4 | 1/4 | 3/4 | | | | | | | | | |

| VEHICLE IDENTIFICATION YEAR MODEL | ADJ. ILL. NO. | CASTER (Degrees) MIN. | PREF. | MAX. | CAMBER (Degrees) MIN. | PREF. | MAX. | TOE-IN (Inches) MIN. | PREF. | MAX. | TOE-IN (Millimeters) MIN. | PREF. | MAX. | TOE-OUT ON TURNS (Degrees) OUTSIDE WHEEL | INSIDE WHEEL | STRG. AXIS INCL. (DEG.) |
|---|---|---|---|---|---|---|---|---|---|---|---|---|---|---|---|---|
| 74 Satellite, Road Runner. . | (7) | | | | | | | | | | | | | | | |
| *W/Power Steering* . . . . . | | - 1/2 | 3/4 | 1 3/4 | | | | 1/16 | 1/8 | 1/4 | 1.6 | 3.2 | 6.4 | 18 | 20 | 8 |
| Left Wheel . . . . . . . . . | | | | | 0 | 1/2 | 1 | | | | | | | | | |
| Right Wheel . . . . . . . . | | | | | - 1/4 | 1/4 | 3/4 | | | | | | | | | |
| *W/Manual Steering* . . . . | (7) | - 1 3/4 | - 1/2 | 1/2 | | | | 1/16 | 1/8 | 1/4 | 1.6 | 3.2 | 6.4 | 18 | 20 | 8 |
| Left Wheel . . . . . . . . . | | | | | 0 | 1/2 | 1 | | | | | | | | | |
| Right Wheel . . . . . . . . | | | | | - 1/4 | 1/4 | 3/4 | | | | | | | | | |
| 74 Valiant, Duster, Barracuda . . . . . . . . . . . . . | | | | | | | | | | | | | | | | |
| *W/Power Steering* . . . . . | (6) | - 1/2 | 3/4 | 1 3/4 | | | | 1/16 | 1/8 | 1/4 | 1.6 | 3.2 | 6.4 | 18 1/2 | 20 | 7 1/2 |
| Left Wheel . . . . . . . . . | | | | | 0 | 1/2 | 1 | | | | | | | | | |
| Right Wheel . . . . . . . . | | | | | - 1/4 | 1/4 | 3/4 | | | | | | | | | |
| *W/Manual Steering* . . . . . | (6) | - 1 3/4 | - 1/2 | 1/2 | | | | 1/16 | 1/8 | 1/4 | 1.6 | 3.2 | 6.4 | 18 1/2 | 20 | 7 1/2 |
| Left Wheel . . . . . . . . . | | | | | 0 | 1/2 | 1 | | | | | | | | | |
| Right Wheel . . . . . . . . | | | | | - 1/4 | 1/4 | 3/4 | | | | | | | | | |
| 73-72 Valiant, Duster, Barracuda, Scamp . . . . . . . | | | | | | | | | | | | | | | | |
| *W/Power Steering* . . . . . | (6) | 1/4 | 3/4 | 1 1/4 | | | | 3/32 | 1/8 | 5/32 | 2.4 | 3.2 | 4.0 | 17 1/2 | 20 | 7 1/2 |
| Left Wheel . . . . . . . . . | | | | | 1/4 | 1/2 | 3/4 | | | | | | | | | |
| Right Wheel . . . . . . . . | | | | | 0 | 1/4 | 1/2 | | | | | | | | | |
| *W/Manual Steering* . . . . | (6) | - 1 | - 1/2 | 0 | | | | 3/32 | 1/8 | 5/32 | 2.4 | 3.2 | 4.0 | 17 1/2 | 20 | 7 1/2 |
| Left Wheel . . . . . . . . . | | | | | 1/4 | 1/2 | 3/4 | | | | | | | | | |
| Right Wheel . . . . . . . . | | | | | 0 | 1/4 | 1/2 | | | | | | | | | |
| 73-72 Fury, VIP . . . . . . . . . . | | | | | | | | | | | | | | | | |
| *W/Power Steering* . . . . . | (6) | 1/4 | 3/4 | 1 1/4 | | | | 3/32 | 1/8 | 5/32 | 2.4 | 3.2 | 4.0 | 18 3/4 | 20 | 7 1/2 |
| Left Wheel . . . . . . . . . | | | | | 1/4 | 1/2 | 3/4 | | | | | | | | | |
| Right Wheel . . . . . . . . | | | | | 0 | 1/4 | 1/2 | | | | | | | | | |
| *W/Manual Steering* . . . . . | (6) | - 1 | - 1/2 | 0 | | | | 3/32 | 1/8 | 5/32 | 2.4 | 3.2 | 4.0 | 18 3/4 | 20 | 7 1/2 |
| Left Wheel . . . . . . . . . | | | | | 1/4 | 1/2 | 3/4 | | | | | | | | | |
| Right Wheel . . . . . . . . | | | | | 0 | 1/4 | 1/2 | | | | | | | | | |
| 73 Satellite, Road Runner. . | | | | | | | | | | | | | | | | |
| *W/Power Steering* . . . . . | (7) | 1/4 | 1/2 | 1 1/4 | | | | 3/32 | 1/8 | 5/32 | 2.4 | 3.2 | 4.0 | 18 1/2 | 20 | 8 |
| Left Wheel . . . . . . . . . | | | | | 1/4 | 1/2 | 3/4 | | | | | | | | | |
| Right Wheel . . . . . . . . | | | | | 0 | 1/4 | 1/2 | | | | | | | | | |
| *W/Manual Steering* . . . . | (7) | - 1 | - 1/2 | 0 | | | | 3/32 | 1/8 | 5/32 | 2.4 | 3.2 | 4.0 | 18 1/2 | 20 | 8 |
| Left Wheel . . . . . . . . . | | | | | 1/4 | 1/2 | 3/4 | | | | | | | | | |
| Right Wheel . . . . . . . . | | | | | 0 | 1/4 | 1/2 | | | | | | | | | |
| 72 Satellite, Road Runner. . | | | | | | | | | | | | | | | | |
| *W/Power Steering* . . . . . | (6) | 1/4 | 3/4 | 1 1/4 | | | | 3/32 | 1/8 | 5/32 | 2.4 | 3.2 | 4.0 | 17 3/4 | 20 | 7 1/2 |
| Left Wheel . . . . . . . . . | | | | | 1/4 | 1/2 | 3/4 | | | | | | | | | |
| Right Wheel . . . . . . . . | | | | | 0 | 1/4 | 1/2 | | | | | | | | | |
| *W/Manual Steering* . . . . | (6) | - 1 | - 1/2 | 0 | | | | 3/32 | 1/8 | 5/32 | 2.4 | 3.2 | 4.0 | 17 3/4 | 20 | 7 1/2 |
| Left Wheel . . . . . . . . . | | | | | 1/4 | 1/2 | 3/4 | | | | | | | | | |
| Right Wheel . . . . . . . . | | | | | 0 | 1/4 | 1/2 | | | | | | | | | |

**FORD MOTOR CO.**

**(A)** Maximum side to side variation; caster ± 3/4 °, camber – 1/2 ° to 1 °.
**(B)** Maximum side to side variation; caster and camber ± 3/4 °.
**(C)** Maximum side to side variation; caster and camber 1 °.
**(D)** Maximum side to side variation; caster 1 ° camber 1/2 °.
**(E)** Caster measurement must be done for **each** wheel regardless of the equipment being used.

| VEHICLE IDENTIFICATION YEAR MODEL | ADJ. ILL. NO. | CASTER (Degrees) MIN. | PREF. | MAX. | CAMBER (Degrees) MIN. | PREF. | MAX. | TOE-IN (Inches) MIN. | PREF. | MAX. | TOE-IN (Millimeters) MIN. | PREF. | MAX. | TOE-OUT ON TURNS (Degrees) OUTSIDE WHEEL | INSIDE WHEEL | STRG. AXIS INCL. (DEG.) |
|---|---|---|---|---|---|---|---|---|---|---|---|---|---|---|---|---|
| **FORD DIVISION** | | | | | | | | | | | | | | | | |
| 81 Escort. . . . . . . . . . . . . (E) | N/A | 29/32 | 1 21/32 | 2 1/32 | | | | 1/32 (in) | 3/32 (out) | 7/32 (out) | .7 (in) | 2.4 (out) | 5.6 (out) | | | |
| Left Wheel . . . . . . . . . . . | | | | | 1 | 1 3/4 | 2 1/2 | | | | | | | 19 31/32 | 20 | 14 21/32 |
| Right Wheel . . . . . . . . . . | | | | | 9/16 | 1 1/32 | 2 1/16 | | | | | | | 17 1/32 | 20 | 15 3/32 |
| 81 LTD . . . . . . . . . . . . . . (B) | (47) | 2 1/4 | 3 | 3 3/4 | - 1/4 | 1/2 | 1 1/4 | 1/16 (out) | 1/16 (in) | 3/16 (in) | 1.6 (out) | 1.6 (in) | 4.8 (in) | 18 1/2 | 20 | 10 31/32 |
| 81 Mustang . . . . . . . . . . . (B) | N/A | 1/4 | 1 | 1 3/4 | - 1/2 | 1/4 | 1 | 1/16 | 3/16 | 5/16 | 1.6 | 4.8 | 8.0 | 19 27/32 | 20 | 15 11/16 |
| 81 Thunderbird . . . . . . . . (B) | N/A | 1/8 | 1 | 1 7/8 | - 1/2 | 3/8 | 1 1/4 | 1/16 | 3/16 | 5/16 | 1.6 | 4.8 | 8.0 | 19 3/4 | 20 | 15 23/32 |
| 81 Fairmont . . . . . . . . . . . . | | | | | | | | | | | | | | | | |
| Sedan. . . . . . . . . . . . . (B) | N/A | 1/8 | 1 | 1 7/8 | - 5/16 | 7/16 | 1 3/16 | 1/16 | 3/16 | 5/16 | 1.6 | 4.8 | 8.0 | 19 27/32 | 20 | 15 23/32 |
| Wagon . . . . . . . . . . . . (B) | N/A | - 1/8 | 3/4 | 1 5/8 | - 1/4 | 1/2 | 1 1/4 | 1/16 | 3/16 | 5/16 | 1.6 | 4.8 | 8.0 | 19 27/32 | 20 | 15 23/32 |
| 81 Granada. . . . . . . . . . . . (B) | N/A | 1/8 | 1 | 1 7/8 | - 5/16 | 7/16 | 1 3/16 | 1/16 | 3/16 | 5/16 | 1.6 | 4.8 | 8.0 | 19 27/32 | 20 | 15 23/32 |
| 80 LTD . . . . . . . . . . . . . . (B) | (47) | 2 1/4 | 3 | 3 3/4 | - 1/4 | 1/2 | 1 1/4 | 1/16 (out) | 1/16 (in) | 3/16 (in) | 1.6 (out) | 1.6 (in) | 4.8 (in) | 18 1/2 | 20 | 10 7/8 |
| 80 Thunderbird . . . . . . . . (A) | N/A | 1/8 | 1 | 1 7/8 | - 1/2 | 3/8 | 1 1/4 | 1/8 | 1/4 | 3/8 | 3.2 | 6.4 | 9.5 | 24 29/32 | 20 | 9 1/2 |
| 80 Fairmont | | | | | | | | | | | | | | | | |
| Sedan. . . . . . . . . . . . . (A) | N/A | 1/8 | 1 | 1 7/8 | - 5/16 | 7/16 | 1 3/16 | 1/16 | 3/16 | 5/16 | 1.6 | 4.8 | 8.0 | 19 3/4 | 20 | 15 1/4 |
| Wagon . . . . . . . . . . . . (A) | N/A | - 1/8 | 3/4 | 1 5/8 | - 1/4 | 1/2 | 1 1/4 | 1/16 | 3/16 | 5/16 | 1.6 | 4.8 | 8.0 | 19 3/4 | 20 | 15 1/4 |
| 80-79 Granada . . . . . . . . (B) | (1)(2) | - 1 1/4 | - 1/2 | 1/4 | - 1/2 | 1/4 | 1 | 0 | 1/8 | 1/4 | 0 | 3.2 | 6.4 | | | 7 1/2 |
| *W/Power Steering* . . . . . | | | | | | | | | | | | | | 18 3/16 | 20 | |
| *W/Manual Steering* . . . . | | | | | | | | | | | | | | 18 7/16 | 20 | |

| VEHICLE IDENTIFICATION<br>YEAR  MODEL | ADJ. ILL. NO. | CASTER (Degrees) | | | CAMBER (Degrees) | | | TOE-IN (Inches) | | | TOE-IN (Millimeters) | | | TOE-OUT ON TURNS (Degrees) | | STRG. AXIS INCL. |
|---|---|---|---|---|---|---|---|---|---|---|---|---|---|---|---|---|
| | | MIN. | PREF. | MAX. | MIN. | PREF. | MAX. | MIN. | PREF. | MAX. | MIN. | PREF. | MAX. | OUTSIDE WHEEL | INSIDE WHEEL | (DEG.) |
| 80-79 Mustang ........ (B) | N/A | 1/4 | 1 | 1 3/4 | − 1/2 | 1/4 | 1 | 3/16 | 5/16 | 7/16 | 4.8 | 8.0 | 11.1 | 19 3/4 | 20 | 15 1/4 |
| 80-79 Pinto Sedan ..... (B) | (8) | 1/4 | 1 | 1 3/4 | − 1/4 | 1/2 | 1 1/4 | 0 | 1/8 | 1/4 | 0 | 3.2 | 6.4 | 18 7/8 | 20 | 10 |
| 80 Pinto Wagon ...... (B) | (8) | − 3/4 | 1/4 | 1 1/4 | − 1/4 | 1/2 | 1 1/4 | 0 | 1/8 | 1/4 | 0 | 3.2 | 6.4 | 18 7/8 | 20 | 10 |
| 79 Pinto Wagon ...... (B) | (8) | − 1/2 | 1/4 | 1 | − 1/4 | 1/2 | 1 1/4 | 0 | 1/8 | 1/4 | 0 | 3.2 | 6.4 | 18 7/8 | 20 | 10 |
| 79 LTD ............. (B) | (47) | 2 1/4 | 3 | 3 3/4 | − 1/4 | 1/2 | 1 1/4 | 1/16 | 3/16 | 5/16 | 1.6 | 4.8 | 8.0 | 18 1/2 | 20 | 11 3/16 |
| 79-78 Fairmont ....... (A) | N/A | 1/8 | 7/8 | 1 5/8 | − 3/8 | 3/8 | 1 1/8 | 3/16 | 5/16 | 7/16 | 4.8 | 8.0 | 11.1 | 19 3/4 | 20 | 15 1/4 |
| 79-77 LTD II ......... (A) | (46) | 3 1/4 | 4 | 4 3/4 | | | | 0 | 1/8 | 1/4 | 0 | 3.2 | 6.4 | 18 | 20 | 9 |
|     Left Wheel .......... | | | | | − 1/4 | 1/2 | 1 1/4 | | | | | | | | | |
|     Right Wheel ......... | | | | | − 1/2 | 1/4 | 1 | | | | | | | | | |
| 79-77 Thunderbird ..... (A) | (46) | 3 1/4 | 4 | 4 3/4 | | | | 0 | 1/8 | 1/4 | 0 | 3.2 | 6.4 | 18 | 20 | 9 1/2 |
|     Left Wheel .......... | | | | | − 1/4 | 1/2 | 1 1/4 | | | | | | | | | |
|     Right Wheel ......... | | | | | − 1/2 | 1/4 | 1 | | | | | | | | | |
| 78-77 Pinto Sedan ..... (B) | (8) | 1/4 | 1 | 1 3/4 | − 1/4 | 1/2 | 1 1/4 | 0 | 1/8 | 1/4 | 0 | 3.2 | 6.4 | 18 7/8 | 20 | 10 |
| 78-77 Pinto Wagon .... (B) | (8) | − 1/2 | 1/4 | 1 | − 1/4 | 1/2 | 1 1/4 | 0 | 1/8 | 1/4 | 0 | 3.2 | 6.4 | 18 7/8 | 20 | 10 |
| 78-77 LTD ........... (A) | (8) | 1 1/4 | 2 | 2 3/4 | | | | 1/16 | 3/16 | 5/16 | 1.6 | 4.8 | 8.0 | 18 3/4 | 20 | 9 3/4 |
|     Left Wheel ......... | | | | | − 1/4 | 1/2 | 1 1/4 | | | | | | | | | |
|     Right Wheel ........ | | | | | − 1/2 | 1/4 | 1 | | | | | | | | | |
| 78-77 Granada ....... (B) | (1)(2) | − 1 1/4 | − 1/2 | 1/4 | − 1/2 | 1/4 | 1 | 0 | 1/8 | 1/4 | 0 | 3.2 | 6.4 | | | 6 3/4 |
|     W/Power Steering ..... | | | | | | | | | | | | | | 18 3/16 | 20 | |
|     W/Manual Steering .... | | | | | | | | | | | | | | 18 7/16 | 20 | |
| 78-75 Mustang ....... (B) | (8) | 1/8 | 7/8 | 5/8 | − 1/4 | 1/2 | 1 1/4 | 0 | 1/8 | 1/4 | 0 | 3.2 | 6.4 | 18 7/8 | 20 | 9 3/4 |
| 77 Maverick ......... (B) | (1)(2) | − 1 1/4 | − 1/2 | 1/4 | − 1/2 | 1/4 | 1 | 0 | 1/8 | 1/4 | 0 | 3.2 | 6.4 | | | 6 3/4 |
|     W/Power Steering ..... | | | | | | | | | | | | | | 18 1/8 | 20 | |
|     W/Manual Steering .... | | | | | | | | | | | | | | 18 3/8 | 20 | |
| 76-75 LTD ........... (A) | (8) | 1 1/4 | 2 | 2 3/4 | | | | 1/16 | 3/16 | 7/16 | 1.6 | 4.8 | 11.1 | 18 3/4 | 20 | 9 3/4 |
|     Left Wheel .......... | | | | | − 1/4 | 1/2 | 1 1/4 | | | | | | | | | |
|     Right Wheel ......... | | | | | − 1/2 | 1/4 | 1 | | | | | | | | | |
| 76-75 Thunderbird...... (A) | (8) | 3 1/4 | 4 | 4 3/4 | | | | 1/16 | 3/16 | 7/16 | 1.6 | 4.8 | 11.1 | 18 1/8 | 20 | 9 |
|     Left Wheel .......... | | | | | − 1/4 | 1/2 | 1 1/4 | | | | | | | | | |
|     Right Wheel ........ | | | | | − 1/2 | 1/4 | 1 | | | | | | | | | |
| 76-75 Granada ....... (B) | (1)(2) | − 1 1/4 | − 1/2 | − 1/4 | − 1/2 | 1/4 | 1 | 0 | 1/8 | 3/8 | 0 | 3.2 | 9.5 | | | 6 3/4 |
|     W/Power Steering ..... | | | | | | | | | | | | | | 18 3/16 | 20 | 6 3/4 |
|     W/Manual Steering .... | | | | | | | | | | | | | | 18 7/16 | 20 | 6 3/4 |
| 76-75 Maverick ...... (B) | (1)(2) | − 1 1/4 | − 1/2 | 1/4 | − 1/2 | 1/4 | 1 | 0 | 1/8 | 3/8 | 0 | 3.2 | 9.5 | | | 6 3/4 |
|     W/Power Steering ..... | | | | | | | | | | | | | | 18 1/8 | 20 | 6 3/4 |
|     W/Manual Steering .... | | | | | | | | | | | | | | 18 3/8 | 20 | 6 3/4 |
| 76-75 Torino, Elite ..... (A) | (8) | 3 1/4 | 4 | 4 3/4 | | | | 0 | 1/8 | 3/8 | 0 | 3.2 | 9.5 | 18 | 20 | 9 |
|     Left Wheel ........ | | | | | − 1/4 | 1/2 | 1 1/4 | | | | | | | | | |
|     Right Wheel ........ | | | | | − 1/2 | 1/4 | 1 | | | | | | | | | |
| 76-75 Pinto Sedan..... (B) | (8) | 1/2 | 1 1/4 | 2 | 0 | 3/4 | 1 1/2 | 1/8 | 1/4 | 3/8 | 3.2 | 6.4 | 9.5 | 18 7/8 | 20 | 10 |
| 76-75 Pinto Wagon .... (B) | (8) | 3/4 | 1 1/2 | 2 1/4 | 0 | 3/4 | 1 1/2 | 1/8 | 1/4 | 3/8 | 3.2 | 6.5 | 9.5 | 18 7/8 | 20 | 10 |
| 74 Ford ............. (A) | (8) | 0 | 2 | 4 | | | | 1/16 | 3/16 | 7/16 | 1.6 | 4.8 | 11.1 | 18 3/4 | 20 | 9 1/2 |
|     Left Wheel ......... | | | | | − 1/2 | 1/2 | 1 1/2 | | | | | | | | | |
|     Right Wheel ......... | | | | | − 3/4 | 1/4 | 1 1/4 | | | | | | | | | |
| 74 Thunderbird ....... (A) | (8) | 1/2 | 2 | 3 1/2 | | | | 1/16 | 3/16 | 7/16 | 1.6 | 4.8 | 11.1 | 18 | 20 | 9 |
|     Left Wheel ......... | | | | | 0 | 1 | 2 | | | | | | | | | |
|     Right Wheel ........ | | | | | − 1/2 | 1/2 | 1 1/2 | | | | | | | | | |
| 74 Mustang ......... (B) | (8) | − 1/4 | 3/4 | 1 3/4 | − 1/2 | 1/2 | 1 1/2 | 0 | 1/8 | 1/4 | 0 | 3.2 | 6.4 | 18 7/8 | 20 | 9 3/4 |
| 74 Maverick ......... (C) | (1)(2) | − 2 1/2 | − 1/2 | 1 1/2 | − 3/4 | 1/4 | 1 1/4 | 1/16 | 3/16 | 3/8 | 1.6 | 4.8 | 9.5 | | | |
|     W/Power Steering ...... | | | | | | | | | | | | | | 18 1/16 | 20 | 6 3/4 |
|     W/Manual Steering .... | | | | | | | | | | | | | | 18 13/32 | 20 | 6 3/4 |
| 74 Torino ........... (C) | (8) | 1/2 | 2 | 3 1/2 | | | | 0 | 1/8 | 3/8 | 0 | 3.2 | 9.5 | 18 1/8 | 20 | 9 |
|     Left Wheel ......... | | | | | − 3/8 | 5/8 | 1 5/8 | | | | | | | | | |
|     Right Wheel ........ | | | | | − 7/8 | 1/8 | 1 1/8 | | | | | | | | | |
| 74 Pinto Sedan/Wagon.. (B) | (8) | − 3/4 | 1 1/4 | 3 1/4 | − 1/4 | 3/4 | 1 3/4 | 1/8 | 1/4 | 3/8 | 3.2 | 6.4 | 9.5 | 18 7/8 | 20 | 10 |
| 73 Ford ............. (D) | (8) | 0 | 2 | 4 | − 1 | 0 | 1 | 1/16 | 3/16 | 7/16 | 1.6 | 4.8 | 11.1 | 18 3/4 | 20 | 7 3/4 |
| 73 Thunderbird ....... (B) | (8) | − 1/2 | 2 | 3 1/2 | − 1/4 | 3/4 | 1 3/4 | 1/16 | 3/16 | 7/16 | 1.6 | 4.8 | 11.1 | 17 3/4 | 20 | 7 3/4 |
| 73 Torino ........... (C) | (8) | − 3/4 | 3/4 | 2 1/4 | − 1/4 | 3/4 | 1 3/4 | 3/16 | 5/16 | 9/16 | 4.8 | 7.9 | 14.3 | 17 3/4 | 20 | NA |
| 73 Pinto Sedan ...... (C) | (8) | − 1 | 1 | 3 | − 1/4 | 3/4 | 1 3/4 | 0 | 1/8 | 1/4 | 0 | 3.2 | 6.4 | 19 | 20 | 8 7/8 |
| 73 Pinto Wagon ...... (C) | (8) | − 1 1/2 | 1/2 | 2 1/2 | − 1/4 | 3/4 | 1 3/4 | 0 | 1/8 | 1/4 | 0 | 3.2 | 6.4 | 19 | 20 | 8 7/8 |
| 73-72 Mustang ....... (B) | (1)(2) | − 2 | 0 | 2 | − 1/2 | 1/2 | 1 1/2 | 1/16 | 3/16 | 3/8 | 1.6 | 4.8 | 9.5 | 17 3/4 | 20 | 6 3/4 |
| 73-72 Maverick ....... (C) | (1)(2) | − 2 1/2 | − 1/2 | 1 1/2 | − 3/4 | 1/4 | 1 1/4 | 1/16 | 3/16 | 3/8 | 1.6 | 4.8 | 9.5 | | | |
|     W/Power Steering ...... | | | | | | | | | | | | | | 18 3/16 | 20 | 6 3/4 |
|     W/Manual Steering .... | | | | | | | | | | | | | | 18 7/16 | 20 | 6 3/4 |
| 72 Ford ............. (D) | (8) | − 1 | 1 | 3 | − 1/2 | 1/2 | 1 1/2 | 1/16 | 3/16 | 7/16 | 1.6 | 4.8 | 11.1 | 19 1/8 | 20 | 7 3/4 |
| 72 Thunderbird ....... (B) | (8) | − 1 | 1 | 3 | − 1/4 | 3/4 | 1 3/4 | 1/16 | 3/16 | 7/16 | 1.6 | 4.8 | 11.1 | 17 3/4 | 20 | 7 3/4 |
| 72 Torino ........... (C) | (8) | − 1 1/4 | 3/4 | 2 3/4 | − 1/4 | 3/4 | 1 3/4 | 1/16 | 3/16 | 7/16 | 1.6 | 4.8 | 11.1 | 17 3/4 | 20 | 7 21/32 |
| 72 Pinto ............ (C) | (8) | − 1/2 | 1 1/2 | 3 1/2 | − 1/4 | 3/4 | 1 3/4 | 1/16 | 3/16 | 7/16 | 1.6 | 4.8 | 11.1 | 19 | 20 | 9 |
| **LINCOLN** | | | | | | | | | | | | | | | | |
| 81 Lincoln, Mark VI ..... (B) | (47) | 2 1/4 | 3 | 3 3/4 | − 1/4 | 1/2 | 1 1/4 | 1/16 (out) | 1/16 (in) | 3/16 (in) | 1.6 (out) | 1.6 (in) | 4.8 (in) | 18 1/2 | 20 | 11 7/8 |
| 80 Continental, Mark VI . (B) | (47) | 2 1/4 | 3 | 3 3/4 | − 1/4 | 1/2 | 1 1/4 | 1/16 (out) | 1/16 (out) | 3/16 (in) | 1.6 (out) | 1.6 (in) | 4.8 (in) | 18 1/2 | 20 | 10 7/8 |

| VEHICLE IDENTIFICATION YEAR MODEL | ADJ. ILL. NO. | CASTER (Degrees) MIN. | PREF. | MAX. | CAMBER (Degrees) MIN. | PREF. | MAX. | TOE-IN (Inches) MIN. | PREF. | MAX. | TOE-IN (Millimeters) MIN. | PREF. | MAX. | TOE-OUT ON TURNS (Degrees) OUTSIDE WHEEL | INSIDE WHEEL | STRG. AXIS INCL. (DEG.) |
|---|---|---|---|---|---|---|---|---|---|---|---|---|---|---|---|---|
| 80-77 Versailles . . . . . . . . (B) | (1)(2) | –1 1/4 | –1/2 | 1/4 | –1/2 | 1/4 | 1 | 0 | 1/8 | 1/4 | 0 | 3.2 | 6.4 | | | |
| *W/Power Steering . . . . .* | | | | | | | | | | | | | | 18 3/16 | 20 | 6 3/4 |
| *W/Manual Steering . . . .* | | | | | | | | | | | | | | 18 7/16 | 20 | 6 3/4 |
| 79-78 Continental . . . . . . (A) | (8) | 1 1/4 | 2 | 2 3/4 | | | | 0 | 1/8 | 1/4 | 0 | 3.2 | 6.4 | 18 1/8 | 20 | 9 1/2 |
| Left Wheel . . . . . . . . . | | | | | –1/4 | 1/2 | 1 1/4 | | | | | | | | | |
| Right Wheel . . . . . . . . | | | | | –1/2 | 1/4 | 1 | | | | | | | | | |
| 79-78 Mark V . . . . . . . . . . (A) | (8) | 3 1/4 | 4 | 4 3/4 | | | | 1/16 | 3/16 | 5/16 | 1.6 | 4.8 | 8.0 | 18 1/8 | 20 | 9 1/2 |
| Left Wheel . . . . . . . . . | | | | | –1/4 | 1/2 | 1 1/4 | | | | | | | | | |
| Right Wheel . . . . . . . . | | | | | –1/2 | 1/4 | 1 | | | | | | | | | |
| 77-75 Continental . . . . . . (A) | (8) | 1 1/4 | 2 | 2 3/4 | | | | 0 | 1/8 | 3/8 | 0 | 3.2 | 9.5 | 18 5/32 | 20 | 9 1/2 |
| Left Wheel . . . . . . . . . | | | | | –1/4 | 1/2 | 1 1/4 | | | | | | | | | |
| Right Wheel . . . . . . . . | | | | | –1/2 | 1/4 | 1 | | | | | | | | | |
| 77 Mark V . . . . . . . . . . . (A) | (8) | 1 1/4 | 2 | 2 3/4 | | | | 1/16 | 3/16 | 5/16 | 1.6 | 4.8 | 8.0 | 18 1/8 | 20 | 9 1/2 |
| Left Wheel . . . . . . . . . | | | | | –1/4 | 1/2 | 1 1/4 | | | | | | | | | |
| Right Wheel . . . . . . . . | | | | | –1/2 | 1/4 | 1 | | | | | | | | | |
| 76 Mark IV. . . . . . . . . . . . (A) | (8) | 1 1/4 | 2 | 2 3/4 | | | | 1/16 | 3/16 | 5/16 | 1.6 | 4.8 | 8.0 | 18 1/8 | 20 | 9 1/2 |
| Left Wheel . . . . . . . . . | | | | | –1/4 | 1/2 | 1 1/4 | | | | | | | | | |
| Right Wheel . . . . . . . . | | | | | –1/2 | 1/4 | 1 | | | | | | | | | |
| 75 Mark IV. . . . . . . . . . . . (A) | (8) | 1 1/4 | 2 | 2 3/4 | | | | 1/16 | 3/16 | 7/16 | 1.6 | 4.8 | 11.1 | 18 1/8 | 20 | 9 |
| Left Wheel . . . . . . . . . | | | | | –1/4 | 1/2 | 1 1/4 | | | | | | | | | |
| Right Wheel . . . . . . . . | | | | | –1/2 | 1/4 | 1 | | | | | | | | | |
| 74-73 Continental . . . . . . (D) | (8) | –1/2 | 1 1/2 | 3 1/2 | –1/2 | 1/2 | 1 1/2 | 0 | 1/8 | 3/8 | 0 | 3.2 | 9.5 | 18 3/4 | 20 | 9 1/2 |
| 74 Mark IV. . . . . . . . . . . . (B) | (8) | 1/4 | 1 3/4 | 3 1/4 | | | | 1/16 | 3/16 | 7/16 | 1.6 | 4.8 | 11.1 | 18 | 20 | 9 |
| Left Wheel . . . . . . . . . | | | | | –1/4 | 3/4 | 1 3/4 | | | | | | | | | |
| Right Wheel . . . . . . . . | | | | | –3/4 | 1/4 | 1 1/4 | | | | | | | | | |
| 73 Mark IV. . . . . . . . . . . . (B) | (8) | –1/2 | 1 1/2 | 3 1/2 | –1/4 | 3/4 | 1 3/4 | 1/16 | 3/16 | 7/16 | 1.6 | 4.8 | 11.1 | 17 3/4 | 20 | 7 3/4 |
| 72 Continental . . . . . . . . (D) | (8) | –1/2 | 1 1/2 | 3 1/2 | –1/2 | 1/2 | 1 1/2 | 0 | 1/8 | 3/8 | 0 | 3.2 | 9.5 | 18 3/4 | 20 | 7 17/32 |
| 72 Mark IV . . . . . . . . . . . (D) | (8) | –1 | 1 1/2 | 3 | –1/4 | 3/4 | 1 3/4 | 1/16 | 3/16 | 7/16 | 1.6 | 4.8 | 11.1 | 17 3/4 | 20 | 7 3/4 |
| **MERCURY** | | | | | | | | | | | | | | | | |
| 81 Lynx . . . . . . . . . . . . . . (E) | N/A | 29/32 | 1 21/32 | 2 1/32 | | | | 1/32 (in) | 3/32 (out) | 7/32 (out) | .7 (in) | 2.4 (out) | 5.6 (out) | | | |
| Left Wheel . . . . . . . . . . . | | | | | 1 | 1 3/4 | 2 1/2 | | | | | | | 19 31/32 | 20 | 14 21/32 |
| Right Wheel . . . . . . . . . . | | | | | 9/16 | 1 1/32 | 2 1/16 | | | | | | | 17 1/32 | 20 | 15 3/32 |
| 81 Mercury . . . . . . . . . . . (B) | 47 | 2 1/4 | 3 | 3 3/4 | –1/4 | 1/2 | 1 1/4 | 1/16 (out) | 1/16 (in) | 3/16 (in) | 1.6 (out) | 1.6 (in) | 4.8 (in) | 18 1/2 | 20 | 10 31/32 |
| 81 Capri . . . . . . . . . . . . . (B) | N/A | 1/4 | 1 | 1 3/4 | –1/2 | 1/4 | 1 | 1/16 | 3/16 | 5/16 | 1.6 | 4.8 | 8.0 | 19 27/32 | 20 | 15 11/16 |
| 81 Cougar XR7 . . . . . . . . (B) | N/A | 1/8 | 1 | 1 7/8 | –1/2 | 3/8 | 1 1/4 | 1/16 | 3/16 | 5/16 | 1.6 | 4.8 | 8.0 | 19 3/4 | 20 | 15 23/32 |
| 81 Cougar. . . . . . . . . . . . (B) | N/A | 1/8 | 1 | 1 7/8 | –5/16 | 7/16 | 1 3/16 | 1/16 | 3/16 | 5/16 | 1.6 | 4.8 | 8.0 | 19 27/32 | 20 | 15 23/32 |
| 81 Zephyr Sedan. . . . . . . (B) | N/A | 1/8 | 1 | 1 7/8 | –5/16 | 7/16 | 1 3/16 | 1/16 | 3/16 | 5/16 | 1.6 | 4.8 | 8.0 | 19 27/32 | 20 | 15 23/32 |
| 81 Zephyr Wagon . . . . . . (B) | N/A | –1/8 | 3/4 | 1 5/8 | –1/4 | 1/2 | 1 1/4 | 1/16 | 3/16 | 5/16 | 1.6 | 4.8 | 8.0 | 19 27/32 | 20 | 15 23/32 |
| 80-79 Capri . . . . . . . . . . . (B) | N/A | 1/4 | 1 | 1 3/4 | –1/2 | 1/4 | 1 | 3/16 | 5/16 | 7/16 | 4.8 | 8.0 | 11.1 | 19 3/4 | 20 | 15 1/4 |
| 80 Zephyr Sedan. . . . . . . (A) | N/A | 1/8 | 1 | 1 7/8 | –5/16 | 7/16 | 1 3/16 | 1/16 | 3/16 | 5/16 | 1.6 | 4.8 | 8.0 | 19 3/4 | 20 | 15 1/4 |
| 80 Zephyr Wagon . . . . . . (A) | N/A | –1/8 | 3/4 | 5/8 | –1/4 | 1/2 | 1 1/4 | 1/16 | 3/16 | 5/16 | 1.6 | 4.8 | 8.0 | 19 3/4 | 20 | 15 1/4 |
| 80 Bobcat Sedan . . . . . . (B) | (8) | 1/4 | 1 | 1 3/4 | –1/4 | 1/2 | 1 1/4 | 0 | 1/8 | 1/4 | 0 | 3.2 | 6.4 | 18 7/8 | 20 | 10 |
| 80 Bobcat Wagon. . . . . . (B) | (8) | –3/4 | 1/4 | 1 1/4 | –1/4 | 1/2 | 1 1/4 | 0 | 1/8 | 1/4 | 0 | 3.2 | 6.4 | 18 7/8 | 20 | 10 |
| 80 Cougar. . . . . . . . . . . . (A) | N/A | 1/8 | 1 | 1 7/8 | –1/2 | 3/8 | 1 1/4 | 1/8 | 1/4 | 3/8 | 3.2 | 6.4 | 9.5 | 24 29/32 | 20 | 15 3/8 |
| 80 Monarch . . . . . . . . . . (B) | (1)(2) | –1 1/4 | –1/2 | 1/4 | –1/2 | 1/4 | 1 | 0 | 1/8 | 1/4 | 0 | 3.2 | 6.4 | | | |
| *W/Power Steering . . . . .* | | | | | | | | | | | | | | 18 3/16 | 20 | 7 1/2 |
| *W/Manual Steering . . . .* | | | | | | | | | | | | | | 18 7/16 | 20 | 7 1/2 |
| 80-79 Mercury . . . . . . . . . (B) | (47) | 2 1/4 | 3 | 3 3/4 | –1/4 | 1/2 | 1 1/4 | 1/16 (out) | 1/16 (in) | 3/16 (in) | 1.6 (out) | 1.6 (in) | 4.8 (in) | 18 1/2 | 20 | 10 7/8 |
| 79-78 Zephyr . . . . . . . . . . (A) | N/A | 1/8 | 7/8 | 1 5/8 | –3/8 | 3/8 | 1 1/8 | 3/16 | 5/16 | 7/16 | 4.8 | 8.0 | 11.1 | 19 3/4 | 20 | 15 1/4 |
| 79-77 Bobcat Sedan . . . . (B) | (8) | 1/4 | 1 | 1 3/4 | –1/4 | 1/2 | 1 1/4 | 0 | 1/8 | 1/4 | 0 | 3.2 | 6.4 | 18 7/8 | 20 | 10 |
| 79-77 Bobcat Wagon . . . (B) | (8) | –1/2 | 1/4 | 1 | –1/4 | 1/2 | 1 1/4 | 0 | 1/8 | 1/4 | 0 | 3.2 | 6.4 | 18 7/8 | 20 | 10 |
| 79-77 Cougar . . . . . . . . . (A) | (46) | 3 1/4 | 4 | 4 3/4 | | | | 0 | 1/8 | 1/4 | 0 | 3.2 | 6.4 | 18 | 20 | 9 |
| Left Wheel . . . . . . . . . . . | | | | | –1/4 | 1/2 | 1 1/4 | | | | | | | | | |
| Right Wheel . . . . . . . . . . | | | | | –1/2 | 1/4 | 1 | | | | | | | | | |
| 79-77 Monarch . . . . . . . . (B) | (1)(2) | –1 1/4 | –1 1/2 | 1/4 | –1/2 | 1/4 | 1 | 0 | 1/8 | 3/8 | 0 | 3.2 | 9.5 | | | |
| *W/Power Steering . . . . .* | | | | | | | | | | | | | | 18 3/16 | 20 | 6 3/4 |
| *W/Manual Steering . . . .* | | | | | | | | | | | | | | 18 7/16 | 20 | 6 3/4 |
| 78-77 Mercury . . . . . . . . . (A) | (8) | 1 1/4 | 2 | 2 3/4 | | | | 1/16 | 3/16 | 5/16 | 1.6 | 4.8 | 8.0 | 18 3/4 | 20 | 9 1/2 |
| Left Wheel . . . . . . . . . | | | | | –1/4 | 1/2 | 1 1/4 | | | | | | | | | |
| Right Wheel . . . . . . . . | | | | | –1/2 | 1/4 | 1 | | | | | | | | | |
| 77 Comet . . . . . . . . . . . . (B) | (1)(2) | –1 1/4 | –1/2 | 1/4 | –1/2 | 1/4 | 1 | 0 | 1/8 | 1/4 | 0 | 3.2 | 6.4 | | | |
| *W/Power Steering . . . . .* | | | | | | | | | | | | | | 18 1/8 | 20 | 6 3/4 |
| *W/Manual Steering . . . .* | | | | | | | | | | | | | | 18 3/8 | 20 | 6 3/4 |
| 76 Bobcat Sedan . . . . . . (B) | (8) | 1/2 | 1 1/4 | 2 | 0 | 3/4 | 1 1/2 | 1/8 | 1/4 | 3/8 | 3.2 | 6.4 | 9.5 | 18 7/8 | 20 | 10 |
| 76 Bobcat Wagon. . . . . . (B) | (8) | 3/4 | 1 1/2 | 2 1/4 | 0 | 3/4 | 1 1/2 | 1/8 | 1/4 | 3/8 | 3.2 | 6.4 | 9.5 | 18 7/8 | 20 | 10 |
| 76-75 Cougar, Montego . (A) | (46) | 3 1/4 | 4 | 4 3/4 | | | | 0 | 1/8 | 3/8 | 0 | 3.2 | 9.5 | 18 | 20 | 9 |
| Left Wheel . . . . . . . . . | | | | | –1/4 | 1/2 | 1 1/4 | | | | | | | | | |
| Right Wheel . . . . . . . . | | | | | –1/2 | 1/4 | 1 | | | | | | | | | |
| 76-75 Monarch . . . . . . . . (B) | (1)(2) | –1 1/4 | –1/2 | 1/4 | –1/2 | 1/4 | 1 | 0 | 1/8 | 3/8 | 0 | 3.2 | 9.5 | | | |
| *W/Power Steering . . . . .* | | | | | | | | | | | | | | 18 3/16 | 20 | 6 3/4 |
| *W/Manual Steering . . . .* | | | | | | | | | | | | | | 18 7/16 | 20 | 6 3/4 |

| VEHICLE IDENTIFICATION YEAR MODEL | ADJ. ILL. NO. | CASTER (Degrees) MIN. | PREF. | MAX. | CAMBER (Degrees) MIN. | PREF. | MAX. | TOE-IN (Inches) MIN. | PREF. | MAX. | TOE-IN (Millimeters) MIN. | PREF. | MAX. | TOE-OUT ON TURNS (Degrees) OUTSIDE WHEEL | INSIDE WHEEL | STRG. AXIS INCL. (DEG.) |
|---|---|---|---|---|---|---|---|---|---|---|---|---|---|---|---|---|
| 76-75 Comet .........(B) | (1)(2) | − 1¼ | − ½ | ¼ | − ½ | ¼ | 1 | 0 | ⅛ | ⅜ | 0 | 3.2 | 9.5 | | | |
| W/Power Steering ..... | | | | | | | | | | | | | | 18 5/32 | 20 | 6¾ |
| W/Manual Steering .... | | | | | | | | | | | | | | 18 13/32 | 20 | 6¾ |
| 76-75 Mercury ........(A) | (8) | 1½ | 2 | 2¾ | | | | 1/16 | 3/16 | 7/16 | 1.6 | 4.8 | 11.1 | 18¾ | 20 | 9½ |
| Left Wheel .......... | | | | | − ¼ | ½ | 1¼ | | | | | | | | | |
| Right Wheel ......... | | | | | − ½ | ¼ | 1 | | | | | | | | | |
| 74 Cougar, Montego.....(B) | (8) | ½ | 2 | 3½ | | | | 0 | ⅛ | ⅜ | 0 | 3.2 | 9.5 | 18⅛ | 20 | 9 |
| Left Wheel .......... | | | | | − ⅜ | ⅝ | 1⅝ | | | | | | | | | |
| Right Wheel ......... | | | | | − ⅞ | ⅛ | 1⅛ | | | | | | | | | |
| 74 Mercury, Meteor.....(D) | (8) | 0 | 2 | 4 | | | | 1/16 | 3/16 | 7/16 | 1.6 | 4.8 | 11.1 | 18¾ | 20 | 9½ |
| Left Wheel .......... | | | | | − ½ | ½ | 1½ | | | | | | | | | |
| Right Wheel ......... | | | | | − ¾ | ¼ | 1¼ | | | | | | | | | |
| 74-72 Comet .....(C) | (1)(2) | − 2½ | − ½ | 1½ | − ¾ | ¼ | 1¼ | 1/16 | 3/16 | ⅜ | 1.6 | 4.8 | 9.5 | | | |
| W/Power Steering ..... | | | | | | | | | | | | | | 18 5/32 | 20 | 6¾ |
| W/Manual Steering .... | | | | | | | | | | | | | | 18 13/32 | 20 | 6¾ |
| 73-72 Cougar .......(B) | (1)(2) | − 2 | 0 | 2 | − ½ | ½ | 1½ | 1/16 | 3/16 | ⅜ | 1.6 | 4.8 | 9.5 | 17¾ | 20 | 6¾ |
| 73 Montego ........(C) | (8) | − ¾ | ¾ | 2¼ | − ¼ | ¾ | 1¾ | 3/16 | 5/16 | 9/16 | 4.8 | 8.0 | 14.3 | 17¾ | 20 | 6¾ |
| 73 Mercury, Meteor.....(D) | (8) | 0 | 2 | 4 | − ¼ | 0 | 1 | 1/16 | 3/16 | 7/16 | 1.6 | 4.8 | 11.1 | 18¾ | 20 | 7¼ |
| 72 Montego ........(C) | (8) | − 1¼ | ¾ | 2¾ | − ¼ | ¾ | 1¾ | 1/16 | 3/16 | 7/16 | 1.6 | 4.8 | 11.1 | 17¾ | 20 | 7 21/32 |
| 72 Mercury, Meteor.....(D) | (8) | − 1 | 1 | 3 | − ½ | ½ | 1½ | 1/16 | 3/16 | 7/16 | 1.6 | 4.8 | 11.1 | 19⅛ | 20 | 7¼ |

## GENERAL MOTORS CORPORATION

### BUICK DIVISION (A)

(A) Maximum side to side variation; caster and camber ½°.
(B) Riviera F.W.D. trim height is measured from the edge of the wheel well opening directly over the center of the wheel to the floor. (See block to right for detail).

### RIVIERA F.W.D. TRIM HEIGHTS

| Model & Year | Front Suspension Inches | MM | Rear Suspension Inches | MM |
|---|---|---|---|---|
| 81-80 | 28⅛ | 726 | 27 5/16 | 694 |
| 79 | 28½ | 724 | 28 | 709 |

| VEHICLE IDENTIFICATION | ADJ. ILL. NO. | CASTER MIN. | PREF. | MAX. | CAMBER MIN. | PREF. | MAX. | TOE-IN (Inches) MIN. | PREF. | MAX. | TOE-IN (Millimeters) MIN. | PREF. | MAX. | TOE-OUT OUTSIDE WHEEL | INSIDE WHEEL | STRG. AXIS INCL. |
|---|---|---|---|---|---|---|---|---|---|---|---|---|---|---|---|---|
| 81-77 Electra, Le Sabre.... | (10) | 2 | 3 | 4 | 0 | 13/16 | 1⅝ | 1/16 | ⅛ | ¼ | 1.6 | 3.2 | 6.4 | NA | NA | 9 9/16 |
| 81-77 Estate Wagon ...... | (10) | 2 | 3 | 4 | 0 | 13/16 | 1⅝ | 1/16 | ⅛ | ¼ | 1.6 | 3.2 | 6.4 | NA | NA | 10¾ |
| 81-79 Riviera ............ | | | | | | | | | | | | | | | | |
| Front .............. | (6) | 1½ | 2½ | 3½ | − 13/16 | 0 | 13/16 | ⅛ (out) | 0 | ⅛ (in) | 3.2 (out) | 0 | 3.2 (in) | NA | NA | 11 |
| Rear .............. | (9) | | Fixed | | | Fixed | | 0 | 5/32 | 5/16 | 0 | 4.0 | 8.0 | | | |
| 81-80 Skylark ......... | (34) | − 2 | 0 | 2 | 0 | ½ | 1 | 0 | 3/32 | 3/16 | 0 | 2.5 | 5.0 | NA | NA | 14½ |
| 81-78 Century, Regal .... | (10) | | | | − 5/16 | ½ | 1 5/16 | 1/16 | ⅛ | ¼ | 1.6 | 3.2 | 6.4 | NA | NA | 8 |
| W/Power Steering ..... | | 2 | 3 | 4 | | | | | | | | | | | | |
| W/Manual Steering .... | | 0 | 1 | 2 | | | | | | | | | | | | |
| 80-76 Skyhawk .......... | (11) | − 1¾ | − ¾ | ¼ | − ½ | ¼ | 1 | -1/16 (out) | 1/16 (in) | 3/16 (in) | 4.8 (out) | 1.6 (in) | 1.6 (in) | NA | NA | 8½ |
| 79-77 Skylark ......... | (10) | | | | 0 | ¾ | 1⅝ | 1/16 | ⅛ | ¼ | 1.6 | 3.2 | 6.4 | NA | NA | 10 |
| W/Power Steering ..... | | 0 | 1 | 2 | | | | | | | | | | | | |
| W/Manual Steering .... | | − 2 | − 1 | 0 | | | | | | | | | | | | |
| 77 Century Regal ........ | (10) | | | | | | | | | | | | | | | |
| W/Radial Tires ........ | | 1 | 2 | 3 | | | | 1/16 (out) | 1/16 (in) | 3/16 (in) | 1.6 (out) | 1.6 (in) | 4.8 (in) | NA | NA | 8 |
| W/Bias Tires.......... | | 0 | 1 | 2 | | | | | | | | | | | | |
| Left Wheel ........ | | | | | ¼ | 1 | 1¾ | | | | | | | | | |
| Right Wheel ........ | | | | | − ¼ | ½ | 1¼ | | | | | | | | | |
| 76 Le Sabre, Electra, Riviera ............. | (10) | ½ | 1½ | 2½ | | | | 1/16 (out) | 1/16 (in) | 3/16 (in) | 1.6 (out) | 1.6 (in) | 4.8 (in) | 18½ | 20 | 9 9/16 |
| Left Wheel ........ | | | | | ¼ | 1 | 1¾ | | | | | | | | | |
| Right Wheel ........ | | | | | − ¼ | ½ | 1¼ | | | | | | | | | |
| 76 Estate Wagon ........ | (10) | ½ | 1½ | 2½ | | | | 1/16 (out) | 1/16 (in) | 3/16 (in) | 1.6 (out) | 1.6 (in) | 4.8 (in) | 18½ | 20 | 10¾ |
| Left Wheel ........ | | | | | ¼ | 1 | 1¾ | | | | | | | | | |
| Right Wheel ........ | | | | | − ¼ | ½ | 1¼ | | | | | | | | | |
| 76 Skylark. ............ | (10) | | | | 0 | ¾ | 1½ | 1/16 (out) | 1/16 (in) | 3/16 (in) | 1.6 (out) | 1.6 (in) | 4.8 (in) | NA | NA | 10 |
| W/Power Steering...... | | 0 | 1 | 2 | | | | | | | | | | | | |
| W/Manual Steering .... | | − 2 | − 1 | 0 | | | | | | | | | | | | |
| 76-75 Century, Regal ..... | (10) | 1 | 2 | 3 | | | | 1/16 (out) | 1/16 (in) | 3/16 (in) | 1.6 (out) | 1.6 (in) | 4.8 (in) | | | 8 |
| Left Wheel ........ | | | | | ¼ | 1 | 1¾ | | | | | | | 18 13/16 | 20 | |
| Right Wheel ........ | | | | | − ¼ | ½ | 1¼ | | | | | | | 19 3/16 | 20 | |
| 75 Skyhawk .......... | (11) | − 1¾ | − ¾ | ¼ | − ¼ | ½ | 1¼ | ⅛ | ¼ | ⅜ | 3.2 | 6.4 | 9.5 | NA | NA | 8½ |
| 75 Electra, Le Sabre, Riviera, Estate Wagon .. | (10) | ½ | 1½ | 2½ | | | | 1/16 (out) | 1/16 (in) | 3/16 (in) | 1.6 (out) | 1.6 (in) | 4.8 (in) | 18½ | 20 | 10½ |
| Left Wheel ........ | | | | | ¼ | 1 | 1¾ | | | | | | | | | |
| Right Wheel ........ | | | | | − ¼ | ½ | 1¼ | | | | | | | | | |

| VEHICLE IDENTIFICATION | ADJ. ILL. NO. | CASTER (Degrees) | | | CAMBER (Degrees) | | | TOE-IN (Inches) | | | TOE-IN (Millimeters) | | | TOE-OUT ON TURNS (Degrees) | | STRG. AXIS INCL. |
| YEAR  MODEL | | MIN. | PREF. | MAX. | MIN. | PREF. | MAX. | MIN. | PREF. | MAX. | MIN. | PREF. | MAX. | OUTSIDE WHEEL | INSIDE WHEEL | (DEG.) |
|---|---|---|---|---|---|---|---|---|---|---|---|---|---|---|---|---|
| 75 Appollo, Skylark ....... | (10) | | | | ¼ | ¾ | 1½ | ¹⁄₁₆ (out) | ¹⁄₁₆ (in) | ³⁄₁₆ (in) | 1.6 (out) | 1.6 (in) | 4.8 (in) | NA | NA | 8¾ |
| W/Power Steering...... | | 0 | 1 | 2 | | | | | | | | | | | | |
| W/Manual Steering .... | | − 2 | − 1 | 0 | | | | | | | | | | | | |
| 74-73 Le Sabre, Electra, Centurion, Riviera, Estate Wagon ....... | (10) | 0 | 1 | 2 | | | | ¹⁄₁₆ (out) | ¹⁄₁₆ (in) | ³⁄₁₆ (in) | 1.6 (out) | 1.6 (in) | 4.8 (in) | 18½ | 20 | 10½ |
| Left Wheel .......... | | | | | ¼ | 1 | 1¾ | | | | | | | | | |
| Right Wheel ......... | | | | | − ¼ | ½ | 1¼ | | | | | | | | | |
| 74 Appollo ........... | (10) | − ½ | ½ | 1½ | − ½ | ¼ | 1 | ¹⁄₁₆ | ³⁄₁₆ | ⁵⁄₁₆ | 1.6 | 4.8 | 8.0 | NA | NA | 8¾ |
| 74 Century, Regal....... | (10) | − 1 | 0 | 1 | | | | ¹⁄₁₆ (out) | ¹⁄₁₆ (in) | ³⁄₁₆ (in) | 1.6 (out) | 1.6 (in) | 4.8 (in) | | | 9⅝ |
| Left Wheel ......... | | | | | ¼ | 1 | 1¾ | | | | | | | 18¹³⁄₁₆ | 20 | |
| Right Wheel ......... | | | | | − ¼ | ½ | 1¼ | | | | | | | 19³⁄₁₆ | 20 | |
| 73 Appollo ........... | (10) | | | | | | | | | | | | | | | |
| W/Power Steering ..... | | − 1 | 0 | 1 | | | | ¹⁄₁₆ (out) | ¹⁄₁₆ (in) | ³⁄₁₆ (in) | 1.6 (out) | 1.6 (in) | 4.8 (in) | NA | NA | NA |
| W/Manual Steering .... | | 0 | 1 | 2 | | | | ¹⁄₁₆ (out) | ¹⁄₁₆ (in) | ³⁄₁₆ (in) | 1.6 (out) | 1.6 (in) | 4.8 (in) | NA | NA | NA |
| Left Wheel ......... | | | | | ¼ | 1 | 1¾ | | | | | | | | | |
| Right Wheel ........ | | | | | − ¼ | ½ | 1¼ | | | | | | | | | |
| 73 Century Regal | | | | | | | | | | | | | | | | |
| W/Power Steering ..... | (10) | − 1 | 0 | 1 | | | | ¹⁄₁₆ (out) | ¹⁄₁₆ (in) | ³⁄₁₆ (in) | 1.6 (out) | 1.6 (in) | 4.8 (in) | | | 9⅝ |
| Left Wheel ........ | | | | | ¼ | 1 | 1¾ | | | | | | | 18¹³⁄₁₆ | 20 | |
| Right Wheel ....... | | | | | − ¼ | ½ | 1¼ | | | | | | | 19³⁄₁₆ | 20 | |
| W/Manual Steering .... | (10) | − 2 | − 1 | 0 | | | | ¹⁄₁₆ (out) | ¹⁄₁₆ (in) | ³⁄₁₆ (in) | 1.6 (out) | 1.6 (in) | 4.8 (in) | | | 9⅝ |
| Left Wheel ....... | | | | | ¼ | 1 | 1¾ | | | | | | | 18¹³⁄₁₆ | 20 | |
| Right Wheel ....... | | | | | − ¼ | ½ | 1¼ | | | | | | | 19³⁄₁₆ | 20 | |
| 72 Centurion, Le Sabre, Riviera, Estate Wagon .. | (10) | 0 | 1 | 2 | − ½ | ½ | 1 | ¹⁄₁₆ | ³⁄₁₆ | ⁵⁄₁₆ | 1.6 | 4.8 | 8.0 | 18½ | 20 | 10½ |
| 72 Skylark, GS, Sport Wagon ........ | | − 1½ | − ½ | ½ | − ¼ | ½ | 1¼ | ¹⁄₁₆ | ³⁄₁₆ | ⁵⁄₁₆ | 1.6 | 4.8 | 8.0 | 18½ | 20 | 9 |

**CADILLAC DIVISION (A) (B)**
(A) Maximum side to side variation: after reset caster and camber 1½ °.
(B) Check suspension height before performing alignment.

| VEHICLE IDENTIFICATION | ADJ. ILL. NO. | CASTER | | | CAMBER | | | TOE-IN (Inches) | | | TOE-IN (Millimeters) | | | TOE-OUT ON TURNS | | STRG. AXIS INCL. |
|---|---|---|---|---|---|---|---|---|---|---|---|---|---|---|---|---|
| | | MIN. | PREF. | MAX. | MIN. | PREF. | MAX. | MIN. | PREF. | MAX. | MIN. | PREF. | MAX. | OUTSIDE WHEEL | INSIDE WHEEL | (DEG.) |
| 81-80 Cadillac except Eldorado and Seville ... | (10) | 2 | 3 | 4 | − ⁵⁄₁₆ | ½ | 1⁵⁄₁₆ | 0 | ⅛ | ¼ | 0 | 3.2 | 6.4 | NA | NA | 10¹⁹⁄₃₂ |
| 81-80 Seville ......... | | | | | | | | | | | | | | | | |
| Front ......., ..... | (6) | 1½ | 2½ | 3½ | − ¹³⁄₁₆ | 0 | ¹³⁄₁₆ | ⅛ (out) | 0 | ⅛ (in) | 3.2 (out) | 0 | 3.2 (in) | NA | NA | 11 |
| Rear ............. | (9) | | Fixed | | | Fixed | | 0 | ⁵⁄₃₂ | ⁵⁄₁₆ | 0 | 4.0 | 8.0 | | | |
| 81-79 Eldorado .......... | | | | | | | | | | | | | | | | |
| Front ............ | (6) | 1½ | 2½ | 3½ | − ¹³⁄₁₆ | 0 | ¹³⁄₁₆ | ⅛ (out) | 0 | ⅛ (in) | 3.2 (out) | 0 | 3.2 (in) | NA | NA | 11 |
| Rear ............. | (9) | | Fixed | | | Fixed | | 0 | ⁵⁄₃₂ | ⁵⁄₁₆ | 0 | 4.0 | 8.0 | | | |
| 79-77 Cadillac except Eldorado and Seville ... | (10) | 2 | 3 | 4 | − ¼ | ½ | 1¼ | ⅛ (out) | 0 | ⅛ (in) | 3.2 (out) | 0 | 3.2 (in) | NA | NA | 10¹⁹⁄₃₂ |
| 79-77 Seville .......... | (10) | 1 | 2 | 3 | − ¾ | 0 | ¾ | ¹⁄₁₆ (out) | ¹⁄₁₆ (in) | ³⁄₁₆ (in) | 1.6 (out) | 1.6 (in) | 4.8 (in) | NA | NA | 10⅝ |
| 76 Seville .......... | (10) | 1 | 2 | 3 | | | | ¹⁄₁₆ (out) | ¹⁄₁₆ (in) | ³⁄₁₆ (in) | 1.6 (out) | 1.6 (in) | 4.8 (in) | NA | NA | 10⅝ |
| Left Wheel ........ | | | | | − ¼ | ½ | 1¼ | | | | | | | | | |
| Right Wheel ........ | | | | | − ½ | ¼ | 1 | | | | | | | | | |
| 76-74 Cadillac except Eldorado, Seville and Fleetwood 75 .......... | (1)(12) | − 1 | 0 | 1 | | | | 0 | ⅛ | ¼ | 0 | 3.2 | 6.4 | NA | NA | 6 |
| Left Wheel ......... | | | | | − ¾ | 0 | ¾ | | | | | | | | | |
| Right Wheel ......... | | | | | − 1 | − ¼ | ½ | | | | | | | | | |
| 76-74 Fleetwood 75 ..... | (1)(2) | − 2 | − 1 | 0 | | | | 0 | ⅛ | ¼ | 0 | 3.2 | 6.4 | NA | NA | 6 |
| Left Wheel ......... | | | | | − ¾ | 0 | ¾ | | | | | | | | | |
| Right Wheel ......... | | | | | − 1 | − ¼ | ½ | | | | | | | | | |
| 76-74 Eldorado ......... | (6) | − 1 | 0 | 1 | | | | ⅛ (out) | 0 | ⅛ (in) | 3.2 (out) | 0 | 3.2 (in) | NA | NA | 11 |
| Left Wheel ......... | | | | | − ¾ | 0 | ¾ | | | | | | | | | |
| Right Wheel ......... | | | | | − 1 | − ¼ | ½ | | | | | | | | | |
| 73 Eldorado ........... | (6) | − 1 | 0 | 1 | − ¾ | 0 | ¾ | ⅛ (out) | 0 | ⅛ (in) | 3.2 (out) | 0 | 3.2 (in) | NA | NA | 11 |

| VEHICLE IDENTIFICATION (YEAR MODEL) | ADJ. ILL. NO. | CASTER (Degrees) MIN. | PREF. | MAX. | CAMBER (Degrees) MIN. | PREF. | MAX. | TOE-IN (Inches) MIN. | PREF. | MAX. | TOE-IN (Millimeters) MIN. | PREF. | MAX. | TOE-OUT ON TURNS (Degrees) OUTSIDE WHEEL | INSIDE WHEEL | STRG. AXIS INCL. (DEG.) |
|---|---|---|---|---|---|---|---|---|---|---|---|---|---|---|---|---|
| 73-72 Cadillac except Eldorado | (1)(12) | −2 | −1 | 0 | | | | 1/16 | 3/16 | 5/16 | 1.6 | 4.8 | 8.0 | NA | NA | 6 |
| Left Wheel | | | | | −3/4 | 0 | 3/4 | | | | | | | | | |
| Right Wheel | | | | | −3/4 | −1/4 | 3/4 | | | | | | | | | |
| 72 Eldorado | (6) | −2 | −1 | 0 | −3/4 | 0 | 3/4 | 1/8 (out) | 0 | 1/8 (in) | 3.2 (out) | 0 | 3.2 (in) | NA | NA | 11 |

**CHEVROLET DIVISION (A)**

(A) Maximum side to side variation after reset caster and camber; All except Chevette 1/2 °, Chevette 2 °.

| VEHICLE IDENTIFICATION (YEAR MODEL) | ADJ. ILL. NO. | CASTER MIN. | PREF. | MAX. | CAMBER MIN. | PREF. | MAX. | TOE-IN (Inches) MIN. | PREF. | MAX. | TOE-IN (Millimeters) MIN. | PREF. | MAX. | TOT OUTSIDE WHEEL | INSIDE WHEEL | STRG. AXIS INCL. |
|---|---|---|---|---|---|---|---|---|---|---|---|---|---|---|---|---|
| 81-77 Chevrolet (full size) | (10) | 2 | 3 | 4 | 0 | 13/16 | 1 5/8 | 1/16 | 1/8 | 1/4 | 1.6 | 3.2 | 6.4 | NA | NA | 9 25/32 |
| 81-78 Malibu, El Camino, Monte Carlo | (10) | | | | −5/16 | 1/2 | 1 5/16 | 1/16 | 1/8 | 1/4 | 1.6 | 3.2 | 6.4 | NA | NA | 7 7/8 |
| W/Power Steering | | 2 | 3 | 4 | | | | | | | | | | | | |
| W/Manual Steering | | 0 | 1 | 2 | | | | | | | | | | | | |
| 81-80 Corvette | | | | | | | | | | | | | | | | |
| Front | (10) | 1 1/4 | 2 1/4 | 3 1/4 | 0 | 3/4 | 1 1/2 | 1/8 | 1/4 | 3/8 | 3.2 | 6.4 | 9.5 | NA | NA | 7 11/16 |
| Rear | (57)(58) | | | | 3/16 | 11/16 | 13/16 | 1/16 | 0 | 1/16 | 0 | 0.8 | 1.6 | | | |
| 81-80 Camaro | (10) | 0 | 1 | 2 | 3/16 | 1 | 1 13/16 | 1/16 | 1/8 | 1/4 | 1.6 | 3.2 | 6.4 | NA | NA | 10 3/8 |
| 81-80 Citation | (34) | −2 | 0 | 2 | 0 | 1/2 | 1 | 0 | 3/32 | 3/16 | 0 | 2.5 | 5.0 | NA | NA | 14 1/2 |
| 81-80 Chevette | (13) | 2 1/2 | 4 1/2 | 6 1/2 | −1/2 | 3/16 | 29/32 | 1/32 (out) | 1/16 (in) | 1/8 (in) | 0.8 (out) | 1.6 (in) | 3.2 (in) | NA | NA | 7 9/16 |
| 80-77 Monza | (11) | −1 13/16 | −13/16 | 3/16 | −5/8 | 3/16 | 1 | 3/16 | 1/16 | 1/16 | 4.8 | 1.6 | 1.6 | NA | NA | 8 9/16 |
| 79-78 Camaro | (10) | 0 | 1 | 2 | 3/16 | 1 | 1 13/16 | 1/16 | 1/8 | 1/4 | 1.6 | 3.2 | 6.4 | NA | NA | 10 3/64 |
| 79-78 Chevette | (13) | 2 1/2 | 4 1/2 | 6 1/2 | −1/2 | 3/16 | 7/8 | 0 | 3/32 | 7/32 | 0 | 2.5 | 5.5 | NA | NA | 7 1/2 |
| 79-78 Nova | (10) | | | | 0 | 13/16 | 1 5/8 | 0 | 1/8 | 1/4 | 1.6 | 3.2 | 6.4 | NA | NA | 10 |
| W/Power Steering | | 0 | 1 | 2 | | | | | | | | | | | | |
| W/Manual Steering | | −2 | −1 | 0 | | | | | | | | | | | | |
| 79 Corvette | | | | | | | | | | | | | | | | |
| Front | (10) | 1 1/4 | 2 1/4 | 3 1/4 | 0 | 3/4 | 1 1/2 | 1/8 | 1/4 | 3/8 | 3.2 | 6.4 | 9.5 | NA | NA | 7 11/16 |
| Rear | (57)(58) | | | | −1 | −1/2 | 0 | 1/16 | 3/32 | 1/8 | 1.6 | 2.4 | 3.2 | NA | NA | |
| 78-77 Corvette | | | | | | | | | | | | | | | | |
| Front | (10) | 1 1/4 | 2 1/4 | 3 1/4 | 0 | 3/4 | 1 1/2 | 1/8 | 1/4 | 3/8 | 3.2 | 6.4 | 9.5 | NA | NA | 7 11/16 |
| Rear | (57)(58) | | | | −1 1/8 | −7/8 | −5/8 | 1/32 (out) | 0 | 1/32 (in) | 0.8 (out) | 0 | 0.8 (in) | NA | NA | |
| 77 Chevette | (13) | 2 1/2 | 4 1/2 | 6 1/2 | −1/2 | 3/16 | 1 | 1/32 (out) | 1/16 (in) | 5/32 (in) | 0.8 (out) | 1.6 (in) | 4.0 (in) | NA | NA | 7 1/2 |
| 77 Chevelle, El Camino | | | | | | | | | | | | | | | | |
| W/Power Steering | (10) | | | | | | | 1/16 (out) | 1/16 (in) | 3/16 (in) | 1.6 (out) | 1.6 (in) | 4.8 (in) | NA | NA | 9 19/32 |
| W/Radial Tires | | 1 | 2 | 3 | | | | | | | | | | | | |
| W/Bias Tires | | 0 | 1 | 2 | | | | | | | | | | | | |
| Left Wheel | | | | | 3/16 | 1 | 1 13/16 | | | | | | | | | |
| Right Wheel | | | | | −5/16 | 1/2 | 1 5/16 | | | | | | | | | |
| W/Manual Steering | (10) | 0 | 1 | 2 | | | | 1/16 (out) | 1/16 (in) | 3/16 (in) | 1.6 (out) | 1.6 (in) | 4.8 (in) | NA | NA | 9 19/32 |
| Left Wheel | | | | | 3/16 | 1 | 1 13/16 | | | | | | | | | |
| Right Wheel | | | | | −5/16 | 1/2 | 1 5/16 | | | | | | | | | |
| 77 Monte Carlo | (10) | 4 | 5 | 6 | | | | 1/16 (out) | 1/16 (in) | 3/16 (in) | 1.6 (out) | 1.6 (in) | 4.8 (in) | NA | NA | 9 19/32 |
| Left Wheel | | | | | 3/16 | 1 | 1 13/16 | | | | | | | | | |
| Right Wheel | | | | | −5/16 | 1/2 | 1 5/16 | | | | | | | | | |
| 77 Camaro | (10) | 0 | 1 | 2 | 3/16 | 1 | 1 13/16 | 1/16 (out) | 1/16 (in) | 3/16 (in) | 1.6 (out) | 1.6 (in) | 4.8 (in) | NA | NA | 10 11/32 |
| 76 Monza | (11) | −1 3/4 | −3/4 | 1/4 | −3/4 | 1/4 | 1 1/4 | 3/16 (out) | 1/16 (out) | 1/16 (in) | 4.8 (out) | 1.6 (out) | 1.6 (in) | NA | NA | 8 9/16 |
| 76-75 Chevrolet (full size) | (10) | | | | | | | 1/16 (out) | 1/16 (in) | 3/16 (in) | 1.6 (out) | 1.6 (in) | 4.8 (in) | NA | NA | 10 |
| W/Radial Tires | | 1/2 | 1 1/2 | 2 1/4 | | | | | | | | | | | | |
| W/Bias Tires | | 0 | 1 | 2 | | | | | | | | | | | | |
| Left Wheel | | | | | 1/4 | 1 | 1 3/4 | | | | | | | | | |
| Right Wheel | | | | | −1/4 | 1/2 | 1 1/4 | | | | | | | | | |
| 76-75 Chevelle, El Camino | | | | | | | | | | | | | | | | |
| W/Power Steering | (10) | 1 | 2 | 3 | | | | 1/16 (out) | 1/16 (in) | 3/16 (in) | 1.6 (out) | 1.6 (in) | 4.8 (in) | NA | NA | 9 5/8 |
| Left Wheel | | | | | 1/4 | 1 | 1 3/4 | | | | | | | | | |
| Right Wheel | | | | | −1/4 | 1/2 | 1 1/4 | | | | | | | | | |
| W/Manual Steering | (10) | 0 | 1 | 2 | | | | 1/16 (out) | 1/16 (in) | 3/16 (in) | 1.6 (out) | 1.6 (in) | 4.8 (in) | NA | NA | 9 5/8 |
| Left Wheel | | | | | 1/4 | 1 | 1 3/4 | | | | | | | | | |
| Right Wheel | | | | | −1/4 | 1/2 | 1 1/4 | | | | | | | | | |
| 77 Nova | (10) | | | | 0 | 13/16 | 1 5/8 | 1/16 (out) | 1/16 (in) | 3/16 (in) | 1.6 (out) | 1.6 (in) | 4.8 (in) | NA | NA | 10 |
| W/Power Steering | | 0 | 1 | 2 | | | | | | | | | | | | |
| W/Manual Steering | | −2 | −1 | 0 | | | | | | | | | | | | |

| VEHICLE IDENTIFICATION | ADJ. ILL. NO. | CASTER (Degrees) | | | CAMBER (Degrees) | | | TOE-IN (Inches) | | | TOE-IN (Millimeters) | | | TOE-OUT ON TURNS (Degrees) | | STRG. AXIS INCL. |
|---|---|---|---|---|---|---|---|---|---|---|---|---|---|---|---|---|
| YEAR MODEL | | MIN. | PREF. | MAX. | MIN. | PREF. | MAX. | MIN. | PREF. | MAX. | MIN. | PREF. | MAX. | OUTSIDE WHEEL | INSIDE WHEEL | (DEG.) |
| 76-75 Nova | (10) | | | | 0 | 3/4 | 1 1/2 | 1/16 (out) | 1/16 (in) | 3/16 (in) | 1.6 (out) | 1.6 (in) | 4.8 (in) | NA | NA | 10 |
| *W/Power Steering* | | 0 | 1 | 2 | | | | | | | | | | | | |
| *W/Manual Steering* | | – 2 | – 1 | 0 | | | | | | | | | | | | |
| 76 Chevette | (13) | 2 1/2 | 4 1/2 | 6 1/2 | – 1/2 | 1/4 | 1 | 1/32 (out) | 1/16 (in) | 5/32 (in) | 0.8 (out) | 1.6 (in) | 4.0 (in) | NA | NA | 7 9/16 |
| 77 Vega | (11) | – 1 13/16 | – 13/16 | 3/16 | 5/8 | 3/16 | 1 | 3/16 (out) | 1/16 (out) | 1/16 (in) | 4.8 (out) | 1.6 (out) | 1.6 (in) | NA | NA | 8 9/16 |
| 76 Vega | (11) | – 1 3/4 | – 3/4 | 1/4 | – 3/4 | 3/4 | 1 1/4 | 3/16 (out) | 1/16 (out) | 1/16 (in) | 4.8 (out) | 1.6 (in) | 1.6 (in) | NA | NA | 8 9/16 |
| 76-75 Camaro | (10) | 0 | 1 | 2 | 1/4 | 1 | 1 3/4 | 1/16 (out) | 1/16 (in) | 3/16 (in) | 1.6 (out) | 1.6 (in) | 4.8 (in) | NA | NA | 10 5/32 |
| 76-73 Monte Carlo | (10) | 4 | 5 | 6 | | | | 1/16 (out) | 1/16 (in) | 3/16 (in) | 1.6 (out) | 1.6 (in) | 4.8 (in) | NA | NA | 9 19/32 |
| Left Wheel | | | | | 1/4 | 1 | 1 3/4 | | | | | | | | | |
| Right Wheel | | | | | – 1/4 | 1/2 | 1 1/4 | | | | | | | | | |
| 76-73 Corvette | (10) | | | | | | | | | | | | | | | |
| *W/Power Steering* | | 1 1/4 | 2 1/4 | 3 1/4 | 0 | 3/4 | 1 1/2 | 1/8 | 1/4 | 3/8 | 3.2 | 6.4 | 9.5 | NA | NA | 7 11/16 |
| *W/Manual Steering* | | 0 | 1 | 2 | | | | 1/8 | 1/4 | 3/8 | 3.2 | 6.4 | 9.5 | NA | NA | 7 11/16 |
| Rear | (57)(58) | | | | – 1 1/8 | – 7/8 | – 5/8 | 1/32 | 1/16 | 3/32 | 0.8 | 1.6 | 2.4 | NA | NA | |
| 75 Vega | (11) | – 1 3/4 | – 3/4 | 1/4 | – 3/4 | 1/4 | 1 1/4 | 1/16 (out) | 1/16 (in) | 3/16 (in) | 1.6 (out) | 1.6 (in) | 4.8 (in) | NA | NA | 8 9/16 |
| 74-73 Chevrolet (full size) | (10) | 0 | 1 | 2 | | | | 1/16 (out) | 1/16 (in) | 3/16 (in) | 1.6 (out) | 1.6 (in) | 4.8 (in) | NA | NA | 9 19/32 |
| Left Wheel | | | | | 1/4 | 1 | 1 3/4 | | | | | | | | | |
| Right Wheel | | | | | – 1/4 | 1/2 | 1 1/4 | | | | | | | | | |
| 74-73 Chevelle, El Camino | | | | | | | | 1/16 (out) | 1/16 (in) | 3/16 (in) | 1.6 (out) | 1.6 (in) | 4.8 (in) | NA | NA | 9 19/32 |
| *W/Power Steering* | (10) | – 1 | 0 | 1 | | | | | | | | | | | | |
| *W/Manual Steering* | (10) | – 2 | – 1 | 0 | | | | | | | | | | | | |
| Left Wheel | | | | | 1/4 | 1 | 1 3/4 | | | | | | | | | |
| Right Wheel | | | | | – 1/4 | 1/2 | 1 1/4 | | | | | | | | | |
| 74-73 Camaro except Z 28 | (10) | – 1 | 0 | 1 | 1/4 | 1 | 1 3/4 | 1/16 | 3/16 | 5/16 | 1.6 | 4.8 | 8.0 | NA | NA | 10 11/32 |
| 74-73 Z 28 only | (10) | – 2 | – 1 | 0 | 0 | 3/4 | 1 1/2 | 1/16 | 3/16 | 5/16 | 1.6 | 4.8 | 8.0 | NA | NA | 10 11/32 |
| 74-73 Nova | (10) | – 1/2 | 1/2 | 1 1/2 | – 1/2 | 1/4 | 1 | 1/16 | 3/16 | 5/16 | 1.6 | 4.8 | 8.0 | NA | NA | 9 |
| 74-72 Vega | (2) | – 1 3/4 | – 3/4 | 1/4 | – 3/4 | 1/4 | 1 1/4 | 1/8 (out) | 1/4 (in) | 3/8 (in) | 3.2 (out) | 6.4 (in) | 9.5 (in) | NA | NA | 8 9/16 |
| 72 Chevrolet (full size) | (10) | 0 | 1 | 2 | – 1/4 | 1/2 | 1 1/4 | 1/16 | 3/16 | 5/16 | 1.6 | 4.8 | 8.0 | NA | NA | 10 |
| 72 Monte Carlo | (10) | – 1 | 0 | 1 | 0 | 3/4 | 1 1/2 | 1/16 | 3/16 | 5/16 | 1.6 | 4.8 | 8.0 | NA | NA | 8 1/4 |
| 72 Corvette | (10) | | | | | | | | | | | | | | | |
| *W/Power Steering* | | 1 1/4 | 2 1/4 | 3 1/4 | 0 | 3/4 | 1 1/2 | 1/8 | 1/4 | 3/8 | 3.2 | 6.4 | 9.6 | NA | NA | 7 |
| *W/Manual Steering* | | 0 | 1 | 2 | 0 | 3/4 | 1 1/2 | 1/8 | 1/4 | 3/8 | 3.2 | 6.4 | 9.6 | NA | NA | 7 |
| Rear | (57)(58) | | | | – 1 1/8 | – 7/8 | – 5/8 | | | | | | | | | |
| 72 Chevelle, El Camino | (10) | – 2 | – 1 | 0 | 0 | 3/4 | 1 1/2 | 1/16 | 3/16 | 5/16 | 1.6 | 4.8 | 8.0 | NA | NA | 8 1/4 |
| 72 Camaro except Z 28 | (10) | – 1 | 0 | 1 | 1/4 | 1 | 1 3/4 | 1/16 | 3/16 | 5/16 | 1.6 | 4.8 | 8.0 | NA | NA | 9 1/2 |
| 72 Z 28 | (10) | – 2 | – 1 | 0 | 0 | 3/4 | 1 1/2 | 1/16 | 3/16 | 5/16 | 1.6 | 4.8 | 8.0 | NA | NA | 9 3/4 |
| 72 Nova | (10) | – 1/2 | 1/2 | 1 1/2 | – 1/2 | 1/4 | 1 | 1/16 | 3/16 | 5/16 | 1.6 | 4.8 | 8.0 | NA | NA | 8 3/4 |

**OLDSMOBILE DIVISION (A)**

**(A)** Maximum side to side variation after reset; caster and camber 1/2 °

**(B)** Toronado F.W.D. trim height is measured between the bottom of the rocker moulding to the floor. The measurement positions are:
81-79 at the front edge of the door and at 71″ (1775 mm) behind the front edge of the door.
78-74 at 6″ (152 mm) behind the front edge of the door, and at 66″ (1676 mm) behind the front edge of the door.
73-72 at 1″ (25.4 mm) behind the front edge of the door, and 3″ (76.2 mm) in front of the rear wheel opening.

| | Front | | Rear | | Maximum Side to Side, Front to Rear Deviation | | |
|---|---|---|---|---|---|---|---|
| Year | Inches | MM | Inches | MM | Year | Inches | MM |
| 81-79 | 9 1/2 | 242 | 9 1/2 | 242 | 81-77 | 3/4 | 19 |
| 78-75 | 9 | 229 | 9 1/4 | 235 | 76-74 | 1/2 | 12.7 |
| 74 | 8 7/8 | 225 | 9 1/4 | 235 | 73-72 | 3/4 | 19 |
| 73-72 std. | 8 7/8 | 225 | 9 3/16 | 233 | | | |
| 73-72 H.D. | 9 3/16 | 233 | 9 1/2 | 242 | | | |

| | | | | | | | | | | | | | | | | |
|---|---|---|---|---|---|---|---|---|---|---|---|---|---|---|---|---|
| 81-80 Omega | (34) | – 2 | 0 | 2 | 0 | 1/2 | 1 | 0 | 3/32 | 3/16 | 0 | 2.5 | 5.0 | NA | NA | 14 1/2 |
| 81-78 Cutlass | (10) | | | | – 5/16 | 1/2 | 1 5/16 | 1/16 | 1/8 | 1/4 | 1.6 | 3.2 | 6.4 | NA | NA | 7 |
| *W/Power Steering* | | 2 | 3 | 4 | | | | | | | | | | | | |
| *W/Manual Steering* | | 0 | 1 | 2 | | | | | | | | | | | | |
| 81-78 88 & 98 Series | (10) | 2 | 3 | 4 | 0 | 3/4 | 1 5/8 | 0 | 1/8 | 1/4 | 0 | 3.2 | 6.4 | NA | NA | 10 1/2 |
| 81-79 Toronado | | | | | | | | | | | | | | | | |
| Front | (6) | 1 1/2 | 2 1/2 | 3 1/2 | 13/16 | 0 | 13/16 | 1/8 (out) | 0 | 1/8 (in) | 3.2 (out) | 0 | 3.2 (in) | NA | NA | 11 |
| Rear | (9) | | Fixed | | | Fixed | | 0 | 5/32 | 5/16 | 0 | 4.0 | 8.0 | | | |
| 80-76 Starfire | (11) | – 1 3/4 | – 3/4 | 1/4 | – 1/2 | 1/4 | 1 | 3/16 (out) | 1/16 (out) | 1/16 (in) | 4.8 (out) | 1.6 (out) | 1.6 (in) | NA | NA | 8 1/2 |

| VEHICLE IDENTIFICATION YEAR MODEL | ADJ. ILL. NO. | CASTER (Degrees) MIN. | PREF. | MAX. | CAMBER (Degrees) MIN. | PREF. | MAX. | TOE-IN (Inches) MIN. | PREF. | MAX. | TOE-IN (Millimeters) MIN. | PREF. | MAX. | TOE-OUT ON TURNS (Degrees) OUTSIDE WHEEL | INSIDE WHEEL | STRG. AXIS INCL. (DEG.) |
|---|---|---|---|---|---|---|---|---|---|---|---|---|---|---|---|---|
| 79-78 Omega | (10) | | | | 0 | 3/4 | 1 5/8 | 1/16 | 1/8 | 1/4 | 1.6 | 3.2 | 6.4 | NA | NA | 10 1/2 |
| *W/Power Steering* | | 0 | 1 | 2 | | | | | | | | | | | | |
| *W/Manual Steering* | | −2 | −1 | 0 | | | | | | | | | | | | |
| 78 Toronado | (6) | −1 | 0 | 1 | | | | 1/8 (out) | 0 | 1/8 (in) | 3.2 (out) | 0 | 3.2 (in) | NA | NA | 11 |
| Left Wheel | | | | | −1/2 | 5/16 | 1 | | | | | | | | | |
| Right Wheel | | | | | −1 | −5/16 | 1/2 | | | | | | | | | |
| 77 88 & 98 Series | (10) | 2 | 3 | 4 | 0 | 3/4 | 1 1/2 | 0 | 1/8 | 1/4 | 0 | 3.2 | 6.4 | NA | NA | 10 1/2 |
| 77-75 Omega | (10) | | | | 0 | 3/4 | 1 1/2 | 1/16 (out) | 1/16 (in) | 3/16 (in) | 1.6 (out) | 1.6 (in) | 4.8 (in) | NA | NA | 10 1/2 |
| *W/Power Steering* | | 0 | 1 | 2 | | | | | | | | | | | | |
| *W/Manual Steering* | | −2 | −1 | 0 | | | | | | | | | | | | |
| 77-75 Cutlass | (10) | 1 | 2 | 3 | | | | 1/16 (out) | 1/16 (in) | 3/16 (in) | 1.6 (out) | 1.6 (in) | 4.8 (in) | NA | NA | 10 1/2 |
| Left Wheel | | | | | 1/4 | 1 | 1 3/4 | | | | | | | | | |
| Right Wheel | | | | | −1/4 | 1/2 | 1 1/4 | | | | | | | | | |
| 77-75 Toronado | (6) | −1 | 0 | 1 | | | | 1/8 (out) | 0 | 1/8 (in) | 3.2 (out) | 0 | 3.2 (in) | NA | NA | 11 |
| Right Wheel | | | | | −1 | −1/4 | −1/2 | | | | | | | | | |
| Left Wheel | | | | | −1/2 | 1/4 | 1 | | | | | | | | | |
| 76-75 88 & 98 Series | (10) | 1/2 | 1 1/2 | 2 1/2 | | | | 1/16 (out) | 1/16 (in) | 3/16 (in) | 1.6 (out) | 1.6 (in) | 4.8 (in) | NA | NA | 10 1/2 |
| Left Wheel | | | | | 1/4 | 1 | 1 3/4 | | | | | | | | | |
| Right Wheel | | | | | −1/4 | 1/2 | 1 3/4 | | | | | | | | | |
| 74 Cutlass except Salon | (10) | 1 | 0 | 1 | | | | 1/16 (out) | 1/16 (in) | 3/16 (in) | 1.6 (out) | 1.6 (in) | 4.8 (in) | NA | NA | 10 1/2 |
| Left Wheel | | | | | 1/4 | 1 | 1 3/4 | | | | | | | | | |
| Right Wheel | | | | | 1/4 | 1/2 | 1 1/4 | | | | | | | | | |
| 74 Cutlass Salon only | (10) | 1 | 2 | 3 | | | | 1/16 (out) | 1/16 (in) | 3/16 (in) | 1.6 (out) | 1.6 (in) | 4.8 (in) | NA | NA | 10 1/2 |
| Left Wheel | | | | | 1/4 | 1 | 1 3/4 | | | | | | | | | |
| Right Wheel | | | | | 1/4 | 1/2 | 1 1/4 | | | | | | | | | |
| 74-73 Omega | (10) | −1/2 | 1/2 | 1 1/2 | −1/2 | 1/4 | 1 | 1/16 | 3/16 | 5/16 | 1.6 | 4.8 | 8.0 | NA | NA | 9 |
| 74-72 Toronado | (6) | −1 | −2 | −3 | | | | 1/8 (out) | 0 | 1/8 (in) | 3.2 (out) | 0 | 3.2 (in) | NA | NA | 11 |
| Left Wheel | | | | | −1/2 | 1/4 | 1 | | | | | | | | | |
| Right Wheel | | | | | −1 | −1/4 | 1/2 | | | | | | | | | |
| 74-73 88 & 98 Series | (10) | 0 | 1 | 2 | | | | 1/16 (out) | 1/16 (in) | 3/16 (in) | 1.6 (out) | 1.6 (in) | 4.8 (in) | NA | NA | 9 5/8 |
| Left Wheel | | | | | 1/4 | 1 | 1 3/4 | | | | | | | | | |
| Right Wheel | | | | | 1/4 | 1/2 | 1 1/4 | | | | | | | | | |
| 73 Cutlass | | | | | | | | | | | | | | | | |
| *W/Power Steering* | (10) | 1 | 0 | 1 | | | | 1/16 (out) | 1/16 (in) | 3/16 (in) | 1.6 (out) | 1.6 (in) | 4.8 (in) | NA | NA | 10 1/2 |
| Left Wheel | | | | | 1/4 | 1 | 1 3/4 | | | | | | | | | |
| Right Wheel | | | | | −1/4 | 1/2 | 1 1/4 | | | | | | | | | |
| *W/Manual Steering* | (10) | −2 | −1 | 0 | | | | 1/16 (out) | 1/16 (in) | 3/16 (in) | 1.6 (out) | 1.6 (in) | 4.8 (in) | NA | NA | 10 1/2 |
| Left Wheel | | | | | 1/4 | 1 | 1 3/4 | | | | | | | | | |
| Right Wheel | | | | | −1/4 | 1/2 | 1 1/4 | | | | | | | | | |
| 72 88 & 98 Series | (10) | 0 | 1 | 2 | | | | 1/8 (out) | 0 | 1/8 (in) | 3.2 (out) | 0 | 3.2 (in) | NA | NA | 9 5/8 |
| Left Wheel | | | | | −1/2 | 1/4 | 1 | | | | | | | | | |
| Right Wheel | | | | | −1 | −1/4 | 1/2 | | | | | | | | | |
| 72 Cutlass, 442 | (10) | −2 1/4 | −1 1/4 | −1/4 | | | | 1/8 (out) | 0 | 1/8 (in) | 3.2 (out) | 0 | 3.2 (in) | NA | NA | 8 |
| Left Wheel | | | | | −1/2 | 1/4 | 1 | | | | | | | | | |
| Right Wheel | | | | | −1 | −1/4 | 1/2 | | | | | | | | | |

**PONTIAC DIVISION (A)**

**(A)** Maximum side to side variation after reset; caster and camber 1/2 °.

| VEHICLE IDENTIFICATION YEAR MODEL | ADJ. ILL. NO. | CASTER MIN. | PREF. | MAX. | CAMBER MIN. | PREF. | MAX. | TOE-IN (Inches) MIN. | PREF. | MAX. | TOE-IN (Millimeters) MIN. | PREF. | MAX. | OUTSIDE WHEEL | INSIDE WHEEL | STRG. AXIS INCL. (DEG.) |
|---|---|---|---|---|---|---|---|---|---|---|---|---|---|---|---|---|
| 81-80 Phoenix | (34) | −2 | 0 | 2 | 0 | 1/2 | 1 | 0 | 3/32 | 3/16 | 0 | 2.5 | 5.0 | NA | NA | 14 1/2 |
| 81-80 Grand Prix and LeMans | | | | | 5/16 | 1/8 | 1 5/16 | 1/16 | 1/8 | 1/4 | 1.6 | 3.2 | 6.4 | NA | NA | 8 |
| *W/Manual Steering* | (10) | 0 | 1 | 2 | | | | | | | | | | | | |
| *W/Power Steering* | (10) | 2 | 3 | 4 | | | | | | | | | | | | |
| 81-78 Firebird | (10) | 0 | 1 | 2 | 3/16 | 1 | 1 13/16 | 1/16 | 1/8 | 1/4 | 1.6 | 3.2 | 6.4 | NA | NA | 10 3/8 |
| 81-78 Catalina and Bonneville | (10) | 2 | 3 | 4 | 0 | 13/16 | 1 5/8 | 1/16 | 1/8 | 1/4 | 1.6 | 3.2 | 6.4 | NA | NA | 10 19/32 |
| 80-78 Sunbird | (10) | −1 13/16 | −13/16 | 3/16 | −5/8 | −3/16 | 1 | 3/16 (out) | 1/16 (out) | 1/16 (in) | 4.8 (out) | 1.6 (out) | 1.6 (in) | NA | NA | 8 9/16 |

| VEHICLE IDENTIFICATION | ADJ. ILL. | CASTER (Degrees) | | | CAMBER (Degrees) | | | TOE-IN (Inches) | | | TOE-IN (Millimeters) | | | TOE-OUT ON TURNS (Degrees) | | STRG. AXIS INCL. |
| YEAR  MODEL | NO. | MIN. | PREF. | MAX. | MIN. | PREF. | MAX. | MIN. | PREF. | MAX. | MIN. | PREF. | MAX. | OUTSIDE WHEEL | INSIDE WHEEL | (DEG.) |
|---|---|---|---|---|---|---|---|---|---|---|---|---|---|---|---|---|
| 79-78 Phoenix | (10) | | | | 0 | 13/16 | 15/8 | 1/16 | 1/8 | 1/4 | 1.6 | 3.2 | 6.4 | NA | NA | 10 |
| W/Power Steering | | 0 | 1 | 2 | | | | | | | | | | | | |
| W/Manual Steering | | −2 | −1 | 0 | | | | | | | | | | | | |
| 79-78 Grand Prix | (10) | | | | −5/16 | 1/2 | 15/16 | 1/16 | 1/8 | 1/4 | 1.6 | 3.2 | 6.4 | NA | NA | 8 |
| W/Power Steering | | 2 | 3 | 4 | | | | | | | | | | | | |
| W/Manual Steering | | 0 | 1 | 2 | | | | | | | | | | | | |
| 79-78 LeMans, Grand AM | (10) | | | | −5/16 | 1/2 | 15/16 | 1/16 | 1/8 | 1/4 | 1.6 | 3.2 | 6.4 | NA | NA | 8 |
| W/Power Steering | | 2 | 3 | 4 | | | | | | | | | | | | |
| W/Manual Steering | | 0 | 1 | 2 | | | | | | | | | | | | |
| 77-76 Phoenix, Ventura | (10) | | | | 0 | 3/4 | 11/2 | 1/16 (out) | 1/16 (in) | 3/16 (in) | 1.6 (out) | 1.6 (in) | 4.8 (in) | 181/2 | 20 | 10 |
| W/Power Steering | | 0 | 1 | 2 | | | | | | | | | | | | |
| W/Manual Steering | | −2 | −1 | 0 | | | | | | | | | | | | |
| 77 Grand Prix | (10) | 4 | 5 | 6 | | | | 1/16 (out) | 1/16 (in) | 3/16 (in) | 1.6 (out) | 1.6 (in) | 4.8 (in) | NA | NA | 103/8 |
| Left Wheel | | | | | 1/4 | 1 | 13/4 | | | | | | | | | |
| Right Wheel | | | | | −1/4 | 1/2 | 11/4 | | | | | | | | | |
| 77 LeMans, Grand LeMans | | | | | | | | | | | | | | | | 103/8 |
| W/Power Steering | (10) | | | | | | | 1/16 (out) | 1/16 (in) | 3/16 (in) | 1.6 (out) | 1.6 (in) | 4.8 (in) | | | |
| W/Belted Tires | | 0 | 1 | 2 | | | | | | | | | | | | |
| W/Radial Tires | | 1 | 2 | 3 | | | | | | | | | | | | |
| Left Wheel | | | | | 1/4 | 1 | 13/4 | | | | | | | 193/16 | 20 | |
| Right Wheel | | | | | −1/4 | 1/2 | 11/4 | | | | | | | 183/16 | 20 | |
| W/Manual Steering | (10) | 0 | 1 | 2 | | | | 1/16 (out) | 1/16 (in) | 3/16 (in) | 1.6 (out) | 1.6 (in) | 4.8 (in) | | | 103/8 |
| Left Wheel | | | | | 1/4 | 1 | 13/4 | | | | | | | 193/16 | 20 | |
| Right Wheel | | | | | −1/4 | 1/2 | 11/4 | | | | | | | 183/16 | 20 | |
| 77 Firebird | (10) | 0 | 1 | 2 | 3/16 | 1 | 13/4 | 1/16 (out) | 1/16 (in) | 3/16 (in) | 1.6 (out) | 1.6 (in) | 4.8 (in) | NA | NA | 103/8 |
| 77-76 Sunbird, Astre | (11) | −13/4 | −3/4 | 1/4 | −9/16 | 3/16 | 1 | 3/16 (out) | 1/16 (out) | 1/16 (out) | 4.8 (out) | 1.6 (out) | 1.6 (out) | NA | NA | 89/16 |
| 77 Catalina, Bonneville, Brougham, Grandville | (10) | 2 | 3 | 4 | 0 | 3/4 | 19/16 | 1/16 | 3/16 | 5/16 | 1.6 | 4.8 | 8.0 | NA | NA | 103/8 |
| 76 Ventura | (10) | | | | 0 | 3/4 | 11/2 | 1/16 (out) | 1/16 (in) | 3/16 (in) | 1.6 (out) | 1.6 (in) | 4.8 (in) | 181/2 | 20 | 10 |
| W/Power Steering | | 0 | 1 | 2 | | | | | | | | | | | | |
| W/Manual Steering | | −2 | −1 | 0 | | | | | | | | | | | | |
| 76-73 Grand Prix | (10) | 2 | 3 | 4 | | | | 1/16 (out) | 1/16 (in) | 3/16 (in) | 1.6 (out) | 1.6 (in) | 4.8 (in) | | | 101/2 |
| Left Wheel | | | | | 1/4 | 1 | 13/4 | | | | | | | 193/16 | 20 | |
| Right Wheel | | | | | −1/4 | 1/2 | 11/4 | | | | | | | 183/8 | 20 | |
| 76-75 LeMans, Grand LeMans | | | | | | | | | | | | | | | | |
| W/Power Steering | (10) | 1 | 2 | 3 | | | | 1/16 (out) | 1/16 (in) | 3/16 (in) | 1.6 (out) | 1.6 (in) | 4.8 (in) | NA | NA | 103/8 |
| Left Wheel | | | | | 1/4 | 1 | 13/4 | | | | | | | 193/16 | 20 | |
| Right Wheel | | | | | −1/4 | 1/2 | 11/4 | | | | | | | 183/16 | 20 | |
| W/Power Steering | (10) | 0 | 1 | 2 | | | | 1/16 (out) | 1/16 (in) | 3/16 (in) | 1.6 (out) | 1.6 (in) | 4.8 (in) | | | 103/8 |
| Left Wheel | | | | | 1/4 | 1 | 13/4 | | | | | | | 193/16 | 20 | |
| Right Wheel | | | | | −1/4 | 1/2 | 11/4 | | | | | | | 183/16 | 20 | |
| 76-75 Firebird | (10) | −1 | 0 | 1 | 1/4 | 1 | 13/4 | 1/16 (out) | 1/16 (in) | 3/16 (in) | 1.6 (out) | 1.6 (in) | 4.8 (in) | NA | NA | 103/8 |
| 76-75 Catalina, Bonneville, Grandville, Brougham | (10) | 1/2 | 11/2 | 21/2 | | | | 1/16 (out) | 1/16 (out) | 3/16 (in) | 1.6 (out) | 1.6 (out) | 4.8 (in) | NA | NA | 89/16 |
| Left Wheel | | | | | 1/4 | 1 | 13/4 | | | | | | | | | |
| Right Wheel | | | | | −1/4 | 1/2 | 11/4 | | | | | | | | | |
| 75 Astre | (11) | −13/4 | −3/4 | 1/4 | −3/4 | 1/4 | 11/4 | 1/8 | 1/4 | 3/8 | 3.2 | 6.4 | 4.8 | NA | NA | 89/16 |
| 75 Ventura | (10) | | | | 0 | 3/4 | 11/2 | 1/16 (out) | 1/16 (in) | 3/16 (in) | 1.6 (out) | 1.6 (in) | | NA | NA | 107/8 |
| W/Power Steering | | 0 | 1 | 2 | | | | | | | | | | | | |
| W/Manual Steering | | −2 | −1 | 0 | | | | | | | | | | | | |
| 74 Ventura | (10) | −1/2 | 1/2 | 11/2 | −1/2 | 1/4 | 1 | 1/16 | 3/16 | 5/16 | 1.6 | 4.8 | 8.0 | NA | NA | 83/4 |
| 74 Firebird | (10) | −1 | 0 | 1 | 1/4 | 1 | 13/4 | 1/16 | 3/16 | 5/16 | 1.6 | 4.8 | 8.0 | 18 | 20 | 103/8 |
| 74-73 LeMans, GTO, Tempest | | | | | | | | | | | | | | | | |
| W/Power Steering | (10) | −1 | 0 | 1 | | | | 1/16 (out) | 1/16 (in) | 3/16 (in) | 1.6 (out) | 1.6 (in) | 4.8 (in) | | | 101/2 |
| Left Wheel | | | | | 1/4 | 1 | 13/4 | | | | | | | 183/16 | 20 | |
| Right Wheel | | | | | −1/4 | 1/2 | 11/4 | | | | | | | 193/16 | 20 | |
| W/Manual Steering | (10) | −2 | −1 | 0 | | | | 1/16 (out) | 1/16 (in) | 3/16 (in) | 1.6 (out) | 1.6 (in) | 4.8 (in) | | | 101/2 |
| Left Wheel | | | | | 1/4 | 1 | 13/4 | | | | | | | 183/16 | 20 | |
| Right Wheel | | | | | −1/4 | 1/2 | 11/4 | | | | | | | 193/16 | 20 | |

| VEHICLE IDENTIFICATION | | ADJ. ILL. NO. | CASTER (Degrees) | | | CAMBER (Degrees) | | | TOE-IN (Inches) | | | TOE-IN (Millimeters) | | | TOE-OUT ON TURNS (Degrees) | | STRG. AXIS INCL. |
|---|---|---|---|---|---|---|---|---|---|---|---|---|---|---|---|---|---|
| YEAR | MODEL | | MIN. | PREF. | MAX. | MIN. | PREF. | MAX. | MIN. | PREF. | MAX. | MIN. | PREF. | MAX. | OUTSIDE WHEEL | INSIDE WHEEL | (DEG.) |
| 74-73 Catalina, Bonneville, Grandville .... | | (10) | 0 | 1 | 2 | | | | 1/16 (out) | 1/16 (in) | 3/16 (in) | 1.6 (out) | 1.6 (in) | 4.8 (in) | 18½ | 20 | 10½ |
| | Left Wheel ........ | | | | | 1/4 | 1 | 1 3/4 | | | | | | | | | |
| | Right Wheel ....... | | | | | − 1/4 | 1/2 | 1 1/4 | | | | | | | | | |
| 73 Ventura ........... | | (10) | − 3/4 | 1/4 | 1 1/4 | − 1/2 | 1/4 | 1 | 1/16 | 3/16 | 5/16 | 1.6 | 4.8 | 8.0 | NA | NA | NA |
| 73 Firebird .......... | | (10) | − 1 1/4 | − 1/4 | 3/4 | 0 | 1 | 2 | 1/16 | 3/16 | 5/16 | 1.6 | 4.8 | 8.0 | NA | NA | 10 3/8 |
| 72 Firebird .......... | | (10) | − 1 | 0 | 1 | 1/4 | 1 | 1 3/4 | 1/16 | 3/16 | 5/16 | 1.6 | 4.8 | 8.0 | 18 | 20 | 8 3/4 |
| 72 Tempest, LeMans, GTO. | | (10) | − 2 1/2 | − 1 1/2 | − 1/2 | − 3/4 | 0 | 3/4 | 1/16 | 1/8 | 5/16 | 1.6 | 3.2 | 8.0 | NA | NA | 9 |
| 72 Grand Prix ........ | | (10) | − 2 1/2 | − 1 1/2 | − 1/2 | − 3/4 | 0 | 3/4 | 1/16 | 1/8 | 5/16 | 1.6 | 3.2 | 8.0 | NA | NA | 9 |
| 72 Ventura ........... | | (10) | − 1/2 | 1/2 | 1 1/2 | − 1/2 | 1/4 | 1 | 1/16 | 3/16 | 5/16 | 1.6 | 4.8 | 8.0 | NA | NA | NA |
| 72 Bonneville, Catalina, Grandville. .......... | | (10) | 0 | 1 | 2 | 0 | 3/4 | 1 1/2 | 1/16 | 3/16 | 5/16 | 1.6 | 4.8 | 8.0 | NA | NA | 8 1/2 |

# U.S. LIGHT TRUCK SECTION

| VEHICLE IDENTIFICATION | | ADJ. ILL. NO. | CASTER (Degrees) | | | CAMBER (Degrees) | | | TOE-IN (Inches) | | | TOE-IN (Millimeters) | | | TOE-OUT ON TURNS (Degrees) | | STRG. AXIS INCL. |
|---|---|---|---|---|---|---|---|---|---|---|---|---|---|---|---|---|---|
| YEAR | MODEL | | MIN. | PREF. | MAX. | MIN. | PREF. | MAX. | MIN. | PREF. | MAX. | MIN. | PREF. | MAX. | OUTSIDE WHEEL | INSIDE WHEEL | (DEG.) |
| **AMERICAN MOTORS:** **JEEP** **(A)** 77 models, 28°-29°; 28°; 75-74 w/std. tires 31°; w/F85 × 15 tires - 34°; 1981 CJ models 31°-32°; 1980-81 except CJ 37°-38°. **(B)** W/F85 × 15 tires - 34°. | | | | | | | | | | | | | | | | | |
| 82 CJ-5 ............... | | (17) | 6 | 6 | 7 | 0 | 0 | 1/2 | 3/64 | | 3/32 | 1.2 | | 2.4 | 29 | | 8 1/2 |
| 82 CJ-7 and Scrambler .... | | (17) | 6 | 6 | 7 | 0 | 0 | 1/2 | 3/64 | | 3/32 | 1.2 | | 2.4 | 32 | | 8 1/2 |
| 82 Cherokee, Wagoneer and Pickup Truck ....... | | (17) | 4 | 4 | 5 | 0 | 0 | 1/2 | 3/64 | | 3/32 | 1.2 | | 2.4 | 36-37 | | 8 1/2 |
| 81 "CJ" Models ......... | | (17) | 6 | 6 | 7 | 1 1/2 | 1 1/2 | 2 | 3/64 | | 3/32 | 1.2 | | 2.4 | (A) 31 | | 8 1/2 |
| 81-80 Cherokee, Wagoneer and Pickup Trucks ...... | | (17) | | 4 | | | 0 | | 3/64 | | 3/32 | 1.2 | | 2.4 | (A) 37 | | 8 1/2 |
| 80-74 "CJ" Models ....... | | (17) | | 3 | | | 1 1/2 | | 3/64 | | 3/32 | 1.2 | | 2.4 | (A) 31 1/2 | | 8 1/2 |
| 79-74 Cherokee, Wagoneer and Pickup Trucks ...... | | (17) | | 4 | | | 1 1/2 | | 3/64 | | 3/32 | 1.2 | | 2.4 | 37 1/2 | | 8 1/2 |
| 73 CJ, DJ, Jeepster, Commando ........... | | (17) | | 3 | | | 1 1/2 | | 3/64 | | 3/32 | 1.2 | | 2.4 | (B) 31 | | 7 1/2 |
| 73 Wagoneer, Trucks ..... | | (17) | | 3 | | | 1 1/2 | | 3/64 | | 3/32 | 1.2 | | 2.4 | 28 | | 7 1/2 |

# GENERAL MOTORS LIGHT DUTY TRUCKS
## CHEVROLET & GMC VERSIONS

| VEHICLE IDENTIFICATION | | ADJ. ILL. NO. | CASTER (Degrees) | | | CAMBER (Degrees) | | | TOE-IN (Inches) | | | TOE-IN (Millimeters) | | | TOE-OUT ON TURNS (Degrees) | | STRG. AXIS INCL. |
|---|---|---|---|---|---|---|---|---|---|---|---|---|---|---|---|---|---|
| **(A)** Vehicle ride heights must be checked and corrected before alignment is performed. **(B)** With JB8 or JF9 add .3°; with R05 subtract .4° for caster. | | | | | | | | | | | | | | | | | |
| 82 S10, S15 ........... | | (10) | 1 | 2 | 3 | 0 | 13/16 | 1 5/8 | 1/16 | 1/8 | 1/4 | 1.6 | 3.2 | 6.4 | NA | NA | NA |

# GENERAL MOTORS LIGHT TRUCKS

## CHEVROLET & GMC VERSIONS

### CHART 1

| VEHICLE IDENTIFICATION CHEVROLET YEAR MODEL | GMC MODEL | ADJ. ILL. NO. | CASTER @ HEIGHT MEASUREMENT (Degrees) Suspension Height Measurement (M) 1 1/2 | 2 | 2 1/2 | 3 | 3 1/2 | 3 3/4 | 4 | CAMBER (Degrees) MIN. | PREF. | MAX. | TOE-IN (Inches) | TOE-IN (Millimeters) |
|---|---|---|---|---|---|---|---|---|---|---|---|---|---|---|
| **1982** | | | | | | | | | | | | | | |
| C-10 | C-1500 | (24) | — | — | 3 5/8 | 3 1/8 | 2 5/8 | 2 3/8 | 2 | 0 | 11/16 | 1 3/8 | 3/16 ± 1/8 | 4.8 ± 3.2 |
| C-20, 30 | C-2500, 3500 | (24) | — | — | 1 1/2 | 15/16 | 5/16 | 1/8 | 0 | − 1/2 | 3/16 | 7/8 | 3/16 ± 1/8 | 4.8 ± 3.2 |
| **1982** | | | | | | | | | | | | | | |
| K-10, 20 | K-1500, 2500 | N/A | — | — | 8 | 8 | 8 | 8 | 8 | 5/16 | 1 | 1 11/16 | 3/16 ± 1/8 | 4.8 ± 3.2 |
| K-30 | K-3500 | N/A | — | — | 8 | 8 | 8 | 8 | 8 | − 3/16 | 1/2 | 1 3/16 | 3/16 ± 1/8 | 4.8 ± 3.2 |
| **1982** | | | | | | | | | | | | | | |
| G-10, 20 | G-1500, 2500 | (24) | 3 1/2 | 3 1/8 | 2 11/16 | 2 3/8 | 2 1/8 | 1 15/16 | 1 7/8 | − 3/16 | 1/2 | 1 3/16 | 3/16 ± 1/8 | 4.8 ± 3.2 |
| G-30 | G-3500 | (24) | 2 7/8 | 2 3/16 | 1 5/8 | 1 | 1/2 | 3/16 | 0 | − 1/2 | 3/16 | 7/8 | 3/16 ± 1/8 | 4.8 ± 3.2 |
| **1982** | | | | | | | | | | | | | | |
| P-10 | P-1500 | (24) | — | — | 2 5/16 | 1 11/16 | 1 3/16 | 15/16 | 5/8 | − 1/2 | 3/16 | 7/8 | 3/16 ± 1/8 | 4.8 ± 3.2 |
| P-20, 30 | P-2500, 3500 | (24) | — | — | 2 15/16 | 2 5/16 | 1 11/16 | 1 3/8 | 1 3/16 | − 1/2 | 3/16 | 7/8 | 3/16 ± 1/8 | 4.8 ± 3.2 |
| **1982** | | | | | | | | | | | | | | |
| P-30 Motor Home | P-3500 Motor Home | (24) | — | — | 5 1/2 | 5 | 4 3/8 | 4 1/8 | 3 7/8 | − 1/2 | 3/16 | 7/8 | 5/16 ± 1/8 | 7.9 ± 3.2 |

### CHART 2

| VEHICLE IDENTIFICATION CHEVROLET YEAR MODEL (A) | GMC MODEL (A) | ADJ. ILL. NO. | CASTER @ HEIGHT MEASUREMENT (Degrees) Suspension Height Measurement (M) 1 1/2 | 2 | 2 1/2 | 3 | 3 1/2 | 4 3/4 | 4 1/2 | CAMBER (Degrees) MIN. | PREF. | MAX. | TOE-IN (Inches) | TOE-IN (Millimeters) |
|---|---|---|---|---|---|---|---|---|---|---|---|---|---|---|
| **1981-80** | | | | | | | | | | | | | | |
| G-10, 20 | G-1500, 2500 | (24) | 3 1/2 | 3 1/8 | 2 11/16 | 2 13/32 | 2 1/8 | 1 13/16 | — | − 3/16 | 1/2 | 1 3/16 | 3/16 ± 1/8 | 4.8 ± 3.2 |
| G-30 | G-3500 | (24) | 2 3/16 | 2 3/16 | 1 5/8 | 1 | 1/2 | 0 | — | − 1/2 | 3/16 | 7/8 | 3/16 ± 1/8 | 4.8 ± 3.2 |
| **1981-79** | | | | | | | | | | | | | | |
| C-10 | C-1500 | (24) | — | — | 2 13/32 | 1 13/16 | 1 3/16 | 11/16 | 3/16 | − 1/2 | 3/16 | 1 7/8 | 3/16 ± 1/8 | 4.8 ± 3.2 |
| C-20, 30 | C-2500, 3500 | (24) | — | — | 1 1/2 | 29/32 | 5/16 | 0 | − 11/16 | − 1/2 | 3/16 | 7/8 | 3/16 ± 1/8 | 4.8 ± 3.2 |
| K-10, 20 | K-1500, 2500 | N/A | — | — | 8 | 8 | 8 | 8 | 8 | 5/16 | 1 | 1 11/16 | 0 ± 1/8 | 0 ± 3.2 |
| K-30 | K-3500 | N/A | — | — | 8 | 8 | 8 | 8 | 8 | − 3/16 | 1/2 | 1 3/16 | 0 ± 1/8 | 0 ± 3.2 |
| P-10 | P-1500 | (24) | — | — | 2 5/16 | 1 11/16 | 1 3/16 | 5/8 | 1/8 | − 1/2 | 3/16 | 7/8 | 3/16 ± 1/8 | 4.8 ± 3.2 |
| P-20, 30 (B) | P-2500, 3500 (B) | (24) | — | 2 29/32 | 2 11/16 | 1 11/16 | 1 3/16 | 5/8 | 3/16 | − 1/2 | 3/16 | 7/8 | 3/16 ± 1/8 | 4.8 ± 3.2 |
| P-30 Motor Home (B) | P-3500 Motor Home (B) | | — | — | 5 1/2 | 5 | 4 13/32 | 3 13/16 | 3 5/16 | − 1/2 | 3/16 | 7/8 | 5/16 ± 1/8 | 7.9 ± 3.2 |
| **1979** | | | | | | | | | | | | | | |
| G-10, 20 | G-1500, 2500 | (24) | 2 29/32 | 2 5/16 | 2 | 1 5/8 | 1 5/16 | 29/32 | — | | N/A | | 3/16 ± 1/8 | 4.8 ± 3.2 |
| G-30 | G-3500 | (24) | 3 13/32 | 2 11/16 | 2 1/8 | 1 1/2 | 1 | 13/32 | — | | N/A | | 3/16 ± 1/8 | 4.8 ± 3.2 |

### CHART 3

| VEHICLE IDENTIFICATION CHEVROLET YEAR MODEL (A) | GMC MODEL (A) | ADJ. ILL. NO. | CASTER @ HEIGHT MEASUREMENT (Degrees) Suspension Height Measurement (M) 2 1/2 | 3 | 3 1/2 | 4 | 4 1/2 | 4 3/4 | 5 | CAMBER (Degrees) MIN. | PREF. | MAX. | TOE-IN (Inches) | TOE-IN (Millimeters) |
|---|---|---|---|---|---|---|---|---|---|---|---|---|---|---|
| **1978-73** | | | | | | | | | | | | | | |
| C-10 | C-1500 | (24) | — | 2 | 1 1/4 | 3/4 | 1/4 | 0 | − 1/2 | | 1/4 | | 3/16 ± 1/8 | 4.8 ± 3.2 |
| **1978-74** | | | | | | | | | | | | | | |
| C-20, 30 | C-2500, 3500 | (24) | 1 1/2 | 1 | 1/2 | 0 | − 1/2 | 3/4 | − 1 | | 1/4 | | 3/16 ± 1/8 | 4.8 ± 3.2 |
| **1978-76** | | | | | | | | | | | | | | |
| K-10, 20, 30 | K-1500, 2500, 3500 | N/A | 8 | 8 | 8 | 8 | 8 | 8 | 8 | | 1/4 | | 0 ± 1/8 | 0 ± 3.2 |
| **1978** | | | | | | | | | | | | | | |
| G-10, 20 | G-1500, 2500 | (24) | 3 1/4 | 2 3/4 | 2 1/2 | 2 | 1 3/4 | 1 1/2 | 1 1/2 | | 1/4 | | 3/16 ± 1/8 | 4.8 ± 3.2 |
| G-30 | G-3500 | (24) | 2 1/4 | 1 1/2 | 1 | 1/2 | 0 | − 1/4 | − 1/2 | | 1/4 | | 3/16 ± 1/8 | 4.8 ± 3.2 |
| **1978-75** | | | | | | | | | | | | | | |
| P-10 | P-1500 | (24) | 2 1/2 | 2 | 1 1/2 | 3/4 | 1/4 | 0 | − 1/4 | | Chart 4 | | 3/16 ± 1/8 | 4.8 ± 3.2 |
| P-20, 30 | P-2500, 3500 | (24) | 2 1/2 | 2 | 1 1/2 | 3/4 | 1/4 | 0 | − 1/4 | | Chart 4 | | 3/16 ± 1/8 | 4.8 ± 3.2 |
| **1977-73** | | | | | | | | | | | | | | |
| G-10, 20, 30 | G-1500, 2500, 3500 | (24) | 2 1/4 | 1 1/2 | 1 | 1/2 | 0 | − 1/4 | − 1/2 | | 1/4 | | 3/16 ± 1/8 | 4.8 ± 3.2 |
| **1976-73** | | | | | | | | | | | | | | |
| K-10, 20 | K-1500, 2500 | N/A | 4 | 4 | 4 | 4 | 4 | 4 | 4 | | 1/2 | | 0 ± 1/8 | 0 ± 3.2 |
| **1974** | | | | | | | | | | | | | | |
| P-10, 20, 30 | P-1500, 2500, 3500 | (24) | — | 2 | 1 1/4 | 3/4 | 1/4 | 0 | − 1/2 | | 1/4 | | 3/16 ± 1/8 | 4.8 ± 3.2 |
| **1973** | | | | | | | | | | | | | | |
| C&P-20, 30 | C&P-2500, 3500 | (24) | 2 | 1 1/4 | 3/4 | 1/4 | − 1/2 | − 1/2 | − 3/4 | | 1/4 | | 3/16 ± 1/8 | 4.8 ± 3.2 |

# CHART 4

| VEHICLE IDENTIFICATION | | ADJ. ILL. NO. | CAMBER @HEIGHT MEASUREMENT (Degrees) Suspension Height Measurement (MM) | | | | | | | | | | |
|---|---|---|---|---|---|---|---|---|---|---|---|---|---|
| CHEVROLET YEAR MODEL | GMC MODEL | | 2½ | 2½ | 2¾ | 3½ | 3¾ | 4 | 4¼ | 4½ | 4¾ | 5 | 5¼ |
| 1978-75 P-10 | P-1500 | (24) | 0 | 0 | ¼ | ¼ | ¼ | ¼ | 0 | 0 | 0 | − ¼ | − ½ |
| P-20, 30 | P-2500, 3500 | (24) | 0 | 0 | ¼ | ¼ | ¼ | ¼ | ¼ | 0 | 0 | − ¼ | − ½ |

NOTE: With vehicle level, measure frame angle with a bubble protractor. Record the suspension height measurement.
a. Subtract an up-in-rear frame angle from a positive caster specification.
b. Subtract a down-in-rear frame angle from a negative caster specification.
c. Add an up-in-rear frame angle to a negative caster specification.
d. Add a down-in-rear frame angle to a positive caster specification.

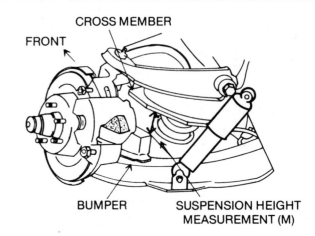

# CHRYSLER CORPORATION LIGHT TRUCKS

| VEHICLE IDENTIFICATION YEAR MODEL | ADJ. ILL. NO. | CASTER (Degrees) MIN. | PREF. | MAX. | CAMBER (Degrees) MIN. | PREF. | MAX. | TOE-IN (Inches) MIN. | PREF. | MAX. | TOE-IN (Millimeters) MIN. | PREF. | MAX. | TOE-OUT ON TURNS (Degrees) OUTSIDE WHEEL | INSIDE WHEEL | STRG. AXIS INCL. (DEG.) |
|---|---|---|---|---|---|---|---|---|---|---|---|---|---|---|---|---|
| **CHRYSLER CORPORATION: DODGE AND PLYMOUTH TRUCKS (A) (B)** | | | | | | | | | | | | | | | | |
| 82-81 B100, 200 300; PB100, 200, 300; MB300, 400 CB300, 400 . . . . . . . . . . . | (8) | 1¼ | 2¼ | 3¼ | 0 | ½ | 1 | 0 | ⅛ | ¼ | 0.0 | 3.2 | 6.0 | NA | NA | NA |
| 82-79 Ramcharger 4 × 2 Trail Duster 4 × 2 D100, 150, 200, 300 . . . . . . | (6) | − ½ | ½ | 1½ | 0 | ¼ | 1 | 0 | ⅛ | ¼ | 0 | 3.2 | 6.4 | 33 | 33 | 8½ |
| 82-79 W150, 200, Ramcharger 4 × 4, Trail Duster 4 × 4 . . . . . . . . | (17) | | 3 | | | 1½ | | ⅛ (out) | | ⅛ (in) | 3.2 (out) | | 3.2 (in) | NA | NA | 8½ |
| 82-79 W200, W250 w/extra equip., W300, W350, W400 . | (17) | | 3 | | | ½ | | ⅛ (out) | | ⅛ (in) | 3.2 (out) | | 3.2 (in) | NA | NA | 8½ |
| 80-77 B100, 200, 300, PB100, 200, 300; MB300, 400; CB300, CB400 . . . . . . . | (8) | 1¼ | 2¼ | 3¼ | 0 | ½ | 1 | 0 | ⅛ | ¼ | 0 | 3.2 | 6.4 | NA | NA | NA |
| w/Heavy Front Axle Load | (8) | ¼ | | 2¾ | − ¼ | ⅜ | 1 | ⅛ (out) | 0 | ⅛ (in) | 3.2 (out) | 0.0 | 3.2 (in) | NA | NA | NA |
| 78 D100, 200, 300 . . . . . . . | (6) | − ½ | ½ | 1½ | | ¼ | | ⅛ (out) | 0 | ⅛ (in) | 3.2 (out) | 0 | 3.2 (in) | NA | NA | NA |
| 78-77 Ramcharger 4 × 2, Trail Duster 4 × 2 . . . . . . . . | (6) | − ½ | ½ | 1½ | 0 | ¼ | 1 | 0 | ⅛ | ¼ | 0 | 3.2 | 6.4 | NA | NA | NA |
| 78-77 Ramcharger 4 × 4, Trail Duster 4 × 4 W100, 200 | (17) | 1½ | 3 | 4½ | 1 | 1½ | 2 | ⅛ (out) | 0 | ⅛ (in) | 3.2 (out) | 0 | 3.2 (in) | NA | NA | 8½ |
| 78-77 W200 w/extra equip., W300, W400 . . . . . | (17) | 1½ | 3 | 4½ | 0 | ½ | 1 | ⅛ (out) | 0 | ⅛ (in) | 3.2 (out) | 0 | 3.2 (in) | NA | NA | 8½ |
| 77 D100, 200, 300 . . . . . . . | (6) | − ½ | ½ | 1½ | 0 | ¼ | 1 | 0 | ⅛ | ¼ | 0 | 3.2 | 6.4 | NA | NA | NA |
| 76-75 Ramcharger 4 × 2, Trail Duster 4 × 2, D100, 200, 300 . . . . . . . . . | (6) | 0 | ½ | 1 | 0 | ¼ | ½ | 1/16 | ⅛ | ⅛ | 1.6 | 3.2 | 3.2 | NA | NA | NA |

| VEHICLE IDENTIFICATION<br>YEAR  MODEL | ADJ. ILL. NO. | CASTER (Degrees) | | | CAMBER (Degrees) | | | TOE-IN (Inches) | | | TOE-IN (Millimeters) | | | TOE-OUT ON TURNS (Degrees)<br>OUTSIDE WHEEL | INSIDE WHEEL | STRG. AXIS INCL. (DEG.) |
|---|---|---|---|---|---|---|---|---|---|---|---|---|---|---|---|---|
| | | MIN. | PREF. | MAX. | MIN. | PREF. | MAX. | MIN. | PREF. | MAX. | MIN. | PREF. | MAX. | | | |
| 76-75 Ramcharger 4 × 4, Trail Duster 4 × 4, W100, 200 | (17) | | 3 | | | 1½ | | 0 | | ⅛ | 0 | | 3.2 | NA | NA | 8½ |
| 76-75 W200 w/extra equip., W300 | (17) | | 3 | | | ½ | | ⅛ (out) | 0 | ⅛ (in) | 3.2 (out) | 0 | 3.2 (in) | NA | NA | 8½ |
| 76-75 B100, 200, 300, PB100, 200, 300, CB300, MB300, 400 | (6) | | | | ¼ | ½ | ¾ | 1/16 | ⅛ | ⅛ | 1.6 | 3.2 | 3.2 | NA | NA | NA |
| W/Power Steering | | 1¾ | 2¼ | 2¾ | | | | | | | | | | | | |
| W/Manual Steering | | 0 | ½ | 1 | | | | | | | | | | | | |
| 74-73 B100, 200, 300, PB100, 200, 300, CB300, MB300, 400 | (6) | | | | 0 | ¼ | ½ | 1/16 | ⅛ | ⅛ | 1.6 | 3.2 | 3.2 | NA | NA | NA |
| W/Power Steering | | 1¾ | 2¼ | 2¾ | | | | | | | | | | | | |
| W/Manual Steering | | 0 | ½ | 1 | | | | | | | | | | | | |
| 74-73 D100, 200, 300 | (6) | 0 | | ½ | 0 | | ¼ | 1/16 | ⅛ | ⅛ | 1.6 | 3.2 | 3.2 | NA | NA | NA |
| 74-73 W100, 200 | (17) | | 3 | | | 1½ | | 0 | | ⅛ | 0 | | 3.2 | NA | NA | 7½ |
| 74-73 W200 w/extra equip., W300 | (17) | | 3½ | | | 1½ | | 0 | | ⅛ | 0 | | 3.2 | NA | NA | 8 |
| 73 P200 | (17) | | 1½ | | | 1½ | | 1/16 | | 3/16 | 1.6 | | 4.8 | NA | NA | NA |
| 73 P300 | (17) | | | | | | | | | | | | | | | |
| W/Power Steering | | | 1½ | | | 1½ | | 1/16 | | 3/16 | 1.6 | | 4.8 | NA | NA | NA |
| W/Manual Steering | | | ½ | | | 2 | | 0 | | ⅛ | 0 | | 3.2 | NA | NA | NA |

**(A)** Engines must be running on vehicles equipped with power steering when centering the steering wheel and setting toe.
**(B)** Caster shown is for unladen vehicles. Caster will vary with increased load. If loaded vehicle wanders, increase caster. If loaded vehicle is hard to steer in corners, decrease caster.

# FORD LIGHT TRUCKS

| YEAR, MODEL AND AXLE PART NUMBER RIDE HEIGHT<br>MIN. | MAX. | ADJ. ILL. NO. | CASTER (Degrees)<br>MIN. | MAX. | CAMBER (Degrees)<br>MIN. | MAX. | TOE-IN (Inches) | TOE-IN (Millimeters) |
|---|---|---|---|---|---|---|---|---|
| **1982**  F-250, 350, 4 × 2 (1)(3) | | N/A | (4) | (4) | (4) | (4) | | |
| 2 | 2¼ | | 5¾ | 9 | −2½ | 0 | 1/32 | 0.8 |
| 2¼ | 2¾ | | 4¾ | 8 | −1½ | 1 | 1/32 | 0.8 |
| 2¼ | 3¼ | | 3¾ | 7 | − ¼ | 1¾ | 1/32 | 0.8 |
| 3¼ | 3¾ | | 2¾ | 6 | ¼ | 2¾ | 1/32 | 0.8 |
| 3½ | 4 | | 1¾ | 5 | 1 | 3½ | 1/32 | 0.8 |
| 4 | 4¼ | | ¾ | 4 | 2 | 4½ | 1/32 | 0.8 |
| **1982-81** F-250, 350 4 × 4 (1)(3) | | N/A | (4) | (4) | | | | |
| 4¾ | 5 | | 3 | 5 | 4¾ | 5 | 1/32 | 0.8 |
| 5 | 5½ | | 3⅛ | 5⅛ | − 1¾ | ¾ | 1/32 | 0.8 |
| 5½ | 6 | | 3⅛ | 5⅛ | − ¾ | 1¾ | 1/32 | 0.8 |
| 6 | 6¼ | | 3¼ | 5¼ | ¼ | 2¾ | 1/32 | 0.8 |
| 6¼ | 6¾ | | 3⅜ | 5⅜ | 1¼ | 4 | 1/32 | 0.8 |
| 6¾ | 7 | | 3½ | 5½ | 2½ | 5 | 1/32 | 0.8 |
| **1982-81** E-100, 150 (1)(3) | | (40) | | | | | | |
| 3¼ | 3½ | | 6¼ | 8 | − 1¾ | − ¼ | 1/32 | 0.8 |
| 3½ | 3¾ | | 5¾ | 7¼ | − 1½ | ¼ | 1/32 | 0.8 |
| 3¾ | 4 | | 5 | 6¾ | − 1 | ¾ | 1/32 | 0.8 |
| 4 | 4¼ | | 4½ | 5¾ | − ½ | 1¼ | 1/32 | 0.8 |
| 4¼ | 4½ | | 4 | 5¼ | 0 | 1¾ | 1/32 | 0.8 |
| 4½ | 4¾ | | 3¼ | 4½ | ½ | 2¼ | 1/32 | 0.8 |
| 4¾ | 5 | | 2½ | 4 | 1 | 2¾ | 1/32 | 0.8 |
| 5 | 5¼ | | 2 | 3¼ | 1½ | 3¼ | 1/32 | 0.8 |
| 5¼ | 5½ | | 1½ | 2¾ | 2 | 3¾ | 1/32 | 0.8 |
| **1982-81** F-100, F-150 4 × 2 (1)(3) | | N/A | (4) | (4) | (4) | (4) | | |
| 2¼ | 2¾ | | 6 | 10 | − 3 | − ½ | 1/32 | 0.8 |
| 2¾ | 3¼ | | 5 | 9 | − 2 | ½ | 1/32 | 0.8 |
| 3¼ | 3½ | | 4 | 8 | − 1¼ | 1¼ | 1/32 | 0.8 |
| 3½ | 4 | | 3 | 7 | − ¼ | 2¼ | 1/32 | 0.8 |
| 4 | 4¼ | | 2 | 6 | ½ | 3 | 1/32 | 0.8 |
| 4¼ | 4¾ | | 1 | 5 | 1½ | 4 | 1/32 | 0.8 |

| YEAR, MODEL AND AXLE PART NUMBER RIDE HEIGHT MIN. | MAX. | ADJ. ILL. NO. | CASTER (Degrees) MIN. | MAX. | CAMBER (Degrees) MIN. | MAX. | TOE-IN (Inches) | TOE-IN (Millimeters) |
|---|---|---|---|---|---|---|---|---|
| **1980** F-100, F-150 4 × 2 (1)(2) | | N/A | (4) | (4) | (4) | (4) | | |
| 2¼ | 2¾ | | 6 | 10 | −3 | −½ | 3/32 | 2.4 |
| 2¾ | 3¼ | | 5 | 9 | −2 | ½ | 3/32 | 2.4 |
| 3¼ | 3½ | | 4 | 8 | −1¼ | 1¼ | 3/32 | 2.4 |
| 3½ | 4 | | 3 | 7 | −¼ | 2¼ | 3/32 | 2.4 |
| 4 | 4¼ | | 2 | 6 | ½ | 3 | 3/32 | 2.4 |
| 4¼ | 4¾ | | 1 | 5 | 1½ | 4 | 3/32 | 2.4 |
| **1982-81** F-150, Bronco 4 × 4 (1)(3) | | (40) | (4) | (4) | | | | |
| 2¾ | 3¼ | | 6 | 9 | −2½ | −¼ | 1/32 | 0.8 |
| 3¼ | 3½ | | 5 | 8 | −1¾ | ½ | 1/32 | 0.8 |
| 3½ | 4 | | 4 | 7 | −¾ | 1½ | 1/32 | 0.8 |
| 4 | 4¼ | | 3 | 6 | 0 | 2¼ | 1/32 | 0.8 |
| 4¼ | 4¾ | | 2 | 5 | 1 | 3¼ | 1/32 | 0.8 |
| 4¾ | 5 | | 1 | 4 | 1¾ | 4 | 1/32 | 0.8 |
| **1980** F-150, Bronco 4 × 4 (1)(2) | | (40) | (4) | (4) | | | | |
| 2¾ | 3¼ | | 6 | 9 | −2½ | −¼ | 3/32 | 2.4 |
| 3¼ | 3½ | | 5 | 8 | −1¾ | ½ | 3/32 | 2.4 |
| 3½ | 4 | | 4 | 7 | −¾ | 1½ | 3/32 | 2.4 |
| 4 | 4¼ | | 3 | 6 | 0 | 2¼ | 3/32 | 2.4 |
| 4¼ | 4¾ | | 2 | 5 | 1 | 3¼ | 3/32 | 2.4 |
| 4¾ | 5 | | 1 | 4 | 1¾ | 4 | 3/32 | 2.4 |
| **1982-80** E-250, 350 (1)(3) | | N/A | | | | | | |
| 3¼ | 3½ | | 9 | 10½ | −1¾ | −¼ | 1/32 | 0.8 |
| 3½ | 3¾ | | 8½ | 9¾ | −1½ | ¼ | 1/32 | 0.8 |
| 3¾ | 4 | | 7⅞ | 9 | −1 | ¾ | 1/32 | 0.8 |
| 4 | 4¼ | | 7⅛ | 8½ | −½ | 1¼ | 1/32 | 0.8 |
| 4¼ | 4½ | | 6½ | 7¾ | 0 | 1¾ | 1/32 | 0.8 |
| 4½ | 4¾ | | 5¾ | 7 | ½ | 2¼ | 1/32 | 0.8 |
| 4¾ | 5 | | 5¼ | 6½ | 1 | 2¾ | 1/32 | 0.8 |
| 5 | 5¼ | | 4⅝ | 6 | 1½ | 3¼ | 1/32 | 0.8 |
| 5¼ | 5½ | | 4 | 5½ | 2 | 3¾ | 1/32 | 0.8 |
| **1980** F-250, 350 4 × 2 (1)(3) | | N/A | (4) | (4) | (4) | (4) | | |
| 2 | 2¼ | | 5¼ | 9 | −2½ | 0 | 1/32 | 0.8 |
| 2¼ | 2¾ | | 4¾ | 8 | −1½ | 1 | 1/32 | 0.8 |
| 2¾ | 2¼ | | 3¾ | 7 | −¾ | 1¾ | 1/32 | 0.8 |
| 3¼ | 3½ | | 2¾ | 6 | ¼ | 2¾ | 1/32 | 0.8 |
| 3½ | 4 | | 1¾ | 5 | 1 | 3½ | 1/32 | 0.8 |
| 4 | 4¼ | | ¾ | 4 | 2 | 4½ | 1/32 | 0.8 |
| **1980** F-250, 350 4 × 4 (1)(2) | | (40) | (4) | (4) | | | | |
| 4¼ | 4¾ | | 3½ | 5¾ | −4 | −1¼ | 3/32 | 2.4 |
| 4¾ | 5 | | 3¼ | 5½ | −2¾ | 0 | 3/32 | 2.4 |
| 5 | 5½ | | 3 | 5¼ | −1½ | 1¼ | 3/32 | 2.4 |
| 5½ | 6 | | 2¾ | 5 | −¼ | 2½ | 3/32 | 2.4 |
| 6 | 6¼ | | 2½ | 4¾ | 1 | 3¾ | 3/32 | 2.4 |
| 6¼ | 6¾ | | 2¼ | 4½ | 2¼ | 5 | 3/32 | 2.4 |
| **1980** E-100, 150 (1)(5) | | N/A | | | | | | |
| 3¼ | 3½ | | 6¼ | 8 | −1¾ | −¼ | ¼ | 6.4 |
| 3½ | 3¾ | | 5¾ | 7¼ | −1½ | ¼ | ¼ | 6.4 |
| 3¾ | 4 | | 5 | 6¾ | −1 | ¾ | ¼ | 6.4 |
| 4 | 4¼ | | 4½ | 5¾ | −½ | 1¼ | ¼ | 6.4 |
| 4¼ | 4½ | | 4 | 5¼ | 0 | 1¾ | ¼ | 6.4 |
| 4½ | 4¾ | | 3¼ | 4½ | ½ | 2¼ | ¼ | 6.4 |
| 4¾ | 5 | | 2½ | 4 | 1 | 2¾ | ¼ | 6.4 |
| 5 | 5¼ | | 2 | 3¼ | 1½ | 3¼ | ¼ | 6.4 |
| 5¼ | 5½ | | 1½ | 2¾ | 2 | 3¾ | ¼ | 6.4 |

(1) All vehicles with normal operating attitude. (2) Nominal toe setting is 2.5 mm (3/32 inch). Range is 0.8 mm (3/32 inch) out to 5.6 mm (7/32 inch) in.
(3) Nominal toe setting is 8 mm (1/32 inch). Range is 2.5 mm (3/32 inch) out to 4 mm (5/32 inch) in. (4) Not Adjustable. (5) Toe range is 0″ to ¼″, ⅛″ nominal.

| | | | | | | | | |
|---|---|---|---|---|---|---|---|---|
| **1979** F-100, F-150, F-250 (D) (6200-6800 GVW) (4 × 2) | | N/A | | | | | | |
| 2¾ | 3 | | 8⅜ | 9⅝ | −2 | −¼ | 3/32 | 2.4 |
| 3 | 3¼ | | 7¾ | 9 | −1½ | ⅛ | 3/32 | 2.4 |
| 3¼ | 3½ | | 7 | 8⅜ | −1⅛ | ½ | 3/32 | 2.4 |
| 3½ | 3¾ | | 6¼ | 7⅝ | −¾ | 1 | 3/32 | 2.4 |
| 3¾ | 4 | | 5⅞ | 7⅛ | −⅜ | 1¼ | 3/32 | 2.4 |
| 4 | 4¼ | | 5⅛ | 6½ | 0 | 1⅝ | 3/32 | 2.4 |
| 4¼ | 4½ | | 4½ | 5⅞ | ⅜ | 2 | 3/32 | 2.4 |
| 4½ | 4¾ | | 3¾ | 5¼ | ¾ | 2⅜ | 3/32 | 2.4 |
| 4¾ | 5 | | 3¼ | 4⅝ | 1¼ | 1¾ | 3/32 | 2.4 |
| 5 | 5¼ | | 2½ | 4 | 1⅞ | 3⅛ | 3/32 | 2.4 |

| YEAR, MODEL AND AXLE PART NUMBER | ADJ. ILL. NO. | CASTER (Degrees) | | CAMBER (Degrees) | | TOE-IN (Inches) | | | TOE-IN (Millimeters) | | |
|---|---|---|---|---|---|---|---|---|---|---|---|
| | | MIN. | MAX. | MIN. | MAX. | | | | | | |

**RIDE HEIGHT** — MIN. / MAX.

**1978 F-250 7800-8000 GVW Regular Cab 6350-7800 Super Cab (4 × 2)** — ADJ. ILL. NO. N/A

| RIDE HT MIN. | RIDE HT MAX. | CASTER MIN. | CASTER MAX. | CAMBER MIN. | CAMBER MAX. | TOE-IN (in) | | | TOE-IN (mm) | | |
|---|---|---|---|---|---|---|---|---|---|---|---|
| 2¾ | 3 | 7⅞ | 8⅞ | − ½ | 1 | 1/32 (out) | 3/32 (in) | 7/32 (in) | 0.8 (out) | 2.4 (in) | 5.6 (in) |
| 3 | 3¼ | 6⅞ | 8⅜ | − 1¼ | 1½ | 1/32 (out) | 3/32 (in) | 7/32 (in) | 0.8 (out) | 2.4 (in) | 5.6 (in) |
| 3¼ | 3½ | 5⅜ | 7¾ | 0 | 1¾ | 1/32 (out) | 3/32 (in) | 7/32 (in) | 0.8 (out) | 2.4 (in) | 5.6 (in) |
| 3¾ | 4 | 5 | 6½ | ¾ | 2½ | 1/32 (out) | 3/32 (in) | 7/32 (in) | 0.8 (out) | 2.4 (in) | 5.6 (in) |
| 4 | 4¼ | 4½ | 5⅞ | 1¼ | 3 | 1/32 (out) | 3/32 (in) | 7/32 (in) | 0.8 (out) | 2.4 (in) | 5.6 (in) |
| 4¼ | 4½ | 3⅜ | 5⅛ | 1¾ | 3¼ | 1/32 (out) | 3/32 (in) | 7/32 (in) | 0.8 (out) | 2.4 (in) | 5.6 (in) |
| 4½ | 4¾ | 3⅛ | 4½ | 2 | 3½ | 1/32 (out) | 3/32 (in) | 7/32 (in) | 0.8 (out) | 2.4 (in) | 5.6 (in) |
| 4¾ | 5 | 2½ | 4 | 2⅜ | 4⅛ | 1/32 (out) | 3/32 (in) | 7/32 (in) | 0.8 (out) | 2.4 (in) | 5.6 (in) |
| 5 | 5¼ | 2 | 3¼ | 2¾ | 4½ | 1/32 (out) | 3/32 (in) | 7/32 (in) | 0.8 (out) | 2.4 (in) | 5.6 (in) |

**1978 F-350 All F-250 8100 GVW Super Cab F-250 Super Cab W/RPO Suspension (4 × 2)** — ADJ. ILL. NO. N/A

| RIDE HT MIN. | RIDE HT MAX. | CASTER MIN. | CASTER MAX. | CAMBER MIN. | CAMBER MAX. | TOE-IN (in) | | | TOE-IN (mm) | | |
|---|---|---|---|---|---|---|---|---|---|---|---|
| 2¾ | 3 | 9¾ | 10⅞ | − ⅜ | − ¼ | 1/32 (out) | 3/32 (in) | 7/32 (in) | 0.8 (out) | 2.4 (in) | 5.6 (in) |
| 3 | 3¼ | 9 | 10¼ | − 1 | − ⅝ | 1/32 (out) | 3/32 (in) | 7/32 (in) | 0.8 (out) | 2.4 (in) | 5.6 (in) |
| 3¼ | 3½ | 8¼ | 9⅝ | − ½ | 1 | 1/32 (out) | 3/32 (in) | 7/32 (in) | 0.8 (out) | 2.4 (in) | 5.6 (in) |
| 3½ | 3¾ | 7⅝ | 8⅝ | − ¼ | 1½ | 1/32 (out) | 3/32 (in) | 7/32 (in) | 0.8 (out) | 2.4 (in) | 5.6 (in) |
| 3¾ | 4 | 6⅞ | 8⅛ | 0 | 1¾ | 1/32 (out) | 3/32 (in) | 7/32 (in) | 0.8 (out) | 2.4 (in) | 5.6 (in) |
| 4 | 4¼ | 6⅜ | 7¼ | ½ | 2¼ | 1/32 (out) | 3/32 (in) | 7/32 (in) | 0.8 (out) | 2.4 (in) | 5.6 (in) |
| 4¼ | 4½ | 5¾ | 7 | ¾ | 2½ | 1/32 (out) | 3/32 (in) | 7/32 (in) | 0.8 (out) | 2.4 (in) | 5.6 (in) |
| 4½ | 4¾ | 5⅛ | 6½ | 1¼ | 3 | 1/32 (out) | 3/32 (in) | 7/32 (in) | 0.8 (out) | 2.4 (in) | 5.6 (in) |
| 4¾ | 5 | 4½ | 5⅞ | 1¾ | 3¼ | 1/32 (out) | 3/32 (in) | 7/32 (in) | 0.8 (out) | 2.4 (in) | 5.6 (in) |
| 5 | 5¼ | 3⅞ | 5⅛ | 2 | 3¼ | 1/32 (out) | 3/32 (in) | 7/32 (in) | 0.8 (out) | 2.4 (in) | 5.6 (in) |

**1978 E-100, 150 (4 × 2)** — ADJ. ILL. NO. N/A

| RIDE HT MIN. | RIDE HT MAX. | CASTER MIN. | CASTER MAX. | CAMBER MIN. | CAMBER MAX. | TOE-IN (in) | | TOE-IN (mm) | |
|---|---|---|---|---|---|---|---|---|---|
| 4 | 4¼ | 3¾ | 6½ | − ¾ | ½ | 0 | ¼ | 0 | 6.4 |
| 4¼ | 4½ | 3¼ | 5¾ | − ½ | ¾ | 0 | ¼ | 0 | 6.4 |
| 4½ | 4¾ | 2½ | 5¼ | 0 | 1¼ | 0 | ¼ | 0 | 6.4 |
| 4¾ | 5 | 2 | 4½ | ½ | 1¾ | 0 | ¼ | 0 | 6.4 |
| 5 | 5¼ | 1¼ | 4 | 1¼ | 2½ | 0 | ¼ | 0 | 6.4 |
| 5¼ | 5½ | ¾ | 3¼ | 1¾ | 3¼ | 0 | ¼ | 0 | 6.4 |
| 5½ | 5¾ | 0 | 2¾ | 1½ | 3¾ | 0 | ¼ | 0 | 6.4 |

**1978 E-250, 350 (4 × 2)** — ADJ. ILL. NO. N/A

| RIDE HT MIN. | RIDE HT MAX. | CASTER MIN. | CASTER MAX. | CAMBER MIN. | CAMBER MAX. | TOE-IN (in) | | | TOE-IN (mm) | | |
|---|---|---|---|---|---|---|---|---|---|---|---|
| 4 | 4¼ | 6¼ | 9 | − 1 | ¾ | 3/32 (out) | 1/32 (in) | 5/32 (in) | 2.4 (out) | 0.8 (in) | 4.0 (in) |
| 4¼ | 4½ | 5¾ | 8¼ | − ½ | 1¼ | 3/32 (out) | 1/32 (in) | 5/32 (in) | 2.4 (out) | 0.8 (in) | 4.0 (in) |
| 4½ | 4¾ | 5¼ | 7¾ | 0 | 1¾ | 3/32 (out) | 1/32 (in) | 5/32 (in) | 2.4 (out) | 0.8 (out) | 4.0 (in) |
| 4¾ | 5 | 4½ | 7¼ | ½ | 2¼ | 3/32 (out) | 1/32 (in) | 5/32 (in) | 2.4 (out) | 0.8 (in) | 4.0 (in) |
| 5 | 5¼ | 4 | 6½ | 1 | 2¾ | 3/32 (out) | 1/32 (in) | 5/32 (in) | 2.4 (out) | 0.8 (in) | 4.0 (in) |
| 5¼ | 5½ | 3¼ | 6 | 1½ | 3¼ | 3/32 (out) | 1/32 (in) | 5/32 (in) | 2.4 (out) | 0.8 (in) | 4.0 (in) |

| YEAR, MODEL AND AXLE PART NUMBER | | ADJ. ILL. NO. | CASTER (Degrees) | | CAMBER (Degrees) | | TOE-IN (Inches) | | | TOE-IN (Millimeters) | | |
|---|---|---|---|---|---|---|---|---|---|---|---|---|
| **RIDE HEIGHT** MIN. | MAX. | | MIN. | MAX. | MIN. | MAX. | | | | | | |
| **1977** F-100, 150, 250 6200-6900 GVW Regular Cab (4 × 2) 2¾ | 3 | N/A | 7½ | 10 | − 1 | ½ | 1/32 (out) | 3/32 (in) | 7/32 (in) | 0.8 (out) | 2.4 (in) | 5.6 (in) |
| 3 | 3¼ | | 7 | 9½ | − ½ | 1 | 1/32 (out) | 3/32 (in) | 7/32 (in) | 0.8 (out) | 2.4 (in) | 5.6 (in) |
| 3¼ | 3½ | | 6½ | 9 | − ¼ | 1½ | 1/32 (out) | 1/32 (in) | 7/32 (in) | 0.8 (out) | 2.4 (in) | 5.6 (in) |
| 3½ | 3¾ | | 5¾ | 8½ | 0 | 1¾ | 1/32 (out) | 3/32 (in) | 7/32 (in) | 0.8 (out) | 2.4 (in) | 5.6 (in) |
| 3¾ | 4 | | 5¼ | 7¾ | ½ | 2¼ | 1/32 (out) | 3/32 (in) | 7/32 (in) | 0.8 (out) | 2.4 (in) | 5.6 (in) |
| 4 | 4¼ | | 4½ | 7 | ¾ | 2½ | 1/32 (out) | 3/32 (in) | 7/32 (in) | 0.8 (out) | 2.4 (in) | 5.6 (in) |
| 4¼ | 4½ | | 4 | 6½ | 1¼ | 3 | 1/32 (out) | 3/32 (in) | 7/32 (in) | 0.8 (out) | 2.4 (in) | 5.6 (in) |
| **1977** F-250 7800-8000 GVW Regular Cab; 6350-7800 Super Cab (4 × 2) 2¾ | 3 | N/A | | | | | | | | | | |
| 3 | 3¼ | | 6½ | 9 | − ¼ | 1½ | 1/32 (out) | 3/32 (in) | 7/32 (in) | 0.8 (out) | 2.4 (in) | 5.6 (in) |
| 3¼ | 3½ | | 5¾ | 8½ | 0 | 1¾ | 1/32 (out) | 3/32 (in) | 7/32 (in) | 0.8 (out) | 2.4 (in) | 5.6 (in) |
| 3½ | 3¾ | | 5¼ | 7¾ | ½ | 2¼ | 1/32 (out) | 3/32 (in) | 7/32 (in) | 0.8 (out) | 2.4 (in) | 5.6 (in) |
| 3¾ | 4 | | 4½ | 7 | ¾ | 2½ | 1/32 (out) | 3/32 (in) | 7/32 (in) | 0.8 (out) | 2.4 (in) | 5.6 (in) |
| 4 | 4¼ | | 4 | 6½ | 1¼ | 3 | 1/32 (out) | 3/32 (in) | 7/32 (in) | 0.8 (out) | 2.4 (in) | 5.6 (in) |
| 4¼ | 4½ | | 3¼ | 6 | 1¾ | 3¼ | 1/32 (out) | 3/32 (in) | 7/32 (in) | 0.8 (out) | 2.4 (in) | 5.6 (in) |
| 4½ | 4¾ | | 2¾ | 5½ | 2 | 3½ | 1/32 (out) | 3/32 (in) | 7/32 (in) | 0.8 (out) | 2.4 (in) | 5.6 (in) |
| **1977** E-100, 150 (4 × 2) 4¼ | 4½ | N/A | 3¼ | 5¾ | − ¾ | ¾ | 1/32 (out) | 3/32 (in) | 7/32 (in) | 0.8 (out) | 2.4 (in) | 5.6 (in) |
| 4½ | 4¾ | | 2½ | 5¼ | − ½ | 1¼ | 1/32 (out) | 3/32 (in) | 7/32 (in) | 0.8 (out) | 2.4 (in) | 5.6 (in) |
| 4¾ | 5 | | 2 | 4½ | 0 | 1¾ | 1/32 (out) | 3/32 (in) | 7/32 (in) | 0.8 (out) | 2.4 (in) | 5.6 (in) |
| 5 | 5¼ | | 1¼ | 4 | ½ | 2½ | 1/32 (out) | 3/32 (in) | 7/32 (in) | 0.8 (out) | 2.4 (in) | 5.6 (in) |
| 5¼ | 5½ | | ¾ | 3¼ | 1½ | 3¼ | 1/32 (out) | 3/32 (in) | 7/32 (in) | 0.8 (out) | 2.4 (in) | 5.6 (in) |
| 5½ | 5¾ | | 0 | 2¾ | 2 | 3¾ | 1/32 (out) | 3/32 (in) | 7/32 (in) | 0.8 (out) | 2.4 (in) | 5.6 (in) |
| **1977** E-250, 350 (4 × 2) 4 | 4¼ | N/A | 6¼ | 9 | − 1 | ¾ | 3/32 (out) | 1/32 (in) | 5/32 (in) | 2.4 (out) | 0.8 (in) | 4.0 (in) |
| 4¼ | 4½ | | 5¾ | 8¼ | − ½ | 1¼ | 3/32 (out) | 1/32 (in) | 5/32 (in) | 2.4 (out) | 0.8 (in) | 4.0 (in) |
| 4½ | 4¾ | | 5¼ | 7¾ | 0 | 1¾ | 3/32 (out) | 1/32 (in) | 5/32 (in) | 2.4 (out) | 0.8 (in) | 4.0 (in) |
| 4¾ | 5 | | 4½ | 7¼ | ½ | 2¼ | 3/32 (out) | 1/32 (in) | 5/32 (in) | 2.4 (out) | 0.8 (in) | 4.0 (in) |
| 5 | 5¼ | | 4 | 6½ | 1 | 2¾ | 3/32 (out) | 1/32 (in) | 5/32 (in) | 2.4 (out) | 0.8 (in) | 4.0 (in) |
| 5¼ | 5½ | | 3¼ | 6 | 1½ | 3¼ | 3/32 (out) | 1/32 (in) | 5/32 (in) | 2.4 (out) | 0.8 (in) | 4.0 (in) |

| YEAR, MODEL AND AXLE PART NUMBER RIDE HEIGHT MIN. | MAX. | ADJ. ILL. NO. | CASTER (Degrees) MIN. | MAX. | CAMBER (Degrees) MIN. | MAX. | TOE-IN (Inches) | | | TOE-IN (Millimeters) | | |
|---|---|---|---|---|---|---|---|---|---|---|---|---|
| **1977** F-350 All F-250 8100 GVW Super Cab (4 × 2) F-250 Super Cab w/RPO Suspension | | N/A | | | | | | | | | | |
| 3½ | 3¾ | | 7 | 9½ | − ¼ | 1½ | 1/32 (out) | 3/32 (in) | 7/32 (in) | 0.8 (out) | 2.4 (in) | 5.6 (in) |
| 3¾ | 4 | | 6½ | 9 | 0 | 1¾ | 1/32 (out) | 3/32 (in) | 7/32 (in) | 0.8 (out) | 2.4 (in) | 5.6 (in) |
| 4 | 4¼ | | 5¾ | 8½ | ½ | 2¼ | 1/32 (out) | 3/32 (in) | 7/32 (in) | 0.8 (out) | 2.4 (in) | 5.6 (in) |
| 4¼ | 4½ | | 5¼ | 7¾ | ¾ | 2½ | 1/32 (out) | 3/32 (in) | 7/32 (in) | 0.8 (out) | 2.4 (in) | 5.6 (in) |
| 4½ | 4¾ | | 4½ | 7 | 1¼ | 3 | 1/32 (out) | 3/32 (in) | 7/32 (in) | 0.8 (out) | 2.4 (in) | 5.6 (in) |
| 4¾ | 5 | | 4 | 6½ | 1¾ | 3¼ | 1/32 (out) | 3/32 (in) | 7/32 (in) | 0.8 (out) | 2.4 (in) | 5.6 (in) |
| 5 | 5¼ | | 3¼ | 6 | 2 | 3¾ | 1/32 (out) | 3/32 (in) | 7/32 (in) | 0.8 (out) | 2.4 (in) | 5.6 (in) |

| VEHICLE IDENTIFICATION YEAR MODEL | ADJ. ILL. NO. | CASTER (Degrees) MIN. | PREF. | MAX. | CAMBER (Degrees) MIN. | PREF. | MAX. | TOE-IN (Inches) MIN. | PREF. | MAX. | TOE-IN (Millimeters) MIN. | PREF. | MAX. | TOE-OUT ON TURNS (Degrees) OUTSIDE WHEEL | INSIDE WHEEL | STRG. AXIS INCL. (DEG.) |
|---|---|---|---|---|---|---|---|---|---|---|---|---|---|---|---|---|
| **FORD MOTOR CO.** | | | | | | | | | | | | | | | | |
| 79 F-250 (4 × 4), F-150 (4 × 4) SC, F-350 (4 × 4) | N/A | 2½ | 4 | 5½ | 0 | 1½ | 3 | | 3/32 | | | 2.4 | | NA | NA | NA |
| 79 F-150 (4 × 4), Bronco | N/A | 6½ | 8 | 9½ | 1 | 1½ | 3 | | 3/32 | | | 2.4 | | NA | NA | NA |
| 78 F-250 (4 × 4) | N/A | 2⅛ | 4 | 5 | ½ | 1 | 1½ | 1/16 | 5/32 | ¼ | 1.6 | 4.0 | 6.4 | NA | NA | NA |
| 78 F-150, (4 × 4), Bronco | N/A | 2¾ | 3½ | 4¼ | 1 | 1½ | 2 | 1/16 | 3/32 | 5/16 | 1.6 | 2.4 | 8.0 | NA | NA | NA |
| 77 F-250 (4 × 4) | N/A | 3 | 4 | 5 | ½ | 1 | 1½ | 1/16 | 5/32 | ¼ | 1.6 | 4.0 | 6.4 | NA | NA | NA |
| 77 F-150 (4 × 4) Bronco | N/A | 2¾ | 3½ | 4¼ | 1 | 1½ | 2 | 1/16 | 5/32 | 5/16 | 1.6 | 4.0 | 8.0 | NA | NA | NA |
| 76-75 E-350 (4 × 2) | N/A | 2½ | | 7½ | ½ | | 3½ | 3/32 (out) | 1/32 (in) | 5/32 (in) | 2.4 (out) | 0.8 (in) | 4.0 (in) | NA | NA | NA |
| 76-75 F-100, 250 (4 × 2) | N/A | ½ | | 8½ | − ½ | | 2½ | 1/32 (out) | 3/32 (in) | 5/32 (in) | 0.8 (out) | 2.4 (in) | 4.0 (in) | NA | NA | NA |
| 76-75 F-350 (4 × 2) | N/A | ½ | | 8½ | 0 | | 3 | 1/32 (out) | 3/32 (in) | 7/32 (in) | 0.8 (out) | 2.4 (in) | 5.6 (in) | NA | NA | NA |
| 76-75 E-100, 150, 250 (4 × 2) | N/A | ½ | | 8½ | ½ | | 3½ | 3/32 (out) | 1/32 (in) | 5/32 (in) | 2.4 (out) | 0.8 (in) | 4.0 (in) | NA | NA | NA |
| 76 F-100, 150, Bronco (4 × 4) | N/A | 2¾ | 4½ | 4¼ | 1 | 1½ | 2 | 1/16 | 5/32 | 5/16 | 1.6 | 4.0 | 8.0 | NA | NA | NA |
| 76-73 F-250 (4 × 4) | N/A | 3½ | 4 | 4½ | 1 | 1½ | 2 | 1/16 | 5/32 | ¼ | 1.6 | 4.0 | 6.4 | NA | NA | NA |
| 75-73 F-100, Bronco (4 × 4) | N/A | 2¾ | 3½ | 4¼ | 1 | 1½ | 2 | 1/16 | 5/32 | ¼ | 1.6 | 4.0 | 6.4 | NA | NA | NA |
| 74 F-100, 250 (4 × 2) | N/A | ½ | | 8½ | − ½ | | 2½ | 1/32 (out) | 3/32 (in) | 7/32 (in) | 0.8 (out) | 2.4 (in) | 5.6 (in) | NA | NA | NA |
| 74 F-350 (4 × 2) | N/A | ½ | | 8½ | 0 | | 3 | 1/32 (out) | 3/32 (in) | 7/32 (in) | 0.8 (out) | 2.4 (in) | 5.6 (in) | NA | NA | NA |
| 74 E-300 (4 × 2) | N/A | 2½ | | 7½ | ½ | | 3½ | 3/32 (out) | 1/32 (in) | 5/32 (in) | 2.4 (out) | 0.8 (in) | 4.0 (in) | NA | NA | NA |
| 74 E-100, 200 (4 × 2) | N/A | ½ | | 8½ | ½ | | 3½ | 3/32 (out) | 1/32 (in) | 5/32 (in) | 2.4 (out) | 0.8 (in) | 4.0 (in) | NA | NA | NA |
| 73 F-100, 250, 350 (4 × 2) | N/A | 3½ | | 8½ | ½ | | 3½ | 1/32 (out) | 3/32 (in) | 7/32 (in) | 0.8 (out) | 2.4 (in) | 5.6 (in) | NA | NA | NA |
| 73 E-100, 200 (4 × 2) | N/A | 3½ | | 8½ | ½ | | 3½ | 3/32 (out) | 1/32 (in) | 5/32 (in) | 2.4 (out) | 0.8 (in) | 4.0 (in) | NA | NA | NA |
| 73 E-300 (4 × 2) | N/A | 2½ | | 7½ | ½ | | 3½ | 3/32 (out) | 1/32 (in) | 5/32 (in) | 2.4 (out) | 0.8 (in) | 4.0 (in) | NA | NA | NA |

If caster or camber are not with specifications, find the true ride height by installing wooden blocks as pictured below. This will aid in determining which wheel is out of specification.

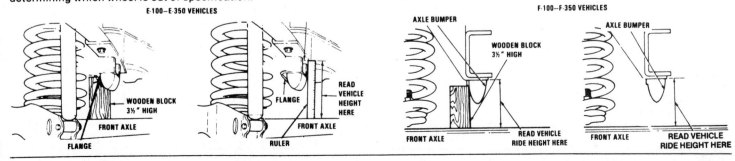

E-100—E-350 VEHICLES     F-100—F-350 VEHICLES

WOODEN BLOCK 3½" HIGH — FRONT AXLE — FLANGE

FLANGE — RULER — FRONT AXLE — READ VEHICLE HEIGHT HERE

AXLE BUMPER — WOODEN BLOCK 3½" HIGH — FRONT AXLE — READ VEHICLE RIDE HEIGHT HERE

AXLE BUMPER — FRONT AXLE — READ VEHICLE RIDE HEIGHT HERE

| VEHICLE IDENTIFICATION — YEAR MODEL | ADJ. ILL. NO. | CASTER (Degrees) MIN. | PREF. | MAX. | CAMBER (Degrees) MIN. | PREF. | MAX. | TOE-IN (Inches) MIN. | PREF. | MAX. | TOE-IN (Millimeters) MIN. | PREF. | MAX. | TOE-OUT ON TURNS (Degrees) OUTSIDE WHEEL | INSIDE WHEEL | STRG. AXIS INCL. (DEG.) |
|---|---|---|---|---|---|---|---|---|---|---|---|---|---|---|---|---|
| **INTERNATIONAL HARVESTER** | | | | | | | | | | | | | | | | |
| 80 Scout, All Models ..... | (17) | | 2½ | | | ½ | | 3/32 | | 5/16 | 1.6 | | 7.9 | NA | NA | 8½ |
| 79 Scout, All Models ..... | (17) | | 0 | | | 1 | | 0 | | 3/16 | 0 | | 5.1 | NA | NA | 8½ |
| 78-75 Scout, All Models ... | (17) | −1 | 0 | 1 | ½ | 1 | 1½ | 0 | 1/16 | 1/8 | 0 | 1.6 | 3.2 | NA | NA | 8½ |
| 74-73 Scout, Scout 800 (4 × 2) ........ | (17) | 0 | | 1 | ½ | 1 | 1½ | 0 | 1/16 | 1/8 | 0 | 1.6 | 3.2 | NA | NA | 8½ |
| 74-73 Scout, Scout 800 (4 × 4) ........ | (17) | −1 | 0 | 1 | ½ | 1 | 1½ | 0 | 1/16 | 1/8 | 0 | 1.6 | 3.2 | NA | NA | 8½ |
| 75-73 1000 Series ........ | (1)(2) | | | | ½ | | 1½ | 1/8 | | 3/16 | 3.2 | | 4.8 | NA | NA | 8 |
| W/Power Steering ..... | | | 0 | | | | | | | | | | | | | |
| W/Manual Steering .... | | | −1 | | | | | | | | | | | | | |
| 75-73 1100 (4 × 2) | (17) | 1 | | 3 | 1 | | 2 | 0 | | 1/8 | 0 | | 3.2 | NA | NA | 4 |
| 75-73 1100 (4 × 4) | (17) | | ½ | | | 1½ | | 1/8 | | 3/16 | 3.2 | | 4.8 | NA | NA | 7½ |
| 75-73 1200 (4 × 2) | (17) | 2½ | | 3½ | 1 | | 2 | 0 | | 1/8 | 0 | | 3.2 | NA | NA | 4 |
| 75-73 1200 (4 × 4) | (17) | | ½ | | | 1½ | | 1/8 | | 3/16 | 3.2 | | 4.8 | NA | NA | 7½ |
| 75-73 1300 (4 × 2) | (17) | 2½ | | 3½ | 1 | | 2 | 0 | | 1/8 | 0 | | 3.2 | NA | NA | 4 |
| 75-73 1300 (4 × 4) | (17) | | 1 9/16 | | | 1½ | | 1/8 | | 3/16 | 3.2 | | 4.8 | NA | NA | 8 |
| 75-73 100, 150 (4 × 4) ..... | (17) | −1 | 0 | 1 | | 1 | | 0 | | 1/8 | 0 | | 3.2 | NA | NA | 8½ |
| 75-73 200 (4 × 4) | (17) | | 1 9/16 | | | 1½ | | 1/32 | | 1/16 | 0.8 | | 1.6 | NA | NA | 8 |
| 75-74 100, 150, 200 (4 × 2) .. | | | | | | | | 0 | | 1/8 | 0 | | 3.2 | NA | NA | NA |
| Measured at 2 13/16 ..... | | 1 | 1½ | 2 | 1/8 L | | 1/8 R | | | | | | | | | |
| 3 3/16 ..... | | ½ | 1 | 1½ | ¼ L | | 0 R | | | | | | | | | |
| 3 5/8 ..... | | 0 | ½ | 1 | ¼ L | | 0 R | | | | | | | | | |
| 4 ..... | | − ½ | 0 | ½ | ¼ L | | 0 R | | | | | | | | | |
| 4 7/16 ..... | | −1 | − ½ | 0 | ¼ L | | 0 R | | | | | | | | | |
| 4 13/16 ..... | | −1½ | −1 | − ½ | 0 L | | ¼ R | | | | | | | | | |
| 5 3/16 ..... | | −2 | −1½ | −1 | − ¼ L | | − ½ R | | | | | | | | | |

R-Right Wheel; L-Left Wheel
Measure suspension height from lower control arm snubber mounting bracket to lower face of front cross member.

# IMPORTED CAR AND LIGHT TRUCK SECTION

| VEHICLE IDENTIFICATION — YEAR MODEL | ADJ. ILL. NO. | CASTER (Degrees) MIN. | PREF. | MAX. | CAMBER (Degrees) MIN. | PREF. | MAX. | TOE-IN (Inches) MIN. | PREF. | MAX. | TOE-IN (Millimeters) MIN. | PREF. | MAX. | TOE-OUT ON TURNS (Degrees) OUTSIDE WHEEL | INSIDE WHEEL | STRG. AXIS INCL. (DEG.) |
|---|---|---|---|---|---|---|---|---|---|---|---|---|---|---|---|---|
| **ALFA ROMEO** | | | | | | | | | | | | | | | | |
| 82-78 Sprint Veloce, Sport Sedan, Spider Veloce..... | N/A | 4 | 4½ | 5 | − 5/16 | 3/16 | 11/16 | 0 | | 5/64 | 0 | | 2.0 | NA | NA | NA |
| 77-76 Alfetta, Alfetta GT ... | N/A | 4 | 4½ | 5 | − 5/16 | 3/16 | 11/16 | 0 | | 5/64 | 0 | | 2.0 | NA | NA | NA |
| 77 Spider Veloce ....... | N/A | 1 | 1½ | 2 | − 5/16 | 3/16 | 11/16 | 0 | | 1/8 | 0 | | 3.0 | NA | NA | NA |
| 76-75 Spider Veloce, GT Veloce ............ | N/A | 1 | ½ | 2 | − 5/16 | 3/16 | 11/16 | 0 | | 1/8 | 0 | | 3.0 | NA | NA | NA |
| 75-73 All Series ......... | N/A | 1 | 1½ | 2 | − 5/16 | 3/16 | 11/16 | 0 | | 1/8 | 0 | | 3.0 | NA | NA | NA |
| **AUDI** | | | | | | | | | | | | | | | | |
| 82-80 4000: Front ........ | (14) | 0 | ½ | 1 | -1 5/32 | − 21/32 | 5/32 | 0 | 5/64 | 5/32 | 0 | 2.0 | 4.0 | NA | NA | NA |
| Rear ........ | N/A | | | | -1 11/32 | −1 | − 21/32 | 0 | 11/64 | 21/64 | 0 | 4.4 | 8.3 | | | |
| 80 5000 up to VIN No. 43A0016066 and all 79-78 5000 ............ | | | | | | | | | | | | | | | | |
| Front ............ | (35) | − 13/16 | − 5/32 | ½ | − 1 | − ½ | 0 | 5/64 (out) | 0 | 3/64 (in) | 2.0 (out) | 0.0 | 1.2 (in) | NA | NA | NA |
| Rear ............... | N/A | | | | − 1 | − ½ | 0 | 5/64 | 11/64 | 13/64 | 2.0 | 4.4 | 5.6 | | | |
| 82-80 5000 after VIN No. 43A0016066; & 82-80 5000 Turbo, All ..... | | | | | | | | | | | | | | | | |
| Front ............ | (35) | ½ | 1 5/32 | 1 13/16 | − 1 | − ½ | 0 | 5/64 (out) | 0 | 3/64 (in) | 2.0 (out) | 0.0 | 1.2 (in) | NA | NA | NA |
| Rear ............... | N/A | | | | − 1 | − ½ | 0 | 5/64 | 11/64 | 13/64 | 2.0 | 4.4 | 5.6 | | | |
| 79-78 Fox: Front ........ | (14) | 0 | ½ | 1 | 0 | ½ | 1 | 5/64 (out) | 5/64 (in) | 5/32 (in) | 2.0 (out) | 2.0 (in) | 4.0 (in) | 18 ¾ | 20 | NA |
| Rear......... | N/A | | | | − 1 5/16 | − 11/16 | 0 | 13/64 (out) | 0 | 13/64 (in) | 5.6 (out) | 0 | 5.6 (in) | | | |
| 77-76 Fox: Front ........ | (14) | 0 | ½ | 1 | 0 | ½ | 1 | 5/64 (out) | 3/64 (in) | 1/8 (in) | 2.0 (out) | 1.2 (in) | 3.2 (in) | 18 ¾ | 20 | NA |
| Rear......... | N/A | | | | − 1 | − ½ | 0 | 3/16 (out) | 0 | 3/16 (in) | 4.8 (out) | 0 | 4.8 (in) | | | |
| 77-75 100 Series ........ | (59) | | | | | | | 1/64 (out) EACH WHEEL | 3/64 | 7/64 (in) | 0.4 (out) EACH WHEEL | 1.2 | 2.8 (in) | 19 27/32 | 20 | NA |
| W/Power Steering ..... Front ............ | | − ¼ | 0 | ¼ | − ½ | 0 | ½ | | | | | | | | | |
| W/Manual Steering .... Front ............ | | − ¾ | − ½ | − ¼ | − ½ | 0 | ½ | | | | | | | | | |
| Rear, All ......... | N/A | | | | − 1 | − ½ | 0 | 1/8 (out) | 0 | 1/8 (in) | 3.4 (out) | 0 | 3.4 (in) | | | |

**100**

| VEHICLE IDENTIFICATION<br>YEAR  MODEL | ADJ. ILL. NO. | CASTER (Degrees) MIN. | PREF. | MAX. | CAMBER (Degrees) MIN. | PREF. | MAX. | TOE-IN (Inches) MIN. | PREF. | MAX. | TOE-IN (Millimeters) MIN. | PREF. | MAX. | TOE-OUT ON TURNS (Degrees) OUTSIDE WHEEL | INSIDE WHEEL | STRG. AXIS INCL. (DEG.) |
|---|---|---|---|---|---|---|---|---|---|---|---|---|---|---|---|---|
| 75-73 Fox: Front ........ | (14) | 0 | 1/2 | 1 | 0 | 1/2 | 1 | 1/32 (out) | 3/64 (in) | 1/8 (in) | 0.8 (out) | 1.2 (in) | 3.4 (in) | 18 3/4 | 20 | NA |
| Rear........ | N/A | | | | − 1 | − 1/2 | 0 | 3/16 (out) | 0 | 3/16 (in) | 4.8 (out) | 0 | 4.8 (in) | | | |
| 74-73 100 Series ....... | (59) | | | | − 1/4 | 1/4 | 3/4 | EACH WHEEL 1/16 (out) | 0 | 1/16 (in) | EACH WHEEL 1.6 (out) | 0 | 1.6 (in) | 19 27/32 | 20 | NA |
| W/Power Steering ..... Front ........... | | 1/16 | 5/16 | 9/16 | | | | | | | | | | | | |
| W/Manual Steering .... Front ........... | | − 7/16 | − 3/16 | 1/16 | | | | | | | | | | | | |
| Rear, All ......... | N/A | | | | − 1 | − 1/2 | 0 | 1/8 (out) | 0 | 1/8 (in) | 3.4 (out) | 0 | 3.4 (in) | | | |

**AUSTIN**

| VEHICLE IDENTIFICATION | ADJ. ILL. NO. | CASTER MIN. | PREF. | MAX. | CAMBER MIN. | PREF. | MAX. | TOE-IN (In.) MIN. | PREF. | MAX. | TOE-IN (mm) MIN. | PREF. | MAX. | OUTSIDE | INSIDE | STRG. AXIS |
|---|---|---|---|---|---|---|---|---|---|---|---|---|---|---|---|---|
| 75-73 Marina .......... | N/A | | 2 | | | 1/4 | | 1/64 | 1/32 | 3/64 | 0.2 | 0.7 | 1.2 | NA | NA | NA |

**BMW (A) (B)**

| VEHICLE IDENTIFICATION | ADJ. ILL. NO. | CASTER MIN. | PREF. | MAX. | CAMBER MIN. | PREF. | MAX. | TOE-IN (In.) MIN. | PREF. | MAX. | TOE-IN (mm) MIN. | PREF. | MAX. | OUTSIDE | INSIDE | STRG. AXIS |
|---|---|---|---|---|---|---|---|---|---|---|---|---|---|---|---|---|
| 82 528e: Front ...... | N/A | 7 3/4 | 8 1/4 | 8 3/4 | − 27/32 | − 11/32 | 5/32 | 1/8 | 11/64 | 7/32 | 3.2 | 4.1 | 5.6 | 18 3/16 | 20 | 12 5/32 |
| Rear........ | N/A | | | | − 2 11/32 | − 1 27/32 | − 1 11/32 | 1/8 | 11/64 | 7/32 | 3.2 | 4.1 | 5.6 | | | |
| 82-77 320i: Front ....... | N/A | 7 13/16 | 8 5/16 | 8 13/16 | − 1/2 | 0 | 1/2 | 1/32 | 1/16 | 3/32 | 0.8 | 1.6 | 2.4 | 19 1/4 | 20 | 10 15/16 |
| Rear...... | N/A | | | | − 2 1/2 | − 2 | − 1 1/2 | 0 | 1/32 | 3/32 | 0 | 0.8 | 2.4 | | | |
| 82-79 528i: Front....... | N/A | 7 3/16 | 7 11/16 | 8 3/16 | − 1/2 | 0 | 1/2 | 1/32 | 1/16 | 3/32 | 0.8 | 1.6 | 2.4 | 18 3/16 | 20 | 8 1/2 |
| Rear ...... | N/A | | | | − 2 1/2 | − 2 | − 1 1/2 | 3/64 | 5/64 | 7/64 | 1.2 | 2.0 | 2.8 | | | |
| 82-78 733i: Front....... | N/A | 8 1/2 | 9 | 9 1/2 | − 1/2 | 0 | 1/2 | 0 | 3/64 | 1/8 | 0 | 1.2 | 3.2 | 18 5/16 | 20 | 11 9/16 |
| Rear ...... | N/A | | | | − 2 | − 1 1/2 | − 1 | 1/8 | 3/16 | 1/4 | 3.2 | 4.8 | 6.4 | | | |
| 82-80 633 CSi: Front ..... | N/A | 7 3/16 | 7 11/16 | 8 3/16 | − 1/2 | 0 | 1/2 | 5/64 | 1/8 | 13/64 | 2.0 | 3.2 | 5.2 | 18 1/2 | 20 | 8 |
| Rear ..... | N/A | | | | − 2 1/2 | − 2 | − 1 1/2 | 3/64 | 5/64 | 7/64 | 1.2 | 2.0 | 2.8 | | | |
| 79-78 530i: Front ..... | N/A | 7 3/16 | 7 11/16 | 8 3/16 | − 1/2 | 0 | 1/2 | 1/32 | 1/16 | 3/32 | 0.8 | 1.6 | 2.4 | 18 3/16 | 20 | 8 1/2 |
| Rear ..... | N/A | | | | − 2 1/2 | − 2 | − 1 1/2 | 0 | 1/32 | 3/32 | 0 | 0.8 | 2.4 | | | |
| 79-77 630 CSi: Front ..... | N/A | 7 5/16 | 7 11/16 | 8 3/16 | − 1/2 | 0 | 1/2 | 1/32 | 1/16 | 3/32 | 0.8 | 1.6 | 2.4 | 18 1/2 | 20 | 8 |
| Rear ..... | N/A | | | | − 2 1/2 | − 2 | − 1 1/2 | 3/64 | 5/64 | 7/64 | 1.2 | 2.0 | 2.8 | | | |
| 76-73 2002, 2002tii: Front.. | N/A | 3 1/2 | 4 | 4 1/2 | 0 | 1/2 | 1 | 1/32 | 1/16 | 3/32 | 0.8 | 1.6 | 2.4 | 19 | 20 | 8 1/2 |
| Rear .. | N/A | | | | − 2 1/2 | − 2 | − 1 1/2 | 0 | 1/16 | 1/8 | 0 | 1.6 | 3.2 | | | |
| 76  3.0 Si: Front ........ | N/A | 7 3/16 | 7 11/16 | 8 3/16 | 0 | 1/2 | 1 | 1/32 | 1/16 | 3/32 | 0.8 | 1.6 | 2.4 | 18 1/2 | 20 | 6 11/32 |
| Rear ........ | N/A | | | | − 2 1/2 | − 2 | − 1 1/2 | 0 | 1/32 | 3/32 | 0 | 0.8 | 2.4 | | | |
| 75  3.0 Series: Front ..... | N/A | 9 3/16 | 9 11/16 | 10 3/16 | 0 | 1/2 | 1 | 1/32 | 1/16 | 3/32 | 0.8 | 1.6 | 2.4 | 18 1/2 | 20 | 6 11/32 |
| Rear ..... | N/A | | | | − 2 1/2 | − 2 | − 1 1/2 | 0 | 1/32 | 3/32 | 0 | 0.8 | 2.4 | | | |
| 77-75 530i: Front ........ | N/A | 7 3/16 | 7 11/16 | 8 3/16 | 0 | 1/2 | 1 | 1/32 | 1/16 | 3/32 | 0.8 | 1.6 | 2.4 | 18 3/16 | 20 | 8 1/2 |
| Rear........ | N/A | | | | − 2 1/2 | − 2 | − 1 1/2 | | 1/32 | | | 0.8 | | | | |
| 74-73 3.0 Series: Front .... | N/A | 9 | 9 1/2 | 10 | − 1/2 | 0 | 1/2 | 1/32 | 1/16 | 3/32 | 0.8 | 1.6 | 2.4 | 18 1/2 | 20 | 6 11/32 |
| Rear .... | N/A | | | | − 2 1/2 | − 2 | − 1 1/2 | 0 | 1/32 | 3/32 | 0 | 0.8 | 2.4 | | | |

**(A)** 320i, 528i, 530 and 733 models aligned with 150 lbs. in each front seat, 150 lbs. in rear seat and 46 lbs. in trunk.

**(B)** 630, 633 aligned with 150 lbs. in each front seat and 30 lbs. in trunk on left side.

**(C)** 2500, 2800, 310, 1600, 2002 aligned with 150 lbs. in each front seat, 150 lbs. in rear seat and 30 lbs. in trunk on left side.

N/A - Not Adjustable.

**BUICK OPEL**

| VEHICLE IDENTIFICATION | ADJ. ILL. NO. | CASTER MIN. | PREF. | MAX. | CAMBER MIN. | PREF. | MAX. | TOE-IN (In.) MIN. | PREF. | MAX. | TOE-IN (mm) MIN. | PREF. | MAX. | OUTSIDE | INSIDE | STRG. AXIS |
|---|---|---|---|---|---|---|---|---|---|---|---|---|---|---|---|---|
| 79 All ................. | (13) | 3 | 4 1/2 | 6 | − 3/4 | 1/4 | 1 1/4 | 1/32 (out) | 1/16 (in) | 5/32 (in) | 0.8 (out) | 1.6 (in) | 4.0 (in) | NA | NA | 7 7/8 |
| 78-76 All ............. | (13) | 3 1/2 | 5 | 6 1/2 | − 1 | 0 | 1 | 1/32 | 1/8 | 7/32 | 0.8 | 3.2 | 5.6 | NA | NA | 7 7/8 |
| 75 All ............ | (16) | 3 | 4 1/2 | 6 | − 1 1/4 | | 1/4 | 3/32 | 1/8 | 5/32 | 2.4 | 3.2 | 4.0 | 19 1/4 | 20 | 8 1/2 |
| 74 All ............ | (16) | 3 | 4 1/2 | 6 | − 1 1/2 | − 1 | − 1/2 | 1/8 | | 3/16 | 3.2 | | 4.8 | 19 1/4 | 20 | 8 1/2 |
| 73 GT ............. | (16) | 2 | 3 | 4 | 1/2 | 1 | 1 1/2 | 1/32 | | 1/8 | 0.8 | | 3.2 | 18 1/2 | 20 | 6 |
| 73  1900 ............. | (16) | 3 1/2 | | 6 1/2 | − 1 1/2 | − 1 | − 1/2 | 1/8 | | 3/16 | 3.2 | | 4.8 | 19 1/4 | 20 | 8 1/2 |

**CAPRI**

| VEHICLE IDENTIFICATION | ADJ. ILL. NO. | CASTER MIN. | PREF. | MAX. | CAMBER MIN. | PREF. | MAX. | TOE-IN (In.) MIN. | PREF. | MAX. | TOE-IN (mm) MIN. | PREF. | MAX. | OUTSIDE | INSIDE | STRG. AXIS |
|---|---|---|---|---|---|---|---|---|---|---|---|---|---|---|---|---|
| 78-76 Capri II .......... | N/A | 1 | 1 1/2 | 2 1/4 | 3/4 | 1 1/2 | 2 1/4 | 0 | | 9/32 | 0 | | 7.1 | 19• | 20 | 8 + |
| 75-74 ......... | N/A | 1/2 | | 1 1/2 | − 1/2 | 0 | 1/2 | 0 | | 1/4 | 0 | | 6.4 | 19• | 20 | 8 + |
| 73 ............. | N/A | 3/4 | | 1 3/4 | − 1/2 | 0 | 1/2 | 0 | | 1/4 | 0 | | 6.4 | 19• | 20 | 8 + |

• Range from 18 1/4 to 19 3/4 acceptable.      + Range from 7 1/2° to 8 1/2° acceptable.

**CHEVROLET LUV**

| VEHICLE IDENTIFICATION | ADJ. ILL. NO. | CASTER MIN. | PREF. | MAX. | CAMBER MIN. | PREF. | MAX. | TOE-IN (In.) MIN. | PREF. | MAX. | TOE-IN (mm) MIN. | PREF. | MAX. | OUTSIDE | INSIDE | STRG. AXIS |
|---|---|---|---|---|---|---|---|---|---|---|---|---|---|---|---|---|
| 82-81 Series 12, 11 (4 × 2) | (10) | 0 | 1/2 | 1 | 0 | 1/2 | 1 | 0 | 1/16 | 1/8 | 0.0 | 2.0 | 4.0 | 33 | 37 | 7 1/2 |
| 82-81 Series 12, 11 (4 × 4) | (18) | − 3/16 | 5/16 | 13/16 | 1/16 | 9/16 | 1 1/16 | 1/16 (out) | 0 | 1/16 (in) | 2.0 (out) | 0 | 2.0 (in) | 33 | 35 | 7 1/2 |
| 80-79 Series 10, 9 (4 × 4) ... | (18) | − 11/16 | 5/16 | 15/16 | − 3/4 | 9/16 | 15/16 | 1/8 (out) | 0 | 1/8 (in) | 3.2 (out) | 0 | 3.2 (in) | 30 | 39 | 7 15/32 |
| 80-77 Series 10,9,8,6 (4 × 2) . | (10) | − 13/16 | 3/16 | 13/16 | − 1/4 | 1/2 | 1 1/4 | 1/8 (out) | 0 | 1/8 (in) | 3.2 (out) | 0 | 3.2 (in) | 30 | 39 | 7 1/2 |
| 76-73 Series 5,4,3,2, (4 × 2) . | (10) | − 11/16 | 5/16 | 15/16 | 1/4 | 1 | 1 3/4 | 0 | 1/8 | 1/4 | 0 | 3.2 | 6.4 | NA | NA | 7 |

## LUV SUSPENSION HEIGHT TABLES

| Model & Year | Measurement at | Front Suspension Inches | Front Suspension MM | Rear Suspension Inches | Rear Suspension MM |
|---|---|---|---|---|---|
| 82-81 (4 × 4) | | 5 | 127 | 7 11/16 | 195 |
| 82-81 (4 × 2) | Base Model, Soft Ride | 4 | 102 | 6 1/8 | 155 |
| | Base Model, Cab & Chassis Comp. | 4 | 102 | 7 1/2 | 190 |
| | Cab & Chassis | 4 | 102 | 8 5/16 | 210 |
| | Long Wheel Base Model | 4 | 102 | 7 1/2 | 190 |
| 80-79 4 × 4 | | 4 13/16 | 122 | 7 11/16 | 195 |
| 80-78 4 × 2 | Base Model, Soft Ride | 4 5/8 | 116.8 | 6 1/8 | 155 |
| | Base Model, Cab & Chassis Comp. | 4 5/8 | 116.8 | 7 1/2 | 190 |
| | Cab & Chassis | 4 5/8 | 116.8 | 8 5/16 | 210 |
| | Long Wheel Base Model | 4 5/8 | 116.8 | 7 1/2 | 190 |
| 77-76 4 × 2 | at Curb | 4 5/8 | 116.8 | 6 | 152.4 |
| | at G.V.W. | 3 5/8 | 92.0 | 5 1/32 | 127.8 |
| 75-73 4 × 2 | at Curb | 4 5/8 | 116.8 | 7 13/32 | 188.0 |
| | at G.V.W. | 3 5/8 | 92.0 | 5 1/32 | 127.8 |

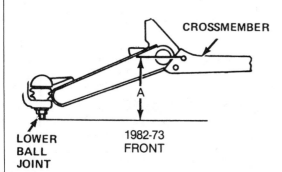

CROSSMEMBER

A

LOWER BALL JOINT

1982-73 FRONT

EDGE OF FRAME

TOP OF AXLE

C

REAR

## TRIM HEIGHTS

| VEHICLE IDENTIFICATION (YEAR MODEL) | ADJ. ILL. NO. | CASTER MIN. | CASTER PREF. | CASTER MAX. | CAMBER MIN. | CAMBER PREF. | CAMBER MAX. | TOE-IN (In.) MIN. | TOE-IN (In.) PREF. | TOE-IN (In.) MAX. | TOE-IN (mm) MIN. | TOE-IN (mm) PREF. | TOE-IN (mm) MAX. | TOE-OUT OUTSIDE WHEEL | TOE-OUT INSIDE WHEEL | STRG. AXIS INCL. (DEG.) |
|---|---|---|---|---|---|---|---|---|---|---|---|---|---|---|---|---|
| **DATSUN** | | | | | | | | | | | | | | | | |
| 82 Nissan Stanza | N/A | | | | | | | | | | | | | | | |
|   Front | | 1 1/16 | 2 3/16 | | − 3/4 | | 3/4 | 0 | | 3/32 | 0.0 | | 2.4 | 18 1/2 | 20 | 14 13/32 |
|   Rear | | | | | 0 | | 1 1/2 | 7/32 (out) | | 5/16 (in) | 5.6 (out) | | 7.9 (in) | | | |
| 82 810: Front | N/A | 2 15/16 | 4 7/16 | | − 5/16 | | 1 3/16 | 1/32 (out) | | 1/32 (in) | 1.0 (out) | | 1.0 (in) | 18 11/16 | 20 | 12 5/16 |
|   Rear | N/A | | | | 15/16 | | 2 7/16 | 1/8 (out) | | 9/32 (in) | 3.2 (out) | | 7.1 (in) | | | |
| 81 810: Front | | 2 15/16 | 4 7/16 | | − 5/16 | | 1 3/16 | 1/32 (out) | | 1/32 (in) | 1.0 (out) | | 1.0 (in) | 18 11/16 | 20 | 12 3/16 |
|   Rear | N/A | | | | | | | 5/32 (out) | | 9/16 (in) | 4.0 (out) | | 14.3 (in) | | | |
| 81-79 510 except Wagon | N/A | 1 1/16 | 2 9/16 | | − 1/4 | | 1 1/4 | 1/32 | | 1/8 | 1.6 | | 3.2 | 19 1/2 | 20 | 8 27/32 |
| 81-78 510 Wagon | N/A | 15/16 | 2 7/16 | | 1/16 | | 1 9/16 | 1/32 | | 1/8 | 1.6 | | 3.2 | 19 1/2 | 20 | 8 5/32 |
| 82-80 200 SX | N/A | 1 3/4 | 3 1/4 | | − 11/16 | | 13/16 | 0 | | 3/32 | 0 | | 2.4 | 18 11/16 | 20 | 8 5/32 |
| 80-77 810: Front | N/A | 1 3/16 | 2 11/16 | | 0 | | 1 1/2 | 0 | | 3/32 | 0 | | 2.4 | 18 29/32 | 20 | 7 29/32 |
|   Rear | N/A | | | | | | | 5/32 | | 9/16 | 4.0 | | 14.3 | | | |
| 82-79 210 except Wagon and Canadian | N/A | 1 11/16 | 3 7/16 | | 0 | | 1 1/2 | 3/64 | | 1/8 | 1.2 | | 3.2 | 19 19/64 | 20 | 8 19/32 |
| 82-79 210 Wagon | N/A | 1 15/16 | 3 7/16 | | 0 | | 1 1/2 | 3/64 | | 1/8 | 1.2 | | 3.2 | 19 19/64 | 20 | 8 19/32 |
| 82-80 210 Canadian 1.2 Litre Mod. | N/A | 1 11/16 | 3 3/16 | | − 1/4 | | 1 3/32 | 0 | | 3/32 | 0 | | 2.4 | 19 19/64 | 20 | 8 13/32 |
| 82-80 280 ZX | N/A | | | | | | | | | | | | | | | |
|   *W/Power Steering* Front | | 4 3/16 | 5 11/16 | | − 9/16 | | 15/16 | 1/16 | | 1/8 | 1.6 | | 3.2 | 18 3/64 | 20 | 9 11/32 |
|   *W/Manual Steering* Front | | 4 3/16 | 5 11/16 | | − 9/16 | | 15/16 | 1/16 | | 1/8 | 1.6 | | 3.2 | 18 45/64 | 20 | 9 11/32 |
|   Rear, All | N/A | | | | − 3/32 | | 1 1/4 | 3/32 | ◆ | 5/32 | 2.4 | ◆ | 4.0 | | | |
| 82 720 Pickup Truck (4 × 2) | (18) | 13/16 | 1 13/16 | | 0 | | 1 | 7/32 | | 9/32 | 5.1 | | 7.1 | 18 | 20 | 9 |
| 82-81 720 Pickup Truck (4 × 4) | (18) | 13/16 | 2 3/16 | | 0 | | 1 | 7/32 | | 9/32 | 5.6 | | 7.1 | 18 | 18 1/2 | 11 |
| 81-80 720 Pickup Truck | (18) | 13/16 | 1 13/16 | | 0 | | 1 | 7/32 | | 9/32 | 5.6 | | 7.1 | 18 | 20 | 9 |

| VEHICLE IDENTIFICATION<br>YEAR MODEL | ADJ. ILL. NO. | CASTER (Degrees) | | | CAMBER (Degrees) | | | TOE-IN (Inches) | | | TOE-IN (Millimeters) | | | TOE-OUT ON TURNS (Degrees) | | STRG. AXIS INCL. |
|---|---|---|---|---|---|---|---|---|---|---|---|---|---|---|---|---|
| | | MIN. | PREF. | MAX. | MIN. | PREF. | MAX. | MIN. | PREF. | MAX. | MIN. | PREF. | MAX. | OUTSIDE WHEEL | INSIDE WHEEL | (DEG.) |
| 82 310: Front | N/A | $7/16$ | | $1^{15}/16$ | $1/4$ | | $1\,3/4$ | 0 | | $3/32$ | 0.0 | | 2.4 | 19■ | 20 | $11^{27}/32$ |
| Rear | N/A | | | | $-\,1/4$ | | $1\,3/4$ | 0 | 0 | 0 | 0.0 | | 0.0 | | | |
| 81-79 310 | N/A | $7/16$ | | $1^{15}/16$ | $1/4$ | | $1\,3/4$ | 0 | | $3/32$ | 0 | | 2.4 | $18^{13}/32$ | 20 | $11^{27}/32$ |
| 79 280 ZX | N/A | | | | | | | | | | | | | | | |
| *W/Power Steering* | | | | | | | | | | | | | | | | |
| Front | | 4 | | $5\,1/2$ | $-\,1/2$ | | 1 | $1/16$ | | $1/8$ | 1.6 | | 3.2 | $26\,1/2$ | 34 • | $9^{11}/32$ |
| *W/Manual Steering* | | | | | | | | | | | | | | | | |
| Front | | 4 | | $5\,1/2$ | $-\,1/2$ | | 1 | $1/16$ | | $1/8$ | 1.6 | | 3.2 | 31 | $35\,1/2$ • | $9^{11}/32$ |
| Rear, All | N/A | | | | $-\,29/32$ | | $1\,1/4$ | $1/32$ | | $1/8$ | 0.8 | | 3.1 | | | |
| 79 Pick up Truck | (18) | $1/2$ | | 2 | $-\,1/4$ | | $1\,1/4$ | $7/32$ | | $9/32$ | 5.6 | | 7.1 | $30\,1/2$ | 35 + | NA |
| 79-78 200 SX | N/A | $1\,1/16$ | | $2\,9/16$ | $5/16$ | | $1^{13}/16$ | $3/32$ | | $5/32$ | 2.4 | | 4.0 | 30 | 35 • | $7^{13}/16$ |
| 78 620 Pick up Truck | (18) | $9/16$ | | $2\,1/16$ | $-\,1/4$ | | $1\,1/4$ | 0 | | $9/32$ | 0 | | 7.1 | $30\,1/2$ | 35 + | NA |
| 78 F10 W/Radial Tires | N/A | $1/4$ | | $1\,3/4$ | $13/16$ | | $2\,1/4$ | 0 | | $3/32$ | 0 | | 2.4 | $32\,1/2$ | $36\,1/2$ | 10 |
| W/Bias Tires | N/A | | | | | | | $1/8$ | | $7/32$ | 3.2 | | 5.6 | $32\,1/2$ | $36\,1/2$ | 10 |
| 78-75 B210 | N/A | 1 | | $2\,1/2$ | $7/16$ | | $1^{15}/16$ | $3/32$ | | $5/32$ | 2.4 | | 4.0 | 32 | 38 + | $8\,1/4$ |
| 78-75 280Z except (2 + 2) | N/A | $2\,1/16$ | | $3^7/16$ | $5/16$ | | $1^{13}/16$ | 0 | | $1/8$ | 0 | | 3.2 | 33 | 34 | 12 |
| Rear | N/A | | | | $-\,3/32$ | | $1\,1/2$ | $7/32$ (out) | | $7/32$ (in) | 5.6 (out) | | 5.6 (in) | | | |
| 78-75 280Z (2 + 2) | N/A | $2\,1/16$ | | $3^7/16$ | $5/16$ | | $1^{13}/16$ | 0 | | $1/8$ | 0 | | 3.2 | $35\,1/2$ | $36\,3/4$ | 12 |
| Rear | N/A | | | | $-\,5/32$ | | $1^{11}/32$ | $7/32$ (out) | | $7/32$ (in) | 5.6 (out) | | 5.6 (in) | | | |
| 77 200 SX | N/A | $1^1/32$ | | $2\,1/4$ | $1/2$ | | $1\,1/2$ | $3/32$ | | $5/32$ | 2.4 | | 4.0 | 30 | 35 • | $7^{13}/16$ |
| 77 620 Pick up Truck | (18) | $1\,1/16$ | $1^{13}/16$ | $2^9/16$ | $1/4$ | $1\,1/4$ | $2\,1/4$ | $3/32$ | | $1/8$ | 2.4 | | 3.2 | 31 | 36 | $6\,1/4$ |
| 77 F10 W/Radial Tires | N/A | $5/16$ | | $1^{13}/16$ | $13/16$ | | $2^5/16$ | 0 | | $3/32$ | 0 | | 2.4 | $32\,1/2$ | $36\,1/2$ | 10 |
| W/Bias Tires | N/A | | | | | | | $7/32$ | | $9/32$ | 5.6 | | 7.1 | $32\,1/2$ | $36\,1/2$ | 10 |
| 77-76 710, 610 | | | | | | | | | | | | | | | | |
| W/Radial Tires | N/A | $1\,1/16$ | | $2^9/16$ | $1\,1/4$ | | $2\,3/4$ | $5/32$ | | $1/4$ | 4.0 | | 6.4 | $30\,1/2$ | $32\,1/2$ + | 7 |
| W/Bias Tires | N/A | | | | | | | $1/4$ | | $5/16$ | 6.4 | | 7.9 | $30\,1/2$ • | $32\,1/2$ + | 7 |
| 76-73 620 Pick up Truck | (18) | $1\,1/16$ | $1^{13}/16$ | $2^9/16$ | $1/4$ | $1\,1/4$ | $2\,1/4$ | $1/32$ | | $3/16$ | 0.8 | | 4.8 | 31 | 36 | $6\,1/4$ |
| 75 710 | N/A | $1^3/16$ | | $2^{11}/16$ | $1^7/16$ | | $2^{15}/16$ | $5/16$ | | $7/16$ | 7.9 | | 11.1 | $30\,1/2$ | $32\,1/2$ + | $6^{13}/32$ |
| 75-74 610 except Wagon | | | | | | | | | | | | | | | | |
| Front | N/A | $1\,1/4$ | | $2\,3/4$ | $1\,1/4$ | | $2\,3/4$ | $7/16$ | | $9/16$ | 11.1 | | 14.3 | $30\,1/2$ | $32\,1/2$ + | $6^{21}/32$ |
| Rear | N/A | | | | $3/4$ | | $2\,1/4$ | $1/8$ | | $1/2$ | 3.2 | | 13.0 | | | |
| 75-74 610 Wagon: Front | N/A | $1\,1/4$ | | $2\,3/4$ | $1\,1/2$ | | 3 | $7/16$ | | $9/16$ | 11.1 | | 14.3 | $30\,1/2$ | $32\,1/2$ + | $6^{21}/32$ |
| 75-74 260Z: Front | N/A | $2^3/16$ | $2^{15}/16$ | $3^{11}/16$ | 0 | $3/4$ | $1\,1/2$ | $3/32$ | | $3/16$ | 2.4 | | 4.8 | $31^{11}/16$ | 33 | $12^5/32$ |
| Rear | N/A | | | | $-\,1/2$ | $1/4$ | 1 | $13/32$ (out) | | $13/32$ (in) | 10.4 (out) | | 10.4 (in) | | | |
| 74 710 | N/A | $1^3/16$ | | $2^{11}/16$ | $1^7/16$ | | $2^{15}/16$ | $9/16$ | | $21/32$ | 14.2 | | 16.6 | $31^{27}/32$ • | $37\,1/2$ + | $6^{13}/32$ |
| 74-73 B210 | N/A | 1 | | $2\,1/4$ | $1\,1/16$ | | $1\,1/16$ | $3/32$ | | $5/32$ | 2.4 | | 4.0 | 32 | 38 + | $8\,1/4$ |
| 73 610 except Wagon | | | | | | | | | | | | | | | | |
| Front | N/A | $3/4$ | | $2\,1/4$ | 1 | | $2\,1/2$ | $1/4$ | | $11/32$ | 6.4 | | 8.7 | $31^{21}/32$ • | $37\,1/2$ + | $7^3/32$ |
| Rear | N/A | | | | $1\,3/4$ | | $2\,1/4$ | $1/8$ | | $1/2$ | 3.2 | | 12.7 | | | |
| 73 610 Wagon: Front | N/A | $15/16$ | | $2^7/16$ | $1^3/16$ | | $2^{11}/16$ | $5/16$ | | $7/16$ | 7.9 | | 11.1 | $31^{21}/32$ • | $37\,1/2$ + | $6^{29}/32$ |
| 73 1200 | N/A | $29/32$ | | $2^{13}/32$ | $11/32$ | | $1^{27}/32$ | $5/32$ | | $1/4$ | 4.0 | | 6.4 | 36 | 43 | $7^{29}/32$ |
| 73 240Z: Front | N/A | $2^3/16$ | $2^{15}/16$ | $3^{11}/16$ | $1/16$ | $13/16$ | $1^9/16$ | $1/16$ | | $7/32$ | 1.6 | | 5.6 | $31\,1/4$ • | $32\,1/2$ + | $12^5/32$ |
| Rear | N/A | | | | $-\,1/2$ | $1/4$ | 1 | $1/16$ (out) | | $1/2$ (in) | 1.6 (out) | | 12.7 (in) | | | |
| 73 PL-510 exc. Wagon | | | | | | | | | | | | | | | | |
| Front | N/A | | $1^9/16$ | | | $7/16$ | | $1/8$ | | $1/4$ | 3.2 | | 6.4 | $31^{21}/32$ | 38 | NA |
| Rear | N/A | | | | | $1^{27}/32$ | | $11/32$ | | $15/32$ | 8.7 | | 11.9 | | | |
| 73 WPL-510 Wagon Std. | N/A | | $1^{11}/16$ | | | $9/16$ | | $5/32$ | | $9/32$ | 4.0 | | 7.2 | $31^{21}/32$ | 38 | NA |
| 73 WPL-510 Wagon W/HD Suspension | N/A | | $1^{11}/16$ | | | $5/16$ | | $1/8$ | | $1/4$ | 3.2 | | 6.4 | $31^{21}/32$ | 38 | NA |

• Plus or minus 2° is considered acceptable.
+ Plus or minus 1° is considered acceptable.
▲ Plus or minus 5½° is considered acceptable.
■ Plus or minus 1½ is considered acceptable.

◆ 1982 280ZX rear toe 0 to $3/32$″, 0 to 2.0 MM

| DODGE (A) (B) | | | | | | | | | | | | | | | | |
|---|---|---|---|---|---|---|---|---|---|---|---|---|---|---|---|---|
| 82-81 Challenger | (19) | $2^3/16$ | $2^{11}/16$ | $3^3/16$ | $11/16$ | $13/16$ | $1^{11}/16$ | 0 | | $9/32$ | 0 | | 7.1 | 32 | 37 | $9\,1/2$ |
| 80 Challenger | (19) | $2^3/16$ | $2^{11}/16$ | $3^3/16$ | $3/4$ | $1\,1/4$ | $1\,3/4$ | $3/32$ | | $11/32$ | 2.4 | | 8.7 | 32 | 37 | $8^{27}/32$ |
| 80 Colt Station Wagon | (19) | $2^3/16$ | $2^{11}/16$ | $3\,1/2$ | $3/4$ | $1\,1/4$ | $1\,3/4$ | $3/32$ | | $11/32$ | 2.4 | | 8.7 | 32 | 37 | $8^{27}/32$ |
| 82-80 D-50 Pick up Truck | (10) | $1\,1/2$ | $2\,1/2$ | $3\,1/2$ | $1/2$ | 1 | $1\,1/2$ | $3/32$ | | $11/32$ | 2.4 | | 8.7 | $30\,1/2$ | 37 | 8 |
| 82-79 FWD Hatch Back Colt | N/A | $1/2$ | $13/16$ | $1\,1/8$ | 0 | $1/2$ | 1 | $3/32$ | | $5/32$ | 2.4 | | 4.0 | $29\,1/4$ | $35^{11}/16$ | $12\,3/4$ |
| | | | | | | | | (out) | | (in) | (out) | | (in) | | | |
| 79 D-50 Pick up Truck | (10) | 2 | 3 | 4 | $1/2$ | 1 | $1\,1/2$ | $3/32$ | | $11/32$ | 2.4 | | 8.7 | $30\,1/2$ | 37 | 8 |
| 79-78 Challenger | (19) | $2^3/16$ | $2^{11}/16$ | $3^3/16$ | 1 | $1\,1/2$ | 2 | $3/32$ | | $11/32$ | 2.4 | | 8.7 | | | |
| *W/Power Steering* | | | | | | | | | | | | | | 31 | 36 | $8^{13}/32$ |
| *W/Manual Steering* | | | | | | | | | | | | | | 32 | 37 | $8^{13}/32$ |

| VEHICLE IDENTIFICATION (YEAR MODEL) | ADJ. ILL. NO. | CASTER (Degrees) MIN. | PREF. | MAX. | CAMBER (Degrees) MIN. | PREF. | MAX. | TOE-IN (Inches) MIN. | PREF. | MAX. | TOE-IN (Millimeters) MIN. | PREF. | MAX. | TOE-OUT ON TURNS (Degrees) OUTSIDE WHEEL | INSIDE WHEEL | STRG. AXIS INCL. (DEG.) |
|---|---|---|---|---|---|---|---|---|---|---|---|---|---|---|---|---|
| 79-78 Colt except Wagon | (19) | 1 9/16 | 2 1/16 | 2 9/16 | 1/2 | 1 | 1 1/2 | 3/32 | | 1/4 | 2.4 | | 6.4 | 30 | 35 | 9 |
| 79 Colt Station Wagon | (19) | 2 3/16 | 2 11/16 | 3 3/16 | 1 | 1 1/2 | 2 | 3/32 | | 11/32 | 2.4 | | 8.7 | 30 1/2 | 35 | 8 27/32 |
| 78 Colt Station Wagon | (19) | 1 5/16 | 2 | 2 11/16 | 1/2 | 1 | 1 1/2 | 3/32 | | 11/32 | 2.4 | | 8.7 | 30 1/2 | 39 | 8 13/32 |
| 77 Colt Hard Top, GT, Station Wagon | (19) | 9/16 | 1 1/16 | 1 9/16 | 3/8 | 7/8 | 1 3/8 | 3/32 | | 1/4 | 2.4 | | 6.4 | 30 | 39 | 9 |
| 77 Colt Coupe & Sedan | (19) | 1 9/16 | 2 1/16 | 2 9/16 | 1/4 | 1 | 1 3/4 | 3/32 | | 1/4 | 2.4 | | 6.4 | 30 | 35 | 9 |
| 76-74 Colt All Models | (19) | 3/4 | 1 1/4 | 1 3/4 | 5/16 | 13/16 | 15/16 | 3/32 | | 1/4 | 2.4 | | 6.4 | 30 1/2 | 39 | 9 |
| 73 Colt All Models | N/A | 3/4 | 1 1/4 | 1 3/4 | 1/2 | 1 | 1 1/2 | 3/32 | | 1/4 | 2.4 | | 6.4 | 32 | 43 | 8 27/32 |

**(A)** Variation between wheels 1/2 °.
**(B)** Side to side variation 1/2 ° or less.

| VEHICLE IDENTIFICATION (YEAR MODEL) | ADJ. ILL. NO. | CASTER (Degrees) MIN. | PREF. | MAX. | CAMBER (Degrees) MIN. | PREF. | MAX. | TOE-IN (Inches) MIN. | PREF. | MAX. | TOE-IN (Millimeters) MIN. | PREF. | MAX. | TOE-OUT ON TURNS (Degrees) OUTSIDE WHEEL | INSIDE WHEEL | STRG. AXIS INCL. (DEG.) |
|---|---|---|---|---|---|---|---|---|---|---|---|---|---|---|---|---|
| **FIAT** | | | | | | | | | | | | | | | | |
| 82-80 Strada: Front | (21) | 1 1/2 | | 2 1/2 | 1 3/16 | | 2 3/16 | 3/16 (out) | | 1/8 (in) | 4.7 (out) | | 3.2 (in) | 31 3/4 | 35 1/2 | NA |
| Rear | (22) | | | | 1/2 | | 1 1/2 | 0 | | 5/32 | 0 | | 4.0 | | | |
| 82-80 X 1/9: Front | (20) | 6 5/16 | | 7 5/16 | −1 | | 0 | 1/16 | | 1/4 | 1.6 | | 6.4 | 28 | 32 11/16 | NA |
| Rear | | | | | −1 3/4 | | −3/4 | 5/32 | | 5/16 | 4.0 | | 7.9 | | | |
| 82-80 Spider 2000 | (23) | 2 11/16 | | 3 11/16 | −5/16 | | 11/16 | 5/32 | | 5/16 | 4.0 | | 7.9 | 28 1/2 | 35 13/16 | 6 |
| 81-77 Brava, 131 | (21) | 3 1/4 | | 4 1/4 | 7/16 | | 1 7/16 | 5/32 | | 5/16 | 4.0 | | 7.9 | 31 | 35 | NA |
| 79 Strada | (21) | 1 1/2 | 2 | 2 1/2 | 1 3/16 | 1 11/16 | 2 3/16 | 3/16 (out) | 1/8 (out) | 3/32 (out) | 4.8 (out) | 3.2 (out) | 2.4 (out) | 31 3/4 | 35 1/2 | NA |
| Rear | (22) | | | | 1/2 | 1 | 1 1/2 | 0 | 1/32 | 3/32 | 0 | 0.8 | 2.4 | | | |
| 79 124 Sport Spider | (23) | 2 11/16 | | 3 11/16 | 5/16 | | 11/16 | 5/32 | | 5/16 | 4.0 | | 7.9 | NA | NA | NA |
| 79-78 128 Sedan & Wagon Front | (21) | 1 1/4 | | 2 1/4 | 1/2 | | 1 1/2 | 1/32 | | 1/8 | 0.8 | | 3.2 | 31 3/4 | 35 | NA |
| Rear | (22) | | | | −3 3/4 | −3 1/4 | −2 3/4 | 3/32 | | 1/4 | 2.4 | | 6.4 | | | |
| 79-75 128-3P | (20) | 1 3/4 | 2 1/4 | 2 3/4 | −1/4 | −3/4 | −1 1/4 | 3/32 (out) | 0 | 3/32 (in) | 2.4 (out) | 0 | 2.4 (in) | 31 | 35 1/4 | NA |
| Rear | (22) | | | | −4 | −3 1/2 | −3 | 1/16 | 5/32 | 1/4 | 1.6 | 4.0 | 6.4 | | | |
| 79 X 1/9 | (20) | 6 1/2 | 7 | 7 1/2 | −1 | | 0 | 5/64 | | 15/64 | 2.0 | | 6.0 | NA | NA | NA |
| Rear | N/A | | | | −2 1/4 | −1 3/4 | −1 1/4 | 5/32 | 1/4 | 5/16 | 4.0 | 6.4 | 8.0 | | | |
| 78-75 X 1/9: Front | (20) | 6 1/2 | 7 | 7 1/2 | −15/16 | −1 | −5/8 | 1/32 | | 3/16 | 0.8 | | 4.8 | 28 | 32 11/16 | NA |
| Rear | N/A | | | | −2 5/16 | −2 | −1 5/8 | 5/32 | | 1/4 | 4.0 | | 6.4 | | | |
| 78-75 124 Sport Coupe/Spider | (23) | 3 | 3 1/2 | 4 | 0 | 1/2 | 1 | 3/32 | 1/8 | 5/32 | 2.4 | 3.2 | 4.0 | 28 1/2 | 35 13/16 | NA |
| 77-75 128 Sedan & Wagon Front | (21) | 1 3/4 | 2 1/4 | 2 3/4 | 1 | 1 1/2 | 2 | 3/32 (out) | 0 | 3/32 (in) | 2.4 (out) | 0 | 2.4 (in) | 31 3/4 | 35 | NA |
| Rear | (22) | | | | −3 3/4 | −3 1/4 | −2 3/4 | 3/32 | | 1/4 | 2.4 | | 6.4 | | | |
| 76-75 131 | (21) | 4 | | 5 | 0 | | 1 | 3/32 | | 5/32 | 2.4 | | 4.0 | 31 | 35 | 6 |
| 74 128 SL: Front | (21) | 1 3/4 | 2 1/4 | 2 3/4 | 1/2 | 1 | 1 1/2 | 3/32 (out) | 0 | 3/32 (in) | 2.4 (out) | 0 | 2.4 (in) | 31 | 35 1/4 | NA |
| Rear | (22) | | | | −4 | −3 1/2 | −3 | 1/16 | 5/32 | 1/4 | 1.6 | 3.2 | 6.4 | | | |
| 74 X 1/9 | (20) | 6 1/2 | 7 | 7 1/2 | −15/16 | −1 | −5/8 | 1/32 | | 3/16 | 0.8 | | 4.8 | 28 | 32 11/16 | NA |
| Rear | N/A | | | | −2 5/16 | −2 | −1 5/8 | 11/32 | | 7/16 | 8.7 | | 11.1 | | | |
| 74 124 Sport Coupe/Spider | (23) | 3 | 3 1/2 | 4 | 0 | 1/2 | 1 | 1/32 | | 7/32 | 0.8 | | 5.6 | NA | NA | 6 |
| 74 128 Sedan & Wagon Front | (21) | 1 3/4 | 2 1/4 | 2 3/4 | 1 1/2 | 2 | 2 1/2 | 3/32 (out) | 0 | 3/32 (in) | 2.4 (out) | 0 | 2.4 (in) | 31 3/4 | 35 | NA |
| Rear | (22) | | | | −3 1/2 | −3 | −2 1/2 | 3/32 | | 1/4 | 2.4 | | 6.4 | | | |
| 74-73 124 Wagon | (23) | 3 | 3 1/2 | 4 | 0 | 1/2 | 1 | 1/32 | | 3/16 | 0.8 | | 4.8 | NA | NA | NA |
| 73 128 SL: Front | (21) | 1 3/4 | 2 1/4 | 2 3/4 | 1 1/2 | 2 | 2 1/2 | 3/32 (out) | 0 | 3/32 (in) | 2.4 (out) | 0 | 2.4 (in) | 31 3/4 | 35 | NA |
| Rear | (22) | | | | 2 1/2 | 3 | 3 1/2 | 1/8 | | 9/32 | 3.2 | | 7.1 | | | |
| 73 124 Sport Coupe/Spider | (23) | 3 5/16 | | 4 | 3/16 | 1/2 | 13/16 | 1/32 | | 1/8 | 0.8 | | 3.2 | NA | NA | NA |
| 73 128 Sedan & Wagon Front | (21) | 1 3/4 | 2 1/4 | 2 3/4 | 1 1/2 | 2 | 2 1/2 | 3/32 (out) | 0 | 3/32 (in) | 2.4 (out) | 0 | 2.4 (in) | 31 3/4 | 35 | NA |
| Rear | (22) | | | | 2 1/2 | 3 | 3 1/2 | 1/8 | | 9/32 | 3.2 | | 7.1 | | | |
| 73 850: Front | (10) | 8 | 9 | 10 | 1 1/4 | | 1 1/2 | 5/32 | | 1/4 | 4.0 | | 6.4 | NA | NA | 4 11/32 |
| Rear | | | | | 1 1/4 | | 1 1/2 | 5/32 | | 15/64 | 4.0 | | 6.0 | | | |
| **FORD MOTOR CO.** | | | | | | | | | | | | | | | | |
| 80-78 Fiesta | N/A | − 7/16 | 5/16 | 15/16 | 1 1/4 | 2 1/4 | 3 1/4 | 15/64 (out) | 13/64 (out) | 5/32 (out) | 6.0 (out) | 2.4 (out) | 4.0 (out) | NA | NA | NA |
| 82-77 Courier | (24) | 3/4 | | 1 1/4 | 1/2 | | 1 1/4 | 0 | | 1/4 | 0 | | 6.4 | 30 11/16 • | 32 1/2 | NA |
| 76-73 Courier | (24) | 3/4 | | 1 1/4 | 1 | | 1 3/4 | 0 | | 1/4 | 0 | | 6.4 | 32 1/2 | 36 | NA |

• 600-14 tire shown; 195 SR-14 tire: Outer wheel 30 11/16 °, Inner wheel 33 15/16 °.

| VEHICLE IDENTIFICATION | ADJ. ILL. | CASTER (Degrees) | | | CAMBER (Degrees) | | | TOE-IN (Inches) | | | TOE-IN (Millimeters) | | | TOE-OUT ON TURNS (Degrees) | | STRG. AXIS INCL. |
| YEAR   MODEL | NO. | MIN. | PREF. | MAX. | MIN. | PREF. | MAX. | MIN. | PREF. | MAX. | MIN. | PREF. | MAX. | OUTSIDE WHEEL | INSIDE WHEEL | (DEG.) |
|---|---|---|---|---|---|---|---|---|---|---|---|---|---|---|---|---|
| **HONDA** | | | | | | | | | | | | | | | | |
| 82 Accord: Front ....... | N/A | 7/16 | 1 7/16 | 2 7/16 | − 1 | 0 | 1 | 1/8 (out) | 0 | 1/8 (in) | 3.2 (out) | 0.0 | 3.2 (in) | NA | NA | 12 1/2 |
| Rear ......... | (45) | | | | | 0 | | 3/64 (out) | 1/32 (in) | 7/64 (in) | 1.2 (out) | 0.8 (in) | 2.8 (in) | | | |
| 81 Accord: Front ....... | N/A | 11/16 | 1 11/16 | 2 11/16 | − 11/16 | 5/16 | 15/16 | 1/8 (out) | 3/64 (out) | 1/32 (in) | 3.2 (out) | 1.2 (out) | 0.8 (in) | NA | NA | 12 1/2 |
| Rear ......... | (45) | | | | | 3/16 | | 1/32 (out) | 5/64 (in) | 1/8 (in) | 0.8 (out) | 1.2 (in) | 3.2 (in) | | | |
| 82-79 Prelude: Front ...... | N/A | 1/2 | 1 1/2 | 2 1/2 | − 1 | 0 | 1 | 1/8 (out) | 0 | 1/8 (in) | 3.2 | 0 | 3.2 | NA | NA | 12 13/16 |
| Rear ...... | (45) | | | | | 0 | | 1/16 | 1/8 | 7/32 | 1.6 | 3.2 | 5.6 | | | |
| 80-79 Accord: Front ...... | N/A | 3/4 | 1 1/4 | 1 3/4 | 0 | 1/2 | 1 | 5/32 (out) | 1/32 (out) | 3/32 (in) | 4.0 (out) | 0.8 (out) | 2.4 (in) | NA | NA | 12 3/16 |
| Rear ...... | (45) | | | | − 1/4 | 1/4 | 3/4 | 1/8 (out) | 1/32 (out) | 1/16 (out) | 3.2 (out) | 0.8 (out) | 1.6 (out) | | | |
| 82 Civic except Wagon.... Front ............ | N/A | 1 1/2 | 2 1/2 | 3 1/2 | − 1 | 0 | 1 | 1/8 (out) | 0 | 1/8 (in) | 3.2 (out) | 0.0 | 3.2 (in) | NA | NA | 12 11/32 |
| Sedan: Rear ........ | N/A | | | | | − 1/2 | | 0 | 5/64 | 5/32 | 0 | 2.0 | 4.0 | | | |
| Hatchback: Rear .... | N/A | | | | | − 1/4 | | 0 | 5/64 | 5/32 | 0 | 2.0 | 4.0 | | | |
| 82 Civic Wagon ........ | N/A | 5/16 | 1 5/16 | 2 5/16 | − 1 | 0 | 1 | 1/8 (out) | 0 | 1/8 (in) | 3.2 (out) | 0 | 3.2 (in) | NA | NA | 12 11/32 |
| 81-80 Civic except Wagon: Front .......... | N/A | 3/4 | 1 3/4 | 2 3/4 | − 1 | 0 | 1 | 1/8 (out) | 0 | 1/8 (in) | 3.2 (out) | 0 | 3.2 (in) | NA | NA | 12 5/16 |
| Rear .......... | N/A | | | | − 1 | 0 | 1 | 5/32 (out) | 0 | (in) | 4.0 (out) | 0 | (in) | | | |
| 81-80 Civic Wagon ....... | N/A | 0 | 1 | 2 | − 1 | 0 | 1 | 1/8 (out) | 0 | 1/8 (in) | 3.2 (out) | 0 | 3.2 (in) | NA | NA | 12 5/16 |
| 79-78 Civic: Front ....... | N/A | 1/4 | 3/4 | 1 1/4 | 0 | 1/2 | 1 | 1/8 (out) | 3/64 (out) | 3/64 (in) | 3.2 (out) | 1.2 (out) | 1.2 (in) | NA | NA | 9 5/16 |
| Rear ........ | N/A | | | | 0 | 1/2 | 1 | | 0 | | | 0 | | | | |
| 79 Civic CVCC except Wagon: Front .......... | N/A | 1/4 | 3/4 | 1 1/4 | 0 | 1/2 | 1 | 1/8 (out) | 3/64 (out) | 3/64 (in) | 3.2 (out) | 1.2 (out) | 1.2 (in) | NA | NA | 9 5/16 |
| Rear ......... | N/A | | | | 0 | 1/2 | 1 | | 0 | | | 0 | | | | |
| 79 Civic CVCC Wagon .... | N/A | 0 | 1/2 | 1 | 0 | 1/2 | 1 | 1/8 (out) | 3/64 (out) | 3/64 (in) | 3.2 (out) | 1.2 (out) | 1.2 (in) | NA | NA | 9 5/16 |
| 78 Civic CVCC except Wagon: Front .......... | N/A | 0 | 1/2 | 1 | 0 | 1/2 | 1 | 1/8 (out) | 1/32 (out) | 1/16 (in) | 3.2 (out) | 0.8 (out) | 1.6 (in) | NA | NA | 9 5/16 |
| Rear .......... | N/A | | | | 0 | 1/2 | 1 | 3/32 (out) | 0 | 3/32 (in) | 2.4 (out) | 0 | 2.4 (in) | | | |
| 78 Civic CVCC Wagon .... | N/A | 0 | 1/2 | 1 | 0 | 1/2 | 1 | 1/32 (out) | 1/32 (in) | 3/32 (in) | 0.8 (out) | 0.8 (in) | 2.4 (in) | NA | NA | 9 5/16 |
| 78 Accord: Front ....... | N/A | 1 | 2 | 3 | − 1/4 | 3/4 | 1 3/4 | 5/32 (out) | 3/64 (out) | 5/64 (in) | 4.0 (out) | 1.2 (out) | 2.0 (in) | NA | NA | 12 3/16 |
| Rear ........ | (45) | | | | − 1/4 | 1/4 | 3/4 | 5/64 (out) | 0 | 5/64 (out) | 2.0 (out) | 0 | 2.0 (in) | NA | NA | |
| 77-73 Civic 1200: Front .... | N/A | 3/4 | 1 3/4 | 2 3/4 | 0 | 1/2 | 1 | 5/32 (out) | 3/64 (out) | 5/64 (in) | 4.0 (out) | 1.2 (out) | 2.0 (in) | NA | NA | 8 15/16 |
| Rear..... | N/A | | | | | 1/2 | | 3/32 (out) | 0 | 3/32 (in) | 2.4 (out) | 0 | 2.4 (in) | | | |
| 77 Civic CVCC Sedan..... Front ............. | N/A | 1 | 1 1/2 | 2 | 0 | 1/2 | 1 | | 3/64 | | | 1.2 | | NA | NA | 9 5/16 |
| Rear ............. | | | | | 0 | 1/2 | 1 | | 0 | | | 0 | | | | |
| 77 Civic CVCC Wagon..... Front ............. | N/A | 1/2 | 1 | 1 1/2 | 0 | 1/2 | 1 | | 3/64 | | | 1.2 | | NA | NA | 9 15/16 |
| 77-76 Accord: Front ...... | N/A | 1 | 2 | 3 | − 1/2 | 1/2 | 1 1/2 | 5/32 (out) | 3/64 (out) | 5/64 (in) | 4.0 (out) | 1.2 (out) | 2.0 (in) | NA | NA | 12 3/16 |
| Rear...... | (45) | | | | | 5/16 | | 1/8 (out) | 3/64 (out) | 3/64 (in) | 3.2 (out) | 1.2 (out) | 1.2 (in) | NA | NA | |
| 76 Civic CVCC Sedan..... Front ............. | N/A | 1 1/2 | 2 | 2 1/2 | 0 | 1/2 | 1 | 1/16 (out) | 1/32 (out) | 0 | 1.6 (out) | 0.8 (out) | 0 | NA | NA | 9 15/16 |
| Rear ............. | N/A | | | | 0 | 1/2 | 1 | 1/32 (out) | 0 | 1/32 (in) | 0.8 (out) | 0 | 0.8 (in) | NA | NA | |

105

| Vehicle Identification (Year / Model) | Adj. Ill. No. | Caster (Deg.) Min. | Pref. | Max. | Camber (Deg.) Min. | Pref. | Max. | Toe-In (In.) Min. | Pref. | Max. | Toe-In (mm) Min. | Pref. | Max. | Toe-Out on Turns (Deg.) Outside Wheel | Inside Wheel | Strg. Axis Incl. (Deg.) |
|---|---|---|---|---|---|---|---|---|---|---|---|---|---|---|---|---|
| **ISUZU** | | | | | | | | | | | | | | | | |
| 82-81 I-Mark | (13) | 3 11/16 | 5 3/16 | 6 3/16 | − 7/8 | 1/8 | 5/8 | 5/64 | 1/8 | 11/64 | 2.0 | 3.0 | 4.0 | 35 | 37 1/2 | 7 7/8 |
| 82-81 Pick up Truck (4 × 2) | (68)(19) | 0 | 1/2 | 1 | 0 | 1/2 | 1 | 0 | 1/16 | 1/8 | 0.0 | 2.0 | 4.0 | 33 | 37 | 7 1/2 |
| 82-81 Pick up Truck (4 × 4) | (18) | − 3/16 | 5/16 | 13/16 | 1/16 | 9/16 | 1 1/16 | 1/16 | 0 | 1/16 | 2.0 | 0.0 | 2.0 | 33 | 35 | 7 1/2 |
| | | | | | | | | (out) | | (in) | (out) | | (in) | | | |
| **JAGUAR** | | | | | | | | | | | | | | | | |
| 80-75 XJ 12: Front | (49) | 3 1/4 | 3 1/2 | 3 3/4 | 1/4 | 1/2 | 3/4 | 1/16 (out) | | 0 (out) | 1.6 (out) | | 0 (out) | NA | NA | 1 1/2 |
| Rear | (51) | | | | − 1 | − 3/4 | − 1/2 | 1/32 (out) | 0 | 1/32 (in) | 0.8 (out) | 0 | 0.8 (in) | | | |
| 82-75 XJ 6: Front | (49) | 2 | 2 1/4 | 2 1/2 | 1/4 | 1/2 | 3/4 | 1/16 (out) | | 1/8 (in) | 1.6 (out) | | 3.2 (in) | NA | NA | 1 1/2 |
| Rear | (51) | | | | − 1 | − 3/4 | − 1/2 | 1/32 (out) | 0 | 1/32 (in) | 0.8 (out) | 0 | 0.8 (in) | | | |
| 82-79 XJS: Front | (49) | 3 1/4 | 3 1/2 | 3 3/4 | 1/4 | 1/2 | 3/4 | 1/16 (out) | | 0 (out) | 1.6 (out) | | 0 (out) | NA | NA | 1 1/2 |
| Rear | (51) | | | | − 1 | − 3/4 | − 1/2 | 1/32 (out) | 0 | 1/32 (in) | 0.8 (out) | 0 | 0.8 (in) | | | |
| 78-75 XJS: Front | (49) | 3 1/4 | 3 1/2 | 3 3/4 | 1/4 | 1/2 | 3/4 | 0 | | 1/8 | 0 | | 3.2 | NA | NA | 1 1/2 |
| Rear | (51) | | | | − 1 | − 3/4 | − 1/2 | 1/32 (out) | 0 | 1/32 (in) | 1.6 (out) | 0 | 1.6 (in) | | | |
| 74-73 XJ 12; XJ 6: Front | (49) | 2 | 2 1/4 | 2 1/2 | 1/4 | 1/2 | 3/4 | 1/16 | | 1/8 | 1.6 | | 3.2 | NA | NA | 1 1/2 |
| Rear | (51) | | | | − 1 | − 3/4 | − 1/2 | | | | | | | | | |
| 75-73 E Type Series 3 | | | | | | | | | | | | | | | | |
| Front | (24)(15) | 2 | 2 1/2 | 3 | 0 | 1/4 | 1/2 | 1/16 | | 1/8 | 1.6 | | 3.2 | NA | NA | NA |
| Rear | (51) | | | | − 1 | − 3/4 | − 1/2 | | | | | | | | | |
| **LANCIA** | | | | | | | | | | | | | | | | |
| 82-80 Beta, Zagato: Front | N/A | 1 3/16 | | 1 13/16 | 3/16 | | 13/16 | 1/8 (out) | | 0 (out) | 3.2 (out) | | 0 (out) | 30 | 36 | NA |
| Rear | N/A | | | | − 1 5/16 | | − 11/16 | 1/32 | | 5/32 | 1.6 | | 4.0 | | | |
| 79-77 Beta, HPE, Zagato | | | | | | | | | | | | | | | | |
| Front | N/A | 1 5/32 | 1 1/2 | 1 27/32 | 5/32 | 1/2 | 27/32 | 5/32 (out) | | 0 (out) | 4.0 (out) | | 0 (out) | 33 1/2 | 36 | 9 3/4 |
| Rear | N/A | | | | − 1 11/32 | − 1 | − 21/32 | 1/64 | | 3/32 | 0.4 | | 4.0 | | | |
| 77-75 Scorpion: Front | N/A | 5 3/4 | 6 | 6 1/4 | − 5/16 | 0 | 5/16 | 5/64 | | 5/32 | 2.0 | | 4.0 | 32 | 32 | 13 1/2 |
| Rear | N/A | | | | 1 11/16 | 2 | 2 5/16 | 5/32 | | 5/16 | 4.0 | | 8.0 | | | |
| 76-75 Beta All Models | | | | | | | | | | | | | | | | |
| Front | N/A | 1 5/32 | 1 1/2 | 1 27/32 | 5/32 | 1/2 | 13/16 | 0 | | 1/8 | 0 | | 3.2 | 31 1/2 | 36 | 9 3/4 |
| Rear | N/A | | | | − 1 5/16 | − 1 | − 11/16 | 1/32 | | 5/32 | 0.8 | | 4.0 | | | |

Add weight to vehicles before performing alignment: Scorpion requires 375 lbs.; Beta Coupe requires 495 lbs.; Beta HPE & Sedan requires 660 lbs.; Zagato requires 495 lbs.

| Vehicle Identification (Year / Model) | Adj. Ill. No. | Caster (Deg.) Min. | Pref. | Max. | Camber (Deg.) Min. | Pref. | Max. | Toe-In (In.) Min. | Pref. | Max. | Toe-In (mm) Min. | Pref. | Max. | Toe-Out on Turns (Deg.) Outside Wheel | Inside Wheel | Strg. Axis Incl. (Deg.) |
|---|---|---|---|---|---|---|---|---|---|---|---|---|---|---|---|---|
| **MAZDA** | | | | | | | | | | | | | | | | |
| 82-81 RX-7 | | | | | 1/2 | 1 | 1 1/2 | 0 | | 1/4 | 0 | | 6.4 | 32 1/4 | 39 11/16 | 10 3/4 |
| Right Wheel | (26) | 3 11/16 | 4 3/16 | 4 11/16 | | | | | | | | | | | | |
| Left Wheel | (26) | 3 3/16 | 3 11/16 | 4 3/16 | | | | | | | | | | | | |
| 82-81 626 | (26) | 2 15/16 | 3 7/16 | 3 15/16 | 3/4 | 1 1/4 | 1 3/4 | 0 | | 1/4 | 0 | | 6.4 | 33 13/16 | 40 | 10 9/16 |
| 82-81 GLC except Station Wagon | N/A | | 1 29/32 | | | 29/32 | | 1/8 (out) | | 1/8 (in) | 3.2 (out) | | 3.2 (in) | NA | NA | 12 5/32 |
| 82-81 GLC Station Wagon | N/A | 3/4 | 1 1/2 | 2 1/4 | 3/4 | 1 1/4 | 1 3/4 | 0 | | 1/4 | 0 | | 6.4 | 31 13/16 | 42 1/2 | 8 1/4 |
| 82-79 B-2000, B2200 | (24) | 11/16 | 1 | 1 5/16 | 7/16 | 3/4 | 1 1/4 | 0 | 1/8 | 1/4 | 0 | 3.2 | 6.4 | 30 11/16 | 32 1/2 | 8 1/4 |
| 80 RX-7 | | | | | 11/16 | 13/16 | 1 11/16 | 0 | 1/8 | 1/4 | 0 | 3.2 | 6.4 | 32 1/4 | 39 11/16 | 10 3/4 |
| Right Wheel | (26) | 4 | 4 1/2 | 5 | | | | | | | | | | | | |
| Left Wheel | (26) | 3 1/2 | 4 | 4 1/2 | | | | | | | | | | | | |
| 80 626 | (26) | | | | 3/4 | 1 1/4 | 1 3/4 | 0 | 1/8 | 1/4 | 0 | 3.2 | 6.4 | 33 13/16 | 40 | 10 11/16 |
| Right Wheel | | 2 29/32 | 3 21/32 | 4 13/32 | | | | | | | | | | | | |
| Left Wheel | | 2 7/16 | 3 3/16 | 3 15/16 | | | | | | | | | | | | |
| 79 RX-7 | (26) | 3 1/4 | 4 | 4 3/4 | 11/16 | 13/16 | 1 11/16 | 0 | 1/8 | 1/4 | 0 | 3.2 | 6.4 | 32 1/4 | 39 11/16 | 10 3/4 |
| 79 626 | (26) | | 3 3/4 | | | 1 1/4 | | 0 | 1/8 | 1/4 | 0 | 3.2 | 6.4 | 33 13/16 | 40 | 10 11/16 |
| 80-79 GLC exc. Wagon | | 15/16 | 1 11/16 | 2 7/16 | 1/4 | 3/4 | 1 1/4 | 0 | 1/8 | 1/4 | 0 | 3.2 | 6.4 | 31 | 42 1/2 | 8 3/4 |
| 80-79 GLC Wagon | N/A | 1 | 1 3/4 | 2 1/2 | 1/2 | 1 | 1 1/2 | 0 | 1/8 | 1/4 | 0 | 3.2 | 6.4 | 31 15/16 | 42 1/2 | 8 1/2 |
| 78 RX-4 Sedan | (26) | 1 1/16 | 1 13/16 | 2 9/16 | 0 | 1 | 2 | 0 | 1/8 | 1/4 | 0 | 3.2 | 6.4 | 30 1/2 | 41 11/16 | 9 3/4 |
| Wagon | (26) | 1 1/16 | 1 13/16 | 2 9/16 | 1/4 | 1 1/4 | 2 1/4 | 0 | 1/8 | 1/4 | 0 | 3.2 | 6.4 | 30 1/2 | 41 11/16 | 9 3/4 |
| 78-77 RX-3 SP | N/A | 1 7/16 | 2 3/16 | 3 15/16 | 1/2 | 1 1/16 | 2 1/16 | 0 | 1/8 | 1/4 | 0 | 3.2 | 6.4 | 32 | 43 | 8 13/16 |
| 78-77 GLC | N/A | 3/4 | 1 9/16 | 2 5/16 | − 5/16 | 11/16 | 1 11/16 | 0 | 1/8 | 1/4 | 0 | 3.2 | 6.4 | 31 | 43 | 8 13/16 |
| 78-77 B-1800 Pick up | (24) | 11/16 | 1 | 1 5/16 | 7/16 | 3/4 | 1 1/4 | 0 | 1/8 | 1/4 | 0 | 3.2 | 6.4 | 30 11/16 | 32 1/2 | 8 1/4 |
| 78-76 Cosmo | | | | | | | | | | | | | | | | |
| W/Manual Steering | (26) | 1 1/16 | 1 13/16 | 2 5/16 | 0 | 1 | 2 | 0 | 1/8 | 1/4 | 0 | 3.2 | 6.4 | 30 1/2 | 41 11/16 | 9 3/4 |
| W/Power Steering | (26) | 1 1/2 | 2 1/4 | 3 | 0 | 1 | 2 | 0 | | 1/4 | 0 | 3.2 | 6.4 | 30 1/2 | 41 11/16 | 9 3/4 |

| VEHICLE IDENTIFICATION (YEAR MODEL) | ADJ. ILL. NO. | CASTER (Degrees) MIN. | PREF. | MAX. | CAMBER (Degrees) MIN. | PREF. | MAX. | TOE-IN (Inches) MIN. | PREF. | MAX. | TOE-IN (Millimeters) MIN. | PREF. | MAX. | TOE-OUT ON TURNS (Degrees) OUTSIDE WHEEL | INSIDE WHEEL | STRG. AXIS INCL. (DEG.) |
|---|---|---|---|---|---|---|---|---|---|---|---|---|---|---|---|---|
| 77 808 (1600) Sedan | N/A | 1 1/16 | 1 13/16 | 2 9/16 | 1/16 | 1 1/16 | 2 1/16 | 0 | 1/8 | 1/4 | 0 | 3.2 | 6.4 | 31 1/2 | 41 | 8 7/16 |
| Coupe | N/A | 1 5/16 | 2 1/16 | 2 13/16 | 1/16 | 1 1/16 | 2 1/16 | 0 | 1/8 | 1/4 | 0 | 3.2 | 6.4 | 31 1/2 | 43 | 8 7/16 |
| Wagon | N/A | 1 1/16 | 1 13/16 | 2 9/16 | 1/4 | 1 1/4 | 2 1/4 | 0 | 1/8 | 1/4 | 0 | 3.2 | 6.4 | 31 1/2 | 43 | 8 1/4 |
| 77-76 Mizer (808-1300) | | | | | | | | | | | | | | | | |
| Sedan | N/A | 1 1/16 | 1 7/16 | 2 3/16 | − 3/16 | 13/16 | 1 13/16 | 0 | 1/8 | 1/4 | 0 | 3.2 | 6.4 | 32 11/16 | 44 1/2 | 8 11/16 |
| Coupe | N/A | 1 | 1 3/4 | 2 1/2 | − 1/16 | 15/16 | 1 15/16 | 0 | 1/8 | 1/4 | 0 | 3.2 | 6.4 | 32 11/16 | 44 1/2 | 8 11/16 |
| Wagon | N/A | 1 1/16 | 1 7/16 | 2 3/16 | 0 | 1 | 2 | 0 | 1/8 | 1/4 | 0 | 3.2 | 6.4 | 32 11/16 | 44 1/2 | 8 9/16 |
| 77-76 Rotary Pick up | (24) | 1 5/8 | 1 15/16 | 2 1/4 | − 1/16 | 1/4 | 9/16 | 0 | 1/8 | 1/4 | 0 | 3.2 | 6.4 | 32 9/16 | 33 5/16 | 8 3/4 |
| 77-76 RX-4 | | | | | | | | | | | | | | | | |
| Sedan & Hard Top | (26) | 1 1/16 | 1 13/16 | 2 9/16 | 0 | 1 | 2 | 0 | 1/8 | 1/4 | 0 | 3.2 | 6.4 | 30 1/2 | 41 11/16 | 9 1/2 |
| Wagon (77 only) | (26) | 1 1/16 | 1 13/16 | 2 9/16 | 1/4 | 1 1/4 | 2 1/4 | 0 | 1/8 | 1/4 | 0 | 3.2 | 6.4 | 30 1/2 | 41 11/16 | 9 1/2 |
| 76 808 (1300) Sedan | N/A | 1 1/16 | 1 7/16 | 2 3/16 | − 3/16 | 13/16 | 1 13/16 | 0 | 1/8 | 1/4 | 0 | 3.2 | 6.4 | 32 11/16 | 44 1/2 | 8 11/16 |
| Coupe | N/A | 15/16 | 1 11/16 | 2 7/16 | − 3/16 | 13/16 | 1 13/16 | 0 | 1/8 | 1/4 | 0 | 3.2 | 6.4 | 32 11/16 | 44 1/2 | 8 11/16 |
| Wagon | N/A | 15/16 | 1 11/16 | 2 7/16 | − 3/16 | 13/16 | 1 13/16 | 0 | 1/8 | 1/4 | 0 | 3.2 | 6.4 | 32 11/16 | 44 1/2 | 8 9/16 |
| 76 808 (1600) Sedan | N/A | 13/16 | 1 9/16 | 2 5/16 | 0 | 1 | 2 | 0 | 1/8 | 1/4 | 0 | 3.2 | 6.4 | 31 1/2 | 43 | 8 7/16 |
| Coupe | N/A | 1 1/16 | 1 13/16 | 2 9/16 | 0 | 1 | 2 | 0 | 1/8 | 1/4 | 0 | 3.2 | 6.4 | 31 1/2 | 43 | 8 7/16 |
| Wagon | N/A | 1 | 1 3/4 | 2 1/2 | 0 | 1 | 2 | 0 | 1/8 | 1/4 | 0 | 3.2 | 6.4 | 31 1/2 | 43 | 8 1/4 |
| 76-75 RX-3 Sedan | N/A | 15/16 | 1 11/16 | 1 13/16 | − 3/16 | 13/16 | 1 13/16 | 0 | 1/8 | 1/4 | 0 | 3.2 | 6.4 | 32 9/16 | 44 9/16 | 8 11/16 |
| Coupe | N/A | 1 13/16 | 1 15/16 | 2 11/16 | − 3/16 | 13/16 | 1 13/16 | 0 | 1/8 | 1/4 | 0 | 3.2 | 6.4 | 32 9/16 | 44 9/16 | 8 11/16 |
| Wagon | N/A | 13/16 | 1 9/16 | 2 5/16 | − 1/16 | 15/16 | 1 15/16 | 0 | 1/8 | 1/4 | 0 | 3.2 | 6.4 | 32 9/16 | 44 9/16 | 8 1/2 |
| 76-73 B-1600 Pick up | N/A | 11/16 | 1 | 1 5/16 | 1 1/8 | 1 7/16 | 1 3/4 | 0 | 1/8 | 1/4 | 0 | 3.2 | 6.4 | 33 1/4 | 35 5/16 | 7 9/16 |
| 75 808 (1600) | N/A | 11/16 | 1 7/16 | 2 3/16 | − 1/4 | 3/4 | 1 3/4 | 0 | 1/8 | 1/4 | 0 | 3.2 | 6.4 | 33 9/16 | 41 1/2 | 8 3/4 |
| 75 Rotary Pick up | (24) | 7/8 | 1 3/16 | 1 1/2 | − 1/16 | 1/4 | 9/16 | 0 | 1/8 | 1/4 | 0 | 3.2 | 6.4 | 32 9/16 | 33 5/16 | 8 3/4 |
| 75-74 RX-4 | (26) | 1 1/4 | 2 | 2 3/4 | 0 | 1 | 2 | 0 | 1/8 | 1/4 | 0 | 3.2 | 6.4 | 29 3/16 | 37 3/16 | 9 3/4 |
| 74 Rotary Pick up | (24) | 1 5/8 | 1 15/16 | 2 1/4 | − 5/16 | 11/16 | 1 11/16 | 0 | 1/8 | 1/4 | 0 | 3.2 | 6.4 | 33 9/16 | 41 1/2 | 8 3/4 |
| 74-73 808 (1600) | N/A | 1/2 | 1 1/4 | 2 | − 5/16 | 11/16 | 1 11/16 | 0 | 1/8 | 1/4 | 0 | 3.2 | 6.4 | NA | NA | 8 3/4 |
| 74-73 RX-3 | N/A | 1/2 | 1 1/4 | 2 | − 5/16 | 11/16 | 1 11/16 | 0 | 1/8 | 1/4 | 0 | 3.2 | 6.4 | NA | NA | 8 3/4 |
| 74-73 RX-2 | N/A | 1/4 | 1 | 1 3/4 | − 1/2 | 1/2 | 1 1/2 | 0 | 1/8 | 1/4 | 0 | 3.2 | 6.4 | 31 | 43 | 8 3/4 |

Maximum variation between wheel, caster 1 1/16°, camber 1/2°.

| VEHICLE IDENTIFICATION (YEAR MODEL) | ADJ. ILL. NO. | CASTER (Degrees) MIN. | PREF. | MAX. | CAMBER (Degrees) MIN. | PREF. | MAX. | TOE-IN (Inches) MIN. | PREF. | MAX. | TOE-IN (Millimeters) MIN. | PREF. | MAX. | TOE-OUT ON TURNS (Degrees) OUTSIDE WHEEL | INSIDE WHEEL | STRG. AXIS INCL. (DEG.) |
|---|---|---|---|---|---|---|---|---|---|---|---|---|---|---|---|---|
| **MERCEDES BENZ** | | | | | | | | | | | | | | | | |
| 82-81 300SD, 380SEL, 380SEC, 380SL, SLC | (54) | 9 1/4 | 9 3/4 | 10 1/4 | − 3/16 | 0 | 3/16 | 9/64 | 5/32 | 7/32 | 3.5 | 4.0 | 5.5 | NA | NA | NA |
| 80-78 300SD | (54) | 9 1/2 | | 10 1/2 | | − 5/16 | | 3/32 | | 5/32 | 2.4 | | 4.0 | 19 | 20 | NA |
| 82-77 300D, 300CD, 300TD | (53) | 8 1/4 | 8 3/4 | 9 1/4 | − 1/16 | 0 | 1/16 | 3/32 | | 5/32 | 2.4 | | 4.0 | 18 13/16 | 20 | NA |
| 81-77 280E, 280CE | (53) | 8 1/4 | 8 3/4 | 9 1/4 | − 3/16 | 0 | 3/16 | 3/32 | | 5/32 | 2.4 | | 4.0 | 18 13/16 | 20 | NA |
| 80-77 280SE | (54) | 9 1/2 | | 10 1/2 | | − 5/16 | | 3/32 | | 5/32 | 2.4 | | 4.0 | 19 | 20 | NA |
| 82-77 240D | (53) | 8 1/4 | 8 3/4 | 9 1/4 | − 1/16 | 0 | 1/16 | 3/32 | | 5/32 | 2.4 | | 4.0 | 18 13/16 | 20 | NA |
| 80-73 450SL, 450SLC | (52) | 3 3/8 | 3 11/16 | 4 | − 5/16 | 0 | 5/16 | 1/32 | | 1/8 | 0.8 | | 3.2 | 19 3/4 | 20 | NA |
| 80-73 450SE, 450SEL | (54) | 9 1/2 | | 10 1/2 | | − 5/16 | | 3/32 | | 5/32 | 2.4 | | 4.0 | 19 | 20 | NA |
| 79-78 6.9 | (54) | 9 1/2 | | 10 1/2 | | − 5/16 | | 3/32 | | 5/32 | 2.4 | | 4.0 | 19 | 20 | NA |
| 79-77 230 | (53) | 8 1/4 | 8 3/4 | 9 1/4 | − 1/16 | 0 | 1/16 | 3/32 | | 5/32 | 2.4 | | 4.0 | 18 13/16 | 20 | NA |
| 76-75 280S | (54) | 9 1/2 | | 10 1/2 | | − 5/16 | | 3/32 | | 5/32 | 2.4 | | 4.0 | 19 | 20 | NA |
| 76-75 300D | | | | | | | | | | | | | | | | |
| W/Power Steering | (52) | 3 3/8 | 3 11/16 | 4 | − 1/16 | 1/4 | 7/16 | 3/32 | | 5/32 | 2.4 | | 4.0 | 19 1/4 | 20 | NA |
| W/Manual Steering | (52) | 2 3/8 | 2 11/16 | 3 | − 1/16 | 1/4 | 7/16 | 3/32 | | 5/32 | 2.4 | | 4.0 | 19 1/4 | 20 | NA |
| 76-74 240D, 230 | | | | | | | | | | | | | | | | |
| W/Power Steering | (52) | 2 3/8 | 3 11/16 | 4 | − 1/16 | 1/4 | 7/16 | 3/32 | | 5/32 | 2.4 | | 4.0 | 19 1/4 | 20 | NA |
| W/Manual Steering | (52) | 2 3/8 | 2 11/16 | 3 | − 1/16 | 1/4 | 7/16 | 3/32 | | 5/32 | 2.4 | | 4.0 | 19 1/4 | 20 | NA |
| 76-73 280, 280C | (52) | 3 7/16 | 3 11/16 | 3 15/16 | − 1/16 | 1/4 | 7/16 | 1/32 | | 1/8 | 0.8 | | 3.2 | 19 1/2 | 20 | NA |
| 73 220/8, 220D/8 | | | | | | | | | | | | | | | | |
| W/Power Steering | (27)(28) | 3 3/4 | 4 | 4 1/4 | − 1/16 | 0 | 7/16 | 3/32 | | 5/32 | 2.4 | | 4.0 | 19 1/4 | 20 | NA |
| W/Manual Steering | (27)(28) | 3 1/4 | 3 1/2 | 3 3/4 | − 1/16 | 0 | 7/16 | 3/32 | | 5/32 | 2.4 | | 4.0 | 19 1/4 | 20 | NA |
| 73 300SEL 4.5 | (27)(28) | 3 3/4 | 4 | 4 1/4 | 0 | 5/16 | 5/16 | 1/32 | | 5/32 | 0.8 | | 4.0 | 19 1/2 | 20 | NA |
| 73 280SE, 280SEL 4.5 | | | | | | | | | | | | | | | | |
| W/Power Steering | (27)(28) | 3 3/4 | 4 | 4 1/4 | 3/16 | 1/2 | 1/2 | 1/32 | | 1/8 | 0.8 | | 3.2 | 19 1/2 | 20 | NA |
| W/Manual Steering | (27)(28) | 3 1/4 | 3 1/2 | 3 3/4 | 3/16 | 1/2 | 1/2 | 1/32 | | 1/8 | 0.8 | | 3.2 | 19 | 20 | NA |

(A) Toe setting taken at wheel rim.

| VEHICLE IDENTIFICATION (YEAR MODEL) | ADJ. ILL. NO. | CASTER (Degrees) MIN. | PREF. | MAX. | CAMBER (Degrees) MIN. | PREF. | MAX. | TOE-IN (Inches) MIN. | PREF. | MAX. | TOE-IN (Millimeters) MIN. | PREF. | MAX. | TOE-OUT ON TURNS (Degrees) OUTSIDE WHEEL | INSIDE WHEEL | STRG. AXIS INCL. (DEG.) |
|---|---|---|---|---|---|---|---|---|---|---|---|---|---|---|---|---|
| **MG** | | | | | | | | | | | | | | | | |
| 81-73 MGB, GT | N/A | 5 | | 7 1/2 | − 1/4 | 1 | 1 1/4 | 1/16 | | 3/32 | 1.6 | | 2.4 | 19 | 20 | 8 |
| 79-73 Midget | N/A | | 3 | | | | 3/4 | 0 | | 1/8 | 0 | | 3.2 | 19 1/4 | 20 | 6 3/4 |
| **NISSAN** | | | | | | | | | | | | | | | | |
| 82 Stanza | | | | | | | | | | | | | | | | |
| Front | N/A | 1 1/16 | | 2 3/16 | − 3/4 | | 3/4 | 0 | | 3/32 | 0.0 | | 2.4 | 18 1/2 | 20 | 14 13/32 |
| Rear | | | | | 0 | | 1 1/2 | 7/32 (out) | | 5/16 (in) | 5.6 (out) | | 7.9 (in) | | | |

| VEHICLE IDENTIFICATION | ADJ. ILL. NO. | CASTER (Degrees) | | | CAMBER (Degrees) | | | TOE-IN (Inches) | | | TOE-IN (Millimeters) | | | TOE-OUT ON TURNS (Degrees) | | STRG. AXIS INCL. |
| YEAR   MODEL | | MIN. | PREF. | MAX. | MIN. | PREF. | MAX. | MIN. | PREF. | MAX. | MIN. | PREF. | MAX. | OUTSIDE WHEEL | INSIDE WHEEL | (DEG.) |
|---|---|---|---|---|---|---|---|---|---|---|---|---|---|---|---|---|
| **PEUGEOT** | | | | | | | | | | | | | | | | |
| 82-80  505: Front . . . . . . . . | N/A | 3 | 3½ | 4 | 0 | ¾ | 1½ | ¼ | 5/16 | 25/64 | 6.0 | 8.0 | 10.0 | NA | NA | 9 1/16 |
| Rear . . . . . . . . | N/A | | | | − 15/16 | − 7/16 | − 1/16 | 7/64 | 9/32 | 21/64 | 3.0 | 7.0 | 9.0 | | | |
| 82-77  604: Front . . . . . . . | N/A | 3 | 3½ | 4 | − ¼ | ½ | 1¼ | 5/32 | 7/32- | 5/16 | 4.0 | 5.6 | 8.0 | NA | NA | 10 |
| Rear . . . . . . . . | N/A | | | | − 2 | − 1½ | − 1 | 5/64 | 5/32 | 5/16 | 2.0 | 4.0 | 8.0 | | | |
| 82-73  504 . . . . . . . . . . . | N/A | 2 5/32 | 2 21/32 | 3 5/32 | ⅛ | ⅝ | 1⅛ | 5/32 | 7/32 | 5/16 | 4.0 | 5.6 | 8.0 | NA | NA | 8 29/32 |
| **PLYMOUTH (A) (B)** | | | | | | | | | | | | | | | | |
| 82-81  Sapporo . . . . . . . . . | (19) | 2 3/16 | 2 11/16 | 3 3/16 | 11/16 | 1 3/16 | 1 11/16 | 0 | | 9/32 | 0 | | 7.1 | 32 | 37 | 9½ |
| 80  Sapporo . . . . . . . . . . . | (19) | 2 3/16 | 2 11/16 | 3 3/16 | ¾ | 1¼ | 1¾ | 5/64 | | 11/32 | 2.0 | | 8.7 | 32 | 37 | 8 27/32 |
| 82-80  Arrow Pickup Truck. . | (10) | 1½ | 2½ | 3½ | ½ | 1 | 1½ | 3/32 | | 11/32 | 2.4 | | 8.7 | 30½ | 37 | 8 |
| 82-79  Champ F.W.D. . . . . . | N/A | ½ | 13/16 | 1⅛ | 0 | ½ | 1 | 3/32 | | 5/32 | 2.4 | | 4.0 | 29¼ | 35 11/16 | 12 21/32 |
| | | | | | | | | (out) | | (in) | (out) | | (in) | | | |
| 80-78  Arrow . . . . . . . . . . | (19) | 1 9/16 | 2 1/16 | 2 9/16 | ½ | 1 | 1½ | 3/32 | | ¼ | 2.4 | | 6.4 | 30 | 35 | 9 |
| 79-78  Sapporo . . . . . . . . . | (19) | 2 3/16 | 2 11/16 | 3 3/16 | 1 | 1½ | 2 | 3/32 | | 11/32 | 2.4 | | 8.7 | | | |
| W/Power Steering . . . . | | | | | | | | | | | | | | 31 | 36 | 8 13/32 |
| W/Manual Steering . . . . | | | | | | | | | | | | | | 32 | 37 | 8 13/32 |
| 79  Arrow Pickup Truck . . . . | (10) | 2 | 3 | 4 | ½ | 1 | 1½ | 3/32 | | 11/32 | 2.4 | | 8.7 | 30½ | 37 | 8 |
| 77  Arrow . . . . . . . . . . . . | (19) | 1 9/16 | 2 1/16 | 2 9/16 | ¼ | 1 | 1¾ | 3/32 | | ¼ | 2.4 | | 6.4 | 30 | 35 | 8 27/32 |
| 76  Arrow . . . . . . . . . . . . | (19) | 1¼ | 2 1/16 | 2¼ | ½ | 1 | 1½ | 3/32 | | ¼ | 2.4 | | 6.4 | 30 | 35 | 8 27/32 |
| 73  Cricket . . . . . . . . . . . | N/A | ½ | 1 | 1½ | ¼ | 1 | 1¾ | 3/64 | 5/32 | 7/32 | 1.2 | | 5.6 | 20 | 20 | 11 |

**(A)** Variation between wheel ½ °.
**(B)** Side to side variation ½ ° or less.

| **PORSCHE (A)** | | | | | | | | | | | | | | | | |
|---|---|---|---|---|---|---|---|---|---|---|---|---|---|---|---|---|
| 82  924, 924 Turbo . . . . . . . | | | | | | | | | | | | | | | | |
| Front . . . . . . . . . . | (29) | 2¼ | 2½ | 3 | − 9/16 | − 5/16 | − 1/16 | ⅛ | 0 | 1/32 | 3.2 | 0 | 0.8 | 19 | 20 | NA |
| | | | | | | | | (out) | | (in) | (out) | | (in) | | | |
| Rear . . . . . . . . . . | (30) | | | | -1 11/16 | − 1 | − 1/16 | 1/32 | 0 | 1/32 | 0.8 | 0 | 0.8 | | | |
| | | | | | | | | (out) | | (in) | (out) | | (in) | | | |
| | | | | | | | EACH WHEEL | | | | | | | | | |
| 81  924, 924 Turbo . . . . . . . | (29) | 2¼ | 2¾ | 3¼ | − 9/16 | − 5/16 | − 1/16 | ⅛ | 0 | 1/32 | 3.2 | 0 | 0.8 | 19 | 20 | NA |
| | | | | | | | | (out) | | (in) | (out) | | (in) | | | |
| Rear Std. Chassis . . . . | (30) | | | | − 1½ | − 1 | − ½ | 1/32 | 0 | 1/32 | 3.2 | 0 | 3.2 | | | |
| | | | | | | | | (out) | | (in) | (out) | | (in) | | | |
| Rear Turbo Chassis . . . . | (30) | | | | − 29/32 | − 13/32 | 3/32 | 1/32 | 0 | 1/32 | 3.2 | 0 | 3.2 | | | |
| | | | | | | | | (out) | | (in) | (out) | | (in) | | | |
| 80-79  924, 924 Turbo . . . . | (29) | 2¼ | 2¾ | 3¼ | − 9/16 | − 5/16 | 1/16 | ⅛ | 0 | 1/32 | 3.2 | 0 | 0.8 | 19 | 20 | NA |
| | | | | | | | | (out) | | (in) | (out) | | (in) | | | |
| Rear Std. Chassis . . . . | (30) | | | | − 1½ | − 1 | − ½ | 1/32 | 0 | 1/32 | 0.8 | 0 | 0.8 | | | |
| | | | | | | | | (out) | | (in) | (out) | | (in) | | | |
| Rear Turbo Chassis . . . . | (30) | | | | − 1½ | − 1 | − ½ | | ⅛ | | | 3.2 | | | | |
| | | | | | | | | | (out) | | | (out) | | | | |
| 82  928: Front . . . . . . . . . | (43) | 3 | 3½ | 4 | − 11/16 | − ½ | − 5/16 | 1/32 | 0 | 1/32 | 0.8 | 0 | 0.8 | 19 | 20 | NA |
| | | | | | | | | (out) | | (in) | (out) | | (in) | | | |
| Rear . . . . . . . . . | (44) | | | | − ⅞ | − 11/16 | − ½ | 1/16 | 3/32 | ⅛ | 1.6 | 2.4 | 3.2 | | | |
| 81-78  928 . . . . . . . . . . . . | (43) | 3¼ | 3½ | 3¾ | − 11/16 | − ½ | − 5/16 | 1/32 | 0 | 1/32 | 0.8 | 0 | 0.8 | 19 | 20 | NA |
| | | | | | | | | (out) | | (in) | (out) | | (in) | | | |
| Rear . . . . . . . . . | (44) | | | | − ⅞ | − 11/16 | − ½ | 1/16 | 3/32 | ⅛ | 1.6 | 2.4 | 3.2 | | | |
| 82-78  911 SC . . . . . . . . . . | (31) | 5 13/16 | 6 1/16 | 6 5/16 | 5/16 | ½ | 11/16 | | 0 | | | 0 | | 19¾ | 20 | 10 15/16 |
| Rear . . . . . . . . . | (32) | | | | − 3/16 | 0 | 3/16 | 0 | 3/32 | 3/16 | 0 | 2.4 | 4.8 | | | |
| 79-77  Turbo Carreara . . . . . | (31) | 5 13/16 | 6 1/16 | 6 5/16 | 5/16 | ½ | 11/16 | | 0 | | | 0 | | 19¾ | 20 | NA |
| Rear . . . . . . . . . | (32) | | | | − ¼ | 0 | ¼ | 0 | 3/32 | 3/16 | 0 | 2.4 | 4.8 | | | |
| 78-76  924 . . . . . . . . . . . . | (29) | 2¼ | 2¾ | 3¼ | − ½ | − 5/16 | ⅛ | 1/32 | 0 | 1/32 | 0.8 | 0 | 0.8 | 19 | 20 | NA |
| | | | | | | | | (out) | | (in) | (out) | | (in) | | | |
| Rear . . . . . . . . . | (30) | | | | − 1½ | − 1 | − ½ | 3/32 | 0 | 3/32 | 2.4 | 0 | 2.4 | | | |
| | | | | | | | | (out) | | (in) | (out) | | (in) | | | |
| 77-75  911 . . . . . . . . . . . . | (31) | 5 13/16 | 6 1/16 | 6 5/16 | − 3/16 | 0 | 3/16 | | 0 | | | 0 | | 19¾ | 20 | 10 15/16 |
| Rear . . . . . . . . . | (32) | | | | 5/16 | ½ | 11/16 | 5/32 | | 5/32 | 4.0 | | 4.0 | | | |
| | | | | | | | | (out) | | (in) | (out) | | (in) | | | |
| 76  Turbo Carreara . . . . . . . | (31) | 5½ | | 6 | 5/16 | ½ | 11/16 | | 0 | | | 0 | | 19¾ | 20 | NA |
| Rear . . . . . . . . . | (32) | | | | − ¼ | 0 | ¼ | 0 | 3/32 | 3/16 | 0 | 2.4 | 4.8 | | | |
| 76  912 . . . . . . . . . . . . . | (31) | 5 13/16 | 6 1/16 | 6 5/16 | 5/16 | ½ | 11/16 | | 0 | | | 0 | | 19¾ | 20 | NA |
| Rear . . . . . . . . . | (32) | | | | | 0 | | 5/32 | | 5/32 | 4.0 | | 4.0 | | | |
| | | | | | | | | (out) | | (in) | (out) | | (in) | | | |
| 75-73  914 . . . . . . . . . . . . | (31) | 5½ | 6 | 6½ | − 5/16 | 0 | 5/16 | | 0 | | | 0 | | 19¾ | 20 | NA |
| Rear . . . . . . . . . . | (32) | | | | − 13/16 | − ½ | − 3/16 | 1/16 | 0 | 1/16 | 1.6 | 0 | 1.6 | | | |
| | | | | | | | | (out) | | (in) | (out) | | (in) | | | |

**(A)** Toe is set with wheels pressed with 33 lbs.

| VEHICLE IDENTIFICATION YEAR MODEL | ADJ. ILL. NO. | CASTER (Degrees) MIN. | PREF. | MAX. | CAMBER (Degrees) MIN. | PREF. | MAX. | TOE-IN (Inches) MIN. | PREF. | MAX. | TOE-IN (Millimeters) MIN. | PREF. | MAX. | TOE-OUT ON TURNS (Degrees) OUTSIDE WHEEL | INSIDE WHEEL | STRG. AXIS INCL. (DEG.) |
|---|---|---|---|---|---|---|---|---|---|---|---|---|---|---|---|---|
| **RENAULT** | | | | | | | | | | | | | | | | |
| 82-79 (late) Le Car: Front... | (41) | 10 | | 13 | 0 | 1/2 | 1 | 0 | 3/64 | 3/32 | 0 | 1.2 | 2.4 | NA | NA | NA |
| Rear ... | N/A | | | | 0 | | 1 1/2 | 0 | | 5/32 | 0 | | 4.0 | NA | NA | NA |
| 82-81 R-18, Fuego ........ | N/A | | | | – 1/2 | 0 | 1/2 | 0 | 3/64 | 3/32 | 0 | 1.2 | 2.4 | NA | NA | 13 1/2 |
| *W/Power Steering* ..... | | 1 1/2 | | 3 | | | | | | | | | | | | |
| *W/Manual Steering* .... | | 1/2 | | 2 | | | | | | | | | | | | |
| 79 (early) -77 Le Car (5 Series): Front ......... | (41) | 10 | | 13 | 0 | 1/2 | 1 | 3/16 (out) | | 3/64 (out) | 4.8 (out) | | 1.2 (out) | NA | NA | 15 1/2 |
| Rear .......... | N/A | | | | 0 | | 1/2 | 0 | | 5/32 | 0 | | 4.0 | | | |
| 79-75 12, 15, 17 & Gordini .. Front ............ | (23) | 4 | 4 | 5 | 1 | 1 1/2 | 2 | 5/32 (out) | | 3/64 (out) | 4.0 (out) | | 1.2 (out) | NA | NA | 18 3/4 |
| Rear ............. | | | | | 0 | | 1/2 | 1/16 (out) | | 0 | 1.6 (out) | | 0 | | | |
| 76 Le Car (5 Series)....... | (41) | 10 | | 13 | 0 | 1/2 | 1 | 3/16 (out) | | 3/64 (out) | 4.8 (out) | | 1.2 (out) | NA | NA | 14 11/16 |
| 74-73 12, 15, 17 & Gordini . Front ............ | (23) | | 4 | | 1 | 1 1/2 | 2 | 1/8 (out) | | 0 | 3.2 (out) | | 0 | NA | NA | NA |
| Rear ............. | | | | | 0 | | 1/2 | 1/16 (out) | | 0 | 1.6 (out) | | 0 | | | |
| **SAAB** | | | | | | | | | | | | | | | | |
| 82-81 99, 900S, Turbo, EMS Front ............ | | | | | 0 | 1/2 | 1 | 3/64 | 5/64• | 1/8 | 1.0 | 2.0• | 3.0 | | | |
| *W/Power Steering* ..... | (10) | 1 1/2 | 2 | 2 1/2 | | | | | | | | | | 20 | 20 3/4 | 11 1/2 |
| *W/Manual Steering* .... | (10) | 1/2 | 1 | 1 1/2 | | | | | | | | | | 20 | 20 3/4 | 11 1/2 |
| Rear ............. | | | | | – 3/4 | – 1/2 | – 1/4 | 1/8 | | 15/64 | 2.0 | | 6.0 | | | |
| 80-75 99, 900 & Turbo .... | | | | | | | | | | | | | | | | |
| *W/Power Steering* ..... | (10) | 1 1/2 | 2 | 2 1/2 | 0 | 1/2 | 1 | 3/64 | 3/32 | 1/8 | 1.2 | 2.4 | 3.2 | 20 | 20 3/4 | 11 1/2 |
| *W/Manual Steering* .... | (10) | 1/2 | 1 | 1 1/2 | 0 | 1/2 | 1 | 3/64 (out) | 0 | 3/64 (in) | 1.2 (out) | 0 | 2.0 (in) | 20 | 20 3/4 | 11 1/2 |
| 74-73 99 Series ......... | (10) | 1/2 | 3/4 | 1 | 1/2 | 3/4 | 1 | 3/64 (out) | 0 | 3/64 (in) | 1.2 (out) | 0 | 2.0 (in) | 20 | 20 1/2 | 11 1/2 |
| 73 95, 96 ............. | (10) | 1 1/2 | 2 | 2 1/2 | 1/2 | 3/4 | 1 | 1/32 | 5/64 | 1/8 | 0.8 | 2.0 | 3.2 | 20 | 22 1/2 | 7 |
| 73 Sonnett (Type 97)...... | (10) | 1 1/2 | 2 | 2 1/2 | – 1/4 | 0 | 1/4 | 0 | 1/32 | 5/64 | 0 | 0.8 | 2.0 | 20 | 22 1/2 | 7 |
| **SUBARU** | | | | | | | | | | | | | | | | |
| 82 exc. 4 W.D., all 2 W.D. models exc. Sta. Wagons . | | | | | | | | | | | | | | | | |
| Front ............ | N/A | – 1 3/16 | | 5/16 | 3/4 | | 2 1/4 | 0 | 3/64 | 5/64 | 0.0 | 1.0 | 2.0 | 35 | 36 1/2 | NA |
| Rear ............. | N/A | | | | – 3/4 | | 3/4 | 1/8 (out) | 0 | 1/8 (in) | 3.2 (out) | | 3.2 (in) | | | |
| 81-80 exc. 4 W.D., all 2 W.D. models exc. Sta. Wagons . | | | | | | | | | | | | | | | | |
| Front ............ | N/A | – 1 3/16 | – 7/16 | 5/16 | 3/4 | 1 1/2 | 2 1/4 | 1/16 | | 5/16 | 2 | | 8 | 35 | 36 1/2 | NA |
| Rear ............. | N/A | | | | – 3/4 | 0 | 3/4 | 1/8 (out) | | 1/8 (in) | 3 (out) | | 3 (in) | | | |
| 82 Station Wagon except 4 W.D.: Front...... | N/A | – 13/16 | | 11/16 | 1 | | 2 1/2 | 0 | 3/64 | 5/64 | 0.0 | 1.0 | 2.0 | 35 | 36 1/2 | NA |
| Rear ...... | N/A | | | | – 3/4 | | 3/4 | 1/8 (out) | 0 | 1/8 (in) | 3.2 (out) | | 3.2 (in) | | | |
| 81-80 Sta. Wagon exc. 4 W.D. : Front ....... | N/A | – 13/16 | – 1/16 | 11/16 | 1 | 1 3/4 | 2 1/2 | 1/16 | | 5/16 | 2 | | 8 | 35 | 36 1/2 | NA |
| Rear ............. | N/A | | | | – 3/4 | 0 | 3/4 | 1/8 (out) | | 1/8 (in) | 3 (out) | | 3 (in) | | | |
| 82 4 W.D. exc. Sta. Wagon and Brat: Front.......... | N/A | – 1 1/4 | | 1/4 | 1 13/16 | | 3 5/16 | 3/64 | 5/64 | 1/8 | 1.2 | 2.0 | 3.2 | 35 | 36 1/2 | NA |
| Rear .......... | N/A | | | | – 3/4 | | 3/4 | 1/8 (out) | 0 | 1/8 (in) | 3.2 (out) | | 3.2 (in) | | | |
| 81-80 4 W.D. exc. Sta. Wagon & Brat: Front ..... | N/A | – 1 1/4 | – 1/2 | 1/4 | 1 13/16 | 2 9/16 | 3 5/16 | 1/4 | | 15/32 | 6.4 | | 12 | 35 | 36 1/2 | NA |
| Rear ...... | N/A | | | | – 3/4 | 0 | 3/4 | 1/8 (out) | | 1/8 (in) | 3 (out) | | 3 (in) | | | |
| 82 4 W.D. Sta. Wagon..... | | | | | | | | | | | | | | | | |
| Front ............ | N/A | – 1 7/16 | | 1/16 | 1 13/16 | | 3 5/16 | 3/64 | 5/64 | 1/8 | 1.2 | 2.0 | 3.2 | 35 | 36 1/2 | NA |
| Rear ............. | N/A | | | | – 7/16 | | 1 1/16 | 1/8 (out) | 0 | 1/8 (in) | 3.2 (out) | | 3.2 (in) | | | |
| 81-80 4 W.D. Sta. Wagon .. | | | | | | | | | | | | | | | | |
| Front ............ | N/A | – 1 7/16 | – 11/16 | 1/16 | 1 13/16 | 2 9/16 | 3 5/16 | 1/4 | | 15/32 | 6.4 | | 1.2 | 35 | 36 1/2 | NA |
| Rear ............. | | | | | – 7/16 | 5/16 | 1 1/16 | 1/8 (out) | | 1/8 (in) | 3 (out) | | 3 (in) | | | |

• At the wheel rim.

| VEHICLE IDENTIFICATION (YEAR MODEL) | ADJ. ILL. NO. | CASTER (Degrees) MIN. | PREF. | MAX. | CAMBER (Degrees) MIN. | PREF. | MAX. | TOE-IN (Inches) MIN. | PREF. | MAX. | TOE-IN (Millimeters) MIN. | PREF. | MAX. | TOE-OUT ON TURNS (Degrees) OUTSIDE WHEEL | INSIDE WHEEL | STRG. AXIS INCL. (DEG.) |
|---|---|---|---|---|---|---|---|---|---|---|---|---|---|---|---|---|
| 82 4 W.D. Brat: Front | N/A | −1 7/16 | | 1/16 | 1 13/16 | | 3 5/16 | 3/64 | 5/64 | 1/8 | 1.2 | 2.0 | 3.2 | 35 | 36 1/2 | NA |
| Rear | N/A | | | | −7/16 | | 1 11/16 | 1/8 | 0 | 1/8 | 3.2 | | 3.2 | | | |
| 81-80 4 W.D. Brat: Front | N/A | −1 9/16 | −13/16 | −1/16 | 1 7/16 | 2 3/16 | 2 15/16 | (out) 1/4 | | (in) 15/32 | (out) 6.4 | | (in) 12 | 35 | 36 | NA |
| Rear | N/A | | | | 9/16 | 15/16 | 2 1/16 | 1/16 | | 1/4 | 2 | | 6.4 | | | |
| 79-77* all exc. Wagon, 4 W.D. & Brat: Front | N/A | −1 19/32 | −27/32 | −3/32 | 1 | 1 3/4 | 2 1/2 | 3/32 | | 21/64 | 2.4 | | 8.3 | 35 | 36 | NA |
| Rear | (42) | | | | −13/32 | | 13/32 | 3/64 | | 13/64 | 1.0 | | 5.0 | | | |
| 79-77* Wagon exc. 4 W.D. Front | N/A | −29/32 | −5/32 | 19/32 | 1 | 1 3/4 | 2 1/2 | 3/32 | | 21/64 | 2.4 | | 8.3 | 35 | 36 | NA |
| Rear | (42) | | | | −13/32 | | 1 29/32 | 5/64 | | 15/64 | 2.0 | | 6.0 | | | |
| 79-77* 4 W.D. except Brat. Front | N/A | −1 19/32 | −27/32 | −3/32 | 1/2 | 1 1/4 | 2 | 15/64 | | 15/32 | 6.0 | | 11.9 | 30 | 31 | NA |
| Rear | (42) | | | | 19/32 | | 23/32 | 5/64 | | 15/64 | 2.0 | | 6.0 | | | |
| 79-77* Brat: Front | N/A | −1 19/32 | −27/32 | −3/32 | 1/2 | 1 1/4 | 2 | 15/64 | | 15/32 | 6.0 | | 11.9 | 35 | 36 | NA |
| Rear | (42) | | | | 19/32 | | 23/32 | 5/64 | | 15/64 | 2.0 | | 6.0 | | | |
| 77-75 All exc. Wagon & 4 W.D.: Front | N/A | 0 | 3/4 | 1 1/2 | 1 | 1 1/2 | 2 | 5/64 | | 5/16 | 2.4 | | 7.9 | 35 | 36 | 12 29/32 |
| Rear | (42) | | | | 1/4 | 1 | 1 1/2 | 3/64 | | 13/64 | 1.2 | | 5.2 | | | |
| 77-75 Wagon: Front | N/A | 0 | 3/4 | 1 1/2 | 1 | 1/2 | 2 | 5/64 | | 5/16 | 2.4 | | 7.9 | 35 | 36 | 12 29/32 |
| Rear | (42) | | | | 1 | 1 1/2 | 2 | 5/64 | | 15/64 | 2.4 | | 6.0 | | | |
| 77-75 4 W.D.: Front | N/A | 0 | 3/4 | 1 1/2 | 2 | 2 1/2 | 3 | 15/64 | | 15/32 | 6.0 | | 11.9 | 30 | 31 | 12 29/32 |
| Rear | (42) | | | | 1 11/32 | 1 27/32 | 2 11/32 | 5/64 | | 15/64 | 2.4 | | 6.0 | | | |
| 74-73 all except 4 W.D. | N/A | 0 | 3/4 | 1 1/2 | 1 | 1 1/2 | 2 | 5/64 | | 5/16 | 2.4 | | 7.9 | 35 | 36 | 12 29/32 |
| Rear | | | | | 1/4 | 1 | 1 1/2 | 3/64 | | 13/64 | 1.2 | | 5.2 | | | |
| 74-73 4 W.D. | N/A | 0 | 3/4 | 1 1/2 | 2 | 2 1/2 | 3 | 15/64 | | 15/32 | 6.0 | | 11.9 | 30 | 31 | 12 29/32 |
| Rear | | | | | | | | 3/64 | | 13/64 | 1.2 | | 5.2 | | | |

*Oblong fender mounted parking lights on 77 Stage II models.

**TOYOTA**

| VEHICLE IDENTIFICATION (YEAR MODEL) | ADJ. ILL. NO. | CASTER MIN. | PREF. | MAX. | CAMBER MIN. | PREF. | MAX. | TOE-IN (in) MIN. | PREF. | MAX. | TOE-IN (mm) MIN. | PREF. | MAX. | OUTSIDE WHEEL | INSIDE WHEEL | STRG. AXIS INCL. |
|---|---|---|---|---|---|---|---|---|---|---|---|---|---|---|---|---|
| 82-81 Starlet | (66) | 1 11/16 | 2 | 2 5/16 | 5/16 | 11/16 | 1 | 3/64 | 5/64 | 1/8 | 1.2 | 2.0 | 3.2 | 34 | 36 27/32 | 9 3/4 |
| 82 Cressida: Sedan | (19) | 1 11/16 | 1 9/16 | 2 1/16 | | 13/16 | | 5/64 | 1/8 | 5/32 | 2.0 | 3.2 | 4.0 | 31 | 38 | 9 3/16 |
| Wagon | (19) | 1 | 1 1/2 | 2 | 5/16 | 13/16 | 15/16 | 5/64 | 1/8 | 5/32 | 2.0 | 3.2 | 4.0 | 31 | 38 | 9 3/16 |
| 81 Cressida: Sedan | (19) | 1 | 1 1/2 | 2 | 5/16 | 13/16 | 15/16 | 5/64 | 1/8 | 5/32 | 2.0 | 3.2 | 4.0 | 31 | 38 | 9 5/32 |
| Wagon | (19) | 1 | 1 1/2 | 2 | 5/16 | 13/16 | 15/16 | 5/64 | 1/8 | 5/32 | 2.0 | 3.2 | 4.0 | 31 | 38 | 9 3/32 |
| 82 Land Cruiser | | | | | | | | | | | | | | | | |
| FJ, BJ, HJ4 Series | N/A | 1/4 | 1 | 1 3/4 | 1/4 | 1 | 1 3/4 | 5/64 | 5/32 | 15/64 | 2.0 | 4.0 | 6.0 | 30 | 30-32 | 9 1/2 |
| FJ, BJ, HJ6 Series | N/A | | | | 1/4 | 1 | 1 3/4 | 5/64 | 5/32 | 15/64 | 2.0 | 4.0 | 6.0 | 30 | 30-32 | 9 1/2 |
| 81 Land Cruiser: FJ40 | N/A | | 1 | | | | 1 | 1/8 | 5/32 | 13/64 | 3.2 | 4.0 | 5.2 | 30 | 32 | 9 1/2 |
| FJ60 | N/A | | 1 3/32 | | | | 1 | 1/8 | 5/32 | 13/64 | 3.2 | 4.0 | 5.2 | 30 | 32 | 9 1/2 |
| 82-80 Hilux (4 × 4) | N/A | 2 3/4 | 3 1/2 | 4 1/4 | 1/4 | 1 | 1 3/4 | | | | | | | 29 | 30 1/2 | 9 1/2 |
| W/Radial Tires | | | | | | | | 0 | 3/64 | 5/64 | 0 | 1.2 | 2.0 | | | |
| W/Bias Tires | | | | | | | | 1/8 | 5/32 | 13/64 | 3.2 | 4.0 | 5.2 | | | |
| 82-80 Tercel: Front | (19) | 1 11/16 | 2 3/16 | 2 11/16 | 0 | 1/2 | 1 | 3/64 | 5/64 | 1/8 | 1.2 | 2.0 | 3.2 | 33 | 35 | 11 5/16 |
| Rear | (45) | | | | −1/2 | 0 | 1/2 | (out) 3/64 | 0 | (in) 3/64 | (out) 1.2 | 0 | (in) 1.2 | | | |
| 82-80 Corolla exc. wagon | (19) | 1 1/4 | 1 3/4 | 2 1/4 | 9/16 | 1 1/16 | 1 9/16 | | | | | | | 32 | 40 | 8 7/16 |
| W/Radial Tires | | | | | | | | 0 | 3/64 | 5/64 | 0 | 1.2 | 2.0 | | | |
| W/Bias Tires | | | | | | | | 5/64 | 1/8 | 5/32 | 2.0 | 3.2 | 4.0 | | | |
| 82-80 Corolla Wagon | (19) | 1 11/16 | 1 9/16 | 2 1/16 | 9/16 | 1 1/16 | 1 9/16 | | | | | | | 32 | 40 | 8 5/16 |
| W/Radial Tires | | | | | | | | 0 | 3/64 | 5/64 | 0 | 1.2 | 2.0 | | | |
| W/Bias Tires | | | | | | | | 5/64 | 1/8 | 5/32 | 2.0 | 3.2 | 4.0 | | | |
| 82 Hilux (4 × 2) Pick up | | | | | | | | | | | | | | | | |
| RN 1/2 Ton | (10) | 1/2 | 1 | 1 1/2 | 9/16 | 1 1/16 | 1 9/16 | | | | | | | 29 | 36 | 7 5/32 |
| W/Bias Tires | | | | | | | | 5/32 | 13/64 | 15/64 | 4.0 | 5.2 | 6.0 | | | |
| W/Radial Tires | | | | | | | | 3/64 | 5/64 | 1/8 | 1.2 | 2.0 | 3.1 | | | |
| RN 3/4 Ton, C & C | (10) | 0 | 1/2 | 1 | 9/16 | 1 1/16 | 1 9/16 | | | | | | | 29 | 36 | 7 5/32 |
| W/Bias Tires | | | | | | | | 5/32 | 13/64 | 15/64 | 4.0 | 5.2 | 6.0 | | | |
| W/Radial Tires | | | | | | | | 3/64 | 5/64 | 1/8 | 1.2 | 2.0 | 3.1 | | | |
| 81-80 Hilux (4 × 2) | (10) | 0 | 1/2 | 1 | 9/16 | 1 1/16 | 1 9/16 | | | | | | | 29 | 36 | 7 3/16 |
| W/Radial Tires | | | | | | | | 3/64 | 5/64 | 1/8 | 1.2 | 2.0 | 3.2 | | | |
| W/Bias Tires | | | | | | | | 5/32 | 13/64 | 1/4 | 4.0 | 5.2 | 6.4 | | | |
| 82 Corona Sedan: | (36) | 1 1/4 | 1 3/4 | 2 1/4 | 1/4 | 1 | 1 3/4 | 0 | 3/64 | 5/64 | 0.0 | 1.2 | 2.0 | 31 1/4 | 38 3/16 | 7 13/16 |
| Wagon | (36) | 1 | 1 1/2 | 2 | 1/4 | 1 | 1 3/4 | 0 | 3/64 | 5/64 | 0.0 | 1.2 | 2.0 | 31 1/4 | 38 3/16 | 7 13/16 |
| 81-80 Corona exc. Wagon | (36) | 1 1/4 | 1 3/4 | 2 1/4 | 1/2 | 1 | 1 1/2 | 0 | 3/64 | 5/64 | 0 | 1.2 | 2.0 | 31 13/16 | 38 3/16 | 7 13/16 |
| 81-80 Corona Wagon | (36) | 1 | 1 1/2 | 2 | 1/2 | 1 | 1 1/2 | 0 | 3/64 | 5/64 | 0 | 1.2 | 2.0 | 31 13/16 | 38 3/16 | 7 13/16 |
| 82 Celica | (19) | 2 13/16 | 3 5/16 | 3 13/16 | 7/16 | 15/16 | 1 7/16 | | | | | | | 32 | 37 | 9 11/32 |
| W/Manual Steering | | | | | | | | 1/8 | 5/32 | 13/64 | 3.2 | 4.0 | 5.2 | | | |
| W/Power Steering | | | | | | | | 5/32 | 13/64 | 15/64 | 4.0 | 5.0 | 5.9 | | | |
| 81-80 Celica | | | | | | | | | | | | | | | | |
| W/Power Steering | N/A | 1 3/16 | 1 11/16 | 2 3/16 | 7/16 | 15/16 | 1 7/16 | 1/8 | 5/32 | 13/64 | 3.2 | 4.0 | 5.2 | 31 1/4 | 38 3/16 | 7 9/16 |
| W/Manual Steering | N/A | 1 3/16 | 1 11/16 | 2 3/16 | 7/16 | 15/16 | 1 7/16 | 0 | 3/64 | 5/64 | 0 | 1.2 | 2.0 | 31 1/4 | 38 3/16 | 7 9/16 |

| VEHICLE IDENTIFICATION (YEAR MODEL) | ADJ. ILL. NO. | CASTER (Degrees) MIN. | PREF. | MAX. | CAMBER (Degrees) MIN. | PREF. | MAX. | TOE-IN (Inches) MIN. | PREF. | MAX. | TOE-IN (Millimeters) MIN. | PREF. | MAX. | TOE-OUT ON TURNS (Degrees) OUTSIDE WHEEL | INSIDE WHEEL | STRG. AXIS INCL. (DEG.) |
|---|---|---|---|---|---|---|---|---|---|---|---|---|---|---|---|---|
| 82 Supra: Front | (19) | 3 11/16 | 4 3/16 | 4 11/16 | 5/16 | 13/16 | 15/16 | 5/64 | 1/8 | 5/32 | 2.0 | 3.2 | 4.0 | 30 3/4 | 37 19/32 | 7 11/16 |
| Rear | (65) | | | | 5 1/2 | 6 | 6 1/2 | 3/64 (out) | 0 | 3/64 (in) | 1.2 (out) | 0 | 1.2 (in) | | | |
| 81-79 1/2 Supra | N/A | 1 1/4 | 1 3/4 | 2 1/4 | 5/16 | 13/16 | 15/16 | 3/64 (out) | 0 | 3/64 (in) | 1.2 (out) | 0 | 1.2 (in) | 31 11/32 | 37 3/32 | 7 11/16 |
| 80-78 Cressida except 78 Wagon | N/A | 3/4 | 1 1/4 | 1 3/4 | 5/16 | 13/16 | 15/16 | 5/64 | 1/8 | 5/32 | 2.0 | 3.2 | 4.0 | 32 | 38 | 7 11/16 |
| 80-79 Land Cruiser | N/A | 1/4 | 1 | 1 3/4 | 1/4 | 1 | 1 3/4 | | | | | | | 30 | 32 | 9 1/2 |
| W/HR78 × 15B | | | | | | | | 3/64 (out) | 0 | 3/64 (in) | 1.2 (out) | 0 | 1.2 (in) | | | |
| W/H78 × 15B | | | | | | | | 7/64 | 5/32 | 13/64 | 2.8 | 4.0 | 5.2 | | | |
| 79 Hilux (4 × 2) | (10) | 0 | 1/2 | 1 | 9/16 | 1 1/16 | 1 9/16 | 5/32 | 13/64 | 15/64 | 4.0 | 5.2 | 6.0 | 29 | 36 | 7 5/32 |
| 79 Hilux (4 × 4) | N/A | | 3 1/2 | | | 1 | | | | | | | | | | 9 1/2 |
| W/HR78 × 15B | | | | | | | | 0 | 3/64 | 5/64 | 0 | 1.2 | 2.0 | 29 | 36 | |
| W/H78 × 15B | | | | | | | | 7/64 | 5/32 | 15/64 | 2.8 | 4.0 | 6.0 | 29 | 30 1/2 | |
| 79 Corona | (36) | 1 1/4 | 1 3/4 | 2 1/4 | 1/2 | 1 | 1 1/2 | 0 | 3/64 | 5/64 | 0 | 1.2 | 2.0 | 31 | 38 | 7 27/32 |
| 79 Celica | N/A | 1 1/4 | 1 3/4 | 2 1/4 | 9/16 | 1 1/16 | 1 9/16 | 0 | 3/64 | 5/64 | 0 | 1.2 | 2.0 | 31 | 38 | 7 7/16 |
| 79-76 Corolla | N/A | 15/16 | 1 15/16 | 2 5/16 | 1/2 | 1 | 1 1/2 | | | | | | | 31 7/16 | 38 | 7 27/32 |
| W/Radial Tires | | | | | | | | 0 | | 5/64 | 0 | | 2.0 | | | |
| W/Bias Tires | | | | | | | | 5/64 | 1/8 | 5/32 | 2.0 | 3.2 | 4.0 | | | 7 27/32 |
| 78 Cressida Wagon | N/A | 1 1/16 | 1 3/16 | 1 11/16 | 5/16 | 13/16 | 15/16 | 5/64 | 1/8 | 5/32 | 2.0 | 3.2 | 4.0 | 32 | 37 1/2 | 7 11/16 |
| 78 Celica | N/A | 1 1/4 | 1 3/4 | 2 1/4 | 7/16 | 15/16 | 1 7/16 | | | | | | | 31 | 38 | 7 9/16 |
| W/Power Steering | | | | | | | | 0 | 1/64 | 1/32 | 0 | 0.4 | 0.8 | | | |
| W/Manual Steering | | | | | | | | 3/64 | 1/16 | 5/64 | 1.2 | 1.6 | 2.0 | | | |
| 78-77 Corona | (36) | 5/16 | | 1 5/16 | 1/16 | | 1 1/16 | | | | | | | | | |
| W/Radial Tires | | | | | | | | 1/64 | | 3/64 | 0.4 | | 1.2 | 31 | 37 1/2 | 7 |
| W/Bias Tires | | | | | | | | 3/64 | | 5/64 | 1.2 | | 2.0 | 31 | 37 1/2 | 7 |
| 78-76 Hilux (4 × 2) | (10) | 0 | 1/2 | 1 | 1/2 | 1 | 1 1/2 | 15/64 | | 9/32 | 6.0 | | 7.1 | | | |
| W/Radial Tires | | | | | | | | | | | | | | 26 1/4 | 31 | 7 1/4 |
| W/Bias Tires | | | | | | | | | | | | | | 27 1/2 | 33 1/4 | 7 1/4 |
| 78-71 Land Cruiser | N/A | | 1 | | | 1 | | 1/8 | 13/64 | | 3.2 | 5.2 | | 30 | 32 | 9 1/2 |
| W/900 × 15 Tires | | | | | | | | 15/64 | 9/32 | | 6.0 | 7.1 | | 24 | 26 | 9 1/2 |
| 77 Celica | N/A | 1 1/4 | | 2 1/4 | 1/2 | | 1 1/2 | 0 | | 1/32 | 0 | | 0.8 | 31 | 38 | 7 9/16 |
| 76 Celica except GT | N/A | 1/2 | | 1 1/2 | 1/2 | | 1 1/2 | | | | | | | | | |
| W/Radial Tires | | | | | | | | 1/64 | | 3/64 | 0.4 | | 1.2 | 31 | 38 | 7 9/16 |
| W/Standard Tires | | | | | | | | 3/64 | | 5/64 | 1.2 | | 2.0 | 31 | 38 | 7 9/16 |
| 76 Celica GT | N/A | 15/16 | | 2 5/16 | 5/16 | | 15/16 | | | | | | | 31 | 38 | 7 9/16 |
| W/Radial Tires | | | | | | | | 1/64 | | 3/64 | 0.4 | | 1.2 | | | |
| W/Standard Tires | | | | | | | | 3/64 | | 5/64 | 1.2 | | 2.0 | | | |
| 76 Mark II: Radial | N/A | 1 3/4 | | 2 3/4 | 9/16 | | 1 9/16 | 0 | | 1/8 | 0 | | 3.2 | 32 1/2 | 36 1/2 | 7 |
| Bias | | 1 3/4 | | 2 3/4 | 9/16 | | 1 9/16 | 5/32 | | 15/64 | 2.0 | | 6.0 | 32 1/2 | 36 1/2 | 7 |
| 75 Corolla: Sedan | N/A | 3/4 | | 1 7/16 | 5/32 | | 15/32 | 0 | 3/64 | 5/64 | 0 | 1.2 | 2.0 | 31 7/16 | 38 | 7 27/32 |
| Wagon | N/A | 3/4 | | 1 7/16 | 5/32 | | 15/32 | 0 | 3/64 | 5/64 | 0 | 1.2 | 2.0 | 31 7/16 | 38 | 7 27/32 |
| Hard Top | N/A | 1 1/2 | | 2 3/16 | 1/4 | | 1 1/4 | 0 | 3/64 | 5/64 | 0 | 1.2 | 2.0 | 31 7/16 | 38 | 7 27/32 |
| 75 Hilux | (10) | -1 1/4 | | 1/4 | 1/4 | | 1 3/4 | 5/64 | | 9/32 | 2.0 | | 7.1 | 31 1/2 | 39 | 7 |
| 75-73 Mark II exc. Wagon | N/A | -5/16 | | 15/16 | 9/16 | | 1 9/16 | 3/16 | | 1/4 | 4.8 | | 6.4 | 32 1/2 | 36 1/2 | 7 |
| 75-73 Mark II Wagon | N/A | 3/16 | | 1 3/16 | 9/16 | | 1 9/16 | 3/16 | | 1/4 | 4.8 | | 6.4 | 32 1/2 | 36 1/2 | 7 |
| 75-73 Celica, Carina | N/A | 1/2 | | 1 1/2 | 1/2 | | 1 1/2 | 13/64 | | 1/4 | 5.2 | | 6.4 | 31 | 38 | 7 9/16 |
| 76-74 Corona | (36) | 5/16 | | 1 5/16 | 1/16 | | 1 1/16 | 1/64 | | 3/64 | 0.4 | | 1.2 | 31 | 38 1/2 | 7 |
| 74 Corolla 1200 (3KC) | N/A | 1 11/16 | | 2 5/16 | 5/16 | | 15/16 | 1/16 | | 3/16 | 1.6 | | 4.8 | 33 | 40 | 8 |
| 74 Corolla 1600 (2TC) | N/A | 15/16 | | 1 | 1/2 | | 1 1/2 | 1/16 | | 3/16 | 1.6 | | 4.8 | 30 | 40 | 7 1/2 |
| 74 Hilux | (10) | -1 1/4 | | 1/4 | 1/4 | | 1 3/4 | 15/64 | | 1/4 | 6.0 | | 6.4 | 31 1/2 | 39 | 7 |
| 73 Hilux (RN22 or 27) | (10) | -1 1/4 | | 1/4 | 1/4 | | 1 3/4 | 3/32 | | 9/32 | 2.4 | | 7.1 | 31 1/2 | 39 | 7 |
| 73 Corolla 1200 (3KC) | N/A | 1 11/16 | | 2 5/16 | 5/16 | | 15/16 | 1/16 | | 3/16 | 1.6 | | 4.8 | 30 | 40 | 7 1/2 |
| 73 Corolla 1600 (2TC) | N/A | 1 9/16 | | 2 1/4 | 1/2 | | 1 1/2 | 1/16 | | 3/16 | 1.6 | | 4.8 | 30 | 40 | 7 1/2 |
| 73 Corona | (10) | -3/16 | | 13/16 | | 15/16 | | 3/16 | | 1/4 | 4.8 | | 6.4 | 31 | 38 1/2 | 7 |
| **TRIUMPH** | | | | | | | | | | | | | | | | |
| 81-75 TR7, TR8 | N/A | 2 1/2 | 3 1/2 | 4 1/2 | -1 1/4 | 1/4 | 3/4 | 0 | | 1/16 | 0 | | 1.6 | NA | NA | 11 1/4 |
| 80-78 Spitfire: Front | (25) | | 4 | | | 3 | | 1/16 | 1/8 | | 1.6 | 3.2 | | NA | NA | 5 3/4 |
| Rear | (60) | | | | | | | 1/16 (out) | 1/8 | | 1.6 (out) | 3.2 | | | | |
| 77-76 Spitfire: Front | (25) | 4 | 4 1/2 | 5 | 1 1/2 | 2 | 2 1/2 | 0 | | 1/16 | 0 | | 1.6 | NA | NA | 6 3/4 |
| Rear | (60) | | | | -4 3/4 | -3 3/4 | -2 3/4 | 0 | | 1/16 | 0 | | 1.6 | | | |
| 76-73 TR6: Front | (25) | 2 1/4 | 2 3/4 | 3 1/4 | -3/4 | -1/4 | 3/4 | 0 | 1/32 | 1/16 | 0 | 0.8 | 1.6 | NA | NA | 9 1/4 |
| Rear | (61) | | | | -1 1/2 | -1 | 1/2 | | | | | | | | | |
| 75 Spitfire: Front | (25) | 4 | 4 1/2 | 5 | 1 1/2 | 2 | 2 1/2 | 0 | | 1/16 | 0 | | 1.6 | NA | NA | 6 3/4 |
| Rear | (60) | | | | -4 1/4 | -3 1/4 | -2 1/4 | 0 | | 1/16 | 0 | | 1.6 | | | |
| 74-73 Spitfire: Front | (25) | 4 | 4 1/2 | 5 | 1 1/2 | 2 | 2 1/2 | 0 | | 1/16 | 0 | | 1.6 | NA | NA | 6 3/4 |
| Rear | (60) | | | | -4 3/4 | -3 3/4 | -2 3/4 | 0 | | 1/16 | 0 | | 1.6 | | | |
| 74-73 Mark IV, 1500: Front | (25) | 3 1/2 | 4 | 4 1/2 | 2 1/2 | 3 | 3 1/2 | 1/16 | | 1/8 | 1.6 | | 3.2 | NA | NA | 5 3/4 |
| Rear | N/A | | | | -2 | -1 | 0 | 3/32 (out) | | 1/32 (out) | 2.4 (out) | | 0.8 (out) | | | |

| VEHICLE IDENTIFICATION | ADJ. ILL. NO. | CASTER (Degrees) | | | CAMBER (Degrees) | | | TOE-IN (Inches) | | | TOE-IN (Millimeters) | | | TOE-OUT ON TURNS (Degrees) | | STRG. AXIS INCL. |
|---|---|---|---|---|---|---|---|---|---|---|---|---|---|---|---|---|
| YEAR MODEL | | MIN. | PREF. | MAX. | MIN. | PREF. | MAX. | MIN. | PREF. | MAX. | MIN. | PREF. | MAX. | OUTSIDE WHEEL | INSIDE WHEEL | (DEG.) |
| 73 GT6 Series: Front...... | (25) | 2 1/2 | 3 1/2 | 4 1/2 | 1 3/4 | 2 3/4 | 3 1/4 | 1/16 | | 1/8 | 1.6 | | 3.2 | NA | NA | 6 |
| Rear...... | (60) | | | | −1 | 0 | 1 | 1/32 (out) | 0 | 1/32 (in) | 0.8 (out) | | 0.8 (in) | | | |
| **VOLKSWAGEN** | | | | | | | | | | | | | | | | |
| 82 Quantum: Front....... | | 0 | 1/2 | 1 | −1 5/32 | −21/32 | −5/32 | 0 | 5/64 | 11/64 | 0.0 | 2.0 | 4.4 | NA | NA | NA |
| Rear....... | | | | | −2 | −1 21/32 | −1 11/32 | 5/64 | 13/64 | 21/64 | 2.0 | 5.2 | 8.3 | | | |
| 82-81 Rabbit Pick-up | | | | | | | | | | | | | | | | |
| Front............... | (4) | 13/16 | 1 5/16 | 1 13/16 | −3/16 | 5/16 | 13/16 | 7/32 (out) | 1/8 (out) | 1/32 (out) | 5.6 (out) | 3.2 (out) | 0.8 (out) | 18 1/2 | 20 | NA |
| Rear............... | N/A | | | | −1 | 0 | 1 | 1/2 (out) | 0 | 1/2 (in) | 12.7 (out) | 0.0 | 12.7 (in) | | | |
| 82-81 Vanagon: Front..... | (62)(63) | 7 | 7 1/4 | 7 1/2 | −1/2 | 0 | 1/2 | 1/16 (out) | 5/64 (in) | 7/32 (in) | 1.5 (out) | 2.0 (in) | 5.5 (in) | NA | NA | NA |
| Rear..... | (64) | | | | −1 5/16 | −13/16 | −5/16 | 5/16 (out) | 0 | 5/16 (in) | 7.9 (out) | 0 | 7.9 (in) | | | |
| 82-80 Rabbit, Scirocco, Jetta: Front............ | (4) | 1 5/16 | 1 13/16 | 2 5/16 | −3/16 | 5/16 | 13/16 | 7/32 (out) | 1/8 (out) | 1/32 (out) | 5.6 (out) | 3.2 (out) | 0.8 (out) | 18 1/2 | 20 | NA |
| Rear............ | N/A | | | | −1 13/16 | −1 1/4 | −11/16 | 0 | 3/16 | 5/16 | 0 | 4.7 | 7.9 | | | |
| 80 Rabbit Pickup: Front.. | (4) | 1 5/16 | 1 13/16 | 2 5/16 | −3/16 | 5/16 | 13/16 | 7/32 (out) | 1/8 (out) | 1/32 (out) | 5.6 (out) | 3.2 (out) | 0.8 (out) | 18 1/2 | 20 | NA |
| Rear... | N/A | | | | −9/16 | 0 | 9/16 | 0 | 5/32 | 5/16 | 0 | 4.0 | 7.9 | | | |
| 80 Vanagon: Front...... | (62)(63) | 7 | 7 1/4 | 7 1/2 | 3/16 | 11/16 | 1 3/16 | 3/16 | 11/32 | 1/2 | 4.8 | 8.7 | 12.7 | NA | NA | NA |
| Rear...... | (64) | | | | −1 5/16 | −13/16 | −5/16 | 3/32 (out) | 0 | 3/32 (in) | 2.4 (out) | 0 | 2.4 (in) | | | |
| 81-78 Dasher: Front...... | (14) | 0 | 1/2 | 1 | 0 | 1/2 | 1 | 1/32 (out) | 3/32 (in) | 7/32 (in) | 0.8 (out) | 2.4 (in) | 5.6 (in) | NA | NA | NA |
| Rear...... | N/A | | | | −1 3/8 | −11/16 | 0 | 13/32 (out) | 0 | 13/32 (in) | 10.3 (out) | 0 | 10.3 (in) | | | |
| 79 Rabbit, Scirocco...... | | | | | | | | | | | | | | | | |
| Front............... | (4) | 1 5/16 | 1 13/16 | 2 5/16 | −3/16 | 5/16 | 13/16 | 7/32 (out) | 1/8 (out) | 1/32 (out) | 5.6 (out) | 3.2 (out) | 0.8 (out) | 18 1/2 | 20 | NA |
| Rear............... | N/A | | | | 11/16 | 1 1/4 | 1 13/16 | 3/32 (out) | 5/32 (in) | 13/32 (in) | 2.4 (out) | 4.0 (in) | 10.3 (in) | | | |
| 79-72 Type 1, Super Beetle. | | | | | | | | | | | | | | | | |
| Front............... | (12) | 1 7/16 | 2 | 2 9/16 | 5/16 | 1 | 1 5/16 | 1/8 | 1/4 | 3/8 | 3.2 | 6.4 | 9.0 | NA | NA | NA |
| Rear............... | (30) | | | | −1 11/16 | −1 | −5/16 | 1/8 (out) | 0 | 1/8 (in) | 3.2 (out) | 6.4 | 3.2 (in) | | | |
| 79-72 Type 2, Bus: Front.. | (12) | 2 5/16 | 3 | 3 11/16 | 3/8 | 11/16 | 1 | 0 | 1/8 | 1/4 | 0 | 3.2 | 6.4 | 17 1/2 | 20 | NA |
| Rear... | (30) | | | | −1 5/16 | −13/16 | −5/16 | 1/16 (out) | 3/32 (in) | 1/4 (in) | 1.6 (out) | 2.4 (in) | 6.4 (in) | | | |
| 78-75 Rabbit, Scirocco.... | | | | | | | | | | | | | | | | |
| Front............... | (4) | 1 5/16 | 1 13/16 | 2 5/16 | −3/16 | 5/16 | 13/16 | 1/4 (out) | 1/8 (out) | 1/32 (out) | 6.4 (out) | 3.2 (out) | 0.8 (out) | 18 1/2 | 20 | NA |
| Rear............... | N/A | | | | | | | | | | | | | | | |
| Rabbit (from chassis #1763241690) and Scirocco (from chassis #5362031722)........... | | | | | −1 9/16 | −1 | −7/16 | 5/32 (out) | 3/32 (in) | 11/32 (in) | 4.0 (out) | 2.4 (in) | 8.8 (in) | | | |
| Rabbit (from chassis #1763241691) and Scirocco (from chassis #5362031722)........... | | | | | −1 13/16 | −1 1/4 | −11/16 | 3/32 (out) | 5/32 (in) | 13/32 (in) | 2.4 (out) | 4.0 (in) | 10.3 (in) | | | |
| 77-74 Dasher: Front...... | (14) | 0 | 1/2 | 1 | 0 | 1/2 | 0 | 1/32 (out) | 3/32 (in) | 7/32 (in) | 0.8 (out) | 2.4 (in) | 5.6 (in) | NA | NA | NA |
| Rear...... | N/A | | | | −1 | −1/2 | 0 | 7/32 (out) | 0 | 7/32 (in) | 5.6 (out) | 0 | 5.6 (in) | | | |
| 77-73 Type 1 Beetle...... | | | | | | | | | | | | | | | | |
| Front............... | (12) | 2 11/16 | 3 5/16 | 4 5/16 | 1/8 | 1/2 | 7/8 | 1/8 | 1/4 | 3/8 | 3.2 | 6.4 | 9.5 | NA | NA | NA |
| Rear............... | (30) | | | | −1 11/16 | −1 | −5/16 | 1/8 (out) | 0 | 1/8 (in) | 3.2 (out) | 0 | 3.2 (in) | | | |
| 74-73 Type 4: Front...... | (39) | 1 3/16 | 1 3/4 | 2 5/16 | 11/16 | 1 3/16 | 1 5/8 | 1/32 | | 5/32 | 0.8 | | 4.0 | NA | NA | NA |
| Rear....... | (38) | | | | −1 1/2 | −1 | 1/2 | 1/32 (out) | 3/32 (in) | 7/32 (in) | 0.8 (out) | 2.4 (in) | 5.6 (in) | NA | NA | NA |
| 73 Type 3: All Front...... | (12) | 1 5/16 | 4 | 4 11/16 | 1 | 1 5/16 | 1 5/8 | 1/8 | | 1/4 | 3.2 | | 6.4 | NA | NA | NA |
| Rear-Sedan... | (30) | | | | −2 | −2 5/16 | −5/8 | 3/32 (out) | 1/32 (in) | 5/32 (in) | 2.4 (out) | 0.8 (in) | 4.0 (in) | | | |
| Rear-Station Wagon.... | (30) | | | | −2 | −1 5/16 | −5/8 | 1/8 (out) | 0 | 1/8 (in) | 3.2 (out) | 0 | 3.2 (in) | | | |

| VEHICLE IDENTIFICATION YEAR MODEL | ADJ. ILL. NO. | CASTER (Degrees) MIN. | PREF. | MAX. | CAMBER (Degrees) MIN. | PREF. | MAX. | TOE-IN (Inches) MIN. | PREF. | MAX. | TOE-IN (Millimeters) MIN. | PREF. | MAX. | TOE-OUT ON TURNS (Degrees) OUTSIDE WHEEL | INSIDE WHEEL | STRG. AXIS INCL. (DEG.) |
|---|---|---|---|---|---|---|---|---|---|---|---|---|---|---|---|---|
| **VOLVO** | | | | | | | | | | | | | | | | |
| 82-80 all models exc. GLT.. | | | | | | | | | | | | | | 20 13/16 | 20 | 12 |
| W/Power Steering ..... | (37) | 3 | | 4 | 1 | | 1½ | 1/16 | 1/8 | 3/16 | 1.6 | 3.2 | 4.8 | | | |
| W/Manual Steering .... | (37) | 2 | | 3 | 1 | | 1½ | 1/8 | 3/16 | 1/4 | 3.2 | 4.8 | 6.4 | | | |
| 82-80 GLT .. | | | | | | | | | | | | | | 20 13/16 | 20 | 12 |
| W/Power Steering ..... | (37) | 3 | | 4 | 1/4 | | 3/4 | 1/16 | 1/8 | 3/16 | 1.6 | 3.2 | 4.8 | | | |
| W/Manual Steering .... | (37) | 2 | | 3 | 1/4 | | 3/4 | 1/8 | 3/16 | 1/4 | 3.2 | 4.8 | 6.4 | | | |
| 79 240, 260 Series ....... | | | | | − 1 | | 0 | | | | | | | 20 13/16 | 20 | 12 |
| W/Power Steering ..... | (37) | 3 | | 4 | | | | 1/16 | 1/8 | 3/16 | 1.6 | 3.2 | 4.8 | | | |
| W/Manual Steering .... | (37) | 2 | | 3 | | | | 1/8 | 3/16 | 1/4 | 3.2 | 4.8 | 6.4 | | | |
| 78-76 240, 260 Series .... | (37) | 2 | | 3 | 1 | | 1½ | | | | | | | 20 13/16 | 20 | 12 |
| W/Power Steering ..... | | | | | | | | 1/16 | 1/8 | 3/16 | 1.6 | 3.2 | 4.8 | | | |
| W/Manual Steering .... | | | | | | | | 1/8 | 3/16 | 1/4 | 3.2 | 4.8 | 6.4 | | | |
| 75 242, 244, 245 ........ | (37) | 2 | | 3 | 1 | | 1½ | | | | | | | 20 13/16 | 20 | 12 |
| W/Power Steering ..... | | | | | | | | 1/16 | 1/8 | 3/16 | 1.6 | 3.2 | 4.8 | | | |
| W/Manual Steering .... | | | | | | | | 1/8 | 3/16 | 1/4 | 3.2 | 4.8 | 6.4 | | | |
| 75-73 164 ........... | (10) | 1½ | | 2½ | 0 | | ½ | 5/64 | | 13/64 | 2.0 | | 5.2 | 22½ | 20 | 7½ |
| 74-73 140 ........... | | | | | | | | | | | | | | 22½ | 20 | 7½ |
| W/Power Steering ..... | (10) | 2 | | 3 | 0 | | ½ | 5/64 | | 13/64 | 2.0 | | 5.2 | | | |
| W/Manual Steering .... | (10) | 1 | | 2 | 0 | | ½ | 5/64 | | 13/64 | 2.0 | | 5.2 | | | |
| 73 1800 ES ........... | (10) | 2 | | 2½ | 0 | | ½ | 0 | | 1/8 | 0 | | 3.2 | 22½ | 20 | 8 |

## ADJUSTMENT ILLUSTRATIONS GROUP I

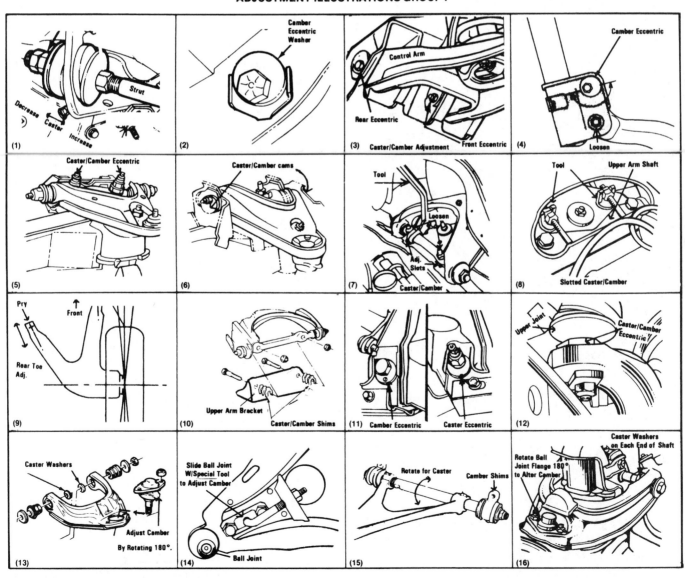

(1) Decrease Caster / Increase / Strut

(2) Camber Eccentric Washer

(3) Control Arm / Rear Eccentric / Caster/Camber Adjustment / Front Eccentric

(4) Camber Eccentric / Loosen

(5) Caster/Camber Eccentric

(6) Caster/Camber cams

(7) Tool / Loosen / Adj. Slots / Caster/Camber

(8) Tool / Upper Arm Shaft / Slotted Caster/Camber

(9) Pry / Front / Rear Toe Adj.

(10) Upper Arm Bracket / Caster/Camber Shims

(11) Camber Eccentric / Caster Eccentric

(12) Upper Joint / Caster/Camber Eccentric

(13) Caster Washers / Adjust Camber By Rotating 180°.

(14) Slide Ball Joint W/Special Tool to Adjust Camber / Ball Joint

(15) Rotate for Caster / Camber Shims

(16) Caster Washers on Each End of Shaft / Rotate Ball Joint Flange 180° to Alter Camber

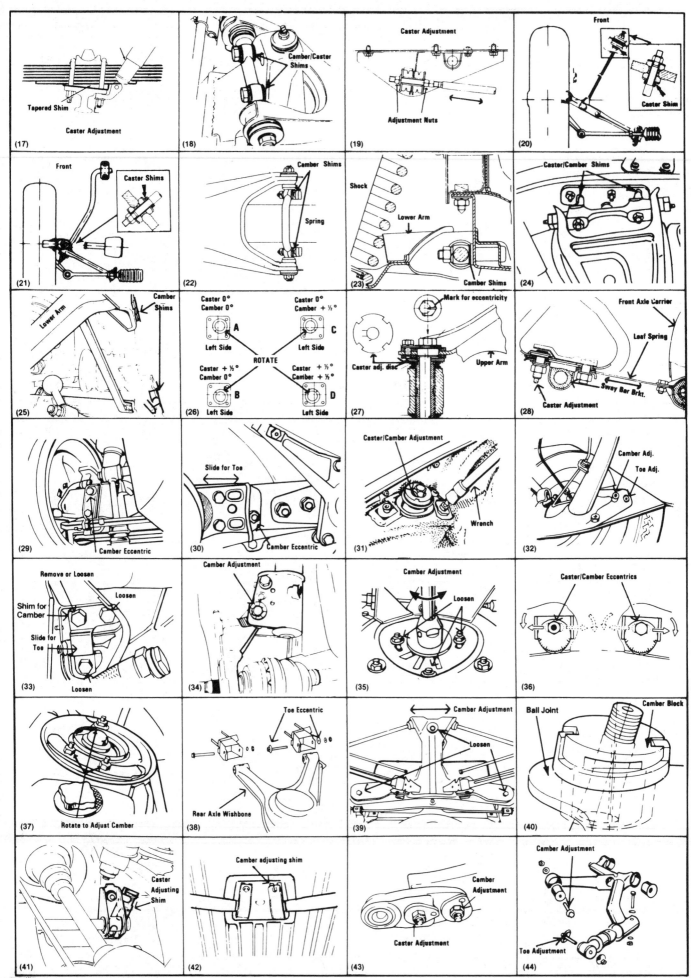

**(17)** Tapered Shim — Caster Adjustment

**(18)** Camber/Caster Shims

**(19)** Caster Adjustment — Adjustment Nuts

**(20)** Front — Caster Shim

**(21)** Front — Caster Shims

**(22)** Camber Shims — Spring

**(23)** Shock — Lower Arm — Camber Shims

**(24)** Caster/Camber Shims

**(25)** Lower Arm — Camber Shims

**(26)** Caster 0° Camber 0° **A** Left Side — Caster 0° Camber + ½° **C** Left Side — ROTATE — Caster + ½° Camber 0° **B** Left Side — Caster + ½° Camber + ½° **D** Left Side

**(27)** Mark for eccentricity — Caster adj. disc — Upper Arm

**(28)** Front Axle Carrier — Leaf Spring — Sway Bar Brkt. — Caster Adjustment

**(29)** Camber Eccentric

**(30)** Slide for Toe — Camber Eccentric

**(31)** Caster/Camber Adjustment — Wrench

**(32)** Camber Adj. — Toe Adj.

**(33)** Remove or Loosen — Loosen — Shim for Camber — Slide for Toe — Loosen

**(34)** Camber Adjustment

**(35)** Camber Adjustment — Loosen

**(36)** Caster/Camber Eccentrics

**(37)** Rotate to Adjust Camber

**(38)** Toe Eccentric — Rear Axle Wishbone

**(39)** Camber Adjustment — Loosen

**(40)** Ball Joint — Camber Block

**(41)** Caster Adjusting Shim

**(42)** Camber adjusting shim

**(43)** Camber Adjustment — Caster Adjustment

**(44)** Camber Adjustment — Toe Adjustment

114

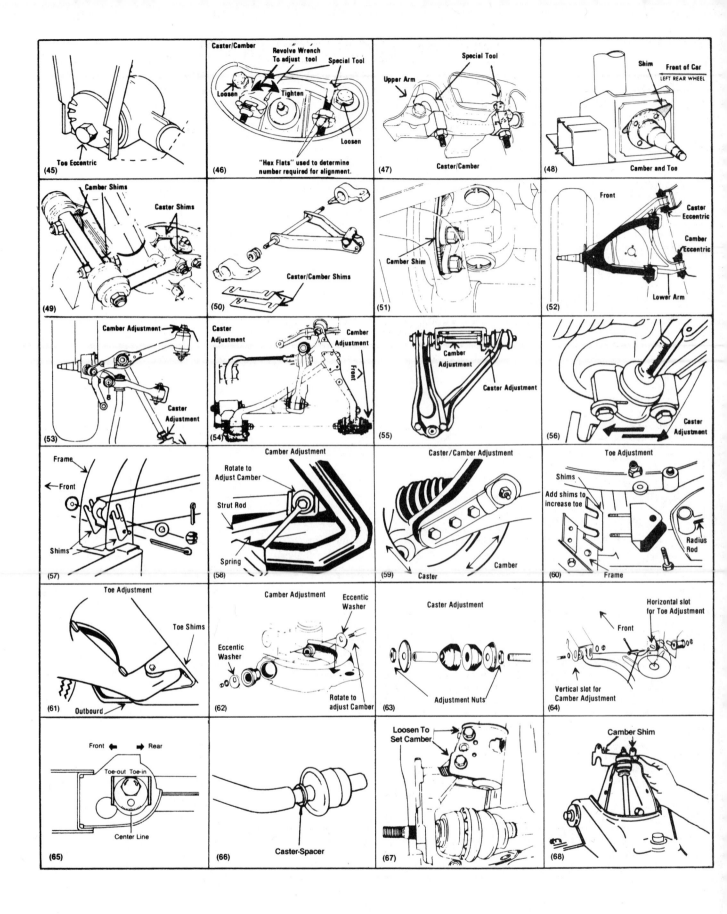

(45) Toe Eccentric

(46) Caster/Camber — Revolve Wrench To adjust tool — Special Tool — Loosen — Tighten — Loosen — "Hex Flats" used to determine number required for alignment.

(47) Upper Arm — Special Tool — Caster/Camber

(48) Shim — Front of Car / LEFT REAR WHEEL — Camber and Toe

(49) Camber Shims — Caster Shims

(50) Caster/Camber Shims

(51) Camber Shim

(52) Front — Caster Eccentric — Camber Eccentric — Lower Arm

(53) Camber Adjustment — Caster Adjustment

(54) Caster Adjustment — Camber Adjustment — Front

(55) Camber Adjustment — Caster Adjustment

(56) Caster Adjustment

(57) Frame — Front — Shims

(58) Camber Adjustment — Rotate to Adjust Camber — Strut Rod — Spring

(59) Caster/Camber Adjustment — Caster — Camber

(60) Toe Adjustment — Shims — Add shims to increase toe — Radius Rod — Frame

(61) Toe Adjustment — Toe Shims — Outbourd

(62) Camber Adjustment — Eccentric Washer — Eccentic Washer — Rotate to adjust Camber

(63) Caster Adjustment — Adjustment Nuts

(64) Horizontal slot for Toe Adjustment — Front — Vertical slot for Camber Adjustment

(65) Front — Rear — Toe-out — Toe-in — Center Line

(66) Caster-Spacer

(67) Loosen To Set Camber

(68) Camber Shim

# CASTER/CAMBER REFRESHER GUIDE

| MAKE AND MODEL | YEAR | MANNER OF CASTER ADJUSTMENT | MANNER OF CAMBER ADJUSTMENT |
|---|---|---|---|
| **American Motors**<br>All models except Pacer<br><br>Pacer | 82-73<br><br>80-75 | Lengthen strut to increase positive caster, shorten strut to decrease positive caster. Rotate rear cam to achieve desired measurement. | Rotate cam to achieve desired measurement. Rotate both eccentric cams to achieve desired measurements. |
| **Chrysler Corporation** · Slotted mounting type adjustment (see illustration 7) | | | |
| Gran Fury, Diplomat<br>New Yorker, Caravelle<br>Full sized Chrysler, Dodge and Plymouth vehicles<br>Cordoba, Mirada<br>LeBaron, Diplomat<br>Volare, Aspen, Caravelle<br>Charger, Coronet, Magnum<br>Satellite, Monaco, Fury<br>St. Regis | 82<br>82<br>81-74<br>82-75<br>81-77<br>81-76<br>80-73<br>78-73<br>81-79 | Increase positive caster by sliding front adjusting bolt away from engine and rear adjusting bolt toward engine. Decrease positive caster by sliding front adjusting bolt toward engine and rear adjusting bolt away from engine. | Increase positive camber by moving both adjusting bolts away from the engine.<br><br>Decrease positive camber by moving both adjusting bolts toward the engine. |
| **Chrysler Corporation** · Cam adjustment type | | | |
| Full sized Chrysler, Dodge, and Plymouth vehicles<br>Challenger, Barracuda<br>Dart, Valiant | 73<br><br>74-73<br>76-73 | Increase positive caster by rotating front cam in the proper direction to move the front of the control arm away from the engine and rotate rear cam to move rear of the control arm toward the engine. Decrease positive caster by reversing the above procedure. | Increase positive camber by rotating both cams to move the control arm away from engine. Decrease positive camber by rotating both cams to move the control arm toward the engine. |
| **Chrysler Corporation** · McPherson strut vehicles. | | | |
| Omni, 024, Horizon, TC3<br>Reliant, Aries<br>400, LeBaron (FWD), Rampage | 82-78<br>82-81<br>82 | Caster is not adjustable. | Turn eccentric cam, which is mounted on the strut attaching bracket, as necessary. |
| **Ford Motor Company** · Models that are not adjustable. | | | |
| Lincoln Continental<br>Zephyr/Fairmont<br>Capri/Mustang<br>Cougar XR7/Thunderbird<br>Escort/Lynx, Granada/Cougar | 82<br>82-78<br>82-79<br>82-80<br>82-81 | Not adjustable. | Not adjustable. |
| **Ford Motor Company** · Models with front coil spring mounted on upper control arm. | | | |
| Maverick/Comet<br>Mustang/Cougar<br>Granada/Monarch<br>Versailles | 77-73<br>73<br>80-75<br>80-77 | Increase positive caster by shortening the lower control arm strut. Decrease positive caster by lengthening the lower control arm strut. | Increase positive camber by rotating lower control arm adjusting cam to move the arm inward. It may be necessary to slightly spread the body bracket at the adjusting cam. Decrease positive camber by reversing the above procedure. |
| **Ford Motor Company** · Models with front coil spring mounted on lower control arm and slotted upper control arm adjustment. | | | |
| Lincoln Town Car<br>Continental Mark VI<br>Full sized Ford/Mercury<br>All Lincoln models<br>Torino/Montego<br>Thunderbird, Meteor<br>Pinto/Bobcat<br>Mustang<br>Cougar/LTD II | 82<br>82<br>82-73<br>81-73<br>76-73<br>79-73<br>80-73<br>78-74<br>79-74 | Increase positive caster by moving front adjusting bolt away from engine, rear adjusting bolt toward engine. Decrease positive caster by moving rear adjusting bolt away from engine, front adjusting bolt toward engine. | Increase positive camber by moving both adjusting bolts away from engine. Decrease positive camber by moving both adjusting bolts toward engine. |
| **GENERAL MOTORS CORPORATION**<br>**Buick Motor Division** · Models with McPherson strut front suspension without strut mounted eccentric cam. | | | |
| Skyhawk (FWD),<br>Skylark (FWD),<br>Century (FWD) | 82 | Caster is not adjustable. | Loosen the lower strut retaining bolts. Grasp the top of the tire firmly and move tire in or out to obtain correct reading. Tighten bolts to final torque specs. (140 ft/lbs.). |
| **Buick Motors Division** Models with shim pack upper control arm adjustment. | | | |
| All Buick models except Skyhawk, 1982-79<br>Riviera and 1982-80 (FWD)<br>Skylark and 1982 Century | 82-73 | Increase positive caster by adding shims to rear shim pack and removing shims from front shim pack. Decrease positive caster by removing shims from rear shim pack and adding shims to front shim pack. | Increase positive camber by removing shims equally from front and rear shim packs. Decrease positive camber by adding shims equally to front and rear shim pack. |
| **Buick Motor Division** · Models with McPherson strut front suspension and strut mounted eccentric cam. | | | |
| Skylark (FWD) | 81-80 | Caster is not adjustable. | Turn eccentric cam on strut mounting bracket as necessary. |
| **Buick Motor Division** · Models with cam type adjustment. | | | |
| Skyhawk | 81-75 | Turn rear cam as necessary. | Turn front cam as necessary. |

| MAKE AND MODEL | YEAR | MANNER OF CASTER ADJUSTMENT | MANNER OF CAMBER ADJUSTMENT |
|---|---|---|---|
| **Buick Motor Division** - Models with upper control arm cam adjustment. | | | |
| Riviera (FWD) | 82-79 | Turn front and rear cams as necessary. | Turn front and rear cams as necessary. |
| **Cadillac Motor Division** - Models with McPherson strut front suspension without strut mounted eccentric cam. | | | |
| Cimarron | 82 | Caster is not adjustable. | Loosen the lower strut retaining bolts. Grasp the top of the tire firmly and move tire in or out to obtain correct reading. Tighten bolts to final torque specs. (140 ft/lbs.) |
| **Cadillac Motor Division** - Models with shim pack upper control arm adjustment. | | | |
| All models except Eldorado, 1981-80 Seville and Cimarron Seville | 82-77 79-76 | Increase positive caster by adding shims to rear shim pack and removing shims from front shim pack. Decrease positive caster by removing shims from rear shim pack and adding shims to front shim pack. | Increase positive camber by removing shims equally from front and rear shim packs. Decrease positive camber by adding shims equally to both front and rear shim packs. |
| **Cadillac Motor Division** - Models with upper control arm cam adjustment. | | | |
| Eldorado (FWD) Seville (FWD) | 82-73 82-80 | Turn front and rear cams as necessary. | Turn front and rear cams as necessary. |
| **Cadillac Motor Division** - Models with eccentric on upper ball joint. | | | |
| All models except Eldorado and Seville | 76-73 | Increase positive caster by shortening lower control arm strut. Decrease positive caster by lengthening lower control arm strut. | Turn ball joint eccentric as necessary. |
| **Chevrolet Motor Division** - Models with shim pack upper control arm adjustment. | | | |
| All models except Vega, Monza, 1982-80 Citation, Celebrity and Cavalier | 82-73 | Increase positive caster by adding shims to rear shim pack and removing shims from front shim pack. Decrease positive caster by removing shim from rear shim pack and adding shims to front shim pack. | Increase positive camber by removing shims equally from front and rear shim packs. Decrease positive camber by adding shims equally to front and rear shim packs. |
| **Chevrolet Motor Division** - Models with cam type adjustment. | | | |
| Vega, Monza | 80-73 | Turn rear cam as necessary. | Turn front cam as necessary. |
| **Chevrolet Motor Division** - Models with McPherson strut front suspension and strut mounted eccentric cam. | | | |
| Citation (FWD) | 81-80 | Caster is not adjustable. | Turn eccentric cam on strut mounting bracket as necessary. |
| **Chevrolet Motor Division** - Models with McPherson strut front suspension without strut mounted eccentric cam. | | | |
| Celebrity, Citation and Cavelier | 82 | Caster is not adjustable. | Loosen the lower strut retaining bolts. Grasp the top of the tire firmly and move tire in or out to obtain correct reading. Tighten bolts to final torque specs. (140 ft/lbs.). |
| **Oldsmobile Motor Division** - Models with McPherson strut front suspension without strut mounted eccentric cam. | | | |
| Ciera, Omega, Firenza | 82 | Caster is not adjustable. | (See Chevrolet Celebrity) |
| **Oldsmobile Division** - Models with shim pack upper control arm adjustment. | | | |
| All models except Starfire, Toronado, 1981-80 Omega (FWD), Ciera, and Firenza | 82-73 | Increase positive caster by adding shims to rear shim pack and removing shims from front shim pack. Decrease positive caster by removing shims from rear shim pack and adding shims to front shim pack. | Increase positive camber by removing shims equally from front and rear shim packs. Decrease positive camber by adding shims equally to front and rear shim packs. |
| **Oldsmobile Division** - Models with cam type adjustment. | | | |
| Starfire | 80-75 | Turn rear cam as necessary. | Turn front cam as necessary. |
| **Oldsmobile Division** - Models with McPherson strut suspension and strut mounted eccentric cam. | | | |
| Omega (FWD) | 81-80 | Caster is not adjustable. | Turn eccentric cam on strut mounting bracket as necessary. |
| **Oldsmobile Division** Toronado | 82-73 | Turn front and rear cams as necessary. | Turn front and rear cams as necessary. |
| **Pontiac Motor Division** - Models with McPherson strut front suspension without strut mounted eccentric cam. | | | |
| Phoenix (FWD), J2000, A6000 | 82 | Caster is not adjustable. | (See Chevrolet Celebrity) |
| **Pontiac Motor Division** - Models with shim pack upper control arm adjustment. | | | |
| All models except Astre, Sunbird, 1981-80 Phoenix (FWD), J2000 and A6000 | 82-73 | Increase positive caster by adding shims to rear shim pack and removing shims from front shim pack. Decrease positive caster by removing shim from rear shim pack and adding shims to front shim pack. | Increase positive camber by removing shims equally from front and rear shim packs. Decrease positive camber by adding shims equally to front and rear shim pack. |
| **Pontiac Motor Division** - Models with McPherson strut front suspension and strut mounted eccentric cam. | | | |
| Phoenix (FWD) | 81-80 | Caster is not adjustable. | Turn eccentric cam on strut mounting bracket as necessary. |
| **Pontiac Motor Division** - Models with cam type adjustment. | | | |
| Astre, Sunbird | 80-73 | Turn rear cam as necessary. | Turn front cam as necessary. |

# BALL JOINT INSPECTION & LIMITS

NOTE: Vehicle weight must be released from load carring ball joints when performing tests.

1. To unload lower ball joint (spring or torsion bar mounted on lower arm) support vehicle weight on lower control arm.
2. To unload upper ball joint (spring mounted on upper control arm) support vehicle weight on frame or cross member.
3. When testing the non-load carrying ball joint, if any measurable movement is detected, replace ball joint. Wheel bearing play is always evident. Do not mistake wheel bearing play for worn ball joint.

CAUTION: Proper safety precautions must be taken when testing ball joints. Safety stands must be used to support vehicle weight. Use proper measuring tools when checking ball joint tolerances.

Axial movement is the vertical (up and down) movement measured when lifting the tire or control arm with pry bar.

Radial movement is the horizontal (in and out) movement of the wheel assembly.

| Make & Model | Year | Max. Axial Tolerance |
|---|---|---|
| **American Motors** | | |
| All | 82-81 | .160 Note 1 |
| All except Pacer | 80-73 | .080 Note 1 |
| Pacer | 79-75 | Note 2 |
| | | |
| **Chrysler Corp.** | | |
| All Models w/RWD | 82-77 | .030 |
| Omni, Horizon, TC3, 024 | | |
| Aries, Reliant, 400, LeBaron | 82-81 | Note 6 |
| Omni, Horizon | 80-78 | Note 7 |
| All except Barracuda, Challenger, Dart, Valiant | 76-74 | .020 |
| Barracuda, Challenger, Dart, Valiant | 75-74 | .070 |
| All except Satellite, Coronet, Charger | 1973 | .070 |
| Satellite, Coronet, Charger | 1973 | Zero |
| Imperial | 73 | Zero |
| | | |
| **Ford Motor Company** | | |
| All except Escort, Lynx, LN7, EXP | 82-81 | Note 3 |
| Escort, EXP, Lynx, LN7 | 82-81 | Note 5 |
| Ford/Mercury, Lincoln Continental, Continental Mark VI, T-Bird/Cougar XR-7 | 1980 | Note 3 |
| Mustang/Capri | 80-79 | Note 3 |
| Fairmont/Zephyr | 80-78 | Note 3 |
| Granada, Monarch, Versailles | 80-75 | Note 4 |
| All except above | 80-73 | .250 |
| | | |
| **General Motors Corp.** | | |
| Buick | | |
| Skyhawk (FWD) Century (FWD) | 82 | Note 5 |
| Skylark (FWD) | 82-80 | Note 5 |
| Riviera | 82-79 | .125 |
| Skyhawk (RWD) | 82-76 | Note 3 |
| Full Size | 82-74 | Note 3 |
| Appollo Skylark | 79-75 | Note 3 |
| Riviera | 78-74 | Note 3 |
| Appollo | 74-73 | .0625 |
| Full Size, Riviera | 1973 | Note 4 |

| Make & Model | Year | Max. Axial Tolerance |
|---|---|---|
| Cadillac | | |
| Cimarron | 82 | Note 5 |
| Seville | 82-80 | .125 |
| All except Eldorado, Seville & Cimarron | 82-74 | Note 3 |
| Eldorado | 82-73 | .125 |
| Seville | 79-76 | Note 3 |
| All except Eldorado | 73 | .0625 |
| Chevrolet | | |
| Citation, Celebrity, Cavalier | 82-80 | Note 5 |
| Chevette | 82-76 | Note 3 |
| Chevelle, Malibu, Monte Carlo, Camaro | 82-74 | Note 3 |
| Full Size | 82-73 | Note 3 |
| Corvette | 82-73 | .0625 |
| Monza | 80-75 | Note 3 |
| Nova | 79-75 | Note 3 |
| Vega | 77-75 | Note 3 |
| Vega | 74-73 | .0625 |
| Nova | 74-73 | .0625 |
| Chevelle, Malibu, Monte Carlo | 1973 | Note 4 |
| Camaro | 73 | Note 4 |
| Oldsmobile | | |
| Omega, Ciera, Firenza | 82-80 | Note 5 |
| Cutlass w/RWD | 82-74 | Note 3 |
| Full Size | 82-73 | Note 3 |
| Toronado | 82-73 | .125 |
| Starfire | 80-75 | Note 3 |
| Omega | 79-75 | Note 3 |
| Omega | 74-73 | .0625 |
| Cutlass | 1973 | Note 4 |
| Pontiac | | |
| Phoenix, J2000, A6000 | 82-80 | Note 5 |
| LeMans, Grand AM, Grand Prix | 82-75 | Note 3 |
| Firebird | 82-74 | Note 3 |
| Full Size | 82-73 | Note 3 |
| Astre, Sunbird | 80-75 | Note 3 |
| Ventura, Phoenix | 79-75 | Note 3 |
| Ventura | 74-73 | .050 |
| Astre | 74-73 | .0625 |
| LeMans, Grand AM, Grand Prix | 1973 | Note 4 |

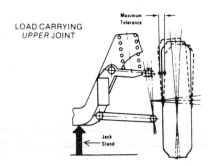

LOAD CARRYING UPPER JOINT

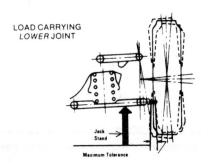

LOAD CARRYING LOWER JOINT

NOTE 1: Maximum radial movement .160″.
NOTE 2: Vehicle level and not raised, remove lube plug and insert rod, scribe mark on rod and measure from lip of rod to scribe mark, if greater than 7/16 replace ball joint.
NOTE 3: Grease filling nipple boss or solid boss should project outside cover surface.
NOTE 4: If there is movement between spindle and control arm of load carrying joint, replace joint.
NOTE 5: If there is movement between lower control arm and spindles, replace lower ball joint.
NOTE 6: With weight of vehicle on tires, try to move grease fitting with fingers. If movement is present, replace joint.
NOTE 7: Maximum of .050″ axial movement between lower leg of steering knuckle and lower control arm.

# CHRYSLER CORPORATION
## FRONT SUSPENSION HEIGHT ADJUSTMENT PROCEDURES AND SPECIFICATIONS

Front suspension height must be set to specifications *before* vehicle alignment is checked or reset. Maximum allowable height difference from side to side is 1/8 inch.

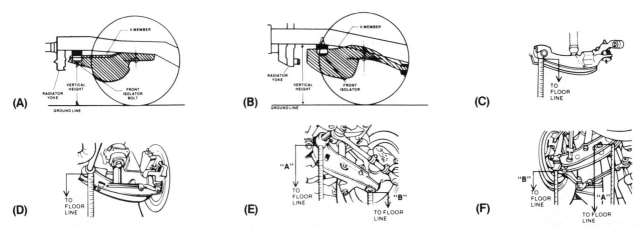

| Year | Vehicle Identification | Pro-cedure | Suspension Height (Inches) | Suspension Height (Millimeters) |
|---|---|---|---|---|
| 1982 | Mirada, Cordoba, Gran Fury, Diplomat, New Yorker, Imperial, Caravelle | A | 12½ ± ¼ | 317.5 ± 6.4 |
| 81-80 | Aspen, Volare, Diplomat, Le Baron, Caravelle, Mirada and Cordoba | A | 12½ ± ¼ | 317.5 ± 6.4 |
| 81-80 | Newport, New Yorker, St. Regis, Gran Fury | B | 16¾ ± ¼ | 425.5 ± 6.4 |
| 1979 | Aspen, Volare, Diplomat, Le Baron and Caravelle | C | 10¼ | 260.4 |
| 1979 | Newport, New Yorker, St. Regis Cordoba, Magnum XE | D | 10¾ | 273.1 |
| 1978 | Aspen, Volare, Diplomat, Le Baron | C | 10¼ | 260.4 |
| 1978 | Monaco, Fury, Charger SE, Magnum XE, Cordoba except Monaco and Fury Wagons | D | 10¾ | 273.1 |
| 1978 | Monaco and Fury Wagons | D | 11¼ | 285.6 |
| 1978 | Newport, New Yorker | D | 10⅛ | 257.2 |
| 1977 | Aspen, Volare, Diplomat, Le Baron | C | 10¼ | 260.4 |
| 1977 | Monaco, Fury, Charger SE, Cordoba except Monaco and Fury Wagons | D | 10¾ | 273.1 |
| 1977 | Monaco and Fury Wagons | D | 11¼ | 285.6 |
| 1977 | Royal Monaco, Gran Fury, Newport, New Yorker | D | 10⅛ | 257.2 |

| Year | Vehicle Identification | Pro-cedure | Suspension Height (Inches) | Suspension Height (Millimeters) |
|---|---|---|---|---|
| 1976 | Valiant, Dart | D | 10¹⁵/₁₆ | 277.8 |
| 1976 | Aspen, Volare | C | 10¼ | 260.4 |
| 1976 | Coronet, Fury, Charge SE, Cordoba except Coronet and Fury Wagons | D | 10¾ | 273.1 |
| 1976 | Coronet and Fury Wagons | D | 11¼ | 285.6 |
| 1976 | Gran Fury, Monaco, Newport, New Yorker | D | 10⅛ | 257.2 |
| 1975 | Valiant, Dart | D | 10¹⁵/₁₆ | 277.8 |
| 1975 | Coronet, Fury, Charger SE, Cordoba except Coronet and Fury Wagons | D | 10¾ | 273.1 |
| 1975 | Coronet and Fury Wagons | D | 11¼ | 285.6 |
| 1975 | Gran Fury, Monaco, Chrysler, Imperial | D | 10⅛ | 257.2 |
| 1974 | Valiant, Dart | F | 1⅞ | 47.6 |
| 1974 | Barracuda, Challenger | F | 1⅛ | 28.6 |
| 1974 | Satellite, Coronet, Charger | E | 1⅞ | 47.6 |
| 1974 | Fury, Polara, Monaco, Chrysler | E | 1 | 25.4 |
| 1974 | Imperial | E | 1 | 25.4 |
| 1973 | Valiant, Dart 4 door only | F | 2⅛ | 54.0 |
| 1973 | Valiant, Dart 2 door only | F | 1⅞ | 47.6 |
| 1973 | Barracuda, Challenger | F | 1⅛ | 28.6 |
| 1973 | Satellite, Coronet, Charger | E | 1⅞ | 47.6 |
| 1973 | Fury, Polara, Monaco | E | 1½ | 38.1 |
| 1973 | Chrysler except Imperial | E | 1¼ | 31.8 |
| 1973 | Imperial | E | 1¾ | 44.5 |

**NOTE:** Before measuring or setting front suspension height, the vehicle should be devoid of cargo and passengers. The tire pressure should be set to specifications. The gasoline tank should be filled. If the gasoline tank cannot be filled, weight should be added to the rear of the vehicle to compensate for the missing gasoline. The front of the car must be jounced vigorously to eliminate friction effects before making car height measurements. The vehicle should be released at the bottom of the downward motion.

**MEASUREMENT PROCEDURE "A"**—Vehicle resting on its tires. Measure from the head of the front suspension cross member front isolator bolt to the floor. (See Ill. "A").

**MEASUREMENT PROCEDURE "B"**—Vehicle resting on its tires. Measure from the bottom of the front frame rail between the radiator yoke and the forward edge of the front suspension cross member to the floor line. (See Ill. "B").

**MEASUREMENT PROCEDURE "C"**—Vehicle resting on its tires. Measure from the lowest point of the lower control arm inner pivot bushing to the floor line indicated by the lowest point of the front tires (See Ill. "C").

**MEASUREMENT PROCEDURE "D"**—Vehicle resting on its tires. Measure from the lowest point of the lower control arm torsion bar anchor at a point 1 inch from the rear face of the anchor to the floor line indicated by the lowest point of the front tires. (See Ill. "D").

**MEASUREMENT PROCEDURE "E"**—Vehicle resting on its tires. Measure the distance from the lowest point of the front torsion bar anchor at the rear of the lower control arm flange to the floor line. Record as distance "A". Measure from the lowest point of the ball joint housing to the floor line and record as distance "B". Subtract distance "B" from distance "A". Answer must equal measurement shown in table or adjustment is required (see Ill. "E"). Measure one side at a time.

**MEASUREMENT PROCEDURE "F"**—Vehicle resting on its tires. Measure the distance from the lowest point of the lower control arm adjusting blade to the floor line. Record as distance "A". Measure the distance from the lowest point of the steering knuckle arm at its center line and record as distance "B". Subtract distance "B" from distance "A". Answer must equal measurement shown in table or adjustment is required. (see Ill. "F"). Measure one side at a time.

# CADILLAC SUSPENSION HEIGHT TABLES

| Model & Year | Front Suspension Inches | Front Suspension MM | Rear Suspension Inches | Rear Suspension MM |
|---|---|---|---|---|
| **Cimarron** | | | | |
| 82 | 1 9/32 | 32.4 | 6 3/4 | 172.0 |
| **Brougham** | | | | |
| 82 Std. (Gas) | 1 7/8-2 5/8 | 47-67 | 5 5/32-5 15/16 | 131-151 |
| 82 Std. (Diesel) | 1 7/8-2 5/8 | 48-68 | 5 5/32-5 15/16 | 131-151 |
| 82 Elc. (Gas) | 1 13/16-2 9/16 | 45-65 | 4 13/16-5 19/32 | 122-142 |
| 82 Elc. (Diesel) | 1 7/8-2 5/8 | 46-66 | 4 13/16-5 19/32 | 122-142 |
| 81-80 w/o ALC (Gas) | 1 13/16-2 19/32 | 45-65 | 5 9/32-6 1/16 | 134-154 |
| 81-80 w/o ALC (Diesel) | 1 7/8-2 5/8 | 47-67 | 5 5/16-6 5/64 | 134-154 |
| 81-80 w/ALC (Gas) | 1 11/16-2 1/2 | 43-63 | 4 3/4-5 17/32 | 120-140 |
| 81-80 w/ALC (Diesel) | 1 3/4-2 9/16 | 45-65 | 4 25/32-5 9/16 | 121-141 |
| 79-77 | 1 7/8-2 5/8 | 48-67 | 5 3/8-6 1/8 | 137-161 |
| 76 | 3 7/8-4 5/8 | 98-118 | 4 7/8-5 5/8 | 124-143 |
| 75 | 3 1/8-3 7/8 | 79-98 | 4 7/8-5 5/8 | 124-143 |
| 74-73 | 3 7/8-4 5/8 | 98-118 | 4 7/8-5 5/8 | 124-143 |
| **DeVille, Calais** | | | | |
| 82 Cpe. Std. (Gas) | 1 7/8-2 5/8 | 47-67 | 5 1/8-5 29/32 | 130-150 |
| 82 Cpe. Std. (Diesel) | 1 7/8-2 5/8 | 48-68 | 5 1/8-5 29/32 | 130-151 |
| 82 Cpe. Elc. (Gas) | 1 13/16-2 9/16 | 45-65 | 4 25/32-5 9/16 | 121-141 |
| 82 Cpe. Elc. (Diesel) | 1 7/8-2 5/8 | 46-66 | 4 25/32-5 9/16 | 121-141 |
| 82 Sdn. Std. (Gas) | 1 7/8-2 5/8 | 47-67 | 5 1/8-5 15/16 | 130-151 |
| 82 Sdn. Std. (Diesel) | 1 7/8-2 5/8 | 48-68 | 5 5/32-5 15/16 | 131-151 |
| 82 Sdn. Elc. (Gas) | 1 13/16-2 9/16 | 45-65 | 4 13/16-5 19/32 | 122-142 |
| 82 Sdn. Elc. (Diesel) | 1 7/8-2 5/8 | 46-66 | 4 13/16-5 19/32 | 122-142 |
| 81-80 w/o ALC (Gas) | 1 13/16-2 19/32 | 45-65 | 5 7/64-5 29/32 | 129-149 |
| 81-80 w/o ALC (Diesel) | 1 7/8-2 5/8 | 46-66 | 5 1/8-5 15/16 | 130-150 |
| 81-80 w/ALC (Gas) | 1 11/16-2 1/2 | 43-63 | 4 3/4-5 17/32 | 120-140 |
| 81-80 w/ALC (Diesel) | 1 3/4-2 9/16 | 45-65 | 4 25/32-5 9/16 | 121-141 |
| 79-77 w/ALC | 2 1/8-2 7/8 | 54-73 | 5 3/8-6 1/8 | 136-156 |
| 79-77 w/o ALC | 2 1/8-2 7/8 | 54-73 | 5 3/4-6 1/2 | 146-170 |
| 76 w/ALC | 3 15/16-4 11/16 | 100-119 | 4 7/8-5 5/8 | 124-143 |
| 76 w/o ALC | 3 15/16-4 11/16 | 100-119 | 5 3/8-6 1/8 | 137-161 |
| 75 w/ALC | 3 3/16-3 15/16 | 81-100 | 4 7/8-5 5/8 | 124-143 |
| 75 w/o ALC | 3 3/16-3 15/16 | 81-100 | 5 13/16-6 1/8 | 148-161 |
| 74-73 w/ALC | 3 15/16-4 11/16 | 100-119 | 4 7/8-5 5/8 | 124-143 |
| 74-73 w/o ALC | 3 15/16-4 11/16 | 100-119 | 5 3/8-6 1/8 | 137-161 |
| **Seville** | | | | |
| 82 (Gas) | 5 3/16-6 | 132-152 | 4 27/32-5 5/8 | 123-143 |
| 82 (Diesel) | 5 1/4-6 1/32 | 133-153 | 4 7/8-5 11/16 | 124-144 |
| 81-80 (Gas) | 5 1/4-6 1/32 (11 5/8-11 7/8) | 133-153 (281-301) | 5 1/16-5 13/16 | 128-148 |
| 81-80 (Diesel) | 5 1/4-6 1/32 | 133-153 | 5-5 25/32 | 127-147 |
| 79-77 w/ALC connected | 2 1/2 | 63 | 3 7/8 | 98 |

| Model & Year | Front Suspension Inches | Front Suspension MM | Rear Suspension Inches | Rear Suspension MM |
|---|---|---|---|---|
| 79-77 w/ALC disconnected | 2 1/2 | 63 | 3 1/2 | 89 |
| 76 w/ALC connected | 1 3/4-2 1/2 | 44-63 | 3 1/2-4 1/4 | 89-107 |
| 76 w/ALC disconnected | 1 3/4-2 1/2 | 44-63 | 3 1/8-3 7/8 | 79-98 |
| **Fleetwood Limo** | | | | |
| 82 Elc. (6.0L) | 2 3/16-2 15/16 | 55-75 | 5 7/32-6 1/32 | 133-153 |
| 81-80 w/ALC | 2 1/8-2 29/32 | 54-74 | 5 9/32-6 1/16 | 134-154 |
| 79-78 w/ALC disconnected | 1 13/16-2 5/8 | 46-76 | 5 3/8-6 1/8 | 136-156 |
| **Fleetwood Formal Limo** | | | | |
| 82 Elc. (6.0L) | 1 31/32-2 3/4 | 50-70 | 5 1/32-5 13/16 | 128-148 |
| 81-80 w/ALC | 1 59/64-2 23/32 | 49-69 | 5 1/8-5 15/16 | 130-150 |
| 79-78 w/ALC disconnected | 1 13/16-2 5/8 | 46-76 | 5 3/8-6 1/8 | 136-156 |
| **Commercial Chassis** | | | | |
| 81-80 | 2 21/64-3 7/64 | 59-79 | 6 3/16-6 31/32 | 157-177 |
| 79-78 w/ALC disconnected | 1 13/16-2 5/8 | 46-76 | 5 3/8-6 1/8 | 136-156 |
| **All 75 Series, Commerical Chassis** | | | | |
| 77 | 1 13/16-2 5/8 | 46-76 | 5 3/8-6 1/8 | 136-156 |
| 76 w/radial tires | 4 5/8-5 5/16 | 118-135 | 6 1/16-6 7/8 | 159-180 |
| 76 w/o radial tires | 4 1/4-4 15/16 | 108-125 | 5 5/8-6 7/16 | 143-169 |
| 75 | 3 5/8-4 3/8 | 92-111 | 5 13/16-6 9/16 | 148-172 |
| 74 w/radial tires | 4 5/8-5 5/16 | 118-135 | 6 1/16-6 7/8 | 159-180 |
| 74 w/o radial tires | 4 1/4-4 15/16 | 108-125 | 5 5/8-6 7/8 | 143-169 |
| 73 | 4 7/16-5 3/16 | 113-132 | 5 9/16-6 5/16 | 141-165 |
| **Eldorado** | | | | |
| 82 (Gas) | 5 3/16-6 | 132-152 | 4 27/32-5 5/8 | 123-143 |
| 82 (Diesel) | 5 1/4-6 1/32 | 133-153 | 4 7/8-5 11/16 | 124-144 |
| 81-80 (Gas) | 5 1/4-6 1/32 (11 5/8-11 7/8) | 133-153 (281-301) | 5-5 25/32 | 127-147 |
| 81-80 (Diesel) | 5 1/4-6 1/32 (11 5/8-11 7/8) | 133-153 (281-301) | 4 31/32-5 3/4 | 126-146 |
| 79 (Note 1) | 5 21/32-6 13/32 (11 1/32-11 25/32) | 148-168 (280-300) | 5 1/8-5 7/8 | 129-149 |
| 78-77 | 8 1/4-8 1/2 | 209-216 | 4 13/16-5 9/16 | 122-141 |
| 76 | 8 3/16-8 7/16 | 208-214 | 4 13/16-5 9/16 | 122-141 |
| 75 | 8 1/4-8 1/2 | 210-216 | 5 1/16-5 5/16 | 129-135 |
| 74 | 8 3/16-8 7/16 | 208-214 | 4 13/16-5 9/16 | 122-141 |
| 73 | 8-8 1/4 | 203-210 | 3 15/16-4 11/16 | 90-119 |

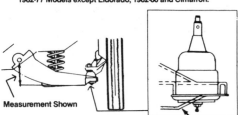

**FRONT SUSPENSION HEIGHT**
1982-77 Models except Eldorado, 1982-80 and Cimarron.

Measurement Shown

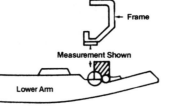

**FRONT SUSPENSION HEIGHT**
1976-73 Models except Eldorado and Seville

Frame
Measurement Shown
Lower Arm

**VERTICAL DISTANCE BETWEEN SPRING SEAT EDGES**
D
CIMARRON, Rear

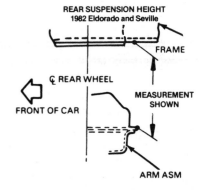

**REAR SUSPENSION HEIGHT**
All Models except Cimarron, Eldorado and Seville

Frame Rail
Measurement Shown
Axle Hsg. or Axle

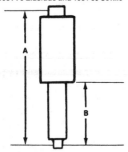

**FRONT SUSPENSION HEIGHT**
1981-73 Eldorado and 1981-80 Seville

A
B

**REAR SUSPENSION HEIGHT**
1982 Eldorado and Seville
FRAME
C REAR WHEEL
MEASUREMENT SHOWN
FRONT OF CAR
ARM ASM

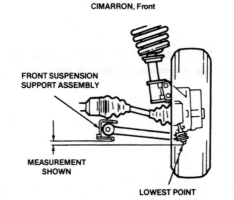

CIMARRON, Front

**FRONT SUSPENSION SUPPORT ASSEMBLY**
MEASUREMENT SHOWN
LOWEST POINT

**Note 1:** Center of bottom shock stud to bottom of shaft shield shown first, (see Ill. measurement B) center to center of both shock studs is shown in parenthesis( ), (see Ill. Measurement A).

ALC: Automatic Leveling Control. ELC: Electric Leveling Control.

# FRONT SUSPENSION—AMC EXCEPT PACER

Except for two types of steering knuckles, service procedures for both two wheel drive and four wheel drive vehicles are very similar. On two wheel drive cars, ball joints can be replaced in a conventional manner. On four wheel drive models, the complete control arm must be replaced.

## Front Wheel Alignment

Front wheel alignment, or steering geometry, refers to the various angles assumed by the components which form the front wheel turning mechanism. There are three adjustable, alignment angles which are caster, camber, and toe-in.

Caster describes the forward or rearward tilt (from vertical) of the steering knuckle. Tilting the top of the knuckle rearward provides positive caster. Tilting the top of the knuckle forward provides negative caster. Caster is a directional stability angle which enables the front wheels to return to a straight-ahead position after turns.

Adjust caster by loosening the strut rod jamnut and turning the rod adjusting nuts in or out to move the lower control arm forward or rearward to obtain the desired caster angle. Tighten adjusting nuts to 65 foot-pounds (88 N·m) torque and jamnut to 75 foot-pounds (102 N·m) torque when adjustment is completed.

Camber describes the inward or outward tilt of the wheel relative to the center of the automobile. An inward tilt of the top of the wheel produces negative camber. An outward tilt produces positive camber.

### Four Wheel Drive Suspension

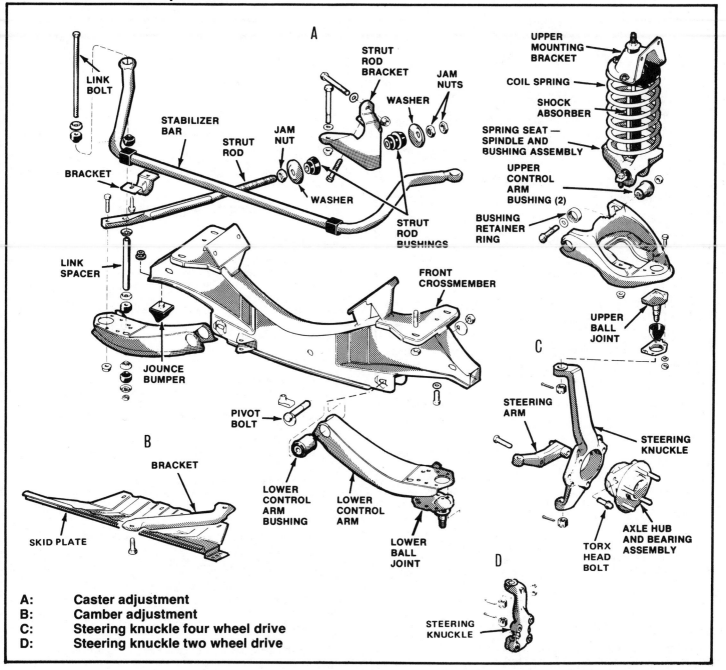

A:     Caster adjustment
B:     Camber adjustment
C:     Steering knuckle four wheel drive
D:     Steering knuckle two wheel drive

Camber greatly affects tire wear. Incorrect camber will cause abnormal wear on the tire, outside or inside edge.

Adjust camber by turning lower control arm inner pivot bolt eccentric. Tighten pivot bolt locknuts to 110 foot-pounds (149 N·m) torque after completing camber adjustment.

Toe-in is a condition that exists when the measured distance at the front of each tire is less than the distance at the rear of the tires. When the distance at the front is less than the rear, the tires are toed-in. Toe-in compensates for normal steering play and causes the tires to roll in a straight-ahead manner. Incorrect toe-in will wear the tires to a feathered edge.

Adjust toe-in by turning tie rod adjuster tubes in or out to shorten or lengthen tie rods to obtain desired toe-in. Place front wheels in straight-ahead position and center steering wheel and gear. Turn tie rod adjusting tubes equally in opposite directions to obtain desired toe-in setting. If steering wheel spoke position was disturbed during toe-in adjustment, correct spoke position by turning tie rod tubes equally in same direction until desired position is obtained.

# Front Wheel Bearings

## TWO WHEEL DRIVE

When repacking and adjusting front wheel bearings, use an EP-type, lithium base wheel bearing lubricant. Pack the bearings with a generous amount of lubricant and place extra lubricant in the rotor hub cavity between the bearings. Always use a new grease seal during assembly.

## Front Wheel Bearing—Two Wheel Drive

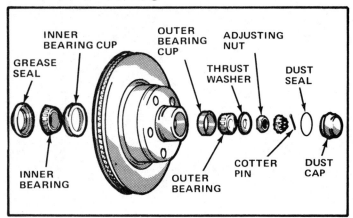

When inspecting, replacing, or repacking bearings, be sure the inner cones of the bearings are free to creep on the spindle. The bearings are designed to creep to allow a constantly changing load contact between the cones and the rollers. Polishing and applying lubricant to the spindle will permit this movement and prevent rust formation.

### Adjustment

1. Raise and support front of automobile.
2. Remove hub cap, grease cap and O-ring, cotter pin and nutlock.
3. On automobiles with styled wheels, remove wheel and hub cap. Install wheel.
4. Tighten spindle nut to 25 foot-pounds (34 N·m) torque while rotating wheel to seat bearings.
5. Loosen spindle nut 1/3-turn. While rotating wheel, tighten spindle nut to 6 inch-pounds (0.7 N·m) torque.

## Axle Hub Assembly—Eagle

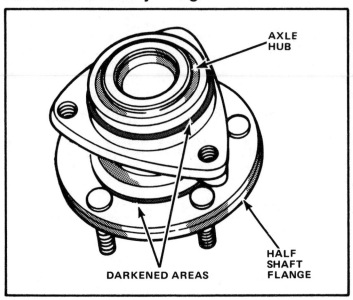

6. Install nutlock on spindle nut so cotter pin holes in nutlock and spindle are aligned.
7. Install replacement cotter pin, grease cap and hub cap.
8. On automobiles with styled wheels, remove wheel, install hub cap and install wheel.

## FOUR WHEEL DRIVE

### Adjustment

Four wheel drive models have a unique front axle hub and bearing assembly. The assembly is sealed and does not require lubrication, periodic maintenance, or adjustment. The hub has ball bearings which seat in races machined directly into the hub. There are darkened areas surrounding the bearing race areas of the hub. These darkened areas are from a heat treatment process, are normal, and should not be mistaken for a problem condition.

### Removal

1. Remove wheel, caliper and rotor.
2. Remove bolts attaching axle shaft flange to halfshaft.
3. Remove cotter pin, nut lock and axle hub nut.
4. Remove halfshaft.
5. Remove steering arm from steering knuckle.
6. Remove caliper anchor plate from steering knuckle.
7. Remove three Torx head bolts retaining hub assembly using too set J-25359.
8. Remove hub assembly from steering knuckle.
9. Clean grease from steering knuckle cavity.

### Installation

1. Partially fill hub cavity of steering knuckle with chassis lubricant and install hub assembly.
2. Tighten hub Torx head bolts to 75 foot-pounds (102 N·m) torque.
3. Install caliper anchor plate and plate retaining bolts.
4. Tighten caliper anchor plate retaining bolts to 100 foot-pounds (136 N·m) torque.
7. Install halfshaft. Install axle flange-to-shaft bolts and install hub nut.
8. Tighten halfshaft-to-flange bolts to 45 foot-pounds (61 N·m) torque.

9. Tighten hub nut to 175 foot-pounds (237 N·m) torque.
10. Install nut lock and cotter pin.
11. Install rotor, caliper and wheel.

## Shock Absorber

### ALL MODELS

#### Removal

1. Remove lower retaining nuts, washers and grommets.
2. Remove upper mounting bracket bolts/nuts from wheelhouse panel.
3. Remove upper bracket and shock absorber from wheelhouse panel.
4. Remove upper retaining nut from shock absorber and remove upper bracket from shock absorber.

#### Installation

1. Install washers, grommets, upper mounting bracket and nut on shock absorber if removed. Tighten nut to 8 foot-pounds (11N·m) torque.

## Shock Absorber

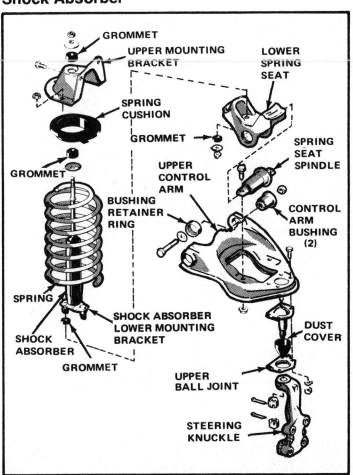

2. Extend shock absorber piston fully.
3. Install grommets on lower mounting studs and position shock absorber in wheelhouse panel.
4. Insert lower mounting studs into lower spring seat and install lower grommets, flat washers, and nuts. Tighten nuts to 15 foot-pounds (20N·m) torque.
5. Install and tighten upper mounting bracket attaching nuts/bolts to 20 foot-pounds (27N·m) torque.

## Upper Ball Joint

### TWO WHEEL DRIVE

#### Inspection

1. Remove upper ball joint lubrication plug and install a dial indicator gauge through the lubrication hole so that you can measure the up and down movement of the ball joint socket.
2. Place a pry bar under tire to load ball joint and raise tire several times to seat gauge tool pin.
3. Pry tire upward to load ball joint and record gauge reading; then release tire to unload ball joint and record gauge reading. Perform this operation several times to ensure accuracy.
4. The difference between load/no-load readings represents ball joint clearance. If clearance is more than 0.080-inch (2.03 mm), ball joint should be replaced.

#### Replacement

1. Install 2×4×5-inch (5.1×10.1×12.7 cm) wood block on frame side sill and under upper control arm.
2. Raise and support front of car.
3. Remove wheel, caliper, and rotor.
4. Remove ball stud cotter pin and retaining nut.
5. Install ball joint remover and loosen ball stud in steering knuckle. Do not remove tool at this time.
6. Place support stand under lower control arm.
7. Remove heads from ball joint attaching rivets using chisel or grinding tool.
8. Drive rivets out of ball joint and control arm using hammer and punch.
9. Disengage ball joint from control arm.
10. Remove tool from ball joint stud and remove ball joint from steering knuckle.
11. Position replacement ball joint in control arm and align bolt holes.
12. Install ball joint attaching bolts (supplied in ball joint replacement kits) and tighten nuts to 25 foot-pounds (34 N·m) torque.
13. Install steering knuckle and retaining nut on ball joint stud. Tighten nut to 75 foot-pounds (102 N·m) torque and install a new cotter pin.
14. Install rotor, caliper and wheel.

### FOUR WHEEL DRIVE

#### Inspection

Inspection procedures are the same as for two wheel drive models.

#### Replacement

If a ball joint is worn (upper and lower) the complete arm assembly must be replaced. Do not attempt to service the ball joint separately.

## Installing Wood Block

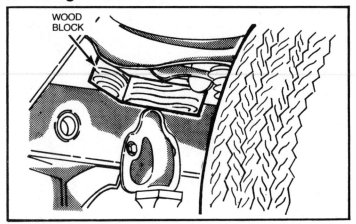

WOOD BLOCK

## Lower Ball Joint

### TWO WHEEL DRIVE

#### Inspection

1. Raise and support front of automobile.
2. Move lower portion of wheel and tire alternately toward and away from center of automobile. Perform this operation several times.
3. Lower ball joint is spring-equipped and preloaded in its socket at all times to minimize looseness and compensate for wear. If lower joint exhibits any lateral movement (shake), ball joint should be replaced.

#### Replacement

1. Install 2×4×5-inch (5.1×10.1×12.7 cm) wood block on frame side sill and under upper control arm.
2. Raise and support front of automobile.
3. Remove wheel, caliper, and rotor.
4. Disconnect strut rod at lower control arm.
5. Disconnect steering arm from steering knuckle.
6. Remove ball stud cotter pin and retaining nut.
7. Install ball joint removal tool and loosen ball stud in steering knuckle. Do not remove tool at this time.
8. Place support stand under lower control arm.
9. Remove heads from ball joint attaching rivets using chisel or grinding tool.
10. Drive rivets out of ball joint and control arm using punch.
11. Disengage ball joint from control arm.
12. Remove ball joint from control arm.
13. Position replacement ball joint on control arm and align bolt holes.
14. Install but do not tighten attaching bolts supplied in replacement ball joint kit.
15. Attach strut rod to lower control arm. Tighten rod attaching bolts to 75 foot-pounds (102 N·m) torque.
16. Tighten ball joint attaching bolts to 25 foot-pounds (34 N·m) torque.
17. Apply chassis grease to steering stops.
18. Install ball joint stud in steering knuckle.
19. Install retaining nut on ball stud. Tighten nut to 75 foot-pounds (102 N·m) torque and install replacement cotter pin.

20. Install steering arm on steering knuckle.
21. Install rotor, caliper, and wheel.

### FOUR WHEEL DRIVE

#### Inspection and Replacement

See the upper ball joint inspection and replacement procedures for four wheel drive models.

# Coil Spring

### ALL MODELS

#### Identification

A plastic identification tag which has the spring part number printed on it is attached to each coil spring. Whenever a spring must be replaced, refer to this part number when ordering a replacement spring.

#### Removal

1. Remove shock absorbers and mounting brackets.
2. Install spring compressor and compress spring approximately 1 inch (25.4 mm).
3. Remove lower spring seat pivot bolt retaining nuts.
4. Raise front of automobile until control arms are free of lower spring seat.

## Lower Spring Seat Position

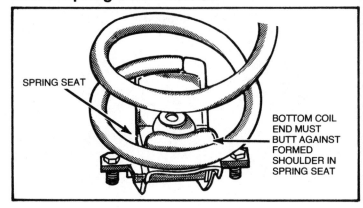

SPRING SEAT

BOTTOM COIL END MUST BUTT AGAINST FORMED SHOULDER IN SPRING SEAT

5. Remove wheel.
6. Pull lower spring seat away from automobile, and guide lower spring seat out and over upper control arm.
7. Remove spring compressor tool and remove lower retainer, spring seat, and spring.

─────────── CAUTION ───────────
*Do not use impact wrench to turn the compression nut. An impact wrench will place unnecessary stress on the compressor tool bolt threads which could result in thread damage or bolt breakage.*
─────────────────────────────────

#### Installation

1. Install spring compressor tool.
2. Install spring upper cushion on top coil of spring. Tape cushion in place to retain it.
3. Install spring in lower spring seat.

NOTE: One side of the lower spring seat has a formed shoulder to help locate the spring properly. Position the spring on the seat so the cut-off end of the spring bottom coil seats against this shoulder. If the spring seat was removed for service, be sure the shouldered end of the spring seat and cut-off end of the spring bottom coil are installed so they face the engine compartment.

4. Position spring in upper seat.
5. Align lower spring seat pivot so that retaining studs will enter upper control arm when spring is in position. Be sure spring lower coil end is properly positioned on seat.
6. Compress spring until lower spring seat pivot studs can be aligned with holes in upper control arm.
7. Turn compression nut counterclockwise and guide spring seat pivot studs into control arm.
8. Install wheel.
9. Remove supports and lower automobile.
10. Install and tighten lower spring seat pivot retaining nuts to 35 foot-pounds (47 N·m) torque.
11. Remove spring compressor tool.
12. Install shock absorber and mounting bracket.

## Upper Control Arm

### ALL MODELS

#### Removal

1. Remove shock absorber and mounting bracket. Install spring compressor tool.
2. Remove lower spring seat pivot retaining nuts, and turn compressor tool until spring is compressed approximately 2 inches (5.03 cm).
3. Raise and support front of automobile.
4. Remove wheel.
5. Remove upper ball joint stud cotter pin and retaining nut.
6. Remove upper ball joint stud from steering knuckle.
7. Remove control arm inner pivot bolts and control arm from wheelhouse panel.

#### Installation

1. Position control arm in wheelhouse panel and install inner pivot bolts.

─────────── CAUTION ───────────
*Do not tighten the pivot bolts until the automobile is resting on the wheels as ride height may be affected.*
──────────────────────────────

2. Install steering knuckle and retaining nut on ball joint stud. Tighten nut to 75 foot-pounds (102 N·m) torque and install a new cotter pin.
3. Turn spring compressor tool compression nut and guide spring seat pivot studs into control arm.

## Lower Control Arm

### TWO WHEEL DRIVE

#### Removal

1. Raise and support front of automobile.
2. Remove wheel, caliper, and rotor.
3. Disconnect steering arm from steering knuckle.
4. Remove lower ball joint stud cotter pin and retaining nut.
5. Remove ball stud from steering knuckle.
6. Disconnect stabilizer bar from control arm, if equipped.
7. Disconnect strut rod from control arm.
8. Remove inner pivot bolt and remove control arm from crossmember.

## Lower Control Arm Components—
## Spirit–AMX–Concord

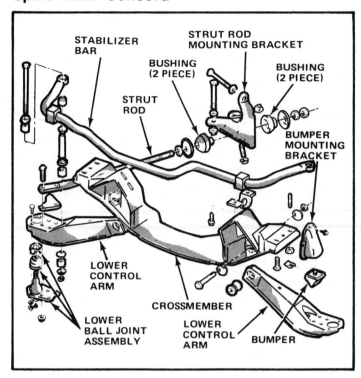

#### Installation

1. Position control arm in crossmember and install inner pivot bolt.

─────────── CAUTION ───────────
*Do not tighten the inner pivot bolt until the automobile weight is supported by the wheels as ride height may be affected.*
──────────────────────────────

2. Install steering knuckle and retaining nut on ball joint stud. Tighten nut to 75 foot-pounds (102 N·m) torque and install replacement cotter pin.
3. Connect strut rod to control arm. Tighten bolts to 75 foot-pounds (102 N·m) torque.
4. Connect stabilizer bar to control arm, if equipped. Tighten bolts to 7 foot-pounds (9 N·m) torque.
5. Connect steering arm to steering knuckle.
6. Install rotor, caliper and wheel.
7. Place a jack under control arm. Raise jack to compress spring slightly and tighten control arm inner pivot bolt to 110 foot-pounds (149 N·m) torque.

### FOUR WHEEL DRIVE

#### Removal

1. Remove cotter pin, nut lock and hub nut.
2. Raise and support front of automobile.
3. Remove wheel, caliper and rotor.
4. Remove lower ball joint cotter pin and retaining nut.
5. Remove ball stud from steering knuckle.
6. Remove halfshaft flange bolts.
7. Remove halfshaft.
8. Remove bolts attaching strut rod to control arm.
9. Disconnect stabilizer bar from control arm.
10. Remove inner pivot bolt and remove control arm.

#### Installation

1. Position control arm in crossmember and install inner pivot bolt.

---

**————— CAUTION —————**

*Do not tighten the inner pivot bolt until the automobile weight is supported by the wheels as ride height may be affected.*

2. Insert ball stud in steering knuckle and install retaining nut on ball joint stud. Tighten nut to 75 foot-pounds (102 N·m) torque and install replacement cotter pin.

3. Connect stabilizer bar to control arm. Tighten lock nut to 7 foot-pounds (9 N·m) torque.

4. Connect strut rod to control arm. Tighten bolts to 75 foot-pounds (102 N·m) torque.

5. Install halfshaft-to-axle flange bolts.

6. Tighten flange bolts to 45 foot-pounds (61 N·m) torque.

**NOTE:** If control arm is worn or bushing is not tight when installed, control arm must be replaced.

## Steering Knuckle and Spindle

### TWO WHEEL DRIVE

#### Removal

1. Raise and support front of automobile.
2. Remove wheel, caliper, and rotor.
3. Remove caliper anchor plate, adapter, steering spindle, and steering arm from knuckle.
4. Remove upper and lower ball joint stud cotter pins and retaining nuts.
5. Remove ball joint studs from steering knuckle.

#### Installation

1. Install steering knuckle and retaining nuts on ball joint studs. Tighten nuts to 75 foot-pounds (102 N·m) torque and install new cotter pins.
2. Install steering arm, spindle, caliper anchor plate, and adapter. Tighten bolts to 55 foot-pounds (75 N·m) torque.
3. Install rotor, caliper, and wheel.

### FOUR WHEEL DRIVE

#### Removal

1. Remove cotter pin, nut lock and hub nut.
2. Remove wheel, caliper and rotor.
3. Remove halfshaft-to-axle flange bolts.
4. Remove halfshaft.
5. Remove steering arm from steering knuckle.
6. Remove caliper anchor plate from steering knuckle.
7. Remove three Torx head attaching bolts remaining front wheel hub assembly.
8. Remove hub assembly from knuckle.
9. Remove rear hub seal from steering knuckle using small pry bar.
10. Remove upper and lower ball joint stud cotter pins and retaining nuts.
11. Remove ball joint studs from steering knuckle using a strike tool to loosen and remove studs from knuckle.

#### Installation

1. Install steering knuckle and ball joint retaining nuts on ball joint studs. Tighten nuts to 75 foot-pounds (102 N·m) torque and install new cotter pins.
2. Install hub rear seal.
3. Partially fill hub cavity of steering knuckle with chassis lubricant and install hub assembly in knuckle.
4. Tighten hub Torx head bolts to 75 foot-pounds (102 N·m) torque.
5. Install caliper anchor plate and retaining bolts.

6. Tighten caliper anchor plate bolts to 100 foot-pounds (136 N·m) torque.

7. Install steering arm and bolts.

8. Tighten steering arm bolts to 100 foot-pounds (136 N·m) torque.

9. Install halfshaft and shaft-to-axle flange bolts.

10. Tighten halfshaft-to-axle flange to 45 foot-pounds (61 N·m) torque.

11. Install rotor, caliper and hub nut.

12. Install wheel.

13. Lower automobile.

14. Tighten hub nut to 180 foot-pounds (244 N·m) torque. Install nut lock and a new cotter pin.

## Strut Rod and Bushing

### ALL MODELS

#### Replacement

1. Raise and support front of automobile.

### Strut Rod Bushings

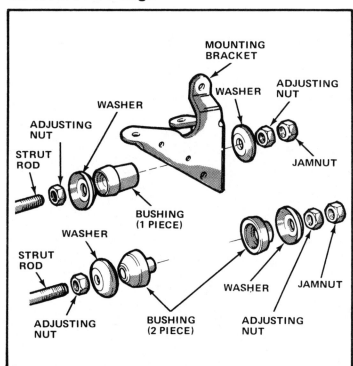

### Control Arm Bushing Replacement

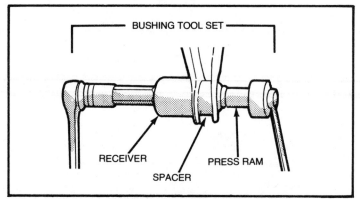

**Press out the old bushing. Press in the new bushing.**

2. Remove jamnut and caster adjustment nut from strut rod.

3. Disconnect strut rod from lower control arm and remove rod, bushings, and washers.

4. On automobiles with one-piece bushing, lubricate the bushing with soapy water and install.

**NOTE: A special tool is required to press the one-piece bushing in and out of the mounting bracket.**

# FRONT SUSPENSION—AMC PACER

## Front Wheel Alignment

Front wheel alignment, or steering geometry, refers to the various angles assumed by the components which form the front wheel turning mechanism. There are three adjustable alignment angles which are caster, camber and toe-in.

Caster describes the forward or rearward tilt (from vertical) of the steering knuckle. Tilting the top of the knuckle rearward provides positive caster. Tilting the top of the knuckle forward provides negative caster. Caster is a directional stability angle which enables the front wheels to return to a straight-ahead position after turns.

Caster is adjusted by turning only the rear pivot bolt eccentric. After adjustment, tighten the pivot bolt locknut to 95 foot-pounds torque.

Camber describes the inward or outward tilt of the wheel relative to the center of the automobile. An inward tilt of the wheel produces negative camber. An outward tilt produces positive camber. Camber greatly affects tire wear. Incorrect camber will cause abnormal wear on the tire outside or inside edge.

Camber is adjusted by turning both the front or both front and rear lower control arm pivot bolt eccentrics as necessary to obtain the desired camber angle. After adjustment, tighten the pivot bolt locknuts to 95 foot-pounds torque.

Toe-in is a condition that exists when the measured distance at the front of each tire is less than the distance at the rear of the tires. When the distance at front is less than the rear, the tires are toed-in. Toe-in compensates the normal steering play and causes the tires to roll in a straight-ahead manner. Incorrect toe-in will wear the tires to a feathered edge.

Adjust toe-in by turning tie rod adjuster tubes in or out to shorten or lengthen tie rods to obtain desired toe-in. Place front wheels in straight-ahead position and center steering wheel and gear. Turn tie rod adjusting tubes equally in opposite directions to obtain desired toe-in setting. If steering wheel spoke position was disturbed during toe-in adjustment, correct spoke position by turning tie rod tubes equally in same direction until desired position is obtained.

## Front Wheel Bearings

When repacking and adjusting front wheel bearings, use an EP-type, lithium base wheel bearing lubricant. Pack the bearings with a generous amount of lubricant and place extra lubricant in the rotor hub cavity between the bearings. Always use a new grease seal during assembly.

When inspecting, replacing, or repacking bearings, be sure the inner cones of the bearings are free to creep on the spindle. The bearings are designed to creep to allow a constantly changing load contact between the cones and the rollers. Polishing and applying lubricant to the spindle will permit this movement and prevent rust from forming.

### Adjustment

1. Raise and support front of automobile.

2. Remove hub cap, grease cap and O-ring, cotter pin and nutlock.

3. On automobiles with styled wheels, remove wheel, remove hub cap, and install wheel.

4. Tighten spindle nut to 25 foot-pounds (34 N·m) torque while rotating wheel to seat bearings.

5. Loosen spindle nut 1/3-turn and, while rotating wheel, tighten spindle nut to 6 inch-pounds (0.7 N·m) torque.

6. Install nutlock on spindle nut so cotter pin holes in nutlock and spindle are aligned.

7. Install replacement cotter pin, grease cap and hub cap.

8. On automobiles with styled wheels, remove wheel, install hub cap and install wheel.

## Shock Absorber

### Removal

1. Remove shock absorber upper locknut, retainer, and grommet.

2. Remove locknuts from shock absorber lower mounting studs and remove shock absorber.

3. Remove lower grommet and jounce bumper retainer from shock absorber piston rod.

### Installation

1. Install jounce bumper retainer and lower grommet on shock absorber piston rod.

2. Extend piston rod to full length.

3. Insert shock absorber through lower control arm and position shock absorber lower mounting bracket on mounting studs. Be sure piston rod is positioned in mounting hole in front crossmember.

4. Install locknuts on lower mounting studs. Tighten nuts to 20 foot-pounds (27.1 N·m) torque.

5. Install upper grommet, retainer, and locknut on shock absorber piston rod. Be sure locating shoulder on grommet seats in hole in crossmember. Tighten upper locknut to 8 foot-pounds (10.8 N·m) torque.

## Front Shock Absorber Assembly

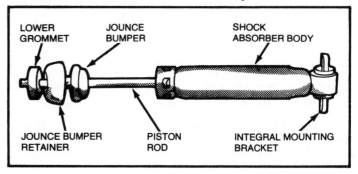

## Upper Ball Joint

### Inspection

1. Position hydraulic jack under lower control arm and raise car until wheel is off floor.

# SUSPENSIONS
## FRONT SUSPENSION—AMC PACER

## Front Suspension—Pacer

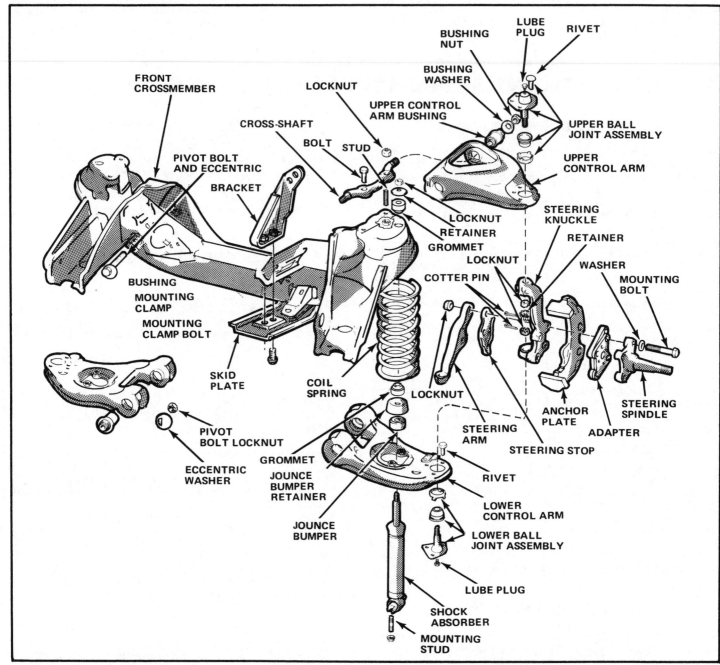

**Caster/camber adjustments :**      **Adjust camber first by moving both eccentric washers. Then adjust caster by moving rear eccentric only.**

2. Move top of tire toward and away from center of car. If ball joint exhibits any looseness or play, replace ball joint.

3. Move upper control arm up and down using pry bar. If ball joint exhibits any looseness or play, replace ball joint.

## Removal

────────── **CAUTION** ──────────
*Always support suspension components in such a way that there is no weight or strain on the component you are working on. This prevents accidents that can cause injury.*

1. Remove wheel and tire assembly.

2. Remove cotter pin and retaining nut from upper ball joint stud.

3. Position a jack under lower control arm and raise arm approximately 1 inch (2.54 cm).

4. Thread ball joint removal tool on ball stud.

5. Remove heads from rivets attaching ball joint to control arm using grinder or chisel.

6. Drive rivets out of ball joint and arm using punch.

7. Lower jack slightly and strike ball joint removal tool with hammer to loosen stud in steering knuckle.

8. Remove tool from ball stud and remove ball joint from knuckle.

## Installation

1. Position replacement ball joint in upper control arm.
2. Install ball joint-to-control arm attaching bolts and nuts supplied in replacement ball joint kit and tighten nuts to 25 foot-pounds (33.9 N·m) torque.

**NOTE: Install the replacement ball joint attaching bolts from the underside of the control arm. The nuts should be on top.**

3. Raise and support lower control arm with a hydraulic jack.
4. Engage upper ball joint stud in steering knuckle and install retaining nut. Tighten nut to 75 foot-pounds (101.6 N·m) torque and install retainer and a new cotter pin.
5. Install wheel and tire assembly.

## Lower Ball Joint

### Inspection

1. Position car on level surface.
2. Remove lube plug from lower ball joint.
3. Fabricate checking tool from a 2- to 3-inch (5.08 to 7.62 cm) length of stiff wire or thin rod.
4. Insert tool into lubrication plug hole until it contacts ball stud. Accurately mark tool with knife or scriber where it is aligned with outer edge of lube plug hole. Distance from the ball stud to outer edge of lubrication plug hole is ball joint clearance.
5. Carefully measure distance from end of tool to mark made in step 4. If distance measured is less than 7/16 inch (11.1 mm) ball joint is serviceable. However, if distance is 7/16 inch (11.1 mm) or more, ball joint should be replaced.

### Removal

1. Remove wheel, tire, caliper, and rotor.
2. Remove bolts attaching steering arm to steering knuckle and move steering arm aside.
3. Disconnect stabilizer bar link bolt at lower control arm, if equipped.
4. Support lower control arm with a jack.
5. Remove cotter pin and retaining nut from lower ball joint stud.
6. Thread ball joint removal tool on to the ball stud; lower the hydraulic jack.
7. Strike tool with hammer to loosen ball stud in knuckle.
8. Raise lower control arm with a jack. Unthread removal tool from ball joint stud and remove ball joint from steering knuckle.
9. Move steering knuckle, steering spindle, adapter, and anchor plate aside to provide working clearance.
10. Remove heads from ball joint attaching rivets using grinder or chisel.
11. Drive rivets out of ball joint and arm using punch.
12. Remove ball joint.

### Installation

1. Position replacement ball joint in lower control arm.
2. Install ball joint attaching bolts and nuts supplied in replacement ball joint kit and tighten bolts to 25 foot-pounds (33.9 N·m) torque.

**NOTE: Install the replacement ball joint attaching bolts from the underside of the control arm. The nuts should be on top.**

3. Remove wire supporting steering knuckle, steering spindle, adapter, and anchor plate.
4. Engage ball joint stud in steering knuckle and install retaining nut on ball stud. Tighten nut to 75 foot-pounds (101.6 N·m) torque and install replacement cotter pin.
5. Install steering arm on steering knuckle and install retaining bolts. Tighten bolts to 55 foot-pounds (74.5 N·m) torque.
6. Connect stabilizer bar link bolt to lower control arm, if equipped. Tighten nut to 7 foot-pounds (9.4 N·m) torque.

## Upper and Lower Ball Joints

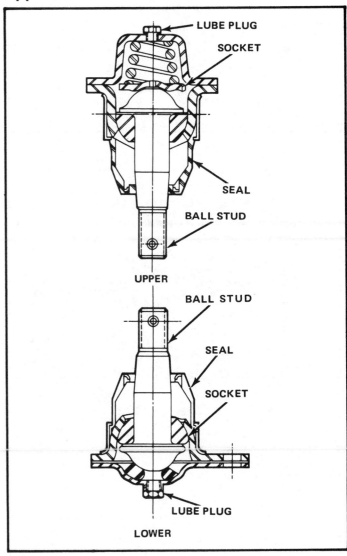

7. Install caliper, rotor, and wheel and tire assembly.
8. Remove supports and jack.
9. Check and adjust front wheel alignment if necessary.

## Coil Spring

### Removal

1. Remove shock absorber.
2. Disconnect stabilizer bar link bolt at lower control arm, if equipped.
3. Remove wheel, tire, caliper and rotor.
4. Remove bolts attaching steering arm to steering knuckle and move steering arm aside.
5. Install spring compressor tool.

———————— **CAUTION** ————————

*Do not use an impact wrench to turn the compressor sleeve. Use a ratchet handle only.*

6. Compress spring evenly until suspension parts are free of strain.
7. Remove cotter pin and nut from lower ball joint stud.

## Coil Spring Removal

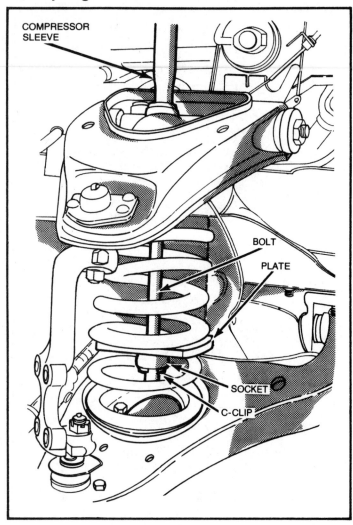

COMPRESSOR SLEEVE

BOLT

PLATE

SOCKET

C-CLIP

8. Thread tool onto stud and strike tool with hammer to loosen stud in steering knuckle.

9. Remove tool from ball joint stud and disengage stud from steering knuckle.

10. Move steering knuckle, steering spindle, and anchor plate aside to provide working clearance.

11. Move lower control arm aside, release spring compressor tool, and remove tool and coil spring.

### Installation

NOTE: The top coil of the spring is flat. This end is installed in the spring pocket of the front crossmember. Pacer model front springs do not use an insulator between the top coil of the spring and the spring pocket.

1. Position upper end of spring in front crossmember spring seat and align cut-off end of bottom coil with formed shoulder in spring seat.

2. Use a jack or support stand to support spring and control arm until spring compressor tool is installed.

3. Install spring compressor tool and compress spring.

4. Insert ball joint stud in steering knuckle and install retaining nut on ball stud. Tighten nut to 75 foot-pounds (101.6 N·m) torque and install a new cotter pin.

5. Install steering arm on steering knuckle and install retaining bolts and nuts. Tighten bolts to 55 foot-pounds (74.5 N·m) torque.

When spring and seat are aligned, release compressor tool, seat spring in control arm, and remove compressor tool.

NOTE: Be sure that the cut-off end of the bottom coil seats against the formed shoulder in the control arm spring seat.

6. Install shock absorber. Tighten lower mounting stud locknuts to 20 foot-pounds (27.1 N·m) torque.

7. Connect stabilizer bar link bolt to lower control arm. Tighten locknut to 7 ft.-lbs. (9.4 N·m) torque.

8. Install rotor, wheel, and tire.

9. Remove supports and lower car.

10. Install shock absorber upper grommet, retainer, and locknut. Be sure locating shoulder on grommet seats in hole in front crossmember. Tighten upper locknut on shock absorber to 8 foot-pounds (10.8 N·m) torque.

# Upper Control Arm

### Removal

1. Raise and support front of car.

2. Remove wheel and tire assembly.

3. Remove cotter pin, retaining nut, and locknut from upper ball joint stud.

4. Thread tool on stud and strike tool with hammer to loosen stud in steering knuckle.

5. Support lower control arm and rotor using hydraulic jack.

6. Remove tool from ball joint stud and disengage stud from steering knuckle.

7. Remove retaining stud locknuts attaching cross-shaft to front crossmember and remove upper control arm assembly.

### BUSHING REPLACEMENT

A special C-clamp tool is required to remove and install upper control arm bushings.

### Installation

1. Position upper control arm assembly on crossmember and install locknuts. Tighten locknuts to 80 foot-pounds (108.4 N·m) torque.

2. Insert upper ball joint stud in steering knuckle. Raise lower control using hydraulic jack to ease installation and relieve spring tension on upper ball joint.

3. Install locknut on upper ball joint stud. Tighten nut to 75 foot-pounds (101.6 N·m) torque and install nut retainer and replacement cotter pin. Do not loosen nut to align slots in nut with hole in ball stud.

4. Tighten cross-shaft bushing nuts to 60 foot-pounds (81.3 N·m) torque.

## Upper Control Arm Removal and Installation

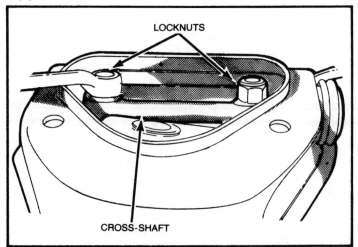

LOCKNUTS

CROSS-SHAFT

## Upper Control Arm Bushing Removal

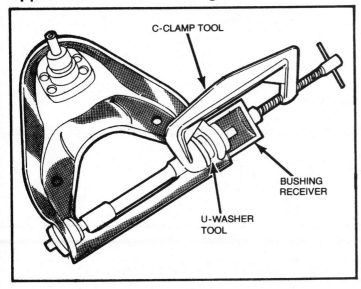

## Lower Control Arm Rear Bushing Installation

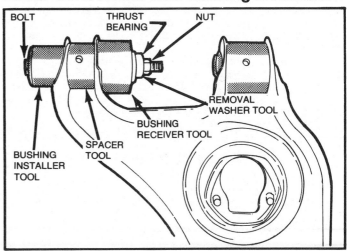

## Lower Control Arm

### Removal

1. Remove shock absorber.
2. Disconnect stabilizer bar link bolt at lower control arm.
3. Remove wheel tire, caliper, and rotor.
4. Remove bolts attaching steering arm to steering knuckle and mobe steering arm aside.
5. Install spring compressor and compress spring evenly until all strain is off lower control arm.
6. Remove cotter pin and nut from lower ball joint stud.
7. Thread tool on stud and strike tool with hammer to loosen stud in steering knuckle.
8. Remove tool from ball joint stud and disengage stud from steering knuckle.
9. Move steering knuckle, steering spindle, adapter, and anchor plate aside to provide working clearance.
10. Remove two pivot bolts attaching lower control arm to front crossmember and remove lower control arm.

——————— CAUTION ———————

*When loosening-tightening the pivot bolts, take care to avoid damaging the steering protective boots. If the boots are cut or torn, the inner tie rod assemblies will be exposed to dirt, road splash, and other debris resulting in premature wear.*

**BUSHING REPLACEMENT**

Do not attempt to remove the bushings without using the spacer tools, to avoid damaging the control arm.

**NOTE: The thrust washer and nut are part of bolt assembly.**

### Installation

1. Position lower control arm in front crossmember and install pivot bolts and locknuts. Tighten locknuts securely but not completely.
2. Insert lower control arm ball joint stud in steering knuckle and install retaining nut on ball stud. Tighten to 75 foot-pounds (101.6 N·m) torque and install a new cotter pin. Do not loosen nut to align slots in nut with hole in ball stud.
3. Install steering arm on steering knuckle. Install retaining bolts and nuts. Tighten bolts to 55 foot-pounds (74.5 N·m) torque.

4. Loosen spring compressor tool until bottom coil of spring rests in control arm. Align bottom coil of spring and spring seat in control arm.

**NOTE: The cut-off end of the bottom coil must seat against the formed shoulder in the spring seat.**

5. Install shock absorber. Tighten lower mounting stud locknuts to 20 foot-pounds (27.1 N·m) torque.
6. Connect stabilizer bar link bolt to lower control arm if equipped. Tighten locknut to 7 foot-pounds (9.4 N·m) torque.
7. Install rotor, caliper, and wheel and tire assembly.
8. Tighten lower control arm pivot bolts to 110 foot-pounds (49.1 N·m) torque.

## Steering Knuckle and Spindle

### Removal

1. Remove wheel, tire, caliper and rotor.
2. Remove bolts attaching steering arm, steering stop, steering spindle, anchor plate, and adapter to steering knuckle and remove these components.
3. Disconnect stabilizer bar link bolt at lower control arm. Support lower control arm using a jack.
4. Remove cotter pins and retaining nuts from upper and lower ball joint studs. Disengage ball joint studs from steering knuckle.
5. Remove steering knuckle.

## Steering Knuckle

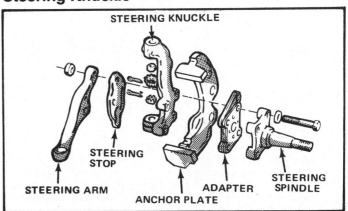

## Installation

1. Install steering knuckle on upper and lower ball joint studs and install stud retaining nuts. Tighten nuts to 75 foot-pounds (101.6 N·m) torque and install new cotter pins. Do not loosen nuts to align nut slots with hole in ball studs.

2. Position steering stop, steering arm, steering spindle, anchor plate, and adapter on steering knuckle and install retaining bolts. Tighten bolts to 85 foot-pounds (115.2 N·m) torque.

3. Connect stabilizer bar link bolt to lower control arm if equipped. Tighten nut to 7 foot-pounds (9.4 N·m) torque.

4. Install wheel, tire, rotor and caliper.

## Steering Arm

### Removal

1. Raise and support front of car.

2. Remove wheel and tire, caliper and rotor.

3. Raise and support lower control arm using hydraulic jack.

4. Remove cotter pin and retaining nut from tie rod end ball stud.

5. Disconnect tie rod end from steering arm.

6. Remove bolts attaching steering arm to steering knuckle; remove steering arm.

### Installation

1. Position steering arm on steering knuckle and install retaining bolts. Tighten bolts to 85 foot-pounds (115.2 N·m) torque.

2. Insert tie rod end ball stud in steering arm (from bottom) and install retaining nut. Tighten nut to 50 foot-pounds (67.7 N·m) torque and install replacement cotter pin.

3. Install rotor, caliper, and wheel and tire.

4. Remove supports and hydraulic jack and lower car.

# RENAULT ALLIANCE—FRONT SUSPENSION

The Renault Alliance front suspension is the MacPherson strut-type with rack and pinion steering. The camber setting is a negative 0.43 inch for the 5 inch steel wheels and 0.28 inch for the aluminum wheels. Caster angle is a plus 2.5 degrees with a toe-oout of from 0.020 to 0.059 inch, per wheel. The caster and camber are not adjustable.

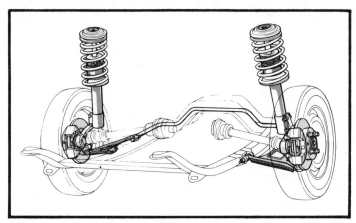

**Renault Alliance front suspension**

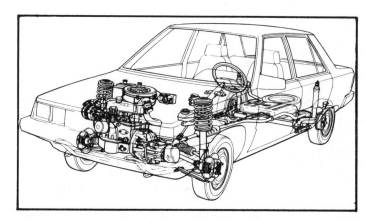

**Phantom view of 1983 Renault Alliance mechanism components**

# REAR SUSPENSION—AMC

## Shock Absorber

### Removal

1. Raise and support rear of automobile and support axle assembly with hydraulic jack.

**NOTE: Support the suspension in such a way that no strain is on shock absorber.**

2. Remove locknut, retainer, and grommet which attach shock absorber lower mounting stud to spring plate.

3. Compress shock absorber by hand and disengage lower mounting stud from spring plate.

4. Remove bolts and lockwashers attaching shock absorber upper mounting bracket to underbody panel and remove shock absorber.

5. Remove locknut, retainer, and grommet which attach mounting bracket to shock absorber upper mounting stud and remove bracket.

6. Remove remaining grommets and retainers from shock absorber upper and lower mounting studs.

### Installation

1. Install retainer and grommet on shock absorber mounting stud. Be sure locating shoulder on grommet faces end of mounting stud.

2. Install mounting bracket on shock absorber upper mounting stud with flat side of bracket facing underbody panel.

3. Install second grommet on mounting stud and install retainer and locknut. Tighten locknut to 8 foot-pounds (11 N·m) torque.

**NOTE: Be sure the locating shoulders on the grommets are centered in the mounting bracket hole before tightening the locknut.**

4. Position assembled mounting bracket and shock absorber on mounting studs in underbody panel. Install lockwashers and bolts. Tighten bolts to 28 foot-pounds (38 N·m) torque on Pacers and 15 foot-pounds (20 N·m) torque on all other models.

## Air Shock System—Spirit and AMX

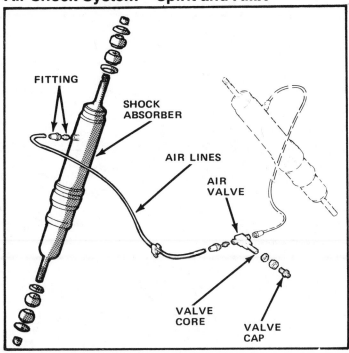

**NOTE: If an adjustable shock absorber is being installed, adjust the ride control setting as necessary before connecting the shock to the spring clip plate.**

5. Engage shock absorber lower mounting stud in spring clip plate.

6. Install second grommet with shoulder of grommet facing spring tie plate and install retainer and locknut. Tighten locknut to 8 foot-pounds (11 N·m) torque.

## Air Shock Inflation Procedure

Do not inflate air shocks until after the automobile has been loaded or had a trailer attached. If the shocks are inflated before loading, the combined force of initial inflation and load weight could exceed shock absorber internal pressure limits and cause damage.

## Leaf Spring

### Removal

1. Raise and support rear of automobile. Support axle assembly with hydraulic jack.
2. Remove shock absorber lower mounting locknut, retainer, and grommet.
3. Remove U-bolts, spring isoclamps, and clamp bracket.
4. Remove pivot bolt and nut from spring front eye.

## Rear Suspension Components

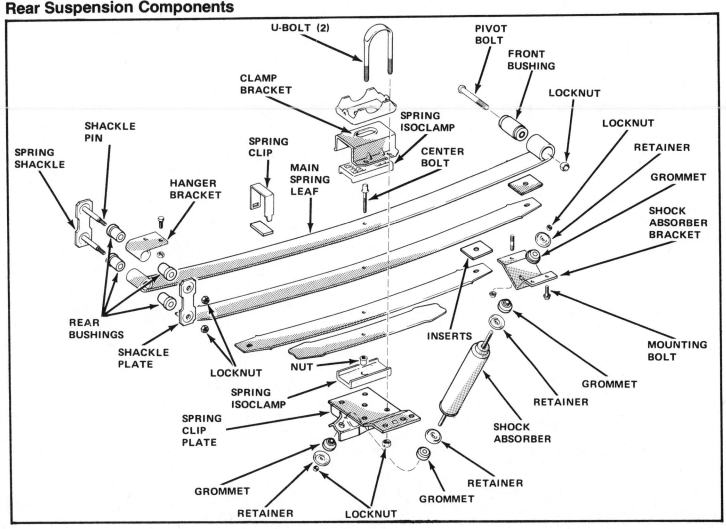

5. On all models except Pacer, remove shackle nuts, shackle plate, and shackle at rear spring eye. Remove spring.

6. On Pacers, remove nuts attaching rear hanger bracket to mounting studs on frame side sill and remove spring. Remove shackle nuts and shackle after spring is removed.

### BUSHING REPLACEMENT

1. Remove bushings from spring eyes using arbor press and suitable size socket or section of pipe.

2. Install replacement bushings in spring eyes using arbor press and suitable size socket or section of pipe. Be sure bushings are centered in spring eyes.

### Installation

1. On all models except Pacer, insert shackle pins into spring rear eye and rear hanger.

2. On Pacers, assemble rear hanger bracket and shackle and install in spring rear eye.

3. On Pacers, position rear hanger bracket on frame side sill and install mounting stud nuts. Tighten nuts to 45 foot-pounds (61 N·m) torque.

4. On all models, position front spring eye in front hanger and install pivot bolt and pivot bolt locknut. Tighten locknut to 110 foot-pounds (149 N·m) torque.

5. Install shackle plate and locknuts on shackle pins. Tighten locknuts to 30 foot-pounds (41 N·m) torque.

6. Install clamp bracket, spring isoclamps, spring plate, and U-bolts.

7. Engage shock lower mounting stud in spring plate and install grommet, retainer, and locknut. Tighten nut to 8 foot-pounds (11 N·m) torque.

## Stabilizer Bar

### Removal

1. Remove nuts and grommets attaching stabilizer bar to connecting links.

2. Remove bolts attaching stabilizer bar mounting clamps to spring clip plates.

3. Remove stabilizer bar.

### Installation

1. Position stabilizer bar and mounting clamps on spring clip plates and install clamp bolts finger tight.

2. Install stabilizer bar on connecting links and install grommets and locknuts.

3. Tighten connecting link locknuts to 7 foot-pounds (9 N·m) torque and tighten stabilizer mounting clamp bolts to 25 foot-pounds (34 N·m) torque.

## Automatic Load Leveling System Operation

The load leveling system automatically adjusts the rear height with changes in vehicle loading.

### COMPRESSOR

The compressor assembly is a positive displacement single piston air pump powered by a 12 volt permanent magnet motor. The compressor head casting contains piston, intake and exhaust valves plus a solenoid operated exhaust valve which releases air from the system when energized.

**NOTE: The compressor is not a serviceable item. If diagnosis indicates the compressor has malfunctioned, replace the compressor as an assembly only. Do not attempt to repair it.**

### Spring to Frame Mounting—Pacer

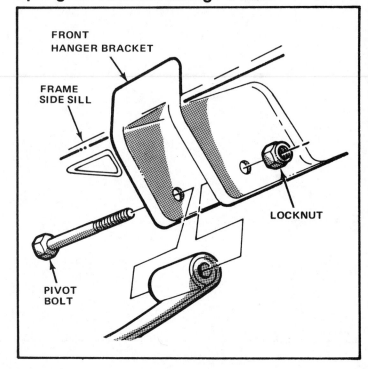

### Rear Spring, Rear Mounting

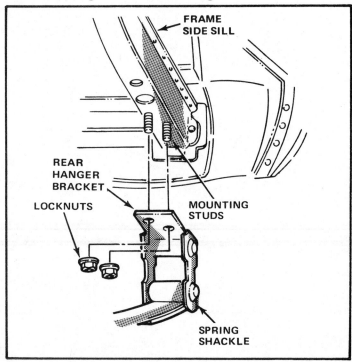

## Rear Stabilizer Bar

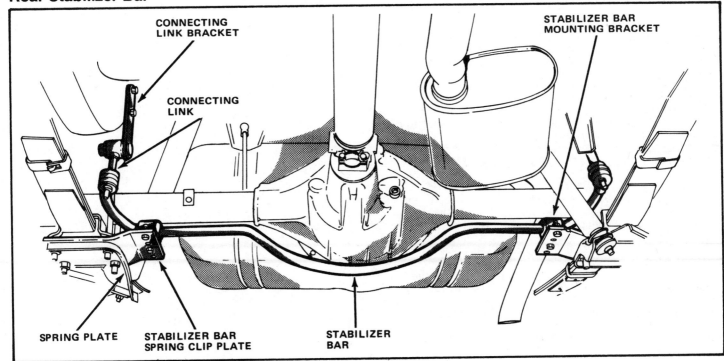

CONNECTING LINK BRACKET

CONNECTING LINK

STABILIZER BAR MOUNTING BRACKET

SPRING PLATE

STABILIZER BAR SPRING CLIP PLATE

STABILIZER BAR

## Automatic Load Leveling System

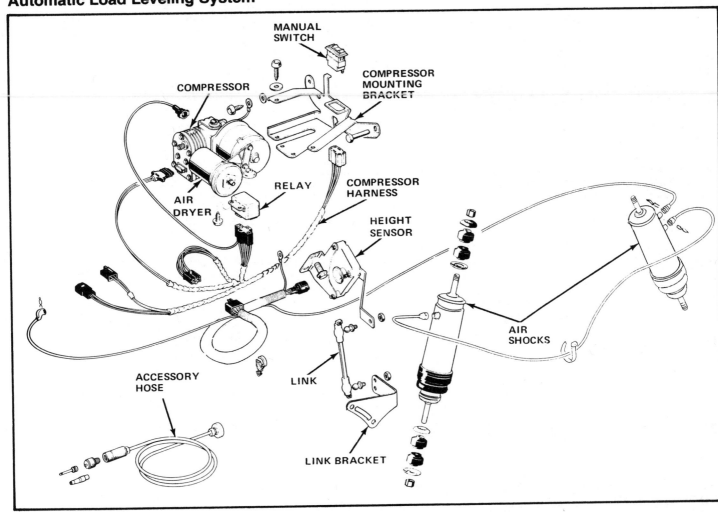

MANUAL SWITCH

COMPRESSOR

COMPRESSOR MOUNTING BRACKET

RELAY

AIR DRYER

COMPRESSOR HARNESS

HEIGHT SENSOR

AIR SHOCKS

ACCESSORY HOSE

LINK

LINK BRACKET

## RAISING THE AUTOMOBILE

When weight is added to the rear of the car, the body is forced downward which causes the height sensor actuating arm to rotate upward. This action causes the height sensor to electrically start the internal time delay circuit. When the time delay (7-15 seconds) has occurred, the sensor then completes the compressor relay circuit to ground. With the relay energized, the 12V(+) circuit to the compressor is complete and the compressor runs, sending air to the air adjustable shock absorbers through air lines. When the body reaches its original trim height (±3/4 inch) the sensor opens the compressor relay circuit, shutting off the compressor.

## LOWERING THE AUTOMOBILE

A high body condition has the effect of rotating the height sensor actuating arm downwad. The height sensor then senses the high condition and starts the time delay circuit. When the time delay (7−15 seconds) has elapsed, the sensor completes the exhaust solenoid circuit to ground. With the exhaust solenoid energized, air escapes from the shocks exiting through the air dryer and exhaust solenoid valve.

As the automobile body lowers, the height sensor actuating arm is rotated toward its original position. When the automobile body reaches its original height (±3/4 inch), the sensor opens the exhaust valve solenoid circuit, which prevents more air from escaping.

A minimum air pressure of 7−14 psi is maintained on the automobile. The minimum pressure provides improved ride characteristics when the automobile has a minimum load. The compressor relief valve is designed to operate at 120−150 psi. See the Diagnosis Chart for troubleshooting procedures.

## HEIGHT SENSOR

### Removal

1. Disconnect connector plug from sensor.
2. Disconnect link from sensor actuating arm.
3. Remove bolts attaching height sensor to underbody and remove sensor.

### Installation

1. Position height sensor on underbody and install sensor attaching bolts.
2. Connect sensor actuating arm to link.
3. Connect wiring harness connector plug to height sensor.

## Compressor Draw Test

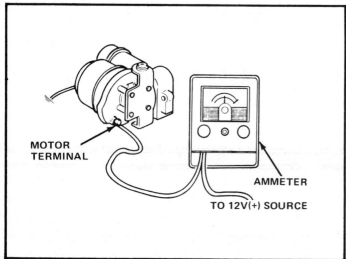

**Current draw should not exceed 14 amps.**

## Height Sensor

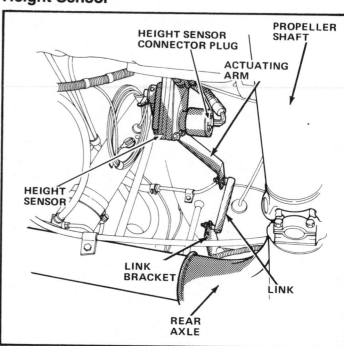

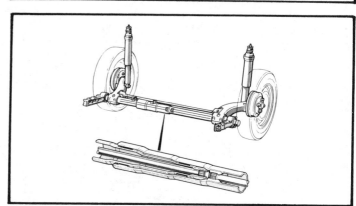

**Renault Alliance rear torsion bar suspension**

## Compressor Pressure Test

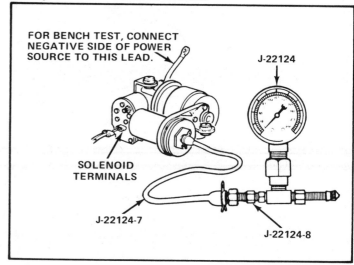

## Wiring Diagram—Automatic Load Leveling System

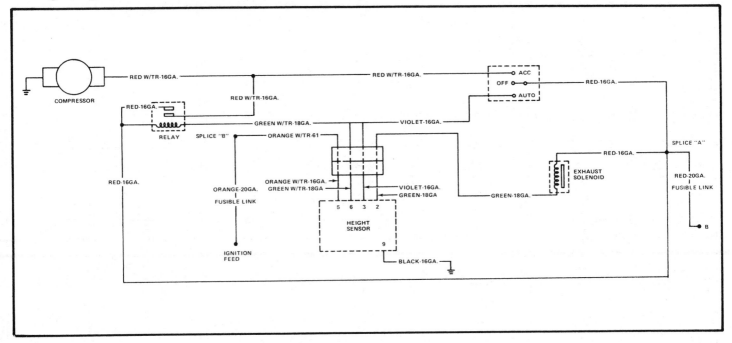

# FRONT SUSPENSION—CHRYSLER FRONT DRIVE

## Alignment

### PRE-ALIGNMENT CHECK

There are six factors which are the foundation to front wheel alignment: Height, caster, camber, toe-in, steering axis inclination and toe-out on turns. Of these six basic factors, only camber and toe are mechanically adjustable.

1. Before any attempt is made to change or correct the wheel alignment, inspection and necessary corrections must be made on those parts which influence the steering of the vehicle.
2. Check and inflate tires to recommended pressures.
3. Check front wheel and tire assembly for radial runout.
4. Check struts (shock absorbers) for extra-stiff, notchy, or spongy operation.
5. Front suspension should be checked only after vehicle has the recommended tire pressures, full tank of fuel, no passenger or luggage compartment load and is on a level floor or alignment rack.
6. To obtain accurate reading, vehicle should be jounced in the following manner just prior to taking measurement. Grasp bumpers at center (rear bumper first) and jounce up and down several times. Always release bumpers at bottom of down cycle after jouncing both rear and front ends an equal number of times.

### Camber Adjustment

1. Loosen cam and through bolts (each side).
2. Rotate upper cam bolt to move upper (knuckle and) wheel in or out to specified camber.
3. Tighten bolts to 115 N·m (85 foot-pounds).

### Toe Adjustment

1. Center steering wheel and hold with steering wheel clamp.
2. Loosen tie rod locknuts. Rotate rods to align toe to specifications.

## Front Wheel Bearings

The vehicle is equipped with permanently sealed front wheel bearings. There is no periodic lubrication, maintenance, or adjustment recommended for these units.

Service repair or replacement of front drive bearing, hub, brake dust shield or knuckle will require assembly removal from the vehicle.

## Strut Damper Assembly

### Removal

1. Remove wheel and tire assembly.
2. Remove cam adjusting bolt, through bolt and brake hose-to-damper bracket retaining screw.
3. Remove strut damper-to-fender shield mounting nut washer assemblies.

### DISASSEMBLY

1. Compress coil with spring compressor tool.
2. Hold strut rod while loosening strut rod nut. Remove nut.
3. Remove retainers and bushings.
4. Remove coil spring.

**NOTE: Mark spring for replacement in original position.**

5. Check retainers for cracks or distortion.
6. Check bearings for binding. Check that they contain an adequate supply of lubricant.

### ASSEMBLY

1. Reassemble and hold strut rod while tightening rod nut to 81 N·m (55 foot-pounds).

**NOTE: Perform step 1 before releasing spring compressor tool.**

2. Release spring compressor tool.

# SUSPENSIONS

## Front Suspension

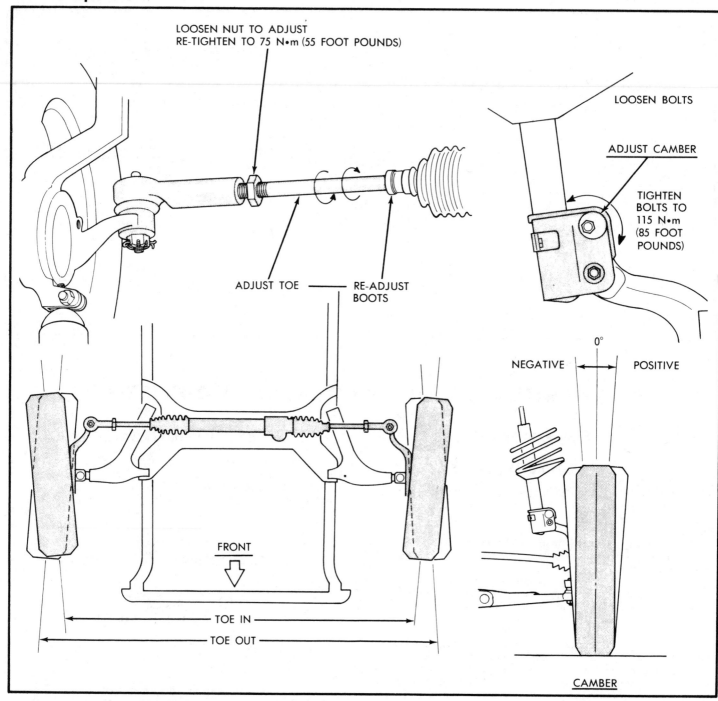

LOOSEN NUT TO ADJUST
RE-TIGHTEN TO 75 N•m (55 FOOT POUNDS)

LOOSEN BOLTS

ADJUST CAMBER

TIGHTEN BOLTS TO 115 N•m (85 FOOT POUNDS)

ADJUST TOE — RE-ADJUST BOOTS

NEGATIVE    0°    POSITIVE

FRONT

TOE IN

TOE OUT

CAMBER

**NOTE: Springs are rated separately for each side of vehicle depending on optional equipment and type of service.**

During reassembly of spring to strut damper, ensure that coil end is seated in strut damper spring seat recess.

### Installation

1. Install unit into fender reinforcement and install retaining nut and washer assemblies. Torque to 20 N•m (27 foot-pounds).

2. Position knuckle leg into strut and install upper (cam) and lower through bolts.

3. Attach brake hose retainer to damper; tighten screw to 12 N•m (10 foot-pounds).

4. Index cam bolt to original mark and tighten bolts to 115 N•m (85 foot-pounds) torque.

5. Install wheel and tire assembly. Tighten wheel nuts to 108 N•m (80 foot-pounds).

## Lower Ball Joint

### Inspection

**1978-80**

1. Raise and support the vehicle.

## Compressing Coil Spring

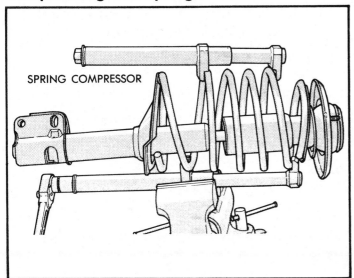

SPRING COMPRESSOR

## Loosening Strut Rod Nut

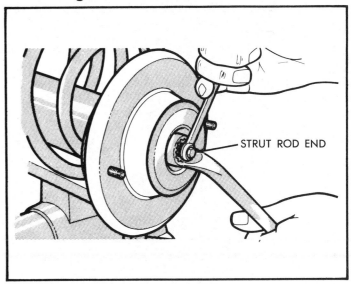

STRUT ROD END

## Upper Spring Retainer Assembly

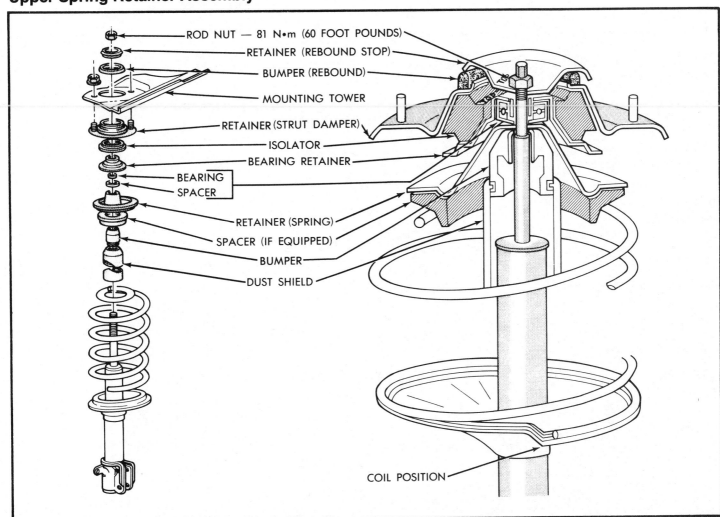

ROD NUT — 81 N•m (60 FOOT POUNDS)
RETAINER (REBOUND STOP)
BUMPER (REBOUND)
MOUNTING TOWER
RETAINER (STRUT DAMPER)
ISOLATOR
BEARING RETAINER
BEARING
SPACER
RETAINER (SPRING)
SPACER (IF EQUIPPED)
BUMPER
DUST SHIELD

COIL POSITION

# SUSPENSIONS

## Strut Assembly

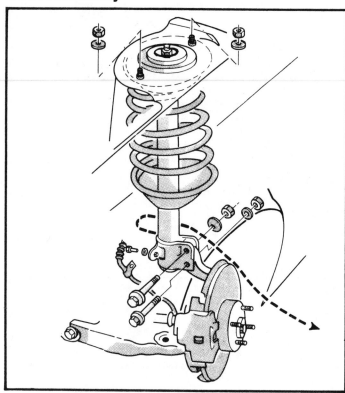

## Checking Ball Joint For Excessive Clearance, Using A Dial Indicator. 1978 And Later Models

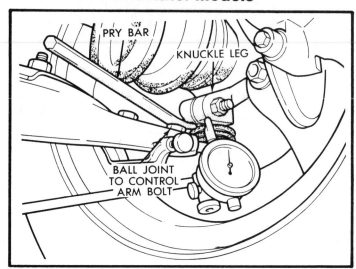

2. With the suspension fully extended (at full travel) clamp a dial indicator to the lower control arm with the plunger indexed against the steering knuckle leg.

3. Zero the dial indicator.

4. Use a stout bar to pry on the top of the ball joint housing-to-lower control arm bolt with the bar tip under the steering knuckle leg.

5. Measure the axial travel of the steering knuckle leg in relation to the control arm by raising the lowering the steering knuckle as in Step 4.

6. If the travel is more than 0.050 in., the ball joint should be replaced.

### 1981 and Later

The lower ball joint is checked at the lube fitting. Try to turn the lube fitting. If it turns or wobbles, the ball joint is worn and should be replaced.

### Removal and Installation

#### 1978

The lower ball joints are permanently lubricated, operate with no free play, and are riveted in place. The rivets must be drilled out and replaced with special bolts.

**NOTE: To avoid damage to the control arm surface adjacent to the ball joint during drilling, the use of a center punch and a drill press are strongly recommended.**

1. Remove the lower control arm.
2. Position the assembly with the ball joint up.
3. Center punch the rivets on the ball joint housing side.
4. Using a drill press with a 1/4 inch bit, drill out the center of the rivet.
5. Using a 1/2 inch bit, drill the center of the rivet until the bit makes contact with the ball joint housing.
6. Using a 3/8 inch bit, drill the center of the rivet. Remove the remainder of the rivet with a punch.

7. Position the new ball joint on the control arm and tighten the bolts to 60 ft. lbs.

8. Install the control arm and tighten the ball joint clamp bolt to 50 ft. lbs.; the pivot bolt to 105 ft. lbs. and the stub strut to 70 ft. lbs.

#### 1979-80

The ball joint housing is bolted to the lower control arm with the joint stud retained in the steering knuckle by a clamp bolt.

1. Raise and support the car.
2. Remove the steering knuckle-to-ball joint stud clamp bolt and separate the stud from the knuckle leg.
3. Remove the 2 bolts holding the ball joint housing to the lower control arm.
4. Remove the ball joint housing.
5. Install a new ball joint housing to the control arm. Torque the retaining bolts to 60 ft. lbs.
6. Install the ball joint stud in the steering knuckle. Tighten the clamp bolt to 50 ft. bls.
7. Lower the car.

## Checking Ball Joint Wear

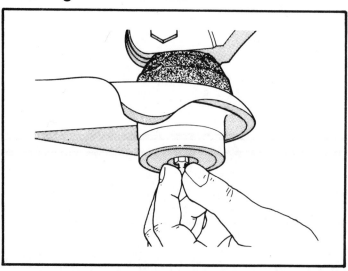

## Ball Joint Replacement

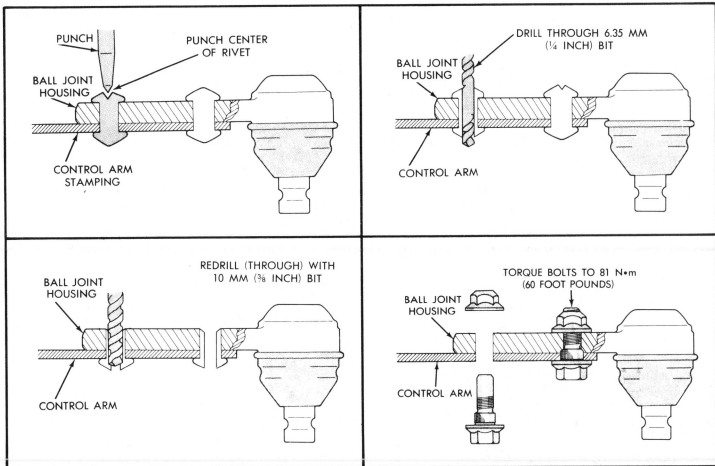

## Ball Joint Bolted To Lower Control Arm. 1978 And Later Models

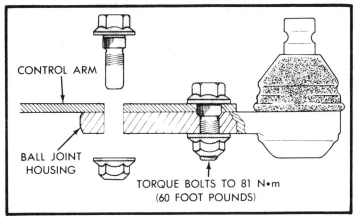

**1981 and Later**

NOTE: On Some 1981 models the front ball joints are welded to the control arms and are not to be pressed out. Those that are welded must be serviced by complete replacement of the control arm and ball joint assembly.

1. Pry off the seal.
2. Position a receiving cup tool C-4699-2 to support the lower control arm.

## Ball Joint Removal

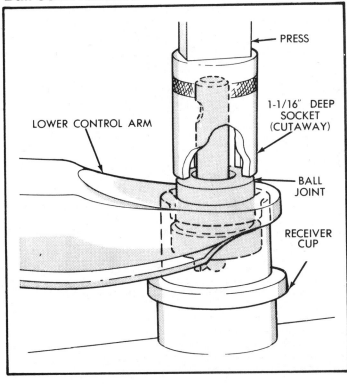

## Ball Joint Installation

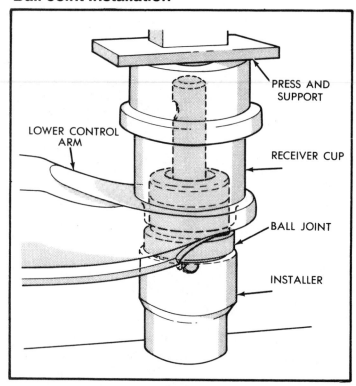

## Ball Joint Seal Installation

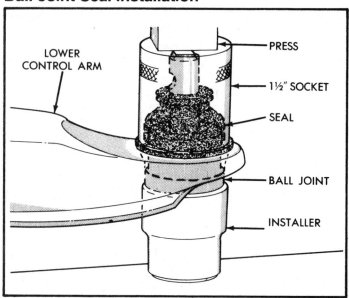

3. Install a 1 1/16" deep socket over the stud and against the joint upper housing.

4. Press the joint assembly from the arm.

5. To install, position the ball joint housing into the control arm cavity.

6. Position the assembly in a press with installer tool C-4699-1 supporting the control arm.

7. Align the ball joint assembly then press it until the housing ledge stops against the control arm cavity down flange.

8. To install a new seal, support the ball joint housing with installing tool C-4699-2 and position a new seal over the stud against the housing.

9. With a 1 1/2" socket, press the seal onto the joint housing with the seat against the control arm.

# Lower Control Arm

### Removal

1. Remove front inner pivot through bolt.
2. Remove rear stub strut nut, retainer and bushing.
3. Remove ball joint-to-steering knuckle clamp bolt.
4. Separate ball joint stud from steering knuckle.

### PIVOT BUSHING REPLACEMENT

1. Position support tool between flanges of lower control arm and around bushing to prevent control arm distortion.

2. Install 1/2 × 2 1/2 inch bolt into bushing.

3. With receiving cup on press base, position control arm inner flange against cup wall to support flange while receiving bushing.

4. Remove bushing by pressing against bolt head.

5. To install, position support tool between flanges of control arm.

6. Install bushing inner sleeve and insulator into installer tool C-4699-1 cavity with the bushing outer shell flange against the tool wall. Position assembly onto press base and align control arm to receive bushing.

7. Position receiving cup tool to support control arm outer flange while receiving bushing.

8. Press bushing into control arm until bushing flange seats against control arm.

### Installation

1. Install retainer, bushing and sleeve on stub strut.

2. Position control arm over sway bar and install rear stub strut and front pivot into crossmember.

3. Install front pivot bolt; install nut but do not tighten yet.

4. Install stub strut bushing and retainer and loosely assemble nut.

5. Install ball joint stud into steering knuckle and install clamp bolt. Tighten clamp bolt to 67 N·m (50 ft.-lbs.).

6. Position sway bar end bushing retainer to control arm. Install retainer bolts and tighten nuts to 30 N·m (22 ft.-lbs.).

7. Lower vehicle. With suspension supporting vehicle (control arm at design height), tighten front pivot bolt to 142 N·m (105 ft.-lbs.) and stub strut nut to 94 N·m (70 ft.-lbs.) torque.

# Knuckle

The front suspension knuckle provides for steering, braking and alignment, and also supports the front (driving) hub (and axle) assembly.

Service repair or replacement of front drive bearing, hub, brake dust shield or knuckle will require removal of the assembly from the vehicle.

### Removal and Installation

1. Remove cotter pin and lock.

2. Loosen hub nut while vehicle is on the floor and brakes applied.

**NOTE: The hub and driveshaft are splined together through the knuckle (bearing) and retained by the hub nut.**

3. Remove wheel and tire assembly.

4. Remove hub nut.

─────── CAUTION ───────

*Ensure that splined driveshaft is "free" to separate from spline in hub during knuckle removal from vehicle. A pulling force on the shaft can separate the inner universal joint. Tap lightly with soft brass punch if required.*

## Lower Control Arm

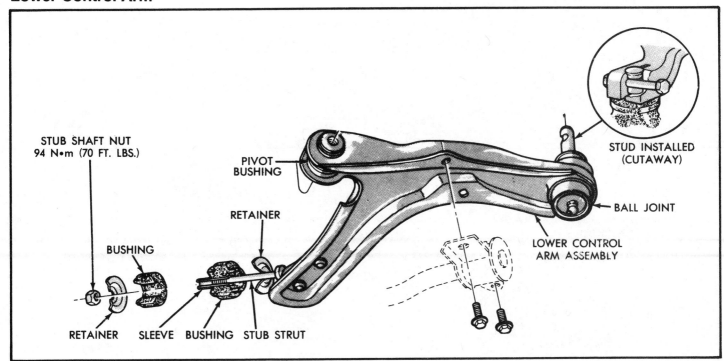

STUB SHAFT NUT
94 N•m (70 FT. LBS.)

PIVOT
BUSHING

BUSHING

RETAINER

RETAINER    SLEEVE   BUSHING   STUB STRUT

STUD INSTALLED
(CUTAWAY)

BALL JOINT

LOWER CONTROL
ARM ASSEMBLY

## Pivot Bushing Removal

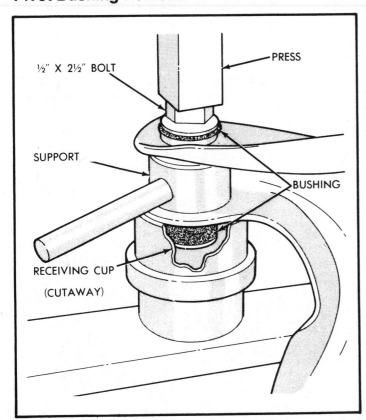

½" X 2½" BOLT

PRESS

SUPPORT

BUSHING

RECEIVING CUP
(CUTAWAY)

## Pivot Bushing Installation

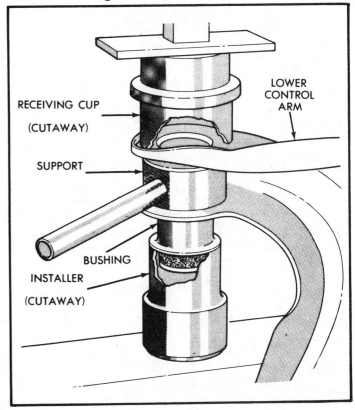

RECEIVING CUP
(CUTAWAY)

SUPPORT

BUSHING

INSTALLER
(CUTAWAY)

LOWER
CONTROL
ARM

# SUSPENSIONS

## Front Suspension Knuckle

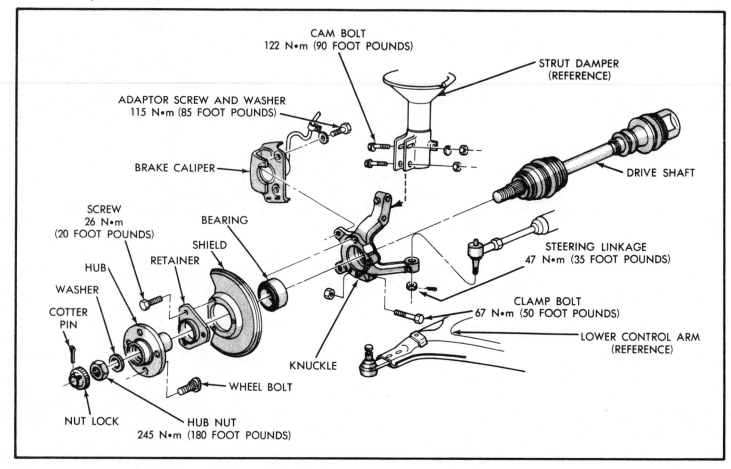

CAM BOLT
122 N•m (90 FOOT POUNDS)

STRUT DAMPER
(REFERENCE)

ADAPTOR SCREW AND WASHER
115 N•m (85 FOOT POUNDS)

BRAKE CALIPER

DRIVE SHAFT

SCREW
26 N•m
(20 FOOT POUNDS)

BEARING

SHIELD

STEERING LINKAGE
47 N•m (35 FOOT POUNDS)

RETAINER

HUB

CLAMP BOLT
67 N•m (50 FOOT POUNDS)

WASHER

COTTER
PIN

LOWER CONTROL ARM
(REFERENCE)

KNUCKLE

WHEEL BOLT

NUT LOCK

HUB NUT
245 N•m (180 FOOT POUNDS)

5. Disconnect tie rod end from steering arm.

6. Disconnect brake hose retainer from strut damper.

7. Remove clamp bolt securing ball joint stud into steering knuckle. Remove brake caliper adaptor screw and washer assemblies.

8. Support caliper with wire hook. Do not allow assembly to hang by brake hose.

9. Remove braking disc (rotor).

10. Mark camber position on upper cam adjusting bolt. Loosen both bolts.

11. To remove assembly from vehicle, support knuckle and remove cam adjusting and through bolts, then move upper knuckle "leg" out of strut damper bracket and lift knuckle off of ball joint stud.

---
**CAUTION**
---

*Support driveshaft during knuckle removal. Do not allow driveshaft to "hang" after separating steering knuckle from vehicle (severe angles will damage inboard universal joint boot).*

12. Installation is the reverse of the removal procedure.

## Hub/Bearing

### Removal and Installation

**NOTE: Do not reuse bearing.**

1. Remove hub (out of bearing) with hub remover tool and fabricated washer shown in illustration.

   a. Place washer and thrust button on hub.

   b. Back out one retainer screw-to-hub, as far as it will go.

## Steering Knuckle and Bearing

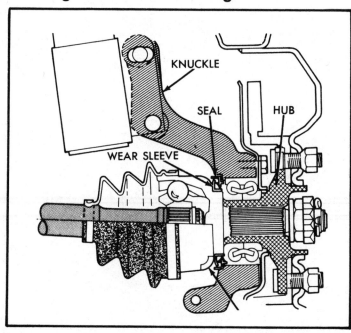

KNUCKLE

SEAL

HUB

WEAR SLEEVE

## Knuckle Assembly Removal

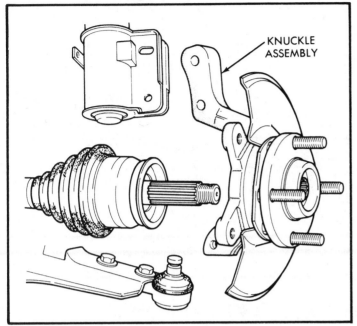

## Hub Removal

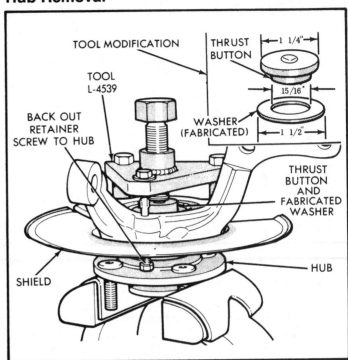

c. Position tool and install two screws firmly into tapped brake adaption extensions and one screw into retaining screw threads.

d. Tighten press screw to remove hub through bearing.

**NOTE: Bearing inner races will separate; outboard race will stay on hub.**

2. Remove bearing outer race from hub with thrust button from tool and universal puller.

3. Remove brake dust shield and bearing retainer.

4. Installation is the reverse of the removal procedure.

## Pressing New Bearing Into Knuckle

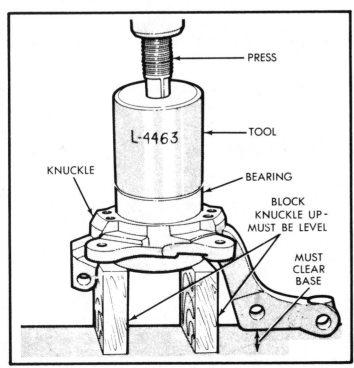

## Pressing Hub Into Knuckle Bearing

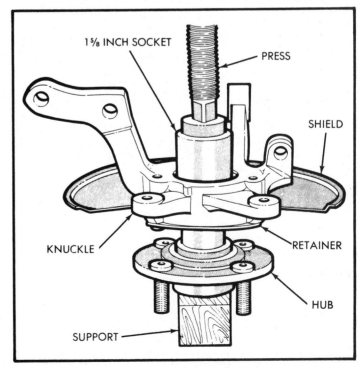

### Outboard Inner Race Removal

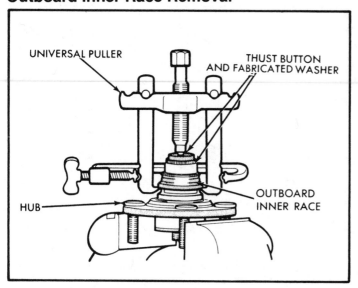

UNIVERSAL PULLER

THUST BUTTON
AND FABRICATED WASHER

HUB

OUTBOARD
INNER RACE

### Press Bearing Out of Knuckle

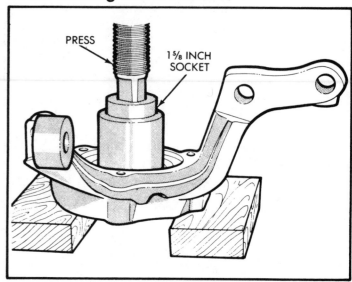

PRESS

1 5/8 INCH
SOCKET

# REAR SUSPENSION—CHRYSLER FRONT DRIVE

## Introduction

Chrysler front drive cars have two types of rear suspension systems. A semi-independent rear suspension is used on "L" body cars (Omni-Horizon); an independent rear suspension is used on "K" cars (Aries/Reliant). The suspensions are similar, although service procedures are different.

## Alignment

### "L" AND "K" BODY CARS

Because of the trailing arm rear suspension of the vehicle, and the incorporation of stub axles or wheel spindles, it is possible to align both the camber and toe of the rear wheels. Alignment is controlled by adding shim stock of .010-inch thickness between the spindle mounting surface and spindle mounting plate.

### L Body Rear Suspension—Horizon/Omni

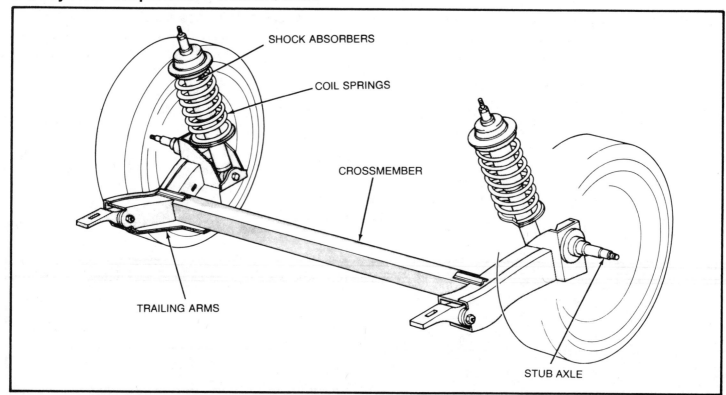

SHOCK ABSORBERS

COIL SPRINGS

CROSSMEMBER

TRAILING ARMS

STUB AXLE

## K Car Flex Arm Rear Suspension

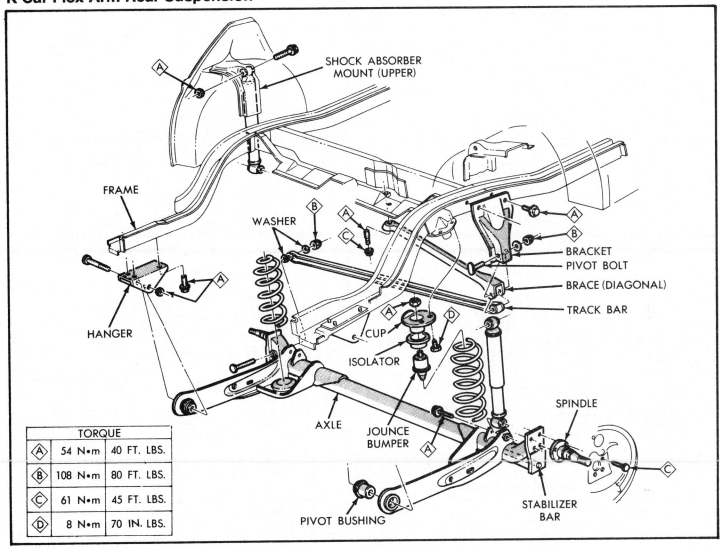

SHOCK ABSORBER MOUNT (UPPER)

FRAME

WASHER

B

A

C

HANGER

A

CUP

ISOLATOR

AXLE

JOUNCE BUMPER

A

PIVOT BUSHING

BRACKET

PIVOT BOLT

BRACE (DIAGONAL)

TRACK BAR

SPINDLE

STABILIZER BAR

C

| TORQUE | | |
|---|---|---|
| A | 54 N•m | 40 FT. LBS. |
| B | 108 N•m | 80 FT. LBS. |
| C | 61 N•m | 45 FT. LBS. |
| D | 8 N•m | 70 IN. LBS. |

## Shim Installation for Negative Camber

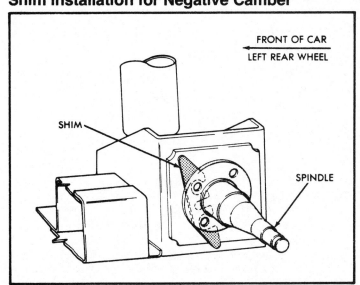

FRONT OF CAR
LEFT REAR WHEEL

SHIM

SPINDLE

## Shim Installation for Toe In

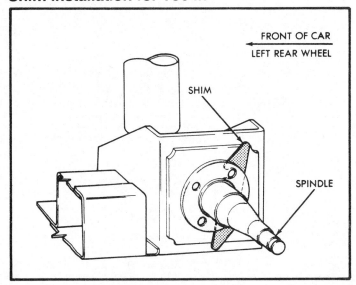

FRONT OF CAR
LEFT REAR WHEEL

SHIM

SPINDLE

## Shim Installation for Toe Out

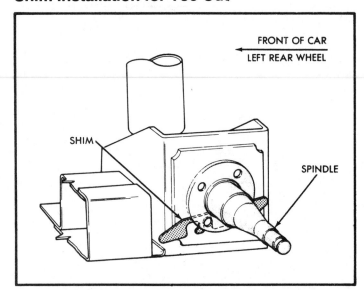

## Shim Installation for Positive Camber

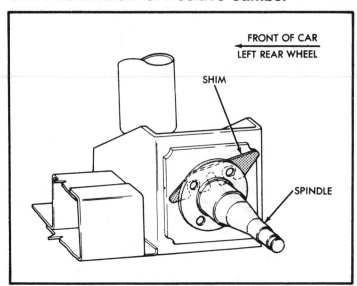

If rear wheel alignment is required, place vehicle on alignment rack and check alignment specifications. Follow equipment manufacturer's recommendations for their equipment. Maintain rear alignment within Chrysler corporation recommendations.

### Installation of Rear Alignment Shims

1. Block front tires so vehicle will not move.
2. Release parking brake.
3. Hoist vehicle so that rear suspension is in full rebound and tires are off the ground.
4. Remove wheel and tire assembly.
5. Pry off grease cap.
6. Remove cotter pin and castle lock.
7. Remove adjusting nut.
8. Remove brake drum.
9. Loosen four (4) brake assembly and spindle mounting bolts enough to allow clearance for shim installation.

**NOTE: Do not remove mounting bolts.**

10. Install shims for desired wheel change.

**NOTE: Wheel alignment changes by 0° 18′ per shim.**

11. Tighten four (4) brake assembly and spindle mounting bolts. Tighten to 60 N·m (45 ft.-lbs.) torque.
12. Install brake drum.
13. Install washer and nut. Tighten adjusting nut to 27-34 N·m (240-300 in.-lbs.) while rotating wheel. Back off adjusting nut with wrench to completely release bearing preload. Finger-tighten adjusting nut.
14. Position nut lock with one pair of slots in line with cotter pin hole. Install cotter pin. The end play should be .025-.076 mm (.001-.003 in.). Clean and install grease cap.
15. Install wheel and tire assembly. Tighten wheel nuts to 108 N·m (80 ft.-lbs.) torque.
16. Lower vehicle.
17. Recheck alignment specifications.

## REAR WHEEL ALLIGNMENT

| | | Acceptable Alignment Range | Preferred Setting |
|---|---|---|---|
| **CAMBER** | | | |
| Horizon-Omni | | −1.5° to −.5° (−1-1/2° to −1/2°) | −1.0° ± .5° (1/2°) |
| Aries-Reliant | | −1.0° to 0° (−1° to 0°) | −.5° ± .5° (1/2°) |
| **TOE**① | | | |
| Horizon-Omni | Specified in Inches | 5/32″ OUT to 11/32″ IN | 3/32″ IN |
| | Specified in Degrees | 0.3° OUT to 0.7° IN | 0.2° IN |
| Aries-Reliant | Specified in Inches | 3/16″ OUT to 3/16″ IN | 0″ ± 1/8″ |
| | Specified in Degrees | .38″ OUT to .38° IN | 0° ± .25° |

①TOE OUT when backed on alignment rack is TOE IN when driving.

## Wheel Bearings

### "L" BODY CARS

#### Lubrication

The lubricant in the wheel bearings should be inspected whenever the drums are removed to inspect or service the brake system, but at least every 22,500 miles (36,000 km). The bearings should be cleaned and repacked with a high temperature multipurpose E.P. grease whenever the brake drums are resurfaced.

**NOTE: Do not add grease to the wheel bearings. Relubricate completely.**

Discard the old seal. Thoroughly clean the old lubricant from the bearings and from the hub cavity. Inspect the rollers for signs of pitting or other surface distress. Light bearing discoloration should be

## Rear Wheel Bearing

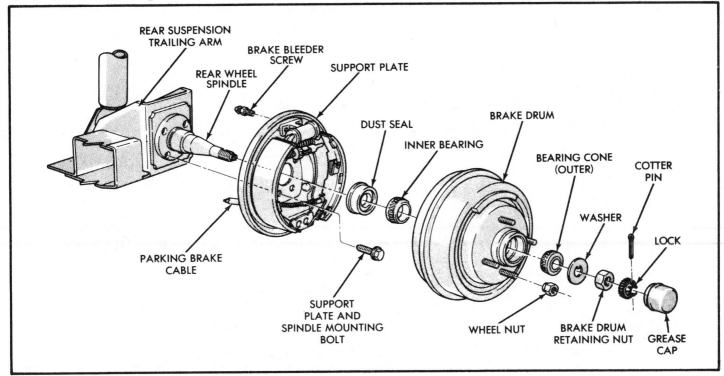

considered normal. Bearings must be replaced if any defects exist. Repack the bearings with a high temperature multipurpose E.P. grease. The use of a bearing packer is recommended. A small amount of new grease should also be added to the hub cavity.

### Adjustment

1. Install hub assembly on spindle.
2. Install outer bearing, thrust washer and nut.
3. Tighten wheel bearing adjusting nut to 240-300 inch-pounds (31 to 38 n·m) while rotating hub.
4. Back off adjusting nut to release all preload, then tighten adjusting nut finger-tight.
5. Position lock on nut with one pair of slots in line with cotter pin hole. Install cotter pin.
6. Install grease cap and wheel and tire assemblies.

## "K" BODY CARS

### Lubrication

The lubricant in the rear wheel bearings should be inspected whenever the drums are removed to inspect or service the brake system, or at least every 48,000 kilometres (30,000 miles). Bearings should be cleaned and repacked with a high temperature multipurpose E.P. grease whenever the brake drums are resurfaced.

**NOTE: Do not add grease to the wheel bearings. Relubricate completely.**

Discard the old seal. Thoroughly clean the old lubricant from the bearings and from the hub cavity.

Inspect the rollers for signs of pitting or other surface distress. Light bearing discoloration should be considered normal. Bearings must be replaced if any defects exist. Repack the bearings with a high temperature multipurpose E.P. grease. Use of a bearing packer is also recommended. A small amount of new grease should also be added to the hub cavity.

## Wheel Bearing Lubrication

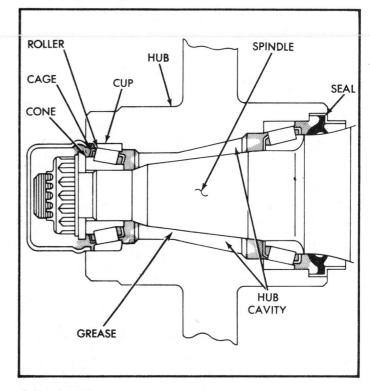

### Adjustment

1. Tighten adjusting nut to 27-34 N·m (240-300 in.-lbs.) torque, while rotating wheel.
2. Stop rotation and back off adjusting nut with wrench to completely release bearing preload.

# SUSPENSIONS
## REAR SUSPENSION—CHRYSLER FRONT DRIVE CARS

## Rear Wheel Components

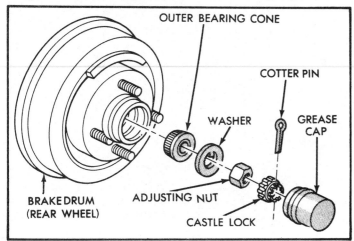

OUTER BEARING CONE

COTTER PIN

WASHER

GREASE CAP

BRAKE DRUM (REAR WHEEL)

ADJUSTING NUT

CASTLE LOCK

3. Finger-tighten adjusting nut.
4. Position nut lock with one pair of slots in line with cotter pin hole.
5. Install cotter pin.
6. The end-play should be .025-.076 mm (.001-.003 in.).
7. Clean and install grease cap.
8. Install wheel and tire assembly.

# Shock Absorber and Coil Spring Assembly

## "L" BODY CARS

### Removal

1. Locate upper shock absorber mounting nut protective cap inside of vehicle at upper rear wheel well area. (On two-door models, remove lower rear quarter trim panel.)
2. Unsnap cap. Use care to retain sound isolation material inside cap.
3. Remove upper shock absorber mounting nut, isolator retainer, and upper isolator. Remove lower shock absorber mounting bolt.
4. Remove shock absorber and coil spring assembly from trailing arm bracket. The shock absorber and coil spring assembly should now be free of vehicle.
5. Place coil spring compressor tool on coil spring and place in vise.

——— CAUTION ———
*Always grip 4 or 5 coils of spring in retractors. Never extend retractors beyond 9 1/4 inches.*

6. Tighten spring retractors *evenly* until pressure is relieved from upper spring seat.
7. Hold flat at end of shock rod. Loosen retaining nut.

——— CAUTION ———
*Be very careful when loosening retaining nut. If coil spring is not compressed enough, serious injury could occur when retaining nut is loosened.*

8. Remove lower isolator, shock rod sleeve, and upper spring seat.
9. Carefully remove shock absorber from coil spring.

### Installation

1. Install lower spring seat on shock absorber. Orient seat recess to centerline of lower bushing.

## Upper Shock Absorber Mounting

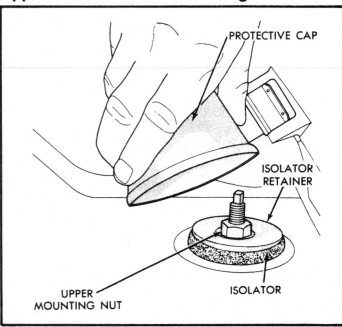

PROTECTIVE CAP

ISOLATOR RETAINER

ISOLATOR

UPPER MOUNTING NUT

## Lower Shock Absorber Mounting Bolt

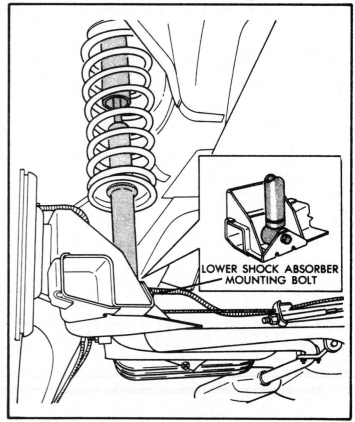

LOWER SHOCK ABSORBER MOUNTING BOLT

2. Install dust shield and jounce bumper on shock absorber.
3. Carefully slip the unit inside the coil spring. Install upper spring seat. Make sure that the leveled surface on both spring seats are in position against the ends of the coil spring.
4. Install sleeve on shock rod. Install retaining nut on end of shock rod. Tighten retaining nut to 27 N·m (20 ft.-lbs.) torque.

## Retract Coil Spring

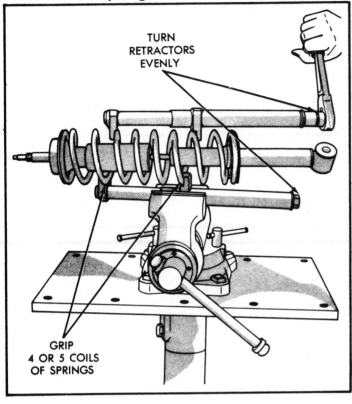

TURN RETRACTORS EVENLY

GRIP
4 OR 5 COILS
OF SPRINGS

## Loosen Retaining Nut

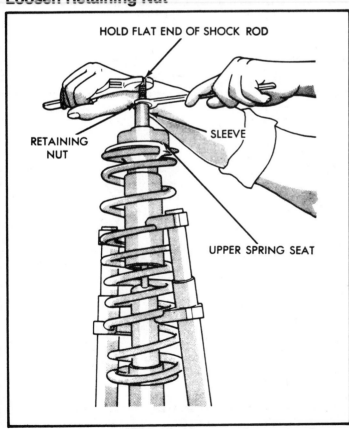

HOLD FLAT END OF SHOCK ROD

RETAINING NUT

SLEEVE

UPPER SPRING SEAT

5. Carefully loosen both coil spring retractors *evenly* and remove retractors from unit.

6. Install lower end of unit in trailing arm bracket. Insert bolt. Finger-tighten only. Make sure that upper end of unit is in proper hole at top of wheel well.

7. Tighten lower shock absorber bolt to 55 N·m (40 ft.-lbs.) torque.

8. Install upper isolator, isolator retainer, and upper mounting nut. Hold shock absorber rod end and tighten nut to 27 N·m (20 ft.-lbs.) torque.

9. Install sound isolation material and snap protective cap on securely.

# Shock Absorbers

## "K" BODY CARS

### Removal

1. Support axle and remove wheel and tire assembly.
2. Remove upper and lower shock absorber fasteners and remove shock absorbers.

### Installation

1. Position shock absorber and install fasteners; loosely assembly lower fastener. Tighten upper fastener to 54 N·m (40 foot-pounds).
2. Install wheel and tire assembly, tighten wheel stud nuts to 108 N·m (80 foot-pounds). Lower vehicle to ground.
3. With suspension supporting vehicle, tighten lower shock absorber fastener to 54 N·m (40 ft.-lbs.).

## K Car Shock Absorber Fasteners Removal and Installation

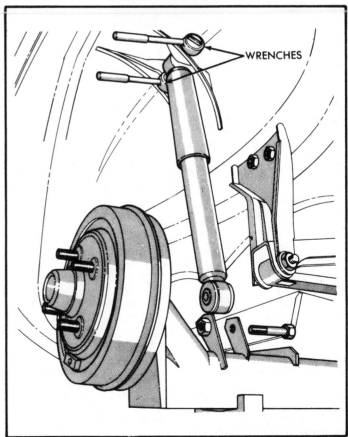

WRENCHES

## K Car Trailing Arms Removal

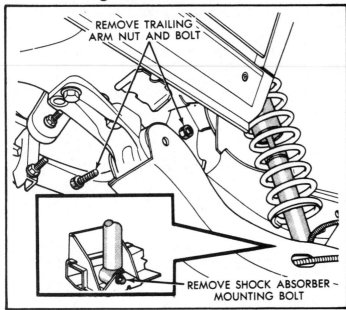

REMOVE TRAILING ARM NUT AND BOLT

REMOVE SHOCK ABSORBER MOUNTING BOLT

# Coil Springs and Jounce Bumper

## "K" BODY CARS

### Removal

1. Lift vehicle.
2. Support axle assembly and remove both lower shock absorber attaching bolts.
3. Lower axle assembly until spring and spring upper isolator can be removed (do not stretch brake hose).
4. Remove two screws holding jounce bumper assembly to rail. Remove jounce bumper assembly.

### Installation

1. Position jounce bumper to rail. Install and tighten attaching screws to 7 N·m (70 in.-lbs.).
2. Install isolator over jounce bumper and install spring.
3. Raise axle and loosely assemble both shock absorber-to-axle screws. Remove axle support and lower vehicle.
4. With suspension supporting vehicle, tighten both shock absorber attaching screws to 54 N·m (40 ft.-lbs.).

# Rear Axle Assembly

## "L" BODY CARS ONLY

### Removal

NOTE: Support the car on the rear crossmember; let the axle hang down.

1. Remove wheel and tire assembly.
2. Remove brake fittings and retaining clips holding flexible brake line.
3. Remove parking brake cable adjusting connection nut.
4. Release both parking brake cables from brackets by slipping ball-end of cables through brake connectors. Pull parking brake cable through bracket.
5. Pry off grease cap.
6. Remove cotter pin and castle lock.

7. Remove adjusting nut. Remove brake drum.
8. Remove four (4) brake assembly and spindle retaining bolts.
9. Set spindle aside and, using a piece of wire, hang brake assembly out of the way.
10. Remove shock absorber mounting bolts.
11. Remove trailing arm-to-hanger bracket mounting bolt.
12. Remove axle assembly.

### Installation

1. Using jacks, raise rear axle assembly into position under vehicle.
2. Install trailing arm-to-hanger mounting bracket; finger-tighten bolts only.
3. Install shock absorber mounting bolts loosely. Remove jacks.
4. Place spindle and brake assembly in position. Install four (4) retaining bolts finger-tight.
5. Tighten the four retaining bolts to 60 N·m (45 ft.-lbs.) torque.
6. Install brake drum.
7. Install washer and nut. Tighten adjusting nut to 27-34 N·m (240-300 in.-lbs.) while rotating wheel. Back off adjusting nut with wrench to completely release bearing pre-load. Finger-tighten adjusting nut.
8. Position nut lock with one pair of slots in line with cotter pin hole. Install cotter pin. The end-play should be .025-.076 mm (.001-.003 in). Clean and install grease cap.
9. Put parking brake cable through bracket.
10. Slip ball-end of parking brake cables through brake connectors on parking brake bracket.
11. Install both retaining clips.
12. Install parking brake cable adjusting connection nut. Tighten until all slack is removed from cables.
13. Install retaining clips and brake tube fittings. Tighten fittings to 12 N·m (110 in.-lbs.).
14. Bleed rear brake system and readjust brakes.
15. Install wheel and tire assembly. Tighten wheel nuts to 108 N·m (80 ft.-lbs.) torque.
16. With vehicle on ground, tighten trailing arm-to-hanger bracket mounting bolts to 55 N·m (40 ft.-lbs.) torque.
17. Tighten shock absorber mounting bolts to 55 N·m (40 ft.-lbs.) torque.

## Disassemble Parking Brake Cables at Rear Crossmember

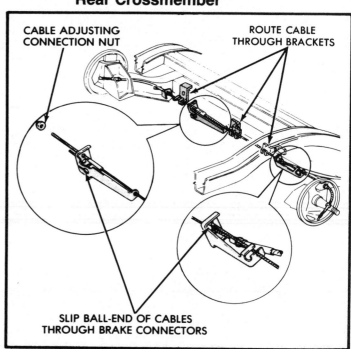

CABLE ADJUSTING CONNECTION NUT

ROUTE CABLE THROUGH BRACKETS

SLIP BALL-END OF CABLES THROUGH BRAKE CONNECTORS

## "K" BODY CARS ONLY

### Removal

1. Raise the vehicle and support safely. Support the rear axle with adjustable jack stands. Remove the wheel assemblies.

2. Separate the parking brake cable at the connector and cable housing at the floor pan bracket.

3. Separate the brake tube assembly from the brake hose at the trailing arm support bracket and remove the lock.

4. Remove the lower shock absorber through bolts and the track bar to axle through bolts. Support the track bar end with wire to keep out of the way.

5. Lower the axle until the spring and isolator assemblies can be removed.

6. Support the pivot bushing end of the trailing arms and remove the pivot bushing hanger bracket to frame screws. Carefully lower the axle assembly and remove from under the vehicle.

### Installation

1. Raise and support the axle assembly on adjustable jack stands.

2. Attach the pivot bushing hanger brackets to the frame rail and tighten the attaching bolts.

3. Install the springs and isolators and carefully raise the axle assembly.

## K Car Spring and Jounce Bumper

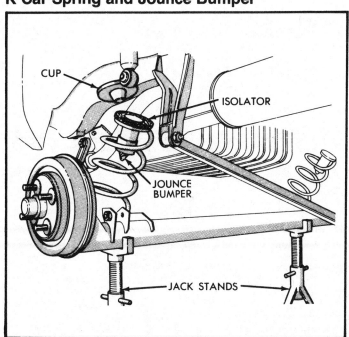

## K Car Flex Arm Rear Suspension

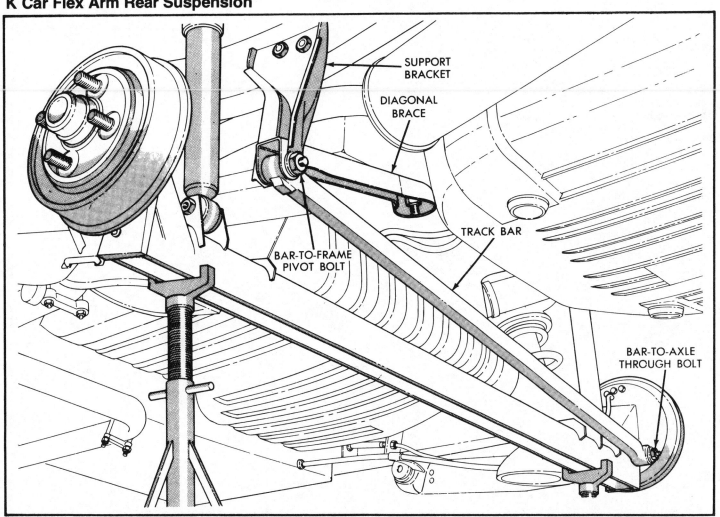

4. Install the shock absorber and track bar through bolts. Do not tighten.

5. Position the spindle and brake support to the axle while routing the parking brake cable through the trailing arm opening and the brake tube over the trailing arm. Install the four retaining bolts loosely, then tighten to 45 ft. lbs. torque.

6. Install the hub and drum, if previously removed.

7. Route the parking brake cable through the fingers in the retaining bracket and lock housing end into the floor pan bracket. Install the cable end into the intermediate connector.

8. Install the brake hose end fitting into the bracket and install the lock. Tighten as required.

9. Install the wheel assemblies and lower vehicle to floor. Tighten the lower shock absorber bolts to 40 ft. lbs. and the track bar bolt to 80 ft. lbs. torque.

10. Bleed the brake system as required.

# FRONT SUSPENSION—CHRYSLER CORPORATION REAR DRIVE

## Alignment

There are six basic factors which are the foundation to front wheel alignment: height, caster, camber, toe-in, steering axis inclination and toe-out on turns. All are mechanically adjustable except steering axis inclination and toe-out on turns. The latter two are valuable in determining if parts are bent or damaged, particularly when the camber and caster adjustments cannot be brought within the recommended specifications.

All adjustments and checks should be made in the following sequence.

1. Front suspension height
2. Caster and camber
3. Toe-in
4. Steering axis inclination (not adjustable)
5. Toe-out on turns (not adjustable)

## HEIGHT

Front suspension heights must be measured with the recommended tire pressures and with no passenger or luggage compartment load. The car should have a full tank of gasoline or equivalent weight compensation. It must be on a level surface.

### Procedure

Rock the vehicle at the center of the front and rear bumpers at least six times to eliminate friction effects before making the vehicle height measurements. Allow the vehicle to settle on its own weight.

For 1976-79 models with longitudinal (front to rear) torsion bars, measure from the lowest point of the lower control arm torsion bar anchor, at a point one inch forward of the rear face of the anchor, to the ground. On models with transverse (across the chassis) torsion bars, measure from the lowest point of the lower control arm inner pivot bushing to the floor.

## Transverse Torsion Bar Suspension

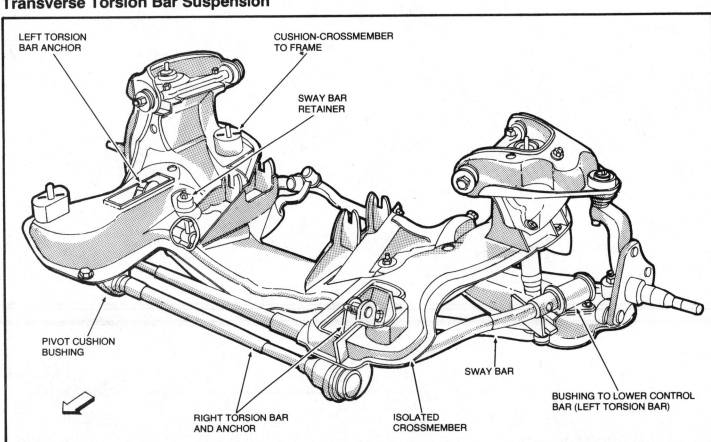

LEFT TORSION BAR ANCHOR

CUSHION-CROSSMEMBER TO FRAME

SWAY BAR RETAINER

PIVOT CUSHION BUSHING

RIGHT TORSION BAR AND ANCHOR

ISOLATED CROSSMEMBER

SWAY BAR

BUSHING TO LOWER CONTROL BAR (LEFT TORSION BAR)

## Longitudinal Torsion Bar Suspension, Typical

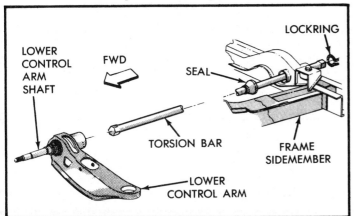

For 1980-and later Gran Fury, St. Regis and Newport/New Yorker, measure from the bottom of the front frame rail, between the radiator yoke and the forward edge of the front suspension crossmember, to the ground. For all other 1980 and later torsion bar front suspension models, measure from the head of the front suspension crossmember front isolator bolt to the ground.

On all models, check the measurements against those given in the Front End Alignment Specifications. Adjust the height by turning the torsion bar adjusting bolt clockwise to raise or counterclockwise to lower. The height should not vary more than 1/8 inch from side to side.

### CAMBER AND CASTER

1. Prepare vehicle for measuring wheel alignment.
2. Determine initial camber and caster readings to confirm variance to specifications before loosening pivot bar bolts.
3. Remove foreign material from exposed threads of pivot bar bolts.
4. Loosen nuts slightly holding pivot (caster/camber) bar. Slightly loosening the pivot bar nuts will allow the upper control arm to be repositioned without slipping to end of adjustment slots.
5. Position claw of tool on pivot bar and pin of tool into holes provided in tower or bracket. Make adjustments by moving pivot bar in or out. Adjust as follows:
*Camber:* Move both ends of upper control arm in or out exactly equal amounts. Camber settings should be held as close as possible to "preferred" setting.
*Caster:* Moving one end of the bar will change caster (and camber). To preserve camber while adjusting caster, move each end of the upper control arm pivot bar exactly equal amounts in opposite directions. For example, to increase positive caster move front of pivot bar away from engine, then move rear of pivot bar towards engine an equal amount. Caster should be held as nearly equal as possible on **both** wheels.

Tighten pivot bar holding bolts to specified torque, as follows. All, except 1976-77 Royal Monaco and Gran Fury, 1976-78 Chrysler models to 150 ft. lbs. (203 N·m). 1976-77 Royal Monaco and Gran Fury, 1976-78 Chrysler models to 175 ft. lbs. (237 N·m).

### TOE

The toe setting should be the final operation of the front wheel alignment adjustments. In all cases, follow equipment manufacturers procedure.

1. Secure steering wheel in "straight-ahead" position. On vehicles equipped with power steering, start engine before centering steering wheel. (Engine should be kept running while adjusting toe).
2. Loosen tie rod clamp bolts.

## Caster/Camber Adjustment

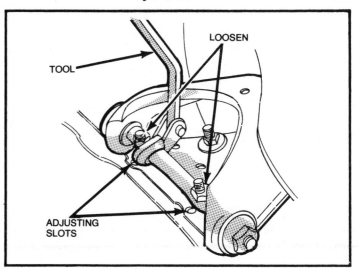

**Loosen pivot bolt nuts**

## Tie Rod Adjustment

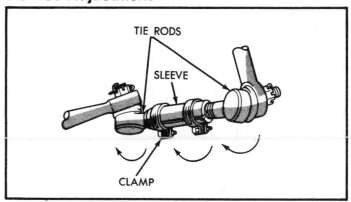

3. Adjust toe by turning tie rod sleeves.

**NOTE: To avoid a binding condition in either tie rod assembly, rotate both tie rod ends in direction of sleeve travel during adjustment. This will ensure that both ends will be in the center of their travel when tightening sleeve clamps.**

4. Shut off engine.
5. Position sleeve clamps so ends do not locate in the sleeve slot, then tighten clamp bolts as specified. Be sure clamp bolts are indexed at or near bottom to avoid possible interference with torsion bars when vehicle is in full jounce.

Upon completion of alignment operations, it is essential that the splash shields, if removed, be correctly reinstalled with all holding clips in place.

## Front Wheel Bearings

### LUBRICATION

Under normal service the lubricant in front wheel bearings should be inspected whenever brake drums or disc brake rotors are removed to inspect or service the brake system, but at least every 30,000 miles (48 000 kilometres).

For severe service vehicles, (such as taxi and police vehicles involving frequent or continuous brake application) wheel bearings

## Front Bearing Hub Cavity

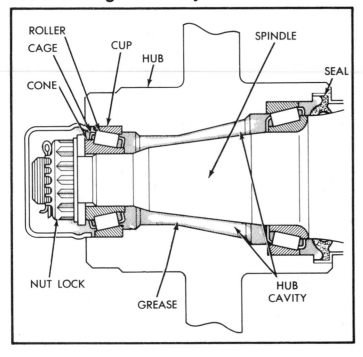

ROLLER
CAGE
CONE
CUP
HUB
SPINDLE
SEAL
NUT LOCK
GREASE
HUB CAVITY

should be inspected whenever the rotors are removed to inspect or service the brake system, or at least every 9,000 miles (14 000 kilometres), whichever occurs first.

Check lubricant to see that it is adequate in quantity and quality. If grease is low in quantity, contains dirt, appears dry or has been contaminated with water to produce a milky appearance, bearings should be cleaned and completely repacked. *Never add grease to wheel bearings.*

When relubrication is required, discard old seal. Thoroughly clean old lubricant from bearings and from hub cavity. Inspect rollers for signs of pitting or other surface distress. Light bearing discoloration should be considered normal. Bearings must be replaced if any defects exist. For all service, repack the bearings with a high temperature wheel bearing grease. Use of a bearing packer is recommended. A small amount of new grease should also be added to hub cavity.

### Adjustment

1. Tighten adjusting nut to 240-300 inch-pounds (27 to 34 N·m) while rotating wheel. Stop rotation and back off adjusting nut with wrench to completely release bearing pre-load. Next, finger-tighten adjusting nut. Position nut lock with one pair of slots in line with cotter pin hole. Install cotter pin. The resulting adjustment should be .0001-.003 inch end play.

2. Clean and install grease cap. Install wheel and tire assembly.

### Removal

1. In the event the bearing cup is found defective during inspection, remove grease cap, cotter pin, nut lock and bearing adjusting nut.

2. Remove the disc brake sliding caliper retaining clips and anti-rattle springs.

3. Slowly slide caliper housing assembly up and away from brake disc and support caliper housing on steering knuckle arm. Do not let caliper housing hang by brake hose, as possible brake hose damage may result.

4. Remove thrust washer and outer bearing cone.

5. Slide wheel hub and disc assembly off the spindle.

6. Carefully drive out inner seal and remove bearing cone with 3/4 inch diameter non-metallic rod.

### Installation

1. Using a bearing drive tool, install new cone. Care must be taken to fully seat new cup against shoulder of hub.

2. Force lubricant between all bearing cone rollers or repack using a suitable bearing packer. A small amount of grease should be added to hub cavity.

3. Install inner cone and a new seal with lip of seal facing inward. Position seal flush with end of hub. The seal flange may be damaged if proper tool is not used.

4. Clean spindle and apply a light coating of wheel bearing lubricant over polished surfaces.

5. Install hub and braking disc assembly on spindle and install outer bearing cone, thrust washer and adjusting nut. Refer to bearing adjustment procedure.

6. Slowly slide caliper housing assembly down on brake disc assembly into position on adaptor. Install caliper retaining clips and anti-rattle springs. Tighten to 180 inch-pounds (20 N·m).

7. Install tire and wheel.

## Shock Absorbers

### Removal

**NOTE: To remove the front shock absorbers on all models, you may find it more convenient to remove the tire and wheel assembly and perform the removal from under the fender.**

1. Loosen and remove nut and retainer from upper end of shock absorber piston rod.

2. Raise car so wheels are clear of floor and remove lower attachment.

3. Compress shock absorber completely by pushing upward. Remove from vehicle by pulling down and out of upper shock absorber mounting bushing.

4. Check appearance of upper shock absorber mounting bushing. If it appears worn, damaged or deteriorated, remove bushing by first pressing out inner sleeve with a suitable tool then prying out or cutting out the rubber bushing. (This bushing will take some set after it has been in service and must be replaced once it has been removed.)

### Installation

1. To install upper rubber bushing, remove inner steel sleeve and immerse bushing in water (do not use oil) and with a twisting motion, start bushing into hole of upper mounting bracket. Tap into position with a hammer. Reinstall steel inner sleeve in bushing.

2. Test and expel air from shock absorber. Compress to its shortest length. Install upper bushing lower retainer and insert rod through upper bushing. Install upper retainer and nut; tighten to 25 foot-pounds (34 N·m).

**NOTE: In each case, install all retainers with the concave side in contact with the rubber.**

3. Position and align lower eye of shock absorber with that of lower control arm mounting holes. Install shock absorber and tighten nut to 50 ft.-lbs. (68 N·m) on bolt-and-nut-type. On suspensions with removal bushings, tighten retainer nut to 35 ft.-lbs. (47 N·m) with full weight of vehicle on the wheels.

**NOTE: When tightening retaining nut, be sure to grip shock absorber at the base area below the weld to avoid reservoir damage.**

## Upper Ball Joint

### Inspection

1. Position jack under the lower control arm and raise wheel clear of floor. Remove wheel cover, grease cap and cotter pin.

2. Tighten bearing adjusting nut enough to remove all play between hub, bearings and spindle.

## Front Shock Absorbers—Transverse Torsion Bar Suspension

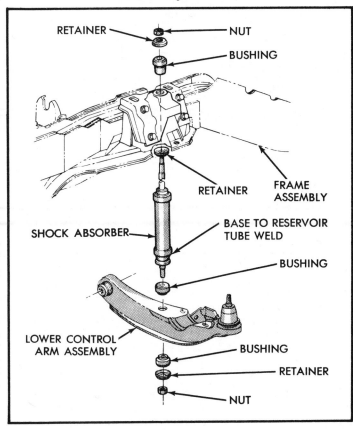

## Front Shock Absorber—Conventional Torsion Bar Suspension

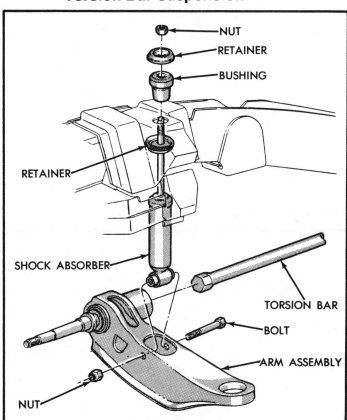

3. Lower jack to allow tire to lightly contact floor (most of vehicle weight relieved from the tire). It is important that the tire have contact with the floor.

4. Grasp the top of the tire and apply force inward and outward. While this force is being applied, have an observer check for any movement at the ball joints between the upper control arm and the knuckle.

5. If any lateral movement is evident, replace the ball joint.

### Removal

1. Place ignition switch in Off or Unlocked position.

2. Raise front of vehicle with hand jack and place short jack stand under lower control arm. Position jack stand as close to wheel as possible. Be sure jack stand is not in contact with brake spash shield. Rubber rebound bumper must not contact frame.

———— CAUTION ————
*Torsion bar will remain in loaded position.*

3. Remove wheel cover, wheel and tire assembly.

4. Remove cotter pins and nuts from upper and lower ball joints to facilitate use of ball joint removal tool.

5. Slide tool on lower ball joint stud allowing tool to rest on knuckle arm. Set tool securely against upper stud.

6. Tighten tool to apply pressure to upper stud and strike knuckle sharply with hammer to loosen stud. Do not attempt to force stud out only with tool.

7. After removing tool, disengage upper ball joint from knukle. Support knuckle and brake assembly to prevent damage to brake hose or lower ball joint.

8. Remove upper ball joint by turning counterclockwise from upper control arm.

## Knuckle Control Arm and Ball Joint

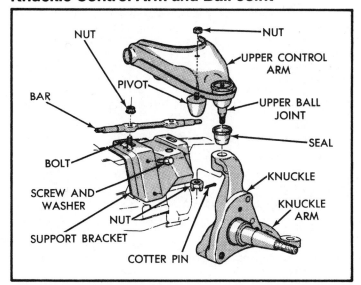

### Installation

1. Screw ball joint squarely into control arm as far as possible by hand. Make certain ball joint threads engage those of control arm correctly if original arm is used. Seals should always be replaced once they have been removed.

2. Tighten ball joint until it bottoms on housing. Tighten to 125 ft.-lbs. (180 N·m)

## Loosening Of Upper Ball Joint Stud From The Steering Knuckle Using A Stud Loosening Tool, Typical

3. Position new seal over ball joint stud and install using tool adapter. Make sure seal is seated on ball joint housing.

4. Position upper ball joint stud in steering knuckle and install nut. Tighten nut to 100 ft.-lbs. (136 N·m).

5. Install lower ball joint stud nut and tighten to 100 ft.-lbs. (136 N·m). Install cotter pin and lubricate upper ball joint.

6. Torsion bar will remain in loaded position.

7. Install wheel and tire assembly with wheel cover.

## Lower Ball Joint

### Inspection

1. Raise the front of vehicle and install safety floor stands under both lower control arms as far outboard as possible. The upper control arms must not contact the rubber rebound bumpers.

2. With the weight of vehicle on the control arm, install dial indicator and clamp assembly to lower control arm.

3. Position dial indicator plunger tip against knuckle arm and zero dial indicator.

4. Measure axial travel of the knuckle arm with respect to the control arm, by raising and lowering the wheel using a pry bar under the center of the tire.

5. If during measurement you find the axial travel of the control arm is .030 inches (0.76 mm) or more, relative to the knuckle arm, the ball joint should be replaced.

### Removal

1. Place ignition switch in Off or Unlocked position.

## Lower Control Arm and Ball Joint

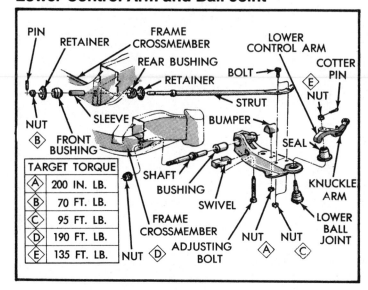

| TARGET TORQUE | |
|---|---|
| Ⓐ | 200 IN. LB. |
| Ⓑ | 70 FT. LB. |
| Ⓒ | 95 FT. LB. |
| Ⓓ | 190 FT. LB. |
| Ⓔ | 135 FT. LB. |

2. Raise vehicle on hoist to place front suspension in rebound. Place jack stands under front frame for additional support.

3. Remove wheel cover, wheel and tire assembly.

4. Remove brake caliper and support with wire hook. Do not hang caliper by brake hose alone.

5. Remove hub and rotor assembly and splash shield. Disconnect shock absorber at lower control arm.

6. Unwind torsion bar.

7. Remove upper and lower ball joint stud cotter pins and nuts. Slide tool over upper stud until tool rests on steering knuckle.

8. Turn threaded portion of tool, locking it securely against lower stud. Tighten tool enough to place lower ball joint stud under pressure, then strike steering knuckle arm sharply with a hammer to loosen stud. Do not attempt to force stud out of knuckle with tool alone.

9. Use tool to press ball joint out of lower control arm.

### Installation

1. Press new ball joint into lower control arm assembly.

2. Place a new seal over ball joint with adapter tool. Press retainer portion of seal down on ball joint housing until it is securely locked in position.

3. Insert ball joint stud into opening in knuckle arm and install stud retaining nuts; tighten as specified. Install cotter pins and lubricate ball joint.

4. Place a load on torsion bar by turning adjusting bolt clockwise.

5. Install wheel, tire and brake assembly and adjust front wheel bearing.

6. Lower vehicle to floor. Adjust front suspension heights.

## Torsion Bars (Longitudinal Type)

Longitudinal torsion bars have a hex formed on each end. One hex end is installed in the lower control arm anchor, the opposite end is anchored in the frame or body crossmember.

Torsion bars are identified for use by length and thickness (depending on carline, body, engine, etc.), and are not interchangeable side for side. The bars are marked either right or left by the letter "R" or "L" stamped on one end of the bar.

### Removal

1. Lift vehicle on hoist to place front suspension in rebound.

2. Release load from torsion bar by turning the anchor adjusting bolt in lower control arm counterclockwise.

## Longitudinal Torsion Bar Removal, Typical

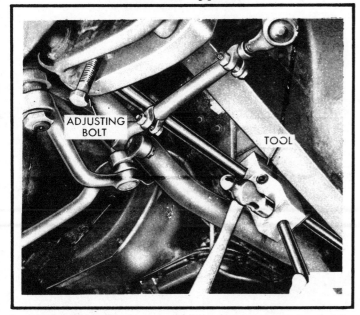

ADJUSTING BOLT

TOOL

3. Remove lock ring from anchor at rear of bar. Install drive tool to remove torsion bar. (If necessary, remove transmission torque shaft to provide clearance.) Place tool toward rear of bar to allow sufficient room for striking pad of tool. Do not apply heat to torsion bar, front anchor or rear anchor.

4. Remove tool and slide bar out through rear anchor. Do not damage balloon seal when removing bar.

### Inspection

1. Inspect torsion bar and seal for damage; replace if damaged.

2. Remove all foreign matter from hex opening(s) in anchors and from hex end(s) of torsion bar.

3. Inspect torsion bar adjusting bolt and swivel and replace if there is corrosion or other damage. Lubricate for easy installation.

### Installation

1. Insert torsion bar through rear anchor.

2. Lubricate inside surface of balloon seal and slide seal over torsion bar (cupped end toward rear of bar).

3. Coat both hex ends of torsion bar with lubricant.

4. Slide torsion bar in hex opening of lower control arm.

**NOTE: If torsion bar hex opening does not index with lower control arm hex opening, loosen lower control arm pivot shaft nut, rotate pivot shaft to index with torsion bar. Install torsion bar. Do not tighten pivot shaft nut while suspension is in rebound.**

5. Install lock ring, making sure it is seated in its groove.

6. Pack rear anchor openigs at lock ring area and area under seals with lubricant. Position lip of seal in groove of anchor.

7. Turn adjusting bolt clockwise to place a load on torsion bar.

8. Lower vehicle to floor and tighten pivot shaft nut to 145 foot-pounds (197 N·m).

9. Adjust front suspension height.

## Torsion Bars (Transverse-Type)

Torsion bars are formed with an angle for transverse mounting. Each bar is hex shaped on the anchor end with a replaceable torsion bar-to-lower control arm bushing on the opposite end and a pivot cushion bushing (permanently attached) midway on the bar creating right and left hand assemblies.

The hex end of the bar is anchored in the crossmember (opposite the affected wheel), extends parallel to the front crossmember, through the pivot cushion bushing (also attached to the crossmember), turns, and attaches to the lower control arm through the torsion bar to lower control arm bushing.

### Removal

1. Raise car on hoist and support vehicle so that front suspension is in full rebound position.

2. Release load on **both** torsion bars by turning anchor adjusting bolts in frame crossmember counterclockwise. Remove anchor adjusting bolt on torsion bar to be removed.

3. Raise lower control arms until clearance between crossmember ledge (at jounce bumper) and torsion bar end bushing is 2⅞ inches (63.0 mm). Support lower control arms at this design height (equal to three passenger position with vehicle on ground). This is necessary to align sway bar and lower control arm attaching points for disassembly and component re-alignment and attachment during reassembly.

## Transverse Torsion Bar Anchor Bolt

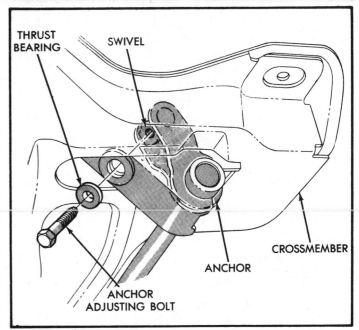

THRUST BEARING

SWIVEL

CROSSMEMBER

ANCHOR

ANCHOR ADJUSTING BOLT

## Correct Anchor And Swivel Installation

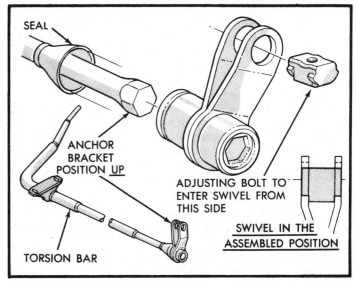

SEAL

ANCHOR BRACKET POSITION UP

TORSION BAR

ADJUSTING BOLT TO ENTER SWIVEL FROM THIS SIDE

SWIVEL IN THE ASSEMBLED POSITION

### Transverse Torsion Bar Front Suspension

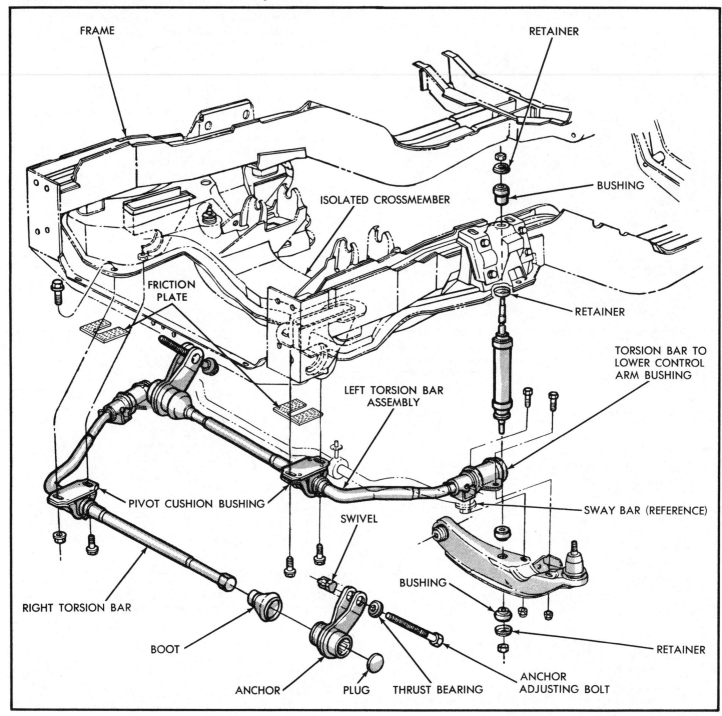

FRAME

RETAINER

ISOLATED CROSSMEMBER

BUSHING

FRICTION PLATE

RETAINER

TORSION BAR TO LOWER CONTROL ARM BUSHING

LEFT TORSION BAR ASSEMBLY

PIVOT CUSHION BUSHING

SWIVEL

SWAY BAR (REFERENCE)

RIGHT TORSION BAR

BUSHING

BOOT

RETAINER

ANCHOR          PLUG     THRUST BEARING

ANCHOR ADJUSTING BOLT

4. Remove sway bar-to-control arm attaching bolt and retainers.

5. Remove two bolts attaching torsion bar end bushing to lower control arm.

6. Remove two bolts attaching torsion bar pivot cushion bushing to crossmember, and remove torsion bar and anchor assembly from crossmember.

7. Carefully separate anchor from torsion bar.

**TORSION BAR-TO-LOWER CONTROL ARM BUSHING REPLACEMENT**

Service replacement bars include pivot cushion bushing and torsion bar to lower control arm bushing.

1. Clamp assembly in vise with rivet head up (hex end of bar down).

— **CAUTION** —

*Never clamp the bar in a vise unless soft vise jaw inserts (brass, aluminum, etc.) are used.*

2. Centerpunch the rivet head and drill a 3/8 inch (9.5 mm) diameter hole approximately 1/2 inch (12.5 mm) deep. A short length of 5/16 inch (8 mm) rod can be used to remove the rivet. It may be necessary to remove flange of rivet head before driving rivet out.

## Measuring Design Height

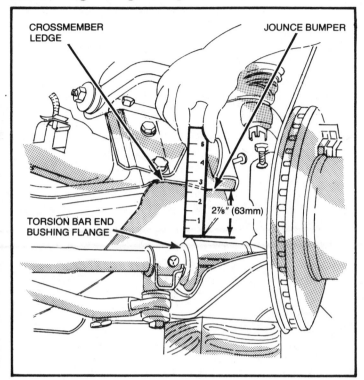

CROSSMEMBER LEDGE

JOUNCE BUMPER

TORSION BAR END BUSHING FLANGE

2⅞" (63mm)

## Transverse Bar Lower Control Arm Mounting

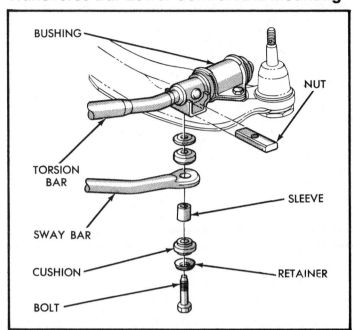

BUSHING

NUT

TORSION BAR

SLEEVE

SWAY BAR

CUSHION

RETAINER

BOLT

---

**———— CAUTION ————**

*Do not enlarge the 7/16 inch (11 mm) diameter hole in the bar.*

---

3. Remove bushing from bar.

4. Install new bushing. Rough area under bushing may be cleaned with sandpaper if necessary for easy assembly. New bushing should go on by hand.

5. Install bushing retaining bolt and tighten nut to 50 foot-pounds (68 N·m).

### Inspection

1. Inspect seal for damage, replace if damaged.

2. Inspect bushing-to-lower control arm and pivot cushion bushing.

**NOTE: Inspect seals on cushion bushing for cuts, tears or severe deterioration that may allow moisture under cushion. If corrosion is evident, the torsion bar assembly should be replaced.**

3. Remove all foreign matter from hex opening(s) in anchors and from hex end(s) of torsion bar.

4. Inspect torsion bar adjusting bolt and swivel and replace if there is any sign of corrosion or other damage. Lubricate for easy installation.

### Installation

1. Carefully slide balloon seal over end of torsion bar (cupped end toward hex).

2. Coat hex end of torsion bar with lubricant.

3. Install torsion bar hex end into anchor bracket. With torsion bar in a horizontal position, the ears of the anchor bracket should be positioned nearly straight up. Position swivel into anchor bracket ears.

4. Place bushing end of bar into position on top of lower control arm. Then, install anchor bracket assembly into crossmember anchor retainer and install anchor adjusting bearing and bolt.

## Bushing Removal, Torsion Bar to Lower Control Arm

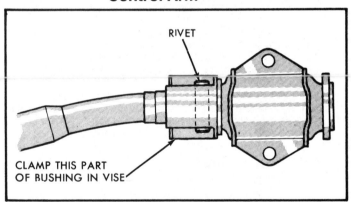

RIVET

CLAMP THIS PART OF BUSHING IN VISE

## Torsion Bar Anchor Assembly

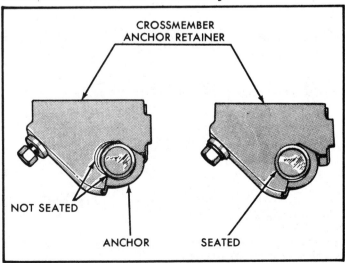

CROSSMEMBER ANCHOR RETAINER

NOT SEATED

ANCHOR

SEATED

**161**

5. Attach pivot cushion bushing to crossmember with two bolt and washer assemblies. Leave bolt and washer assemblies loose enough to install friction plates.

6. With lower control arms at "design height," install two bolt and nut assemblies attaching torsion bar bushing to lower control arm. Tighten to 70 foot-pounds (95 N·m).

7. Ensure that torsion bar anchor bracket is fully seated in crossmember. Install friction plates between crossmember and pivot cushion bushing with open end of slot to rear and bottomed out on mounting bolt. Tighten cushion bushing bolts to 85 foot-pounds (115 N·m).

8. Position balloon seal over anchor bracket.

9. Reinstall bolt, through sway bar, retainer cushions and sleeve, and attach to lower control arm end bushing. Tighten bolt to 50 foot-pounds (68 N·m).

10. Load torsion bar by turning anchor adjusting bolt clockwise.

11. Lower vehicle and adjust torsion bar height to specifications.

## Upper Control Arm

### Removal

1. Place ignition switch in Off or Unlocked position.

2. Raise front of vehicle with hand jack and remove wheel cover, wheel and tire assembly.

3. Position short jack stand under lower control arm near splash shield and lower hand jack. Observe that jack stand does not contact shield and rebound bumpers are under no load.

4. On some models, remove brake caliper and set aside to provide clearance for ball joint remover tool.

5. Remove cotter pin and nut from upper and lower ball joints to facilitate use of tool to free ball joint.

6. Slide spreader tool over lower ball joint stud to allow tool to rest on steering knuckle arm. Tighten tool to apply pressure to upper ball joint stud and strike steering knuckle boss sharply with hammer to loosen stud. Do not attempt to force stud out of knuckle with tool alone.

7. After removing tool, support brake and knuckle assembly to prevent damage to brake hose or lower ball joint, then disengage upper ball joint from knuckle.

8. From under hood, remove engine splash shield to expose upper control arm pivot bar.

## Upper Control Arm And Knuckle Assembly

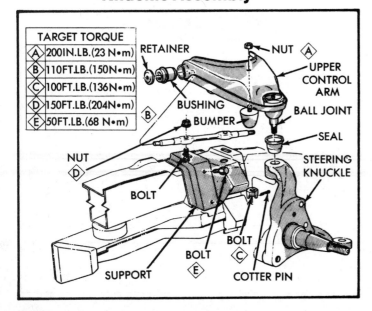

| TARGET TORQUE | |
|---|---|
| Ⓐ | 200 IN.LB. (23 N·m) |
| Ⓑ | 110 FT.LB. (150 N·m) |
| Ⓒ | 100 FT.LB. (136 N·m) |
| Ⓓ | 150 FT.LB. (204 N·m) |
| Ⓔ | 50 FT.LB. (68 N·m) |

### Upper Control Arm Bushing Tool

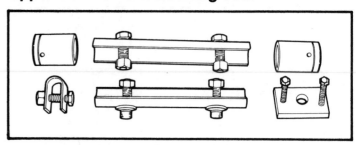

9. Scribe a line on support bracket along inboard edge of pivot bar (to re-establish suspension alignment during reassembly).

10. Remove pivot bolts or nuts and lift upper control arm with ball joint and pivot bar assembly from bracket.

**DISASSEMBLY (BUSHINGS)**

1. Place upper control arm in vise and remove pivot bar nuts and bushing retainers.

2. Bolt support tool C-4253-1 to pivot bar.

3. Place puller tool C-4253-2 over end of pivot bar and reinstall nut. Snug bolts against arm.

4. Screw bolts equally until bushing is free in arm and remove tool and bushing.

**ASSEMBLY (BUSHINGS)**

1. With control arm in vise, put pivot bar in arm and attach support bracket spacer tool.

2. Slip bushings over each end of pivot bar and pilot into holes in arm.

3. Install bushing cups over both bushings and press bushings together until both bushings are fully seated in arm. Pound bushings in place at the same time or use an arbor press. Bushing flange must be bottomed on control arm extrusion.

4. Install retainers and nuts on pivot bar. Snug nuts against retainers.

**NOTE: Pivot bar bushing retainer nuts are to be tightened to specifications AFTER suspension (upper control arm) is at design height.**

### Installation

1. Place upper control arm with ball joint and pivot bar on bracket. Install and snug attaching bolts against arm.

2. Set inboard edge of pivot bar on mounting bracket. Tighten bolts to 150 ft.-lbs. (204 N·m).

3. Replace engine splash shield.

4. Install ball joint stud through steering knuckle and install upper and lower ball joint nuts, tighten to specifications and install cotter pins.

5. With vehicle at design height tighten pivot bar nuts to 110 ft.-lbs. (150 N·m).

## Lower Control Arm (Longitudinal-Type Torsion Bars)

### Removal

1. Place ignition switch in Off or Unlocked position.

2. Remove rebound bumper.

3. Raise vehicle on hoist to place front suspension in rebound. Place jack stands under front frame for additional support.

4. Remove wheel cover, wheel and tire assembly.

5. Remove brake caliper and set aside. Do not hang caliper by brake hose alone. Disconnect shock absorber lower bolt.

6. Remove hub and rotor assembly, splash shield and lower shock mounting nut. Remove bolt and nut.

## Longitudinal Torsion Bar Suspension

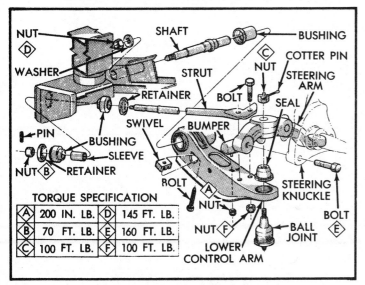

| TORQUE SPECIFICATION | | | |
|---|---|---|---|
| A | 200 IN. LB. | D | 145 FT. LB. |
| B | 70 FT. LB. | E | 160 FT. LB. |
| C | 100 FT. LB. | F | 100 FT. LB. |

7. Remove two (2) strut bar attaching bolts from lower control arm.

8. Remove automatic transmission gear shift torque shaft assembly if required for tool clearance.

9. Measure torsion bar anchor bolt depth into lower control arm before unwinding torsion bar. Unwind bar.

10. Remove torsion bar.

11. Separate lower ball joint from knuckle arm.

12. Remove lower control arm shaft nut from control arm shaft and push out shaft from frame crossmember. Strike threaded end of shaft with soft hammer to loosen if necessary. Remove lower control arm and shaft as an assembly from the vehicle.

13. In the event the shaft bushing indicates wear or deterioration, replacement is recommended.

### DISASSEMBLY (BUSHINGS)

1. Place lower control arm in vise and remove torsion bar adjusting bolt and swivel.

2. Place lower control arm assembly in arbor press with torsion bar hex opening up and with a support around anchor on bottom end.

3. Place a brass drift into hex opening and press shaft out of lower control arm. The bushing inner shell will remain on shaft.

4. Cut and remove rubber portion of bushing from control arm or shaft. Remove bushing outer shell by cutting with a chisel. Use care not to cut into control arm.

5. Remove bushing inner shell with pivot shaft.

### ASSEMBLY (BUSHINGS)

1. Position new bushing on shaft (flange end of bushing first). Press shaft into inner sleeve until bushing seats on shoulder of shaft.

2. Press shaft and bushing assembly into lower control arm using an arbor press.

3. Install torsion bar adjusting bolt and swivel.

### Installation

1. Position lower control arm with shaft in crossmember. Install lower control arm shaft nut and finger-tighten nut.

2. Position lower ball joint stud into knuckle arm and tighten nut to 100 ft.-lbs. (136 N·m).

3. Install torsion bar into lower control arm.

4. Replace transmission gear shaft torque shaft if removed.

5. Position strut bar with two attaching bolts to lower control arm. Tighten to 100 foot-pounds (136 N·m).

6. Attach brake splash shield and secure lower shock mounting bolt to lower control arm.

7. Attach hub and rotor and install brake caliper.

## Lower Bushing Remover Tool

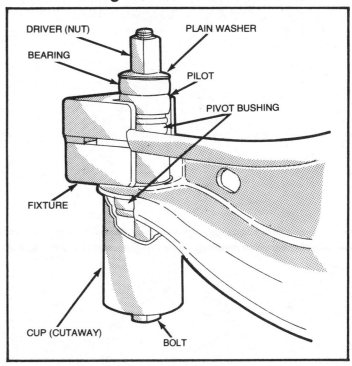

8. Install wheel and tire assembly.

9. Lower vehicle to floor and adjust front suspension heights. Tighten lower control arm pivot shaft nut to 145 foot-pounds (197 N·m). Install rebound bumper and tighten to 200 in-lbs. (23 N·m).

10. Adjust wheel alignment.

## Lower Control Arm (Transverse-Type Torsion Bars)

### Removal

1. Raise car on hoist and remove wheel and tire assembly.

2. Remove brake caliper retaining screws, clips and anti-rattle springs and remove caliper from adaptor and support caliper assembly on wire hook. (Do not hang caliper by brake hose.)

3. Remove hub and rotor assembly and splash shield.

4. Remove shock absorber lower nut, retainer and insulator.

5. Release load on both torsion bars by turning anchor adjusting bolts counterclockwise. Releasing both torsion bars is required because of sway bar reaction from the opposite torsion bar.

6. Raise lower control arm until clearance between crossmember ledge (at jounce bumper) and torsion bar to lower control arm bushing is 2⅞ inches (73 mm). Support control arm at this "design height" and remove two bolts attaching torsion bar end bushing to lower control arm.

7. Separate lower ball joint from knuckle arm.

8. Remove lower control arm pivot bolt and lower control arm.

### DISASSEMBLY (BUSHINGS)

1. Place lower control arm in vise and install bushing removal tool.

2. Place support fixture between flanges of control arm and around bushing. Proper fixture position is required to prevent control arm distortion during bushing removal.

3. Position cup over flanged bushing end with bolt through cup and bushing.

4. Install pilot, thrust washer, plain washer and nut on through bolt.

## Lower Control Arm With Transverse Torsion Bar Suspension

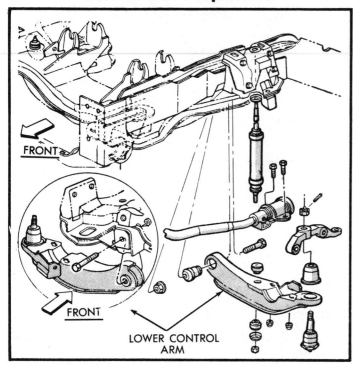

LOWER CONTROL ARM

## Lower Control Arm Bushing Installation

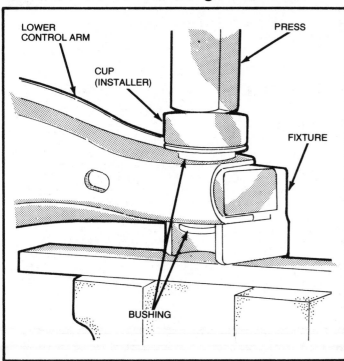

LOWER CONTROL ARM

CUP (INSTALLER)

PRESS

FIXTURE

BUSHING

5. Press bushing out of lower control arm by holding bolt on cup end while turning nut on pilot end.

### ASSEMBLY (BUSHINGS)

1. Place support fixture on lower control arm flanges and position assembly on base of suitable press. Proper fixture position is required to prevent control arm distortion during bushing installation.

2. Position flange end of new bushing into cup squarely and press bushing into control arm until bushing flange seats on arm.

### Installation

1. Position lower control arm in crossmember, install pivot bolt and finger-tighten flanged nut.

2. Position lower ball joint stud into steering knuckle arm and tighten nut to 100 foot-pounds (136 N·m). Insert cotter key.

3. Install torsion bar into lower control arm.

4. Load torsion bar by returning torsion bar adjusting bolt depth to original position before removal.

5. Position strut bar with two attaching bolts to lower control arm. Tighten to 100 foot-pounds (136 N·m).

6. Attach brake splash shield and secure lower shock mounting bolt to lower control arm.

7. Attach hub and rotor and install brake caliper.

8. Install wheel and tire assembly.

9. Lower vehicle to floor and adjust front suspension heights. Tighten lower control arm pivot shaft nut to 145 foot-pounds (197 N·m). Install rebound bumper and tighten to 200 in-lbs. (23 N·m).

10. Adjust wheel alignment.

# Steering Knuckle Arm

### Removal

1. Place ignition switch in Off or Unlocked position.

2. Remove rebound bumper.

3. Raise vehicle on hoist to place front suspension in rebound. Use jack stands under front frame for additional support.

4. Remove wheel cover, wheel and tire assembly.

5. Remove brake caliper and hang out of way with wire hook during this operation to prevent damage to brake hose.

6. Remove hub and brake disc assembly.

7. Remove brake splash shield from steering knuckle.

8. Unload torsion bars, by turning anchor adjusting bolt counterclockwise.

9. Disconnect tie rod from steering knuckle arm. Use care not to damage seals.

10. Remove lower ball joint stud from knuckle arm.

11. Separate knuckle arm from steering knuckle by remove two (2) nuts and two (2) attaching bolts.

12. Remove steering knuckle arm.

### Installation

1. Attach steering knuckle arm to knuckle and install two bolts and nuts. Tighten to 160 ft.-lbs. (217 N·m).

2. Attach lower ball joint stud to knuckle arm. Tighten nut to 100 foot-pounds (136 N·m) and install cotter key.

3. Attach tie rod end to steering knuckel arm and inside nut. Tighten to 40 ft.-lbs. 54 N·m) and install cotter pin.

4. Load torsion bar by turning adjusting bolt on lower control arm clockwise.

## Tie Rod End Puller

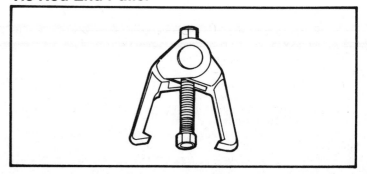

5. Install brake splash shield onto steering knuckle.

6. Install hub and disc assembly. Adjust wheel bearings. Install caliper.

7. Install wheel and tire assembly and attach wheel cover.

8. Lower vehicle to floor, adjust front suspension heights and wheel alignment as necessary.

# Sway Bar (Longitudinal-Type Torsion Bar)

## Removal

1. Place ignition switch in Off or Unlocked position.

2. Raise car on hoist to place front suspension in rebound.

3. Remove wheel cover, wheel and tire assembly.

4. Remove nut and bolt on each end of bar attaching sway bar to strut clamp. Remove nut and bolt from both sway bar link straps to free sway bar from links.

5. Remove sway bar by pulling unit out through frame crossmember openings in direction of area where wheel has been removed.

6. In the event strut cushions and sway bar bushings show excessive wear or deterioration of rubber, replacement is recommended.

## Installation

1. On side where wheel assembly has been removed, install sway bar with center offset in downward position. (Color code on bar is always on driver's side.)

2. Attach sway bar with bolt and nut to strut retainer clamp on each side of bar and tighten to 35 foot-pounds (47 N·m).

3. Lower vehicle to floor and attach both sway bar frame link straps. Tighten to 35 ft.-lbs. (47 N·m).

# Sway Bar (Transverse-Type Torsion Bar)

## Removal

1. Raise car on hoist.

**NOTE: Sway bar-to-lower control arm attaching points are aligned ONLY when lower control arms are at "design height" (equal to three passenger position with vehicle on ground). If frame contact or twin post hoist is used, release load on torsion bar by turning adjuster bolts counterclockwise. Raise lower control arms until clearance between crossmember ledge (at jounce bumper) and torsion bar to lower control arm bushing is 2⅞ inches (73 mm). Support lower control arms with jack stand during sway bar removal and installation.**

2. With lower control arms supported as described in note above, remove sway bar-to-torsion bar bushing attaching bolts, retainers, cushions and sleeves.

3. Remove retainer assembly strap bolts and retainer straps. Remove sway bar.

4. Inspect cushions and bushings for excessive wear or deterioration and replace if required.

## Installation

1. Position sway bar bushings against retainers and install retainer straps. Loosely assemble retainer bolts.

2. Reinstall bolt through sway bar retainer, cushions and sleeve, and attach to torsion bar lower control arm bushing. Tighten bolt to 50 foot-pounds (68 N·m) torque.

3. Tighten sway bar retainer and strap bolts to 30 foot-pounds (41 N·m).

4. Load torsion bar by turning anchor adjusting bolt in crossmember clockwise.

5. Lower vehicle and adjust torsion bar height to specifications.

## Longitudinal Type Torsion Bar

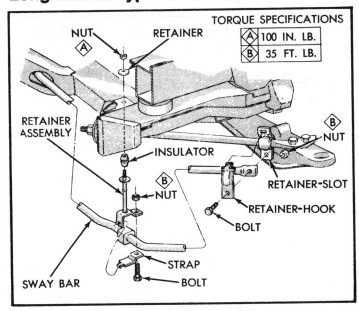

| TORQUE SPECIFICATIONS | |
|---|---|
| A | 100 IN. LB. |
| B | 35 FT. LB. |

## Sway Bar—Transverse Type Torsion Bar

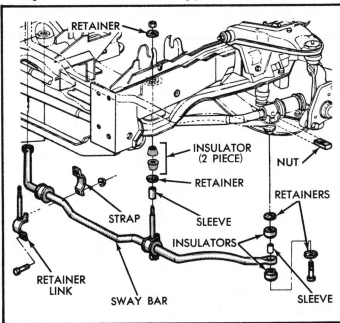

# REAR SUSPENSION—CHRYSLER REAR DRIVE

## Shock Absorber

### Removal

1. Raise axle to relieve load on shock absorber.
2. Remove shock absorber lower end, as follows: Loosen and remove nut, retainer and bushing from spring plate. *When loosening retaining nut grip shock absorber at the base (below the base to reservoir tube weld) to avoid reservoir damage.*
3. Loosen and remove nut and bolt from upper shock absorber mounting, and remove shock absorber.

### Installation

1. Expel air from new shock absorber.

## Expelling Air From New Shock Absorber

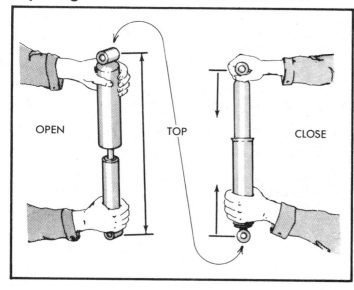

2. Position and align upper eye of shock absorber with mounting holes in crossmember and install bolt and nut. Do not fully tighten.
3. Install shock absorber lower end, as follows: Install upper bushing on shock absorber stud and pull stud through spring plate mounting hole. Install lower bushing, cupped washer and nut. Tighten as specified.
4. Lower vehicle until full weight of vehicle is on the wheels. Tighten upper nut 70 foot-pounds (95 N·m). Tighten lower nut to 35 foot-pounds (47 N·m).

## Springs

### MEASURING SPRING HEIGHT

When measuring rear spring heights, place vehicle on a level floor, have correct front suspension height on both sides, correct tire pressures, no passenger or luggage compartment load and a full tank of fuel.

1. Jounce car several times (front bumper first). Release bumpers at same point in each cycle.
2. Measure shortest distance from top of axle housing to the rail at side of rear axle bumper strap (at rear of bumper).
3. Measure both right and left sides.

If these measurements vary by more than 3/4 inch (side to side), it is an indication that one of the rear springs may need replacing.

### Removal

1. Using floor stands under axle assembly, raise axle assembly to relieve weight on rear spring.
2. Disconnect rear shock absorber at spring plate. Lower axle assembly, permitting rear springs to hang free. (On vehicles so equipped, disconnect rear sway bar links.)
3. Loosen and remove U-bolt nuts and remove U-bolts and spring plate.
4. Loosen and remove the nuts holding front spring hanger to body mounting bracket.
5. Loosen and remove rear spring hanger bolts and let spring drop far enough to pull front spring hanger bolts out of body mounting bracket holes.

## Rear Suspension Cars With Transverse Front Suspension

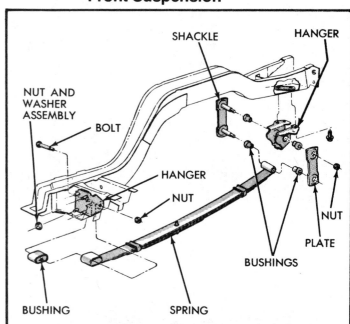

## Rear Suspension Models With Longitudinal Torsion Bar Suspension

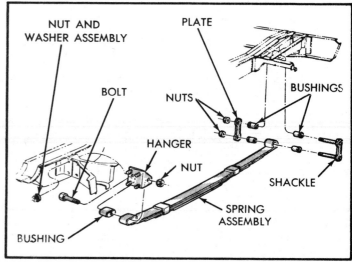

## Rear Spring Isolators

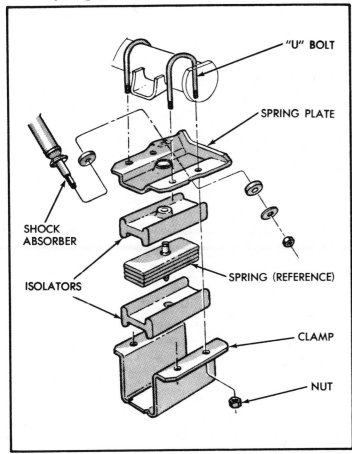

## Pivot Bushing Removal and Installation

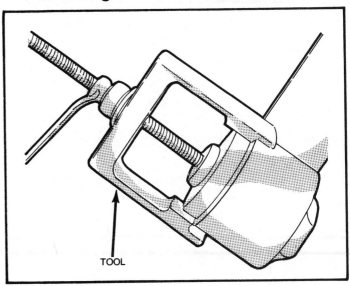

## Bending Bushing Flanges

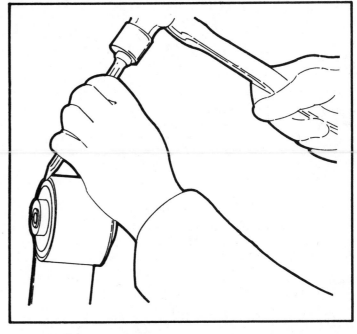

6. Loosen and remove front pivot bolt from front spring hanger.

7. Loosen and remove shackle nuts and remove shackle from rear spring.

### BUSHING REPLACEMENT

It is recommended that the spring assembly be removed from the vehicle for bushing replacement on the bench.

1. Bend two locking tabs away from spring eye on opposite side and remove bushing.

2. Press old bushing out.

3. Press new bushing in.

### Installation

1. Assemble shackle to spring. Do not fully tighten bolt nut.

2. Install front spring hanger and insert pivot bolt and nut; do not fully tighten.

3. Install rear spring hanger-to-body bracket.

4. Raise spring and insert spring hanger mounting bracket bolts. Tighten to 30-35 ft.-lbs. (42-46 N·m).

5. Align axle assembly with spring center bolt. Position center bolt over lower spring plate. Insert U-bolt and nut. Tighten to 45 ft.-lbs. (60 N·m).

6. Connect shock absorbers.

7. Lower car. Tighten pivot bolts to 105 ft.-lbs. (142 N·m). Tighten shackle nuts to 35 ft.-lbs. (46 N·m).

# FRONT SUSPENSION—FORD FRONT DRIVE

## Description

The front wheel drive front suspension is a MacPherson strut design with cast steering knuckles. The shock absorber strut assembly includes a rubber top mount and a coil spring insulator, mounted on the shock strut.

The entire strut assembly is attached at the top by two bolts, which retain the top mount of the strut to the body side apron. The lower end of the assembly is attached to the steering knuckle. A forged lower arm assembly is attached to the underbody side apron and steering knuckle. A stabilizer bar connects the outer end of lower arm to the engine mount bracket. The drive shaft outer stub shaft and wheel hub

# SUSPENSIONS
## FRONT SUSPENSION—FORD FRONT DRIVE CARS

### Front Wheel Drive Suspension

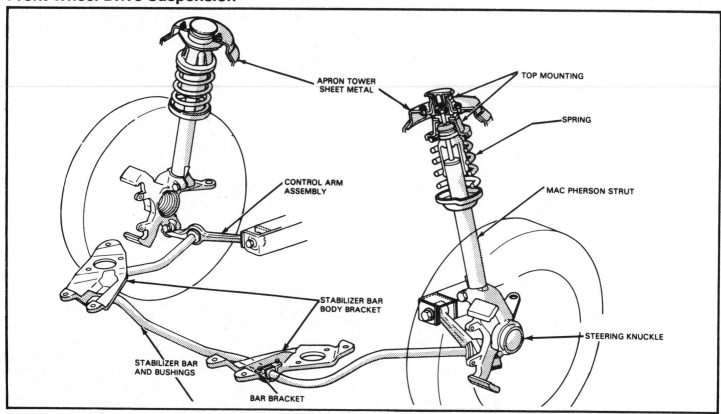

APRON TOWER SHEET METAL

TOP MOUNTING

SPRING

CONTROL ARM ASSEMBLY

MAC PHERSON STRUT

STABILIZER BAR BODY BRACKET

STEERING KNUCKLE

STABILIZER BAR AND BUSHINGS

BAR BRACKET

### Tie Rod Adjustment

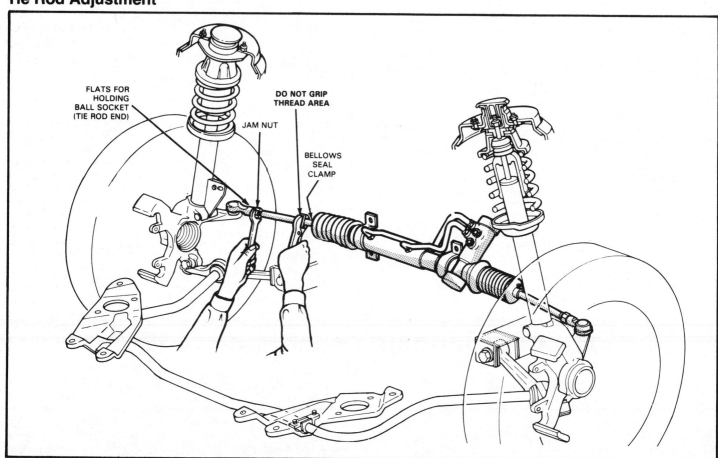

FLATS FOR HOLDING BALL SOCKET (TIE ROD END)

DO NOT GRIP THREAD AREA

JAM NUT

BELLOWS SEAL CLAMP

are attached inside the steering knuckle hub by a pressed fit of mating splines. The assembly is secured by a staked nut on the end of the stub shaft. The hub rotates on two non-adjustable tapered roller bearings which seat against cups in the steering knuckle.

## Wheel Alignment

### TOE

Toe is the difference in distance between the front and the rear of front wheels.

1. Start the engine (power steering only) and move the steering wheel back and forth several times until it is in the straight ahead or centered position.
2. Turn the engine off (power steering only) and lock the steering wheel in place using a steering wheel holder. Loosen and slide off small outer clamp from boot prior to starting toe adjustment to prevent boot from twisting.
3. Adjust left and right tie rods until each wheel has one-half of the desired total toe specification.

**NOTE: When jam nuts are loosened for toe adjustment, the nuts must be tightened to specifications. Attach boot clamp after setting is completed and make sure boot is not twisted.**

### CASTER AND CAMBER

Caster and camber angles of this suspension system are preset at the factory and cannot be adjusted. Measurement procedures are for diagnostic purposes.

**NOTE: Caster measurements must be made on the left side by turning the left wheel through the prescribed angle of sweep and on the right side by turning the right wheel through the prescribed angle of sweep.**

### FRONT WHEEL TURNING ANGLE

When the inside wheel is turned 20 degrees, turning angle of outside wheel should be as specified. The turning angle cannot be adjusted directly, because it is a result of the combination of caster, camber and toe adjustments and should, therefore, be measured only after the toe adjustment has been made.

**NOTE: If the turning angle does not measure to specification, check the knuckle or other suspension or steering parts for a bent condition.**

## Wheel Bearings

Front wheel bearings are located in the front knuckle, not the rotor. The bearings are protected by inner and outer grease seals and an additional inner grease shield immediately inboard of the inner grease seal. The wheel hub is installed with an interference fit to the constant velocity universal joint outer race shaft. The hub nut and washer are installed and tightened to 240-270 N·m (180-200 ft.-lbs.). The rotor fits loosely on the hub assembly and is secured when the wheel and wheel nuts are installed.

### Adjustment

The front wheel bearings have a set-right design that requires no scheduled maintenance. The bearing design relies on component stack-up and deformation/torque at assembly to determine bearing setting. Therefore, bearings cannot be adjusted. In addition to maintaining bearing adjustment, the hub nut torque of 240-270 n·m (180-200 ft.-lbs.) restricts bearing/hub relative movement and maintains axial position of the hub. Due to the importance of the hub nut torque/tension relationship, certain precautions must be taken during service.

## Hub Nut Installation

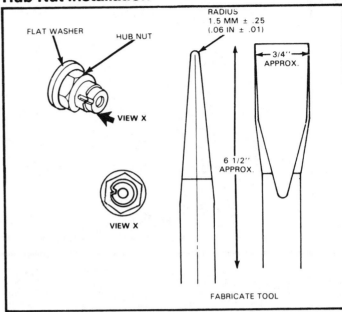

**Tighten nut to 240–270 N·m (180–200 lb-ft) after hub is seated, using special hub installer tool T81P-1104-A. Do not use impact wrench for removal or installation of hub nut. After nut is torqued, deform nut collar into slot of driveshaft using tool as indicated. The nut must not split or crack when staked. If nut is split or cracked after staking, it must be removed and replaced with new unused nut. Removing or moving a hub nut after staking requires the nut be replaced with a new nut. The staking tool can be fabricated from an existed hardened chisel. The correct radius on the chisel tip will prevent improper staking. Do not attempt to stake with a sharp edged tool.**

1. The hub nut must be replaced with a new nut whenever the nut is backed off or removed after the nut has been staked. Never re-use the nut.
2. The hub nut must not be backed off after reaching the required torque of 240-270 N·m (180-200 ft.-lbs.) during installation.
3. The hub nut collar must be staked into the outboard constant velocity joint slot with the proper tool to make sure the required torque is maintained during vehicle operation. The nut collar must not split or crack when staked. If the collar splits or cracks, the nut must be replaced.
4. Impact type tools must not be used to tighten the hub nut or bearing damage will result.
5. The hub and constant velocity joint splines have an interference fit requiring special tools for removal and assembly. The hub nut must not be used to accomplish assembly.
6. To remove the hub nut, apply sufficient torque to the nut to overcome the prevailing torque feature of the crimp in the nut collar. Do not use tools such as a screwdriver or chisel to remove the crimp.

## Front Hub

### Removal

1. Remove hub retaining nut and washer by applying sufficient torque to the nut to overcome prevailing torque feature of the crimp in nut collar. Do not use tools such as a screwdriver or chisel to remove

## Steering Knuckle

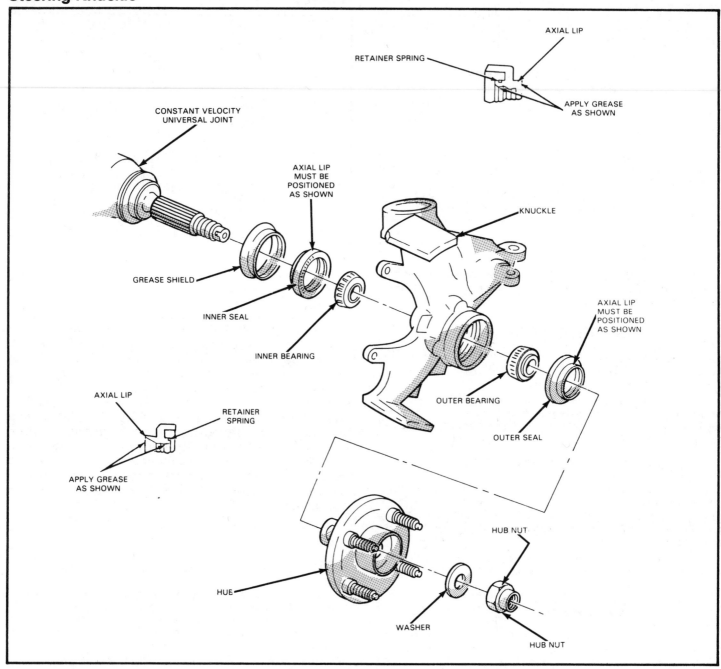

the crimp or use an impact-type tool to remove the hub nut. The hub nut must be discarded after removal.

2. Remove brake caliper. Do not remove caliper pins from the caliper assembly. Lift caliper off the rotor and hang it free of the rotor. Do not allow caliper assembly to hang from the brake hose. Support caliper assembly.

3. Remove rotor from hub by pulling it outboard off the hub bolts.

4. Install hub remover/installer tool, T81P-1104-A with T81P-1104-C and adaptors T81P-b or equivalent, and remove the hub. If outer bearing is seized on the hub, use a puller to remove the bearing. Be careful not to damage bearing if it is being re-used and not to raise burrs on the hub journal diameter. If bearings are being re-used, carefully inspect both bearing cone and rollers, bearing cups

and lubrication for any signs of damage or contamination. If damage or contamination exists, replace all bearing components including cups and seals. In the event the bearings are acceptable, clean and repack bearing components. Inner and outer grease retainers and hub nut must be replaced whenever bearings are inspected.

5. Remove front suspension knuckle.

### Installation

1. Place front knuckle and bearing assembly in a vise so that the inner knuckle bore faces upward (to prevent inner bearing from falling out of the knuckle). Start hub into outer knuckle bore and push the hub by hand through outer and inner wheel bearings as far as possible.

## Wheel Bearing and Caliper Assembly

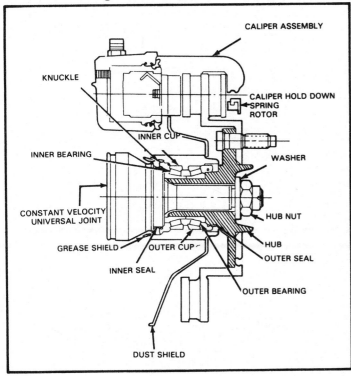

## Hub Removal

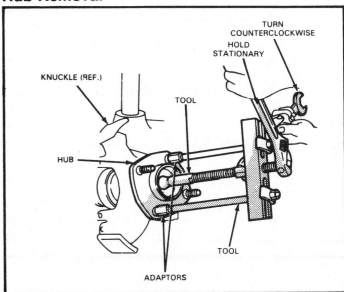

Remove hub from constant velocity universal joint splined stub shaft as shown.

NOTE: Crocus cloth may be used to remove burrs, score marks and rust from the hub barrel.

2. With the hub fully seated in the bearings, position hub and knuckle assembly to front strut. Attach the knuckle to the strut.

3. Lightly lubricate the constant velocity joint stub shaft splines using S.A.E. 30 motor oil.

4. Using hand pressure only, insert splines of the constant velocity joint stub shaft into knuckle/hub assembly as far as possible. Install hub installer tool T81P-1104-C-B-A to the hub and stub shaft.

─── CAUTION ───

*Care must be taken during installation to prevent hub from backing out of bearing assembly. Otherwise, it will be necessary to again reassemble hub through bearings.*

─── CAUTION ───

*Prior to assembly, remove burrs, nicks, score marks, foreign material (rust, dirt, etc.) from hub bearing journal. Due to the very close tolerance 0.0121 mm (0.0005 in) between the wheel bearing inside diameter and the hub assembly, it is important to install hub completely through inner and outer wheel bearings. Hand pressure only is essential to this procedure. Forcing or jamming bearing race (cone) on the hub barrel will cause burrs that can prevent proper installation. Do not strike hub with any type of tool.*

## Hub Installation

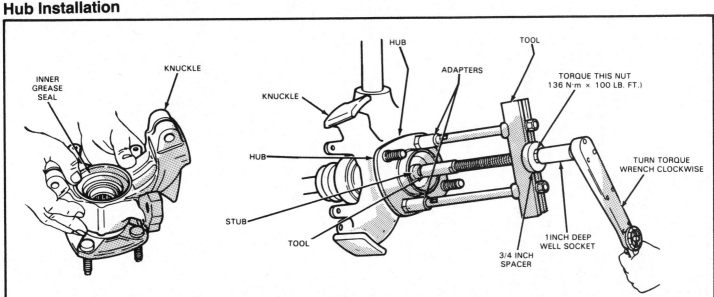

Step 1: Install hub to knuckle after bearing installation. Knuckle must be positioned as shown and hub must be inserted through bearings using hand pressure only. Step 2: Install hub to constant velocity universal joint splined stub shaft. Tighten tool nut to 150 N·m (110 lb-ft) using torque wrench to seat hub.

5. Tighten hub installer tool to 163 N·m (120 ft.-lbs.) torque to make sure that the hub is fully seated. Remove tool and install washer and hub nut. Tighten the hub nut finger tight.

6. Install disc brake rotor and brake caliper.

7. Lower vehicle and block wheels to prevent rolling.

8. Tighten wheel nuts to 109[142 N·m (80[105 ft.-lbs.) torque.

## Strut, Spring and Upper Mount

### Removal

1. Raise front of vehicle and place jack stands under frame jack pads, rearward of the wheels.

2. Remove tire and wheel assembly.

3. Remove brake line flex hose clip from strut.

4. Place a floor jack under lower control arm and raise strut as far as possible without lifting vehicle.

5. Install spring compressor tool by placing top jaw on second coil from top and bottom jaw so as to grip a total of five coils. Compress spring until there is about 1/8 inch between any two coils.

--- CAUTION ---
*Use hand wrenches (no impact wrenches).*

6. Using a pry bar slightly spread knuckle-to-strut pinch joint.

7. Place a piece of wood, 2 inches by 4 inches, and 7½ inches long, against shoulder on the knuckle. Using a short pry bar between wood block and lower spring seat, separate the strut from the knuckle.

8. Remove two top mounting nuts.

9. Remove strut, spring and top mount assembly from vehicle.

10. Place an 18 mm deep socket on strut shaft nut. Insert a 6 mm allen wrench into shaft end and then clamp mount into a vise. Remove top shaft mounting nut from shaft while holding allen wrench with vise grips or a suitable extension.

11. Remove strut top mount components and spring.

### Installation

1. Position compressed spring in lower spring seat. Be sure that:

## Separate Shock Absorber Strut From Knuckle

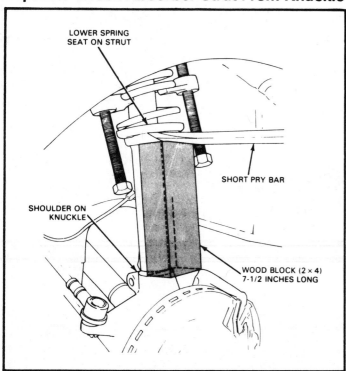

LOWER SPRING SEAT ON STRUT

SHORT PRY BAR

SHOULDER ON KNUCKLE

WOOD BLOCK (2 × 4) 7-1/2 INCHES LONG

## Compressing Spring

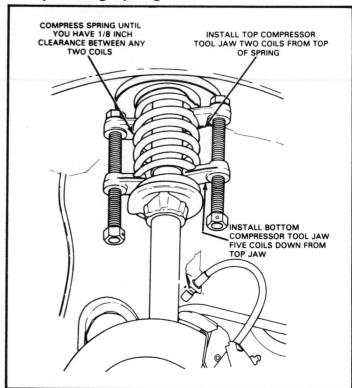

COMPRESS SPRING UNTIL YOU HAVE 1/8 INCH CLEARANCE BETWEEN ANY TWO COILS

INSTALL TOP COMPRESSOR TOOL JAW TWO COILS FROM TOP OF SPRING

INSTALL BOTTOM COMPRESSOR TOOL JAW FIVE COILS DOWN FROM TOP JAW

## Raise Strut But Not Vehicle

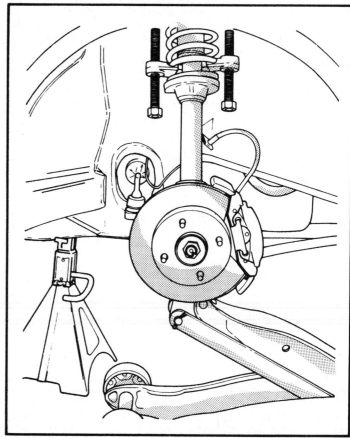

## Position Spring Compressor Tool

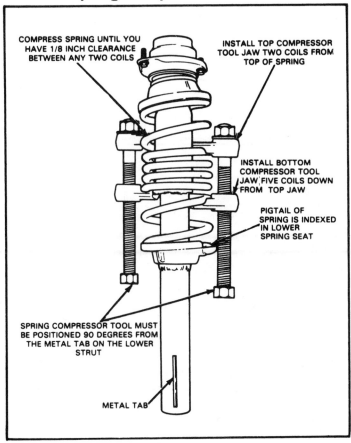

COMPRESS SPRING UNTIL YOU HAVE 1/8 INCH CLEARANCE BETWEEN ANY TWO COILS

INSTALL TOP COMPRESSOR TOOL JAW TWO COILS FROM TOP OF SPRING

INSTALL BOTTOM COMPRESSOR TOOL JAW FIVE COILS DOWN FROM TOP JAW

PIGTAIL OF SPRING IS INDEXED IN LOWER SPRING SEAT

SPRING COMPRESSOR TOOL MUST BE POSITIONED 90 DEGREES FROM THE METAL TAB ON THE LOWER STRUT

METAL TAB

a. Pigtail of spring is indexed in seat.

b. Spring compressor tool is positioned 90 degrees from metal tab on lower part of strut.

2. Using a new nut, assemble top mount components to strut.

### —— CAUTION ——

*Be sure that the correct assembly sequence and proper positioning of bearing and seal assembly is followed. If bearing and seal assembly is out of position, damage to bearing will result.*

3. Tighten shaft nut to torque of 65-85 N·m (48-62.5 ft.-lbs.).

4. Install strut, spring, upper mount and spring compressor into the vehicle as an assembly.

5. Position two top mounts attaching studs through holes in apron and start two new nuts. Do not tighten nuts at this time.

6. Install strut into steering knuckle pinch point.

7. Install a new pinch bolt in the steering knuckle and tighten to torque of 90-110 N·m (68-81 ft.-lbs.).

8. Tighten two upper mount attaching nuts to torque of 30-40 N·m (22-29 ft.-lbs.).

9. Remove spring compressor from the vehicle. As the compressor is loosened, be sure spring ends are indexed in upper and lower spring seats.

10. Install brake line flex hose clip to strut.

11. Install tire and wheel assembly.

12. Remove jack stands and lower vehicle.

## Steering Knuckle

### Removal

1. Raise vehicle on a hoist.

2. Remove tire and wheel assembly.

## Top Shaft Mounting Nut Removal

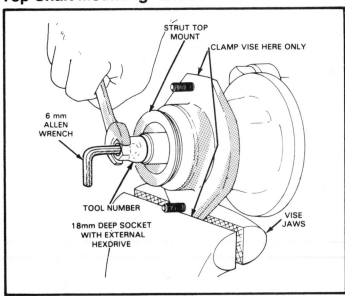

STRUT TOP MOUNT

CLAMP VISE HERE ONLY

6 mm ALLEN WRENCH

TOOL NUMBER

18mm DEEP SOCKET WITH EXTERNAL HEXDRIVE

VISE JAWS

## Top Mount Components

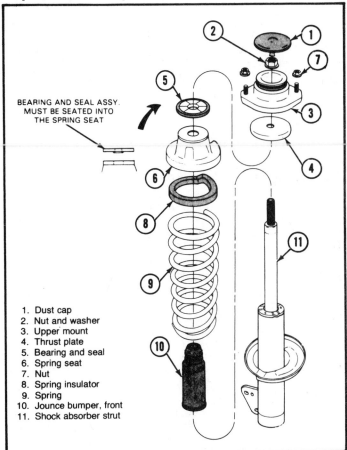

BEARING AND SEAL ASSY. MUST BE SEATED INTO THE SPRING SEAT

1. Dust cap
2. Nut and washer
3. Upper mount
4. Thrust plate
5. Bearing and seal
6. Spring seat
7. Nut
8. Spring insulator
9. Spring
10. Jounce bumper, front
11. Shock absorber strut

3. Remove cotter pin from the tie rod end stud and remove slotted nut.

4. Remove tie rod end from knuckle.

5. Remove the brake caliper.

6. Remove the hub from the driveshaft.

# SUSPENSIONS

## Front Suspension Fasteners

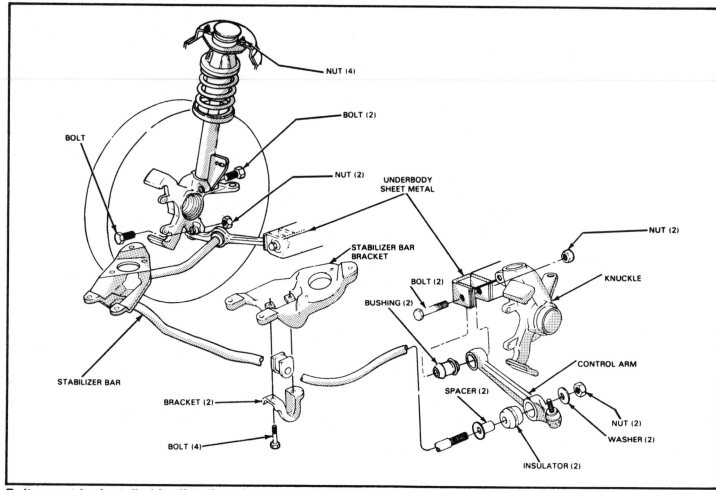

**Bolts must be installed in direction shown.**

7. Remove lower arm-to-steering knuckle pinch bolt and nut. (A drift punch may be used to remove production bolt.) Using a screwdriver, slightly spread the knuckle-to-lower arm pinch joint and remove lower arm from steering knuckle.

**NOTE:** Be sure steering column is in unlocked position, and do not use a hammer to separate ball joint from knuckle.

8. Remove shock absorber strut-to-steering knuckle pinch bolt. Using a pry bar, slightly spread knuckle-to-strut pinch joint.
9. Remove steering knuckle from the shock absorber strut.
10. Place assembly on a bench and remove the seals and bearings.
11. Remove rotor splash shield from knuckle.

### Installation

1. Install the rotor splash shield.
2. Install bearings and seals.
3. Install steering knuckle onto shock absorber strut and install a new pinch bolt in knuckle to retain strut. Tighten nut to torque of 90-110 N·m (66-81 ft.-lbs.).
4. Install hub on the driveshaft.
5. Install lower control arm to knuckle, ensuring that ball stud groove is properly positioned. Install a new nut and bolt. Tighten to torque of 50-60 N·m (37-44 ft.-lbs.).
6. Install the brake caliper.
7. Position tie rod end into knuckle, install a new slotted nut and tighten to torque of 31-47 N·m (23-35 ft.-lbs.). If necessary, advance nut to align slot and install a new cotter pin.
8. Install tire and wheel assembly.

## Lower Ball Joint Check

1. Raise vehicle on a frame contact hoist or by floor jacks placed beneath the underbody until wheels fall to the full down position.
2. Ask an assistant to grasp the lower edge of the tire and move wheel and tire assembly in and out.
3. As wheel is being moved in and out, observe the lower end of the knuckle and the lower control arm. Any movement between the lower end of knuckle and lower arm indicates abnormal ball joint wear.
4. If any movement is observed, install a new lower control arm assembly.

## Lower Control Arm

### Removal

1. Raise vehicle on a hoist.
2. Remove nut from the stabilizer bar. Pull off large dished washer.
3. Remove lower control arm inner pivot bolt and nut.
4. Remove lower control arm ball joint pinch bolt. Slightly spread the knuckle pinch joint and separate control arm from the steering knuckle. A drift punch may be used to remove the bolt.

**NOTE:** Be sure steering column is in unlocked position, and do not use a hammer to separate ball joint from knuckle.

## Checking Lower Ball Joint

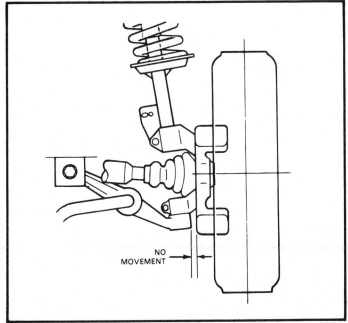

As wheel is being moved in and out, observe the lower end of the knuckle and the lower control arm. Any movement between lower end of the knuckle and the lower arm indicates abnormal ball joint wear.

### Installation

1. Assemble lower control arm ball joint stud to the steering knuckle, insuring that the ball stud groove is properly positioned.

2. Insert a new pinch bolt and nut. Tighten nut to torque of 50-60 N·m (37-44 ft.-lbs.).

3. Position lower control arm onto stabilizer bar and then position lower control arm to the inner underbody mounting. Install a new nut and bolt. Tighten bolt to torque of 60-75 N·m (44-55.3 ft.-lbs.).

4. Assemble stabilizer bar, dished washer and a new nut to stabilizer bar. Tighten nut to torque of 80-100 N·m (59-73 ft.-lbs.).

## Stabilizer Bar and/or Insulators

### Removal

1. Raise vehicle.
2. Remove nut from stabilizer bar at each lower control arm and pull off large dished washer.
3. Remove stabilizer bar insulator mounting bracket bolts and remove stabilizer bar assembly.
4. Cut worn insulators from bar.

### Installation

1. Coat bar and insulators with tire mounting lubricant or soapy water. Slide new insulators onto bar and position in approximate final location.

2. Install washer spacers onto bar ends and push mounting brackets over insulators.

3. Insert ends of bar into lower control arms. Using new bolts, attach bar and insulator mounting brackets to body. Tighten to 68-81 N·m (50-60 ft.-lbs.).

4. Using new nuts and original dished washers, attach bar to lower control arms. Tighten nuts to 80-100 N·m (59-73 ft.-lbs.).

## Stabilizer Bar Components

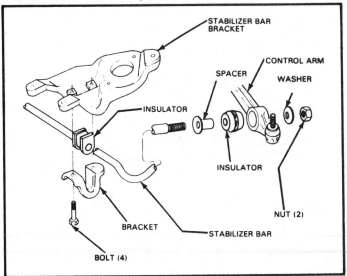

## Stabilizer Bar Bushing Removal

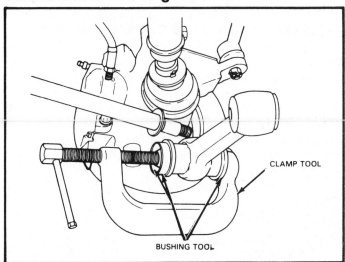

## Stabilizer Bar-to-Control Arm Insulator

### Removal

1. Raise vehicle on a hoist.
2. Remove stabilizer bar-to-control arm nut and dished washer.
3. Remove control arm inner pivot nut and bolt and pull arm down from the underbody and away from stabilizer bar.
4. Using Tool T81P-5493-A and T74P-3044-A1 or equivalent, remove old insulator bushing from the control arm.

### Installation

1. Saturate new insulator bushing and lower arm with vegetable oil such as Mazola® oil or an equivalent oil.

**NOTE:** Use only vegetable oil. Any mineral- or petroleum-based oil or brake fluid will deteriorate the rubber bushing.

## Stabilizer Bar Bushing Installation

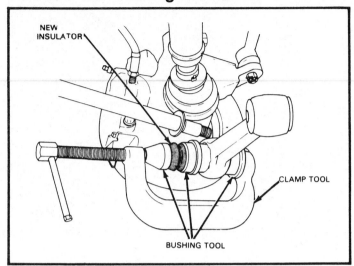

2. Using Tool T81P-5493-A and T74P-3044-A1, install new insulator bushing in lower control arm by tightening the C-clamp very slowly until bushing pops in place.

# Lower Arm Inner Pivot Bushing

### Removal

1. Raise vehicle on a hoist.
2. Remove stabilizer bar to control arm nut and dished washer.
3. Remove control arm inner pivot nut and bolt and pull arm down from underbody and away from stabilizer bar.
4. Using a sharp knife, carefully cut away retaining lip of bushing prior to its removal.
5. Using Tool T81P-5493-B and T74P-3044-A1 or equivalent, remove old bushing from control arm.

**NOTE: This operation can be done in vehicle without removing arm from knuckle.**

### Installation

1. Saturate new bushing and lower arm with vegetable oil such as Mazola® or equivalent.

**NOTE: Use only vegetable oil. Any mineral- or petroleum-based oil or brake fluid will deteriorate the rubber bushing.**

2. Using Tool T81P-5493-B and T74P-3044-A1 or equivalent, install new bushing in lower control arm.
3. Position control arm onto stabilizer bar. Be sure washer is in place.
4. Position control to underbody and install a new nut and bolt. Tighten to torque of 60-75 N·m (44-55.3 ft.-lbs.).
5. Install a new nut and the original dished washer on stabilizer bar. Tighten nut to torque of 80-100 N·m (59-73 ft.lbs.).

# Rear Suspension Usage

All Comets, Mavericks, Monarchs, Versailles, and 1975-80 Granadas use a leaf-spring rear suspension. A pair of leaf springs support the axle housing, which is secured to the springs by two U-bolts and retaining plates. Each spring is suspended from the underbody side rails by a hanger at the front and a shackle at the rear. The shock absorbers are mounted between the leaf spring retaining plates and brackets bolted to the crossmember.

## Inner Pivot Bushing Removal

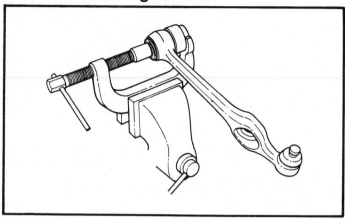

## Inner Pivot Bushing Installation

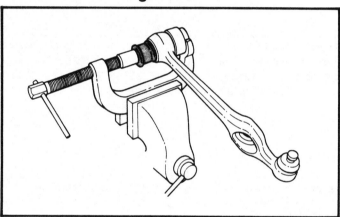

All Torinos, Montegos, Elites, LTD IIs, 1977-79 Thunderbirds, and 1975-79 Cougars utilize a coil spring rear suspension. The axle housing is suspended from the frame by an upper and lower trailing arm, and a shock absorber at each side of the vehicle. These arms pivot in the frame members and the rear axle housing brackets. Each coil spring is mounted between a lower seat which is welded to the axle housing and an upper seat integral with the frame. The shock absorbers are bolted to the spring upper seats at the top and brackets mounted on the axle housing at the bottom. A rear stabilizer bar, attached to the frame side rail brackets and the two axle housing brackets, is available as optional equipment.

All Fairmonts, Zephyrs, Mustangs, Capris, 1980 and later Thunderbirds and Cougar XR-7s, and 1981 and later Cougars and Granadas, Futura and 1982 and later Lincoln Continental have a four bar link coil spring suspension. The lower links are parallel to the frame, and serve to locate the lower end of the coil springs. The upper links are angled 45° toward the differential housing. Shock absorbers are mounted vertically at the outside of the frame rails. The rear stabilizer bar, optional on some models, mounts to the two lower links.

All Escorts/Lynx, EXP/LN7 have an independent trailing arm type rear suspension with a modified MacPherson strut and endent trailing arm type rear suspension with a modified MacPherson strut and coil spring mounted on a lower control arm.

# REAR SUSPENSION—FORD FRONT DRIVE

## Wheel Bearings

### Adjustment

Tighten adjusting nut "A" to 23-24 N·m (17-25 ft.-lbs.) while rotating hub and drum assembly. Back off adjusting nut approximately 100 degrees. Position nut retainer "B" over adjusting nut so slots are in line with cotter pin hole without rotating adjusting nut. Install cotter pin.

NOTE: The spindle has a prevailing torque feature that prevents adjusting the nut by hand.

## Component Replacement

The following applies regarding components that are replaced individually or as an assembly.

1. The shock absorber strut upper mounting is separately serviceable.

## Independent Rear Suspension, Typical Of Escort/Lynx, EXP/LN7 Models

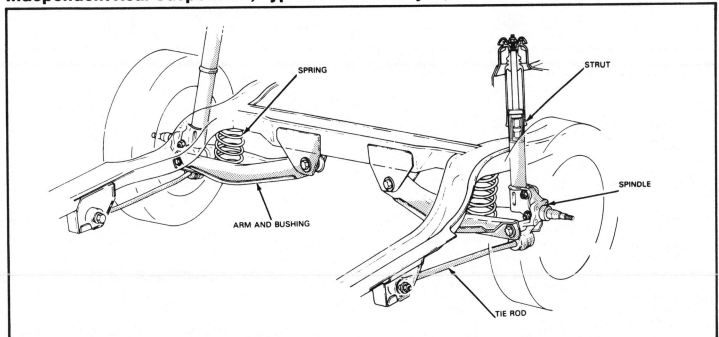

## Rear Bearing

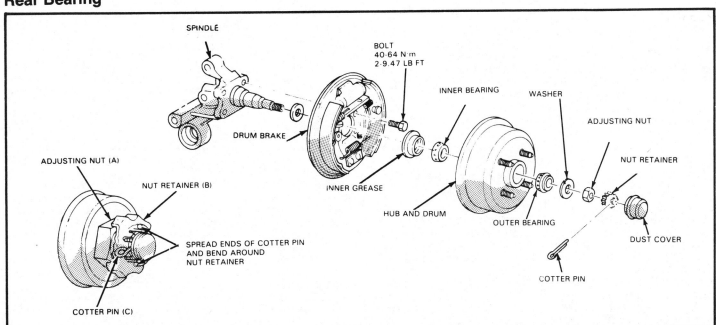

2. The shock absorber strut is not repairable and must be replaced as an assembly.

3. Lower control arm bushings are not serviceable. They must be replaced with a lower control arm and bushing assembly.

4. Tie-rod bushings can be serviced separately at both the forward and rearward locations. However, if the tie-rod requires replacement, new bushings must be installed in the spindle at the same time.

5. Coil springs are serviceable. If a rear coil spring is replaced, the upper spring insulator must also be replaced.

## Shock Absorber Strut

### Removal

1. Remove rear compartment access panels. Four-door model requires removal of quarter trim panel.

2. Loosen but do not remove top shock absorber attaching nut using an 18 × 18 × 43 mm deep socket with an external hex (Tool D81P-18045-A3) while holding the strut rod with a 6 mm Allen wrench.

**NOTE: If the shock absorber is to be re-used, do not grip the shock absorber shaft with pliers or vise grips, as this will damage the shaft surface finish.**

3. Raise vehicle on a hoist, and remove the tire and wheel assembly.

**NOTE: If a frame contact hoist is used, support the lower control arm with a floor jack. If a twin post hoist is used, support the body with floor jacks on lifting pads forward of the tie-rod body bracket.**

4. Remove clip retaining the brake flexible hose to the rear shock and carefully move hose aside.

5. Loosen two nuts and bolts retaining shock to the spindle. DO NOT remove bolts at this time.

6. Remove top mounting nut, washer and rubber insulator.

7. Remove two bottom mounting bolts and remove the shock from the vehicle.

### Installation

1. Extend shock absorber to its maximum length.

2. Install a new lower washer and insulator assembly, using tire lubricant to ease insertion into the quarter panel shock tower. (Use of a lubricant other than ESA-M1B6-B or soapy water is not recommended as it may damage the rubber insulator.)

3. Position upper part of shock absorber shaft into shock tower opening in the body and push slowly on lower part of the shock until mounting holes are lined up with mounting holes in the spindle.

4. Install new lower mounting bolts and nuts. DO NOT tighten at this time.

**NOTE: The heads of both bolts must be to the rear of the vehicle.**

5. Place a new upper insulator and washer assembly and nut on the upper shock absorber shaft. Tighten nut to torque of 81-95 N·m (60-70 ft.-lbs.), using the 18 × 18 × 43 mm deep socket with external hex (Tool D81P-18045-A3) while holding the strut shaft with a 6 mm allen wrench. Do not grip the shaft with pliers or vise grips.

6. Tighten two lower mounting bolts to torque of 122-135 N·m (90-100 ft.-lbs.).

7. Install brake flex hose and retaining clip.

8. Install wheel and tire assembly.

9. Install quarter trim panels on four-door models and access panels on other models.

## Top Mount Components

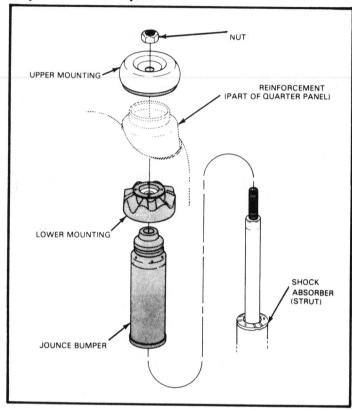

NUT

UPPER MOUNTING

REINFORCEMENT (PART OF QUARTER PANEL)

LOWER MOUNTING

SHOCK ABSORBER (STRUT)

JOUNCE BUMPER

## Lower Control Arm

### Removal

**NOTE: The lower control arm is replaced as a unit. The bushing is not serviceable.**

1. Remove tire and wheel assembly.

2. Place a floor jack under the lower control arm between spring and spindle end mounting.

3. Remove nuts from control arm-to-body mounting and control arm-to-spindle mounting. Do not remove bolts at this time.

4. Remove spindle end mounting bolt. Slowly lower floor jack until spring and spring insulator can be removed.

5. Remove bolt from the body end and remove control arm from the vehicle.

### Installation

1. Attach lower control arm to body bracket using a new bolt and a new nut. DO NOT tighten at this time. Install this bolt with bolt head to the front of the vehicle.

2. Place spring in spring pocket in lower control arm. Be sure the spring pigtail is in the proper index in lower control arm and the insulator is at the top of the spring, properly seated and indexed. Insulator must be replaced on the spring before spring is placed in position.

3. Using a floor jack, raise lower control arm until it comes in line with mounting hole in the spindle.

4. Install lower control arm to spindle using a new nut, new bolt and new washers. DO NOT tighten at this time. Install this bolt with the bolt head to the front of the vehicle.

5. Using a floor jack, raise lower control arm to its curb height.

6. Tighten control arm-to-spindle bolt to torque of 122-135 N·m (90-100 ft.-lbs.).

## Rear Suspension Fasteners

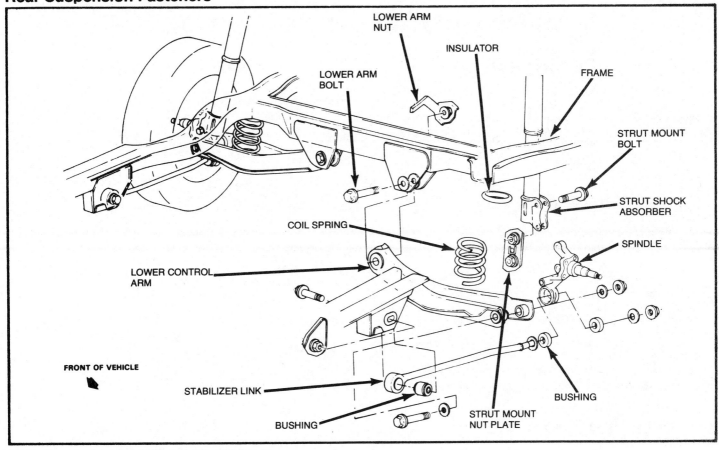

LOWER ARM NUT

INSULATOR

FRAME

LOWER ARM BOLT

STRUT MOUNT BOLT

STRUT SHOCK ABSORBER

COIL SPRING

SPINDLE

LOWER CONTROL ARM

BUSHING

FRONT OF VEHICLE

STABILIZER LINK

BUSHING

STRUT MOUNT NUT PLATE

**All bolts must be installed in direction shown.**

## Position Floor Jack Under Control Arm

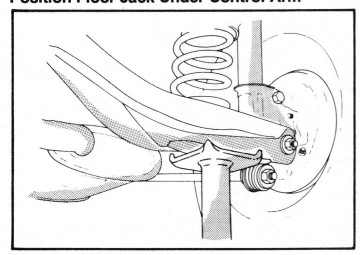

## Snap In Insulator

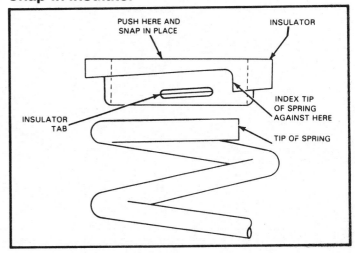

PUSH HERE AND SNAP IN PLACE

INSULATOR

INSULATOR TAB

INDEX TIP OF SPRING AGAINST HERE

TIP OF SPRING

# FRONT SUSPENSION—FORD, LINCOLN, MERCURY REAR DRIVE CARS, SINGLE ARM DESIGN

## Description

The design utilizes shock struts with coil springs mounted between the lower arm and a spring pocket in the crossmember. The shock struts are non-repairable, and they must be replaced as a unit. The ball joints and lower suspension arm bushings are not separately serviced, and they also must be replaced as a suspension arm, bushing, and ball joint assembly.

## Wheel Alignment

Caster and camber angles of this suspension are set at the factory, and cannot be adjusted in the field. Toe is adjustable.

### TOE

Start the engine and move the steering wheel back and forth several times until it is in the straight ahead or centered position. Turn the engine off, and lock the steering wheel in place using a steering wheel holder.

Adjust the left and right spindle connecting rod sleeves until each wheel has one-half of the desired total toe specification.

**NOTE: For all vehicles, whenever the jam nuts are loosened for toe adjustment, the nuts must be tightened to 48-67 N·m (33-50 ft.-lbs.).**

## Wheel Bearings

### Replacement and Lubrication

1. Raise the vehicle until the tire clears the floor, and remove wheel from hub and rotor.
2. Remove the caliper from the spindle, and wire it to the underbody to prevent damage to the brake hose.
3. Remove the grease cap from the hub. Remove the cotter pin, nut lock, adjusting nut, and flatwasher from the spindle. Remove the outer bearing cone and roller assembly.
4. Pull the hub and rotor assembly off the spindle.
5. Using tool 1175-AC or equivalent, remove and discard the grease retainer. Remove the inner bearing cone and roller assembly from the hub.
6. Clean the inner and outer bearing cups with solvent. Inspect the cups for scratches, pits, excessive wear, and other damage. If the cups are worn or damaged, replace.

### Adjustment

If the wheel is loose on the spindle or does not rotate freely, adjust the front wheel bearings as follows:
1. Raise the vehicle until the tire clears the floor.
2. Remove the wheel cover. Remove the grease cap from the hub.

## Single Arm Front Suspension

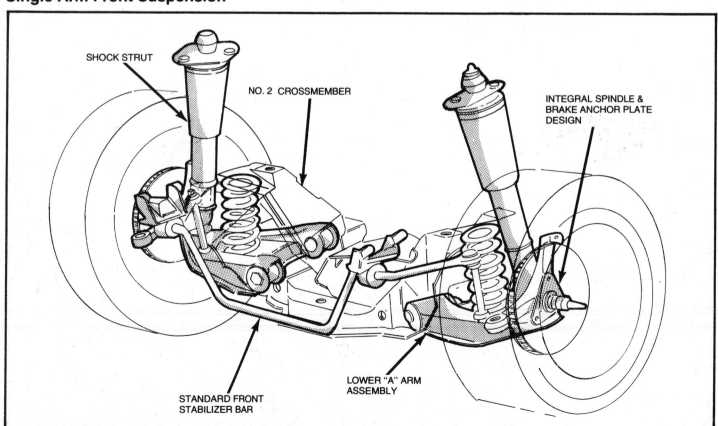

Thunderbird/XR-7, Fairmont/Zephyr, Granada/Cougar, Mustang/Capri

3. Wipe the excess grease from the end of the spindle. Remove the cotter pin and nut lock.

4. Loosen the adjusting nut three turns. Rock the wheel, hub, and rotor assembly in and out several times to push the shoe and linings away from the rotor.

5. While rotating the wheel, hub, and rotor assembly, tighten the adjusting nut to 24-33 N·m (17-25 ft.-lbs.), to seat the bearings.

6. Loosen the adjusting nut one-half turn, then retighten 1.2-1.6 N·m (10-15 in.-lbs.), using a torque wrench.

7. Place the nut lock on the adjusting nut, so the castellations on the lock are in line with the cotter pin hole in the spindle.

8. Install a new cotter pin, and bend the ends around the castellated flange of the nut lock.

9. Check the front wheel rotation. If the wheel rotates properly, reinstall the grease cap and wheel cover. If rotation is noisy or rough, follow the inspection, lubrication, and replacement procedures.

10. Before driving the vehicle, pump the brake pedal several times to restore normal brake pedal travel.

## Lower Ball Joint

### Inspection

1. Support the vehicle in normal driving position with both ball joints loaded.

2. Wipe the wear indicator and ball joint cover checking surface, so they are free of dirt and grease.

3. The checking surface should project outside the cover. If the checking surface is inside the cover, replace the lower arm assembly.

## Shock/Strut

### Removal

1. Place the ignition key in the unlocked position to permit free movement of front wheels.

2. From the engine compartment, remove the one 16 mm strut to upper mount attaching nut. A screwdriver in the slot will hold the rod stationary while removing the nut.

3. Raise the front of the vehicle by the lower control arms, and position safety stands under the frame jacking pads, rearward of the wheels.

4. Remove the tire and wheel assembly.

5. Remove brake caliper and rotate out of position.

6. Remove the two lower nuts and bolts attaching the strut to the spindle.

7. Lift the strut up from the spindle to compress the rod, then pull down and remove the strut.

## Toe Adjustment

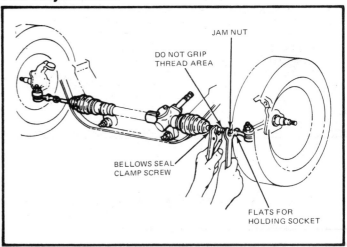

## Front Wheel Bearing Adjustment

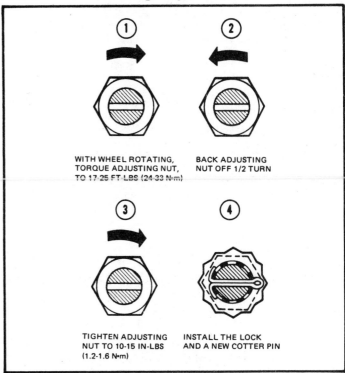

## Checking Lower Ball Joint—Fairmont/Zephyr, Mustang/Capri

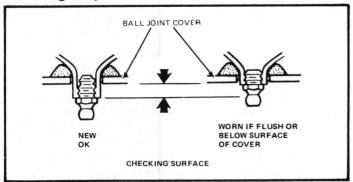

## Checking Lower Ball Joint— Thunderbird/XR-7, Granada/Cougar

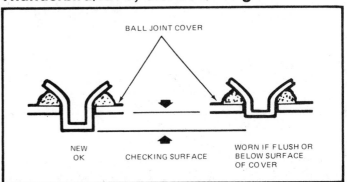

### Installation

1. With the rod half extended, place the rod through the upper mount; and hand start with a new 16 mm nut, engaging as many nut threads as possible.
2. Extend the strut, and position into the spindle.
3. Install two new lower mounting bolts, and hand start nuts.
4. Tighten the new 16 mm strut to upper mount attaching nut, inside the engine compartment to 81-102 N·m (60-75 ft.-lbs.). A screwdriver in the slot will hold the rod stationary while the nut is being tightened.
5. Remove the suspension load from the lower control arms by lowering the hoist, and tighten the lower mounting nuts to 203-244 N·m (150-180 ft.-lbs.).

## Lower Suspension Arm

### Removal

1. Raise the front of the vehicle, and position safety stands under both sides at the jack pads just behind the lower arms.
2. Remove the wheel and tire assembly.
3. Disconnect the stabilizer bar link from the lower arm.
4. Remove the disc brake caliper, rotor, and dust shield.
5. Remove the steering gear bolts, and position steering gear out of way.
6. Remove the cotter pin from the ball joint stud nut, and loosen the ball joint nut one or two turns.
7. Tap the spindle boss to relieve the stud pressure.
8. Remove the tie-rod end from the spindle with tool 3290-C or equivalent. Place floor jack under lower arm, supporting arm at both bushings. Remove both lower arm bolts, lower jack, and remove coil spring.
9. Remove ball joint nut, and remove arm assembly.

### Installation

1. Place the new arm assembly into spindle, and tighten ball joint nut to 108-163 N·m (80-120 ft.-lbs.). Install a new cotter pin.

## Spring Compressor Tool in Position Showing Marks on Spring Where Upper and Lower Plates Are Located

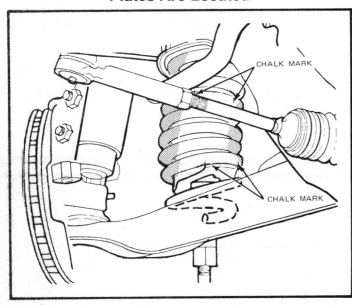

CHALK MARK

CHALK MARK

2. Position the coil spring in the upper spring pocket. Be sure the insulator is on top of the spring and the lower end (pigtail) is properly positioned (between two holes) in the depression of the lower arm.
3. Carefully raise the lower arm with the floor jack until the bushings are properly positioned in the crossmember.
4. Install the lower arm bolts and nut (finger-tight only).
5. Install the steering gear bolts, and tighten to 122-136 N·m (90-100 ft.-lbs.).
6. Connect the tie-rod end, and install the nut. Tighten to 47-64 N·m (35-41 ft.-lbs.).
7. Connect the stabilizer link bolt and nut. Tighten to 11-16 N·m (8-12 ft.-lbs.).
8. Install the brake dust shield, rotor, and caliper.
9. Install the wheel and tire assembly.
10. Remove the safety stands, and lower the vehicle. After the vehicle has been lowered to the floor and the vehicle is at curb height, tighten the lower arm nuts to 271-298 N·m (200-220 ft.-lbs.).

## Spring

### Removal

1. Raise the front of the vehicle, and position safety stands under both sides at the jack pads just back of the lower arms. Remove the wheel and tire assembly.
2. Disconnect the stabilizer bar link from the lower bar.
3. Remove the steering gear bolts, and move the steering gear out of the way.
4. Disconnect the tie rod from the steering spindle.
5. Using the spring compressor tool D78P-5310-A or equivalent, install one plate with the pivot ball seat down into the coils of the spring. Rotate the plate, so that it is fully seated into the lower suspension arm spring seat.
6. Install the other plate with the pivot ball seat up into the coils of the spring. Insert the ball nut through the coils of the spring, so it rests in the upper plate.
7. Insert compression rod into the opening in the lower arm through the lower and upper plate. Install upper ball nut on the rod, and return the securing pin.

**NOTE: This pin can only be inserted one way into the upper ball nut because of a stepped hole design.**

8. With the upper ball nut secured, turn the upper plate, so it walks up the coil until it contacts the upper spring seat.
9. Install the lower ball nut, thrust bearing and forcing nut on the compression rod.
10. Rotate the nut until the spring is compressed enough, so it is free in its seat.
11. Remove the two lower control arm pivot bolts and nuts and disengage the lower arm from the frame crossmember. Remove the spring assembly.
12. If a new spring is to be installed, mark the position of the upper and lower plates on the spring with chalk. Measure the compressed length of the spring as well as the amount of spring curvature to assist in the compression and installation of a new spring.
13. Loosen the nut to relieve spring tension, and remove the tools from the spring.

### Installation

1. Assemble the spring compressor tool and locate tool through spring.
2. Before compressing the coil spring, be sure the upper ball nut securing pin is inserted properly.
3. Compress the coil spring.
4. Position the coil spring assembly into the lower arm.

**NOTE: Be sure that the lower end (pigtail) of the coil spring is properly positioned between the two holes in the lower arm spring pocket depression.**

5. Install coil spring. Reverse removal procedures.

## Coil Spring Removal Tool

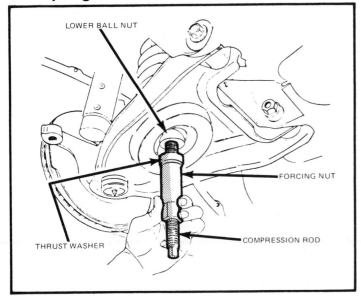

## Spring Compressed for Removal

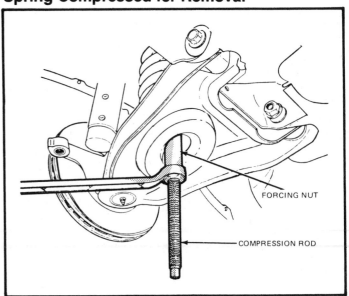

## Spindle Assembly

### Removal

1. Raise the front of the vehicle, and position safety stands under both sides at the jacking pads just behind the lower arms.
2. Remove the wheel and tire assembly.
3. Remove the brake caliper, rotor, and dust shield.
4. Remove the stabilizer link from the lower arm assembly.
5. Remove the tie-rod end from the spindle.
6. Remove the cotter pin from the ball joint stud nut, and loosen the ball joint nut one or two turns.

—————— CAUTION ——————
*Do not remove the nut from the ball joint stud at this time.*

7. Tap the spindle boss to relieve the stud pressure.
8. Place a floor jack under the lower arm, compress the coil spring and remove the stud nut.
9. Remove the two bolts and nuts attaching the spindle to the shock strut. Compress the shock strut until working clearance is obtained.
10. Remove the spindle assembly.

### Installation

1. Place the spindle on the ball joint stud, and install the stud nut. Do not tighten at this time.
2. Lower the shock strut until the attaching holes are in line with the holes in the spindle. Install two new bolts and nuts.
3. Tighten the ball joint stud nut to 108-163 N·m (80-120 ft.-lbs.), and install a new cotter pin.
4. Tighten the shock strut-to-spindle attaching nuts to 203-244 N·m (150-180 ft.-lbs.).
5. Lower the floor jack from under the suspension arm, and remove the jack.
6. Install the stabilizer link and tighten the attaching bolt and nut to 11-16 N·m (8-12 ft.-lbs.).
7. Attach the tie-rod end, and tighten the retaining nut to 47-64 N·m (35-47 ft.-lbs.).
8. Install the disc brake dust shield, rotor, and caliper.
9. Install the wheel and tire assembly.

## Stabilizer Bar Link Insulators

### Removal

To replace the link insulators on each stabilizer link, use the following procedure:
1. Raise the vehicle on a hoist.
2. Remove the nut, washer, and insulator from the upper end of the stabilizer bar attaching link bolt.
3. Remove the bolt and the remaining washers, insulators, and spacer.

### Installation

1. Install the stabilizer bar link insulators by reversing the above steps.
2. Tighten the attaching nuts to 11-16 N·m (8-12 ft.-lbs.).

## Stabilizer Bar and/or Insulator

### Removal

1. Raise the vehicle on a hoist.
2. Disconnect the stabilizer from each stabilizer link and both stabilizer insulator attaching clamps. Remove the stabilizer bar assembly.
3. Cut the worn insulators and plastic sleeves from the stabilizer bar.

### Installation

1. Coat the necessary parts of the stabilizer bar with D9AZ-19583-A or an equivalent lubricant; install new plastic sleeves with the flange inboard, and slide insulators onto the stabilizer bar and over sleeves. Be sure the insulator is fully seated against the flange.
2. Using a new nut and bolt, secure each end of the stabilizer bar to the lower suspension arm. Tighten these nuts to 11-16 N·m (8-12 ft.-lbs.).
3. Using new fasteners, clamp the stabilizer bar to the attaching brackets on the side rail. Tighten these bolts to 27-33 N·m (20-25 ft.-lbs.).

## Single Arm Front Suspension

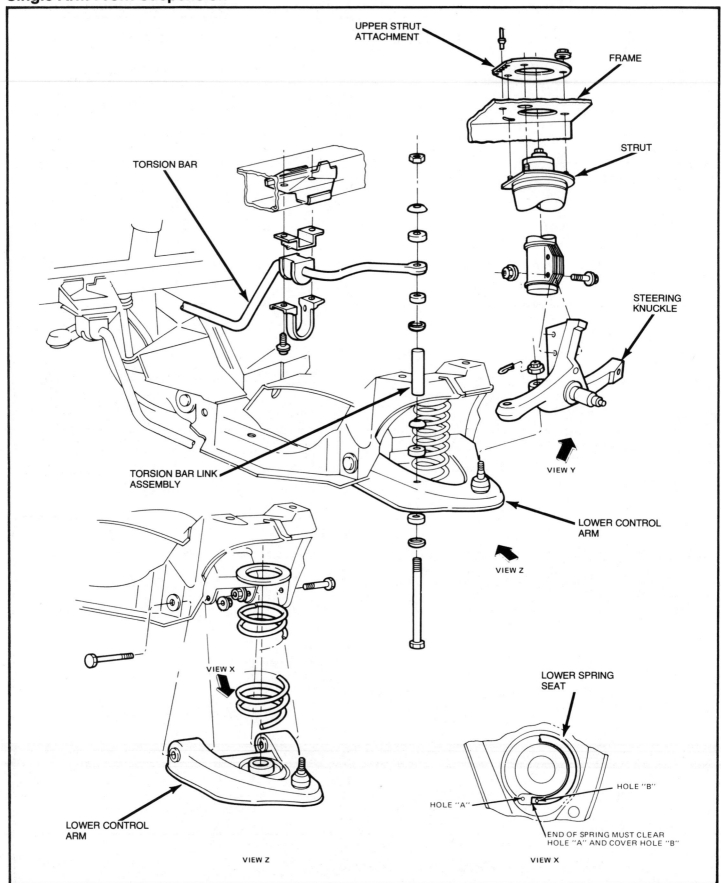

UPPER STRUT ATTACHMENT

FRAME

STRUT

TORSION BAR

STEERING KNUCKLE

VIEW Y

TORSION BAR LINK ASSEMBLY

LOWER CONTROL ARM

VIEW Z

VIEW X

LOWER SPRING SEAT

LOWER CONTROL ARM

HOLE "B"

HOLE "A"

END OF SPRING MUST CLEAR HOLE "A" AND COVER HOLE "B"

VIEW Z

VIEW X

# FRONT SUSPENSION—FORD, LINCOLN, MERCURY REAR DRIVE CARS, SPRING ON UPPER ARM DESIGN

## Wheel Alignment

### ADJUSTMENTS

NOTE: Two types of caster-camber adjustment locations are used on front suspensions with the spring on upper arm. One, both caster and camber are controlled by adjustments on the lower control arm. Two, both caster and camber are controlled by adjustments on the upper control arm.

### ADJUSTMENTS ON LOWER ARM

**GRANADA, MONARCH AND VERSAILLES—THROUGH 1980**

#### Caster

Caster is controlled by the front suspension strut. To obtain positive caster, loosen the strut rear nut and tighten the strut from nut against the bushing. To obtain negative caster, loosen the strut front nut and tighten the strut rear nut against the bushing.

#### Camber

Camber is controlled by the eccentric cam located at the lower arm attachment to the side rail. To adjust the camber, loosen the camber adjustment bolt nut at the rear of the body bracket. Spread the body bracket at the camber adjustment bolt area just enough to permit lateral travel of the arm when the adjustment bolt is turned. Rotate the bolt and eccentric as required to increase or decrease camber.

After the caster and camber have been adjusted to specification, torque the lower arm eccentric bolt nut and the strut front nut to specification.

### ADJUSTMENTS ON UPPER ARM

#### Type A

NOTE: Camber and caster on the Fairmont, Zephyr, Capri, 1979 and later Mustang, 1980 and later Cougar and Thunderbird, 1981-82 Granada, and Monarch, and 1982 Escort, Lynx, EXP and LN7 is permanently set at the factory. Only toe-in can be adjusted.

1. Working from inside the front wheel housing, install the special service tool T74P-3000, one at each end of the upper arm inner shaft. Turn special tool bolts inward until the bolt ends contact the body metal.
2. Loosen the two upper arm inner shaft-to-body attaching nuts. The upper arm inner shaft will then move inboard until stopped by the special tool bolt ends.

## Camber and Caster Adjusting Tools

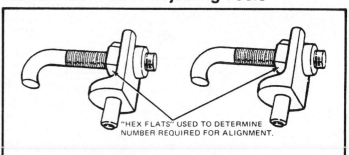

"HEX FLATS" USED TO DETERMINE NUMBER REQUIRED FOR ALIGNMENT.

## Caster/Camber Adjustment At Lower Control Arm And Strut

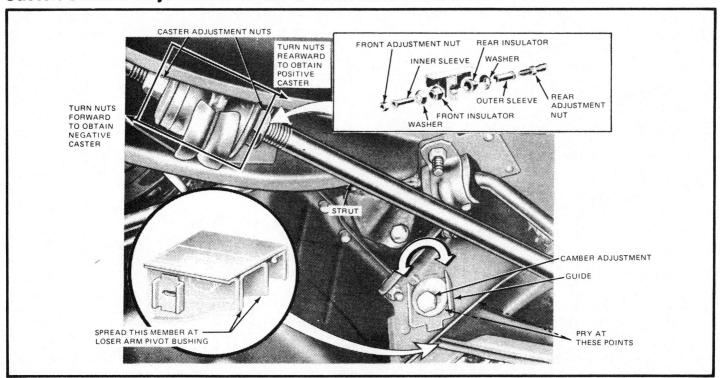

CASTER ADJUSTMENT NUTS

TURN NUTS REARWARD TO OBTAIN POSITIVE CASTER

TURN NUTS FORWARD TO OBTAIN NEGATIVE CASTER

FRONT ADJUSTMENT NUT — REAR INSULATOR
INNER SLEEVE — WASHER
OUTER SLEEVE — REAR ADJUSTMENT NUT
FRONT INSULATOR
WASHER

STRUT

SPREAD THIS MEMBER AT LOSER ARM PIVOT BUSHING

CAMBER ADJUSTMENT
GUIDE
PRY AT THESE POINTS

## Caster And Camber Adjustment, Typical Of Suspension Using Upper Arm Inner Shaft

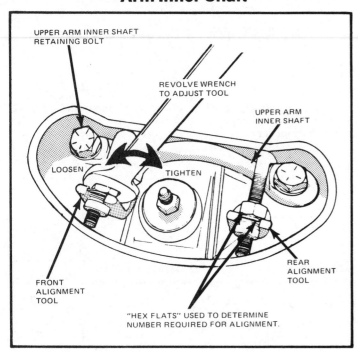

## Caster And Camber Adjustment, Pinto And Bobcat

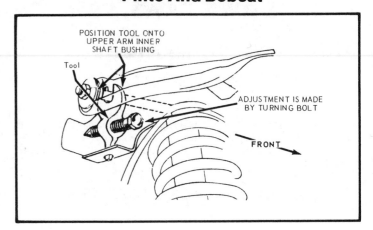

### Left-hand Sleeve

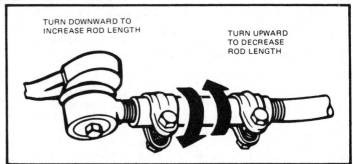

### Right-hand Sleeve

3. Turn the special tool bolt(s) inward or outward until caster and camber are within specifications. Tightening the special tool bolts forces the arm outward; loosening the adjustment bolts on the special tool permits the arm and inner shaft to move inboard due to weight force.

### Type B

#### PINTO AND BOBCAT

Position one Ford tool T74P-3000 or its equivalent at each end of the upper control arm, pivot shaft with the leg of the tools through the holes in the sheet metal (see illustration). Turn the adjusting bolts until they are solidly contacting sheet metal, and loosen the pivot shaft retaining bolts.

Caster is adjusted by turning the front and rear adjusting bolts in the opposite direction. Camber is adjusted by turning both bolts an equal amount in the same direction. Following the adjustments, tighten the pivot shaft retaining bolts, remove the adjusting tools, and recheck caster and camber.

4. When caster and camber specifications are attained, tighten the upper arm inner shaft-to-body attaching nuts to (95-120 ft.-lbs.). 129-162 N·m.

5. Loosen the special tool bolts and remove tools.

## TOE (ALL MODELS)

Start the engine and move the steering wheel back and forth several times until it is in the straight ahead or centered position. Turn the engine off and lock the steering wheel in place using a steering wheel holder. Align the straight ahead marks at the base of the steering wheel and the head of the steering column.

Adjust the left and right spindle connecting rod sleeves until each wheel has one-half of the desired total toe specifications. Tighten jam nuts to 35-50 ft.-lbs.

## Front Wheel Bearing

### Adjustment

If the wheel is loose on the spindle or does not rotate freely, adjust the front wheel bearings as follows.

1. Raise the vehicle until the tire clears the floor.

2. Remove the wheel cover. Remove the grease cap from the hub.

3. Wipe the excess grease from the end of the spindle. Remove the cotter pin and nut lock.

4. Loosen the adjusting nut three turns. Rock the wheel, hub, and rotor assembly in and out several times to push the shoe and linings away from the rotor.

5. While rotating the wheel, hub and rotor assembly, tighten the adjusting nut to 23-34 N·m (17-25 ft.-lbs.), to seat the bearings.

6. Loosen the adjusting nut one-half turn, then retighten 1.1-1.7 N·m (18-15 in.-lbs.), using a torque wrench.

7. Place the nut lock on the adjusting nut so the castellations on the lock are in line with the cotter pin hole in the spindle.

8. Install a new cotter pin and bend the ends around the castellated flange of the nut lock.

9. Check front wheel rotation. If the wheel rotates properly, reinstall the grease cap and wheel cover. If rotation is noisy or rough, inspect, lubricate and replace.

10. Before driving the vehicle, pump the brake pedal several times to restore normal brake pedal travel.

## Replacement and Lubrication

1. Raise the vehicle until the tire clears the floor. Remove wheel from hub and rotor.

2. Remove the caliper from the spindle and wire it to the underbody to prevent damage to the brake hose.

3. Remove the grease cap from the hub. Remove the cotter pin, nut lock, adjusting nut and flatwasher from the spindle. Remove the outer bearing cone and roller assembly.

4. Pull the hub and rotor assembly off the spindle.

5. Using tool 1175-AC or equivalent, remove and discard the grease retainer. Remove the inner bearing cone and roller assembly from the hub.

6. Clean the inner and outer bearing cups with solvent. Inspect the cups for scratches, pits, excessive wear and other damage. If the cups are worn or damaged, remove them with tools D80L-927-A and T77F-1102-A or equivalent and replace them.

# Shock Absorber

## Removal

1. Raise the hood and remove three shock absorber upper mounting bracket-to-spring tower attaching nuts. Position wood blocks.

2. Raise the front of the vehicle and place safety stands under the lower arms.

3. Remove the shock absorber lower attaching nuts, washers and insulators.

4. Lift the shock absorber and upper bracket from the spring tower and remove the bracket from the shock absorber. Remove the insulators from the lower attaching studs.

## Installation

1. Install the upper mounting bracket on the shock absorber and torque to specification. Install the insulators on the lower attaching studs.

2. Position the shock absorber and upper mounting bracket in the spring tower, making sure the shock absorber lower studs are in the pivot plate holes.

3. Install the two nuts on the shock absorber lower studs, and torque to specification.

4. Install the three shock absorber upper mounting bracket-to-spring tower attaching nuts, and torque to specification. Remove the safety stands and lower the vehicle. Remove wood blocks.

# Front Spring

## Removal

1. Remove the shock absorber and upper mounting bracket as an assembly. Position wood blocks.

2. Raise the vehicle on a hoist, install safety stands, and remove the wheel cover or hub cap.

3. Remove the grease cap from the hub. Remove the cotter pin, nut lock, adjusting nut, and outer bearing from the hub.

4. Remove the wheel, tire, rotor and caliper assembly from the spindle.

5. Install a spring compressor tool and compress the spring.

## Shock Absorber Upper Attachment

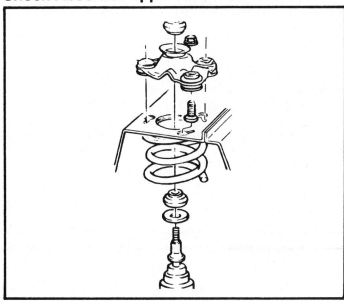

## Shock Absorber Lower Attachment

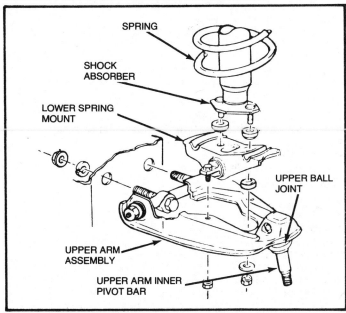

6. Remove two upper arm-to-spring tower attaching nuts and swing the upper arm outboard from the spring tower.

7. Release the spring compressor tool and remove the tool from the spring. Remove the spring from the vehicle.

## Installation

1. Place the spring upper insulator on the spring and secure in place with tape.

2. Position the spring in the spring tower. Install the spring compressor and compress the spring.

3. Swing the upper arm inward and insert the bolts through the holes in the side of the spring tower. Install the attaching nuts and lockwasher and torque them to specification.

4. Release the spring pressure and guide the spring into the upper arm spring seat.

**NOTE:** The end of the spring must be no more than ½ inch from the tab on the spring seat.

# SUSPENSIONS

## Position of Wood Block

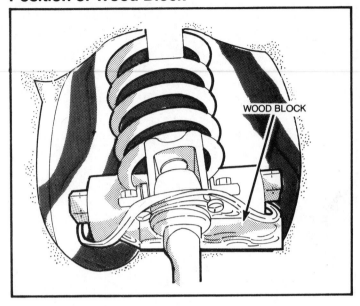

WOOD BLOCK

5. Remove the spring compressor and position the wheel, tire, and hub and drum on the spindle.

6. Install the bearing, washer, adjusting nut, and lock nut. Adjust the wheel bearing. Install the cotter pin, grease cap, and hub cap or wheel cover.

7. Lower the vehicle and install the shock absorber and upper mounting bracket. Remove wood blocks.

## Lower Arm

### Removal

**NOTE: If equipped with camber adjuster bolt, match-mark cam washer to frame before removal.**

1. Position wood blocks.
2. Raise the vehicle, position safety stands, and remove the wheel and tire.
3. Remove the stabilizer bar and link attaching nut. Disconnect the bar from the link and remove the link bolt.
4. Remove the strut-to-lower arm attaching nuts and bolts.
5. Remove the cotter pin from the lower ball joint stud nut.
6. Loosen the lower ball joint stud nut one to two turns.

**NOTE: Do not remove the nut from the stud at this time.**

7. Install tools T71P-3006-A and T60K-3006-A between upper and lower ball joint studs, against ends of studs, not nuts.
8. With a wrench, turn the adapter screw until the tool places the stud under compression. Tap the spindle near the lower stud with a hammer to loosen the stud in the spindle.

**NOTE: Do not loosen the stud from the spindle with tool pressure only.**

9. Remove the nut from the lower ball joint stud and lower the arm.
10. Remove the lower arm to underbody cam bolt, nut, and lock washer. Remove the lower arm.

### Installation

1. Position the lower arm to the underbody bracket and install the cam bolt, lock washer, and nut loosely.
2. Raise the lower arm, guide the lower ball joint stud into the spindle bore, and install the stud attaching nut loosely.

## Spring Compressor Tool Installation

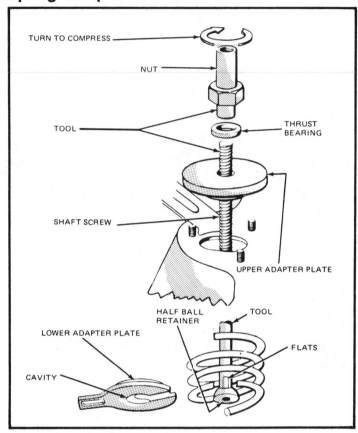

TURN TO COMPRESS

NUT

TOOL

THRUST BEARING

SHAFT SCREW

UPPER ADAPTER PLATE

HALF BALL RETAINER

TOOL

LOWER ADAPTER PLATE

FLATS

CAVITY

## Lower Arm Attachment

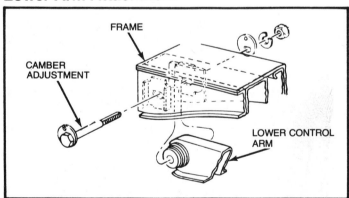

FRAME

CAMBER ADJUSTMENT

LOWER CONTROL ARM

3. Install the stabilizer link bolt, washer, bushings, and spacer. Connect the stabilizer bar to the link. Install the attaching nut and torque to specification.

4. Position the strut to the lower arm. Install the attaching bolts and nuts, and torque to specification.

5. Torque the lower ball joint stud nut to specification, continue to tighten the nut until the next slot aligns with the cotter pin hole, and install a new cotter pin.

6. While holding the head of the bolt with a wrench, torque the lower arm-to-underbody pivot bolt and nut to specification.

7. Remove the safety stands and lower the vehicle.

8. Remove the wood blocks.

9. Adjust caster, camber, and toe to specification.

## Front Suspension—Granada/Monarch, Versailles Similar

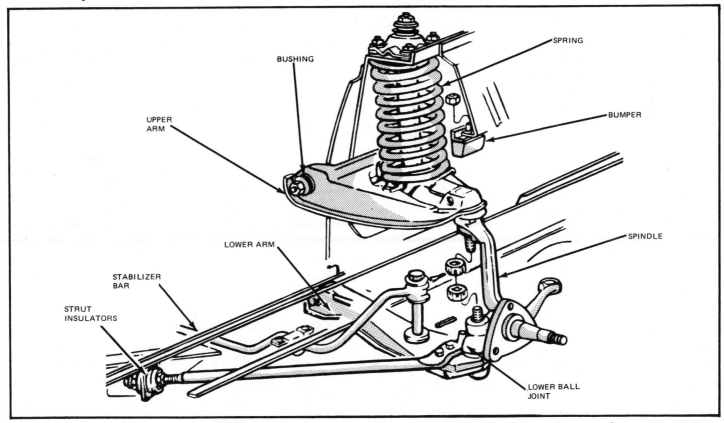

BUSHING

SPRING

UPPER
ARM

BUMPER

LOWER ARM

SPINDLE

STABILIZER
BAR

STRUT
INSULATORS

LOWER BALL
JOINT

**Ball joints must not be replaced. Upper or lower suspension arms should be replaced as a unit.**

## Upper Arm

NOTE: **Upper arm shaft and bushing may not be disassembled on these vehicles. The entire upper arm assembly must be replaced as a unit.**

### Removal

1. Position wood blocks. Raise the vehicle, position safety stands under the frame and lower the vehicle slightly.
2. Remove the wheel and tire.
3. Remove the shock absorber lower attaching nuts and washers.
4. Remove the shock absorber upper mounting bracket attaching nuts and remove the shock absorber and bracket as an assembly. On all 8-cylinder vehicles, remove the air cleaner to obtain access for tool installation.
5. Install the spring compressor tool and compress the spring.
6. Position a safety stand under the lower arm.
7. Remove the cotter pin from the nut on the upper ball joint stud, and loosen the nut one or two turns.

NOTE: **Do not remove the nut from the stud at this time.**

8. Position the appropriate ball joint remover tool between the upper and lower ball joint studs. The tool should seat firmly against the ends of both studs and not against the stud nuts.
9. Turn the tool with a wrench until the tool places the studs under considerable compression, then hit the spindle sharply near the upper stud with a hammer to break the stud loose in the spindle.

NOTE: **Do not loosen the stud in the spindle with tool pressure only.**

10. Remove the nut from the upper stud and lift the stud out of the spindle.

## Upper Arm Attachment

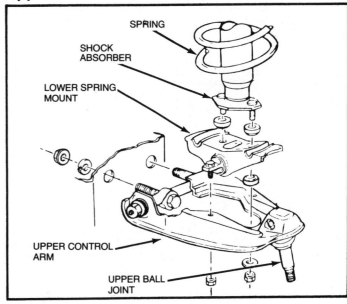

SPRING

SHOCK
ABSORBER

LOWER SPRING
MOUNT

UPPER CONTROL
ARM

UPPER BALL
JOINT

11. Remove the upper arm inner shaft attaching nuts from the engine compartment and remove the upper arm.

### Installation

1. Position the upper arm and spring seat assembly to the spring tower and install the nuts on the two inner shaft attaching bolts. Torque the nuts to specification.

2. Position the upper ball joint stud in the top of the wheel spindle, and install the stud nut. Torque the nut to specification, and continue to tighten it until the next slot aligns with the cotter pin hole. Install a new cotter pin.

3. Release the coil spring, remove the tool, and install the front shock absorber and the wheel and tire.

4. Remove safety stands and lower vehicle. Remove wood blocks.

5. If the upper arm is being replaced due to accidental damage, check caster, camber and tow and adjust as required.

## Wheel Spindle

### Removal

1. Position wood blocks between the upper arm and frame, then raise the vehicle and position safety stands.

2. Remove the two bolts attaching the caliper to the spindle. Remove the caliper from the rotor and wire it to the underbody to prevent damage to the brake hose.

3. Remove the grease cap from the hub, then remove the cotter pin, nut lock, adjusting nut, washer, and outer bearing cone and roller assembly.

4. Pull the hub and rotor assembly off the wheel spindle.

5. Remove the three caliper shield attaching bolts and remove the shield.

6. Disconnect the spindle connecting rod (tie-rod) end from the spindle arm.

7. Remove the cotter pins from both ball joint stud nuts, and loosen the nuts one or two turns.

**NOTE: Do not remove the nuts from the studs at this time.**

8. Position the appropriate ball joint remover tool between the upper and lower ball joint studs. The tool should seat firmly against the ends of both studs and not against the stud nuts.

9. Turn the tool with a wrench until the tool places the studs under considerable tension, and, with a hammer, hit the spindle near the studs to break them loose in the spindle.

## Spindle Attachments

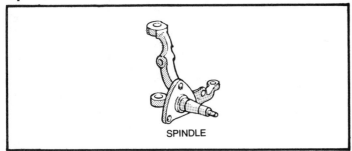

SPINDLE

**NOTE: Do not loosen the studs in the spindle with tool pressure alone.**

10. Position a floor jack under the lower arm.

11. Remove the upper and lower ball joint stud nuts, lower the jack and remove the spindle.

### Installation

1. Position the spindle on the lower ball joint stud and install the stud nut. Torque the nut to specification. Continue tightening until the next slot aligns with the cotter pin hole and install a new cotter pin.

2. Raise the lower arm, and guide the upper ball joint stud into the spindle. Install the stud nut.

3. Torque the nut to specifications and install a new cotter pin. Remove the floor jack.

4. Connect the spindle connecting rod (tie-rod) end to the spindle arm and install the attaching nut. Torque the nut to specification and install a new cotter pin.

5. Position the caliper splash shield to the spindle and install the attaching bolts and nuts. Torque the nuts and bolts to specification.

6. Install the hub and rotor on the spindle. Position the caliper to the rotor and spindle and install the attaching bolts. Torque the bolts to specification. Install the wheel and tire on the hub and adjust the wheel bearing.

7. Install the hub cap or wheel cover.

8. Remove the safety stands, and lower the vehicle.

9. Remove the wood blocks from between the upper arm and frame. Check and adjust caster, camber and tow as required.

## Stabilizer Bar End Bushing

### Removal

To replace the end bushings on each stabilizer link, use the following procedure:

1. Raise the vehicle on a hoist.

2. Remove the nut, washer, and insulator from the lower end of the stabilizer bar attaching bolt.

3. Remove the bolt and the remaining washers, insulators, and the spacer.

### Installation

1. Assemble a cup washer and a new insulator on the bolt.

2. Insert the bolt through the stabilizer bar, then install a new insulator and cup washer.

3. Install the spacer, cup washer, and another new insulator on the bolt.

4. Insert the bolt through the lower arm and install a new insulator and cup washer. Install and torque the attaching nut to specification.

## Strut-To-Frame Attachment

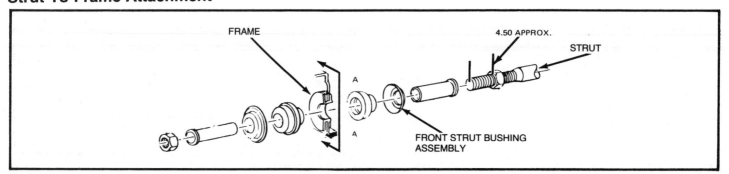

FRAME

4.50 APPROX.

STRUT

A

A

FRONT STRUT BUSHING ASSEMBLY

## Lower Arm Strut and/or Insulators

### Removal

1. Position wood blocks under the upper arm.
2. Raise the vehicle, position safety stands.
3. Remove the nut from the front of the strut.
3. Remove the nut from the front of the strut.
4. Remove the two nuts attaching the strut to the lower arm. Tap the strut upward to loosen the bolt serrations in the lower arm. Remove the strut.
5. With two pry bars approximately 18 inches long, one at each side at the rear of the front washer, pry it forward to separate the inner sleeve from the outer sleeve. Remove insulators, washers, inner and outer sleeves from the No. 1 crossmember.

### Installation

1. Install outer sleeve, rear washer (large I.D.) and rear insulator bushing on the forward end of the strut rod.
2. Position the strut into the crossmember and to the lower suspension arm. Install the strut-to-arm attaching bolts and nuts, and torque them to specification.
3. Install the forward insulator washer (small ID), inner sleeve, and adjustment nut on the forward end of the strut.

**NOTE: Remove 1/4 inch from the inner sleeve prior to assembly.**

## Strut and Stabilizer Attachment to Lower Arm

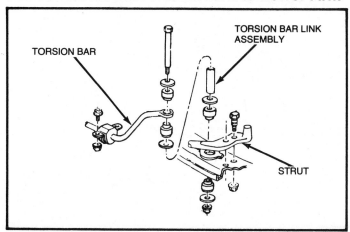

**Bolt must be installed with head up. Do not press into hole. Bolt head must be held from turning when tightening nut.**

4. Remove the safety stands, and lower the vehicle. Remove the wood blocks from under the upper arm.
5. Adjust the caster and camber to specification.

# FRONT SUSPENSION—FORD, LINCOLN, MERCURY REAR DRIVE CARS, SPRING ON LOWER ARM DESIGN

## Wheel Alignment

### ADJUSTMENTS

#### Caster and Camber

Special tools must be used in order to accurately adjust caster and camber.

Using these special tools, caster and camber adjustments are made in a single operation.

1. Check suspension with the front wheels in the straight-ahead position. Run the engine so that the power steering control valve will be in the center (neutral) position (if equipped).
2. Check caster and camber and record the readings.
3. Compare camber and caster readings with specifications to determine if adjustment is required to bring vehicle to nominal setting.
4. If adjustment is required, insert alignment tools into frame holes and "snug" the tool hooks finger-tight against the upper arm inner shaft. Then tighten hex nut of each alignment tool 1 additional "hex flat."
5. Loosen upper arm inner shaft-to-frame attaching bolts so that lockwashers on bolts are unloaded. Then firmly tap bolt heads to assure loosening of the lower assemblies.
6. Adjust camber and caster on each wheel.
7. Torque upper arm inner shaft-to-frame attaching bolts to 136-190 N·m (100-140 ft.-lbs.). It is not necessary to recheck caster and camber after this adjustment procedure is performed.
8. Check toe-in and steering wheel spoke position and adjust both (as required) at the same time.

#### Toe and Steering Wheel Spoke Position

After adjusting caster and camber, check the steering wheel spoke

## Special Tools for Camber and Caster Adjustment

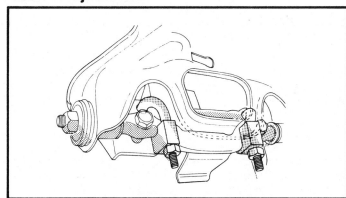

## Spindle Connecting Rod Adjustment

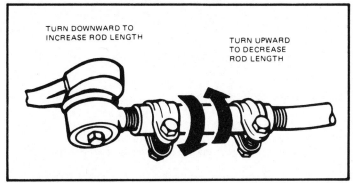

TURN DOWNWARD TO INCREASE ROD LENGTH

TURN UPWARD TO DECREASE ROD LENGTH

**Left-hand sleeve**

# SUSPENSIONS

## Strut Bar Suspension

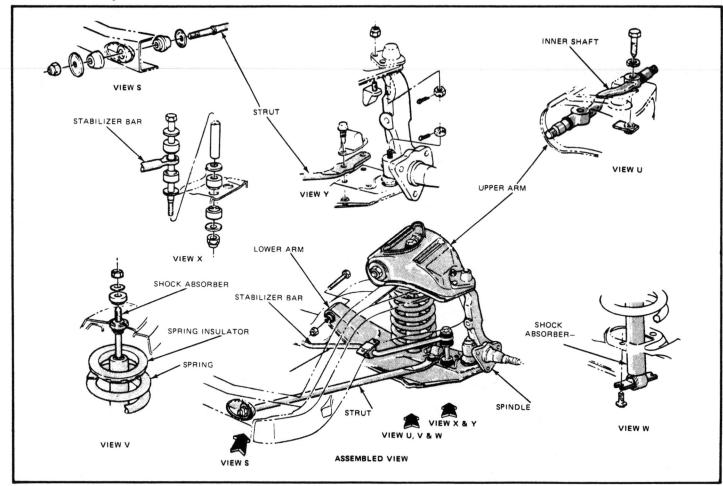

position with the front wheels in straight-ahead position. If the spokes are not in their normal position, they can be properly adjusted while toe is being adjusted.

1. Loosen the two clamp bolts on each spindle connecting rod sleeve.

2. Adjust toe. If the steering wheel spokes are in their normal position, lengthen or shorten both rods equally to obtain correct toe.

3. If the steering wheel spokes are not in their normal position, make the necessary rod adjustments to obtain correct toe and steering wheel spoke alignment.

4. When toe and the steering wheel spoke position are both correct, lubricate clamp, bolts, and nuts and tighten the clamp bolts on both connecting rod sleeves and specification. The sleeve position should not be changed when the clamp bolts are tightened for proper clamp bolt orientation.

## Wheel Bearings

### Adjustment and Inspection

If the wheel is loose on the spindle or does not rotate freely, adjust the front wheel bearings as follows:

1. Raise the vehicle until the tire clears the floor.

2. Remove the wheel cover. Remove the grease cap from the hub.

3. Wipe the excess grease from the end of the spindle. Remove the cotter pin and nut lock.

## Positioning Clamp Bolts

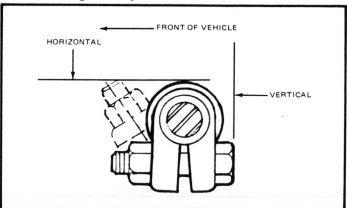

**After toe setting, the two clamp bolts on each side of vehicle must be positioned within limits shown, with threaded end of bolts toward the front of the vehicle.**

4. Loosen the adjusting nut three turns and rock the wheel, hub and rotor assembly in and out several times to push the shoe and linings away from the rotor.

5. While rotating the wheel, hub and rotor assembly, torque the adjusting nut to 23-34 N·m (17-25 ft.-lbs.) to seat the bearings.

## Steering Wheel Spoke Alignment

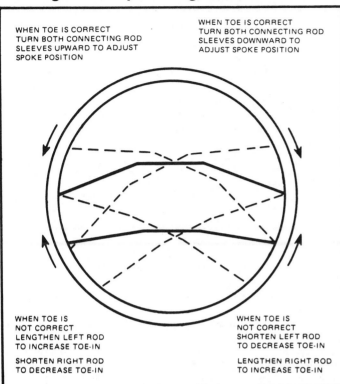

WHEN TOE IS CORRECT
TURN BOTH CONNECTING ROD
SLEEVES UPWARD TO ADJUST
SPOKE POSITION

WHEN TOE IS CORRECT
TURN BOTH CONNECTING ROD
SLEEVES DOWNWARD TO
ADJUST SPOKE POSITION

WHEN TOE IS
NOT CORRECT
LENGTHEN LEFT ROD
TO INCREASE TOE-IN

SHORTEN RIGHT ROD
TO DECREASE TOE-IN

WHEN TOE IS
NOT CORRECT
SHORTEN LEFT ROD
TO DECREASE TOE-IN

LENGTHEN RIGHT ROD
TO INCREASE TOE-IN

**Adjust both rods equally to maintain normal spoke position.**

TURN
DOWNWARD
TO DECREASE
ROD LENGTH

TURN UPWARD TO
INCREASE ROD LENGTH

### Right-hand sleeve

6. Loosen the adjusting nut one-half turn, then retighten to 1.2-1.6 N·m (10-15 in.-lbs.) using a torque wrench.

7. Place the nut lock on the adjusting nut so the castellations on the lock are in line with the cotter pin hole in the spindle.

8. Install a new cotter pin and bend the ends around the castellated flange of the nut lock.

9. Check front wheel rotation. If the wheel rotates properly, reinstall the grease cap and wheel cover. If rotation is noisy or rough, remove wheel hub and check for bearing problems.

10. Before driving the vehicle, pump the brake pedal several times to restore normal brake pedal travel.

### Replacement and Lubrication

1. Raise the vehicle until the tire clears the floor. Remove wheel from hub and rotor.

2. Remove the caliper from the spindle and wire it to the underbody to prevent damage to the brake hose.

3. Remove the grease cap from the hub. Remove the cotter pin, nut lock, adjusting nut and flatwasher from the spindle. Remove the outer bearing cone and roller assembly.

## Front Wheel Bearings Adjustment Procedure

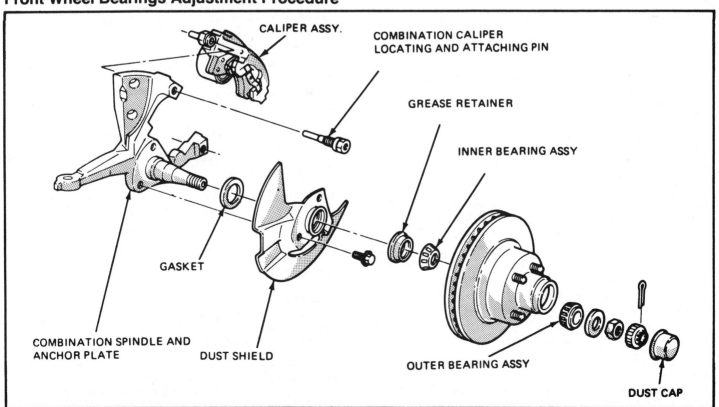

CALIPER ASSY.

COMBINATION CALIPER
LOCATING AND ATTACHING PIN

GREASE RETAINER

INNER BEARING ASSY

GASKET

COMBINATION SPINDLE AND
ANCHOR PLATE

DUST SHIELD

OUTER BEARING ASSY

DUST CAP

# SUSPENSIONS

## Front Wheel Bearing Adjustment

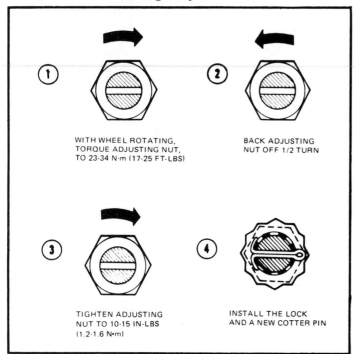

① WITH WHEEL ROTATING, TORQUE ADJUSTING NUT, TO 23-34 N·m (17-25 FT-LBS)

② BACK ADJUSTING NUT OFF 1/2 TURN

③ TIGHTEN ADJUSTING NUT TO 10-15 IN-LBS (1.2-1.6 N•m)

④ INSTALL THE LOCK AND A NEW COTTER PIN

4. Pull the hub and rotor assembly off the spindle.

5. Using tool 1175-AC or equivalent, remove and discard the grease retainer. Remove the inner bearing cone and roller assembly from the hub.

6. Clean the inner and outer bearing cups with solvent. Inspect the cups for scratches, pits, excessive wear, and other damage. If the cups are worn or damaged, remove them with tools D80L-927-A and T77F-1102-A or equivalent and replace them.

## Shock Absorber

### Removal

1. Remove the nut, washer, and bushing from the shock absorber upper end.

2. Remove the two thread-cutting screws attaching the shock absorber to the lower arm and remove the shock absorber. Lightly wire brush the shock studs to free of rust, oil or corrosion.

### Installation

1. Place a washer and bushing on the shock absorber top stud and position the shock absorber inside the front spring. Install the thread-cutting screws and torque to specifications. If the threads in the lower arm become stripped or damaged, the removed thread cutting screws should be re-used, along with 5/16-18 lock nuts. Torque to the same specifications as when thread cutting screws are secured directly to the lower arm.

2. Remove the safety stands and lower the vehicle.

3. Place a bushing and wather on the shock absorber top stud and install nut.

## Spring

### Removal

1. Raise the vehicle. Remove the wheel and tire assembly.

2. Disconnect the stabilizer bar link from the lower arm.

## Coil Spring Removal

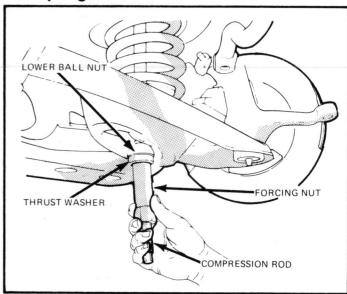

LOWER BALL NUT

THRUST WASHER

FORCING NUT

COMPRESSION ROD

3. Remove the two bolts attaching the shock absorber to the lower arm assembly.

4. Remove the upper nut, retainer and grommet from the shock absorber and remove the shock.

5. Remove the steering center link from the pitman arm.

6. Support the vehicle with safety stands under the jacking pads and lower the hoist for working room.

7. Using the spring compressor tool, install one plate with the pivot ball seat facing downward into the coils of the spring. Rotate the plate so that it is flush with the upper surface of the lower arm.

8. Install the other plate with the pivot ball seat facing upward into the coils of the spring. Insert the upper ball nut through the coils of the spring so that nut resta in the upper plate.

9. Insert the compression rod into the opening in the lower arm, through the upper and lower plate and upper ball nut. Insert the securing pin through the upper ball nut and compression rod.

**NOTE: This pin can only be inserted one way into the upper ball nut because of a stepped hole design.**

10. With the upper ball nut secured, turn the upper plate so that it walks up the coil until it contacts the upper spring seat, then back off one half turn.

11. Install the lower ball nut and thrust washer on the compression rod, and screw on the forcing nut.

12. Tighten the forcing nut until the spring is compressed enough so that it is free in its seat.

13. Remove the two lower arm pivot bolts and disengage the lower arm from the frame crossmember; remove the spring assembly.

14. If a new spring is to be installed, mark the position of the upper and lower plates on the spring with chalk, and measure the compressed length of the spring and amount of spring curvature to assist in compression and installation of a new spring.

15. Loosen the forcing nut to relieve spring tension and remove the tools from the spring.

### Installation

1. Assemble the spring compressor and locate in the same position as indicated in step 14 of Spring Removal.

2. Before compressing the coil spring, be sure that the upper ball nut securing pin is inserted properly.

3. Compress the coil spring until the spring height reaches the dimension obtained in step 14 of Spring Removal.

4. Position the coil spring assembly into the lower arm.

5. To install coil spring, reverse removal procedures.

## Ball Joints

### Inspection

The checking surface should project outside the cover. If the checking surface is inside the cover, replace the lower arm assembly, or install ball joint kit.

### Replacement

The manufacturer recommends replacing the complete arm if the ball joint is worn. However, aftermarket suppliers have ball joint kits available to replace worn ball joints without replacing the complete arm.

## Upper Arm

### Removal

1. Raise the front of the vehicle and position safety stands under both sides of the frame just back of lower arm.
2. Remove the wheel and tire.
3. Remove the cotter pin from the upper ball joint stud nut.
4. Loosen the upper ball joint stud nut one or two turns.

——— CAUTION ———

*Do not remove the nut from the stud at this time.*

5. Insert ball joint press tool between the upper and lower ball joint studs with the adapter screw on top.
6. With a wrench, turn the adapter screw until the tool places the stud under compression. Tap the spindle near the upper stud with a hammer to loosen the stud in the spindle.

**NOTE: Do not loosen the stud from the spindle with tool pressure only. Do not contact the boot seal with the hammer.**

7. Remove the tool from between the ball joint studs and place a floor jack under the lower arm.
8. Remove the upper arm attaching bolts, and remove the upper arm assembly.

### Installation

1. Transfer the bumper from the old arm to the new arm.
2. Position the upper arm in new shaft to the frame bracket, and install the two attaching bolts and washers to a snug fit.
3. Connect the upper ball joint stud to the spindle and install the attaching nut. Torque the nut to specification and continue to tighten the nut until the cotter pin hole in the stud is in line with the nut slots, then install the cotter pin.
4. Install the wheel and tire adjust the wheel bearing.
5. Remove the safety stands and lower the front of the vehicle.
6. Adjust caster, camber and toe-in to specification.

## Upper Arm Bushings

### Replacement (With Arm Removed)

1. Remove the nuts and washer from both ends of the upper arm shaft.

**NOTE: Use the existing C-clamp tool part number T74P-3044-A-1 or equivalent and adapters to remove the bushings.**

2. Position the shaft and new bushings to the upper arm and install the bushings and shaft to the upper arm.

**NOTE: The front bushing is a larger diameter than the rear, requiring that adapter part number T79P-3044-A2 or equivalent, is used when installing rear bushings.**

3. Make certain that the inner shaft is positioned so that the serrated side contacts the frame.

## Lower Ball Joint Check

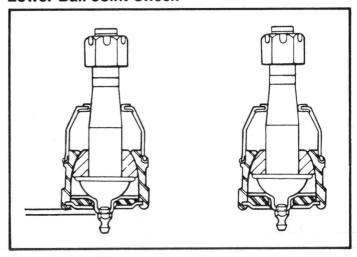

## Special Tool to Break Ball Joint Loose

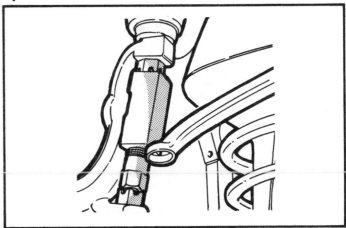

**The tool should be seated firmly against the ends of both studs and not against the nuts or lower stud cotter pin.**

## Upper Arm Bushing Installation

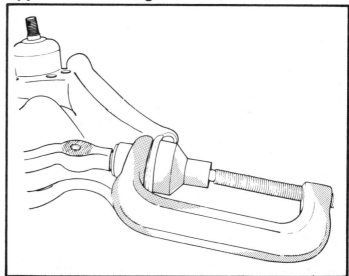

## Upper Arm Bushing Removal

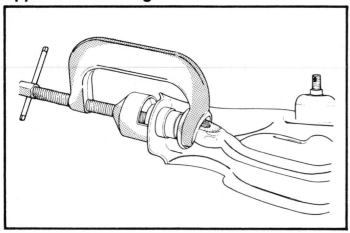

4. Install an inner washer, rear bushing only, and two outer washers and new nuts on each side of the inner shaft. Torque to specification.

# Lower Arm

### Removal

1. Raise the front of the vehicle and position safety stands under both sides of the frame just back of the lower arms.
2. Remove the wheel and tire.
3. Remove the brake caliper and rotor, and dust shield as outlined in brake section.
4. Remove the shock absorber.
5. Disconnect the stabilizer bar link from the lower arm.
6. Disconnect the steering center link from the pitman arm.
7. Remove the cotter pin from the lower ball joint nut.
8. Loosen the lower ball joint stud nut one or two turns.
9. Install ball joint press tool between the upper and lower ball joint studs.
10. Install the coil spring compression tools outlines in "Spring Removal" and remove the spring.
11. Remove the ball joint nut, and remove arm assembly.
12. With a wrench, turn the adapter screw until the tool places the stud under compression. Tap the spindle near the lower stud with a hammer to loosen the stud in the spindle.
13. Remove the ball joint press tool.
14. Place a floor jack under the lower arm.
15. Gently lower the arm until all tension is relieved.
16. Remove lower arm center bolt and remove arm.

### Installation

1. Position the arm assembly ball joint stud into the spindle and install the nut. Torque to specification and install a new cotter pin.
2. Position the coil spring into the upper spring pocket; raise the lower arm and align the holes in the arm with the holes in the crossmember. Install bolts and nuts. Do not tighten at this time.

**NOTE: Be sure that the pigtail of the lower coil of the spring is in the proper location of the seat of the lower arm, between the two holes.**

3. Remove the spring compressor tool.
4. Connect the steering center link at the pitman arm, install the nut and tighten to specification. Install a new cotter pin.
5. Install the shock absorber and torque fasteners to specifications.
6. Install the jounce bumper and torque nut to specification.
7. Install the dust shield, rotor and caliper.
8. Position the stabilizer link to the lower arm and install the bolt and attaching nut.

## Front Spring/Lower Control Arm Removal And Installation

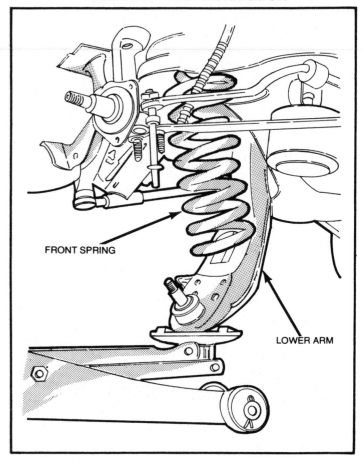

FRONT SPRING

LOWER ARM

9. Install the wheel and tire.
10. Remove the safety stands and lower the vehicle. After the vehicle has been lowered to floor and at curb height, torque the lower pivot bolts and nuts to 136-189 N·m (100-140 ft.-lbs.).

# Wheel Spindle

### Removal

1. Raise the front of the vehicle and position safety stands under both sides of the frame just back of the lower arm.
2. Remove the wheel and tire.
3. Remove the brake rotor, caliper and dust shield.
4. Disconnect the tie-rod end from the spindle.
5. Remove the cotter pins from both ball joint stud nuts and loosen the nuts one or two turns.

**NOTE: Do not remove the nuts at this time.**

6. Position the ball joint remover tool between the upper and lower ball joint studs.
7. Turn the tool with a wrench until the tool places the studs under compression. With a hammer, sharply hit the spindle near the studs to break it loose in the spindle.
8. Position a floor jack under the lower arm at the lower ball joint area.
9. Remove the upper and lower ball joint stud nuts, lower the jack *carefully* and remove the spindle.

## Spindle Mounting

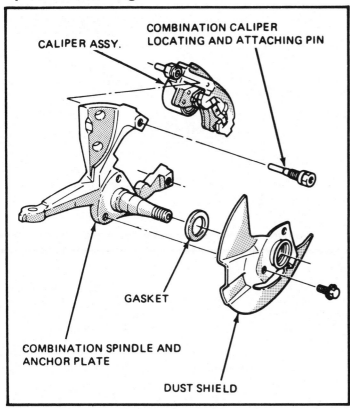

CALIPER ASSY.

COMBINATION CALIPER
LOCATING AND ATTACHING PIN

GASKET

COMBINATION SPINDLE AND
ANCHOR PLATE

DUST SHIELD

### Installation

1. Position the spindle on the lower ball joint stud and install the stud nut. Torque the nut to specification and install a new cotter pin.
2. Raise the lower arm and guide the upper ball joint stud into the spindle. Install the stud nut.
3. Torque the nut to specifications and install a new cotter pin. Remove the floor jack.
4. Connect the tie-rod to the spindle. Install the nut and torque to specifications. Install a new cotter pin.
5. Install the brake dust shield, caliper and rotor.

6. Install the wheel and tire assembly.
7. Remove the safety stands and lower the vehicle.
8. Check caster, camber and toe-in and adjust as required.

## Stabilizer Bar End Bushing

### Replacement

1. Raise the vehicle on a hoist.
2. Remove the nut, washer and insulator from the lower end of the stabilizer bar attaching bolt.
3. Remove the bolt and the remaining washers, insulators, and the spacer.
4. Assemble a cup washer and new insulator on the bolt.
5. Insert the bolt through the stabilizer bar, then install new insulator and cup washer.
6. Install the spacer, cup washer, and another new insulator on the bolt.
7. Insert the bolt through the lower arm and install a new insulator and cup washer. Install and torque the attaching nut to specification.

## Stabilizer Bar and/or Insulator

### Removal

1. Raise the vehicle on a hoist, and place jack stands under the lower arm.
2. Disconnect the stabilizer from each stabilizer link and both stabilizer insulator attaching clamps. Remove the stabilizer bar assembly.
3. Cut the worn insulators from the stabilizer bar.

### Installation

1. Coat the necessary parts of the stabilizer bar with Ruglyde or an equivalent lubricant; slide insulators onto the stabilizer bar.
2. Using a new nut and bolt, secure each end of the stabilizer bar to the lower suspension arm, making sure the bolt head is at the bottom. Tighten nuts to 9-16 N·m (6-12 ft.-lbs.).
3. Using new fasteners, clamp the stabilizer bar to the attaching brackets on the side rail. Tighten bolts to 19-35 N·m (14-26 ft.-lbs.).

# REAR SUSPENSION—FORD, LINCOLN, MERCURY REAR DRIVE CARS, FOUR-BAR LINK DESIGN

## Shock Absorber

### Removal

1. Remove the attaching nut, washer, and insulator from the shock absorber's upper stud.
2. Raise the vehicle on a hoist, and support the rear axle.
3. From underneath the vehicle, compress the shock absorber to clear it from the hole in the upper shock tower.
4. Remove the lower shock absorber bolt, washer, and nut from the axle bracket.
5. Remove the shock absorber.

### Installation

1. Expel all air from the new shock absorber.
2. Compress the shock absorber and position the shock's mount-ing eye to the axle bracket mounting hole. Place a new load bearing washer between the shock eye and axle bracket. Install a new Torx drive belt or equivalent through the shock eye, washer, and axle bracket, then hand start the bolt into a new self-wrenching nut. Do not tighten at this time.
3. After compressing the shock absorber, place the absorber's lower mounting eye between the ears of the lower shock mounting bracket. Then insert the bolt. The bolt head must seat on the inboard side of the shock bracket, through the shock bracket and the shock absorber mounting eye. Install the prevailing torque attaching nut. Do not tighten the nut at this time.
4. Place the inner washer and insulator on the upper attaching stud.
5. Extend the shock absorber's upper stud, and position it through the mounting hole in the shock tower.
6. **Fairmont/Zephyr, Mustang/Capri, Granada/Cougar:** While holding the shock absorber in position, tighten the lower attaching

bolt to 94.9 N·m (70 ft.-lbs.) using tool number D80P-2100-T55 or equivalent. Allow the self-wrenching nut to rotate freely, so that the wrenching tab seats on the outboard leg of the axle bracket. Do not restrain the nut using any other method.

7. **Thunderbird/XR-7:** While holding the shock absorber in position, tighten the lower shock cross bolt to 94.9 N·m (70 ft.-lbs.).

## Rear Shock Lower Installation— Fairmont/Zephyr, Mustang/Capri, Granada/Cougar

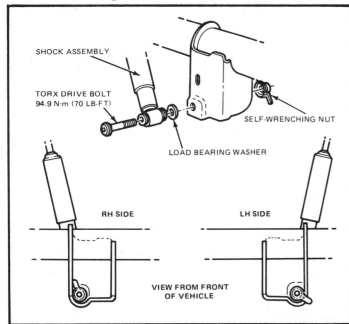

SHOCK ASSEMBLY

TORX DRIVE BOLT 94.9 N·m (70 LB·FT)

SELF-WRENCHING NUT

LOAD BEARING WASHER

RH SIDE

LH SIDE

VIEW FROM FRONT OF VEHICLE

**Allow the self-wrenching nut to rotate freely, so that the wrenching tab seats on the outboard leg of the axle bracket. Do not restrain the nut using any other method.**

## Protective Cover Installation— Thunderbird/XR-7

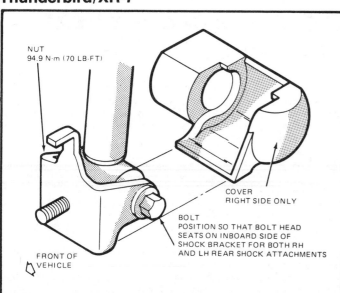

NUT 94.9 N·m (70 LB·FT)

COVER RIGHT SIDE ONLY

BOLT POSITION SO THAT BOLT HEAD SEATS ON INBOARD SIDE OF SHOCK BRACKET FOR BOTH RH AND LH REAR SHOCK ATTACHMENTS

FRONT OF VEHICLE

8. **Thunderbird/XR-7:** Install the protective cover (only one is required) to the right hand shock absorber. This is done by inserting the bolt point and nut into the cover's open end, sliding the cover over the bolt head. Properly installed, the cover will completely conceal the bolt point, nut, and bolt head. The rounded or closed end of the cover should be pointing inboard.

9. Lower the vehicle. Install the insulator, outer washer, and a new nut to the upper shock stud, and tighten. Install the rubber cap on the shock stud. Install the inside panel trim covers.

## Spring

### Removal

**NOTE: If vehicle is equipped with a rear stabilizer bar, remove the bar.**

1. Raise the vehicle and support the body at the rear body crossmember.
2. Lower the hoist until the rear shocks are fully extended.

**NOTE: The axle must be supported by the hoist, a transmission jack, or jack stands.**

3. Place a transmission jack under the lower arm axle pivot bolt, and remove the bolt and nut. Lower the transmission jack slowly until the coil spring load is relieved.
4. Remove the coil spring and insulators from the vehicle.

### Installation

1. Place the upper spring insulator on top of the spring. Tape in place if necessary.
2. Place the lower spring insulator on the lower arm (except Mustang/Capri). Install the internal damper into the spring (except Thunderbird/XR-7.)
3. Position the coil spring on the lower arm spring seat, so that the pigtail on the lower arm is at the rear of the vehicle and pointing toward the left hand side of the vehicle. Slowly raise the transmission jack until the arm is in position. Insert a new rear pivot bolt and nut with the nut facing outwards. Do not tighten at this time.
4. Lower the transmission jack. Raise the axle to curb height. Tighten the lower arm pivot bolt to 135 N·m (100 ft.-lbs.).
5. If vehicle was equipped with a rear stabilizer bar, install bar.
6. Remove crossmember supports, and lower the vehicle.

## Lower Arm

### Removal

1. Raise the vehicle and support body at the rear body crossmember.
2. Lower the hoist until the rear shocks are fully extended.

**NOTE: The axle must be supported by the hoist, a jack, or stands.**

3. Place the transmission jack under the lower arm rear pivot bolt, and remove the bolt and nut.
4. Lower the jack slowly until the coil spring can be removed.
5. Remove the lower arm assembly.

### Installation

1. Position the lower arm assembly into the front arm bracket, and insert a new front pivot bolt and nut with nut facing outwards. Do not tighten at this time.
2. Install coil spring. Holding the spring in position, use the jack under the rear of the lower arm. Raise the jack until the arm is in position. Insert a new rear pivot bolt and nut with nut facing outwards. Do not tighten at this time.
3. Lower the jack. Raise the axle with the hoist to curb height. Tighten the lower arm front bolt and the rear pivot bolt heads to 135 N·m (100 ft.-lbs.).

## Four-Bar Link Coil Rear Suspension

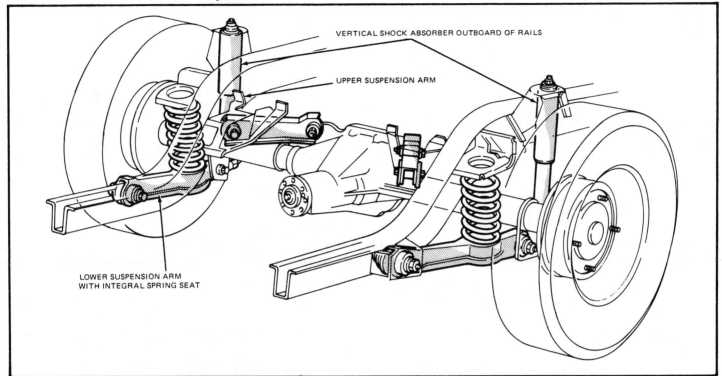

VERTICAL SHOCK ABSORBER OUTBOARD OF RAILS

UPPER SUSPENSION ARM

LOWER SUSPENSION ARM WITH INTEGRAL SPRING SEAT

Vehicle application: Thunderbird/XR-7, Fairmont/Zephyr, Granada/Cougar, Mustang/Capri.

## Rear Suspension—Thunderbird/XR-7

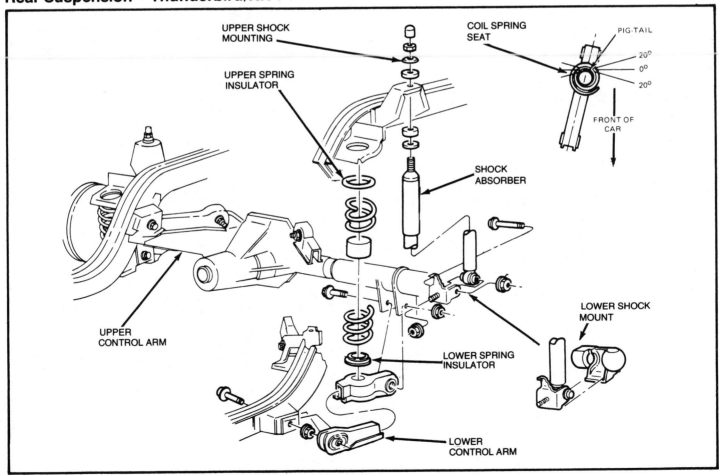

UPPER SHOCK MOUNTING

UPPER SPRING INSULATOR

COIL SPRING SEAT

PIG-TAIL

20°
0°
20°

FRONT OF CAR

SHOCK ABSORBER

UPPER CONTROL ARM

LOWER SHOCK MOUNT

LOWER SPRING INSULATOR

LOWER CONTROL ARM

4. If vehicle is equipped with a rear stabilizer bar, install bar.

5. Remove crossmember supports, and lower vehicle.

## Upper Arm and Axle Bushing

### Replacement

1. Remove upper arm rear pivot bolt and nut.

### Bushing Removal and Installation

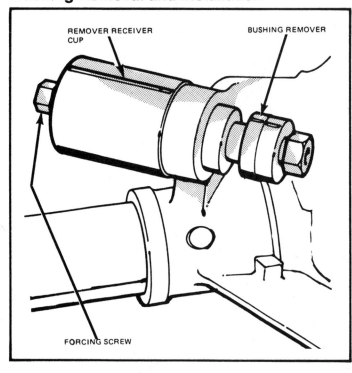

REMOVER RECEIVER CUP

BUSHING REMOVER

FORCING SCREW

2. Remove front pivot bolt and nut. Remove upper arm from vehicle.

3. Place the upper arm rear bushing remover tool in position, and remove the bushing assembly.

4. Using the installer tool, install the bushing assembly into the bushing ear of the rear axle.

## Stabilizer Bar

### Removal

1. Raise vehicle on hoist.

2. Remove four bolts attaching stabilizer bar to brackets in lower arms.

3. Remove stabilizer bar from vehicle.

### Installation

NOTE: Make sure bar is not installed upside down. A color code is provided on stabilizer bar (passenger side only) as an aid for proper orientation.

1. Align four holes in stabilizer bar with holes in brackets in lower arms.

2. Install four new bolts, and tighten nuts to 27 N·m (20 ft.-lbs.).

3. Visually inspect installation to insure adequate clearance between stabilizer bar and lower arm.

## Stabilizer Bar Brackets

### Removal

1. Raise vehicle on hoist, and support body at rear crossmember.

2. Remove stabilizer bar.

3. Disconnect the shock absorbers at the lower shock bracket.

4. Slowly lower the suspension until the front bracket-to-arm bolt clears the body side rail.

NOTE: Do not stress the brake hose when lowering the suspension.

### Rear Stabilizer Bar Installation

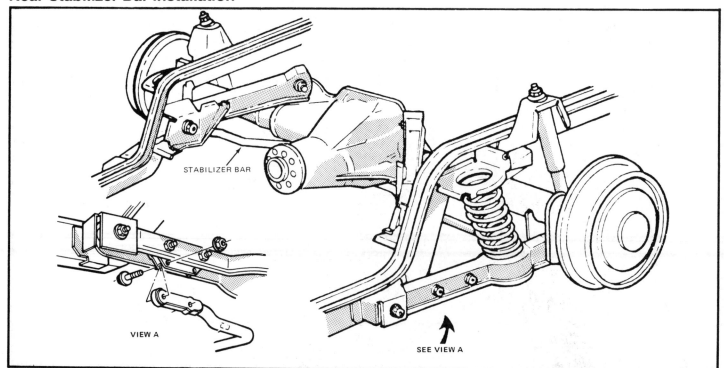

STABILIZER BAR

VIEW A

SEE VIEW A

5. Remove both bracket to arm bolts and nuts, and remove bracket from the arm.

### Installation

1. Insert the bracket into the arm and align holes in arm and

bracket. Install new bolts and nuts. Tighten the nut to 94 N·m (70 ft.-lbs.).

2. Raise the suspension and reassemble the rear shock absorber lower attachment using a new attaching nut.

3. Install stabilizer bar.

4. Remove crossmember supports, and lower the vehicle.

# REAR SUSPENSION—FORD, LINCOLN, MERCURY REAR DRIVE CARS, LEAF SPRING

## Rear Shock Absorber

### Removal and Installation

1. Remove the lower end of the shock absorber from the U-bolt plate.

2. Remove the nut that fastens the upper end of the shock absorber to the mounting bracket.

3. Compress the shock absorber and remove it from the vehicle.

4. Transfer the washers and bushings to the new shock absorber. Insert the stud on the upper end of the shock absorber through the

mounting bracket. Install the new attaching nut fingertight.

5. Compress the shock absorber and assemble the lower end to the U-bolt plate together with the washers and bushings. Tighten the new attaching nut.

## Rear Spring

### Removal

1. Raise the vehicle on a hoist and place supports beneath the underbody and under the axle.

## Exploded View Of Leaf Spring Rear Suspension

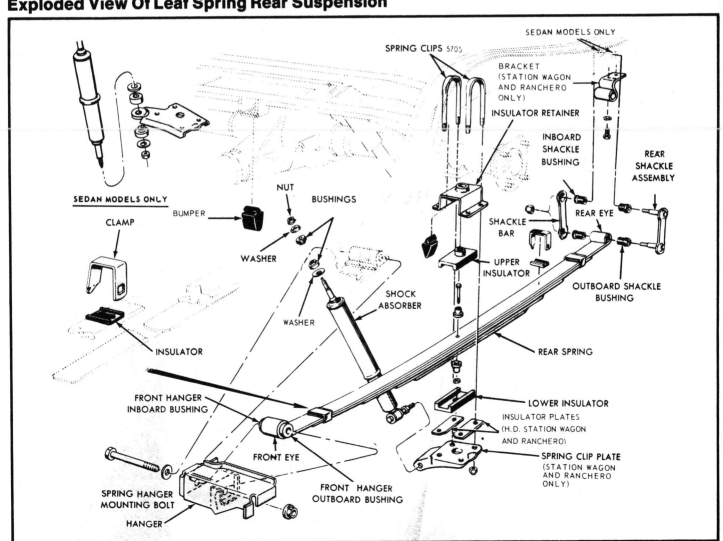

2. Disconnect the lower end of the shock absorber from the U-bolt plate and push the shock out of the way. Remove the supports from under the axle.

3. Remove the U-bolt plate nuts from the U-bolts. Then remove the plate. Raise the rear axle just enough to remove the weight from the spring.

4. Remove the two attaching nuts, rear shackle bar and the two shackle inner bushings.

5. Remove the rear shackle assembly and the two outer bushings.

6. Remove the front hanger bolt and nut from the eye at the forward end of the spring. Lift out the spring assembly.

**Installation**

1. Position the spring under the rear axle. Insert the shackle assembly into the rear hanger bracket and the rear eye of the spring.

2. Install the shackle inner bushings, after lubricating them. Install the shackle plate and the locknuts. Tighten the locknuts fingertight.

3. Position the spring front eye in the front hanger. Install retainer and insulators on spring (if so equipped). Lubricate isoclamp insulators with a lubricant or soap and water solution prior to installation. From the inboard side, insert the bolt through the hanger.

## Spring Bushing Removal and Installation

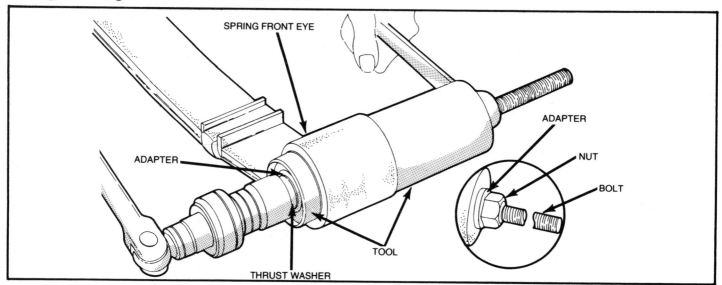

## GM BODY IDENTIFICATION

### FRONT SUSPENSION—GENERAL MOTORS "A" AND "X" BODY CARS

### BUICK

| Model | Body Code |
| --- | --- |
| Regal 1976-81 | A |
| Regal 1982 and later | G |
| Century 1976-81 | A |
| Century 1982 and later ① | A |
| LeSabre 1976 and later ② | B |
| Electra 1976 and later | C |
| Riviera 1976, 1979 and later | E |
| Riviera 1977-78 | B |
| Skyhawk 1976-81 | H |
| Skyhawk 1982 and later ① | J |
| Skylark 1976-79 | X |
| Skylark 1980 and later ① | X |

① Front Wheel Drive
② 1980 and later Electra Estate Wagon-B

### CHEVROLET

| Model | Body Code |
| --- | --- |
| Malibu, Monte Carlo 1976-81 ① | A |
| Malibu, Monte Carlo 1982 and later | G |
| Impala, Caprice 1976 and later | B |
| Camaro 1976 and later | F |
| Vega 1976-77 | H |
| Monza 1976-80 | H |
| Chevette 1976 and later | T |
| Nova 1976-79 | X |
| Citation 1980 and later | X |
| Celebrity 1982 and later | A |
| Cavalier 1982 and later | J |
| Corvette 1976 and later | Y |

① Includes Chevelle
② Spring of 1981

## CADILLAC

| Model | Body Code |
|---|---|
| Fleetwood Brougham 1976 and later | C |
| Calais 1976 | C |
| DeVille 1976 and later | C |
| Fleetwood 1976 and later | D |
| Eldorado 1976 and later | E |
| Seville 1976 and later | K |
| Cimarron 1982 and later ① | J |
| Commercial Chassis 1976-81 | Z |

① Spring of 1981

## OLDSMOBILE

| Model | Body Code |
|---|---|
| Cutlass 1976-81 | A |
| Cutlass 1982 and later | G |
| Cutlass Ciera 1982 and later ① | A |
| Delta 1976 and later | B |
| Ninety Eight 1976 and later | C |
| Toronado 1976 and later | E |
| Starfire 1976-80 | H |
| Omega 1976 and later ② | X |
| Firenza 1982 and later | J |

① Front wheel drive
② 1980 and later—front wheel drive

## PONTIAC

| Model | Body Code |
|---|---|
| LeMans 1976-81 | A |
| Grand Am 1978-80 | A |
| Catalina 1976-81 | B |
| Bonneville 1976-81 | B |
| Bonneville 1982 and later | G |
| Firebird 1976 and later | F |
| Grand Prix 1976-80 | G |
| Grand Prix 1981 | A |
| Astre 1976-77 | H |
| Sunbird 1976-80 | H |
| Ventura 1976-77 | X |
| Phoenix 1977-81 | X |
| T1000 1981 and later | T |
| 6000 1982 and later ① | A |
| J2000 1982 and later ② | J |

① Front wheel drive
② Spring of 1981

# FRONT SUSPENSION—GENERAL MOTORS A AND X BODY CARS—FRONT WHEEL DRIVE

## Wheel Alignment

Front alignment consists of the camber adjustment and toe setting. The caster setting is built into the vehicle with no provisions for adjustment.

Two bolts clamp the lower end of the MacPherson strut assembly to the upper arm of the steering knuckle. The lower of the two bolts has an eccentric washer at the head providing the camber adjustment. These special high tensile bolts with the loose nuts are torqued to 270 N·m (210 ft.-lbs.). The camber setting is plus .5 degree with a ± .5 degree tolerance.

The toe adjustment is conventional, with adjusting sleeves at the tie rod ends held in place with locking jam nuts. The toe setting is plus .1 degree with a tolerance of ± .1 degree.

## Wheel Bearings

The front wheel bearing is a double row ball bearing design. It is a prelubricated sealed bearing and requires no regular maintenance. The bearing is a loose fit in the steering knuckle. The drive axle outer joint shaft is a splined fit through the bearing. The hub nut and washer are used to pre-load the bearing.

### DIAGNOSIS

Check for proper drive axle nut torque, 250 N·m. Clean threads, remove nut, install new nut and torque to proper specification. Free the shoes from the disc or remove calipers. Reinstall two wheel nuts to secure disc to bearing. Mount dial indicator. Grasp disc and use a

203

# SUSPENSIONS

## Front Suspension

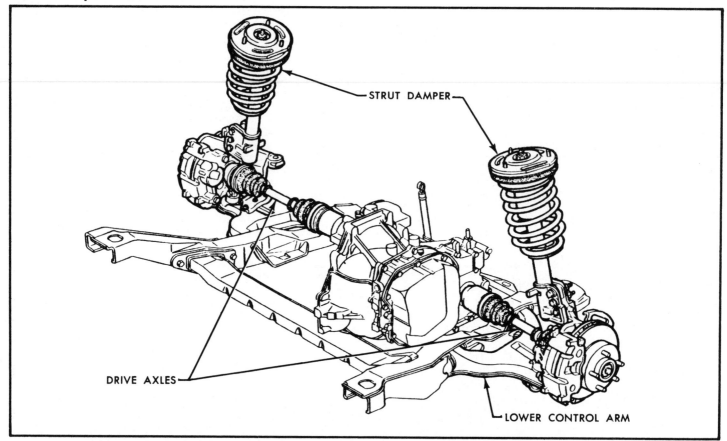

DRIVE AXLES

STRUT DAMPER

LOWER CONTROL ARM

## Knuckle/Strut Mounting Bolts For Camber Adjustment

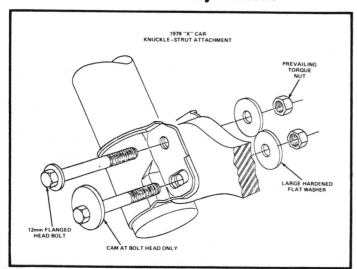

1979 "X" CAR
KNUCKLE–STRUT ATTACHMENT

PREVAILING TORQUE NUT

LARGE HARDENED FLAT WASHER

12mm FLANGED HEAD BOLT

CAM AT BOLT HEAD ONLY

## Front Wheel Bearing

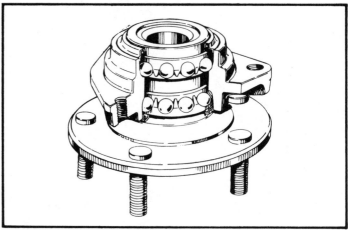

**If looseness exceeds .005 in. (.5080 mm) replace bearing.**

push-pull movement. Do not rock disc as this will give a false reading. If looseness exceeds .5080 mm (.005″) replace bearing.

### Removal

1. Break hub nut.
2. Raise car and remove wheel.
3. Remove hub nut and discard.
4. Remove brake caliper.

5. Remove three hub and bearing attaching bolts. If old bearing is being re-used, mark attaching bolts and corresponding holes for installation.

6. Install tool J-28733 and remove bearing. If excessive corrosion is present make sure bearing is loose in knuckle before using tool J-28733.

**NOTE: A boot protector should be installed whenever servicing front suspension components to prevent damage to the drive axle boot.**

## Dial Indicator Mounting

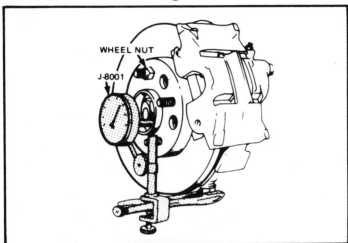

### Installation

1. Clean and inspect bearing mating surfaces and steering knuckle bore for dirt, knicks and burrs.
2. If installing steering knuckle seal, use tool J-28671. Apply grease to seal and knuckle bore.
3. Push bearing on axle shaft.
4. Torque new hub nut until bearing is seated.
5. Install brake caliper.
6. Lower car.
7. Apply final torque to hub nut, 215 ft.-lbs. (305 N·m).

### TORQUES
Top strut nut—68 ft.-lbs. (90 N·m)
Top mount nuts—18 ft.-lbs. (24 N·m)
Lower strut bolts—140 ft.-lbs. (190 N·m)

# MacPherson Strut

### Removal

1. Support the car so that there is no weight on the lower control arm.
2. Remove wheel.
3. Clean up and mark the camber adjusting cam.
4. Remove the brake hose clip.
5. Remove the top three bolts from the lower strut bolts.
6. Remove the strut assembly, and take a sample to work bench.

——————— CAUTION ———————
*A reliable spring compressor tool is essential to disassemble and assemble strut bumper to avoid personal injury.*

7. Compress spring with compressor until there is no pressure on the upper spring seat.
——————— CAUTION ———————
*Do not compress spring until it bottoms.*

8. Remove the top nut from the strut shaft and remove the bumper shaft and the top mounting assembly.
9. Remove the spring from the strut assembly.

### Installation

1. Install the new strut assembly into the spring and attach the mounting components on to the strut assembly.
2. Tighten the top strut nut 68 ft.-lbs. (90 N·m) and remove the spring compressor.
3. Install the spring and strut assembly, first in the upper spring seat, then connect the low end of the strut to the lower control arm.

## MacPherson Strut

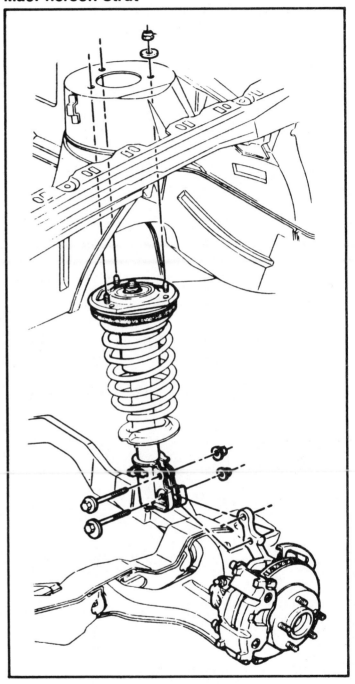

4. Install brake caliper and wheel.

# Steering Knuckle

### Removal

1. Remove wheel and wheel bearing.
2. Mark and remove lower strut bolts.
3. Remove tie-rod end and ball joint.
4. Remove steering knuckle.

### Installation

1. Install knuckle to ball joint and tighten.

# SUSPENSIONS

## Upper Strut Mounting

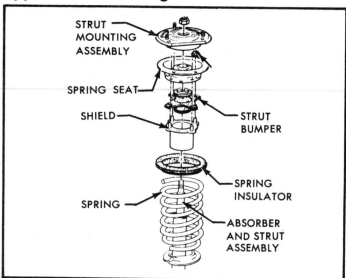

STRUT MOUNTING ASSEMBLY

SPRING SEAT

SHIELD

STRUT BUMPER

SPRING

SPRING INSULATOR

ABSORBER AND STRUT ASSEMBLY

## Steering Knuckle and Components

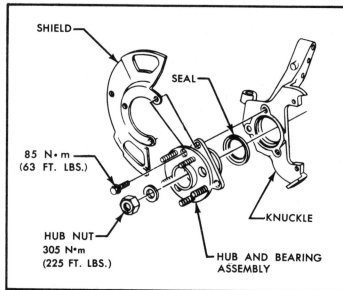

SHIELD

SEAL

85 N•m (63 FT. LBS.)

HUB NUT 305 N•m (225 FT. LBS.)

KNUCKLE

HUB AND BEARING ASSEMBLY

2. Loosely install knuckle to strut.
3. Install front wheel bearing.
4. Jack control arm into position and install tie-rod end.
5. Tighten cam bolts.
6. Reset steering camber and toe.
7. Install brake caliper and wheel.

## Lower Control Arm Ball Joint

### Inspection

1. Raise front suspension by placing jack or lift under the cradle.
2. Grasp the wheel at top and bottom and shake top of wheel in an in-and-out motion. Observe for any horizontal movement of the knuckle relative to the control arm. Replace ball joint if such movement is noted.
3. If the ball stud is disconnected from the knuckle and any looseness is detected, or if the ball stud can be twisted in its socket using finger pressure, replace the ball joint.

### Removal

1. Raise car and remove wheel.

## Lower Strut Bolts

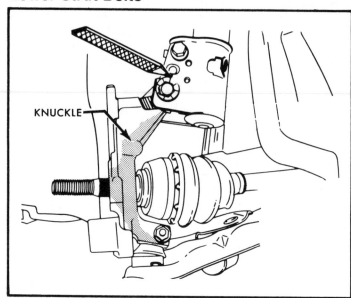

KNUCKLE

## Front Stabilizer and Bushing Removal

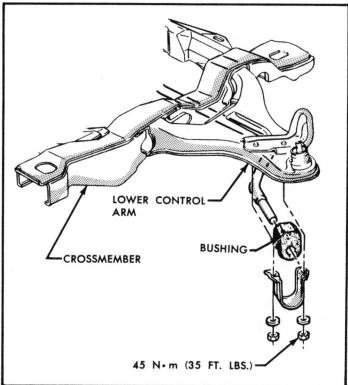

LOWER CONTROL ARM

CROSSMEMBER

BUSHING

45 N•m (35 FT. LBS.)

**Do not remove studs from control arm. End of stabilizer should be an equal distance from bushing on both sides.**

2. Remove parts.
3. Remove ball joint from knuckle.

### Installation

1. Install ball joint to knuckle.
2. Install parts.
3. Install wheel and lower car.
4. Check toe-in setting. Adjust as required.

## Ball Joint Removal and Installation

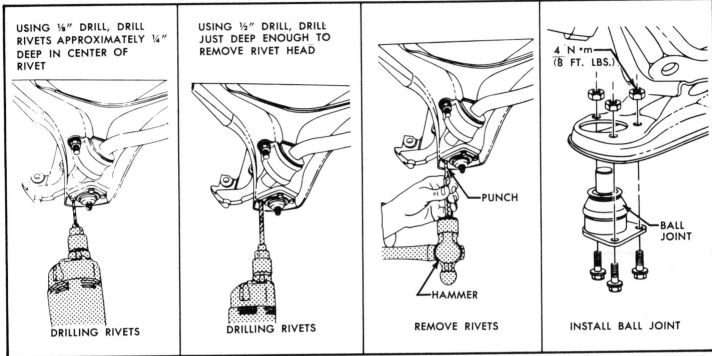

USING ⅛" DRILL, DRILL RIVETS APPROXIMATELY ¼" DEEP IN CENTER OF RIVET

DRILLING RIVETS

USING ½" DRILL, DRILL JUST DEEP ENOUGH TO REMOVE RIVET HEAD

DRILLING RIVETS

PUNCH

HAMMER

REMOVE RIVETS

4 N•m (8 FT. LBS.)

BALL JOINT

INSTALL BALL JOINT

**Ball joint inspection: vertical movement .000 and horizontal movement .000. If ball joint shows any movement replace. It is not necessary to remove control arm to replace joint.**

## Lower Control Arm and/or Bushings

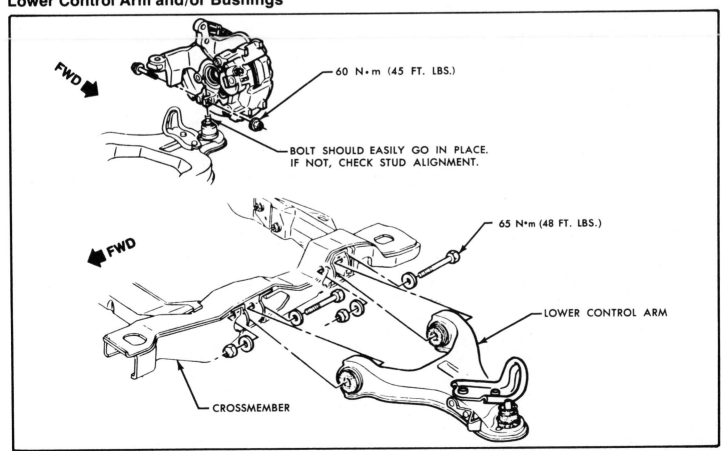

FWD

60 N•m (45 FT. LBS.)

BOLT SHOULD EASILY GO IN PLACE. IF NOT, CHECK STUD ALIGNMENT.

FWD

65 N•m (48 FT. LBS.)

LOWER CONTROL ARM

CROSSMEMBER

# REAR SUSPENSION—GENERAL MOTORS A AND X BODY CARS

## Wheel Bearings

The rear wheel bearing is a double row ball bearing. It is pre-lubricated and sealed at the factory. The bolt on bearing should be replaced if the looseness exceeds recommendations.

### Removal

1. Remove wheel and brake drum.

NOTE: **Do not hammer on brake drum as damage to the bearing could result.**

2. Remove four hub and bearing assembly-to-rear axle attaching bolts and remove hub and bearing assembly from axle.

## Bolt-on Wheel Bearings

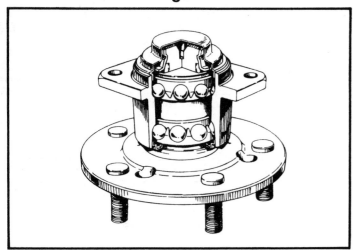

## Rear Wheel Stud Removal

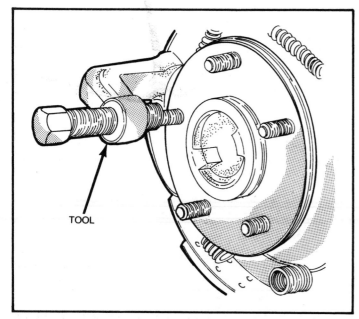

TOOL

**Press stud out. Do not hammer.**

NOTE: **If the studs must be removed from the hub, do not remove with a hammer as damage to bearing will result.**

### Installation

1. Install hub and bearing assembly to rear axle. Tighten bolts to 55 N·m (35 ft.-lbs.).

## Shock Absorber

### Removal

1. Open deck lid, remove trim cover and remove upper shock attaching nut.
2. Raise car on hoist and support rear axle assembly.
3. Remove lower attaching bolt and nut and remove shock.

### Installation

1. Install shock at lower attachment, feed bolt through holes and loosely install nut.
2. Lower car enough to guide upper stud through body opening and install nut loosely.
3. Torque lower nut to 47 N·m (34 ft.-lbs.).
4. Lower car all the way and torque upper nut. Torque to 10 N·m (7 ft.-lbs.).

## Track Bar

### Removal

1. Raise car on hoist and support rear axle.
2. Remove nut and bolt at both the axle and body attachments and remove track bar.

### Installation

1. Position track bar in left hand reinforcement and loosely install bolt and nut. The open side of the bar and nut must face rearward.
2. Place other end of track bar in body reinforcement and install bolt and nut (nut must be at the rear of reinforcement of both attachments). Torque nut at axle bracket to 45 N·m (33 ft.-lbs.) and torque nut at body reinforcement to 47 N·m (34 ft.-lbs.).

## Spring

### Removal and Installation

NOTE: **Do not use a twin-post type hoist. The swing arc tendency of the rear axle when some fasteners are removed may cause it to slip from the hoist.**

1. Raise the car. Support the rear axle while removing the brake line brackets, the track bar and the shock absorber lower mounts.
2. Lower rear axle and remove springs.
3. When installing, position springs correctly.

## Control Arm Bushing

### Removal

1. Raise car on hoist and support rear axle under front side of spring seat.
2. If removing right bushing, disconnect parking brake cable from hook guide.

## Rear Suspension

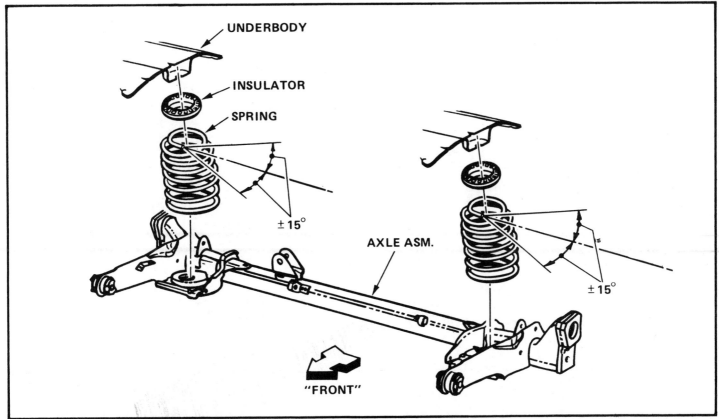

Position leg of upper coil on springs parallel to axle assembly and towards left-hand side of vehicle within limits shown. When removing rear springs, do not use a twin-post type hoist. The swing arc tendency of the rear axle assembly when certain fasteners are removed may cause it to slip from the hoist. Perform operation on floor if necessary.

3. Remove dual parking brake cables from bracket attachment and pull out of way.

4. Disconnect brake line bracket attachment from frame.

5. Remove shock lower attaching nut and bolt and pull spring out of way.

6. Remove four control arm bracket-to-underbody attaching bolts and allow control arm to rotate downward.

7. Remove nut and bolt from bracket attachment and remove bracket.

8. Press bushing out of control arm.

### Installation

1. Press bushing into control arm.

NOTE: Cut-outs on rubber portion of bushing must face front and rear.

2. Install bracket to control arm and torque nut to 47 N·m (34 ft.-lbs.). Bracket must be at a 45-degree angle.

3. Raise control arm into position and install four control arm bracket-to-underbody attaching bolts. Torque to 27 N·m (20 ft.-lbs.).

4. Replace spring and insulator and install shock lower attaching nut and bolt. Torque nut to 47 N·m (34 ft.-lbs.).

5. Install brake line bracket to frame and torque screw to 11 N·m (8 ft.-lbs.).

6. On right side only, reconnect brake cables to bracket, and reinstall brake cable to hook. Adjust cable as necessary.

## Control Arm Bushing Tool

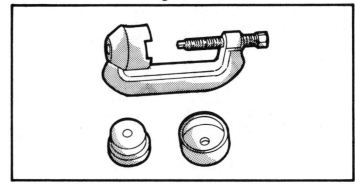

## Rear Axle Assembly

### Removal

NOTE: When removing rear axle assembly, do not use a twin-post type hoist. The swing arc tendency of the rear axle assembly when certain fasteners are removed may cause it to slip from the hoist.

1. Remove wheel and brake drum.

NOTE: Do not hammer on brake drum as damage to the bearing could result.

## Rear Suspension

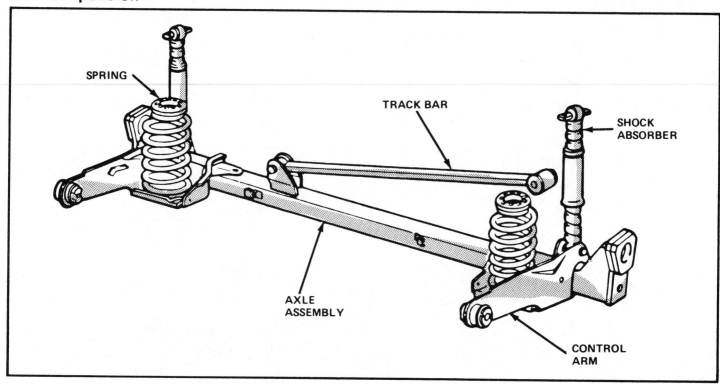

2. Disconnect parking brake cable from hook connection.

3. Remove brake line brackets from frame.

4. Remove shock lower attaching bolts and nuts at axle and disconnect shocks from axle.

5. Remove track bar attaching nut and bolt at axle and disconnect track bar.

6. Lower rear axle and remove coil springs and insulators.

7. Disconnect brake lines from control arm attachments.

8. Remove brake cable from rear axle attachments.

9. Remove hub attaching bolts and remove hub and bearing assembly. Move backing plate out of way.

10. Remove control arm bracket-to-underbody attaching bolts (four per side) and lower axle down to bench. This may require two people to steady axle.

11. Remove control arm brackets from control arms.

### Installation

1. Install control arm brackets to control arms. Torque nuts to 47 N·m (34 ft.-lbs.). Brackets must be at a 45-degree angle.

2. Place axle assembly on transmission jack and raise into position. Attach control arms to underbody with four bolts per side. Torque bolts to 27 N·m (20 ft.-lbs.).

3. Install backing plate and hub and bearing assembly to rear axle. Torque bolts to 55 N·m )35 ft.-lbs.).

4. Install brake line connections to frame.

5. Attach brake cable to rear axle assembly.

6. Position coil springs and insulators in seats and raise rear axle. Leg of upper coil on springs must be parallel to axle assembly and face outboard on both sides.

7. Install shock absorbers to rear axle and torque nuts to 47 N·m (34 ft.-lbs.).

8. Install track bar to rear axle and torque nut to 45 N·m (33 ft.-lbs.).

9. Install brake line brackets to control arm brackets and torque screws to 11 N·m (8 ft.-lbs.).

10. Connect parking brake cable to guide hook and adjust as necessary.

11. Install brake drums and wheels. Torque lug nuts to 140 N·m (103 ft.-lbs.).

12. Remove transmission support and lower car.

13. Bleed brake system and refill reservoir.

## Superlift Shock Absorbers

The Superlift system is an assist-type leveling device which the driver controls manually by varying air pressure in the system. The leveling unit is a combination of a pliable neoprene boot and air cylinder built around a hydraulic shock absorber.

## PRECAUTIONS

To insure satisfactory functioning of the Superlift system, observe the following precautions:

1. Maintain a minimum of 70 kPa (10 psi) for best ride characteristics with an empty car.

2. Vary pressure up to a maximum of 620 kPa (90 psi) to level the car with loads.

# FRONT SUSPENSION—GENERAL MOTORS 1982 F BODY CARS—REAR WHEEL DRIVE

## Wheel Alignment

### CASTER AND CAMBER ADJUSTMENTS

Caster and camber can be adjusted by moving the position of the upper strut mount assembly. Moving the strut mount forward/rearward adjusts the caster while moving the strut mount inward and outward, adjusts the camber.

The position of the strut mount can be changed after loosening the three retaining nuts. The weight of the vehicle will normally cause the strut assembly to move to the full inboard position.

Install special tool J-29724 or its equivalent between the strut mount and a fender bolt and tighten the tool's turnbuckle until the proper camber reading is obtained. If an adjustment of caster is required, tap the strut mount either forward or rearward with a rubber mallet until the caster reading is obtained. Tighten the three mount screws to 20 ft. lbs. (28 N·m).

Remove the tool from the strut mount to fender bolt and re-install the fender bolt in place.

## Suspension Description

Each wheel is independently connected to the frame by a steering knuckle, strut assembly, ball joint, and lower control arm. The steering knuckles move in a prescribed three dimensional arc. The front wheels are held in proper relationship to each other by two tie rods which are connected to the steering knuckles and to a relay rod assembly.

## Front End Alignment Procedure

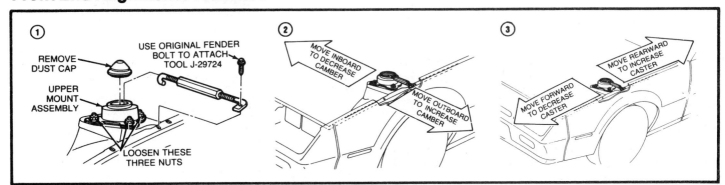

## Exploded View Of 1982 And Later F Body Front Suspension

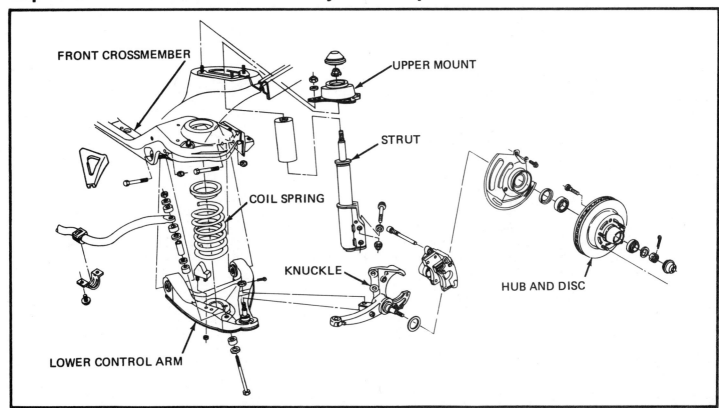

Coil chassis springs are mounted between the spring housings on the front crossmember and the lower control arms. Ride control is provided by double, direct acting strut assemblies. The upper portion of each strut assembly extends through the fender well and attaches to the upper mount assembly with a nut.

Side roll of the front suspension is controlled by a spring steel stabilizer shaft. It is mounted in rubber bushings which are held to the frame side rails by brackets. The ends of the stabilizer are connected to the lower control arms by link bolts isolated by rubber grommets.

The inner ends of the lower control arm have pressed-in bushings. Bolts, passing through the bushings, attach the arm to the suspension crossmember. The lower ball joint assembly is a press fit in the arm and attaches to the steering knuckle with a torque prevailing nut.

## WHEEL BEARINGS

The proper functioning of the front suspension cannot be maintained unless the front wheel tapered roller bearings are correctly adjusted. The bearings must be a slip fit on the spindle and the inside diameter of the bearings should be lubricated to insure proper operation. The spindle nut must be a free-running fit on the threads.

### Adjustment

1. Remove dust cap from hub.
2. Remove cotter pin from spindle and spindle nut.
3. Tighten the spindle nut to 16 N·m (12 ft.-lbs.) while turning the wheel assembly forward by hand to fully seat the bearings. This will remove any grease or burrs which could cause excessive wheel bearing play later.
4. Back off the nut to the "just loose" position.
5. Hand tighten the spindle nut. Loosen spindle nut until either hole in the spindle lines up with a slot in the nut. Not more than 1/2 flat.
6. Install new cotter pin. Bend the ends of the cotter pin against nut, cut off extra length to ensure ends will not interfere with the dust cap.

7. Measure the looseness in the hub assembly. There will be from .03 to .13 mm (.001 to .005 inches) end play when properly adjusted.
8. Install dust cap on hub.

## STRUT ASSEMBLY

### Removal and Installation

1. Raise vehicle.
2. Remove wheel-and-tire.
3. Support lower control arm with jackstand.
4. Remove brake hose bracket.
5. Remove two strut-to-knuckle bolts.
6. Remove cover from upper mount ass'y.
7. Remove nut from upper end of strut.
8. Remove strut and shield.
9. Reverse order of removal to replace strut.

## LOWER BALL JOINT

### Removal

1. Raise car, support with floor stands under frame.
2. Remove tire and wheel assembly.

---
### — CAUTION —
*Floor jack must remain under control arm spring seat during removal and installation to retain spring and control arm in position.*

---

4. Remove cotter pin, and loosen castellated nut. Use too J-24292A or equivalent, to break ball joint loose from knuckle. Remove tool, and separate joint from knuckle.
5. Guide lower control arm out of opening in splash shield with a putty knife or similar tool.
6. Remove grease fittings, and install tools as shown below. Press ball joint out of lower control arm.

## Removal And Installation Of Lower Control Arm Ball Joint

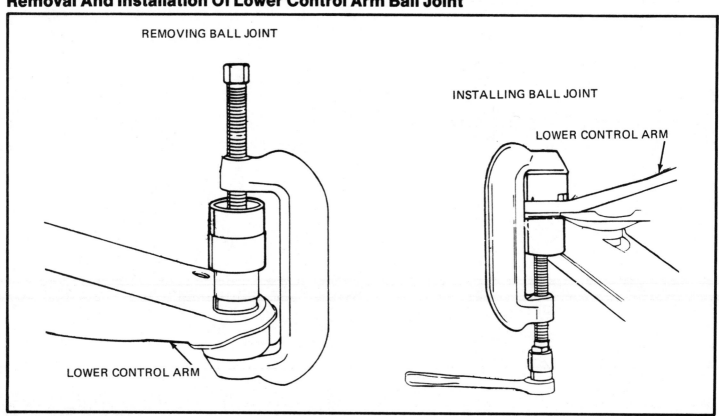

REMOVING BALL JOINT

LOWER CONTROL ARM

INSTALLING BALL JOINT

LOWER CONTROL ARM

## Lower Control Arm And Spring Assembly

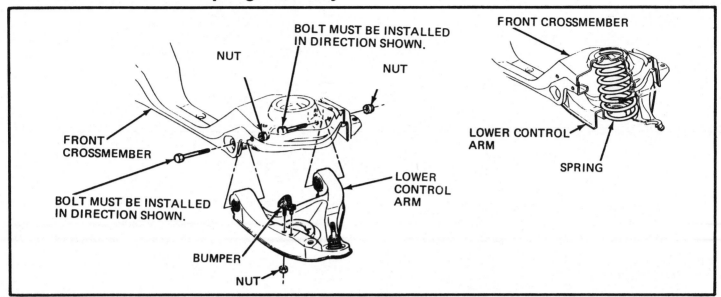

### Inspection

Inspect the tapered hole in the steering knuckle. Remove any dirt. If out-of-roundness, deformation or damage is noted, the knuckle MUST be replaced.

### Installation

1. Position ball joint into lower control arm and press in until it bottoms on the control arm, using tools as illustrated below. Grease purge on seal must be located facing inboard.
2. Place ball joint stud in steering knuckle.
3. Torque ball stud nut to 120 N·m (90 lb.-ft.). Then tighten an additional amount enough to align slot in nut with hole in stud. Install cotter pin.
4. Install and lubricate ball joint fitting until grease appears at the seal.
5. Install tire and wheel assembly.
6. Check front alignment.

## COIL SPRING/LOWER CONTROL ARM

### Removal and Installation

1. Raise vehicle using a hoist.
2. Remove wheel and tire.
3. Remove stabilizer link and bushings at lower control arm.
4. Remove pivot bolt nuts. DO NOT remove pivot bolts at this time.
5. Install tool J23028 adaptor or equivalent, to jack and place into position with tool J-23028 or equivalent, supporting bushings.
6. Install jackstand under outside frame rail on opposite side of vehicle.
7. Raise tool J-23028 or equivalent, enough to remove both pivot bolts.
8. Lower tool J-23028 or equivalent, carefully, as shown below.
9. Remove spring and insulator tape insulator to spring.
10. Remove ball joint from knuckle using tool J-24292A or equivalent, as outlined earlier.
11. Replace bushings in lower control arm.
12. Install parts in reverse order of removal.

**NOTE: After assembly, end of spring coil must cover all or part of one inspection drain hole. The other hole must be partly exposed or completely uncovered.**

## Knuckle, Hub And Disc Assembly

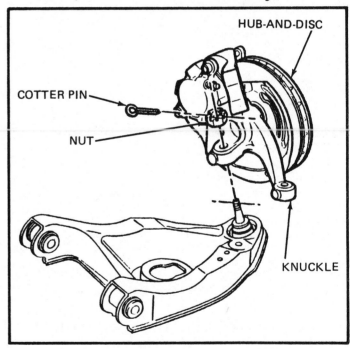

## KNUCKLE, HUB AND DISC

### Removal and Installation

1. Siphon master cylinder, to avoid fluid leakage.
2. Raise vehicle.
3. Remove wheel-and-tire.
4. Remove brake hose from strut.
5. Remove caliper support safely.
6. Remove hub-and-disc.
7. Remove splash shield.
8. Disconnect tie rod from knuckle.
9. Support lower control arm.
10. Disconnect ball joint from knuckle, using tool J-24292A or equivalent.

11. Remove two bolts attaching strut to knuckle, and remove knuckle.

12. Reverse order of removal to install.

## STABILIZER SHAFT

**Removal and Installation**

1. Raise vehicle on hoist.

2. Remove link bolt, nut, grommets, spacer and retainers.
3. Remove insulators and brackets.
4. Remove stabilizer shaft.
5. Install parts in reverse order of removal.
6. Hold stabilizer shaft at approx. 55.0 mm from bottom of side rail when tightening stabilizer shaft insulators.
7. Lower vehicle.

# REAR SUSPENSION—GENERAL MOTORS 1982 F BODY CARS—REAR WHEEL DRIVE

## Suspension Description

The rear axle assembly is attached to the vehicle through a link type suspension system. The axle housing is connected to the body by two lower control arms and a track bar. A single torque arm is used in place of upper control arms and is rigidly mounted to the rear axle housing at the rear and through a rubber bushing to the transmission at the front. Coil springs are used to support the weight of the car and ride control is provided by shock absorbers mounted to the rear of the axle housing. A stabilizer shaft is optional.

## SHOCK ABSORBERS

**Removal**

1. Hoist car and support rear axle.
2. From above, pull back carpeting and remove shock absorber upper mounting nut.

**NOTE: Axle assembly must be supported before removing upper shock absorber nut to avoid possible damage to brake lines, track bar and prop shaft.**

3. Loosen and remove shock absorber lower mounting nut from shock absorber and remove shock.

**Installation**

1. Position shock absorber through body mounting hole and loosely install the lower shock absorber mounting nut.
2. From above, install the upper shock absorber retainer and nut and torque nut to 17 N·m (13 ft.-lbs.).
3. Torque lower shock absorber nut to 95 N·m (70 ft.-lbs.).
4. Remove rear axle support and lower car.

## COIL SPRINGS AND INSULATORS

**Removal**

1. Hoist car on non-twin post-type joist and support rear axle assembly with an adjustable lifting device.
2. Remove track bar mounting bolt at axle assembly and loosen track bar bolt at body brace.
3. Disconnect rear brake hose clip at underbody to allow additional axle drop.
4. Remove right and left shock absorber lower attaching nuts.
5. Remove prop shaft on vehicles equipped with 4-cylinder engines.
6. Carefully lower rear axle and remove spring(s) and or insulator(s).

**NOTE: DO NOT suspend rear axle by brake hose. Damage to hose could result.**

**Installation**

1. Position springs and insulators in spring seats and raise rear axle until rear axle supports weight of vehicle at normal curb height position.
2. Install shocks to rear axle and torque nuts to 95 N·m (70 ft.-lbs.).
3. Thoroughly clean track bar to axle assembly bolt and nut.
4. Reinstall track bar mounting bolt at axle and torque nut to 125 N·m (93 ft.-lbs.). Torque track bar to body bracket nut to 78 N·m (58 ft.-lbs.).
5. Install brake line clip to underbody.
6. Install prop shaft on 4-cylinder engine equipped cars.
7. Remove adjustable lifting device from beneath axle and lower car.

## TRACK BAR

**Removal**

1. Hoist car and support rear axle, at curb height position.
2. Remove track bar mounting bolt and nut at rear axle and at body bracket.
3. Remove track bar.

**Installation**

1. Position track bar in body bracket and loosely install bolt and nut.
2. Thoroughly clean track bar to axle assembly bolt and nut.
3. Position track bar at axle and install bolt and nut, torque bolt to 125 N·m (93 ft.-lbs.).
4. Torque track bar nut at body bracket to 78 N·m (58 ft.-lbs.).
5. Remove rear axle support and lower car.

## TRACK BAR BRACE

**Removal**

1. Hoist car and support rear axle.
2. Remove heat shield screws from track bar brace.
3. Remove three track bar brace to body brace screws.
4. Remove nut and bolt at body bracket and remove track bar brace.

**Installation**

1. Position track bar brace and loosely install nut and bolt at body bracket.
2. Position other end of track bar brace at body bracket and install three screws. Torque screws to 47 N·m (34 ft.-lbs.).
3. Torque track bar nut at body brace to 78 N·m (58 ft.-lbs.).
4. Install heat shield screws to track bar brace.
5. Remove rear axle support and lower car.

## Exploded View Of 1982 And Later F Body Rear Suspension Assembly

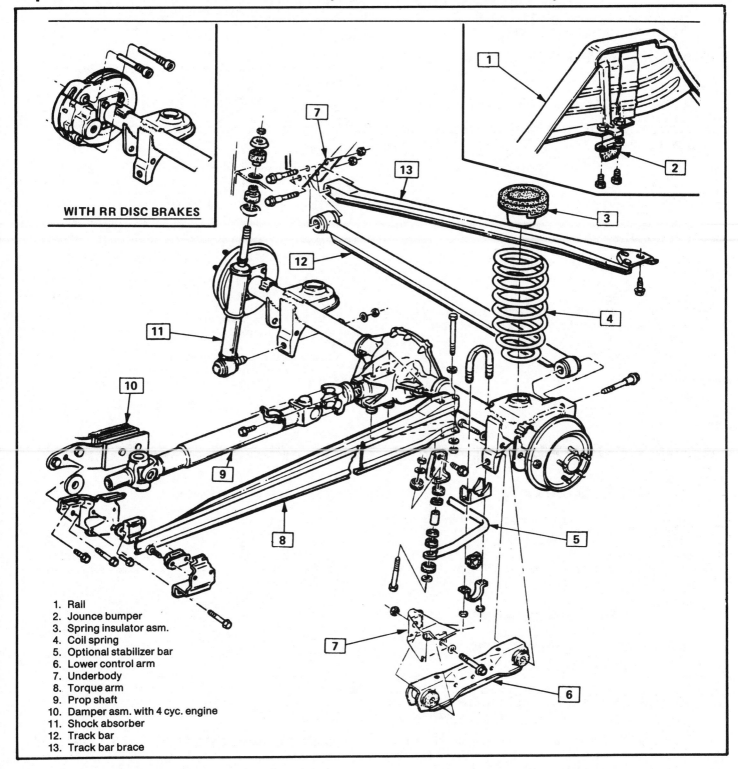

**WITH RR DISC BRAKES**

1. Rail
2. Jounce bumper
3. Spring insulator asm.
4. Coil spring
5. Optional stabilizer bar
6. Lower control arm
7. Underbody
8. Torque arm
9. Prop shaft
10. Damper asm. with 4 cyc. engine
11. Shock absorber
12. Track bar
13. Track bar brace

## REAR LOWER CONTROL ARM

**NOTE: If both control arms are being replaced, remove and replace one control arm at a time to prevent the axle from rolling or slipping sideways making replacement difficult.**

**Removal**

1. Hoist car and support rear axle at curb height position.
2. Remove lower control arm to axle housing bolt and control arm to underbody bolt.
3. Remove control arm.

### Installation

1. Position control arm and install front and rear nuts and bolts.
2. Torque front and rear bolts to 93 N·m (68 ft.-lbs.).
3. Remove rear axle support and lower car.

## BUSHING (REAR LOWER CONTROL ARM)

### Removal

1. Remove control arm as specified in Rear Lower Control Arm Removal Procedure.
2. Place receiver J-25317-2 over flanged side of bushing.
3. Use an arbor press to force the bushing out of the arm, using large O.D. of a driver such as J-21465-8 contacting O.D. of bushing outer sleeve.

### Installation

To install the bushing, reverse the tool and push bushing into position. Connect the rear control arms as outlined in Rear Lower Control Arm Installation procedure.

## TORQUE ARM

### Removal

**NOTE: Coil springs must be removed before removing torque arm to avoid rear axle forward twist which may cause vehicle damage.**

1. Hoist car on a non-twin post-type hoist and support rear axle assembly with an adjustable lifting device.
2. Remove track bar mounting bolt at axle assembly and loosen track bar bolt at body brace.

3. Disconnect rear brake hose clip at underbody to allow additional axle drop.
4. Remove right and left shock absorber lower attaching nuts.
5. Remove prop shaft on vehicles equipped with 4-cylinder engines.
6. Carefully lower rear axle and remove coil springs.

**NOTE: Do not suspend rear axle by brake hose. Damage to hose could result.**

7. Remove torque arm rear attaching bolts.
8. Remove front torque arm outer bracket and remove torque arm.

### Installation

1. Position torque arm and loosely install torque arm bolts.
2. Install front torque arm bracket and torque nuts to 27 N·m (20 ft.-lbs.).
3. Torque rear torque arm nuts to 135 N·m (100 ft.-lbs.).
4. Position springs and insulators in spring seats and raise rear axle until rear axle supports weight of axle of vehicle at normal curb height position.
5. Install shocks to rear axle and torque nuts to 95 N·m (70 ft.-lbs.).
6. Thoroughly clean track bar to axle assembly bolt and nut as outlined under Recommendations for Reuse of Prevailing Torque Fasteners in section 10.
7. Reinstall track bar mounting bolt at axle and torque nut to 125 N·m (93 ft.-lbs.), torque track bar to body bracket nut to 78 N·m (58 ft.-lbs.).
8. Install brake line clip to underbody.
9. Install prop shaft on 4-cylinder engine equipped cars.
10. Remove adjustable lifting device and lower car.

# FRONT SUSPENSION—GENERAL MOTORS E AND K SERIES

## General Description

The front suspension consists of control arms, stabilizer bar, shock absorber and a right and left torsion bar. Torsion bars are used instead of the conventional coil springs. The front end of the torsion bar is attached to the lower control arm. The rear of the torsion bar is mounted into an adjustable arm at the torsion bar crossmember. The trim height of the car is controlled by this adjustment.

## Wheel Alignment

### HEIGHT ADJUSTMENT (TORSION BAR SUSPENSION MODELS)

The standing height must be checked, and adjusted if necessary, before performing the front end alignment procedure. The standing height is controlled by the adjustment setting of the torsion bar adjusting bolt.

Clockwise rotation of the bolt increases the front height; counter-clockwise decreases the front height.

Car must be on a level surface, gas tank full, or a compensating weight added, front seat all the way to the rear, and front and rear tires inflated to the proper pressures. Doors, hood and trunk must be closed and no passengers or additional weight should be in car or trunk.

These tolerances are production specifications on bumper height. If there is more than 1 inch (25 mm) difference, side to side, at the wheel well opening, corrective measures may need to be implemented on a case by case basis. These are curb height dimensions which include a full tank of fuel.

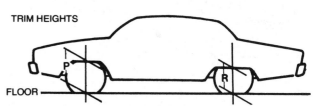

TRIM HEIGHTS

FLOOR

P = FRONT WHEEL OPENING TO FLOOR
R = REAR WHEEL OPENING TO FLOOR

|  | Trim | Height | Tolerance |
|---|---|---|---|
| Front (P) | mm | 726 | +32 |
|  | Inches | 28.6 | +1.25 |
| Rear (R) | mm | 694 | +32 |
|  | Inches | 27.3 | +1.25 |

## CAMBER AND CASTER ADJUSTMENTS

These adjustments can be made either from under car or under hood, as desired. If under hood approach is used, however, be sure to recheck alignment after all operations are completed. Change in weight distribution caused by opened hood is sufficient to disturb final alignment settings.

1. Loosen nuts on upper suspension arm front and rear cam bolts.
2. Note camber reading and rotate front bolt to correct for 1/2 of incorrect reading or as near as possible.

## Front Suspension

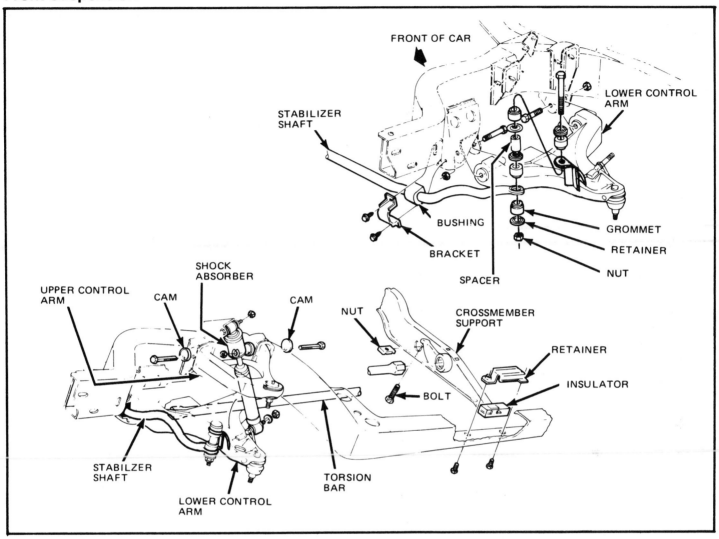

FRONT OF CAR

STABILIZER SHAFT

LOWER CONTROL ARM

BUSHING

BRACKET

GROMMET

RETAINER

NUT

SPACER

UPPER CONTROL ARM

CAM

SHOCK ABSORBER

CAM

NUT

CROSSMEMBER SUPPORT

RETAINER

INSULATOR

BOLT

STABILZER SHAFT

LOWER CONTROL ARM

TORSION BAR

3. Rotate rear cam bolt to bring camber reading to 0°.

NOTE: Do not use a socket to adjust rear cam bolt on left side as brake pipes could be damaged. An offset box end wrench is recommended at this adjustment point.

4. Tighten front and rear bolts and check caster. If caster requires adjustment, proceed with step 5; if not, move to step 8.

5. Loosen front and rear cam bolt nuts.

6. Using camber scale on alignment equipment, rotate front bolt so that the camber changes an amount equal to 1/4 of the desired caster change. (A caster change-to-camber change ratio of about 2 to 1 is inherent to the Eldorado and Seville suspension system. That is, when one cam is rotated sufficiently to change camber 1°, caster reading will change about 2°.)

If adjusting to correct for excessive negative caster, rotate front bolt to increase positive camber. If adjusting to correct for excessive positive caster, rotate front bolt to increase negative camber.

7. Rotate the rear bolt until camber setting returns to its corrected position (step 3).

8. Tighten upper suspension arm cam nuts to 130 N·m (95 ft.-lbs.). Hold head of bolt securely; any movement of the cam will affect final setting and will require a recheck of the camber and caster adjustments.

## Caster/Camber Adjustments

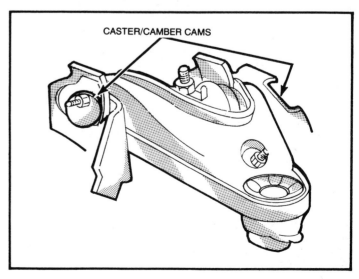

CASTER/CAMBER CAMS

217

## E Series Tie Rod Clamp and Sleeve Positioning

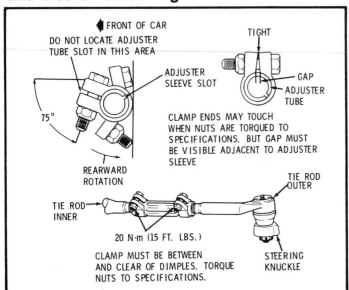

**Bolts must be installed in direction shown. Rotate both inner and outer tie rod housings rearward to the limit of ball joint travel before tightening clamps. With this same rearward rotation, all bolt centerlines must be between angles shown after tightening clamps.**

## Bearing Adjustment

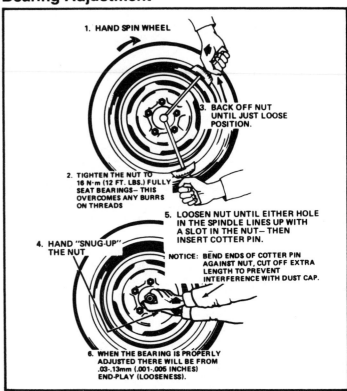

## TOE ADJUSTMENT

Before checking toe-in, make certain that the intermediate rod height is correct.

Toe-in is adjusted by turning the tie-rod adjuster tubes at the outer ends of each tie-rod after loosening clamp bolts. The readings should be taken only when the front wheels are in a straight ahead position so that the steering gear is on its high spot.

1. Center steering wheel, raise car, and check wheel runout.
2. Loosen tie-rod adjuster nuts and adjust tie-rods to obtain proper toe setting.
3. Position tie-rod adjuster clamps so that openings of clamps are facing up. Interference with front suspension components could occur while turning if clamps are facing down.

## Wheel Bearings (Tapered Roller Bearings)

### Lubrication

For normal application, clean and repack front wheel bearings with a high melting point wheel bearing lubricant at each front brake lining replacement or 30,000 miles (48 000 km), whichever comes first. For heavy duty application, clean and repack front wheel bearings at each front brake lining replacement or 15,000 miles (24 000 km) whichever comes first. Use wheel bearing lubricant; "long fiber" or "viscous" type lubricant should not be used. Do not mis wheel bearing lubricants. Be sure to thoroughly clean bearings and hubs of all old lubricant before repacking.

**NOTE: Tapered roller bearings have a slightly loose feel when properly adjusted. They must never be overtightened (pre-loaded) or severe bearing damage may result.**

### Adjustment

The proper functioning of the front suspension cannot be maintained unless the front wheel tapered roller bearings are correctly adjusted. Cones must be a slip fit on the spindle and the inside diameter of cones should be lubricated to insure that the cones will creep. Spindle nut must be a free-running fit on threads.

1. Remove cotter pin from spindle and spindle nut.
2. Tighten the spindle nut to 16 N·m (12 ft.-lbs.) while turning the wheel assembly forward by hand to fully seat the bearings. This will remove any grease or burrs which could cause excessive wheel bearing play later.
3. Back off the nut to the "just loose" position.
4. Hand tighten the spindle nut. Loosen spindle nut until either hole in the spindle lines up with a slot in the nut (not more than 1/2 flat).
5. Install new cotter pin. Bend the ends of the cotter pin against nut. Cut off extra length to ensure ends will not interfere with the dust cap.
6. Measure the looseness in the hub assembly. There will be from .03 to .13 mm (.001 to .005 inches) end play when properly adjusted.
7. Install dust cap on hub.

## Sealed Wheel Bearing

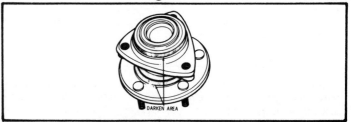

## Wheel Bearings (Bolt On-Type Bearings)

Starting in 1979, the E and K Series have front and rear sealed wheel bearings. The bearings are pre-adjusted and require no lubrication maintenance or adjustment. There are darkened areas on the bearing assembly. These darkened areas are from a heat treatment process and do not require bearing replacement.

### Shock Absorber Removal and Installation

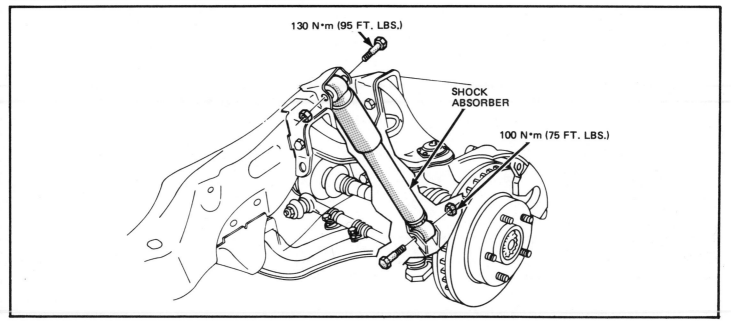

130 N•m (95 FT. LBS.)

SHOCK ABSORBER

100 N•m (75 FT. LBS.)

### Check Ball Joint

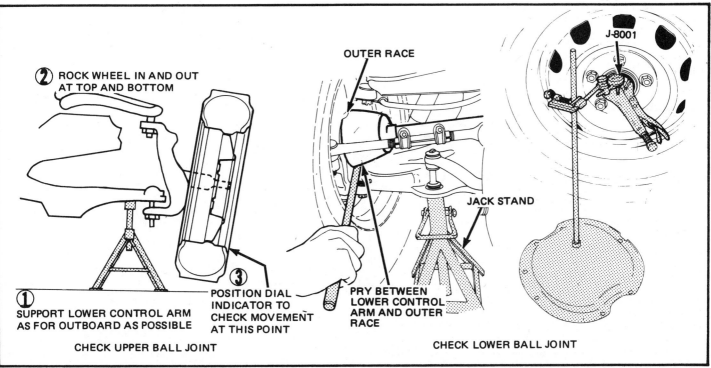

OUTER RACE

J-8001

② ROCK WHEEL IN AND OUT AT TOP AND BOTTOM

JACK STAND

① SUPPORT LOWER CONTROL ARM AS FOR OUTBOARD AS POSSIBLE

③ POSITION DIAL INDICATOR TO CHECK MOVEMENT AT THIS POINT

PRY BETWEEN LOWER CONTROL ARM AND OUTER RACE

CHECK UPPER BALL JOINT

CHECK LOWER BALL JOINT

**Check ball joint as shown. If dial indicator reading exceeds 3.2 mm (.125 in.) or if ball stud is disconnected from knuckle and any looseness is detected or ball stud can be twisted in its socket with fingers, replace ball joint.**

# SUSPENSIONS

## Lower Ball Joint Removal and Installation

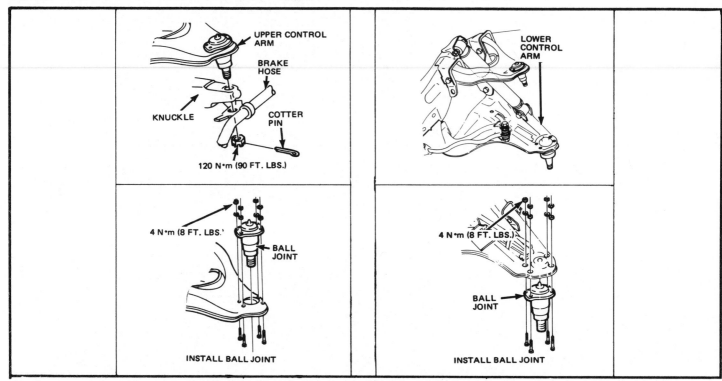

To remove: hoist car and remove wheel. Remove knuckle. Remove hub and bearing assembly, knuckle and knuckle seal. Remove parts as shown. To install: install parts as shown. Install steering knuckle. Install hub and bearing assembly, knuckle and knuckle seal. Install wheel and lower hoist.

## Upper Ball Joint Removal and Installation

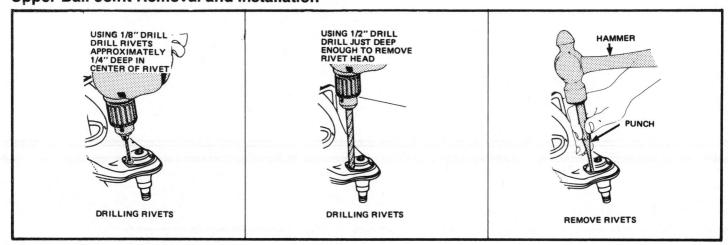

To remove: hoist car and remove wheel, disconnect upper ball joint, and remove parts as shown. To install: install parts as shown, install wheel, and lower hoist.

## Upper Control Arm and/or Bushings Removal and Installation

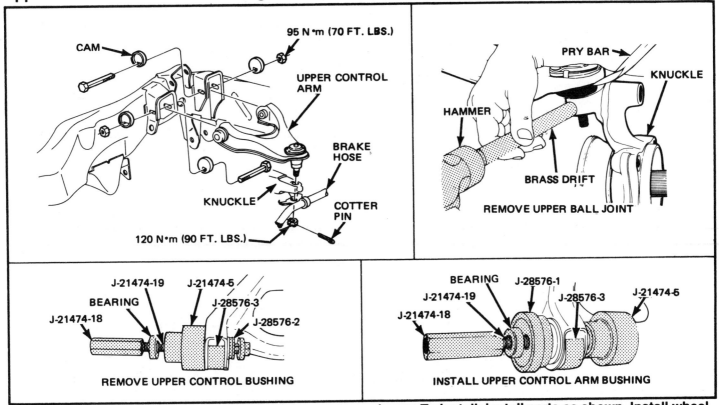

To remove: hoist car, remove wheel and remove parts as shown. To install: install parts as shown. Install wheel and lower hoist. Check alignment and adjust as necessary.

## Lower Control Arm Removal and Installation

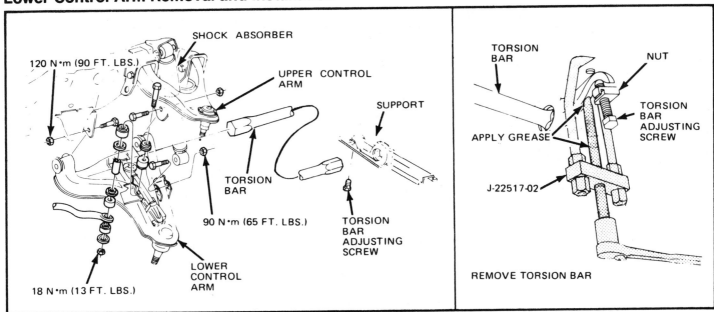

To remove: hoist car and remove wheel, remove knuckle, and remove parts as shown. To install: install parts as shown. Install knuckle, install wheel and lower hoist. Adjust trim height.

# SUSPENSIONS

## Lower Control Arm Bushing Removal and Installation

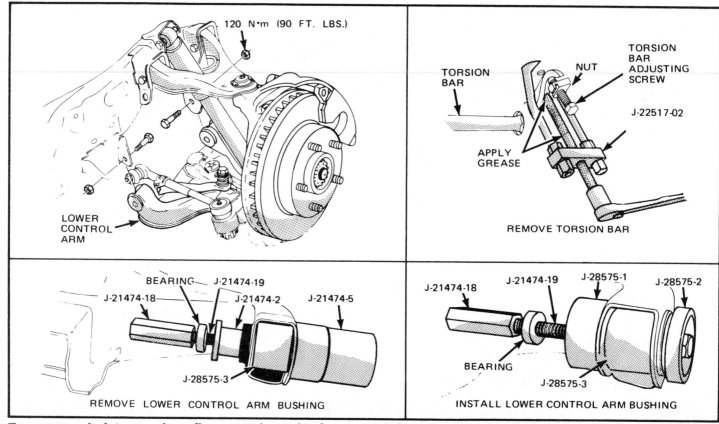

To remove: hoist car, place floor stands under frame, and then remove wheel. Remove parts as shown. To install: install parts as shown. Raise front hoist under lower control arm and remove floor stands. Install wheel and lower hoist. Adjust trim height.

## Stabilizer Shaft Removal and Installation

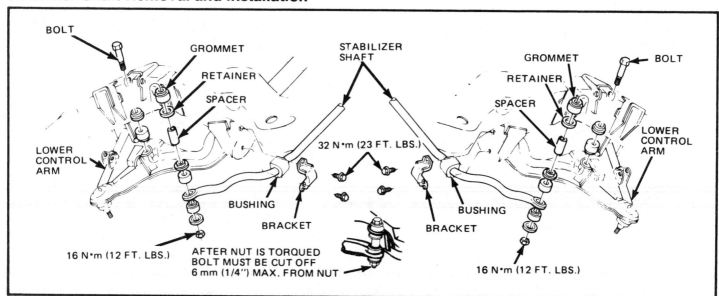

To remove: hoist car and remove parts as shown. To install: install parts as shown and lower hoist.

## Torsion Bar and/or Support Removal and Installation

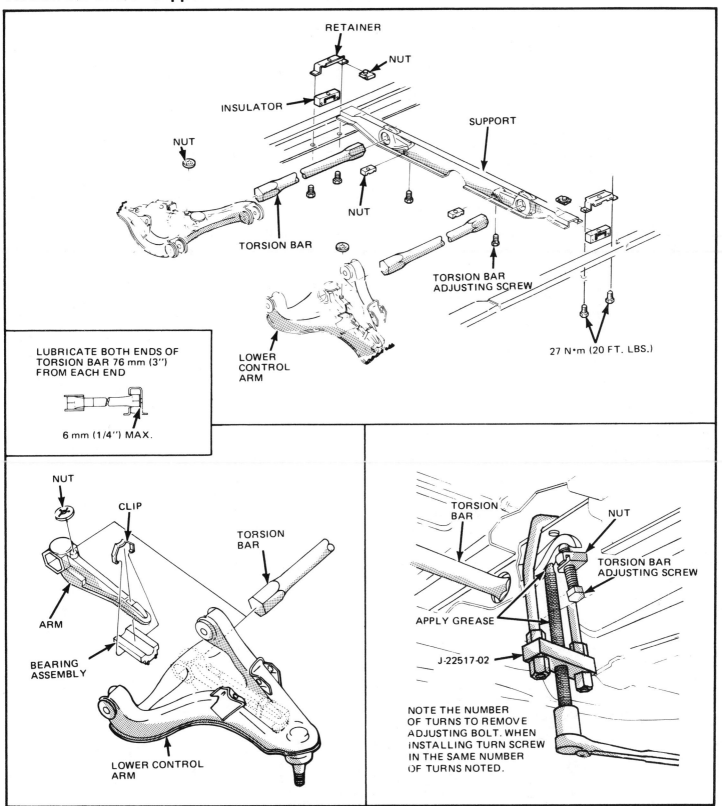

RETAINER

NUT

INSULATOR

NUT

SUPPORT

TORSION BAR

NUT

TORSION BAR ADJUSTING SCREW

LOWER CONTROL ARM

27 N•m (20 FT. LBS.)

LUBRICATE BOTH ENDS OF TORSION BAR 76 mm (3") FROM EACH END

6 mm (1/4") MAX.

NUT

CLIP

TORSION BAR

ARM

BEARING ASSEMBLY

LOWER CONTROL ARM

TORSION BAR

NUT

TORSION BAR ADJUSTING SCREW

APPLY GREASE

J-22517-02

NOTE THE NUMBER OF TURNS TO REMOVE ADJUSTING BOLT. WHEN INSTALLING TURN SCREW IN THE SAME NUMBER OF TURNS NOTED.

**To remove: hoist car and remove parts as shown. To remove torsion bar(s) only: remove torsion bar adjusting screw as shown. Slide torsion bar forward in lower control arm until torsion bar clears support. Then pull down on bar and remove from control arm. To install: install parts as shown. Lower hoist and adjust trim height.**

## Bolt on Sealed Wheel Bearing Diagnosis

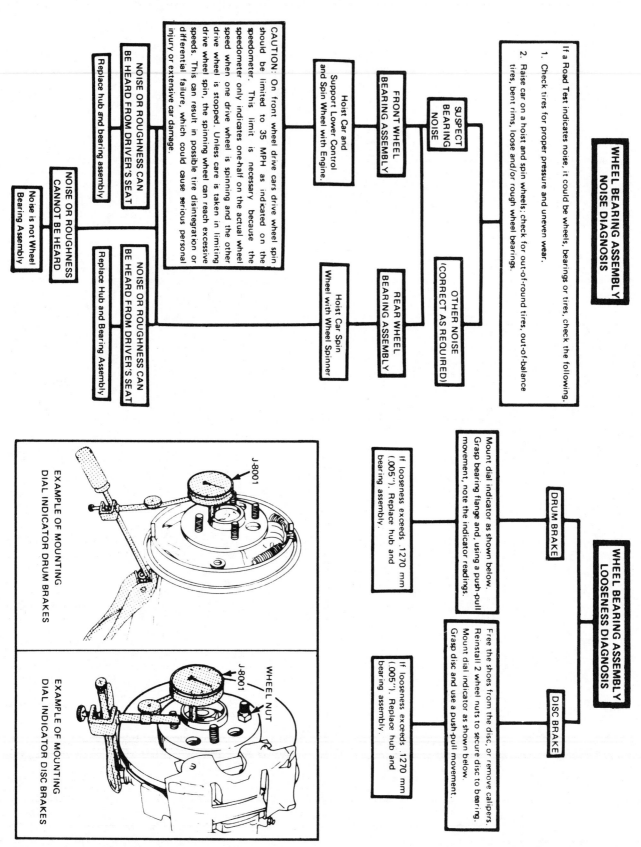

**WHEEL BEARING ASSEMBLY NOISE DIAGNOSIS**

If a Road Test indicates noise, it could be wheels, bearings or tires, check the following.

1. Check tires for proper pressure and uneven wear.
2. Raise car on a hoist and spin wheels; check for out-of-round tires, out-of-balance tires, bent rims, loose and/or rough wheel bearings.

**SUSPECT BEARING NOISE**

**OTHER NOISE (CORRECT AS REQUIRED)**

**FRONT WHEEL BEARING ASSEMBLY**

Hoist Car and Support Lower Control and Spin Wheel with Engine.

CAUTION: On front wheel drive cars drive wheel spin should be limited to 35 MPH as indicated on the speedometer. This limit is necessary because the speedometer only indicates one-half on the actual wheel speed when one drive wheel is spinning and the other drive wheel is stopped. Unless care is taken in limiting drive wheel spin, the spinning wheel can reach excessive speeds. This can result in possible tire disintegration or differential failure, which could cause serious personal injury or extensive car damage.

**NOISE OR ROUGHNESS CAN BE HEARD FROM DRIVER'S SEAT**

Replace hub and bearing assembly

**NOISE OR ROUGHNESS CANNOT BE HEARD**

Noise is not Wheel Bearing Assembly

**REAR WHEEL BEARING ASSEMBLY**

Hoist Car Spin Wheel with Wheel Spinner

**NOISE OR ROUGHNESS CAN BE HEARD FROM DRIVER'S SEAT**

Replace Hub and Bearing Assembly

**WHEEL BEARING ASSEMBLY LOOSENESS DIAGNOSIS**

**DRUM BRAKE**

Mount dial indicator as shown below. Grasp bearing flange and, using a push-pull movement, note the indicator readings.

If looseness exceeds .1270 mm (.005"). Replace hub and bearing assembly.

**DISC BRAKE**

Free the shoes from the disc, or remove calipers. Reinstall 2 wheel nuts to secure disc to bearing. Mount dial indicator as shown below. Grasp disc and use a push-pull movement.

If looseness exceeds .1270 mm (.005"). Replace hub and bearing assembly.

EXAMPLE OF MOUNTING DIAL INDICATOR DRUM BRAKES

J-8001

EXAMPLE OF MOUNTING DIAL INDICATOR DISC BRAKES

J-8001

WHEEL NUT

## Hub and Bearing Assembly, Knuckle and Knuckle Seal Removal and Installation

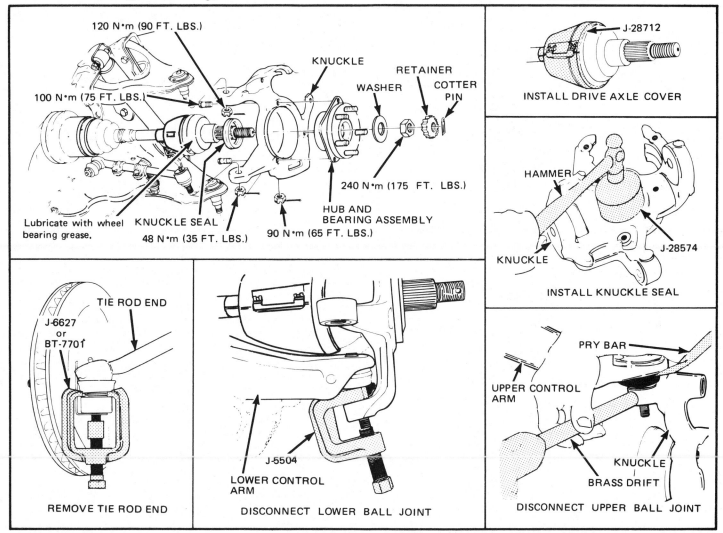

To remove: hoist car and remove wheel, remove disc, and remove parts as shown. To install: be sure that bearing surfaces are clean and free of burrs. Install parts as shown. Install disc, install wheel and lower hoist.

# REAR SUSPENSION—GENERAL MOTORS E AND K SERIES

## Description

The E and K Series have a semi-trailing arm-type rear suspension system with a relatively long control arm for a minimum camber change. The system consists of boxed control arms, coil springs, super-lift shock absorbers and a stabilizer bar.

The control arms are welded together. The hub and bearing attachment plane is machined for precise suspension alignment.

The hub and wheel bearing is a unit assembly which eliminates the need for wheel bearing adjustments and does not require periodic maintenance.

## Operation

The left and right rear wheel suspensions, being independent of each other, permit the vertical movement of one wheel without affecting the wheel on the opposite side of the car.

This independent wheel movement is obtained by an A frame control arm. The control arm is hinged at the frame to provide the up and down movement of the wheel. The solid stabilizer bar forces the wheel to travel in through a controlled arc.

The control arm also carries the rear brake mounting bracket and hub and bearing assembly.

## Alignment

Satisfactory operation may occur over a wide range of rear wheel alignment settings. Nevertheless, should settings vary beyond certain tolerances, readjustment of alignment is advisable. The specifications stated in column 1 of the charts should be used as guidelines.

These specifications provide an acceptable all-around operating range in that they prevent abnormal tire wear caused by wheel alignment.

In the event the actual settings are beyond the specifications set

## Rear Suspension

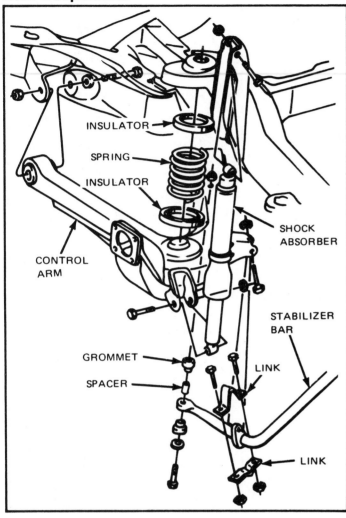

forth in column 1 or 2 (whichever is applicable), or whenever for other reasons the alignment is being reset, the factory recommends that the specifications given in column 3 of the charts be used.

Rear wheel alignment should be checked and adjusted as necessary in the following procedure.

1. Check front and rear trim heights.
2. Check electronic level control for proper operation.
3. Using an alignment machine, use one of the following procedures.

Preferred method

  a. If machine does not have guide line, place tape on floor from wheel plate to rear of car to use as a guide for lining up car on machine.

  b. Back car onto alignment machine placing rear wheel on wheel plates.

  c. Place a straightedge at same rib of tire at front and rear and measure distance from inside edge of straight edge and edge of guide line. The measurement from the guide line to the straight edge must be greater at the rear tire by 16 mm (5/8 in.) ± 6 mm (1/4 in.).

Alternate method

  a. Place a one inch tape on the floor along the righthand side of the center line between the wheel plates of the alignment machine.

  b. Back car onto alignment machine making sure car is as straight as possible.

  c. Hanging plumbs on the front and rear crossmembers at gage holes will give guide lines for near perfect centering of the car.

4. Attach alignment mirrors to rear wheel and take toe and camber readings.

**NOTE: With car backed onto alignment machine, toe-in and toe-out are reversed. Toe-in will be read as toe-out. It is very important that the readings be made and understood properly.**

5. Toe adjustments are made at the inner pivot bushings. Loosening the nut and bolt at the inner bushing will enable the toe to be moved in or out as necessary.

6. Tighten bushing mounting nut to 135 N·m (97 ft.-lbs.) and recheck toe for correct setting. It may be necessary to use a pry bar to move the control arm. Moving the control arm rearward increases toe-in; moving it forward increases toe-out.

7. Check camber. There are not camber adjustment provisions; ses Diagnosis.

## Shock Absorber Removal and Installation

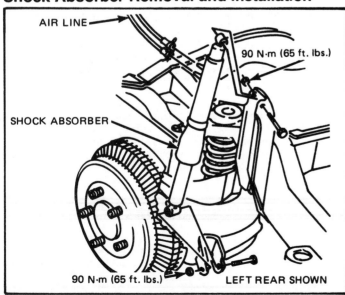

To remove: hoist car and support control arm, remove parts as shown. To install: install parts as shown, remove support and lower hoist.

## Rear Wheel Alignment Diagnosis

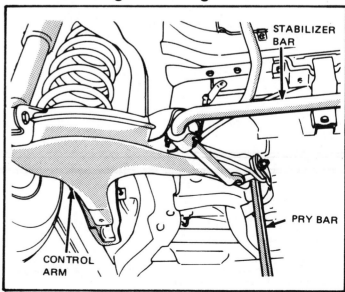

## Rear Wheel Alignment

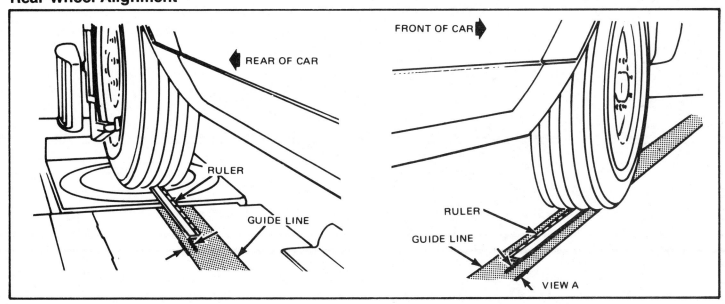

## Rear Wheel Alignment

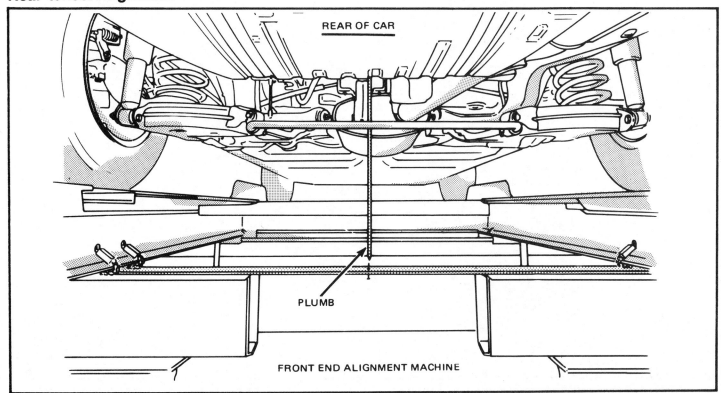

# SUSPENSIONS

## Rear Suspension Components

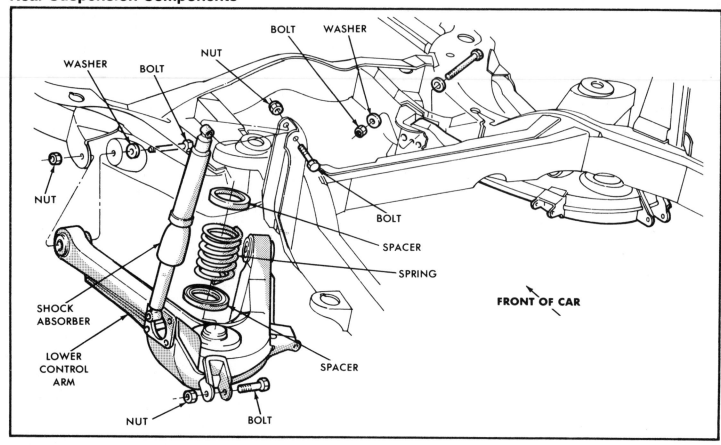

WASHER

WASHER

BOLT

BOLT

NUT

BOLT

WASHER

NUT

BOLT

SPACER

SPRING

FRONT OF CAR

SHOCK ABSORBER

LOWER CONTROL ARM

SPACER

NUT

BOLT

## Stabilizer Bar and/or Bushing Removal and Installation

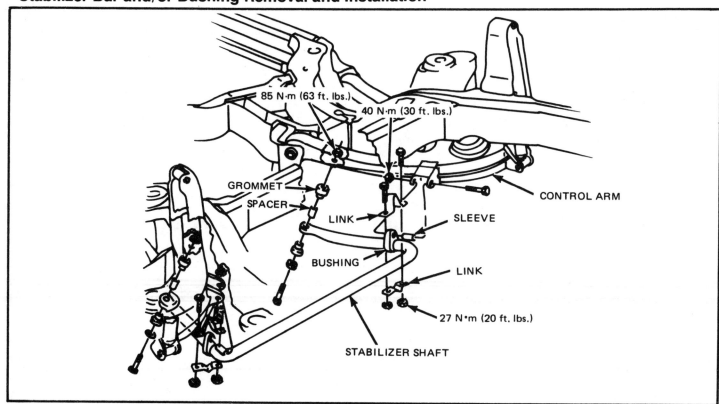

85 N·m (63 ft. lbs.)

40 N·m (30 ft. lbs.)

CONTROL ARM

GROMMET

SPACER

LINK

SLEEVE

BUSHING

LINK

27 N·m (20 ft. lbs.)

STABILIZER SHAFT

To remove: hoist car and remove parts as shown. To install: install parts as shown, and lower hoist.

## Rear Control Arm, Spring and/or Bushing Removal and Installation

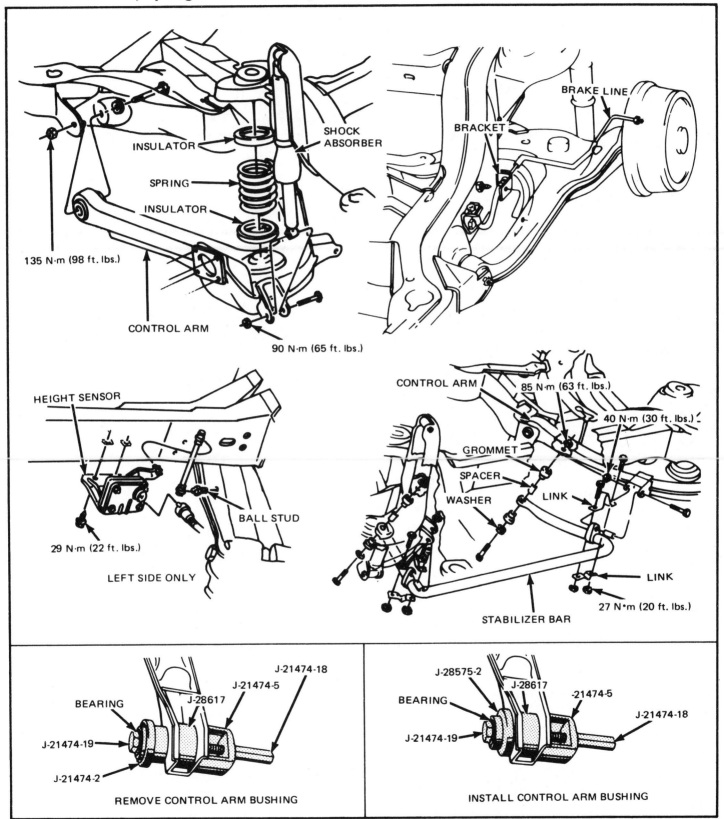

INSULATOR

SHOCK ABSORBER

SPRING

INSULATOR

135 N·m (98 ft. lbs.)

CONTROL ARM

90 N·m (65 ft. lbs.)

BRAKE LINE

BRACKET

HEIGHT SENSOR

29 N·m (22 ft. lbs.)

LEFT SIDE ONLY

BALL STUD

CONTROL ARM

85 N·m (63 ft. lbs.)

40 N·m (30 ft. lbs.)

GROMMET

SPACER

WASHER

LINK

LINK

27 N·m (20 ft. lbs.)

STABILIZER BAR

BEARING

J-21474-19

J-21474-2

J-28617

J-21474-5

J-21474-18

REMOVE CONTROL ARM BUSHING

J-28575-2

J-28617

BEARING

J-21474-19

-21474-5

J-21474-18

INSTALL CONTROL ARM BUSHING

**To remove: hoist car and remove wheel. Remove hub and bearing assembly. Remove parts as shown. To install: install parts as shown. Install hub and bearing assembly. Bleed brakes. Install wheel and lower hoist.**

# SUSPENSIONS

## Hub and Bearing Assembly Removal and Installation

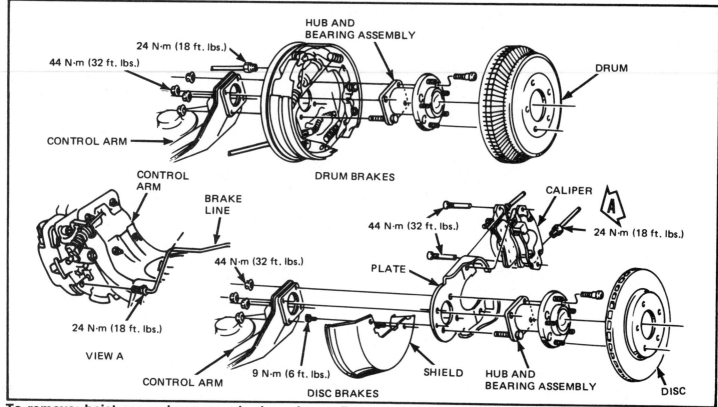

To remove: hoist car and remove wheel as shown. Remove parts as shown. To install: be sure that bearing surfaces are clean and free of burrs. Install parts as shown. If equipped with disc brakes, bleed brakes. Install wheel and lower hoist.

## TORQUE SPECIFICATIONS

|  | N·m | Ft.Lbs. |
|---|---|---|
| **Stabilizer** | | |
| Stabilizer Link Nut | 18 | 13 |
| Stabilizer Bar Brkt. to Frame Bolts & Nuts | 33 | 24 |
| **Shock Absorber** | | |
| Shock Absorber Upper Attaching Nut | 130 | 95 |
| Shock Absorber to Control Arm Bolts | 100 | 75 |
| **Control Arms** | | |
| Upper Control Arm to Frame Attaching Nuts | 95 | 70 |
| Lower Control Arm to Frame Attaching Nuts | 120 | 90 |
| **Ball Joints** | | |
| Service Ball Joints to Upper Control Arm | 11 | 8 |
| Lower | 90 | 65 |
| Upper | 120 | 90 |
| **Front Wheel Drive** | | |
| Drive Axle Nut | 240 | 175 |
| Hub and Bearing to Knuckle Bolts | 100 | 75 |
| Torsion Bar Crossmember Retainer Bolts | 27 | 20 |
| Drive Axle to Output Shaft Bolts | 80 | 60 |
| Tie Rod to Knuckle Nut | 54 | 40 |

## REAR SUSPENSION DIAGNOSIS

| PROBLEM | CAUSE | CORRECTION |
|---|---|---|
| Toe not adjustable within specifications | Lower control arm bent<br>Frame bent<br>Car not properly centered on machine | Replace control arm<br>Bring frame within specification<br>Center car on machine |
| Camber out of specification | Control arm bent<br>Frame bent<br>Hub and bearing assembly not seated on mounting surface | Replace control arm<br>Bring frame within specifications<br>Properly mount Hub and bearing assembly |

## REAR WHEEL ALIGNMENT DIAGNOSIS

| PROBLEM | CAUSE | CORRECTION |
|---|---|---|
| Toe not adjustable within specifications | Lower control arm bent<br>Frame bent<br>Car not properly centered on alignment machine<br>Bearing mounting flange bent<br>Wheel bent | Replace control arm<br>Bring frame within specification<br>Center car on alignment machine<br>Replace bearing assembly<br>Replace wheel |
| Camber out of specification | Control arm bent<br>Frame bent<br>Spindle-bearing | Replace control arm<br>Bring frame within specifications<br>Properly mount |

## REAR WHEEL ALIGNMENT

| | Specifications for Diagnosis for Warranty Repair or Customer Paid Service | Specifications for Periodic Motor Vehicle Inspection | Specification for Alignment Resetting |
|---|---|---|---|
| Camber (measure only) | −1.3° to +0.3° | −1.5° to +0.5° | Refer to Rear Suspension Diagnosis |
| Toe-in per wheel | 0.00° to +0.30°<br>(0″ to +⁵/₃₂″) | −0.25° to +0.55°<br>(−⅛″ to +⁹/₃₂″) | +0.15° ±0.15°<br>(+³/₃₂″ ±³/₃₂″) |
| Toe-in both wheels | 0.00° to +0.60°<br>(0″ to +⁵/₁₆″) | −0.50° to +1.10°<br>(−¼″ to +⁹/₁₆″) | +0.30° ±0.30°<br>(+⁵/₃₂″ ±⁵/₃₂″) |

NOTE: It is important that toe-in be measured per wheel. If equipment is not available to measure each wheel, measure toe-in both wheels. When resetting be sure that toe-in on each wheel is the same.

CAUTION: With car backed onto alignment machine, toe-in and toe-out are reversed. Toe-in will be read as toe-out. It is very important that the readings be made and understood properly.

# FRONT SUSPENSION—GENERAL MOTORS J BODY CARS

## Description

The front suspension is a MacPherson strut design. The lower control arms pivot from the lower side rails. Rubber bushings are used for the lower control arm pivots. The upper end of the strut is isolated by a rubber mount which contains a non-serviceable bearing for wheel turning. The tie-rods connect to the steering arm on the strut, just below the spring seat. The lower end of the wheel steering knuckle pivots on a ball stud for wheel turning. The ball stud is riveted in the lower control arm and is fastened to the steering knuckle with a castellated nut and cotter pin.

## Wheel Alignment

### TOE

Toe is controlled by the tie-rod position. To adjust toe setting, loosen the clamp bolts at the outer end of the tie rod. Rotate adjuster to align toe to specifications. Tighten bolts to 20 N·m (15 ft.-lbs.).

**Adjustment**

1. Loosen clamp bolts at outer tie-rod.

# SUSPENSIONS

## Front Suspension

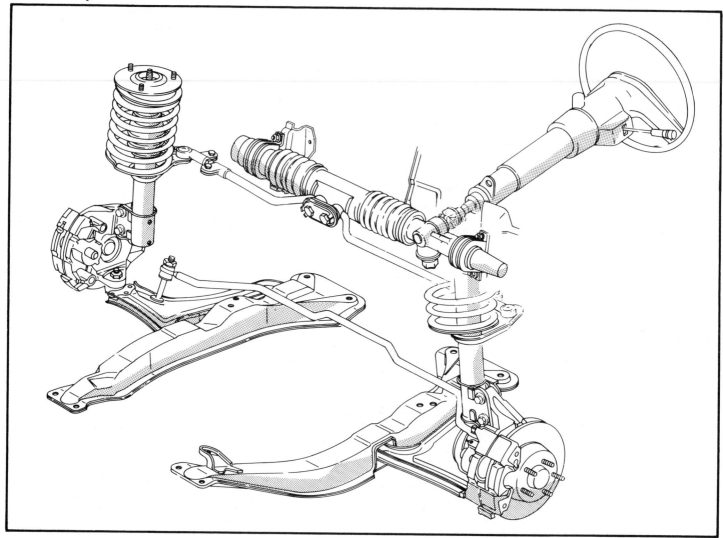

## Toe Adjustment

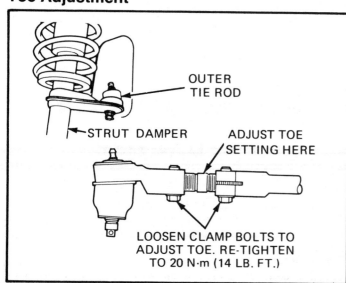

OUTER TIE ROD

STRUT DAMPER

ADJUST TOE SETTING HERE

LOOSEN CLAMP BOLTS TO ADJUST TOE. RE-TIGHTEN TO 20 N·m (14 LB. FT.)

2. Square the vehicle.
3. Rotate adjuster to set toe to specifications.
4. Tighten clamp bolts.

## CAMBER

In special circumstances when camber adjustment becomes necessary, refer to the following procedure for instructions on modifying the front suspension strut assembly.

### Adjustment

1. Position the car on the alignment equipment. Follow the manufacturer's instructions to obtain the camber reading.
2. Use appropriate extensions to reach around both sides of the tire. Loosen both strut-to-knuckle bolts just enough to allow movement between the strut and the knuckle. Remove tools.
3. Grasp the top of the tire firmly, and move the tire inboard or outboard until the correct camber reading is obtained.
4. Carefully reach around the tire with appropriate extensions and tighten both bolts enough to hold the correct camber while the wheel and tire is removed to allow final torque.
5. With wheel and tire removed, torque both bolts to specifications.
6. Reinstall wheel and tire. Tighten nuts to specifications.

## MacPherson Strut

### Removal

1. Raise hood and disconnect upper strut-to-body nuts.
2. Hoist car, allowing front suspension to hang free.
3. Remove wheel and tire.
4. Install drive axle cover.
5. Disconnect tie-rod from strut.
6. Remove both strut-to-knuckle bolts.
7. Remove strut.

### Installation

1. Install strut by reversing removal steps 1-6.
2. Place flats on both mounting bolts in a horizontal position.
3. Torque all fasteners to specifications.

### STRUT MODIFICATION (ONLY FOR ADJUSTMENT OF CAMBER SETTING)

1. Place strut in vise. (It is not necessary to remove the strut from the car. Filling can be accomplished by disconnecting strut from knuckle.)
2. File the holes in outer flanges to enlarge the bottom holes until they match the slots in the inner flanges.

## Strut Damper Removal And Installation, Showing Modification Procedure For Camber Adjustment

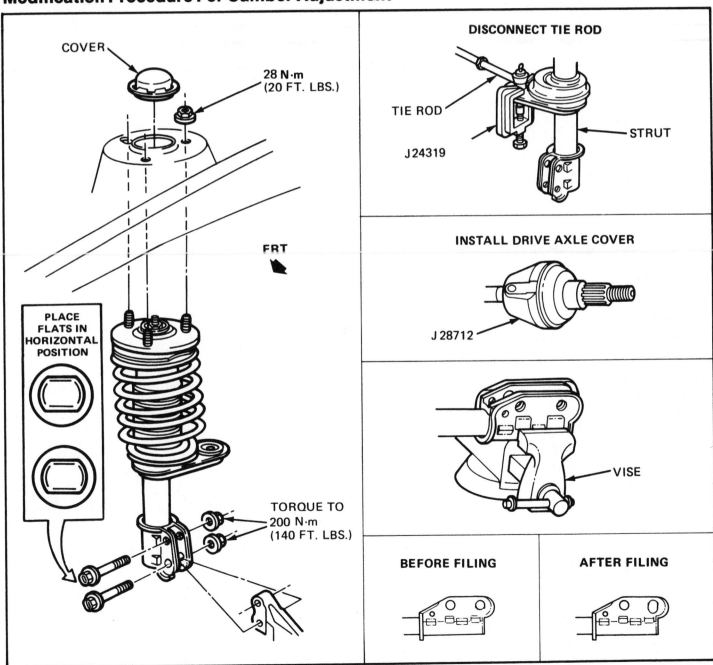

233

## Disassembly and Reassembly of Strut Assembly

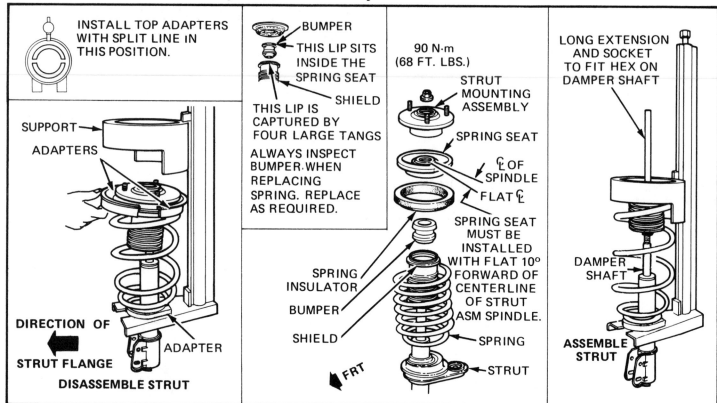

INSTALL TOP ADAPTERS WITH SPLIT LINE IN THIS POSITION.

SUPPORT
ADAPTERS

DIRECTION OF
STRUT FLANGE
ADAPTER
DISASSEMBLE STRUT

BUMPER
THIS LIP SITS INSIDE THE SPRING SEAT
SHIELD
THIS LIP IS CAPTURED BY FOUR LARGE TANGS
ALWAYS INSPECT BUMPER WHEN REPLACING SPRING. REPLACE AS REQUIRED.

SPRING INSULATOR
BUMPER
SHIELD

90 N·m
(68 FT. LBS.)
STRUT MOUNTING ASSEMBLY
SPRING SEAT
C OF SPINDLE
FLAT C
SPRING SEAT MUST BE INSTALLED WITH FLAT 10° FORWARD OF CENTERLINE OF STRUT ASM SPINDLE.
SPRING
STRUT
FRT

LONG EXTENSION AND SOCKET TO FIT HEX ON DAMPER SHAFT
DAMPER SHAFT
ASSEMBLE STRUT

### STRUT DISASSEMBLY

1. Mount strut compressor in vise.
2. Place strut assembly in bottom adapter of compressor and install J 26584-86 (make sure adapter captures strut and that locating pins are engaged).
3. Rotate strut assembly to align top mounting assembly lip with strut compressor support notch.
4. Insert *both* J 26584-88 top adapters *between* the top mounting assembly and the top spring seat. Position top adapters so that the split line is in the 9 o'clock-3 o'clock position.
5. Using a ratchet with 1-inch socket, turn compressor forcing screw clockwise until top support flange contacts the J 26584-88 top adapters. Continue turning the screw, compressing the strut spring approximately 1/2 inch (four complete turns).

**NOTE: Never bottom spring or strut damper rod.**

6. The top nut can now be removed from the strut damper shaft and the top mounting assembly (containing bearing) can be lifted off the strut assembly.
7. Turn strut compressor forcing screw counterclockwise until the strut spring tension is relieved. Remove top adapters, bottom adapter, then remove components.

### STRUT ASSEMBLY

----- **CAUTION** -----

*Never place a hard tool such as pliers or a screwdriver against the polished surface of the damper shaft. The shaft can be held up from the top end with your fingers, or with the extension, to prevent it from receding into the strut assembly, while the spring is being compressed.*

1. Clamp strut compressor body J 26584 in vise.
2. Place strut assembly in bottom adapter of compressor and install J 26584-86 (make sure adapter captures strut and locating pins are engaged).
3. Rotate strut assembly until mounting flange is facing out, directly opposite the compressor forcing screw.

4. Position spring on strut making sure spring is properly seated on bottom spring plate.
5. Install strut spring seat assembly on top of spring.
6. Place *both* J 26584-88 top adapters over spring seat assembly.
7. Turn compressor forcing screw until compressor top support just contacts top adapters (do not compress spring at this time).
8. Install a long extension with a socket to fit the hex on the damper shaft through the top spring seat. Use the extension to guide the components during reassembly.
9. Compress spring by turning screw clockwise until approximately 1½ inch of damper shaft extends through the top spring plate.

**NOTE: Do not compress spring until it bottoms.**

10. Remove extension and socket, position top mounting assembly over damper shaft and install nut.
11. Turn forcing screw counterclockwise to back off support, remove top adapters and bottom adapter, and remove strut assembly from compressor.

**NOTE: Special tool J-26584 must be used to disassemble and assemble strut damper, or damage could occur.**

### REPLACE STRUT CARTRIDGE

The internal piston rod, cylinder assembly, and fluid can be replaced utilizing a service cartridge and nut. Internal threads are located immediately below a cut line groove.

1. Mount strut in vise. Do not overclamp! Excessive clamping may damage tube and/or bracket.
2. Locate cut line groove. It is important to locate groove as accurately as possible because mislocation will result in thread damage. Cut around groove with a pipe cutter until reservoir tube is completely cut through.
3. Remove and discard end cap, cylinder, and piston rod assembly. Remove strut from vise and discard fluid.
4. Reclamp strut in vise. A flaring cup tool is provided in service package to flare and deburr cut edge of reservoir tube to accept service

nut. Place flaring cup on open end of reservoir tube. Strike flaring cup with a mallet or hammer until flaring cup's flat outer surface rests on reservoir tube. Remove the flaring cup tool and discard. At this time, try nut to assure positive start and smooth threading into reservoir tube threads. Remove nut after this check. Flaring cup must be placed in contact with tube so there is not gap between cup and tube when struck.

5. Place strut cartridge in reservoir tube. Turn cartridge until it settles into indentations at base of tube so cartridge cannot be easily turned. Place nut over cartridge.

6. Using tool J 29778 for 53 mm hex nut and a torque wrench, tighten to 190-230 N·m (140-170 ft.-lbs.) in upright mounting position. Stroke the piston rod once or twice to check for proper operation.

## Brake Rotor

### Removal

1. Hoist car and remove wheel.
2. Remove allen head mounting bolts.
3. Remove caliper from rotor and suspend with a wire hook from the frame.
4. Remove rotor.

## Strut Cartridge Replacement

## Upper Ball Joint Check

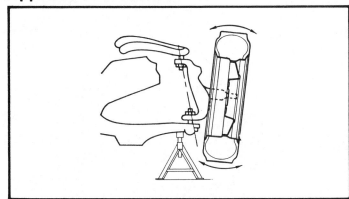

**Support lower control arm as far outboard as possible. Position dial indicator to check movement at point shown. Rock wheel in and out at top and bottom.**

### Installation

1. Install rotor and caliper.
2. Install mounting bolts and torque to 28-47 N·m (21-35 ft.-lbs.).
3. Install wheels and lower car.

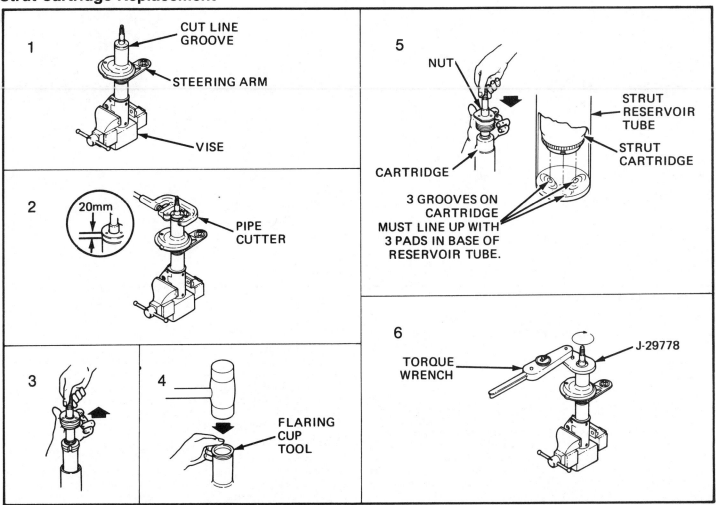

## Caliper Removal and Installation

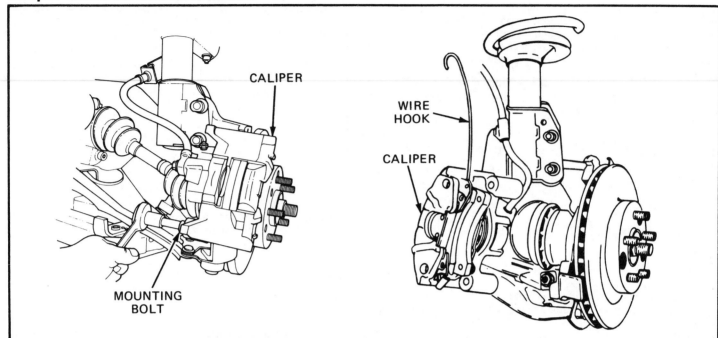

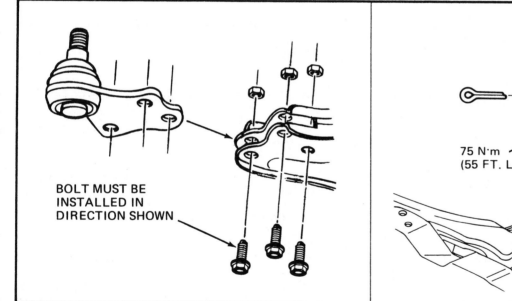

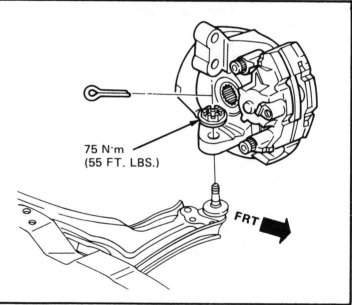

Install ball joint to control arm.

# Ball Joints

## Inspection

Car must be supported by the wheels so weight of car will properly load the ball joints.

The lower ball joint is inspected for wear by visual observation alone. Wear is indicated by the protrusion of the 12.7 mm (1/2 in.) diameter nipple into which the grease fitting is threaded. This round nipple projects 1.27 mm (.050 in.) beyond the surface of the ball joint cover on a new, unworn joint. Normal wear will result in the surface of this nipple retreating very slowly inward.

## Removal

1. Raise car and remove wheel and tire.
2. If no countersink is found on the lower side of the rivets, carefully locate the center of the rivet body and mark with a punch.
3. Use the proper sequence to drill out rivets.
4. Use tool J 29330 to separate joint from knuckle.
5. Disconnect stabilizer from control arm.
6. Remove ball joint.

## Installation

1. Install new ball joint using three bolts.
2. Reverse removal steps 1-5 to install. Tighten all fasteners to specifications.
3. Check toe setting. Adjust as required.

## Lower Control Arm Support and Bushings Removal and Installation

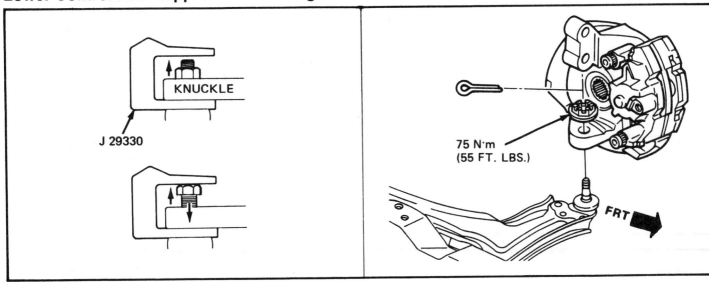

Place J29330 into position as shown. Loosen nut and back off until the nut contacts the tool. Continue backing off the nut until the nut forces the ball stud out of the knuckle.

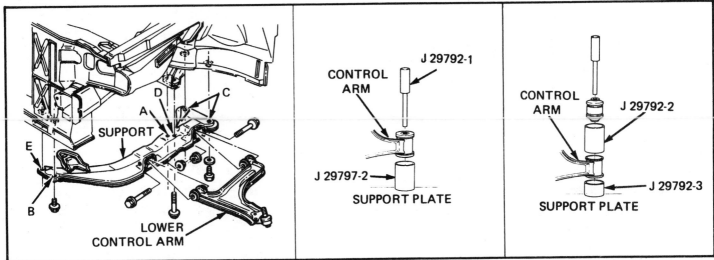

Front suspension support assembly attaching bolt/screw sequence: loosely install center bolt into hole (A). Loosely install tie bar bolt into outboard hole (B). Install both rear bolts into holes (C). Torque rear bolts. Install bolt into center hole (D), then torque. Torque bolt in hole (A). Install bolt into front hole (D), then torque. Torque bolt in hole (A). Install bolt into front hole (E), then torque. Torque bolt in hole (B). Torque support-to-body bolts 90 N·m (63 ft. lbs.) and LCA pivot bolts 95 N·m (67 ft. lbs.).

To remove, insert J29792-1 into bushing, support control arm on J29792-2. Press as shown.

To install, support control arm on J29792-3. Place bushing into J29792-2 and press bushing into control arm using J29792-1. Lubricate bushing.

# SUSPENSIONS

## Control Arm

### Removal

1. Raise car and remove wheel and tire.
2. Disconnect stabilizer bar from control arm and/or support.
3. Separate knuckle from ball joint using tool J 29330.
4. Remove control arm/support.

### Installation

1. Install control arm/support.
2. When installing support, install the center bolts first.
3. Install ball joint to knuckle.
4. Install wheel and tire.
5. Lower car. Check toe setting, adjust as required.

## Hub and Bearing

### Removal

**NOTE:** The car must not be moved while the driveshaft is out of the hub-and-bearing, nor until the hub nut is installed to final torque.

1. Loosen hub nut.
2. Raise car. Remove wheel and tire.
3. Install boot cover J-28712.
4. Remove hub nut.
5. Remove caliper and rotor.
6. Remove hub-and-bearing mounting bolts. Remove shield. If bearing assembly is to be re-used, mark attaching bolt and corresponding hole for installation in the same position.

## Hub and Bearing Assembly Removal and Installation

Install drive axle cover

Using long bolt for installing hub nut

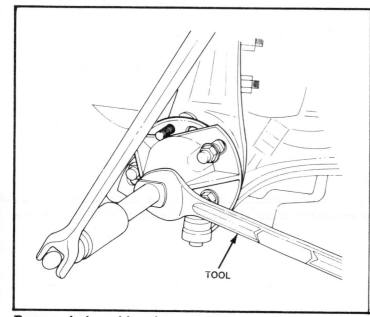

Remove hub and bearing assembly

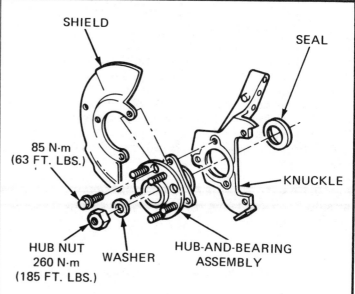

## Removing Axle Stub From Knuckle Assembly

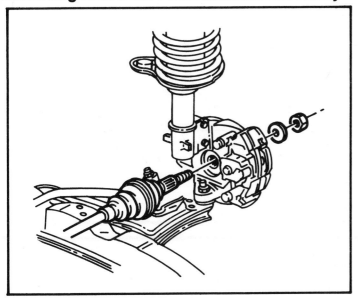

## Strut Bolt Location On Knuckle Assembly

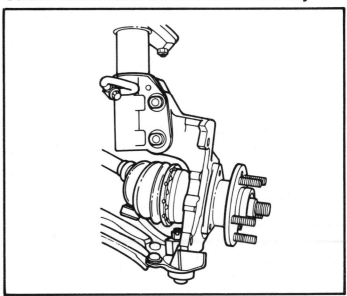

7. Install tool J-28733, and turn bolt to press the hub-and-bearing assembly off of the drive shaft. If excessive corrosion is present, make sure the hub-and-bearing is loose in the knuckle before using J-28733.

8. If installing a new bearing assembly, replace the steering knuckle seal, using tool J-22388.

### Installation

1. Clean and inspect bearing mating surfaces and steering knuckle bore for dirt, knicks and burrs.

2. If installing knuckle seal, apply grease to seal and to bore of knuckle.

3. Replace parts in reverse order of removal. When attaching hub-and-bearing mounting bolts, use one long bolt to extend through cut-out. This will serve as a reaction point to allow enough torque on hub nut to seat axle shaft into bearing. After tightening hub nut to 100 N·m (70 ft.-lbs.), remove long bolt and replace with normal bolt.

4. Lower car. Apply final torque to hub nut, 260 N·m (185 ft.-lbs.).

## Steering Knuckle

### Removal

1. Hoist car. Remove wheel and tire.
2. Remove front wheel hub-and-bearing.
3. Disconnect ball joint from knuckle, using tool J-29330.
4. Remove both strut-to-knuckle mounting bolts. Remove steering knuckle.

### Installation

1. Install both strut-to-knuckle mounting bolts loosely.
2. Install knuckle to ball joint. Torque ball joint nut to 75 N·m (55 ft.-lbs.). Install cotter pin.
3. Tighten mounting bolts to 200 N·m (140 ft.-lbs.).
4. Install remaining components.

# REAR SUSPENSION—GENERAL MOTORS J BODY CARS

## Description

## Shock Insulator

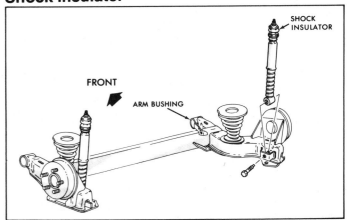

This vehicle has a semi-independent rear suspension which consists of an axle with trailing arms and twisting cross beam, two coil springs, two shock absorbers, two upper spring insulators, and two spring compression bumpers. The axle assembly attaches to the underbody through a rubber bushing located at the front of each control arm. The brackets are integral with the underbody side rails. The axle structure itself maintains the geometrical relationship of the wheels relative to the body. A serviceable stabilizer bar is available as an option. It is attached to the inside of the axle beam and to the lower surface of the control arms as a subassembly of the axle.

The two coil springs support the weight of the car in the rear. Each spring is retained between a seat in the underbody and a seat welded to the top of the control arm. A rubber cushion is used to isolate the coil spring upper end from the underbody seat, while the lower end sits on a combination compression bumper and spring insulator.

# SUSPENSIONS

### Shock Absorber and Stabilizer Bar Attachment

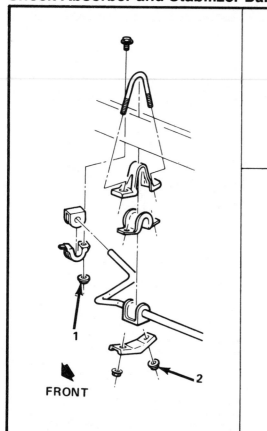

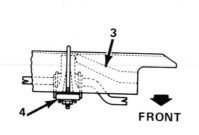

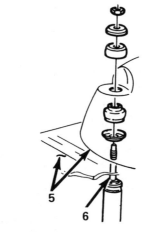

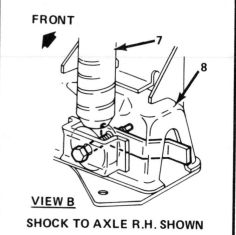

VIEW B

SHOCK TO AXLE R.H. SHOWN

1. 17 N·m 13 ft. lb.
2. 14 N·m 10 ft. lb.
3. Brace
4. Spacer must contact brace when spacer and parts are installed.
5. Underbody
6. Shock stud
7. Shock
8. Axle assembly

The double-acting rear shock absorbers are filled with a calibrated amount of fluid, and sealed during production. They are non-adjustable, non-refillable, and cannot be disassembled. The only service the shock absorbers require is replacement is they have lose their resistance, are damaged, or are leaking fluid.

The lower ends of the shock absorbers are attached to the axle assembly, with bolts and paddle nuts. The upper ends are attached to the body in the wheelhouse area with conventional insulators, washers and nuts.

A single unit hub-and-bearing assembly is bolted to both ends of the rear axle assembly. This hub-and-bearing assembly is a sealed unit. The bearing is not replaceable as a separate unit.

## Shock Absorber

### Removal

--- CAUTION ---
*Do not remove both shock absorbers at one time as suspending rear axle at full length could result in damage to brake lines and hoses.*

1. Open deck lid, remove trim cover and remove upper shock attaching nut. Remove one shock at a time when both shocks are being replaced.
2. Raise vehicle on hoist and support rear axle assembly. When lifting vehicle with body hoist it will be necessary to support rear axle with adjustable jack stands. When lifting vehicle with suspension hoist care should be taken to align axle on the hoist prior to lifting.
3. Remove lower attaching bolt and nut and remove shock.

### Installation

1. Install shock absorbers at lower attachment, feed bolt through holes and loosely install paddle nut.

2. Lower vehicle enough to guide upper stud through body opening and install nut loosely.
3. Torque lower bolt to 55 N·m (41 ft.-lbs.).
4. Remove axle support and lower car all the way and torque upper nut. Torque to 17 N·m (13 ft.-lbs.).
5. Replace rear trim cover.

## Stabilizer Bar

### Removal

1. Raise vehicle on hoist and support body with jack stands.
2. Remove nuts and bolts at both the axle and control arm attachments and remove bracket, insulator and stabilizer bar.

### Installation

1. Install U-bolts, upper clamp, spacer and insulator and trailing axle. Position stabilizer bar in insulators and loosely install lower clamp and nuts.
2. Attach the end of stabilizer bar to control arms and torque all nuts to 17 N·m (13 ft.-lbs.).
3. Torque axle attaching nut to 14 N·m (10 ft.-lbs.).
4. Lower vehicle and remove from hoist.

## Springs and Insulators

### Removal

--- CAUTION ---
*When removing rear springs do not use a twin-post type hoist. The swing arc tendency of the rear axle assembly when certain fasteners are removed may cause it to slip from the hoist. Perform operation on floor if necessary.*

## Coil Spring and Brake Line Bracket Attachment

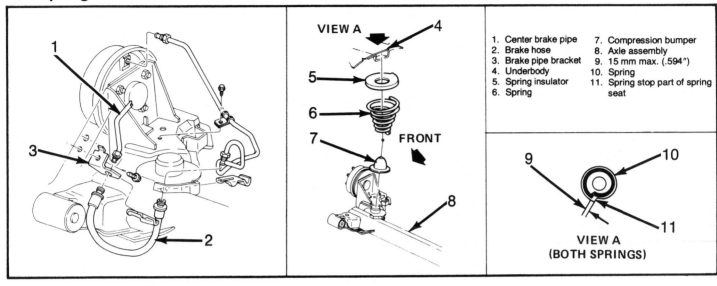

1. Center brake pipe
2. Brake hose
3. Brake pipe bracket
4. Underbody
5. Spring insulator
6. Spring
7. Compression bumper
8. Axle assembly
9. 15 mm max. (.594")
10. Spring
11. Spring stop part of spring seat

VIEW A

FRONT

VIEW A
(BOTH SPRINGS)

1. Raise vehicle using frame contact type hoist if possible and support rear control arms with jack stands. If necessary to lift vehicle with twin post hoist, lift by tires and support the control arms or body with jack stands.
2. Remove wheel and tire assembly.
3. Remove right and left brake line bracket attaching screws from body and allow brake line to hang free.
4. Remove right and left shock absorber lower attaching bolts.

— CAUTION —
*Do not suspend rear axle by brake hoses. Damage to hoses could result.*

5. Lower rear axle and remove spring(s) and/or insulator(s).

### Installation

1. Position springs and insulators in seats and raise axle. The ends of the upper coil on the spring must be positioned in the seat of the body. Prior to installing spring it will be necessary to install upper insulators to the body with adhesive to keep it in position while raising the axle assembly and springs.
2. Connect shocks to rear axle and torque bolts to 55 N·m (41 ft.-lbs.). It will be necessary to bring the axle assembly to standing height prior to torquing bolts on the shocks.
3. Install brake line brackets to body and torque screws to 11 N·m (8 ft.-lbs.).

4. Install wheel and tire assembly. Torque lug nuts to 140 N·m (102 ft.-lbs.).
5. Remove jack stands and lower vehicle.

## Control Arm Bushing

### Removal

1. Raise vehicle on hoist.
2. Remove wheel and tire assembly and support body with jack stands.
3. If removing right bushing, disconnect brake line from body. If left bushing is being removed, disconnect brake line bracket from body, and parking brake cable from hook guide on the body. Replace only one bushing at a time.
4. Remove nut, bolt and washer from the control arm and bracket attachment, and rotate control arm downward.
5. Remove bushing as follows:
   a. Install J 29376-1 on control arm over bushing and tighten attaching nuts until tool is securely in place.
   b. Install J 21474-19 bolt through plate J 29376-7 and install into J 29376-1 receiver.
   c. Place J 29376-6 remover into position on bushing and install nut J 21474-18 onto J 21474-19 bolt.

## Control Arm Bushing Installation

1. Receiver J-29376-1
2. Plate J-29376-7
3. Nut J-21474-18
4. Installer J-29376-4
5. Bushing must be indexed in installer, and installer arrow must align with arrow on receiver for proper bushing installation.

d. Remove bushing from control arm by turning bolt.

### Installation

1. Install bushing on bolt and position onto housing. Align bushing installer J 29376-4 arrow with arrow on receiver for proper indexing of bushing.
2. Install nut J 21474-18 onto bolt J 21474-19.
3. Press bushing into control arm by turning bolt. When bushing is in proper position the end flange will be flush against the face of the control arm.
4. Use a screw type jack stand to position control arm into bracket and install bolt and nut. Do not torque bolt at this time. It will be necessary to torque the bolt of the control arm with vehicle at standing height.
5. Install brake line bracket to frame and torque screw to 11 N·m (8 ft.-lbs.).
6. If left side was disconnected, reconnect brake cables to bracket, and reinstall brake cable to hook. Adjust cable as necessary.
7. While supporting vehicle at standing height, tighten control arm bolt to 90 N·m (67 ft.-lbs.).
8. Remove jack stands and install wheel assembly and lower vehicle from hoist.

## Hub and Bearing

### Removal

1. Raise vehicle on hoist.
2. Remove wheel and tire assembly and brake drum.

---
**CAUTION**
---
*Do not hammer on brake drum as damage to the assembly could result.*

3. Remove hub-and-bearing assembly-to-rear axle attaching bolts and remove hub and bearing assembly from axle. The top rear attaching bolt will not clear the brake shoe when removing the hub-and-bearing assembly. Partially remove hub-and-bearing assembly prior to removing this bolt.

### Installation

1. Position top rear attaching bolt in hub-and-bearing assembly prior to the installation in the axle assembly.
2. Install remaining bolts and nuts. Torque bolts to 52 N·m (39 ft.-lbs.).
3. Install brake drum, and wheel and tire assembly. Torque lug nuts to 140 N·m (103 ft.-lbs.).
4. Lower vehicle and remove from hoist.

### Wheel Stud Removal

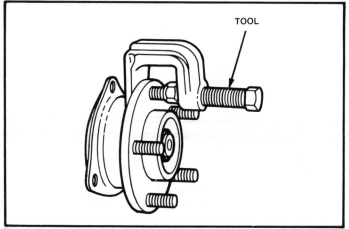

TOOL

**Do not remove stud with hammer, as damage to bearing will result.**

## Hub and Bearing Assembly

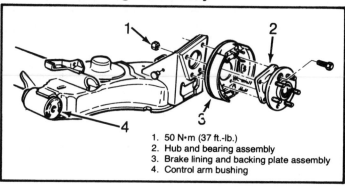

1. 50 N·m (37 ft.-lb.)
2. Hub and bearing assembly
3. Brake lining and backing plate assembly
4. Control arm bushing

## Rear Axle Assembly

### Removal

1. Raise vehicle on hoist and support assembly with jack stands under the control arms.
2. Remove stabilizer bar from axle assembly, if so equipped.
3. Remove wheel and tire assembly and brake drum.

---
**CAUTION**
---
*Do not hammer on brake drum as damage to the bearing could result.*

4. Remove shock absorber lower attaching bolts and paddle nuts at axle and disconnect shocks from control arm.
5. Disconnect parking brake cable from the axle assembly.
6. To insure that axle assembly is not suspended by brake lines, disconnect brake line at the brackets from axle assembly.
7. Lower rear axle and remove coil springs and insulators.
8. Remove control arm bolts from underbody bracket and lower axle.
9. Remove hub attaching bolts and remove hub, bearing and packing plate assembly. Be careful not to drop hub and bearing assembly, as damage to the bearing could result.

### Installation

1. Install backing plate, and hub-and-bearing assembly to rear axle. Hold nuts and torque bolts to 52 N·m (39 ft.-lbs.).
2. Install stabilizer bar, if so equipped, by attaching nut and bolts to axle assembly and at the end to the control arms.
3. Place axle assembly on transmission jack and raise into position. Attach control arms to underbody bracket with bolts and nuts. Do not torque bolts at this time. It will be necessary to torque the bolt of the control arm at standing height.
4. Install brake line connections to axle assembly.
5. Attach brake cable to rear axle assembly.
6. Position coil springs and insulators in seats and raise rear axle. The end of upper coil on the springs must be parallel to axle assembly and seated in pocket.
7. Install shock absorber lower attachment bolts and paddle nuts to rear axle, torque bolt to 55 N·m (41 ft.-lbs.).
8. Connect parking brake cable to guide hook and adjust as necessary.
9. Install brake drums, wheels and tire assembly. Torque lug nuts to 140 N·m (103 ft.-lbs.).
10. Bleed brake system and refill reservoir.

# FRONT SUSPENSION—GENERAL MOTORS REAR DRIVE EXCEPT H AND T BODY CARS

## Wheel Alignment

Front wheel alignment factors are caster, camber, toe-in, toe-out, and trim height. Before any corrections are made, the car must be on a level surface with a full gas tank and the front seat to the rear. All doors must be closed with no passengers or excess weight in the car.

### CASTER AND CAMBER ADJUSTMENTS

To adjust caster and camber, loosen the upper control arm shaft-to-frame nuts, add or subtract shims as required, and retorque nuts.

A normal shim pack will leave at least two threads of the bolt exposed beyond the nut. The difference between front and rear shim packs must not exceed .40 inches.

If these requirements cannot be met in order to reach specifications, check for damaged control arms and related parts. Always tighten the nut on the thinner shim pack first, for improved shaft-to-frame clamping force and torque retention.

### TOE-IN ADJUSTMENT

Toe-in can be increased or decreased by changing the length of the tie-rods. A threaded sleeve is provided for this purpose.

When the tie-rods are mounted ahead of the steering knuckle, they must be decreased in length in order to increase toe-in.

Loosen the clamp bolts at each end of the steering tie-rod adjustable sleeves. With steering wheel set in straight ahead position, turn tie-rod adjusting sleeves to obtain proper toe-in adjustment.

**NOTE: Before locking clamp bolts on the rods, make sure that the tie-rod ends are in alignment with their ball studs by rotating both tie-rod ends in the same direction as far as they will go. Then tighten adjuster tube clamps to specified torque. Make certain that adjuster tubes and clamps are positioned correctly.**

### TOE-OUT

Toe-out on turns refers to the difference in angles between the front wheels and the car frame during turns. Toe-out on turns is non-adjustable.

## Tie Rod Clamp and Sleeve Positioning

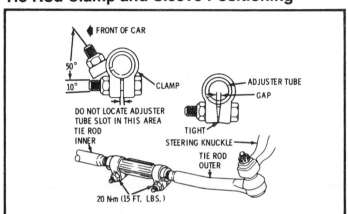

**Bolts must be installed in direction shown. Rotate both inner and outer tie rod housings rearward to the limit of ball joint travel before tightening clamps. With this same rearward rotation all bolt centerlines must be between angles shown after tightening clamps. Clamp ends may touch when nut is torqued to specifications, but gap must be visible adjacent to adjuster sleeve. Clamp must be between and clear of dimples. Torque nuts to specification.**

### TRIM HEIGHT ADJUSTMENT

When checking trim height, the car should be parked on a level surface, full tank of gas, front seat rearward, doors closed and the tire pressure as specified.

If there is more than 24 mm (1 inch) difference side to side at the wheel well opening, corrective measures should be taken to make the car level.

1. Check tire sizes.
2. Check tire wear.

## Caster and Camber Adjustment

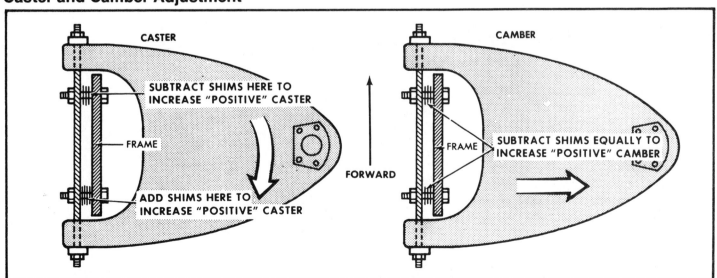

**Pivot shaft inboard of frame**

# SUSPENSIONS

SUSPENSION—GENERAL MOTORS REAR DRIVE, EXCEPT "H" AND "T" BODY CARS

## Front Suspension

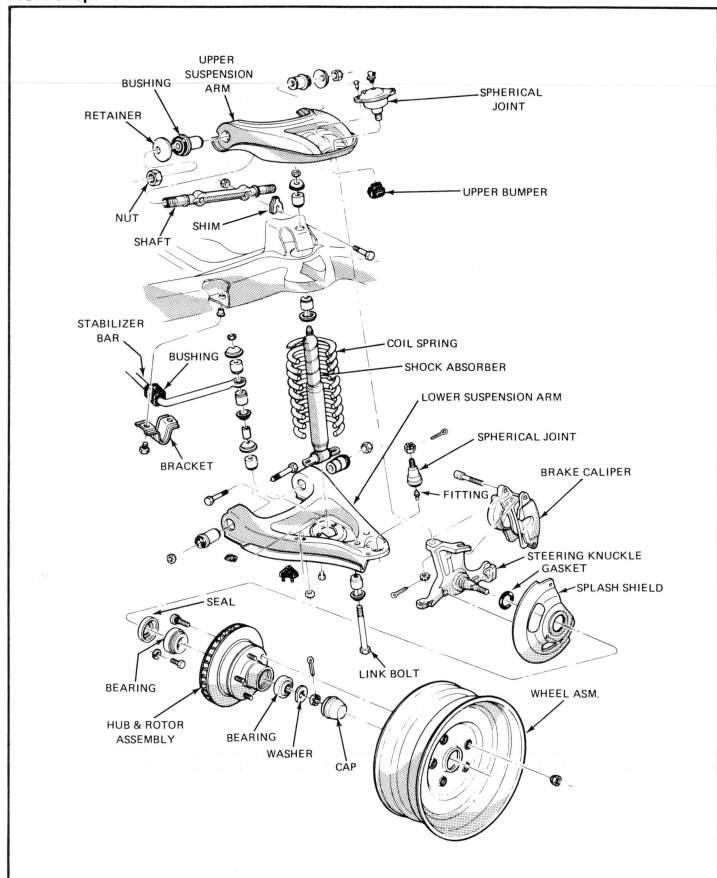

BUSHING

UPPER SUSPENSION ARM

RETAINER

SPHERICAL JOINT

NUT

UPPER BUMPER

SHIM

SHAFT

STABILIZER BAR

BUSHING

COIL SPRING

SHOCK ABSORBER

LOWER SUSPENSION ARM

SPHERICAL JOINT

BRACKET

BRAKE CALIPER

FITTING

STEERING KNUCKLE

GASKET

SPLASH SHIELD

SEAL

LINK BOLT

BEARING

WHEEL ASM.

HUB & ROTOR ASSEMBLY

BEARING

WASHER

CAP

3. Check coil spring height.

4. Check for worn suspension parts.

# Wheel Bearings

For normal use, clean and repack front wheel bearings with a high melting point wheel bearing lubricant at each front brake lining replacement or 30,000 miles (48,000 km), whichever comes first. For heavy duty application such as police cars and taxi cabs, clean and repack front wheel bearings at each front brake lining replacement or 15,000 miles (24,000 km) whichever comes first.

"Long fiber" or "viscous" type lubricants should not be used. Do not mix wheel bearing lubricants. Be sure to thoroughly clean bearings and hubs of all old lubricant before repacking.

## Wheel Bearing Adjustment

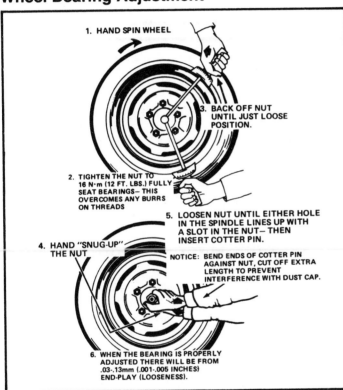

**NOTE: Tapered roller bearings used in these cars have a slightly loose feel when properly adjusted. They must never be over-tightened (pre-loaded) or severe bearing damage may result.**

## Adjustment

The proper functioning of the front suspension cannot be maintained unless the front wheel taper roller bearings are correctly adjusted. Cones must be a slip fit on the spindle and the inside diameter of cones should be lubricated to insure that the cones will creep. Spindle nut must be a free-running fit on threads.

1. Remove dust cap from hub.

2. Remove cotter pin from spindle and spindle nut.

3. Tighten the spindle nut to 16 N·m (12 ft.-lbs.) while turning the wheel assembly forward by hand to fully seat the bearings. This will remove any grease or burrs which could cause excessive wheel bearing play later.

4. Back off the nut to the "just loose" position.

5. Hand-tighten the spindle nut. Loosen spindle nut until either hole in the spindle lines up with a slot in the nut (not more than 1/2 flat).

6. Install new cotter pin. Bend the ends of the cotter pin against nut. Cut off extra length to ensure ends will not interfere with the dust cap.

7. Measure the looseness in the hub assembly. There will be from .03 to .13 mm (.001 to .005 inches) end-play when properly adjusted.

8. Install dust cap on hub.

# Shock Absorbers

### Removal

1. Raise car on hoist, and with an open end wrench hold the shock absorber upper stem from turning. Remove the upper stem retaining nut, retainer and rubber grommet.

## Shock Absorber Attachment

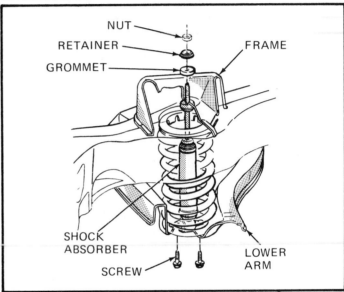

**Upper nut torque is 19 N·m (14 ft. lbs.). Lower studs torque is 30 N·m (22 ft. lbs.).**

2. Remove the two bolts retaining the lower shock absorber pivot to the lower control arm and pull the shock absorber assembly out from the bottom.

### Installation

1. With the lower retainer and rubber grommet in place over the upper stem, install the shock absorber (fully extended) up through the lower control arm and spring so that the upper stem passes through the mounting hole in the upper control arm frame bracket.

2. Install the upper rubber grommet, retainer and attaching nut over the shock absorber upper stem.

3. With an open end wrench, hold the upper stem from turning and tighten the retaining nut.

4. Install the retainers attaching the shock absorber lower pivot to the lower control arm, torque and lower car to floor.

# Upper Ball Joint

### Inspection

1. Raise the car and position floor stands under the left and right lower control arm as near as possible to each lower ball joint. Car must be stable and should not rock on the floor stands. Upper control arm bumper must not contact frame.

# SUSPENSIONS

2. Position dial indicator against the wheel rim.

3. Grasp front wheel and push in on bottom of tire while pulling out at the top. Read gauge, then reverse the push-pull procedure. Horizontal deflection on dial indicator should not exceed 3.18 mm (1.25 in.).

4. If dial indicator reading exceeds 3.18 mm (.125 in.), or if ball stud has been disconnected from knuckle assembly and any looseness is detected, or the stud can be twisted in its socket with your fingers, replace ball joint.

## Disconnecting Upper Ball Joint

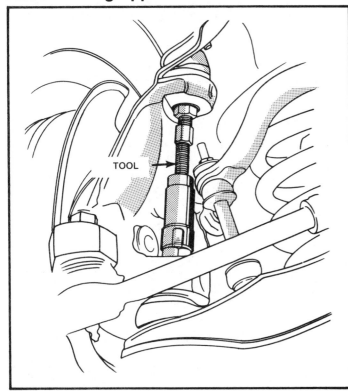

## Upper Ball Joint Removal

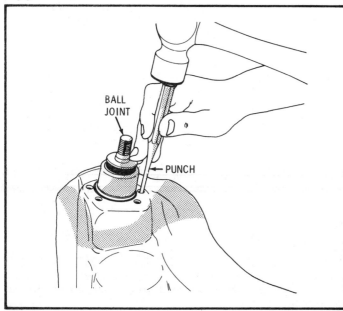

## Removal

1. Raise front of car and support lower control arm with floor stands.

------ CAUTION ------

*Floor jack or stand must remain under control arm spring seat during removal and installation to retain spring and control arm in position.*

Since the weight of the car is used to relieve spring tension on the upper control arm, the floor stands must be positioned between the spring seats and ball joints of the lower control arms for maximum leverage.

2. Remove wheel, then loosen the upper ball joint from the steering knuckle as follows:

   a. Remove upper ball joint nut and install push tool.

   b. Apply pressure on stud by expanding the tool until the stud breaks loose.

3. Remove tool and upper ball joint nut, then pull stud from knuckle. Support the knuckle assembly to prevent weight of the assembly from damaging the brake hose.

4. With control arm in the raised position, drill four rivets 1/4 in. deep using a 1/8 in. diameter drill.

5. Drill off rivet heads using a 1/2 in. diameter drill.

6. Punch out rivets using a small punch, and remove ball joint.

### Installation

1. Position new ball joint in control arm and install four attaching bolts. Torque nuts to 11 N·m (8 ft.-lbs.).

2. Connect ball joint to steering knuckle. Torque nut to 40 N·m (30 ft.-lbs.).

## Lower Ball Joint

### Inspection

Car must be supported by the wheels so weight of car will properly load the ball joints.

The lower ball joint is inspected for wear by visual observation alone. Wear is indicated by protrusion of the 12.7 mm (1/2 in.) diameter nipple into which the grease fitting is threaded. This round nipple projects 1.27 mm (.050 in.) beyond the surface of the ball joint cover on a new, unworn joint. Normal wear will result in the surface of this nipple retreating slowly inward.

To inspect for wear, wipe grease fitting and nipple free of dirt and grease as for a grease job. Observe or scrape a scale, screwdriver or fingernail across the cover. If the round nipple is flush or inside the cover surface, replace the ball joint.

### Removal

1. Raise the car, support with floor stands under frame.

2. Remove tire and wheel assembly.

3. Place floor jack under control arm spring seat.

------ CAUTION ------

*Floor jack must remain under control arm spring seat during removal and installation to retain spring and control arm in position.*

4. To disconnect the lower control arm ball joint from the steering knuckle. Remove the cotter pin from ball joint stud and remove stud nut. Tool J-8806 can be used to break the ball joint loose from knuckle after stud breaks loose.

5. Guide lower control arm out of opening in splash shield with a putty knife or similar tool.

6. Block knuckle assembly out of the way by placing a wooden block between frame and upper control arm.

7. Remove ball joint seal by prying off retainer with a pry bar or driving off with a chisel.

8. Remove grease fittings and install special tool to remove lower ball joint from lower control arm.

## Lower Ball Joint Wear Indicator

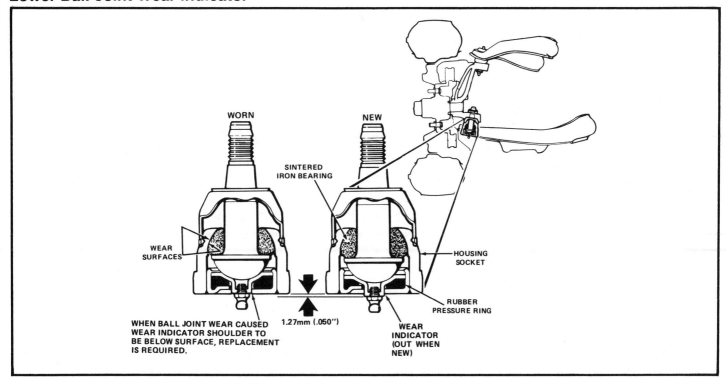

## Guiding Lower Control Arm Past Shield

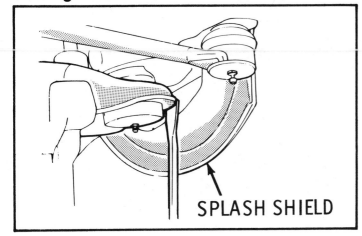

### Installation

1. Position the lower ball joint into the lower control arm and with special press tools, press the ball joint in until it bottoms in the control arm.

**NOTE: The grease purge on the seal must be located facing inboard.**

2. Place the ball joint stud into the bottom hole of the steering knuckle. Force stud into tapered hole to a torque of 40 ft.-lbs. (55 N·m).

3. Install the stud nut and torque to 90 ft.-lbs. (120 N·m).

**NOTE: Some replacement ball joints will use cotter pins. Be sure they are in place when required.**

4. Install the grease fitting and lubricate the ball joint until the grease appears at the seal.

5. Install the wheel and lower the vehicle.

## Disconnecting Lower Ball Joint

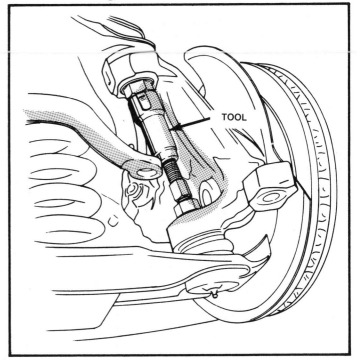

## Coil Spring

### Removal

1. Place transmission in neutral so steering wheel is unlocked.

2. Clean shock upper threads; oil, then remove nut, washer, and grommet.

# SUSPENSIONS

## Lower Ball Joint Removal

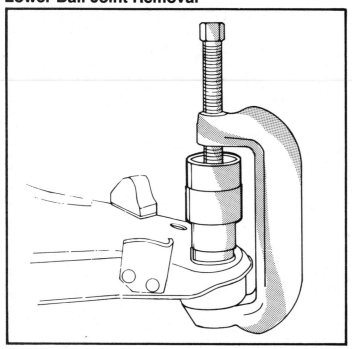

3. Hoist car. Remove wheel and tire.
4. Remove stabilizer link nut, grommets washers, and bolt.
5. Support car with floor stands.
6. Remove shock.

## Compressed Coil Spring

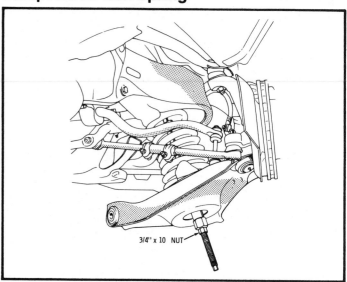

3/4" x 10 NUT

7. Install coil spring tool. Make sure tool is fully seated into lower control arm spring seat.
8. Rotate nut until spring is compressed enough so that it is free in its seat.
9. Remove the two lower control arm pivot bolts and disengage lower control arm from frame. Rotate arm with spring rearward and remove spring from arm.

## Coil Spring Positioning

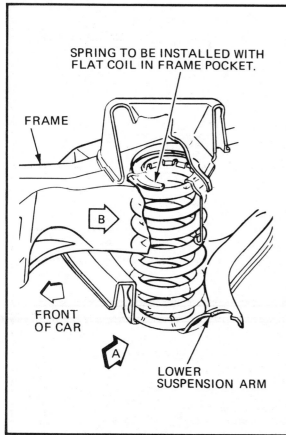

SPRING TO BE INSTALLED WITH FLAT COIL IN FRAME POCKET.

FRAME

B

FRONT OF CAR

A

LOWER SUSPENSION ARM

AFTER ASSEMBLY, END OF SPRING COIL MUST COVER ALL OR PART OF ONE INSPECTION DRAIN HOLE. THE OTHER HOLE MUST BE PARTLY EXPOSED OR COMPLETELY UNCOVERED.

LOWER SUSPENSION ARM

VIEW A

WHEN COMPRESSING A PORTION OF THE SPRING, DO NOT COMPRESS TO GAP BETWEEN ACTIVE COILS OF LESS THAN 6mm (.22 INCHES)—C-CAR

VIEW B

IF ENTIRE SPRING IS COMPRESSED, THE OVERALL DIMENSION MUST NEVER BE LESS THAN 244mm (9.62 INCHES)—C-CAR

### Installation

1. Install spring on bench.
2. Insert compressed spring into place.
3. Twist spring into proper position.
4. Carefully lift lower control arm and attach the lower control arm pivot bolts. Tighten nuts to 120 N·m (90 ft.-lbs.).

## Upper Control Arm

### Removal

1. Raise front of car and support lower control arm with floor stands.

— CAUTION —

*Floor jack must remain under control arm spring seat during removal and installation to retain spring and control arm in position.*

NOTE: Since the weight of the car is used to relieve spring tension on the upper control arm, the floor stands must be positioned between the spring seats and ball joints of the lower control arms for maximum leverage.

2. Remove wheel, then loosen the upper ball joint from the steering knuckle as follows:
   a. Remove the upper ball joint nut.
   b. Apply pressure on stud by expanding the tool until the stud breaks loose.
3. Remove tool and upper ball joint nut, then pull stud free from knuckle. Support the knuckle assembly to prevent weight of the assembly from damaging the brake hose.
4. Remove the upper control arm attaching bolts to allow clearance to remove upper control arm assembly.
5. Remove upper control arm from the car.

**PIVOT SHAFT BUSHING REPLACEMENT**
1. Remove upper control arm assembly from the car.
2. Remove nuts from ends of pivot shaft.
3. Position control arm assembly and tools and push bushing out of control arm.
4. To install bushings, place pivot shaft in control arm and push new bushing into control arm and over end of pivot.

### Installation

1. Position the upper control arm attaching bolts loosely in the frame and install the pivot shaft on the attaching bolts.
2. Install the inner pivot bolt with the heads to the front on the front bushing and to the rear on the rear bushing.
3. Install the alignment shims between the pivot shaft and frame on their respective bolts. Torque the nuts to 73 ft.-lbs. on the B and 1976-81 F body models and 45 ft.-lbs. on the rear drive A body models.
4. Remove the temporary support from the hub assembly and connect the ball joint to the steering knuckle.
5. Install the wheel, check front end alignment and adjust as required.

## Lower Control Arm

### Removal

1. Place transmission in neutral so steering wheel is unlocked.
2. Clean shock upper threads; oil, then remove nut, bolt, washer and grommet.
3. Hoist car. Remove wheel and tire.
4. Remove stabilizer link nut, grommets washers, and bolt.
5. Support car with floor stands and lower hoist. Remove shock.
6. Loosen the lower ball joint nut and use tool J-8806 or equivalent. Apply pressure on stud by expanding the tool until the stud breaks loose.

### Upper Control Arm Pivot Shaft Bushing Installation Clearance

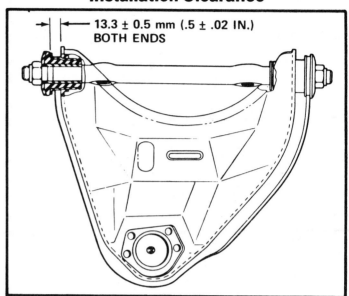

13.3 ± 0.5 mm (.5 ± .02 IN.) BOTH ENDS

### Upper Control Arm Bushing Removal

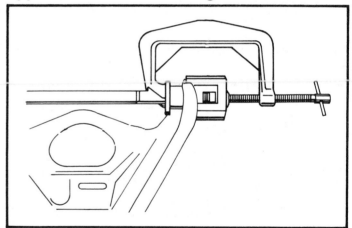

### Upper Control Arm

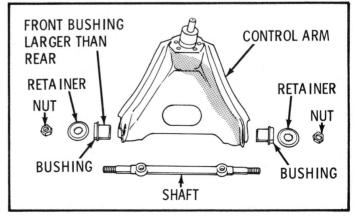

FRONT BUSHING LARGER THAN REAR
RETAINER NUT
BUSHING
CONTROL ARM
RETAINER NUT
BUSHING
SHAFT

## Rear Bushing Removal

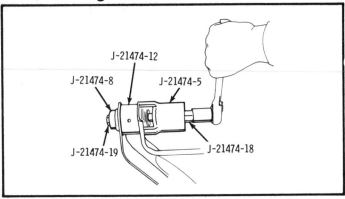

**Press out old bushing. Press in new bushing**

## Bushing Flare Angle Degree Of Between 40 - 45 Degrees

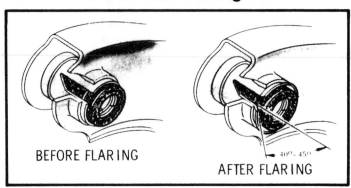

BEFORE FLARING     AFTER FLARING

7. Install spring compressor in through front spring. Compress spring until all tension is off lower control arm.

8. Remove pivot bolts and ball joint.

9. Remove complete control arm.

### Installation

1. Install the ball joint stud into the lower hole of the steering knuckle. Install the nut loosely.

2. Position the spring on the lower arm and safety chain it to the lower arm to prevent personal injury.

3. Position a jack under the rear of the arm and raise the arm rear pivot bushings into position on the crossmember brackets. Install the pivot bolts.

4. Remove the jack and safety chain. Lower the vehicle to the ground and torque the bolts as noted in the front spring removal and installation section.

### LOWER CONTROL ARM BUSHINGS

#### Removal and Installation

The removal and installation of the lower control arm bushings require the use of a press tool. A flare is found on the bushings used except with the rear wheel drive A models. When the bushings are installed and the flare is required, the tool must be capable of flaring the metal flange on the bushing to its proper angle.

## Steering Knuckle

It is recommended that the car be raised and supported so that the front coil spring remains compressed, yet the wheel and steering knuckle assembly remain accessible. If a frame hoist is used, support the lower suspension arm with an adjustable jack stand to retain spring in the curb height position.

### Removal

1. Raise car on hoist and support lower suspension arm.

2. Remove wheel and tire assembly.

3. Remove tie-rod end from steering knuckle.

4. Remove brake caliper and hub and rotor assembly. Use a piece of wire to attach caliper to upper suspension arm.

**NOTE: Never allow caliper to hang from brake hose, as hose may be damaged.**

5. Remove splash shield.

6. Remove upper and lower ball joint studs.

7. Remove studs from steering knuckle.

### Installation

1. Place steering knuckle into position and insert upper and lower ball joint studs into knuckle bosses.

2. Install stud nuts and torque upper and lower nuts to 55 N·m (40 ft.-lbs.).

## Steering Knuckle and Hub Assembly

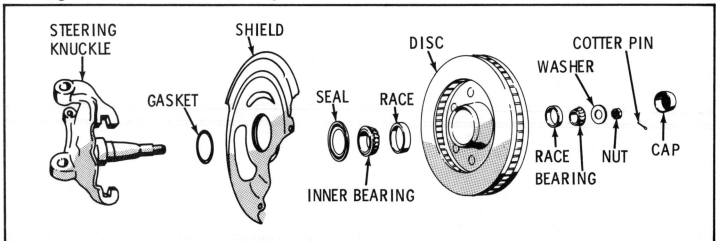

3. Install splash shield. Torque screws to 9.5 N·m (85 in.-lbs.).
4. Install hub and rotor assembly.
5. Install outer bearing, spindle washer and nut. Adjust bearing.
6. Install brake caliper.
7. Install wheel and tire assembly. Tighten nuts to 140 N·m (100 ft.-lbs.).
8. Lower car to floor.
9. Check front wheel alignment.

## Stabilizer Shaft

### Removal

1. Hoist car.
2. Disconnect each side of stabilizer linkage by removing nut from link bolt. Pull bolt from linkage and remove retainers, grommets and spacer.
3. Remove bracket-to-frame or body bolts and remove stabilizer shaft, rubber bushings and brackets. Some models require a special tool to remove stabilizer shaft bolt.

### Installation

To replace, reverse sequence of operations, being sure to install with the identification forming on the right side of the car. The rubber bushings should be positioned squarely in the brackets with the slit in the bushings facing the front of car. Torque stabilizer link nut to 18 N·m (13 ft.-lbs.) and bracket bolts to 33 N·m (24 ft.-lbs.).

## Stabilizer Shaft

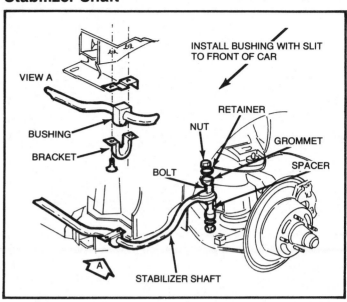

# REAR SUSPENSION—GENERAL MOTORS REAR DRIVE SOLID AXLE EXCEPT H AND T BODY CARS

## Description

### COIL SPRING SYSTEM

The rear axle assembly is attached to the frame through a link type suspension system. Two rubber bushed lower control arms mounted between the axle assembly and the frame maintain fore and aft relationship of the axle assembly to the chassis. Two rubber bushed upper control arms, angularly mounted with respect to the centerline of the car, control driving and braking torque and sideways movement of the axle assembly. The rigid axle holds the rear wheels in proper alignment.

The upper control arms are shorter than the lower arms, causing the differential housing to "rock" or tilt forward on compression. This rocking or tilting lowers the rear propeller shaft to make possible the use of a lower tunnel in the rear floor pan area.

The rear upper control arms control drive forces, side sway and pinion nose angle. Pinion angle adjustment can greatly affect car smoothness and must be maintained as specified.

The rear chassis springs are located between brackets on the axle tube and spring seats in the frame. The springs are held in the seat pilots by the weight of the car and by the shock absorbers which limit axle movement during rebound.

Ride control is provided by two identical direct doublt acting shock absorbers angle-mounted between brackets attached to the axle housing and the rear spring seats.

### LEAF SPRING SYSTEM

The rear axle assembly is attached to multi-leaf springs by "U" bolts. The spring front eyes are attached to the frame at the front hangers, through rubber bushings. The rear ends of the springs are attached to the frame by the use of shackles which allow the spring to "change its length" while the vehicle is in motion. Control arms are not used with leaf springs.

## Shock Absorbers

### Removal

Raise car and support rear axle to prevent stretching of brake hose. The lower end has a stud which is an integral part of the shock. Remove the nut and tap shock free from bracket. To disconnect the shock at the top, on all models, remove the two bolts, nuts and lockwashers.

### Installation

Loosely attach shock at both ends. Tighten upper bolts and nuts to 26 N·m (20 ft.-lbs.). Tighten lower nut to 90 N·m (65 ft.-lbs.).

## Coil Springs

### Removal

1. Hoist rear of car on axle housing and support at frame rails with floor stands. Do not lower hoist at this time.

**NOTE: Do not allow the rear brake hose to become kinked or stretched.**

2. Disconnect brake line at axle housing.
3. Disconnect upper control arms at axle housing.
4. Remove shock at lower mount.
5. Lower hoist at rear axle.
6. Remove spring.

### Installation

1. Install coil spring.
2. Raise hoist at rear axle.
3. Install shock at lower mount.

## Rear Suspension

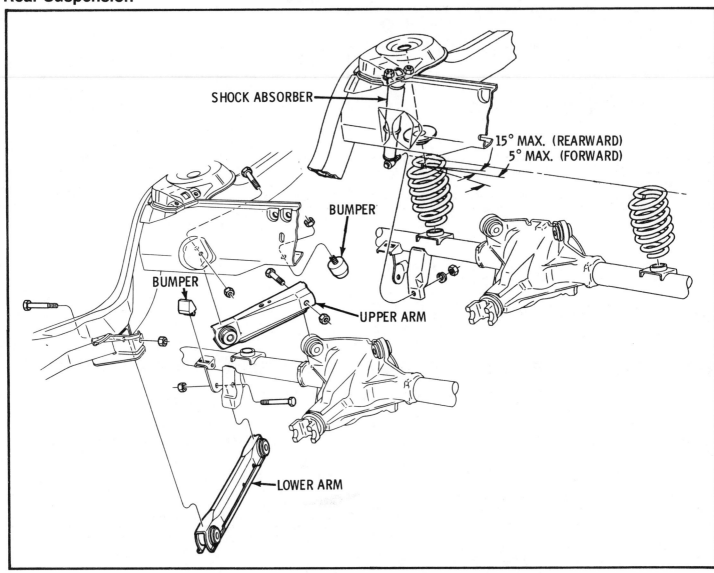

SHOCK ABSORBER

15° MAX. (REARWARD)
5° MAX. (FORWARD)

BUMPER

BUMPER

UPPER ARM

LOWER ARM

## Coil Spring Mounting

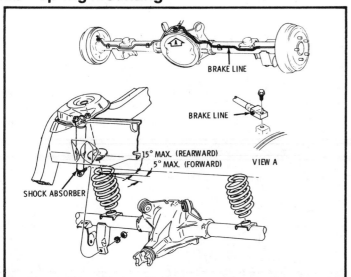

BRAKE LINE

BRAKE LINE

15° MAX. (REARWARD)
5° MAX. (FORWARD)

VIEW A

SHOCK ABSORBER

4. Install upper control arm bolts at axle housing.
5. Connect brake line at axle housing.
6. Remove jack stands and lower car.

# Upper Control Arm

### —CAUTION—
*If both control arms are to be replaced, remove and replace one control arm at a time to prevent the axle from rolling or slipping sideways. This might occur with both upper control arms removed, making replacement difficult.*

### Removal

1.  Remove nut from rear arm to rear axle housing bolt and while rocking rear axle, remove the bolt. On some cars disconnecting lower shock absorber stud will provide clearance. Use support under rear axle nose to aid in bolt removal.
2.  Remove front and rear arm attaching nuts and bolts.
3.  Remove suspension arm and inspect bushings.

## Typical Upper Control Arm

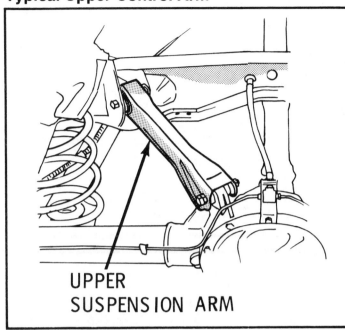

UPPER
SUSPENSION ARM

## Differential Housing Bushings

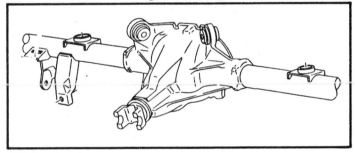

### Installation

To install, reverse removal procedure. Torque nuts with car resting at normal trim height.

## Lower Control Arm

—————— CAUTION ——————

*If both control arms are to be replaced, remove and replace one control arm at a time to prevent the axle from rolling or slipping sideways. This might occur with both lower control arms removed, making replacement difficult.*

### Removal

1. Raise car and support under axle housing.
2. Remove rear arm-to-axle housing bracket bolt.
3. Remove front arm-to-bracket bolts and remove lower control arm.

### Installation

To replace arm, reverse the removal sequence of operations. Torque arm attaching nuts with the weight of the car on the rear springs.

## Bushing Removal (Rear Axle Housing)

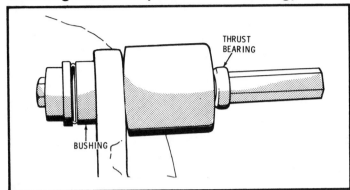

THRUST BEARING

BUSHING

**Press out old bushing. Press in new bushing.**

## Control Arm Bushing Removal

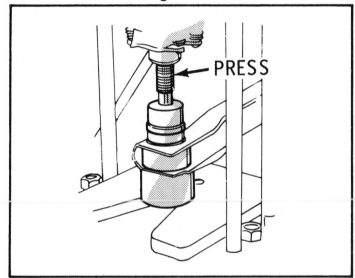

PRESS

## Control Arm Bushing Installation

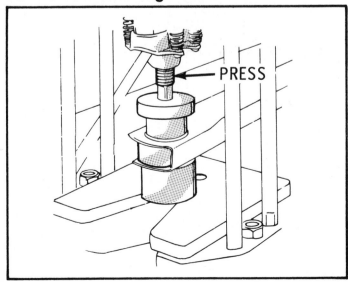

PRESS

## Leaf Spring Rear Suspension

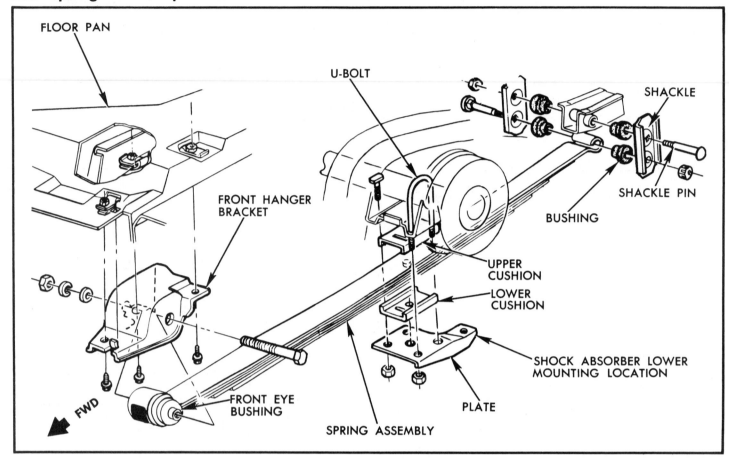

# Leaf Spring

## Removal

1. Raise vehicle on hoist. Support vehicle so axle can be raised and lowered.

2. Raise axle assembly so that all tension is removed from spring.

3. Disconnect shock absorber lower attaching mount.

4. Loosen the spring front eye-to-bracket retaining bolt.

5. Remove the screws securing the spring front bracket to the underbody.

6. Lower axle assembly sufficiently to permit access to spring front bracket and remove bracket from spring.

7. The spring eye bushing can be replaced without completely removing the spring from the vehicle.

8. Pry parking brake cable out of the retainer bracket mounted on the spring mounted plate.

9. Remove lower spring plate-to-axle bracket retaining nuts.

10. Remove upper and lower rubber spring pads and spring plate.

11. Support spring, then remove lower bolt from spring rear shackle. Separate shackle and withdraw spring from vehicle.

12. Remove rear spring shackle upper bolt and withdraw shackle bushings from frame.

## Installation

1. Position spring front mounting bracket to spring front eye.

Spring attaching bolt must be installed so that head of bolt is toward center of vehicle.

2. Position spring shackle upper bushings in frame. Position shackles to bushings and loosely install bolt and nut.

3. Install bushing halves in spring rear eye, place spring in shackles and loosely install shackle lower bolt and nut. When installing spring, make sure spring is positioned so that parking brake cable is on underside of spring.

4. Raise front end of spring and position bracket to underbody. Guide spring into position so that it will index in the axle bracket and also make sure that the tab on spring bracket is indexed in slot provided in the underbody.

5. Loosely install spring-to-underbody bracket.

6. Position spring upper cushion between spring and axle bracket so that spring cushion ribs align with axle bracket locating ribs.

7. Place lower spring cushion on spring so that cushion is indexed on locating dowel. Upper cushion and lower cushion will be aligned if installation is correct.

8. Place lower mounting plate over locating dowel on spring lower pad and loosely install retaining nuts.

9. If new mounting plate was installed, transfer parking brake cable retaining bracket to new plate.

10. Attach shock absorber to spring mounting plate.

11. Position parking brake cable in retaining bracket and securely clamp bracket to retain cable.

12. Tighten all affected parts with vehicle weight on suspension components.

# REAR SUSPENSION—GENERAL MOTORS REAR DRIVE INDEPENDENT SUSPENSION EXCEPT H AND T BODY CARS

## Description

The rear suspension features a transverse spring mounted on a fixed differential carrier. Each rear wheel is mounted by a three-link independent suspension. These three links are made up of wheel drive shaft, a camber control strut rod and a wheel spindle support arm.

## Rear Wheel Alignment

To align the rear suspension, "back" the car onto the machine normally used to align front suspension. Camber will now be read in the normal manner. However, with the vehicle "backed" in, toe-in will now read as toe-out, while toe-out will be read as toe-in.

**NOTE: Check condition of strut rods. They should be straight. Rear wheel alignment could be affected if they are bent.**

### REAR WHEEL ALIGNMENT

| | |
|---|---|
| Camber | 0° ± ½° |
| Toe-in (Per Wheel) | .06° ± .06° |

NOTE Each wheel must be adjusted independently.

### CAMBER

Wheel camber angle is obtained by adjusting the eccentric cam and bolt assembly located at the inboard mounting of the strut rod. Place rear wheels on alignment machine and determine camber angle. To adjust, loosen cam bolt nut and rotate cam and bolt assembly until specified camber is reached. Tighten nut securely and torque to specifications.

### TOE-IN

Wheel toe-in is adjusted by inserting shims of varying thickness inside the frame side member on both sides of the torque control arm pivot bushing. Shims are available in thickness of .40 mm (1/64 in.), 79 mm (1/32 in.), 3.18 mm (1/8 in.) and 6.35 mm (1/4 in.).

To adjust toe-in, loosen torque control arm pivot bolt. Remove cotter pin retaining shims and remove shims. Position torque control

## Type B Axle Bearing

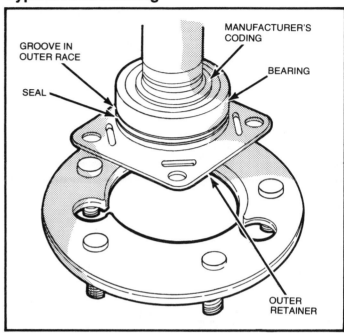

**Axle bearing is held by outer plate.**

arm to obtain specified toe-in. Shim the gap toward vehicle centerline between torque control arm bushing and frame side inner wall. Do not use thicker shim than necessary, and do not use undue force when shimming inner side of torque control arm. To do so may cause toe setting to change.

Shim outboard gap as necessary to obtain solid stackup between torque control arm bushing and inner wall of frame side member. After correct shim stack has been selected, install cotter pin (with loop outboard) through shims. Torque nut to specifications, and install cotter pin. If specified torque does not permit cotter pin insertion, tighten nut to next flat.

## Rear Wheel Camber Adjustment

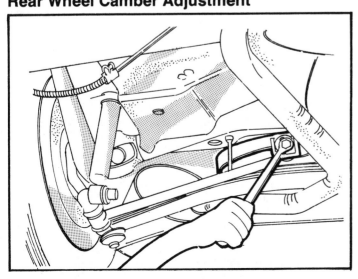

## Toe-in Adjusting Shim Location

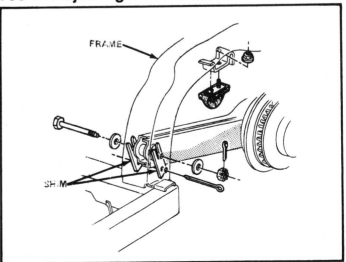

## Spindle Bearing End Play Check

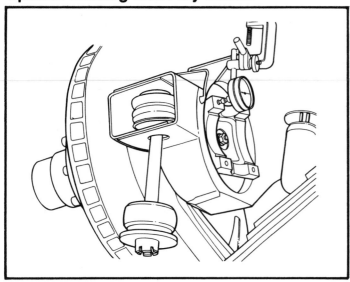

## Drive Spindle Removal

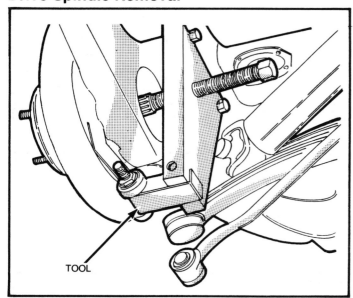

TOOL

## Spindle

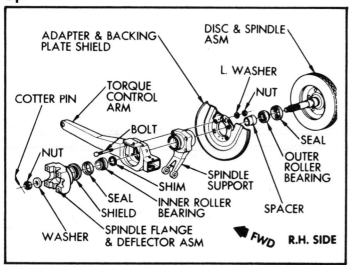

ADAPTER & BACKING PLATE SHIELD

DISC & SPINDLE ASM

TORQUE CONTROL ARM

L. WASHER

COTTER PIN

NUT

BOLT

NUT

SEAL

OUTER ROLLER BEARING

SPINDLE SUPPORT

SEAL SHIELD

SHIM

INNER ROLLER BEARING

SPACER

WASHER

SPINDLE FLANGE & DEFLECTOR ASM

FWD  R.H. SIDE

# Wheel Bearing

## END-PLAY CHECK

The tapered-roller spindle bearings should have end play of .003 to .20 mm (.001 in. to .008 in.). During inspection, check end play and, when necessary, adjust as outlined in this section.

1. Raise vehicle on hoist, being careful not to bend the strut rods.
2. Disengage bolt lock tabs and disconnect outboard end of axle drive shaft from wheel spindle flange.
3. Mark camber cam in relation to bracket. Loosen and turn camber bolt until strut rod forces torque control arm outward. Position loose end of axle drive shaft to one side for access to spindle.
4. Remove wheel and tire assembly. Mount dial indicator on torque control arm adjacent surface and rest pointer on flange or spindle end.
5. Grasp brake disc and move axially (in and out) while reading movement on dial indicator. If end movement is within the .003 to .20 mm (.001 in. to .008 in.) limit, bearings do not require adjustment. If not within .003 to .20 mm (.001 in. to .008 in.) limit, record reading for future reference and adjust bearings.

### Spindle Removal

1. Apply parking brake to prevent spindle from turning and remove cotter pin and nut from spindle.
2. Release parking brake and remove drive spindle flange from splined end of spindle. It may be necessary to use tool J-8614-01 to remove flange from spindle.
3. Remove brake caliper.
4. Install thread protector J-21859-1 over spindle threads. Remove drive spindle from spindle support, using tool J-22602.

When using tool J-22602 to remove drive spindle, make sure puller plate is positioned vertically in the torque control arm before applying pressure to the puller screw.

5. When the spindle is removed, the outer bearing will remain on the spindle. The inner bearing, tubular spacer, end-play adjustment shim and both outer races will remain in the spindle support.
6. Remove bearing, spacer and shim. Record shim thickness for later use.

### Bearing Replacement and Adjustment, and Spindle Installation

1. With the spindle assembly on a bench, place the two halves of J-24489-1 into position between the outer bearing and the oil seal.
2. Mount J-8433-1 to J-24489-1 and draw bearing off spindle.
3. Remove outer oil seal from spindle shaft and inspect for damage. Replace if necessary.
4. Remove the outer races from the spindle support and install new ones, using J-7817 for reinstallation.
5. Pack new bearings with EPB-2 bearing lubricant, or equivalent.
6. Check bearing end play as measured in step 5 of Wheel Bearing End Play Check. Use the same adjusting shim thickness as the original. If end play was *not* within limits, use the following steps to determine the proper shim thickness:
   a. If end play was greater than .20 mm (.008 in.), it will be necessary to reduce shim thickness to bring end play within limits.
   b. For example, if end-play reading was .33 mm (.013 in.), and the shim measured 3.66 mm (.144 in.), you will have to decrease the shim thickness. Reducing the shim by .25 mm (.010 in.), from 3.66 to 3.40 mm (.144 to .134 in.), will also reduce end play by .25 mm (.010 in.), from .33 to .08 mm (.013 to .003 in.).
   c. If no end play was found on inspection, add .08 mm (.003 in.) to the original shim as a starting point.
7. To check bearing end-play before final installation, use J-24626 as follows:
   a. Mount the outer bearing onto the large shoulder, with the large end of the bearing against the flange.

b. Place the tubular spacer, with the large end against the outer bearing, and the shim selected in step 6 onto J-24626.

c. Place the tool into position in the spindle support and install inner bearing, large washer and nut.

d. Tighten nut to 140 N·m (100 ft.-lbs.) to simulate actual installed conditions.

e. Mount a dial indicator and check bearing end-play.

f. After shim thickness as necessary to obtain end-play from .03 to .20 mm (.001 to .008 in.). Shims are available in thicknesses from 2.46 to 3.68 mm (.097 to .145 in.).

g. Remove J-24626 from spindle support.

8. Install outer bearing into outer race. Install outer oil seal into bore of spindle support, making sure it is firmly seated.

9. Carefully install spindle assembly through the outer oil seal (being careful not to dislodge seal from the bore) and through the outer bearing.

10. Place the tubular spacer and the shim selected in step 7 onto the spindle shaft.

11. Place the inner bearing onto the spindle shaft.

12. Thread tool J-24490-1 onto the spindle shaft, then install sleeve J-24490-2, and washer and nut. Tighten nut against sleeve. Spindle shaft will now be drawn through the bearings to its final installed position.

13. Remove J-24490-1 and J-24490-2.

14. Position drive flange over spindle, making sure flange is aligned with spindle splines. Install washer and nut on spindle, then tighten nut to specifications and install cotter pin. If specified torque does not permit cotter pin insertion, tighten nut to next flat.

15. Install caliper onto disc.

16. Install axle drive shaft, wheel and tire assembly, adjust camber cam to original position and torque all components to specifications.

## Spindle Support

### Removal

1. Remove wheel spindle as outlined previously.

2. Disconnect parking brake cable from actuating lever.

3. Remove four nuts securing spindle support to torque control arm and withdraw brake backing plate. Position it out of the way.

4. Disconnect shock absorber lower eye from strut rod mounting shaft. It may be necessary to support spring outer end before disconnecting shock absorber, as shock absorber has internal rebound control.

5. Remove cotter pin and nut from strut rod mounting shaft, then pull shaft from support and strut rod.

6. Separate support from torque control arm.

### Installation

1. Position support over torque arm bolts with strut rod fork toward center of vehicle and downward.

2. Place backing plate in position; install four nuts and torque to specifications.

3. Install strut rod and shock absorber mounting shaft onto support arm. Install shock absorber. Torque to specifications.

4. Connect parking brake cable to actuating lever.

5. Install drive spindle assembly.

## Shock Absorber

### Removal

1. Raise vehicle on hoist.

2. Disconnect shock absorber upper mounting bolt.

3. Remove lower mounting nut and lock washer.

4. Slide shock upper eye out of frame bracket and pull lower eye and rubber grommets off strut and mounting shaft.

5. Inspect grommets and shock absorber upper eye for excessive wear.

## Sectional View of Spindle

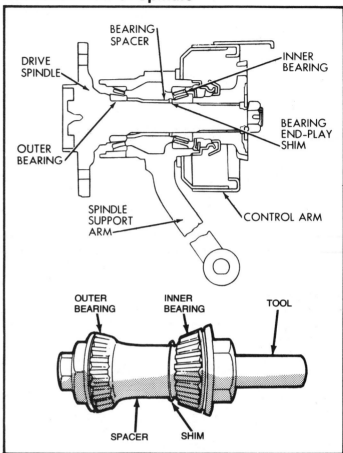

## Shock Absorber Mounting

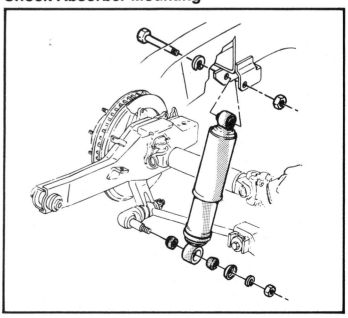

### Installation

1. Slide upper mounting eye into frame mounting bracket and install bolt, lock washer and nut.
2. Place rubber grommet, shock lower eye, inboard grommet, washer and nut over strut rod shaft. Install washer with curve pointing inboard (away from grommet).
3. Torque nuts to specifications.
4. Lower vehicle and remove from hoist.

## Strut Rod and Bracket

### Removal

1. Raise vehicle on hoist.
2. Disconnect shock absorber lower eye from strut rod shaft.
3. Remove strut rod shaft cotter pin and nut. Withdraw shaft by pulling toward front of vehicle.
4. Mark relative position of camber adjusting cam and bracket, so they may be reassembled in same location.
5. Loosen camber bolt and nut. Remove four bolts, lock washers and flat washers securing strut rod bracket to carrier and lower bracket.
6. Remove cam bolt nut and cam and bolt assembly. Pull strut down out of bracket and remove bushing caps.
7. Inspect strut rod bushings for wear and replace where necessary.

## Marking Camber Cam and Bracket

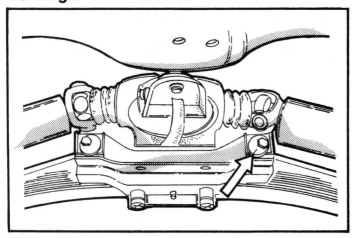

## Strut Rod Mounting

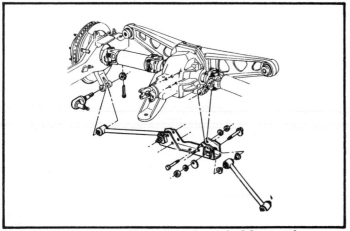

**Install strut with outboard end angled forward**

### Installation

1. Place bushing caps over inboard bushing and slide rod into bracket. Install cam and bolt assembly and adjust cam to line up with mark of bracket. Tighten nut but do not torque at this point.
2. Raise bracket and assemble to carrier lower mounting surface. Be sure both flat washer and lock washer are between bolt and bracket. Torque bolts to specifications.
3. Raise outboard end of strut rod into fork so that flat on shaft lines up with corresponding flat in spindle fork. Install retaining nut, but do not torque.
4. Place shock absorber lower eye and bushing over strut shaft, install washer and nut and torque to specifications.
5. With weight on wheels torque camber cam nut and strut rod shaft nut to specifications. Then install cotter pin through rod bolt.
6. Check rear wheel camber and adjust where necessary.
7. Lower vehicle and remove from hoist.

## Transverse Spring

### Removal

1. Raise vehicle on hoist allowing axle to hang. Remove wheels and tires.
2. Install a C clamp on spring approximately 23 cm. (9 in.) from one end. Tighten securely.
3. Place adjustable lifting device under spring with lifting pad of jack inboard of link bolt near the C clamp. Place a suitable piece of wood between jack pad and C clamp screw. The C clamp is merely acting as a stop so the jack will not slip when the spring is released. The wood block is used to protect the clamp threads from distortion due to contact with the jack pad.
4. Raise jack until all load is off link. Remove link cotter key and link nut. Remove cushion. Do not grip shank of spring link bolt with Vise Grips. Use new bolt if the bolt surface is scored or damaged.

### Clamp Plate Removal

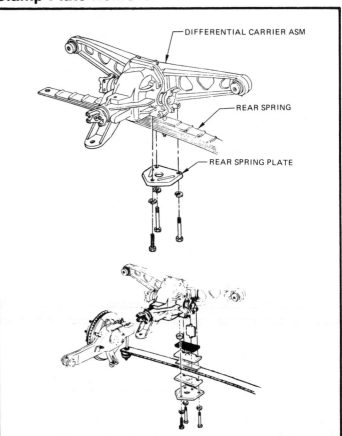

5. Carefully lower jack until spring tension is released.

6. Repeat steps 2-5 for other side.

7. Remove four bolts and washers securing spring center clamp plate.

8. Slide spring out from under vehicle.

## Installation

1. Place spring on carrier cover mounting surface, indexing center bolt head with hole in cover.

2. Place center clamp plate in position and install bolts and washers. Snug bolts to position spring and torque to specifications.

3. Install C clamp as in step 2 of removal procedure.

4. Place adjustable lifting device inboard of link bolt near C clamp. Add wooden block as in step 3 of removal procedure.

5. Raise spring outer end until spring is nearly flat, aligning torque arm with spring end.

6. Install new attaching parts. Whenever servicing spring or removing spring attaching parts, always install new link bolts, rubber cushions, retainers, nuts and cotter pins.

7. Lower jack making sure cushions remain indexed in retainers. Remove C clamp.

8. Remove jack and repeat for other side.

9. Place vehicle weight on wheels and torque center clamp bolts to specifications.

## Torque Control Arm

### Removal

1. Disconnect spring on side torque arm is to be removed. Follow steps 1-5 of the spring removal procedure. If vehicle is so equipped, disconnect stabilizer shaft from torque arm.

2. Remove shock absorber lower eye from strut rod shaft.

3. Disconnect and remove strut rod shaft and swing strut rod down.

4. Remove four bolts securing axle drive shaft to spindle flange and disconnect driveshaft. It may be necessary to force torque arm outboard to provide clearance to lower driveshaft.

5. Disconnect brake line at caliper and from torque arm. Disconnect parking brake cable.

6. Remove torque arm pivot bolt and toe-in shims and pull torque arm out of frame. Tape shims together and identify for correction reinstallation.

### Installation

1. Place torque arm in frame opening.

2. Install pivot bolt. Place toe-in shims in original position on both sides of torque arm. Install cotter pin retaining shims with loop of pin pointed outboard. Do not tighten pivot bolt nut at this time.

3. Raise axle driveshaft into position and install to drive flange. Torque bolts to specifications.

4. Raise strut rod into position and insert strut rod shaft so that flat lines up with flat in spindle support fork. Install nut and torque to specifications.

5. Install shock absorber lower eye and tighten nut to specifications.

6. Connect spring end as outlined under spring installation, step 3-6. If vehicle is so equipped, connect stabilizer shaft to torque arm.

## Transverse Spring Mounting

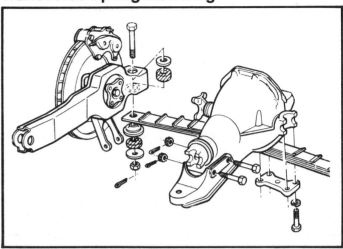

## Stabilizer Shaft Installation

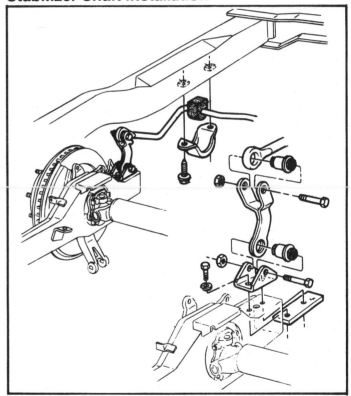

7. Install brake line at caliper and torque arm. Bleed brakes.

8. Install wheel and tire. Torque the torque arm pivot bolt to specifications and install cotter pin with weight on wheels.

# FRONT SUSPENSION—GENERAL MOTORS REAR DRIVE H BODY

## Description

The front suspension is of the A frame type with short and long control arms. The upper control arm is bolted to the front end sheet metal at each inner pivot point. Rubber bushings are used for mounting.

The lower control arm attaches to the front end sheet metal, with two cam type bolts, through rubber bushings. The cam bolts adjust camber and caster. The front cam bolts adjust camber and the rear cam bolts adjust caster.

# SUSPENSIONS

## Front Suspension

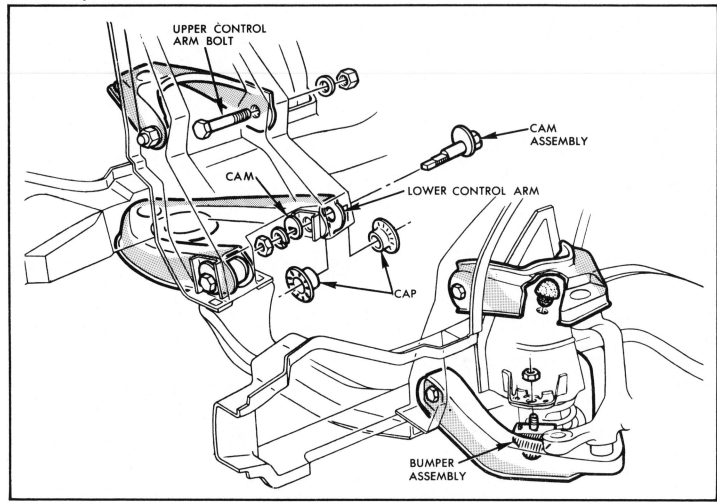

The upper ball joint is riveted to the upper control arm and the lower ball joint is pressed into the lower control arm. The lower ball joints use the "wear indicator" feature.

The coil springs are mounted between the lower control arms and the shock absorber tower. The shock absorber towers have helical spring seats which are used for spring indexing.

Shock absorbers mount to the lower control arm and run through the center of the coil spring to the shock absorber tower where the upper stem mounts.

The steering knuckle is nodular iron with an integral steering arm. The wheel hub is an integral part of the brake disc and mounts to the wheel spindle in the conventional manner with inner and outer wheel bearings.

## Wheel Alignment

Front end alignment consists of three adjustments which should be done in order: camber (first), caster (second), toe-in (third).

## CASTER AND CAMBER

Caster and camber adjustments are made by rotating or changing the position of the cam bolts. Use the front cam to adjust the camber setting and the rear cam to adjust the caster setting. The lower control arm is designed so that the camber setting should be made first.

The front cam tends to move the control arm in or out with respect to the vehicle. This movement will change camber.

### Adjustment

Camber angle is adjusted by loosening the front lower control arm pivot nut and rotating the cam until proper setting is reached. This eccentric cam action will move the lower control arm in or out, thereby varying the camber. Hold the cam bolt head while tightening the nut.

Caster angle is adjusted by loosening the rear lower control arm pivot nut and rotating the cam until proper setting is reached. This eccentric cam action will tend to move the lower control arm fore or aft thereby varying the caster. Hold the cam bolt head while tightening the nut. Recheck camber after setting caster.

## TOE-IN

Toe-in can be increased or decreased by changing the length of the tie rods. A threaded sleeve is provided for this purpose. The tie rods are mounted ahead of the steering knuckle and must be decreased in length in order to increase toe-in. Toe-in adjustment must be checked after camber and caster adjustment.

### Adjustment

Toe-in is the difference in the distance measured between the front and rear of the front wheels. The wheels must in the straight ahead position when adjusting toe-in. Check position of tie rod clamps (frame clearance) after setting toe-in.

## Wheel Bearings

NOTE: Tapered roller bearings are used on all series vehicles and they have a slightly loose feel when properly adjusted. A design feature of front wheel tapered roller bearings is that they must never be pre-loaded. Damage can result from pre-loading.

Cones must be a slip fit on the spindle and the inside diameter of cones should be lubricated to insure that the cones will creep. Spindle nut must be a free-running fit on threads.

### Inspection

1. Raise vehicle and support at front lower control arm.
2. Spin wheel to check for unusual noise or roughness.
3. If bearings are noisy, tight, or excessively loose, they should be cleaned, inspected and relubricated prior to adjustment.

To check for tight or loose bearings, grip the tire at the top and bottom and move the wheel assembly in and out on the spindle. Measure movement of hub assembly. If movement is less than .001 in. or greater than .005 in., adjust bearings.

### Adjustment

1. Remove hub cap or wheel disc from wheel.
2. Remove dust cap from hub.
3. Remove cotter pin from spindle and spindle nut.
4. Tighten the spindle nut to 12 ft.-lbs. while turning the wheel assembly forward by hand to fully seat the bearings. This will remove any grease or burrs which could cause excessive wheel bearing play later.
5. Back off the nut to the "just loose" position.
6. Hand tighten the spindle nut. Loosen spindle nut until either hole in the spindle lines up with a slot in the nut. (Not more than 1/2 flat).
7. Install new cotter pin. Bend the ends of the cotter pin against nut, cut off extra length to ensure ends will not interfere with the dust cap.
8. Measure the looseness in the hub assembly. There will be from .001 to .005 inches end play when properly adjusted.
9. Install dust cap on hub.
10. Replace the wheel cover or hub cap.

## Shock Absorber

### SPIRAL GROOVE RESERVOIR

If this type of shock has been stored or allowed to lay in a horizontal position for any length of time, an air void will develop in the pressure chamber of the shock absorber. If this air void is not purged, a technician may diagnose the shock as defective. To purge the air from the pressure chamber, proceed as follows:

1. Holding the shock in its normal vertical position (top end up), fully extend shock.
2. Hold the top end of the shock down and fully collapse the shock.
3. Repeat pumping action at least five times to assure air is purged.

## Upper Ball Joint

### Inspection

Raise car and move front wheel vertically and horizontally by hand. If there is any free play, the ball joint must be replaced.

### Removal

The ball joint is riveted to the control arm. Replace by drilling out rivets and installing new ball joint with bolts and nuts.

## Shock Absorber

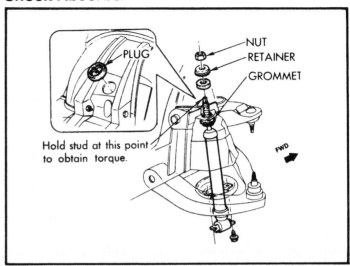

Lower studs 59 N·m (43 in. lbs.). Top nut 10 N·m (7 in. lbs.).

## Drilling Upper Ball Joint Rivet

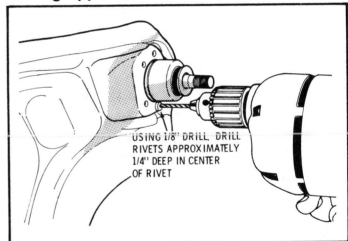

Using 1/8 in. drill, drill rivets approximately 1/4 in. deep in center of rivet. Using 1/2 in. drill, drill just deep enough to remove rivet head.

1. Raise the car with the suspension hanging in full rebound, and place a support under the lower control arm.
2. Separate the ball joint from the knuckle, and remove the control arm.

### Installation

1. Install ball joint into control arm with nuts and bolts.
2. Install control arm.
3. Install ball joint stud to knuckle. Install a new cotter pin.

## Lower Ball Joint

### Inspection

Wear indicators are incorporated in the lower ball joints.

## Ball Joint Replacement

Upper ball pivot bolt 64 N·m (47 in. lbs.). Lower ball
pivot bolt 66 N·m (49 in. lbs.).

## Lower Ball Joint Installation

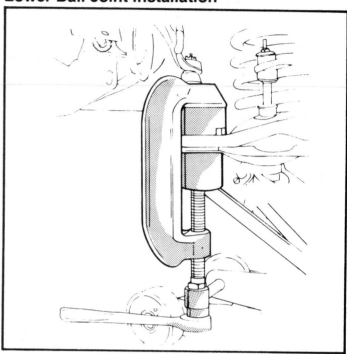

## Control Arm Bushings Replacement

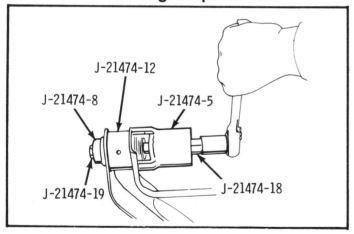

J-21474-12
J-21474-8
J-21474-5
J-21474-19
J-21474-18

## Built-in Wear Indicator for Lower Ball Joint

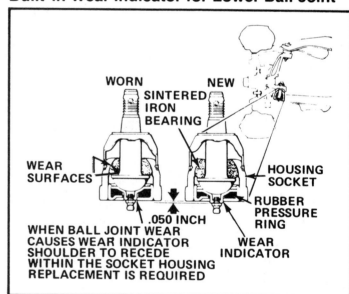

WORN   NEW
SINTERED IRON BEARING
WEAR SURFACES
HOUSING SOCKET
RUBBER PRESSURE RING
WEAR INDICATOR
.050 INCH
WHEN BALL JOINT WEAR CAUSES WEAR INDICATOR SHOULDER TO RECEDE WITHIN THE SOCKET HOUSING REPLACEMENT IS REQUIRED

## Removal and Installation

The ball joint is a press fit in the control arm.

1. Raise the car with the suspension hanging in full rebound, and support the lower control arm at the outer end.

2. Free the ball joint from the knuckle, and press it out of the control arm.

## Lower Control Arm

### Removal

1. Raise the vehicle with suspension hanging in full rebound. Remove the shock absorber and coil spring and separate the ball joint from the knuckle.

2. Remove the inner pivot cam nuts and bolts after marking them for installation reference.

### BUSHING REPLACEMENT

1. Remove lower arm from vehicle.
2. Press out old bushing.
3. Press in new bushing.

**NOTE: An arbor press can be used, providing appropriate supports are used to prevent distorting control support arms.**

### Installation

1. When installing, the front cam bolt must be installed with the head toward the vehicle front and the rear cam bolt with the head toward the vehicle rear.

2. Check front alignment after installation.

## Coil Spring

### Removal

1. Remove the shock absorber and stabilizer bar, and raise the vehicle with suspension hanging in full rebound.

2. Support the lower control arm with a jack, separate the ball joint and tie-rod end from the knuckle, and lower the control arm until the spring can be removed.

### Installation

1. Properly position spring in spring tower and on control arm, and lift control arm with hydraulic jack. Be sure the insulator is indexed with the closed end located on the high point in the spring seat.

2. Guide the lower control arm ball stud into the steering knuckle and install nut. Torque nut to specifications and insert cotter pin. Remove the safety chain.

Do not back off nut to insert cotter pin. Advance nut to the next slot that lines up with the hole in the stud.

3. Install the tie rod end to the steering arm.
4. Install the shock absorber.
5. Install the stabilizer bar if removed.
6. Install the wheel and tire assembly.
7. Remove the vehicle from the jackstands and lower to the floor. Install the upper end of the shock absorber.

## Front Coil Spring Positioning

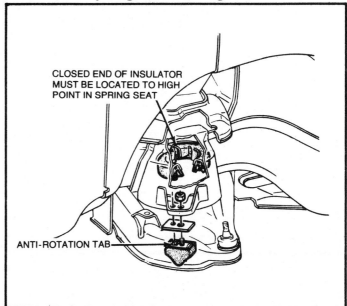

CLOSED END OF INSULATOR MUST BE LOCATED TO HIGH POINT IN SPRING SEAT

ANTI-ROTATION TAB

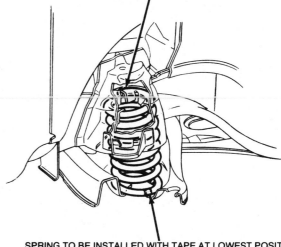

HELIX END OF SPRING (TOP) MUST BE SEATED IN UPPER SPRING SEAT. VISUALLY CHECK THROUGH SHOCK HOLE IN LOWER CONTROL ARM

SPRING TO BE INSTALLED WITH TAPE AT LOWEST POSITION. TOP OF SPRING IS COILED HELICAL AND BOTTOM OF SPRING IS COILED FLAT WITH A GRIPPER NOTCH NEAR END OF THE WIRE

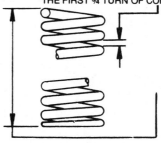

WHEN A PORTION OF THE ACTIVE COILS ARE COMPRESSED, THE COIL-TO-COIL GAP MUST NEVER BE LESS THAN .44 IN. THE FIRST ¾ TURN OF COIL ON EACH END IS NOT ACTIVE

WHEN ENTIRE SPRING IS COMPRESSED, THE OVERALL DIMENSION MUST NEVER BE LESS THAN 7.30 IN.

# REAR SUSPENSION—GENERAL MOTORS REAR DRIVE H BODY

## Description

The rear suspension and axle assembly is attached to the vehicle through a link type suspension arrangement. The major components are:

1. A single upper control arm, (torque arm) rigidly mounted to the differential housing and mounted to the transmission through rubber bushings.
2. Two conventional rubber-bushed lower control arms.
3. Two coil springs.
4. Two conventional shock absorbers.
5. A stabilizer bar.
6. A tie rod, mounted between the axle and the underbody of the vehicle.

### Torque Arm Suspension

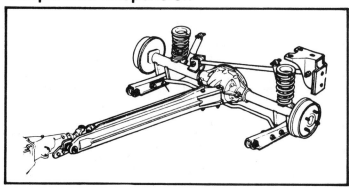

### Rear Shock Mounting

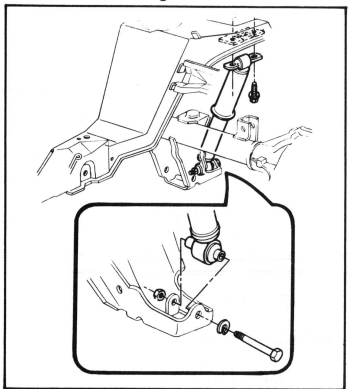

## Shock Absorber

### Removal

1. Remove upper attaching bolts and lower attaching bolt and nut.
2. Remove shock absorber.

### Installation

1. Place shock into installed position and install upper retaining bolts. Torque to specifications.
2. Install bolt and nut onto lower shock attachment. Torque to specifications.

## Coil Spring

### Removal

1. Raise vehicle on a hoist.
2. Support rear axle with an adjustable lifting device.
3. Disconnect both shock absorbers from lower brackets.
4. Lower the axle and remove the springs, insulators, and retainer/bumper assemblies. One or both springs may be removed. When lowering the axle, do not stretch the brake hose running from frame to axle.

### Installation

1. Install retainer/bumper assembly onto top of spring. Place the insulator onto the retainer.
2. Raise the axle into the proper position, being sure the spring assembly is properly oriented to the underbody.
3. Reconnect the shock absorbers and torque to specifications.

## Torque Arm Attachments

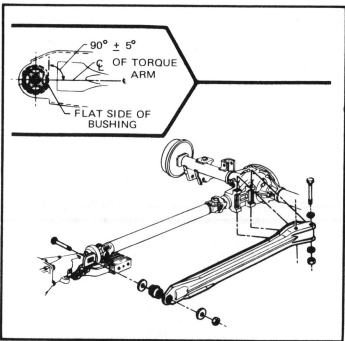

## Torque Arm to Support Attachment

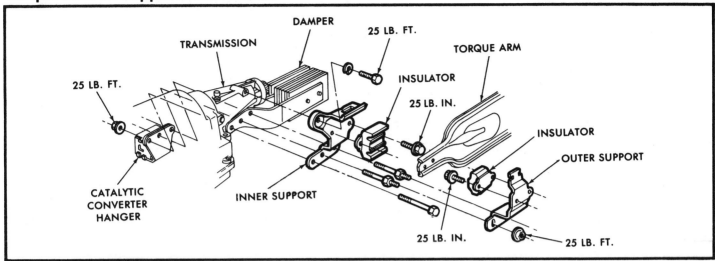

## Coil Springs

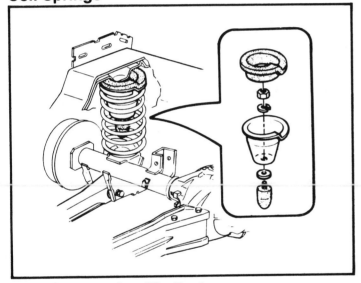

# Tie-Rod

### Replacement

A tie-rod is mounted to the rear of the axle. It is important to use shims to position the axle assembly so that equal clearances exist between tire and wheelhouse on each side of the vehicle.

1. Raise vehicle on a hoist, and support the rear axle.
2. Remove bolt at underbody end of track rod.
3. Remove bolt at axle bracket.
4. Remove tie-rod from vehicle.
5. Place new tie-rod into position and install bolt at each bracket.

## Rear Suspension Tie Rod

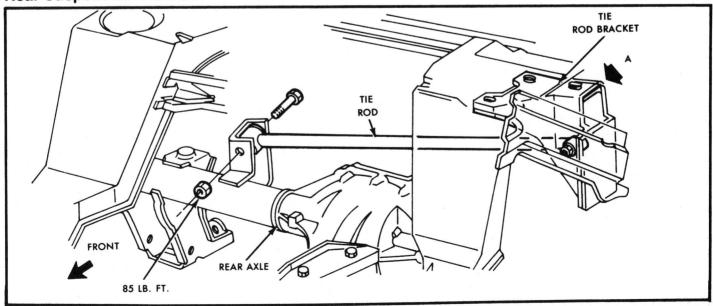

# FRONT SUSPENSION—GENERAL MOTORS REAR DRIVE T BODY

## Description

The front suspension system uses conventional long and short arm design and coil springs. The control arms attach to the vehicle with bolts and bushings at the inner pivot points, and to the steering knuckle/front wheel spindle assembly at the outer pivot points. The lower ball joints use the wear indicator feature used on other General Motor original equipment ball joints.

## Wheel Alignment

### CAMBER

Camber angle can be increased approximately one degree. Remove the upper ball joint, rotate it one-half turn and reinstall it with the flat of the upper flange on the inboard side of the control arm.

### CASTER

Shims placed between the upper control arm and legs control caster. Always use two washers totalling 12 mm thickness, placing one washer at each end of the locating tube.

## Front Suspension

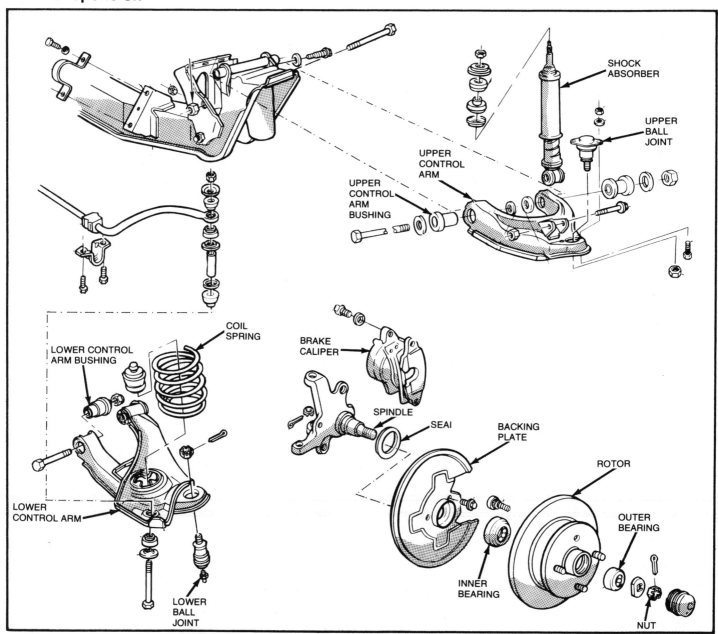

## Camber Adjustment

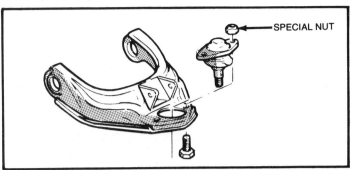

**To increase camber, disconnect upper ball joint and rotate 180° to position "flat" of flange inboard. Then reconnect ball joint.**

### TOE

Adjust by changing tie-rod position. Loosen the nuts at the steering knuckle end of the tie rod and the rubber cover at the other end. Rotate tie-rod to change adjustment.

# Wheel Bearings

NOTE: Tapered roller bearings are used on all series vehicles and they have a slightly loose feel when properly adjusted. A design feature of front wheel tapered roller bearings is that they must never be pre-loaded. Damage can result from pre-loading.

The proper functioning of the front suspension cannot be maintained unless the front wheel taper roller bearings are correctly adjusted. Cones must be a slip fit on the spindle and the inside diameter of cones should be lubricated to insure that the cones will creep. Spindle nut must be a free-running fit on threads.

### Inspection
1. Raise vehicle and support at front lower control arm.
2. Spin wheel to check for unusual noise or roughness.
3. If bearings are noisy, tight, or excessively loose, they should be cleaned, inspected, and relubricated prior to final adjustment. If it is necessary to inspect bearings, movement should be from 0.025 mm to 0.127 mm (.001 — .005 in.). If movement is not in this range, adjust bearings.

### Adjustment
1. Remove hub cap or wheel disc from wheel.
2. Remove dust cap from hub.
3. Remove cotter pin from spindle and spindle nut.
4. Tighten the spindle nuto to 16 N·m (12 ft.-lbs.) while turning the wheel assembly forward by hand to fully seat the bearings. This will remove any grease or burrs which could cause excessive wheel bearing play later.
5. Back off the nut to the "just loose" position.
6. Hand-tighten the spindle nut. Loosen spindle nut until either hol in the spindle lines up with a slot in the nut (not more than 1/2 flat).
7. Install new cotter pin. Bend the ends of the cotter pin against nut, cut off extra length to ensure ends will not interfere with the dust cap.
8. Measure the looseness in the hub assembly. There will be from 0.025 to 0.127 mm (.001 to .005 in.) end-play when properly adjusted.
9. Install dust cap on hub.
10. Replace the wheel cover or hub cap.

## Front Shock Absorber

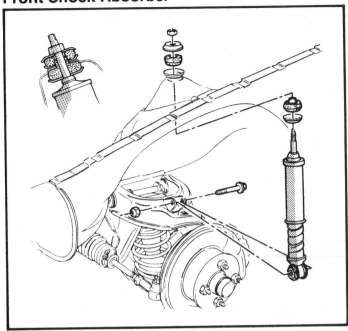

## Removal of Ball Joints From Knuckle

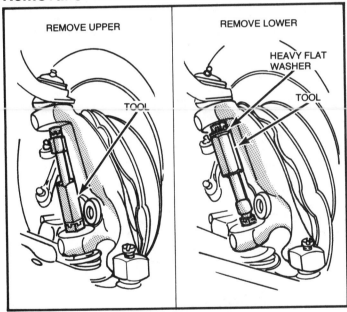

# Shock Absorbers

### Removal
1. Hold the shock absorber upper stem and remove the nut, upper retainer and rubber grommet.
2. Raise vehicle on a hoist.
3. Remove the bolts from the lower end of the shock absorber.
4. Lower the shock absorber from the vehicle.

### Installation
1. With the lower retainer and rubber grommet in position, extend the shock absorber stem and install the stem through the wheelhouse opening.

2. Install the lower bolts. Torque to 48-70 N·m (35-50 ft.-lbs.).

3. Lower the vehicle to the floor.

4. Install the upper rubber grommet, retainer and nut to the shock absorber stem.

5. Hold the stem and tighten the nut to 7-13 N·m (60-120 kn.-lbs.). Torque is obtained by running nut to unthreaded portion of stud.

## Upper Ball Joint

### Removal

1. Raise the vehicle on a hoist.

2. Remove the tire and wheel assembly.

3. Support the lower control arm with a floor jack.

4. Remove upper ball stud nut. Reinstall nut finger-tight.

5. Install spreader tool and push stud loose from knuckle.

6. Remove tool and remove nut from ball stud.

7. Remove two nuts and bolts attaching ball joint to upper control arm, then remove ball joint.

### Installation

Inspect the tapered hole in the steering knuckle. Remove any dirt and if any out-of-roundness, deformation, or damage is noted, the knuckle *must* be replaced.

1. Install bolts and nuts attaching ball joint to upper control arm, then mate the upper control arm ball stud to the steering knuckle. The ball joint studs use a special nut which *must be discarded* whenever loosened and removed. On reassembly, use a standard nut to draw the ball joint into position on the knuckle. Torque the standard nut to 30 N·m (22 ft.-lbs.), then remove that nut and *install a new special nut* for final installation.

2. Install the ball stud nut and torque to 39-49 N·m (29-36 ft.-lbs.).

3. Install the tire and wheel assembly.

## Lower Ball Joint

### Removal

1. Raise vehicle on hoist.

2. Remove the tire and wheel assembly.

3. Support the lower control arm with a hydraulic floor jack.

## Correct Spring Position

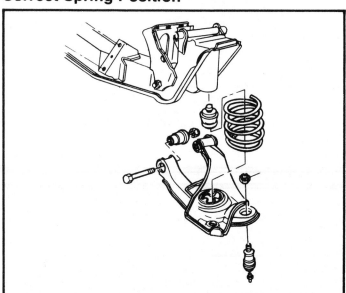

4. Remove lower ball stud nut, then reinstall nut finger-tight.

5. Install spreader tool and push the ball joint stud until it is free of the steering knuckle.

6. Remove tool and remove nut from ball stud.

7. Remove ball joint from lower control arm.

### Installation

Inspect the tapered hole in the steering knuckle. Remove any dirt and if any out-of-roundness, deformation, or damage, is noted, the knuckle *must* be replaced.

1. Mate the ball stud through the lower control arm and into the steering knuckle. The ball joint studs use a special nut which *must be discarded* whenever loosened and removed. On reassembly, use a standard nut to draw the ball joint into position on the knuckle. Torque the standard nut then remove that nut and *install a new special nut* for final installation.

2. Install the ball stud nut and torque to 56-73 N·m (41-54 ft.-lbs.).

3. Install the tire and wheel assembly.

## Front Spring/Lower Control Arm

### Removal

1. Remove wheel and tire assembly.

2. Disconnect stabilizer from lower control arm. Disconnect tie-rod from steering knuckle.

3. Support lower control arm with a jack.

4. Remove the nut from the lower ball joint. Install spreader tool and push the ball joint stud loose in the steering knuckle.

5. Swing the knuckle-and-hub out of the way, and attach securely with wire.

6. Loosen lower control arm pivot bolts.

7. Install chain through coil spring as a safety precaution.

---
**CAUTION**

*The coil spring is under load. Be sure to install a chain and to slowly lower the jack.*

---

8. Slowly lower the jack.

9. When the spring is extended as far as possible, use a prybar to carefully lift the spring over the lower control arm seat. Remove the spring.

10. Remove pivot bolts, and then remove lower control arm.

### Installation

1. Install lower control arm and pivot bolts to underbody brackets.

2. Position spring and install spring into upper pocket. Use tape to hold insulator onto spring.

3. Install spring lower end onto lower control arm. It may be necessary to have an assistant help you compress the spring far enough to slide it over the raised area of the lower control arm seat.

4. Use a jack to raise the lower control arm and compress the coil spring.

5. The ball joint studs uses a special nut which *must be discarded* whenever loosened and removed. On reassembly, use a standard nut to draw the ball joint into position on the knuckle, then remove that nut and *install a new special nut* for final installation. Install the ball joint through the lower control arm and into the steering knuckle. Install nut to ball joint stud and torque to 56-73 N·m (41-53 ft.-lbs.).

6. Connect stabilizer bar and tie-rod. Install wheel and tire assembly. Torque to specifications.

## Upper Control Arm

### Removal

1. Raise vehicle on a hoist.

2. Remove the tire and wheel assembly.

3. Support the lower control arm with a floor jack.

4. Remove upper ball joint from steering knuckle.

5. Remove control arm pivot bolts and remove control arm from vehicle.

### Installation

1. Install upper control arm and pivot bolt to vehicle. The inner pivot bolt must be installed with the bolt head toward the front.

2. Install the pivot bolt nut.

3. Position the control arm in a horizontal plane and torque the nut to 59-68 N·m (43-50 ft.-lbs.).

4. The ball joint studs use a special nut which *must be discarded* whenever loosened and removed. On reassembly, use a standard nut to draw the ball joint into position on the knuckle, then remove that nut and *install a new special nut* for final installation. Install ball joint to upper control arm and to steering knuckle, as described earlier. Install nut; tighten to specifications.

5. Install wheel and tire; torque to specifications.

6. Lower vehicle to floor.

## Steering Knuckle

### Removal

1. Raise vehicle and support the lower control arm with a jackstand.

—————————— CAUTION ——————————
*This keeps the coil spring compressed. Use care to support safely.*
—————————————————————————————

2. Remove the tire and wheel assembly.

3. Remove the disc brake caliper. Do not allow the caliper to hang by the brake hose. Insert a piece of wood between the shoes to hold the piston in the caliper bore. The block of wood should be about the same thickness as the brake disc.

4. Remove the hub and disc.

5. Remove the splash shield.

6. Remove the tie-rod end from the steering knuckle.

7. Loosen both ball stud nuts. Using a spreader tool, push both the upper and lower ball studs from the steering knuckle.

8. Remove ball stud nuts and remove the steering knuckle.

### Installation

1. Place steering knuckle in position and insert the upper and lower ball studs into knuckle bosses.

### Stabilizer Bar Attachement

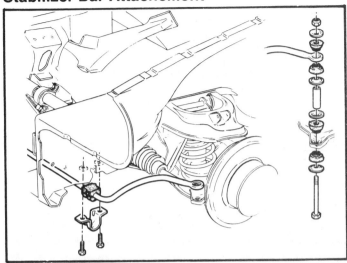

2. The ball joint studs use a special nut which *must be discarded* whenever loosened and removed. On reassembly, use a standard nut to draw the ball joint into position on the knuckle. Torque the standard nut then remove that nut and *install a new special nut* for final installation. Install ball stud nuts and tighten to specifications.

## Stabilizer Bar

### Removal

1. Raise the vehicle on a hoist.

2. Remove stabilizer bar nut and bolt from lower control arm.

3. Remove stabilizer bar bracket from body.

### Installation

1. Hold stabilizer bar in place and install the body bushings and brackets.

2. Install the retainers, grommets and spacers to the lower control arm and install nuts.

3. Lower the vehicle to the floor.

4. Torque nut to 16-24 N·m (12-18 ft.-lbs.). Torque is obtained by running nut to unthreaded portion of link bolt.

# REAR SUSPENSION—GENERAL MOTORS REAR DRIVE T BODY

## Description

The solid rear axle is attached to the body through two tubular lower control arms, a straight track rod, two shock absorbers and a bracket at the front end of the axle extension. Variable rate coil springs mount between the axle and body.

Two rubber bushed lower control arms mounted between the axle assembly and the frame maintain fore and aft relationship of the axle assembly to the chassis. The rigid axle holds the rear wheels in proper alignment.

The rear chassis springs are located between brackets on the axle tube and spring seats in the frame. The springs are held in the seat pilots by the weight of the car and by the shock absorbers which limit axle movement during rebound.

Ride control is provided by two identical direct double acting shock absorbers angle-mounted between brackets attached to the axle housing and the rear spring seats.

## Shock Absorber

### Removal

1. Support rear axle assembly.

2. Remove upper attaching bolts and lower attaching bolt and nut.

3. Remove shock absorber.

# SUSPENSION

## Rear Suspension

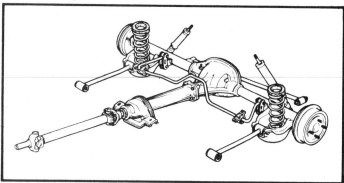

### Installation

1. Install retainer and rubber grommet onto shock.
2. Place shock into installed position and install upper retaining bolts. Torque to specifications.
3. Install bolt and nut onto lower shock attachment. Torque to specifications.
4. Lower vehicle and remove from hoist.

## Coil Spring

### Removal

1. Raise vehicle on hoist.
2. Support rear axle with an adjustable lifting device.
3. Disconnect both shock absorbers from lower brackets.
4. Disconnect rear axle extension bracket.

——— CAUTION ———

*Be sure to use caution when disconnecting extension assembly. Be sure to support assembly safely.*

5. Lower axle and remove springs and spring insulators. One or both springs may be removed at this point.

——— CAUTION ———

*When lowering axle, do not stretch brake hose running from frame to axle or damage to the brake line may result.*

### Installation

1. Install insulators on top and bottom of springs then position spring between upper and lower seats.
2. Raise axle and reconnect shock absorbers. Torque nut to specifications.
3. Remove lifting device from axle.
4. Lower vehicle and remove from hoist.

## Lower Control Arm and Tie-Rod

### Removal

——— CAUTION ———

*If both control arms are to be replaced, remove and replace one control arm at a time to prevent the axle from rolling or slipping sideways.*

1. Raise the car.
2. Support the rear axle.
3. Disconnect the stabilizer bar.

## Rear Shock Absorber

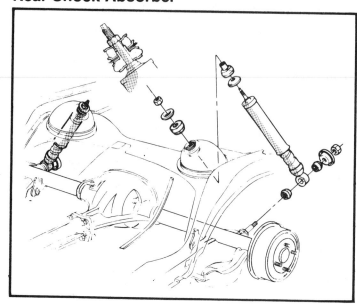

## Coil Springs

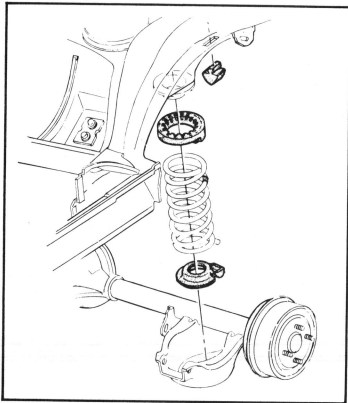

4. Remove the control arm front and rear attaching bolts and remove the control arm.
5. Remove the track rod attaching bolts and remove the track rod.

## BUSHING REPLACEMENT

1. Use appropriate tools to press bushings out of control arm/tie rod.
2. Inspect for distortion, burrs, etc.
3. Press bushing into place.

### Installation

1. Place control arm into position and install front and rear bolts. Torque to specifications.
2. Place tie-rod into position; torque bolts to specifications. Car must be at curb height when tightening pivot bolts. Tighten pivot bolts to 45 N·m (33 ft.-lbs.).
3. Reattach stabilizer bar.

## Stabilizer Bar

### Removal

1. Raise vehicle on hoist.
2. Remove bolts securing brackets to body and link to axle and remove bar.

### Installation

1. Place stabilizer into position. Install bolts and nuts. Torque to 20 N·m (15 ft.-lbs.).

### Lower Control Arm and Tie Rod

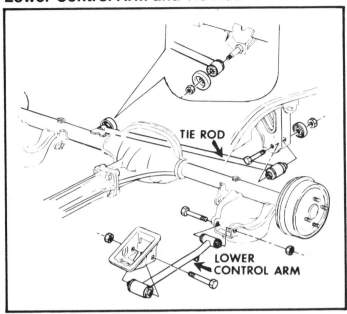

# GENERAL MOTORS ELECTRONIC LEVEL CONTROL

## Description

The electronic level control (ELC) system automatically adjusts the rear height with varying car loads. The system is activated when weight is added to, or removed from, the rear of the car.

## Components

The electronic level control system consists of the following components:
1. Compressor
2. Air adjustable shock absorbers
3. Electronic height sensor
4. Compressor relay (two with E series)
5. Exhaust solenoid
6. Air dryer
7. Wiring and air tubing
8. Pressure regulator (E series only)

The E and K series front drive cars (torsion bar front suspension) have a pressure limiter valve added to the system. This valve is located in the engine compartment in the pressure line which runs from the compressor to the shocks. The limiter allows a maximum of 85 psi (586 kPa) ± 5 psi (34 kPa) to reach the rear shocks.

### COMPRESSOR RELAY

This relay is a single pole single throw type that completes the 12V(+) circuit to the compressor motor when energized. The compressor relay is located on the compressor mounting bracket.

### COMPRESSOR

The basic compressor assembly is a positive displacement single piston air pump powered by a 12 volt DC permanent magnet motor. The compressor head casting contains piston intake and exhaust valves plus a solenoid.

### AIR DRYER

The air dryer is attached externally to the compressor output and provides a dual function.

1. It contains a dry chemical that absorbs moisture from the air before it is delivered to the shocks and returns the moisture to the air when it is being exhausted. This action provides a long chemical life.

2. The air dryer also contains a valving arrangement that maintains 8-15 pounds minimum air pressure in the shock absorbers (except the E and K series which have 14-20 lb. retention.

### EXHAUST SOLENOID

The exhaust solenoid is located in the compressor head assembly and provides two functions.

### Air Dryer

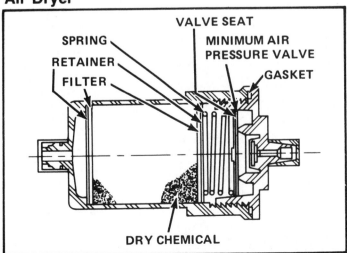

## Compressor Assembly

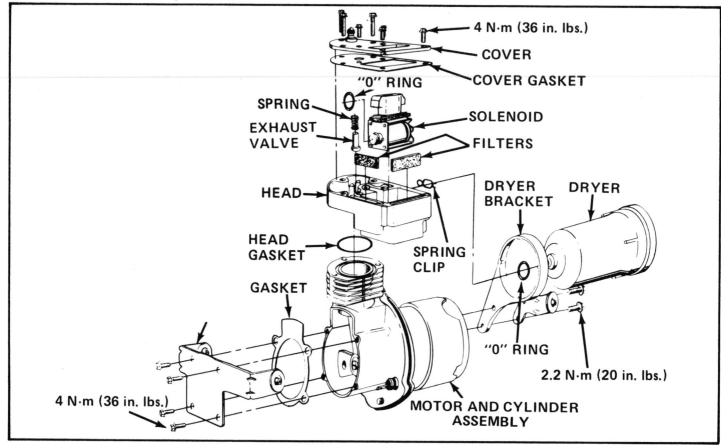

4 N·m (36 in. lbs.)
COVER
COVER GASKET
"O" RING
SPRING
SOLENOID
EXHAUST VALVE
FILTERS
HEAD
DRYER BRACKET
DRYER
HEAD GASKET
SPRING CLIP
GASKET
"O" RING
2.2 N·m (20 in. lbs.)
4 N·m (36 in. lbs.)
MOTOR AND CYLINDER ASSEMBLY

1. It exhausts air from the system when energized. The height sensor controls this function.

2. It acts as a blow off valve to limit maximum pressure output of the compressor.

### HEIGHT SENSOR

The height sensor is an electronic device that controls two basic circuits.

1. Compressor relay coil ground circuit.
2. Exhaust solenoid coil ground circuit.

To prevent falsely actuating the compressor relay or exhaust solenoid circuits during normal ride motions, the sensor circuitry provides an 8-14 second delay before either circuit can be completed.

In addition, the sensor electronically limits compressor run time or exhaust solenoid energized time to a maximum of 3½ minutes. This time limit function is necessary to prevent continuous compressor operation in case of a solenoid malfunction. Turning the ignition "off" and "on" resets the electronic timer circuit to renew the 3½ minute maximum run time. The height sensor is mounted to the frame crossmember in the rear. The sensor actuator arm is attached to the rear upper control arm by a link.

## Air Lines and Fittings

**NOTE: While the lines are flexible for easy routing and handling, care should be taken not to kink them and to keep them from coming in contact with the exhaust system.**

When the air line is attached to the shock absorber fittings or compressor dryer fitting the retainer clip snaps into a groove in the fitting locking the air line in position. To remove the air line, spread the retainer clip, release it from the groove and pull on the air line.

### Exhaust Solenoid

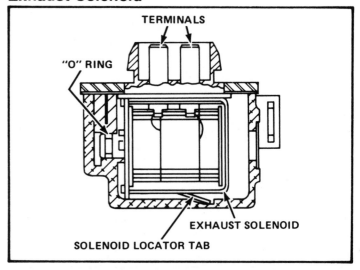

TERMINALS
"O" RING
EXHAUST SOLENOID
SOLENOID LOCATOR TAB

## System Operation Check

**NOTE: When certain tests require raising the car on a hoist, the hoist should support the rear wheels or axle housing. When a frame type hoist is used, two additional jack stands should be used to support the rear axle housing in its normal curb weight position.**

1. Select a suitable location at rear wheelhouse opening and measure distance to floor.

2. Start engine momentarily. Leave switch "ON".

3. Apply load to rear of car (two people or approximately 300-350 pounds).

a. There should be 8-14 second delay before compressor turns on and the car begins to raise.

b. Car should raise to within 3/4 in. (19 mm) of measurement made in step 1 by the time the compressor shuts off. If car does not raise, refer to the diagnosis chart.

**NOTE: Failure of car to return to within 3/4 in. (19 mm) of unloaded dimension can be caused by unusually heavy loading in the trunk which exceeds the capacity of the system. If this type of loading is encountered, remove it and repeat test.**

4. Remove load applied in step 3.

a. There should be 8-14 second delay before car begins to lower.

b. Car should lower to within 3/4 in. (19 mm) of measurement made in step 1 in less than 3½ minutes.

## COMPRESSOR/DRYER DIAGNOSIS CHART

| Malfunction | Correction |
|---|---|
| 1. Current draw exceeds 14 amps. | 1. Replace motor cylinder assembly. |
| 2. Compressor Inoperative. | 2. Replace motor cylinder assembly. |
| 3. Pressure build up OK but leaks down below 90 psi before holding steady. | 3. Replace solenoid exhaust valve assembly. |
| 4. Compressor pressure leaks down to 0 psi. | 4. Leak test compressor/dryer assembly. |
| 5. Compressor output less than 110 psi and current draw normal. | 5. Perform compressor/dryer leak test. If no leak is found, replace motor/cylinder assembly. |

## Height Sensor

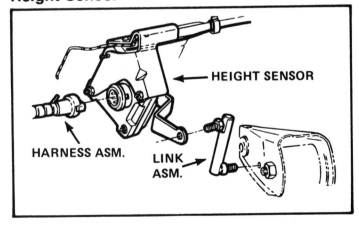

HEIGHT SENSOR

HARNESS ASM.

LINK ASM.

## Compressor/Dryer Performance Test

### COMPRESSOR CURRENT DRAW, PRESSURE OUTPUT AND LEAK DOWN TEST

1. Disconnect wiring from compressor motor and exhaust solenoid terminals.

## Airline Retainer Clip

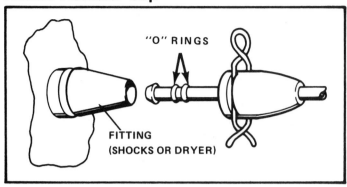

"O" RINGS

FITTING (SHOCKS OR DRYER)

2. Disconnect existing pressure line from dryer and attach pressure gauge to dryer fitting.

3. Connect ammeter to 12V source and to compressor.

a. Current draw should NOT exceed 14 amp.

b. When gauge reads 110-120 psi SHUT COMPRESSOR OFF and observe if pressure leaks down.

**NOTE: If compressor is permitted to run until it reaches its maximum output pressure, the solenoid exhaust valve will act as a relief valve. The resulting leak down when compressor is shut off will indicate a false leak.**

c. Leak down pressure should not drop below 90 psi when compressor is shut off.

## RESIDUAL AIR CHECK

1. Remove air line from dryer fitting and attach it to gauge. Attach gauge air line to dryer fitting.

2. Turn ignition "ON" and perform system check, to inflate shocks.

## COMPRESSOR/DRYER DIAGNOSIS CHART

| Malfunction | Correction |
|---|---|
| 1. Current draw exceeds 14 amps. | 1. Replace motor cylinder assembly. |
| 2. Compressor Inoperative. | 2. Replace motor cylinder assembly. |
| 3. Pressure build up OK but leaks down below 90 psi before holding steady. | 3. Replace solenoid exhaust valve assembly. |
| 4. Compressor pressure leaks down to 0 psi. | 4. Leak test compressor/dryer assembly. |
| 5. Compressor output less than 110 psi and current draw normal. | 5. Perform compressor/dryer leak test. If no leak is found, replace motor/cylinder assembly. |

## Height Sensor Operational Check/Adjustment

### OPERATIONAL CHECK

1. Turn ignition switch "ON" and raise car on hoist. If frame hoist is used, rear wheels or axle must be supported. Jacks should be adjusted upward until axle housing and/or wheels reach trim/curb weight position.

2. Compare neutral position of the height sensor metal arm with position of sensor arm being tested. (Shocks should have minimum air pressure.) If neutral position varies more than 3-4° check for correct sensor and/or link, sensor mounting bolts tight, sensor mounting bracket not bent. Make necessary corrections as required.

3. Disconnect link from height sensor arm.

4. Disconnect and reconnect wiring to height sensor to assure resetting the sensor time limit function. Failure to do this can result in erroneous diagnosis.

5. Move sensor metal arm upward approximately 1½-2 in. above neutral position. There should be 8-15 seconds delay before compressor turns "ON". As soon as shocks noticeably inflate move sensor arm down slowly and note arm position where compressor stops. This position should be very close to the neutral position.

6. Move arm down approximately 1½ in. below the point where the compressor stopped. There should be 8-15 seconds delay before shocks start to deflate. Allow shocks to deflate until only the retention pressure is left in the shocks (approximately 8-15 lbs.).

### TRIM ADJUSTMENT

**NOTE: Link should be attached to metal arm when making the adjustment.**

1. Loosen lock nut that secures metal arm to height sensor plastic arm.

2. To increase car trim heiggt move white plastic actuator arm upward and tighten lock nut.

**NOTE: If all adjustment is used up, check trim height.**

3. To lower car trim height, follow step 1 and move plastic arm down.

4. If adjustment cannot be made, check for correct height sensor.

### Height Sensor Operation Check

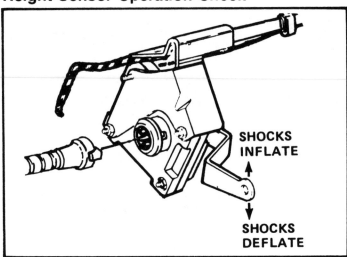

**SHOCKS INFLATE**

**SHOCKS DEFLATE**

### Height Sensor Adjustment

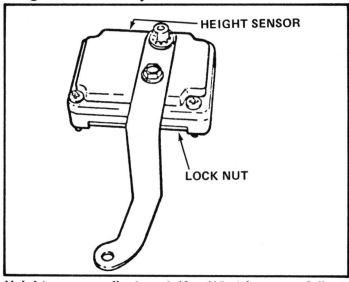

HEIGHT SENSOR

LOCK NUT

Height sensor adjustment 1° = ¼″ at bumper. Adjustment of 5° total.

## Front Drive

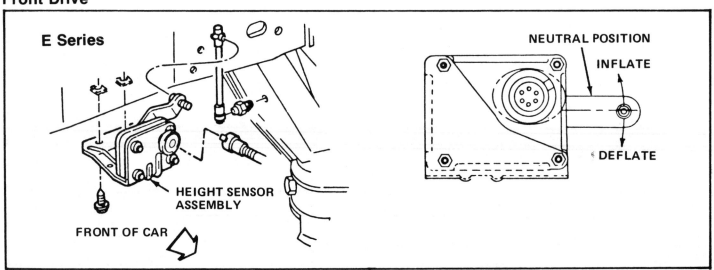

E Series

HEIGHT SENSOR ASSEMBLY

FRONT OF CAR

NEUTRAL POSITION

INFLATE

DEFLATE

## Rear Drive

### B-C Series

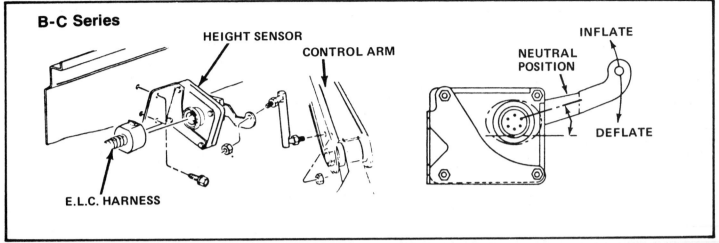

Neutral position may vary from 3 to 5°

### A Series

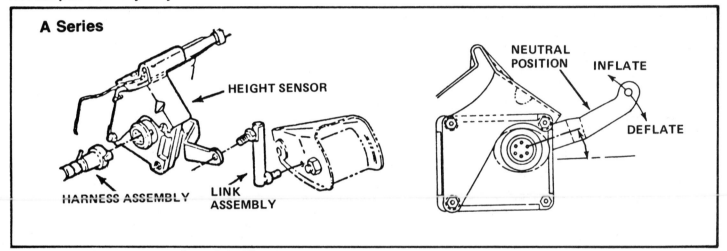

Neutral position may vary from 3 to 5°

## Automatic level control system component locations except Eldorado and Seville

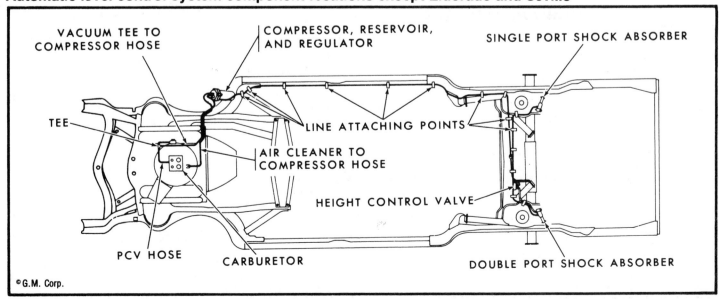

©G.M. Corp.

**Automatic level control system component locations (front drive cars)**

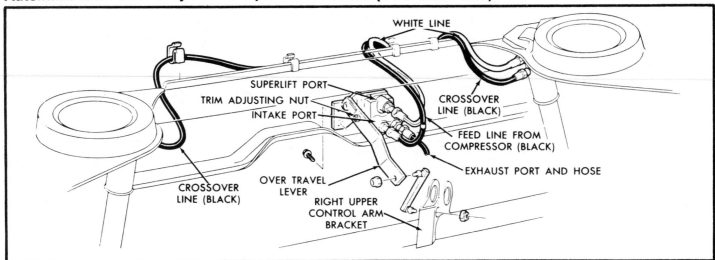

WHITE LINE

SUPERLIFT PORT

TRIM ADJUSTING NUT

INTAKE PORT

CROSSOVER LINE (BLACK)

FEED LINE FROM COMPRESSOR (BLACK)

EXHAUST PORT AND HOSE

CROSSOVER LINE (BLACK)

OVER TRAVEL LEVER

RIGHT UPPER CONTROL ARM BRACKET

**Automatic level control compressor mounting (front drive cars)**

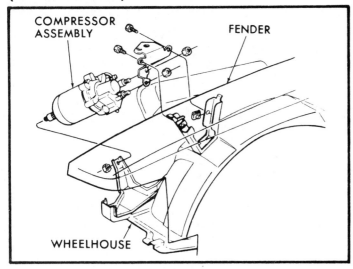

COMPRESSOR ASSEMBLY

FENDER

WHEELHOUSE

2. Install mounting brackets to compressor assembly and torque screws to 4 N·m (36 in.-lbs.).

3. Connect solenoid and motor connectors.

4. Install two radiator support to compressor bracket screws and torque to 6 N·m (48 in-lbs.).

5. Install support bracket screws and torque to 10 N·m (7 ft.-lbs.).

6. Install two compressor relay attaching screws.

7. Rotate clip on high pressure line until clip snaps in groove, then connect high pressure line at air dryer.

8. Cycle ignition switch and test for system operation and leaks at air dryer.

# Compressor and Bracket

### Removal

1. Remove negative battery cable.
2. Deflate system through service valve.
3. Disconnect high pressure line at air dryer by revolving spring clip 90° while holding connector end and removing tube assembly.
4. Remove two relay-to-compressor bracket screws and allow relay to hang to one side.
5. Remove support bracket screws.
6. Remove two radiator support to compressor bracket screws.
7. Disconnect solenoid and motor connectors.
8. Remove compressor and bracket assembly.
9. Remove three compressor mounting bracket screws then remove bracket.
10. If replacing compressor assembly remove dryer, dryer bracket, and compressor cylinder housing bracket and gasket.

### Installation

1. If compressor was replaced install dryer and bracket and torque to 2.2 N·m (20 in-lbs.).

# Air Dryer

### Removal

1. Deflate system through service valve.
2. Disconnect high pressure line at air dryer by revolving spring clip and removing tube assembly.
3. Disconnect air dryer from compressor by revolving spring clip and sliding air dryer assembly away from compressor head through its bracket. Remove "O" ring from compressor head.

### Installation

Lubricate dryer O-ring with Vaseline or equivalent before installing dryer in head casting.

1. Reverse removal procedure.
2. Check for leaks.

# Air Line Repair

The air lines used on the superlift shock absorbers and the electronic level control systems can be repaired by splicing in a coupling at the leaking area.

# STRUT OVERHAUL

## Strut Service and Repair

MacPherson struts are appearing on the front (and rear) wheels of more and more cars. The strut design takes up less room in the engine compartment, compared to a conventional upper and lower arm with shock absorber arrangement. The trend toward smaller, lighter and more efficient packaging mandates the use of a strut suspension to permit more room for engine accessories and front wheel drive components.

## Strut Suspension Design

In a conventional front suspension, the wheel is attached to a spindle, which is in turn, connected to upper and lower control arms through upper and lower ball joints. A coil spring between the control arms (sometimes on top of the upper arm) supports the weight of the vehicle and a shock absorber controls rebound and dampens oscillations.

In a strut type suspension, the strut performs a shock dampening function, like a shock absorber, but unlike a conventional shock absorber, the strut is a structural part of the vehicle's suspension.

The strut assembly usually contains a spring seat to retain the coil spring that supports the vehicle's weight. The shock absorber is built into the body of the strut housing. The strut is normally attached at the bottom to the lower control arm and at the top to the car body. The upper mount usually features a bearing that permits the coil spring to rotate as the wheels turn for

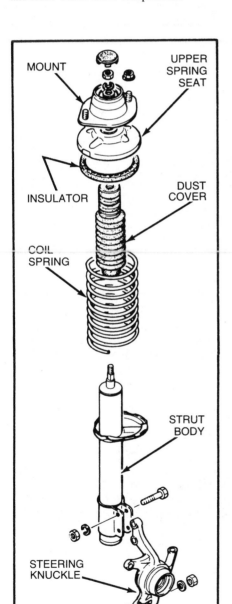

**Exploded view of a typical strut assembly**

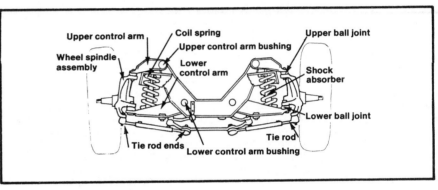

**Conventional upper and lower arm suspension**

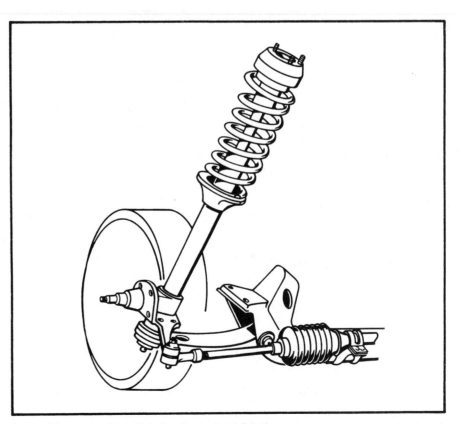

**Strut with concentric coil spring (rear wheel drive)**

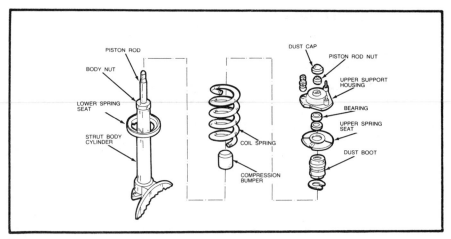

**Exploded view of a typical strut**

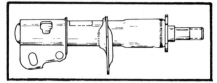

**A sealed strut has no body nut and is serviceable by replacement**

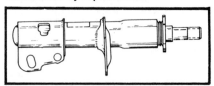

**Serviceable struts have a removeable body nut to allow replacement of the strut cartridge**

smoother steering. The entire design eliminates the need for the upper control arm, upper ball joint and many of the conventional suspension bushings. The lower ball joint is no longer a load carrying unit, because it is isolated from the weight of the vehicle.

Domestic struts have taken 2 forms—a concentric coil spring around the strut itself and a spring located between the lower control arm and the frame. GM and Chrysler (except for '82 and later Camaro and Firebird) use the traditional concentric coil spring around the strut. Ford (except the Escort and Lynx) and '82 Camaros and Firebirds use the spring off the strut between the lower control arm and frame. The location of the spring on the lower control arm instead of on the strut, allows minor road vibrations to be absorbed through the chassis rather than be fed back to the driver through the steering system.

## Serviceability

Struts fall into 2 broad categories—serviceable and sealed units. A sealed strut is designed so that the top closure of the strut assembly is permanently sealed. There is no access to the shock absorber cartridge inside the strut housing and no means of replacing the cartridge. It is necessary to replace the entire strut unit.

A serviceable strut is designed so that the cartridge inside the housing, that provides the shock absorbing function, can be replaced with a new cartridge. Serviceable struts use a threaded body nut in place of a sealed cap to retain the cartridge.

The shock absorber device inside a serviceable strut is generally "wet". This means that the shock absorber contains oil that contacts and lubricates the inner wall of the strut body. The oil is sealed inside the strut by the body nut, O-ring and piston rod seal.

Servicing a "wet" strut with the equivalent components involves a thorough cleaning of the inside of the strut body, absolute cleanliness and great care in reassembly.

Cartridge inserts were developed to simplify servicing "wet" struts. The insert is a factory sealed replacement for the strut shock absorber. The replacement cartridge is simply substituted for the original shock absorber cartridge and retained with the body nut, avoiding the near laboratory-like conditions required to service a "wet" strut with "wet" service components.

Most OEM domestic struts are serviced

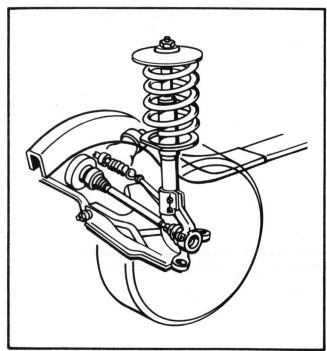

**Strut with concentric coil spring (front wheel drive)**

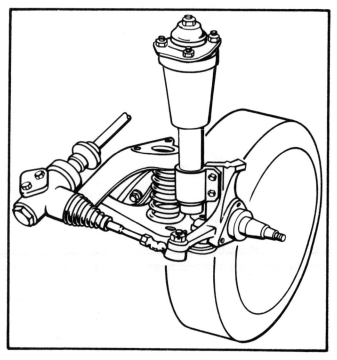

**Modified MacPherson strut design with coil spring on the lower arm**

by replacement of the entire unit. There is no strut cartridge to replace. Exceptions to this general rule are the struts used on GM front wheel drive J-cars and A-cars, which feature an internally threaded housing, accessible by removing the OEM cap from the housing. Once the old cartridge is removed, a new cartridge can be threaded into the housing.

Sealed, OEM units can also be serviced by replacement with an aftermarket unit, that will permit future servicing by cartridge replacement.

## WHEEL ALIGNMENT

It is not always necessary to re-align the wheels after struts are serviced. If care is taken matchmarking affected components and in reassembling, alignment may be unaffected. However, if wheels were not in proper alignment prior to service, or if the entire strut assembly was replaced, a wheel alignment check should be made. Generally, only camber is adjustable, and then only within a narrow range.

**NOTE: Do not attempt to bend components to correct wheel alignment.**

Since the majority of OEM struts are serviced by replacement, most manufacturers recommend wheel alignment following strut replacement.

## Tools

Without the right tools, a strut job will take longer than necessary and can be dangerous.

A normal selection of hand tools such as open end and box wrenches, sockets, pliers, screwdrivers and hammers are necessary to work on struts. Extensions and universal joints will help reach tight spots. Be sure to have both metric and inch-sized wrenches on hand. Two big time-savers are ''crowsfeet'' and ratcheting box wrenches in assorted sizes. Torx fasteners are also showing up more and more in chassis fasteners.

In addition to the normal handtools, some sort of spanner is necessary to remove the body nut on serviceable struts. Sometimes a pipe wrench can be used successfully.

Strut and cartridge replacement requires a spring compressor.

— CAUTION —
*Makeshift tools for compressing coil springs—threaded rod, chains, wire or other methods—should never be used. The coil spring is under tremendous compression and can fly off causing personal injury and damage to equipment. Use only a good quality spring compressor such as described below.*

Economy, or manual, spring compressors are the least expensive but more time consuming to use. Angle hooks grasp the

## MAINTAINING WHEEL ALIGNMENT

The location and method of adjusting wheel alignment determines the components that must be match-marked to maintain wheel alignment. There are 4 basic methods of adjusting wheel alignment. Almost all cars use one of these or a slight variation.

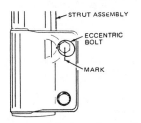

Mark the eccentric (camber adjusting bolt) relative to the clevis mounting bracket.

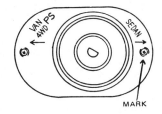

Mark the mounting stud that faces the front of the vehicle. This type of bracket is reversible for varying applications.

Mark the upper support housing relative to the inner fender before removing the strut from the upper mount.

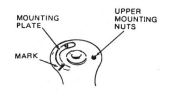

Mark the location of the mounting plate relative to the location on the inner fender.

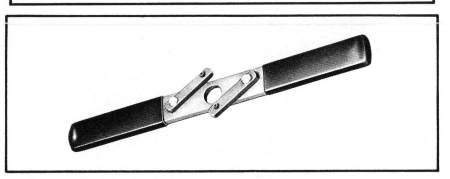

**A simple spanner wrench designed for use with body nuts equipped with recessed lugs. A pipe wrench is a frequent substitute.**

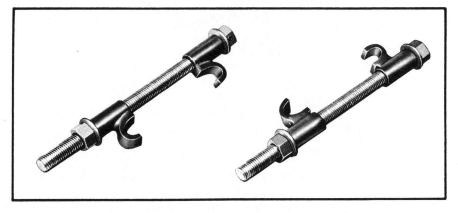

**An economical manual spring compressor**

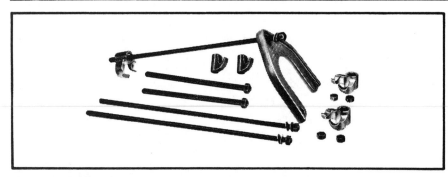

A manual spring compressor with plates or hooks for servicing virtually any strut.

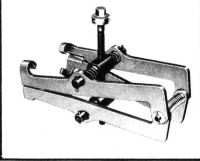

Lightweight, air operated, portable spring compressor can be used on or off the vehicle. Extra shoes are available to handle all strut applications

"Jaws" type spring compressor

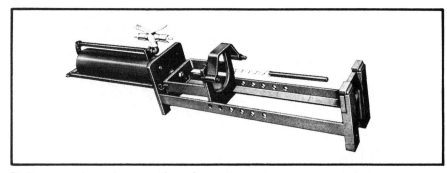

Stationary, universal pneumatic spring compressor

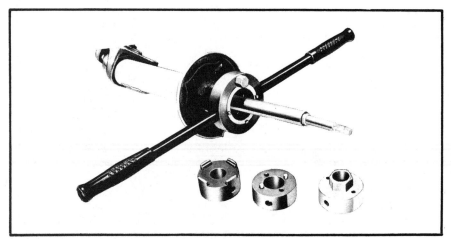

Spanner wrench with adaptor inserts for various applications of body nuts. This type of spanner can be used with a torque wrench for retorqueing the body nut.

spring coils and must be compressed with a wrench. For those who service struts infrequently, this is probably the wisest investment for purchase.

Other manual spring compressors (jaws type) are faster to operate, have a more positive gripping action and can be used on or off the car. These types are probably not cost effective for the do-it-yourselfer, but can be rented from auto supply stores for single-time use.

For volume work, compressors that are pneumatically or hydraulically operated are best. Air operated compressors are suitable for all types of struts (through use of adaptors), are lightweight and can be used on or off the vehicle. Bench mounted hydraulically operated units are probably the safest, but are also the most expensive and require that the strut be removed from the vehicle, which means separating brake lines and other connections which can be time consuming.

There are also universal kits that fit all struts in either the manual or air operated types.

Regardless of what type of spring compressor you're using, GM front wheel drive A-, J-, and X-cars as well as Chrysler Corp. Omni, Horizon and K-cars, require the use of a special spring compressor with self-leveling plates to grasp the spring seats as the spring is compressed. Likewise, the portable, pneumatic units have extra wide shoe sets suitable for these cars. The shoes are also epoxy coated to avoid scratching the coated springs on these models.

GM front wheel drive A-, J- and X-cars also make use of a camber assist tool, that makes camber adjustment a one man job.

A tube cutter is necessary on GM J-cars to cut the welded top from the strut housing for cartridge replacement.

Spring compressor for GM and Chrysler product applications

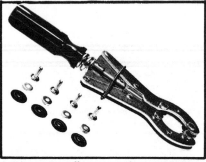

A tube cutter allows opening of the GM J-car struts for cartridge replacement

A camber assist tool makes GM cars a one-man job

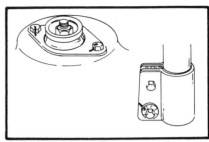

Mark the position of the attachments that control wheel alignment. See Maintaining Wheel Alignment earlier in this section

## Repair Tips

1. Make sure you have all the tools you'll need. NEVER IMPROVISE A SPRING COMPRESSOR.

2. Normally both front struts should be repaired or replaced at the same time.

3. The easiest way to work on most struts is to remove the entire unit from the vehicle, unless you have access to an air operated spring compressor. Some struts, however, can, and should, be repaired while installed on the vehicle.

4. Always read the instructions packaged with any replacement parts. In particular, note whether the body nut is supplied new or re-used.

5. Mark the position(s) of any bearing plate nuts or cam bolts to assure proper alignment after installation.

6. Be sure to protect the rubber boot on the drive axle of front wheel drive cars.

7. If necessary to remove the brake caliper, do not let the caliper hang by the brake hose. Suspend the caliper from a wire hook or rope.

8. Be careful in clamping a strut in a vise. Special fixtures are available to hold struts in a vise, but are not necessary if care is used to be sure the housing is not crushed or dented. A block of soft wood on either side of the housing will prevent most damage.

9. Use a spring compressor to relieve tension from the spring. Be sure to clean and lubricate the screw threads, particularly on hand operated (manual) spring compressors.

**NOTE: Some springs have a special coating that should not be scuffed.**

10. If you are replacing the strut cartridge, clean the inside of the strut housing and the body nut threads before replacing the oil and installing a new cartridge.

11. Be sure to use OEM quality fasteners any time a fastener is replaced.

## STRUT OVERHAUL (OFF-CAR)

Following is a typical overhaul procedure of a serviceable MacPherson strut, after having removed the strut from the vehicle. The vehicle should be firmly supported. If it is necessary, to separate the brake line from the strut for strut removal, the brakes will have to be bled after reinstallation. See the manufacturer's car section for specific MacPherson strut removal and installation procedures.

Photos Courtesy Gabriel Div., Maremont Corp.

Examine the strut assembly for damage, dented strut body, spring seat, broken or missing strut mounting parts. Any of these will require replacement of the complete assembly. Also inspect other suspension components for wear or damage.

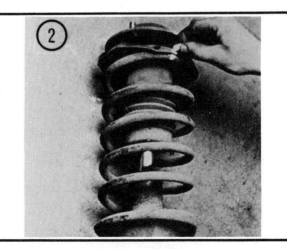

Matchmark the upper end of the coil spring and bearing plate to avoid confusion during reassembly.

To make servicing easier, clamp the strut in a strut vise. The strut vise is designed to clamp the strut tight without damage to strut cylinder. It is very handy for strut work and can be used in your shop vise or mounted to any bench

Before using the manual spring compressor, lubricate both sides of the thrust washers and the threads with a light coat of grease

Install the compressor hooks on opposite sides of the coil spring with the hooks attached to the upper-most and lower-most spring coils. To avoid possible slippage, use tape or small hose clamps on either side of the compressor hooks

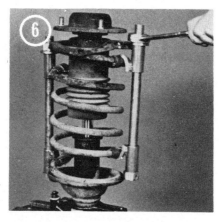

Alternately tighten the bolts a few turns at a time until all tension is removed from the spring seat

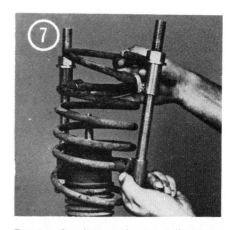

Remove the piston rod nut and disassemble the upper mounting parts, keeping them in order for reassembly. Remove the coil spring. There is no need to remove the compressor from the coil spring

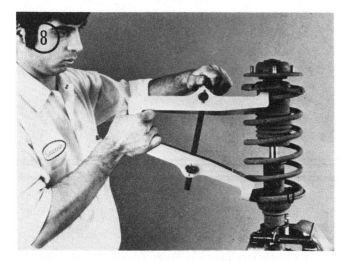

An alternative to the manual compressor is the "jaws" type. Turn the load screw to open or close the compressor until the maximum number of spring coils can be engaged

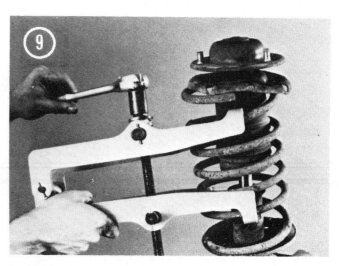

Tighten the load screw until the coil spring is loose from the spring seats. There is no need to compress the spring any further

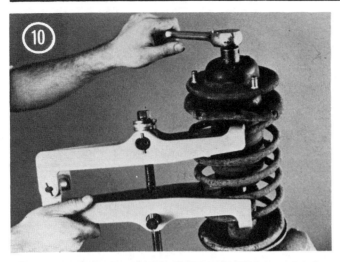

Remove the piston rod nut and disassemble the upper mounting parts

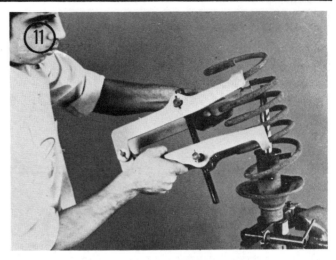

Like the manual compressor, there is no need to remove the compressor from the coil spring. Remove the coil spring and compressor

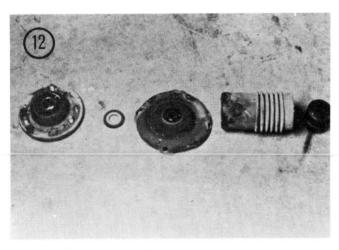

Keep the upper mounting parts in order of their removal. They'll be re-assembled in reverse order

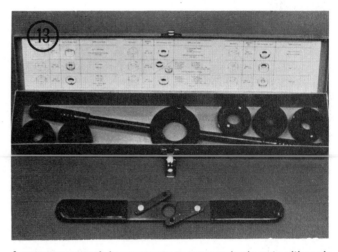

A spanner wrench is necessary to remove body nuts although a pipe wrench will do the job

Use the spanner wrench or pipe wrench to loosen the body nut

Remove the body nut and discard if a new body nut came with the replacement cartridge. If not, save the body nut

Use a scribe or suitable tool to remove the O-ring from the top of the housing

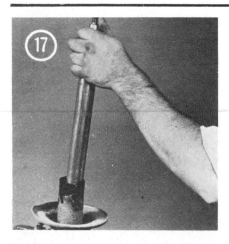

Grasp the piston rod and pull cartridge out of the housing. Remove it slowly to avoid splashing oil. Be sure all pieces come out of the housing

Pour all of the strut fluid into a suitable container, clean the inside of the strut cylinder, and inspect the cylinder for dents and to insure that all loose parts have been removed from inside of strut body

Refill the cylinder with one ounce (a shot glass) of the original oil or fresh oil. The oil helps dissipate internal cartridge heat during operation and results in a cooler running, longer lasting unit. Do not put too much oil in—otherwise the oil may leak at the body nut after it expands when heated

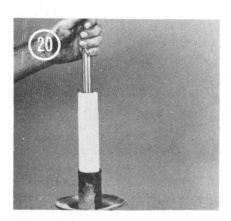

Insert the new replacement cartridge into the strut body

Push the piston rod *all* the way down, to avoid damage to the piston rod if the spanner wrench slips, and start the body nut by hand. Be sure it is not cross-threaded

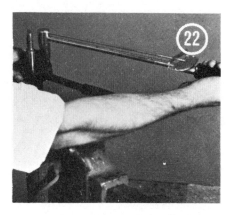

Tighten the body nut securely

Inspect the loose parts prior to re-assembly. Note the chalk mark location for proper seating of the upper spring seat

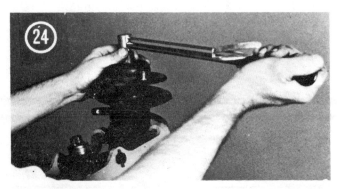

Re-assemble the coil spring and upper mounting parts in reverse order. Tighten the piston rod nut and remove the spring compressor. Install the dust cap. Install the strut in the vehicle. See the car section for details

## STRUT OVERHAUL

Most domestic car OEM MacPherson struts are sealed units and not repairable. The exceptions are GM front wheel drive A- and J-cars, which use replaceable cartridges. All other cars must use aftermarket struts to be serviceable at a future date. The following procedures cover disassembly of the strut, installation of a serviceable strut, reassembly and cartridge replacement on GM front wheel drive A- and J-models. Consult the applicable manufacturer's car section for removal and installation procedures.

Photos Courtesy Gabriel Div., Maremont Corp.

Most domestic cars are serviced initially by replacing the entire strut rather than by using a replacement cartridge. This is necessary because the original equipment struts are sealed shut and cannot be serviced with a replacement cartridge. Aftermarket struts are designed with serviceable threaded body nuts which means they can be serviced in the future by installing a replacement cartridge, using normal cartridge service methods, rather than by replacing the entire strut.

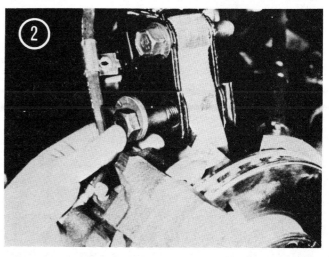

An X-Car is shown, but the lower mount on the Citation is typical of many vehicles. They all have two bolt clevis mounts and the position of the strut determines the camber adjustment. This means that if you are replacing a sealed strut, front end alignment is necessary because the original alignment is eliminated when you change the strut. If the car has a serviceable strut, you can retain the alignment by marking the position of the mounting bolt relative to the strut.

GM has made a running change on the lower mount of their X-Car. The earlier type had an eccentric bolt for camber adjustment. Camber on the latest type is adjusted by pushing or pulling on the wheel with the bolts loosened slightly, but the eccentric can be installed on later cars.

A special type spring compressor is required for the GM cars and Chrysler K and L cars. A compressor should be used that does not damage the protective coating on the coil spring. Virtually any compressor can be used on other car lines/models

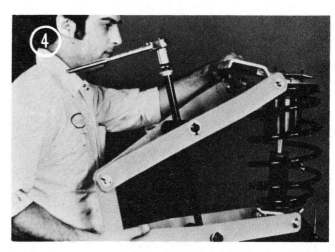

Secure the strut in the strut vise, turn the load screw counter-clockwise until the lower plate can be fitted under the lower spring seat and the upper plate can be fitted between the upper spring seat and support housing

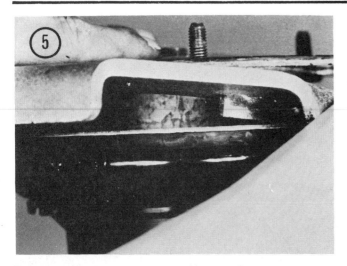

Make sure that the crescent shaped bars on the upper compression plate are located inside the upper spring seat

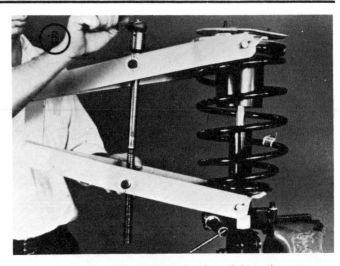

Turn the load screw clockwise enough to tighten the compression plates on the spring seats. Stop and make sure that the coil spring will not arch, and that the pivot points are aligned with the center-line of the coil spring

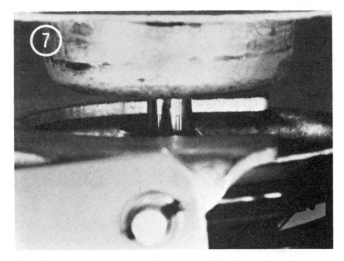

Continue to tighten the load screw until the upper support housing can be pulled up to expose about ½ inch of piston rod. This assures that the spring load has been removed from upper spring seat

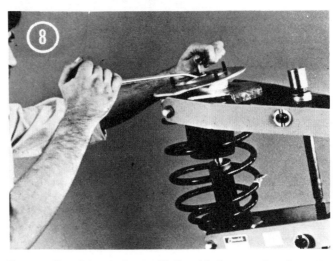

Remove the piston rod nut with the aid of a wrench to keep the piston rod from turning and remove upper support housing

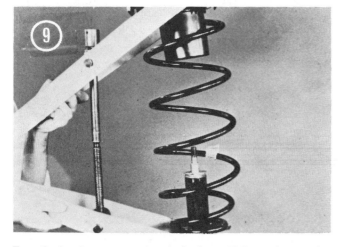

Turn the load screw counter-clockwise until the spring tension is completely relieved. Remove the compressor, coil spring and upper support housing from the strut

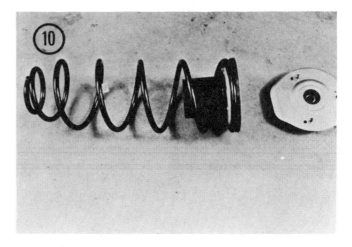

Assemble the upper mounting parts in order of their removal. They'll be re-assembled in reverse order

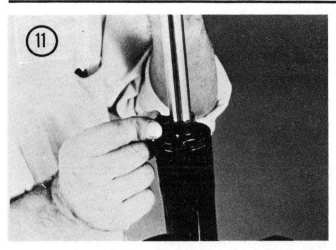

Place the new strut in the vise and extend piston rod fully and install clip (spring type clothes pin will do) as shown. This keeps the piston rod extended while assembling the spring and upper mounting parts.

Install the coil spring and upper spring seat on the new strut

Make sure that the spring helix is aligned with the lower spring seat.

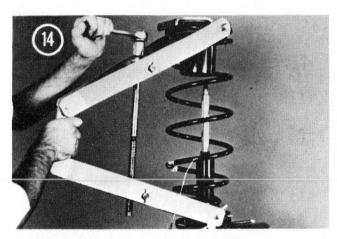

Locate upper and lower compression plates on spring seats

Make sure that the crescent shaped bars on the upper compression plate are located on the upper spring seat as shown. Turn the load screw clockwise enough to tighten the compression plates on the upper and lower spring seats. Stop. Again, to assure that the coil spring will not arch, make sure that the pivot points are aligned with the centerline of the coil spring. Then continue turning the load screw clockwise until about 1-½ inches of piston rod is showing above the upper spring seat

Install upper support housing on piston rod. Tighten the piston rod nut and remove the compressor from the strut and the strut from the vise. Install the strut. See the car section for details

## GM J- AND A-CARS ONLY

Place the strut assembly in a vise, and compress the coil spring. Remove the piston rod nut, upper support housing, spring seat and coil spring. If the universal pneumatic spring compressor is used, an adaptor provided with the compressor should be fastened to the strut under the steering arm. The ears of the adaptor should be aligned with steering arm. The adaptor provides a square seating surface for the strut while it is being compressed

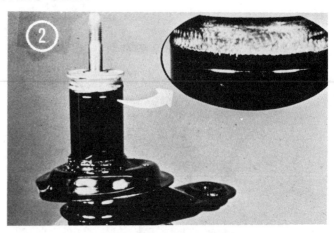

J and A-car struts have a welded upper closure, but the strut is designed so the damping mechanism can be replaced with a cartridge insert. Just below the spin weld there is a cut-line scribed in the strut body

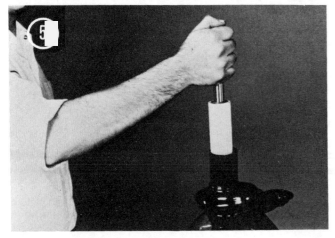

Using a pipe cutter, cut open the strut body at the scribed line. Note: *It is important that the cut be made on the cut-line*

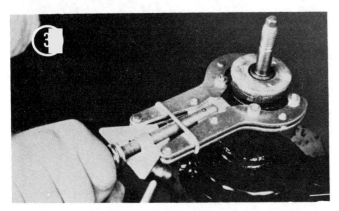

Remove the cartridge and oil from the strut. Note the threads on the inside of the strut. Deburr the top of the strut body if necessary

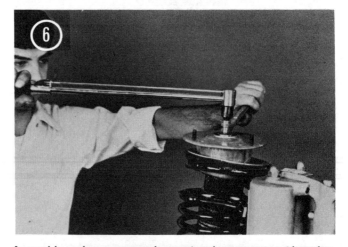

Pour about one ounce of oil into the strut body and insert the replacement cartridge. Push the piston rod down and start the body nut by hand. Tighten the nut securely.

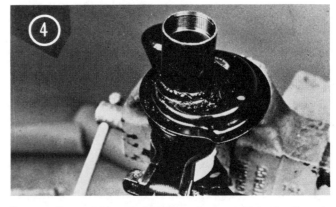

Assemble spring, upper spring seat and upper support housing on the strut and tighten the new piston rod nut. The renewed strut is now ready to install on the vehicle. Release the spring tension

## MACPHERSON STRUT PROBLEM DIAGNOSIS

Problems with MacPherson struts generally fall into 3 main categories: suspension, tire wear and steering. In general, the symptoms encountered are not significantly different from those encountered on conventional suspensions.

### Suspension

#### Sag

Vehicle "sag" is a visible tilt of the car from one side to the other or one end to the other while parked on a level surface.

Weak or damaged strut springs could cause this condition and should be repaired immediately.

Sag will also cause steering and tire wear problems to be more pronounced and vehicle instability on rough roads. Front wheel alignment will not solve the problem.

Weak strut springs increase vehicle sag. See "Tire Cupping".

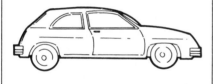

#### Cartridge Leaks

Strut cartridge leaks (not seepage) indicate the need for cartridge or strut replacement. Be sure the leakage is coming from the strut, and not from elsewhere on the vehicle.

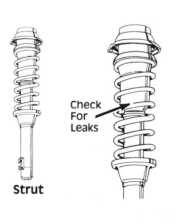

Check For Leaks

**Strut**

### Abnormal Tire Wear

#### Wear on One Side

One sided tire wear indicates incorrect camber. Check the causes in the accompanying illustration and be sure the wheel alignment is correct.

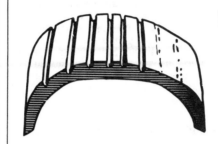

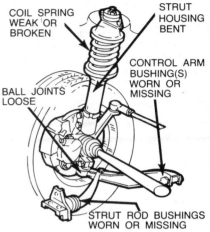

COIL SPRING WEAK OR BROKEN — STRUT HOUSING BENT — CONTROL ARM BUSHING(S) WORN OR MISSING — BALL JOINTS LOOSE — STRUT ROD BUSHINGS WORN OR MISSING

#### Tire "Cupping"

Cupped tires indicate any or all of the following problems.

1. A weak strut cartridge can be verified by bouncing each corner of the car vigorously and letting go. The car should not bounce more than once, if the shock absorber cartridges are good.

2. Weak strut springs allow sag to increase with only a slight amount of downward pressure. A visual inspection will reveal any broken springs or shiny spots.

3. Check for loose or worn wheel bearings with the weight of the car off of the wheel.

4. Check the wheel balance.

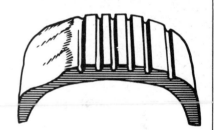

#### Tread Edge Wear

Wear along tread edges (feathering) indicates a suspension or steering system problem.

1. Strut rod bushings are worn or missing.

2. Tie rod end wear can be determined by grabbing the tie rod end firmly and forcing it up, down or sideways to check for lost motion.

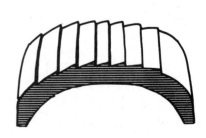

## MACPHERSON STRUT PROBLEM DIAGNOSIS

Problems with MacPherson struts generally fall into 3 main categories: suspension, tire wear and steering. In general, the symptoms encountered are not significantly different from those encountered on conventional suspensions.

### Steering

### Tires

Both front tires should match and both rear tires should match. Be sure air pressure is correct.

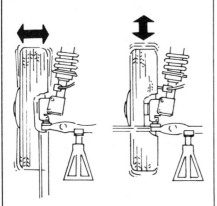

### Ball Joints

Support the car under the frame or crossmember so that the jack does not interfere with the control arm. Rock the tire in and out and up and down. Excessive movement means that both ball joints should be replaced.

Struts with lower weight-carrying ball joints should be supported at the outer edge of the lower control arm. These vehicles usually have wear indicating ball joints that can be checked visually.

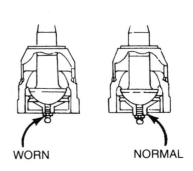

WORN        NORMAL

### Stabilizer Bar Bushings

Check for worn bushings or lost motion with the vehicle level and the weight evenly distributed on all wheels.

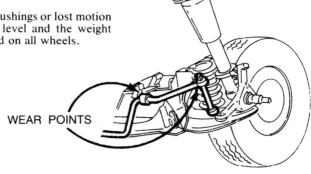

WEAR POINTS

### Strut Rod Bushings

Grasp the strut rod and shake it. Any noticeable play indicates excessive wear and need for parts replacement.

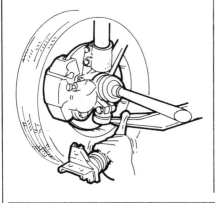

### Control Arm Bushings

Support the car under the frame or body and remove the weight from the wheel and control arm. Check for free-play in the bushings at the pivot point, using a pry bar.

**NOTE: Some control arm bushings are serviceable only by replacing the entire arm.**

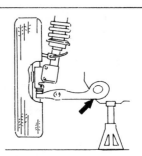

### Strut Assembly

Check the strut assembly for cracks or dents in the housing. Look for worn, bent or loose piston rods or dents that will inhibit piston rod movement.

### Steering Gear

Check for worn steering gear or loose or worn mounting bolts and bushings.

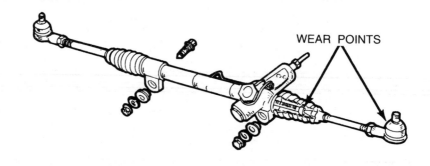

WEAR POINTS

## ROAD TEST TROUBLESHOOTING

Following are possible solutions to common potential problems which might be noticed during the road test after strut service is completed. Many are not exclusively strut service related.

| Problem | Correction |
|---------|-----------|
| Brake pedal low or soft | Bleed brakes<br>Check for leaks<br>    Brake lines<br>    Wheel cylinder<br>    Caliper piston seal |
| Erratic steering | Check upper support housing components for proper assembly<br>Check spring assembly right side up<br>Check for spring helix riding correctly on spring seat<br>Check wheel alignment |
| Noises and rattles | Check torques<br>    Piston rod nut<br>    Upper support housing nuts & bolts<br>    Lower mounting nuts & bolts<br>    Body nut<br>Check cartridge assembly in the body<br>    Spacer used<br>    Centering collar used |

# AMC MANUAL RACK AND PINION-PACER

## DESCRIPTION

The manual steering gear rack and pinion design combines the steering gear and steering linkage into one compact assembly.

### Steering Gear Inspection

1. After removing assembly from car, place in a vise using protective jaws.

——————— CAUTION ———————
*Do not clamp any part of tube in vise clamp housing only in vise.*

2. Cut and remove large diameter boot clamp on housing end of gear, and slide boot away from housing end of gear.
3. Turn flexible coupling to expose as many rack teeth as possible.
4. Clean and check rack teeth for signs of chipped, cracked, broken, excessive wear or tooth flaking.
5. If any of these signs appear the gear assembly must be replaced.

**NOTE: Do not replace steering gear if rack teeth have machining marks on them or appear excessively bright or shiny. These are normal conditions.**

6. If teeth are in good condition, remove flexible coupling pinch bolt and separate coupling from pinion shaft.
7. Remove adjuster plug.
8. Remove pinion shaft from housing by pulling up and rotating counterclockwise.
9. Clean and inspect pinion shaft, if teeth are chipped cracked broken or excessively worn, steering gear assembly must be replaced.

### Disassembly

1. Remove contraction plug from housing, using 1/4 inch diameter brass rod, insert rod through upper and lower pinion bushings and tap on rod to dislodge plug.
2. Remove lower pinion bushing and preload spring from housing using brass rod.
3. Move rack to center position in tube housing.
4. Install pinion shaft and adjuster plug in housing, hand tighten adjuster plug.
5. Loosen adjuster tube clamp nuts and remove adjuster tubes and tie rod assemblies from inner tie rods.

**NOTE: Mark position of adjuster tubes on inner tie rods for assembly reference.**

6. Mark location of breather tube on tube and housing assembly.

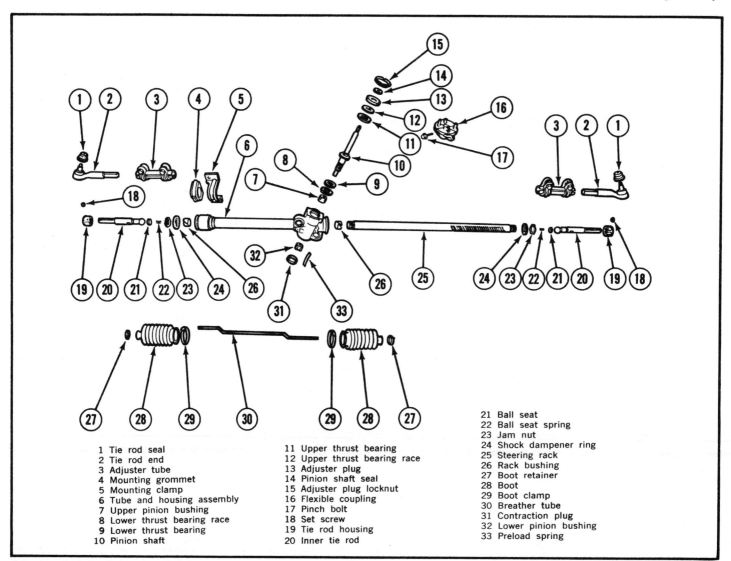

| | |
|---|---|
| 1 Tie rod seal | 11 Upper thrust bearing |
| 2 Tie rod end | 12 Upper thrust bearing race |
| 3 Adjuster tube | 13 Adjuster plug |
| 4 Mounting grommet | 14 Pinion shaft seal |
| 5 Mounting clamp | 15 Adjuster plug locknut |
| 6 Tube and housing assembly | 16 Flexible coupling |
| 7 Upper pinion bushing | 17 Pinch bolt |
| 8 Lower thrust bearing race | 18 Set screw |
| 9 Lower thrust bearing | 19 Tie rod housing |
| 10 Pinion shaft | 20 Inner tie rod |

| |
|---|
| 21 Ball seat |
| 22 Ball seat spring |
| 23 Jam nut |
| 24 Shock dampener ring |
| 25 Steering rack |
| 26 Rack bushing |
| 27 Boot retainer |
| 28 Boot |
| 29 Boot clamp |
| 30 Breather tube |
| 31 Contraction plug |
| 32 Lower pinion bushing |
| 33 Preload spring |

**Pacer manual steering rack and pinion**

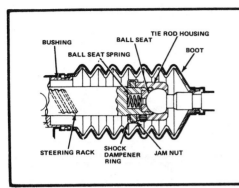

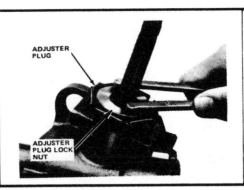

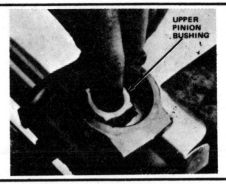

**Cross section of tie rod housing and inner tie rod**

**Removing the adjuster plug**

**Removing the upper pinion shaft bushing**

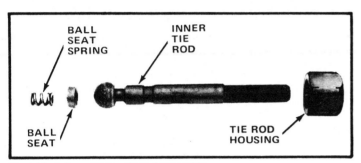

**Exploded view of inner tie rod assembly**

Boots and breather tube must be installed in same position to ensure proper sealing.

7. Cut and remove all boot clamps and remove boots.
8. Remove breather tube and shock dampener rings.
9. Clamp inner tie rod housing in vise and loosen jam nut.

### CAUTION

*Tie rod housing must be held securely when loosening or tightening jamnut to prevent damage to internal components of steering gear.*

10. Loosen tie rod housing set-screws and remove inner tie rod housing, ball seats, springs, jam nuts and shock dampener rings from rack.
11. Remove adjuster plug and pinion shaft from housing.
12. Remove pinion shaft from housing by pulling upward and rotating counterclockwise. Remove lower thrust bearing and race.
13. Remove upper pinion bushing from housing with fingers.
14. Pull steering rack from tube and housing assembly.
15. Remove rack bushings by slipping knife blade under bushings, use needlenose pliers and pull straight out.
16. Remove pinion shaft seal by threading adjuster plug into housing and prying seal out with screwdriver.

### Assembly

1. Install a new pinion shaft seal in the adjuster plug using a suitable sized socket. Press the seal in until it is flush with the face of the adjuster plug. Do not press on the lip of the seal.
2. Replace the rack bushings, if removed. Compress the leading end and force them into the housing or tube opening. Once they are past the lip of the tube or housing, the bushing will snap back to its original shape.
3. Coat the rack teeth with lithium based chassis grease and install the rack in the housing. Install the upper pinion bushing in the housing.
4. Lubricate the pinion shaft lower race and thrust bearing with the same type grease and install the race and thrust bearing in the housing with the flanged edge of the race facing up.
5. Center the steering rack in the housing. Set the distance between the end of the steering rack and the inner lip of the housing at 4 in.
6. Start the pinion shaft into the housing and rack with the flat on the splined end of the pinion shaft at about the 10 o'clock position. Turn the pinion shaft counterclockwise and push down until the pinion shaft race is bottomed on the thrust bearing.
7. Reset the distance between the end of the rack and the housing (step 5). The flat on the pinion shaft should be at the 3 o'clock position now. Be sure the pinion race is bottomed in the housing. If the flat on the pinion shaft is not at the 3 o'clock position with the rack set at 4 in., start over again at step 4.

**NOTE: The rack must be centered; otherwise, the steering wheel travel from left to right will be unequal.**

8. Install the adjuster plug using a spanner type tool that fits in the two holes in the top of the plug. Tighten the plug until it bottoms. Mark the adjuster plug and housing at a spanner hole. Back off the adjuster plug (counterclockwise) until the hole marked is 3/16 in to 1/4 in. past (counterclockwise) the reference mark made on the housing. Install and tighten the locknut to 50 ft./lbs.

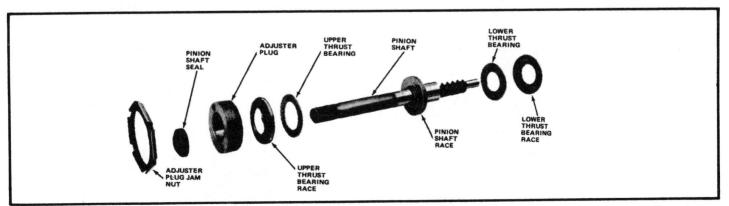

**Exploded view of adjuster plug and pinion shaft assembly**

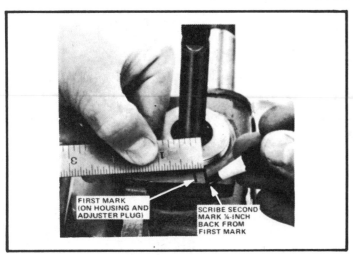

Marking the housing for the adjustment of the adjusting plug

Backing off the adjusting plug to the proper adjustment

9. Turn the assembly over and mount it in a vise. Fill the space around the pinion shaft with the same type of grease (step 3). Do not overfill because the pinion bushing and spring have to be installed yet.

10. Install the preload spring in the housing with the center hump of the spring bearing against the housing. Allow the spring to extend about 1/4 in. from the end of the housing.

11. Hold the top of the preload spring against the housing with needlenose pliers and install the busing in the housing with the chamfered end facing downward.

12. Seat the contraction plug in the housing using a brass rod or a suitable size socket.

13. Install the shock-dampener rings on each end of the steering rack with the open ends facing out and install both jamnuts.

14. Liberally apply some lithium based chassis lubricant to all of the inner tie rod assembly wear surfaces. Pack the tie rod housings with the same type grease.

15. Assemble and install the inner tie rod assemblies to the rack. Tighten the tie rod housing to 75 in./lbs. while rocking the inner tie rod to relieve grease lock, loosen 1/2 turn and retighten the housing to 50 in./lbs.

16. Tighten the housing setscrews to 60 in./lbs.

17. Clamp the tie rod housings in the vise and tighten the jamnuts to 100 ft./lbs. using a crow-foot adapter on the end of a torque wrench.

18. Slip the shock dampener rings over the jamnuts.

19. Install the mounting clamp and grommet on the tube using the alignment marks made during disassembly.

20. Install the boot on the mounting bracket side of the tube and housing in position so that the hole in the boot aligns with the hole in the mounting grommet. Slide the short end of the tube through the grommet and boot breather tube holes. The long end of the tube lies against the tube and housing.

**NOTE: The breather tube transfers air from one boot to the other during the turning of the front wheels. If the tube is blocked in any way, dust and water could be drawn into the inner tie rod assemblies.**

21. Install the opposite side boot with the hole in the boot aligned with the breather tube. The boot lip must fit into the housing flange to seat the tube.

22. Slide the small outer collars of the boots over the inner tie rod grooves. Install the small diameter boot clamps on the boots and tighten the clamps.

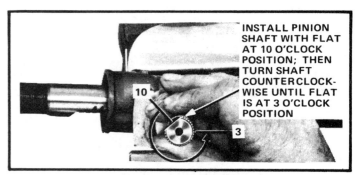

Installing the pinion shaft and engaging the rack

INSTALL PINION SHAFT WITH FLAT AT 10 O'CLOCK POSITION; THEN TURN SHAFT COUNTERCLOCKWISE UNTIL FLAT IS AT 3 O'CLOCK POSITION

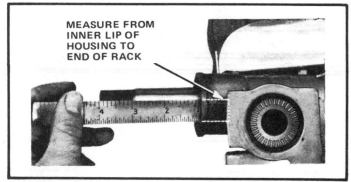

MEASURE FROM INNER LIP OF HOUSING TO END OF RACK

Centering the steering rack in the tube and housing

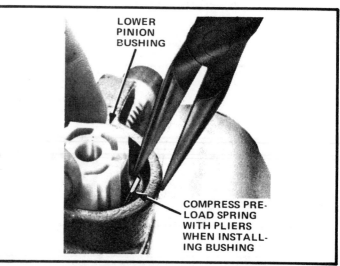

LOWER PINION BUSHING

COMPRESS PRELOAD SPRING WITH PLIERS WHEN INSTALLING BUSHING

Installing the lower pinion bushing

23. Install the adjuster tubes and tie rod ends on the inner tie rods. Align the tubes and tie rods using the marks made during disassembly. At least 3 threads should be visible at both ends of the adjuster tubes. The number of threads per side should not differ by more than 3.

24. Install the flexible coupling on the pinion shaft, flat to flat, and install the pinch bolt. Tighten the pinch bolt to 30 ft./lbs.

25. Install the steering gear in the car and check the toe-in adjustment.

# CHRYSLER MANUAL RECIRCULATING BALL TYPE

## Disassembly and Assembly

1. Attach the steering gear assembly to a holding fixture and put the holding fixture in a bench vise. Thoroughly clean the outside surface before disassembly.

2. Loosen the cross-shaft (sector shaft) adjusting screw locknut, and back out the adjusting screw about two turns to relieve the mesh load between the ball nut rack and the sector gear teeth. Remove the cross-shaft seal as given in the procedure for cross-shaft seal replacement.

3. Position the steering gear worm shaft in a straight ahead position.

4. Remove the attaching bolts from the cross-shaft cover and slowly remove the cross-shaft while sliding arbor tool into the housing. Remove the locknut from the adjusting screw and remove the screw from the cover by turning screw clockwise. Slide the adjustment screw and its shim out of the slot in the end of the cross-shaft.

5. Loosen the worm shaft bearing adjuster locknut with a brass drift (punch) and remove the locknut. Hold the worm shaft steady while unscrewing the adjuster. Slide the worm adjuster off the shaft.

––––––––– CAUTION –––––––––

*Handle the adjuster carefully to avoid damaging the aluminum threads. Also, do not run the ball nut down to either end of the worm shaft to avoid damaging the ball guides.*

6. Carefully remove the worm shaft and ball nut assembly. This assembly is serviced as a complete assembly only and is not to be disassembled or the ball return guides removed or disturbed.

7. Remove the cross-shaft needle bearing by placing the gear housing in an arbor press; insert tool in the lower end of the housing and press both bearings through the housing. The cross-shaft cover assembly, including a needle bearing or bushing, is serviced as an assembly.

8. Remove the worm shaft oil seal from the worm shaft bearing adjuster by inserting a blunt punch behind the seal and tapping alternately on each side of the seal until it is driven out of the adjuster.

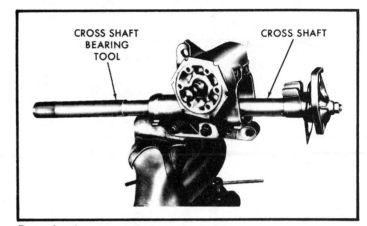

**Removing the cross shaft**

9. Remove the worm shaft upper bearing cup in the same manner as that given in step 8. *Be careful not to cock the bearing cup and distort the adjuster counter bore.*

10. Remove the lower cup if necessary by placing the locking head jaws of remover tool C-3868 behind the bearing cup and expanding the remover head by pressing down on the center plunger of the tool. Pull the bearing cup out by turning the remover screw clockwise while holding the center screw steady.

11. Wash all parts in clean solvent and dry thoroughly. Inspect all parts for wear, scoring, pitting, etc. Test operation of the worm shaft and ball nut assembly. If ball nut does not travel smoothly and freely on the worm shaft or if there is binding, replace the assembly.

**NOTE: Extreme care must be taken when handling the aluminum worm bearing adjuster to avoid thread damage. Also, be careful not to damage the threads in the gear housing. Always lubricate the worm bearing adjuster before screwing it into the housing.**

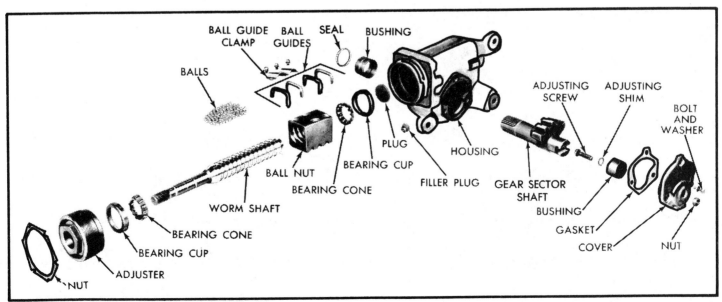

**Exploded view of manual steering gear - Chrysler Corp.**

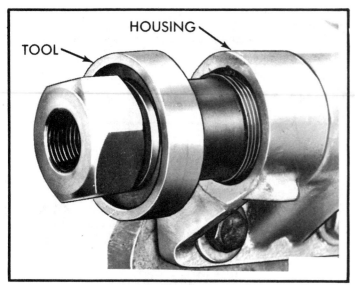

**Removing cross shaft oil seal**

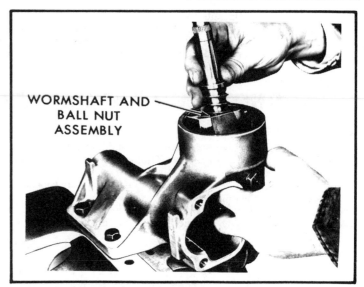

**Removing worm and ball nut assembly**

12. Inspect the cross-shaft for wear and check the fit of the shaft in the housing bearings. Inspect the fit of the shaft pilot in cover bearing. Be sure the worm shaft is not bent or damaged. The cross-shaft and wormshaft oil seals should be replaced when the unit is reassembled.

13. Install the cross-shaft lower needle bearing. Press the bearing into the housing about 7/16 in. below the end of the bore to leave space for the new oil seal.

14. Install the upper needle bearing in the same manner and press it into the inside of the housing bore flush with the inside end of the bore surface.

15. Install the worm shaft bearing cups (upper and lower) by placing them and their spacers in the adjuster nut and in the housing. Then press them into place.

16. Install the worm shaft oil seal by placing them and their spacers in the adjuster nut and press them into place.

17. Install the worm shaft oil seal by placing the seal in the worm shaft adjuster with the metal seal retainer up. Drive the seal into place with a suitable sleeve until it is just below the end of the bore in the adjuster.

**NOTE: Apply a coating of steering gear lubricant to all moving parts during assembly. Also, put lubricant on and around oil seal lips.**

18. Clamp the holding fixture and housing in a bench vise with the bearing adjuster opening upward. Place a thrust bearing in the lower cup in the housing.

19. Hold the ball nut from turning and insert the worm shaft and ball nut assembly into the housing with the end of the worm shaft

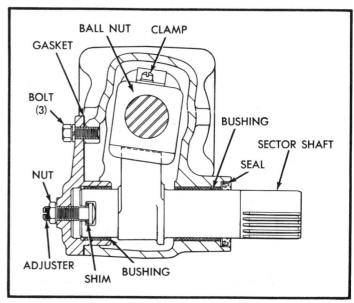

**Sectional view of sector shaft**

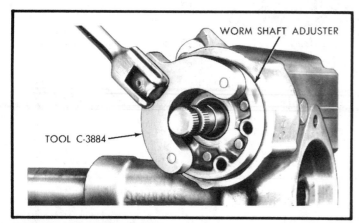

**Removing wormshaft adjuster**

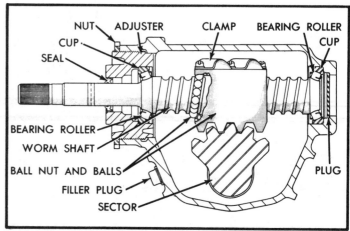

**Sectional view of worm shaft**

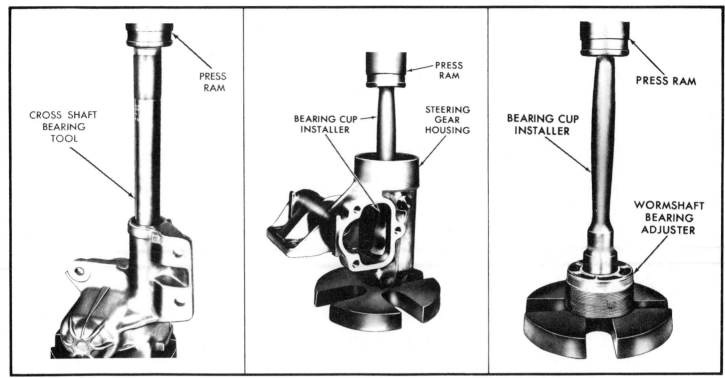

**Removing cross shaft inner and outer bearings**

**Installing lower bearing cup**

**Installing upper bearing cup**

resting in the thrust bearing. Place the upper thrust bearing on the worm shaft. Thoroughly lubricate the threads on the adjuster and the threads in the housing.

20. Place a protective sleeve of tape over the splines on the worm shaft to avoid damaging the seal. Slide the adjuster assembly over the shaft.

21. Thread the adjuster into the housing and, with Tool wrench C-3884 and the splined nut set, tighten the adjuster to 50 ft./lbs. while rotating the worm shaft to seat the bearings.

22. Loosen the adjuster so no bearing preload exists. Tighten the adjuster for a worm shaft bearing preload of $1\frac{1}{8}$ to $4\frac{1}{2}$ in./lbs. Tighten the bearing adjuster locknut and recheck the preload.

# TROUBLESHOOTING CHRYSLER MANUAL STEERING

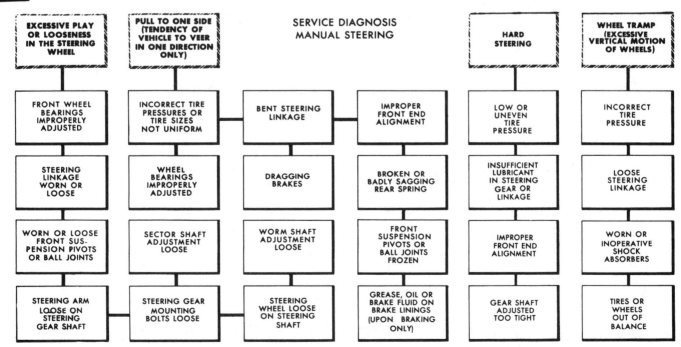

SERVICE DIAGNOSIS
MANUAL STEERING

**EXCESSIVE PLAY OR LOOSENESS IN THE STEERING WHEEL**
- FRONT WHEEL BEARINGS IMPROPERLY ADJUSTED
- STEERING LINKAGE WORN OR LOOSE
- WORN OR LOOSE FRONT SUSPENSION PIVOTS OR BALL JOINTS
- STEERING ARM LOOSE ON STEERING GEAR SHAFT

**PULL TO ONE SIDE (TENDENCY OF VEHICLE TO VEER IN ONE DIRECTION ONLY)**
- INCORRECT TIRE PRESSURES OR TIRE SIZES NOT UNIFORM
- WHEEL BEARINGS IMPROPERLY ADJUSTED
- SECTOR SHAFT ADJUSTMENT LOOSE
- STEERING GEAR MOUNTING BOLTS LOOSE

- BENT STEERING LINKAGE
- DRAGGING BRAKES
- WORM SHAFT ADJUSTMENT LOOSE
- STEERING WHEEL LOOSE ON STEERING SHAFT

- IMPROPER FRONT END ALIGNMENT
- BROKEN OR BADLY SAGGING REAR SPRING
- FRONT SUSPENSION PIVOTS OR BALL JOINTS FROZEN
- GREASE, OIL OR BRAKE FLUID ON BRAKE LININGS (UPON BRAKING ONLY)

**HARD STEERING**
- LOW OR UNEVEN TIRE PRESSURE
- INSUFFICIENT LUBRICANT IN STEERING GEAR OR LINKAGE
- IMPROPER FRONT END ALIGNMENT
- GEAR SHAFT ADJUSTED TOO TIGHT

**WHEEL TRAMP (EXCESSIVE VERTICAL MOTION OF WHEELS)**
- INCORRECT TIRE PRESSURE
- LOOSE STEERING LINKAGE
- WORN OR INOPERATIVE SHOCK ABSORBERS
- TIRES OR WHEELS OUT OF BALANCE

23. Before installing the cross-shaft, pack the worm shaft cavities in the housing above and below the ball nut with steering gear lubricant. A good grade of multi-purpose lubricant may be used if steering gear lubricant is not available. *Do not use gear oil.* Pack enough lubricant into the worm cavities to cover the worm.

24. Slide the cross-shaft adjusting screw and shim into the slot in the end of the shaft. Check the end clearance for no more than 0.004 in. clearance. If the clearance is not within the limit, remove old shim and install a new shim, available in three different thicknesses, to get the proper clearance.

25. Start the cross-shaft and adjuster screw into the bearing in the housing cover. Using a screwdriver through the hole in the cover, turn the screw counterclockwise to pull the shaft into the cover. Install the adjusting screw locknut, but do not tighten at this time.

26. Rotate the worm shaft to center the ball nut.

27. Place a new gasket on the housing cover and install the cross-shaft and cover assembly into the steering gear housing. *Be sure to coat the cross-shaft and sector teeth with steering gear lubricant before installing the cross-shaft in the housing.* Allow some lash between the cross-shaft sector teeth and the ball nut rack. Install and tighten the cover bolts to 25 ft./lbs.

28. Place the cross-shaft seal on the cross-shaft with the lip of the seal facing the housing. Press the seal in place.

29. Turn the worm shaft about 1/4 turn away from the center of the high spot position. Using a torque wrench and a 3/4 in. socket on the worm shaft spline, check the torque needed to rotate the shaft through the high spot. The reading should be between 8 and 11 in./lbs. Readjust the cross-shaft adjusting screw until the proper reading is obtained. Tighten the locknut to 24 ft./lbs. and recheck cross-shaft torque.

## CROSS-SHAFT OIL SEAL

### Replacement

1. Remove the steering gear arm retaining nut and lockwasher.
2. Remove seal with a seal puller or other appropriate tool.
3. Place a new oil seal onto the splines of the cross-shaft with the lip of the seal facing the housing.
4. Remove the tool, and install the steering gear arm, lockwasher, and retaining nut. Tighten the nut to 180 ft./lbs. torque.

# CHRYSLER CORP. MANUAL STEERING RACK/PINION ASSEMBLY

## Reliant, Aries, LeBaron, 400, Horizon, Omni

The manual steering rack and pinion gear cannot be adjusted or serviced. Should a malfunction occur, the complete rack and pinion assembly must be replaced.

# FORD RECIRCULATING BALL TYPE

## Disassembly and Assembly

1. Rotate the steering shaft to the center position.
2. Remove the sector shaft adjusting screw locknut and the housing cover bolts and remove the sector shaft with the cover. Remove the cover from the shaft by turning the screw clockwise. *Keep the shim with the screw.*
3. Loosen the worm bearing adjuster locknut and remove the adjuster assembly and wormshaft upper bearing.
4. Carefully pull the wormshaft and ball nut from the housing,

and remove the wormshaft lower bearing. *Do not run the ball nut to either end of the worm gear to prevent damaging the ball return guides. Disassemble the ball nut only if there are signs of binding or tightness.*

5. To disassemble the ball nut, remove the ball return guide clamp and the ball return guides from the ball nut. *Keep ball nut clamp side up until ready to remove the balls.*

6. Turn the ball nut over and rotate the worm shaft from side to side until all the balls have dropped out into a clean pan. With all balls removed, the ball nut will slide off the wormshaft.

7. Remove the upper bearing cup from the bearing adjuster and the lower cup from the housing. It may be necessary to tap the housing or the adjuster on a wooden block to jar the bearing cups loose.

8. If the inspection shows bearing damage, the sector shaft bearing and the oil seal should be pressed out.

9. If the sector shaft bearing and oil seal have been removed, press a new bearing and oil seal into the housing. Do not clean, wash, or soak seals in cleaning solvent. Apply steering gear lubricant to the housing and seals.

10. Install a bearing cup in the lower end of the housing and in the adjuster. This is a clearance fit not a press fit.

11. Install a new seal in the bearing adjuster if the old seal was removed.

12. Insert the ball guides into the holes in the ball nut, lightly tapping them if necessary to seat them.

13. Insert half of the balls into the hole in the top of each ball guide. If necessary, rotate the shaft slightly to distribute the balls evenly in the circuit.

14. Install the ball guide clamp, tighten the screws to 42-70 in./lbs.

**Exploded view of manual steering - Ford Motor Co.**

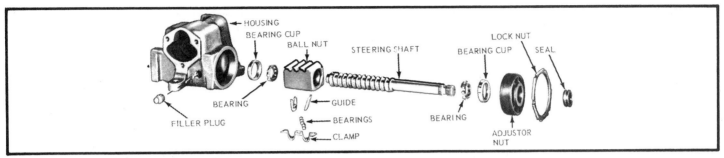

**Ball nut and steering shaft assembly - typical**

for Maverick and Comet, and 18-42 in./lbs. on larger models. Check that the wormshaft rotates freely.

15. Coat the threads of the steering shaft bearing adjuster, the housing cover bolts, and the sector adjusting screw with a suitable oil-resistant sealing compound. Do not apply sealer to female threads. *Do not get sealer on the steering shaft bearings.*

16. Coat the worm bearings, sector shaft bearings, and gear teeth with steering gear lubricant.

17. Clamp the housing in a vise, with the sector shaft axis horizontal, and place the wormshaft lower bearing in its cup. Place the wormshaft and ball nut assemblies in the housing.

18. Position the wormshaft upper bearing on top of the worm gear and install the wormshaft bearing adjuster, adjuster nut, and the bearing cup. Leave the nut loose.

19. Adjust the worm bearing preload according to the instructions given earlier.

20. Position the sector adjusting screw and adjuster shim, and check for a clearance of not more than 0.002 in. between the screw head and the end of the sector shaft. If the clearance exceeds 0.002 in., add enough shims to reduce the clearance to under 0.002 in. clearance.

21. Start the sector shaft adjusting screw into the housing cover. Install a new gasket on the cover.

22. Rotate the steering shaft until the ball nut teeth mesh with the sector gear teeth, tilting the housing so the ball will tip toward the housing cover opening.

23. Lubricate the sector shaft journal and install the sector shaft and cover. With the cover moved to one side, fill the gear with steering gear lubricant. Push the cover and the sector shaft into place, and install the two top housing bolts. Do not tighten the bolts until checking to see that there is some lash between the ball nut and the

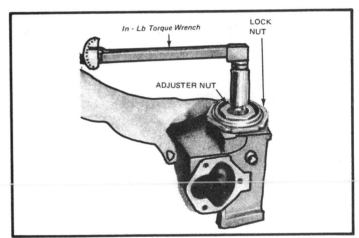

**Checking steering shaft bearing preload**

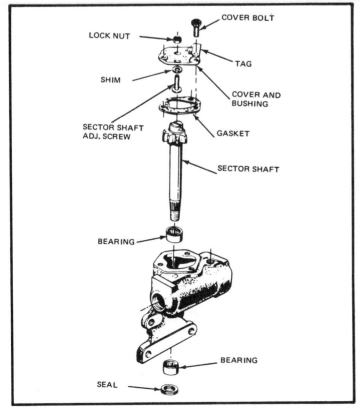

**Sector shaft and housing - typical**

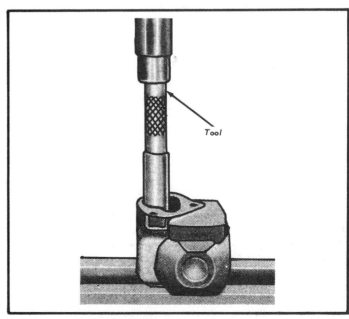

**Removing oil seal and bearing**

sector gear teeth. Hold or push the cover away from the ball nut and tighten the bolts to 17-25 ft./lbs. on Maverick and Comet, and 30-40 ft./lbs. on larger models.

24. Loosely install the sector shaft adjusting screw locknut and adjust the sector shaft mesh load as given earlier. Tighten the adjusting screw locknut.

# FORD/AMC MANUAL RACK AND PINION TYPE

## Disassembly

1. Clean exterior of gear and place in a bench-mounted holding fixture.
2. Remove the yoke cover, shims, gasket, yoke spring and yoke.
3. Remove tie rods and sockets from the ends of the rack by:
   a. Remove tie rod ends and jam nuts.
   b. Remove four bellows clamps, drain the lubricant and remove the bellows.
   c. A special tool is available for drilling the lock pin. Install the tool on the ball socket. Position the fixture so that the pin in the ball socket is lined up with the drill guide.
   d. Drill out the lock pin.
   e. Remove the tie rods and ball sockets.
4. Move the rack to either lock and note the position of the flat on the input shaft.
5. Remove the pinion cover bolts, pinion cover, gasket, shaft, spacer, shims and upper bearing.
6. Remove the rack from the housing.
7. Remove the lower bearing through the pinion shaft bore.

## Assembly

1. Install the lower bearing in the bottom of the housing.
2. Install the rack in the housing.
3. Install the pinion shaft making sure the pinion shaft gear end is engaged in the lower bearing ID.
4. Install the upper bearing, shims, spacer, gasket and pinion cover. Torque cover bolts to 15-20 ft./lbs.
5. Assemble the yoke and spring. Install a new gasket and, if necessary, adjust the yoke-to-rack spring tension.
6. Assemble the shim pack and cover.
7. Coat the cover bolts with sealant and torque to 15-20 ft./lbs.

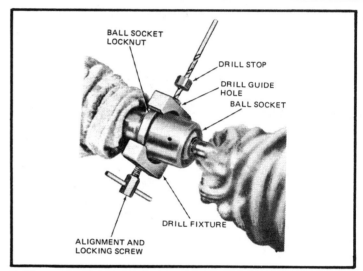

**Drilling out retaining pin**

8. Install a new spring in rack end and assemble ball socket to tie rod.
9. Install a new inner thrust bearing in the ball socket.
10. Apply lubricant to the spring, thrust bearing, tie rod ball and ball socket.
11. Thread the ball joint locknut on the rack end.
12. Thread the ball socket onto the rack until the tie rod movement stiffens.
13. Rotate tie rod about 10 times to check movement.

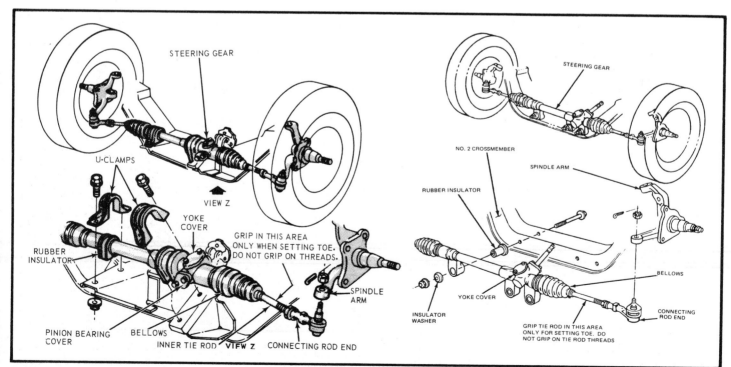

**Typical attaching methods for the manual rack and pinion assembly to the cars**

14. Install tie rod ends.

15. Place hook of pull scale through hole in threaded area of the ball joint. With the tie rod in mid-position, parallel to the rack adjust the position of the ball socket on the rack until the pull effort is 4-6 lbs. Secure the unit by torquing the ball socket lock nut to 35-40 ft./lbs.

16. Install the drill fixture on the ball socket. Drill a hole on the line of contact between the lock nut and the ball socket.

**NOTE: A new hole must be drilled even if the halves of the old hole align. A total of only two drilled holes are allowed on one end of the rack.**

17. Insert the retaining pin flush in the drilled hole. Stake the pin. Clean rack and housing bores.

18. Install bellows and clamps. Add seven ounces of lubricant to either bellows.

19. Install jam nuts.

# GM SAGINAW RECIRCULATING BALL TYPE

## A.M.C. Cars Except Pacer, All G.M. Corp. Cars

### WORM BEARING PRELOAD

#### Adjustment

— CAUTION —
*Do not turn steering wheel hard against stops as damage to ball nut assembly may result.*

1. Disconnect the ball stud from the pitman arm, and retighten the pitman arm nut.

2. Loosen the pitman shaft adjusting screw locknut and back off adjusting screw a few turns.

3. Attach spring scale to the steering wheel and measure the pull needed to move the steering wheel when off the high point. The pull should be between 1/8 and 3/8 lbs.

4. To adjust the worm bearing, loosen the worm bearing adjuster locknut with a brass drift and turn the adjuster screw until the proper pull is obtained. When adjustment is correct, tighten the adjuster locknut, and recheck with the spring scale again.

### SECTOR AND BALL NUT BACKLASH

#### Adjustment

1. After the worm bearing preload has been adjusted correctly, loosen the pitman shaft adjusting screw locknut and turn the pitman shaft adjusting screw clockwise until a pull of 3/4 to 1⅛ lbs. is shown on the spring scale. When the adjustment is correct, tighten the pitman shaft adjusting screw locknut and recheck the adjustment.

**NOTE: A torque wrench calibrated in in./lbs. may be substituted for the spring scale in adjusting steering gear.**

2. Turn the steering wheel to the center of its turning limits (pitman arm disconnected). If the steering wheel is removed, the mark on the steering shaft should be at top center.

3. Connect the ball stud to the pitman arm, tightening the attaching nut to 115 ft./lbs. (Vega—93 ft./lbs.).

### STEERING GEAR

#### Disassembly and Assembly

1. After removing the steering gear from the car, place the steering gear assembly in a bench vise.

**NOTE: Worm seal may be replaced without disassembling gear. Be careful not to damage shaft or housing when removing seal.**

2. Rotate the worm shaft until it is centered with the mark facing upward. Remove three cover attaching screws and the adjusting screw locknut. Remove the cover and gasket by turning adjusting screw clockwise through the cover.

3. Remove the adjusting screw with its shim from the slot in the

**Exploded view of manual recirculating ball type steering - GM Saginaw**

PITMAN SHAFT NUT
LOCK WASHER
SEAL RETAINER
SEAL
BUSHING
HOUSING
SEAL
ADJUSTING SCREW
SHIM
WITH GEAR CENTERED MARK ON SHAFT TO BE UP
RETAINER
LOWER WORM BEARING
SEAT
UPPER WORM BEARING
WORM SHAFT
BALL NUT
BALL BEARINGS
RETURN GUIDES
CLAMP
WORM BEARING ADJUSTER
SEAT
LOCK NUT
PITMAN SHAFT
GASKET
SIDE COVER
SILL ADAPTER
LOCK NUT

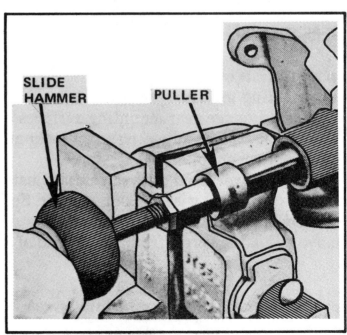

**Pitman shaft bushing removal**

SLIDE HAMMER  PULLER

# MANUAL STEERING
## GM SAGINAW RECIRCULATING BALL TYPE

## ☲ TROUBLESHOOTING GM MANUAL STEERING

| CONDITION | POSSIBLE CAUSE | CORRECTION |
|---|---|---|
| **Hard or Erratic Steering** | (1) Incorrect tire pressure | (1) Inflate tires to recommended pressures |
| | (2) Insufficient or incorrect lubrication | (2) Lubricate as required (refer to Maintenance Section |
| | (3) Suspension, or steering linkage parts damaged or misaligned | (3) Repair or replace parts as necessary |
| | (4) Improper front wheel alignment | (4) Adjust incorrect wheel alignment angles |
| | (5) Incorrect steering gear adjustment | (5) Adjust steering gear |
| | (6) Sagging springs | (6) Replace springs |
| **Play or Looseness in Steering** | (1) Steering wheel loose | (1) Inspect shaft splines and repair as necessary. Tighten attaching nut and stake in place |
| | (2) Steering linkage or attaching parts loose or worn | (2) Tighten, adjust, or replace faulty components |
| | (3) Pitman arm loose | (3) Inspect shaft splines and repair as necessary. Tighten attaching nut and stake in place |
| | (4) Steering gear attaching bolts loose | (4) Tighten bolts |
| | (5) Loose or worn wheel bearings | (5) Adjust or replace bearings |
| | (6) Steering gear adjustment incorrect or parts badly worn | (6) Adjust gear or replace defective parts |
| **Wheel Shimmy or Tramp** | (1) Improper tire pressure | (1) Inflate tires to recommended pressures |
| | (2) Wheels, tires, or brake rotors or drums out-of-balance or out-of-round | (2) Inspect and replace out-of-balance parts |
| | (3) Inoperative, worn, or loose shock absorbers or mounting parts | (3) Repair or replace shocks or mountings |
| | (4) Loose or worn steering or suspension parts | (4) Tighten or replace as necessary |
| | (5) Loose or worn wheel bearings | (5) Adjust or replace bearings |
| | (6) Incorrect steering gear adjustments | (6) Adjust steering gear |
| | (7) Incorrect front wheel alignment | (7) Correct front wheel alignment |
| **Tire Wear** | (1) Improper tire pressure | (1) Inflate tires to recommended pressures |
| | (2) Failure to rotate tires | (2) Rotate tires |
| | (3) Brakes grabbing | (3) Adjust or repair brakes |
| | (4) Incorrect front wheel alignment | (4) Align incorrect angles |
| | (5) Broken or damaged steering and suspension parts | (5) Repair or replace defective parts |
| | (6) Wheel runout | (6) Replace faulty wheel |
| | (7) Excessive speed on turns | (7) Make driver aware of condition |
| **Car Leads to One Side** | (1) Improper tire pressures | (1) Inflate tires to recommended pressures |
| | (2) Front tires with uneven tread depth, wear pattern, or different cord design (i.e., one bias ply and one belted tire on front wheels) | (2) Install tires of same cord construction and reasonably even tread depth and wear pattern |
| | (3) Incorrect front wheel alignment | (3) Align incorrect angles |
| | (4) Brakes dragging | (4) Adjust or repair brakes |
| | (5) Pulling due to uneven tire construction | (5) Replace faulty tire |

end of the pitman shaft. Remove the pitman shaft from the housing being careful not to damage the seal in the housing.

4. Loosen the worm bearing adjuster locknut with a brass drift and remove the adjuster and bearing. Remove the bearing retainer with a screwdriver.

5. Remove the worm steed shaft assembly with the ball nut assembly and bearing. Remove the ball nut return guide clamp by removing screws. Remove the guides, turn ball nut over, and remove

the steel balls by rotating the shaft from side to side. After all steel balls have been removed, take the ball nut off the worm shaft.

6. Clean all parts in solvent. Inspect all bearings, bearing cups, bushings, seals, worm groove, and gear teeth for signs of wear, scoring, pitting, etc. If the pitman shaft bushings or seal, steering shaft seal, or upper and lower bearing cups need replacement, see the replacement procedures given below.

7. Remove the pitman shaft seal with a screwdriver or punch. If

there is leakage around the threads of the bearing adjuster, apply a non-hardening sealer.

8. Remove faulty bushings from the pitman shaft with Puller and Slide Hammer. Install new bushings, seating the inner end of the bushing flush with the inside surface of the housing.

9. Remove the steering shaft seal with a punch or screwdriver. Tap new seal in place, using a section of tubing to seat the seal.

10. Remove the upper or lower bearing cup from the worm bearing adjuster or steering gear housing using Puller and Slide Hammer. Install the new bearing cups.

11. Lubricate all seals, bushings, and bearings before installing into the steering gear assembly.

12. Position the ball nut on the worm shaft. Install the steel balls in the return guides and the ball nut, placing an equal number in each circuit of the ball nut. Install the return guide clamp and screws.

---
**CAUTION**

*Do not rotate the worm shaft while installing the steel balls since the balls may enter the crossover passage between the circuits, causing incorrect operation of the ball nut.*

---

13. Place bearing on shaft above the worm gear, center ball nut on worm gear; then, slide the steering shaft, bearing, and ball nut into the housing. *Do not damage the steering shaft seal in the housing.*

14. Place the bearing in the worm adjuster, install the bearing retainer, and install the adjuster and locknut on the housing, tightening it just enough to hold the bearing in place.

15. Install the pitman shaft adjusting screw and selective shim in the pitman shaft. Be sure there is no more than 0.002 in. of end play of the screw in the slot. If the end-play is more than 0.002 in., install a new selective shim to get the proper clearance. Shims are available in four thicknesses: 0.063 in., 0.065 in., 0.067 in., and 0.069 in.

16. Install the pitman shaft and adjusting screw with the sector and ball nut positioned as shown.

17. Install the cover and gasket on the adjusting screw, turning screw counterclockwise until it extends through the cover from 5/8 to 3/4 in. Install the cover attaching screws and torque to 35 ft./lbs. (Vega—18 ft./lbs.).

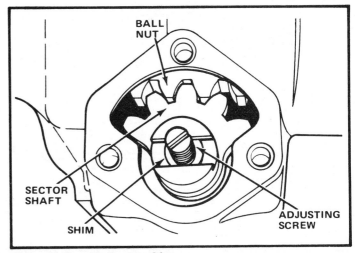

**Pitman shaft and ball nut position**

18. Tighten the pitman shaft adjusting screw so that the teeth on the shaft and the ball nut engage but do not bind. Final adjustment must be made later.

19. Wrap the pitman shaft splines with tape to protect the seal and install the seal.

20. Fill steering gear with a good quality steering gear lubricant. Turn the steering gear from one extreme to the other to make sure it does not bind. *Do not allow the ball nut to strike the ends of the ball races on the worm gear to avoid damaging the ball return guides.*

21. Install the steering gear as described previously. Perform the final adjustments on the worm bearing preload and the sector and ball nut backlash adjustments.

# GM MANUAL RACK AND PINION-CHEVETTE, T1000

### Disassembly

1. Position assembly in vise, clamping housing near center. Use soft jaws to prevent damage to housing.

2. Loosen jam nuts. Remove outer tie rod.

---
**CAUTION**

*Hold housing while loosening nuts so as not to damage internal gear components.*

---

3. Remove inner boot clamp by cutting. Remove the outer clamp by relieving tension in clamp. Remove boot by pulling. Repeat procedure for other end.

4. Position rack in soft jaw vise, and remove inner tie rod assemblies (both ends).

---
**CAUTION**

*To prevent internal gear damage when removing housing, turn housing counterclockwise until assembly separates from rack.*

---

5. Remove adjuster plug locknut, adjuster plug, and spring.

6. Remove rack bearing from housing.

7. Clean surface at seal. Pierce seal at one of the two round spots on surface. Pry out seal.

8. Using snap ring pliers, remove retaining ring from bore.

9. Position end of shaft in soft jaw vise. Tap housing to separate pinion assembly from housing.

---
**CAUTION**

*With pinion separated, rack may slide from housing and be damaged.*

---

10. Remove rack from housing.

11. The rack and pinion assembly is now disassembled. Clean all components, except inner tie rod assemblies, with an approved solvent. Air dry and inspect. Replace any seals which are cut or badly worn. If the pinion seal is removed, it must be replaced.

**NOTE: Check major wear areas for cracking, chipping, etc. Replace as required.**

## MOUNTING GROMMETS

Do not remove grommets unless replacement is required. Replace both grommets if either requires replacement.

Cut through grommet and remove.

Lube inside of seals lightly with chassis lube. Start with left seal first and force it past the right side (smaller inside diameter) boss. Start right hand grommet and seat. Remove housing from vise and slide grommet to left hand mounting. Assemble grommet to housing.

## GUIDE BUSHINGS

No attempt to replace the guide bushing should be made unless it is damaged or broken. If this occurs, replace the housing.

## RACK BUSHING

The rack bushing should only be replace if evidence of heavy wear is observed.

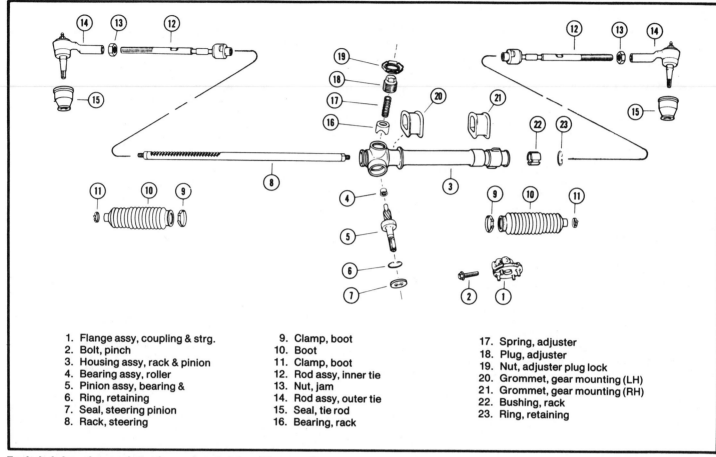

1. Flange assy, coupling & strg.
2. Bolt, pinch
3. Housing assy, rack & pinion
4. Bearing assy, roller
5. Pinion assy, bearing &
6. Ring, retaining
7. Seal, steering pinion
8. Rack, steering
9. Clamp, boot
10. Boot
11. Clamp, boot
12. Rod assy, inner tie
13. Nut, jam
14. Rod assy, outer tie
15. Seal, tie rod
16. Bearing, rack
17. Spring, adjuster
18. Plug, adjuster
19. Nut, adjuster plug lock
20. Grommet, gear mounting (LH)
21. Grommet, gear mounting (RH)
22. Bushing, rack
23. Ring, retaining

**Exploded view of manual steering rack and pinion - Chevette, T1000**

Remove retaining ring. Using a suitable size socket and extension, drive the bushing out of the housing. If a puller is available, position fingers of puller behind bushing and remove bushing using slide hammer.

Using a suitable size socket, press new rack bushing into housing until it bottoms. Install retaining ring.

## ROLLER BEARING ASSEMBLY

Check condition of pinion pilot. If scored or badly worn, replace pinion and roller bearing assembly.

Press or tap out bearing using drift and press or hammer.

Using a suitable size socket, press or drive new bearing into housing until it bottoms.

## BEARING AND PINION ASSEMBLY

Inspect roller bearing pilot, pinion teeth, and rotor bearing assembly. If pilot is scored, teeth are chipped, or is loose on pinion shaft, the bearing and pinion assembly should be replaced.

## INNER TIE TODS

The inner tie rod assemblies cannot be serviced. If the pivot is loose, replace the tie rod assembly. If the joint rocking or turning torque exceeds 17 N/m (150 in./lbs.), replace the inner tie rod assembly.

### Assembly

1. Install rack with teeth facing pinion into housing. The flat on the teeth should be parallel with pinion shaft. Measure and set 68.5 mm (2.70 in.) from lip of housing to end of rack.

**NOTE: Insert pinion with flat at 75° from vertical. Tap on pinion shaft with soft hammer until pinion seats. Reset 68.5 mm (2.70 in.) dimension of rack position. Flat should now be vertical. If flat is at plus or minus 30° from vertical, restart procedure.**

—— **CAUTION** ——
*Rack must be centered as described. If not, the steering wheel cannot travel fully, causing unequal turning radii.*

2. Install retaining ring using too J-4245. Beveled edge of retaining ring should be up.

3. Liberally coat top of pinion bearing with anhydrous calcium grease, then seat pinion seal flush with housing. Seal can be seated by tapping on alternate sides with hammer.

4. Install rack bearing. Coat bearing with lithium based grease.

5. Coat both ends of preload spring and threads of adjuster plug with lithium based grease.

6. Assemble adjuster plug and spring assembly into housing. Turn adjuster plug clockwise until it bottoms, then counterclockwise 45° to 60°. Torque required to turn pinion should be between 0.9-1.1 N/m (8-10 in./lbs.). Turn plug in or out to adjust as required. Tighten lock nut to 68 N/m (50 ft./lbs.).

7. Lube both ends of rack with lithium based grease. Fill rack teeth with lube. Move rack back and forth several times by turning pinion shaft, adding grease to rack teeth each time.

8. Install inner tie rod assemblies to rack. Turn inner tie rod assemblies until they bottom out.

—— **CAUTION** ——
*Support rack in vise or with another wrench to avoid internal gear damage.*

9. Use wood block or vise support and stake tie rod housing to rack flat. Stake both sides.

10. Position one of the large clamps on the housing. Place boot lip into position over undercut. Position clamp over boot at undercut and secure using side cutter type pliers or tool J-22610.

11. Slip end of boot into rod undercut. Do not assemble clamp over boot until toe adjustment is made. Straighten boots if twisted before assembling clamps.

12. Thread jam nuts (both sides) onto tie rods.

13. Thread on tie rod ends. Do not tighten jam nuts until toe adjustment is made. Then tighten to 67 N/m (50 ft./lbs.).

14. Slip on coupling assembly. Flat on inside diameter of coupling mates with flat on pinion shaft. Install pinch bolt, but do not tighten until vehicle installation, then tighten to 41 N/m (30 ft./lbs.).

## TORQUE FIGURES: GM MANUAL RACK & PINION—CHEVETTE, T1000

| | | |
|---|---|---|
| **Rack and Pinion Steering** | | |
| Inner Tie Rod Rocking or Turning Torque (Not to Exceed) | 5 N·m | (45 in. lbs.) |
| Tie Rod Jam Nuts | 68 N·m | (50 ft. lbs.) |
| Pinion Turning Torque | 0.7 to 1.5 N·m | (6-13 in. lbs.) |
| Adjuster Plug Lock Nut | 68 N·m | (50 ft. lbs.) |
| Rack and Pinion Assembly Clamp Bolts | 19 N·m | (14 ft. lbs.) |
| **Steering Column** | | |
| Flexible Coupling Pinch Bolt | 41 N·m | (30 ft. lbs.) |
| Intermediate Shaft to Flexible Coupling Bolts | 24 N·m | (18 ft. lbs.) |
| Steering Column to Dash Cover Screws | 2 N·m | (18 in. lbs.) |
| Column to Toe Pan Screw | 27 N·m | (20 ft. lbs.) |
| Column to I.P. Bracket Bolts | 30 N·m | (22 ft. lbs.) |
| Column Bracket to I.P. Nuts | 27 N·m | (20 ft. lbs.) |
| Column Shroud to Housing Screws | 2 N·m | (18 in. lbs.) |
| Column Housing & Shroud to Jacket | 7 N·m | (60 in. lbs.) |
| Ignition Switch to Jacket Screws | 4 N·m | (35 in. lbs.) |
| Turn Signal Switch Screws | 4 N·m | (35 in. lbs.) |

# G.M. MANUAL STEERING RACK AND PINION—A BODY CARS

## Celebrity, 6000, Ciera, Century 82-83

### OUTER TIE ROD

#### Removal

1. Loosen the jam nut and remove the tie rod from the steering knuckle. Count the number of turns needed to remove.

2. Remove the outer tie rod.

#### Installation

1. Install the outer tie rod by screwing it on the inner tie rod the same amount of turns as was needed to remove the outer tie rod.

2. Adjust the toe-in/out by turning the inner tie rod. Tighten the lock nut to 50 ft./lbs.

3. Install the outer boot clamp and secure.

### BOOT SEAL

#### Removal and Installation

To remove the boot seal from either side of the steering assembly, the outer tie rod must be removed. Remove the lock nut and the boot clamps. Remove the boot by sliding it off the inner tie rod.

To install the boot, reverse the removal procedure. Secure the boot clamps.

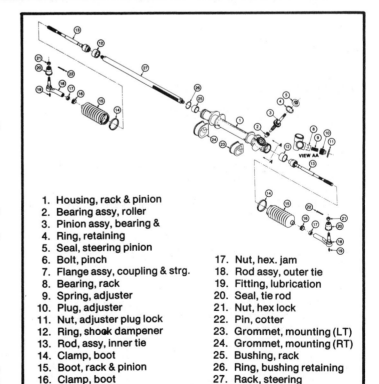

1. Housing, rack & pinion
2. Bearing assy, roller
3. Pinion assy, bearing &
4. Ring, retaining
5. Seal, steering pinion
6. Bolt, pinch
7. Flange assy, coupling & strg.
8. Bearing, rack
9. Spring, adjuster
10. Plug, adjuster
11. Nut, adjuster plug lock
12. Ring, shock dampener
13. Rod, assy, inner tie
14. Clamp, boot
15. Boot, rack & pinion
16. Clamp, boot
17. Nut, hex. jam
18. Rod assy, outer tie
19. Fitting, lubrication
20. Seal, tie rod
21. Nut, hex lock
22. Pin, cotter
23. Grommet, mounting (LT)
24. Grommet, mounting (RT)
25. Bushing, rack
26. Ring, bushing retaining
27. Rack, steering

**Exploded view of manual steering rack and pinion - GM front wheel A body cars**

## INNER TIE ROD

### Removal

1. The steering assembly must be out of the vehicle.
2. Position a wrench on the rack and hold it as the inner tie rod is unscrewed.

### Installation

1. Screw the inner tie rod into the rack and with a wrench holding the rack to avoid teeth damage, tighten the inner tie rod to 70 ft./lbs.
2. Stake the housing on both sides.

**NOTE: Be sure the tie rod rocks freely in the housing before staking.**

3. When staking is completed, a 0.010 inch feeler gauge must not pass between the rack and the housing stakes. Check both sides.

## RACK BEARING

### Removal

1. Remove the adjuster plug lock nut, the adjuster plug, spring and the rack bearing.

### Installation

1. Lubricate the metal parts before installation, install then in the housing in the reverse order of their removal.
2. Turn the adjuster plug in until it bottoms and then back off approximately 40 to 60 degrees.
3. Check the torque on the pinion by turning it with a torque wrench. The correct pinion torque is 8 to 10 inch pounds.
4. Tighten the lock nut to 50 ft./lbs. while holding the adjuster plug.

## PINION SEAL

### Removal

1. Pierce the seal in one or two round spots and pry it from the housing.

### Installation

1. Lubricate the seal and seat the seal flush with the housing.

## PINION SHAFT ASSEMBLY

### Removal

1. Remove the seal and the retaining ring from the housing assembly.
2. Place the pinion shaft in a soft jawed vise and tap on the housing to separate the two.

— CAUTION —

*With the pinion removed from the housing, the rack can slide from the housing and be damaged.*

### Installation

1. Lubricate and slide the rack into the housing.
2. Position the rack that 63.5 mm protrudes from the pinion shaft end of the housing.
3. Position the pinion so that the center of the flat is facing the 4:30 position and install it into the housing.
4. When the pinion is seated properly, the center of the flat will be facing the 9:00 position.
5. Install the retainer ring and install the seal.

**NOTE: The distance between the holes in the retaining ring should be 7.0 mm apart.**

## ROLLER BEARING

### Removal and Installation

The pinion must be out of the housing. Drive the bearing from the housing and press the new bearing in place. Complete the pinion installation.

## RACK BUSHING

### Removal

1. Remove the pinion and rack assembly from the housing.
2. Remove the retaining ring from the housing and with a special long legged puller, remove the bushing from the housing.

### Installation

1. Press a new bushing into the housing until it is firmly seated.
2. Install the retaining ring and complete the rack and pinion installation.

# GM MANUAL STEERING RACK AND PINION—J BODY CARS

## Cavalier, J-2000, Skyhawk 82-83, Firenza

### OUTER TIE ROD

#### Removal

1. With the steering assembly from the vehicle, loosen the outer rod pinch bolt and turn the tie rod from the adjuster stud, counting the number of turns until the tie rod separates from the adjuster stud.

#### Installation

1. Turn the tie rod onto the adjuster stud the same number of turns as was needed to remove.
2. Tighten the pinch bolt until the toe-out can be verified. Re-loosen and adjust as required.

## INNER TIE ROD AND INNER PIVOT BUSHING

### Removal

1. Bend back the lock plate tabs and loosen the inner tie rod bolt and remove.
2. Remove the inner tie rod by sliding it out between the bolt support plate and rack/pinion boot.

**NOTE: If both inner tie rods are to be removed, re-install the inner tie rod bolt in the first tie rod retaining bolt hole to keep the rack and pinion boot and other parts aligned, while tie rods are out.**

3. With the tie rod disconnected, the pivot bushings can be pressed out and new ones pressed in.

### Installation

1. Be sure the center housing cover washers are fitted into the rack and pinion boot, before rod installation.

2. Remove the locating bolt from the rack and position one inner tie rod assembly in place over the rack. Place the bolt through the lock plate and the tie rod. Place the second inner tie rod in place and install the bolt through the lock plate and the tie rod.

3. Tighten the inner tie rod bolts to 65 ft./lbs. and bend the lock tabs against the flats of the inner tie rod bolts after torquing.

## RACK AND PINION BOOT, RACK GUIDE, BEARING GUIDE, MOUNTING GROMMET OR HOUSING END COVER

### Removal

1. Separate right-hand mounting grommet and remove. Left-hand mounting grommet need not be removed unless replacement is required.
2. Cut both boot clamps and discard.
3. Using constant pressure, slide rack and pinion boot over boot retaining bushing and off housing.
4. The boot retaining bushing on housing tube end need not be removed unless damaged.
5. Remove housing end cover only if damaged.

### Installation

1. Remove boot retaining bushing from pinion end of boot.
2. Slide new boot clamp on boot. Install bushing into boot.
3. Install new bearing guide on rack guide if necessary.
4. Install new boot retaining bushing on housing if necessary.
5. Install rack guide on rack.
6. Coat inner lip of boot retaining bushing lightly with grease for ease of assembly.
7. Install boot on housing.
8. Be sure center housing washers are in place on boot.
9. For ease of assembly, install inner tie rod bolts through cover washers and boot. Screw into rack lightly. This will keep rack, rack guide and boot in proper alignment.
10. Slide boot and boot retaining bushing until seated in bushing groove at pinion end of housing. Crimp new boot clamp.
11. Slide other end of boot onto boot retaining bushing in housing at tube end. Crimp new boot clamp.

## FLANGE AND STEERING COUPLING ASSEMBLY

### Removal

1. Loosen and remove the pinch bolt.
2. Remove the coupling.

**NOTE: The dash seal can be removed or installed with the coupling off the steering assembly.**

### Installation

1. Install the flange and steering coupling assembly on the pinion shaft.
2. Install the pinch bolt and torque to 29 ft./lbs.

## RACK BEARING

### Removal

1. Remove the adjuster plug lock nut.
2. Remove the adjuster plug, the spring, O-ring and rack bearing.

### Installation

1. Install the parts in the reverse order of the removal procedure.

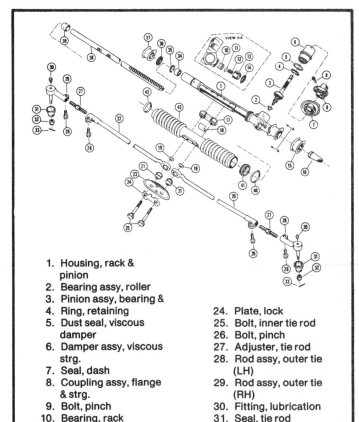

1. Housing, rack & pinion
2. Bearing assy, roller
3. Pinion assy, bearing &
4. Ring, retaining
5. Dust seal, viscous damper
6. Damper assy, viscous strg.
7. Seal, dash
8. Coupling assy, flange & strg.
9. Bolt, pinch
10. Bearing, rack
11. Seal, O-ring
12. Spring, adjuster
13. Plug, adjuster
14. Nut, adjuster plug lock
15. Grommet, mounting (LH)
16. Cover, housing end
17. Guide, rack
18. Guide, bearing
19. Washer, center hsg cover
20. Rod, inner tie (LH)
21. Bushing, inner pivot
22. Rod, inner tie (RH)
23. Plate, bolt support
24. Plate, lock
25. Bolt, inner tie rod
26. Bolt, pinch
27. Adjuster, tie rod
28. Rod assy, outer tie (LH)
29. Rod assy, outer tie (RH)
30. Fitting, lubrication
31. Seal, tie rod
32. Nut, hex slotted
33. Pin, cotter
34. Bushing, rack
35. Ring, internal retaining
36. Bushing, boot retaining
37. Grommet, mounting (RH)
38. Rack, steering
39. Cover, housing end
40. Clamp, boot
41. Bushing, boot retaining
42. Boot, rack & pinion
43. Clamp, boot

**Exploded view of manual steering rack and pinion - GM J body cars**

2. With the rack centered, tighten the adjuster plug to a torque of 6 to 11 ft./lbs. Back off adjuster plug to 50 to 70 degrees.

3. Assemble lock nut and tighten while holding the adjuster plug stationary. Tighten to 50 ft./lbs.

4. Rotate the pinion with an inch pound torque wrench and socket. The turning torque should be 8 to 20 inch pounds. Adjust as required.

## VISCOUS STEERING DAMPER ASSEMBLY

### Removal

1. Using a three-fingered wheel puller on flange of viscous damper, remove damper.

**NOTE: Three finger puller must be used or damage to damper will occur.**

2. Replace dust seal if necessary.

# MANUAL STEERING
## GM RACK & PINION TYPE

### Installation

1. Remove retaining shield from damper.
2. Be sure dust seal is installed on damper.
3. Line up flat on damper with adjuster plug.
4. Using a press, press on inner hub of damper with suitable pipe. Be sure tabs on damper line up with slots in housing. Press until seated on pinion shaft.
5. Using suitable pipe, press on outer housing of damper until fully seated in rack and pinion housing.
6. Reinstall retaining shield.

## PINION SHAFT ASSEMBLY

### Removal

1. Turn the pinion shaft until the rack guide is equal distance from both sides of the housing opening.
2. Mark the location of the stub shaft flat on the housing. Remove the retaining ring.
3. Remove the pinion by placing it in a soft jawed vise and tapping on the housing with a soft faced hammer.

### Installation

1. Measure the rack guide so it is equal distance on both sides of the housing opening.
2. Install the pinion assembly so when the pinion is fully seated, the pinion shaft flat and the mark on the housing line up and the rack guide is centered in the housing opening.
3. Install the retaining ring.

## RACK

### Removal

1. With the pinion out of the steering housing, thread an inner tie rod bolt into the rack.
2. Slide the rack back and forth until the housing end cover is forced from the end of the tube.
3. Unthread the bolt and slide the rack from the steering housing.

### Installation

1. Slide rack into the housing and seat the end cover into the end of the housing tube.

## PINION SHAFT ROLLER BEARING

### Removal and Installation

The roller bearing is pressed out and the new one pressed into the housing.

## RACK BUSHING

### Removal and Installation

With the rack out of the housing, remove the internal retaining ring from the tube. A long legged puller is used to remove the bushing from the tube housing. A press is used to install the new bushing. Press the bushing into position and install the retaining ring.

# AMC POWER RACK AND PINION-PACER

## DESCRIPTION

This type of power steering unit combines the steering gear and linkage into one assembly.

The power steering gear consists of an internal tube and housing assembly which contains the steering rack and piston, the pinion shaft and valve body assembly and the adjuster plug assembly. The tube and housing are permanently connected during manufacture by a plastic injection-bonding process.

### Disassembly

1. Remove the steering gear from the vehicle and mount the unit in a vise, clamping only the housing.
2. Cut and remove the two boot clamps from the housing end of the steering gear.

3. Slide the boot away from the housing to expose the rack teeth.
4. Turn the flexible coupling to move the rack toward the housing end of the steering gear and expose as many teeth as possible.
5. Wipe the rack teeth clean with a clean cloth and inspect the rack for chipped, cracked, broken, flaking, or excessively worn teeth. If any of the above conditions exist, the steering gear assembly must be replaced. Machining marks or shiny and bright rack teeth are normal conditions.
6. Remove the flexible coupling pinch bolt and separate the coupling from the stub shaft.
7. Remove the adjuster plug lock-nut and remove the adjuster plug with a spanner type tool that fits into the two holes in to top of the adjuster plug.
8. Remove the valve body assembly by pulling straight up on the stub shaft. Do not disassemble the valve body.
9. Remove the pinion shaft from the housing using pliers. Grip the pinion shaft at the drive tang and rotate the shaft clockwise while pulling up.

**Removing the adjusting plug lock nut and plug**

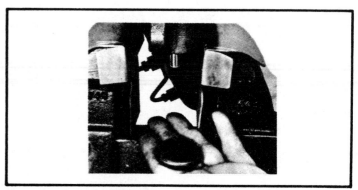
**Removing the contraction plug**

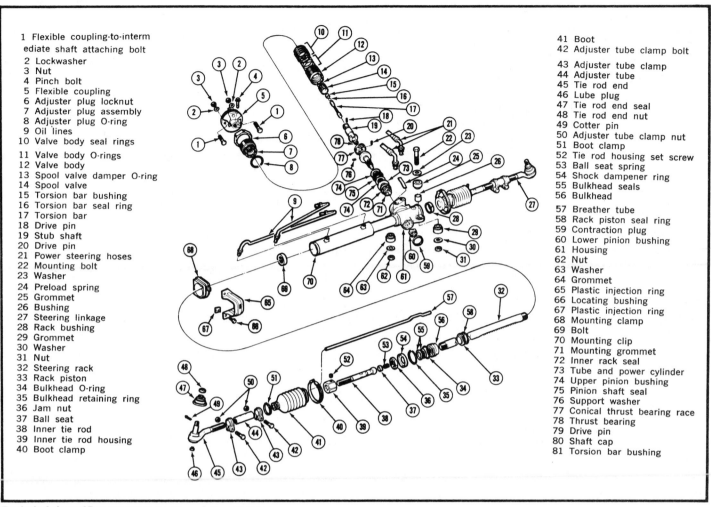

1 Flexible coupling-to-interm
ediate shaft attaching bolt
2 Lockwasher
3 Nut
4 Pinch bolt
5 Flexible coupling
6 Adjuster plug locknut
7 Adjuster plug assembly
8 Adjuster plug O-ring
9 Oil lines
10 Valve body seal rings
11 Valve body O-rings
12 Valve body
13 Spool valve damper O-ring
14 Spool valve
15 Torsion bar bushing
16 Torsion bar seal ring
17 Torsion bar
18 Drive pin
19 Stub shaft
20 Drive pin
21 Power steering hoses
22 Mounting bolt
23 Washer
24 Preload spring
25 Grommet
26 Bushing
27 Steering linkage
28 Rack bushing
29 Grommet
30 Washer
31 Nut
32 Steering rack
33 Rack piston
34 Bulkhead O-ring
35 Bulkhead retaining ring
36 Jam nut
37 Ball seat
38 Inner tie rod
39 Inner tie rod housing
40 Boot clamp

41 Boot
42 Adjuster tube clamp bolt
43 Adjuster tube clamp
44 Adjuster tube
45 Tie rod end
46 Lube plug
47 Tie rod end seal
48 Tie rod end nut
49 Cotter pin
50 Adjuster tube clamp nut
51 Boot clamp
52 Tie rod housing set screw
53 Ball seat spring
54 Shock dampener ring
55 Bulkhead seals
56 Bulkhead
57 Breather tube
58 Rack piston seal ring
59 Contraction plug
60 Lower pinion bushing
61 Housing
62 Nut
63 Washer
64 Grommet
65 Plastic injection ring
66 Locating bushing
67 Plastic injection ring
68 Mounting clamp
69 Bolt
70 Mounting clip
71 Mounting grommet
72 Inner rack seal
73 Tube and power cylinder
74 Upper pinion bushing
75 Pinion shaft seal
76 Support washer
77 Conical thrust bearing race
78 Thrust bearing
79 Drive pin
80 Shaft cap
81 Torsion bar bushing

**Exploded view of Pacer power steering rack and pinion**

10. Clean and inspect the pinion shaft. If any of the conditions mentioned in step 5 exist, replace the steering gear.

11. Remove the contraction plug from the housing using a 1/4 in. diameter brass rod. Insert the rod through the upper and lower pinion bushings and tap on the rod to remove the plug.

12. Remove the lower pinion bushing and preload spring from the housing with the brass rod.

13. Move the rack to the centered position in the tube and housing.

14. Install the pinion shaft in the housing. Be sure the pinion is fully seated.

15. Install the valve body and adjuster plug in the housing and hand tighten the plug only.

**NOTE: Make sure that the valve body is seated in the housing. Do not press on the stub shaft to seat the valve body. Press directly on the valve body with only your thumbs.**

16. Loosen, but do not remove, the adjuster tube clamp nuts and remove the tie rod ends and adjuster tubes as assemblies. Mark the position of the tubes and tie rods for reference during assembly. Use penetrating oil if the threads are corroded. Hold the inner tie rod with a 9/16 in. wrench while removing the toe rods and tubes.

17. Remove the remaining boot clamps and boot retainers.

18. Remove both protective boots.

19. Slide the shock dampener rings back off the jamnuts by rotating and pushing them back.

20. Remove the steering gear from the vise, and clamp the inner tie rod housing. Loosen the jamnut using 1½ in. wrench. Loosen the jamnut at the opposite end of the rack also.

**NOTE: The inner tie rod housings must be held in a vise while loosening the jamnuts to prevent damage to the rack, pinion, stub shaft and valve body.**

21. Loosen but do not remove the set screw in each inner tie rod housing.

22. Reposition the steering gear in the vise as before, and remove the inner tid rod housings, inner tie rods, ball seats, ball seat springs, jamnuts, and shock dampener rings.

23. Remove the adjuster plug. Pull straight up and out on the stub shaft and remove the assembly.

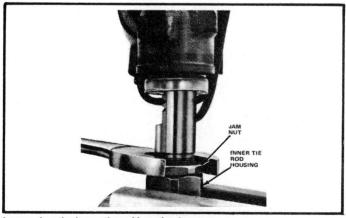

**Loosening the inner tie rod housing jamnuts**

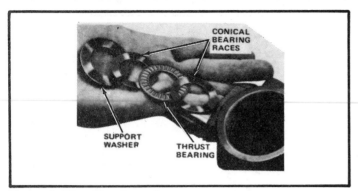

**Removal of thrust bearing and races**

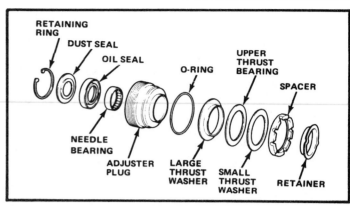

**Exploded view of the pinion shaft and adjusting plug assemblies**

24. Remove the pinion shaft from the housing with a pair of pliers. Grip the pinion shaft at the drive tang and rotate the shaft clockwise while pulling upward to remove the pinion shaft.

25. Remove the pinion thrust bearing conical thrust bearing races, and support washer from the housing with either your fingers, an external-type snap ring pliers, or a magnet.

26. Mark the position of the breather tube, mounting clamp and grommet on each end of the tube and housing for reference during assembly. The breather tube, grommet and mounting clamp must be reassembled in their original positions to ensure proper seating and sealing.

27. Remove the breather tube and the grommet and mounting clamp by pulling and twisting at the same time.

28. Remove the bulkhead retaining ring from the end of the tube by inserting a pin punch through the access hole in the end of the tube to force the retaining ring out of its groove. Then, place the blade of a screwdriver behind the ring and pry it out.

**NOTE: Be careful not to scratch the bore of the tube when removing the retaining ring. Remove any scratches with crocus cloth.**

29. Pull the steering rack and bulkhead out of the tube and housing assembly.

**NOTE: Do not remove the steering rack unless a new inner rack seal is available as the seal is rendered useless when the rack is removed. As the rack is removed, the rack piston will force the bulkhead out of the tube at the same time.**

30. Remove the bulkhead from the steering rack.

31. Remove the plastic rack bushing from the housing. Insert a knife blade under the bushing and pry up. Grasp the bushing with a needlenose pliers and pull the bushing out of the housing.

32. Remove the inner rack seal by driving it out of the tube with a hammer and a brass rod 12 to 14 in. long.

33. Turn the steering gear over in the vise and remove the upper pinion bushing and pinion shaft seal with a 5/8 in. socket and extension. When the bushing separates from the housing, it will force the seal out also.

34. Remove the outer O-ring and two inner lip-type seals from the bulkhead. Note their positions for correct reassembly.

35. Using the blade of a small screwdriver or knife, carefully remove the seal from the rack piston. Be careful not to scratch the piston.

36. To disassemble the adjuster plug:

   a. Remove the thrust bearing retainer with a screwdriver and discard the retainer. Be careful not to damage the needle bearing bore.

   b. Remove the thrust bearing spacer, bearing, and bearing races.

   c. Remove the adjuster plug O-ring seal and discard the seal.

   d. Remove the stub shaft seal retaining ring with snap ring pliers.

   e. Remove the stub shaft dust seal and oil seal by prying them out with a small screwdriver.

   f. Remove the needle bearing.

37. To disassemble the valve body and stub shaft assembly:

**NOTE: Do not disassemble the valve for any other reason than to replace the seals. If replacement of any valve part other than the seal rings or O-rings is necessary, replace the complete valve body assembly.**

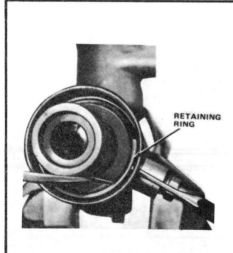

**Removing the bulkhead retaining ring**

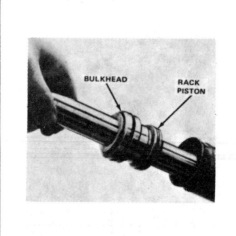

**Removing the steering rack and bulkhead**

**Removing the valve body assembly**

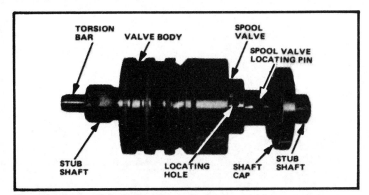

**The valve body, stub shaft and spool valve assembly**

a. Hold the valve body in your hand with the stub shaft pointing downward. Tap the stub shaft lightly against the workbench until the shaft cap is free of the valve body.

b. Pull the stub shaft until the shaft cap clears the valve body by about 1/4 in. Do not pull the stub shaft out any further than 1/4 in. because the spool may become cocked in the body.

c. Carefully remove the spool valve locating pin from the spool valve and remove the stub shaft.

d. Remove the spool valve from the valve body by pushing and rotating the valve. If the valve becomes cocked, carefully align the valve, then remove it.

e. Remove the damper O-ring from the spool valve and discard it.

f. Carefully cut and remove the four seal rings and O-rings from the valve body.

38. To replace the housing hose connector seats:

**NOTE: The hose connector seats do not need to be replaced unless they are damaged and/or do not provide proper sealing.**

a. Insert a no. 4 screw extractor into the seat and turn counter-clockwise to remove the seat.

b. Install a new seat by driving it into place with a brass rod. Be sure the seat is bottomed in the housing and is not cocked.

39. Wash all of the parts, except the rubber boots and nylon bushings in a suitable solvent and dry them with compressed air. Make sure you do not wash off the alignment marks for the breather tube and mounting clamp.

40. Inspect all of the parts for the conditions mentioned in step 4a.

**NOTE: If the tube and housing or steering rack and pinion shaft need replacing, replace the complete steering gear assembly, less the steering linkage components.**

Bushings, thrust bearings, steering linkage components, the breather tube and boots, and the flexible coupling can all be replaced individually.

The adjuster plug assembly can be replaced separately.

The valve body and stub shaft assembly must be replaced as a unit if found to be defective.

## Assembly

1. Install the split-type nylon rack bushing in the housing. Compress the leading edge of the bushing with your fingers and insert it into the housing. The bushing will snap into place once it is past the lip of the housing.

2. Install the inner rack seal on the steering rack by first cutting a 2½ in. × 4 in. section of a manila envelope or similar cardboard-type paper. Form the paper over the rack teeth. The paper will protect the seal from the rack teeth during installation. Dip the seal in power steering fluid and slide it over the rack and onto the paper. Install the seal on the rack with the seal lip facing the rack piston and the metal surface of the seal facing away from the rack piston. Remove the paper protector when the seal is over the rack teeth.

3. Lightly coat the outside diameter of the inner rack seal and bulkhead retaining ring groove in the tube with chassis grease.

4. Coat the rack teeth liberally with lithium base chassis lubricant.

5. Dip the rack piston seal ring in power steering fluid and install the seal ring on the rack piston. Be careful not to overstretch or twist the seal ring when installing it.

6. Carefully insert the steering rack into the tube. Push the steering rack into the tube as far as possible so that the rack piston will start the inner rack seal into its seat in the end of the tube. Bottom the seal in its seat. Install the upper pinion bushing with the chamfered side down, and start the pinion shaft seal, with the seal lip facing the bore, into the seat in the housing.

7. Place the support washer on top of the pinion shaft seal.

8. Using a 1¼ in. socket with an extension, lightly tap on the support washer until the pinion shaft seal and support washer are fully seated in the housing.

9. Lubricate the pinion shaft thrust bearing and races with petroleum jelly and install them on the pinion shaft. The bearing is installed between the two races.

10. Position the rack teeth parallel to the housing bore and set the end of the rack 4 in. from the machined inner face of the housing.

11. Install the pinion shaft into the housing bore with the drive pin located between the 3 and 4 o'clock position. Push the pinion down until it bottoms in the housing.

12. Center the steering rack to the 4 in. setting. With the rack centered, the pinion shaft drive pin should now be located at the 12 o'clock position. If the drive pin is positioned incorrectly either at the 11 o'clock or 1 o'clock position, remove the pinion shaft and start over again at step 10.

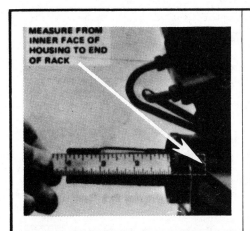

**Centering the steering rack**

**Installing the stub shaft in the valve body**

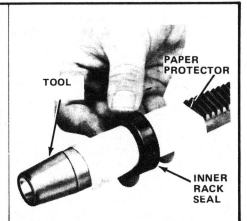

**Installing the inner rack seal**

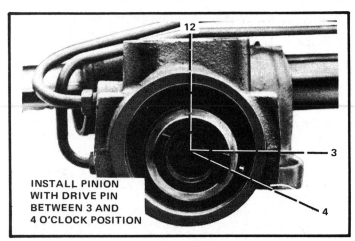

**Centering the pinion shaft in relation to the rack**

13. Assemble the valve body and stub shaft in the following manner:

a. If the valve body O-rings and teflon rings were removed, install new O-rings in the oil ring grooves and lubricate them with power steering fluid.

b. Lubricate the 4 teflon seal rings with power steering fluid and install them in the grooves over the O-rings. Do not be concerned if the teflon rings appear to be distorted; the heat of the fluid during operation will straighten them.

c. Lubricate the spool valve dampener O-ring with power steering fluid and install it over the spool valve.

d. Lubricate the spool valve and valve body with power steering fluid and slide the spool valve into the valve body. Rotate the spool valve while pushing it into the valve body. Push the spool valve on through the body until the shaft pin hole is visible from the opposite end. The spool valve should be flush with the shaft cap end of the valve body.

e. Lubricate the stub shaft assembly with power steering fluid and carefully install it into the spool valve until the shaft pin can be placed into the spool valve.

f. Align the notch in the shaft cap with the pin in the valve body and press the spool valve and shaft assembly into the valve body. Make sure that the notch in the shaft cap mates with the valve body pin.

14. Align the notch in the valve body with the drive pin in the pinion shaft and install the valve body in the housing. Be sure the drive lugs on the pinion shaft fully engage the slots in the stub shaft. When the valve body is correctly installed, the fluid return hole in the housing will be exposed. If the hole is not visible, either the pinion shaft is not seated, the spool valve locating pins are misaligned, or the valve body stub shaft locating pins are misaligned.

**NOTE: Do not press on the stub shaft to seat the valve body. Press only with your thumbs directly on the valve body.**

15. Assemble the adjuster plug components as follows:

a. If you are replacing the needle bearing, drive the new bearing into the plug bore with a soft drift until it bottoms. The bearing identification number faces up.

b. Lubricate the new stub shaft oil seal and install the seal far enough to allow clearance for the dust seal and retaining ring.

c. Lubricate the new dust seal and install it in the plug with the identification number facing outward.

d. Install the retaining snap ring with snap ring pliers.

e. Lubricate the new O-ring seal with petroleum jelly and install it into the groove on the adjuster plug.

f. Assemble the large thrust bearing race, bearing, small bearing race and spacer on the plug. Press the bearing retainer into the needle bearing bore with a brass or wooden drive tool, being very careful not to damage the dimples. Radial location of the dimples is not important.

**Backing off the adjuster plug to the proper adjustment**

16. Install the adjuster plug in housing using a spanner type tool that fits into the two holes on the adjuster plug face. Tighten the plug until it is fully seated.

17. Adjust the thrust bearing preload by first measuring back (counterclockwise) 3/16 to 1/4 in. from one of the adjuster plug holes and making a mark. Back the adjuster plug off (counterclockwise) so that the hole used for reference is opposite the mark. Install the adjuster plug locknut and tighten it to 80 ft./lbs.

18. Install the O-ring and lip-type seals in the bulkhead. The lips of the seals must face the interior of the tube.

19. Slide the bulkhead into the tube and bottom it against the counterbore in the tube by tapping it with a brass rod.

20. Install the bulkhead retaining ring. The opening in the ring should be 1/4 in. away from the access hole in the tube.

21. Turn the steering gear over in the vise and place some lithium base chassis grease in the housing bore. Do not overfill.

22. Install the preload spring in the housing with the center hump of the spring against the housing. The end of the spring must enter the upper pinion bushing. Allow about 1/4 in. of spring to extend past the end of the housing.

23. Hold the preload spring against the housing with needlenose pliers, and install the lower pinion bushing. The chamfered end of the bushing faces inward toward the pinion shaft. Tap lightly on the spring and bushing until they are both seated.

24. Install the contraction plug in the housing and seat it using a brass rod or suitable size socket.

25. Install the mounting clamp and grommet on the tube. Position them according to the marks made during disassembly.

26. Install the shock dampener ring on each end of the steering rack with the open end facing out. Thread the jamnuts on the rack fully.

27. Liberally apply lithium bas chassis grease to all wear surfaces of the inner tie rod assembly. Pack the tie rod housings with the same type grease.

28. Assemble the tie rod housing, inner tie rods, ball seats, ball seat springs, and install the assemblies on the steering rack. Tighten the inner tie rod housings to 75 in./lbs. while rocking the inner tie rods to prevent grease lock. Loosen the tie rod housings 1/2 of a turn and retighten the tie rod housings to 50 in./lbs.

29. Tighten the tie rod housing set-screws to 60 in./lbs.

30. Clamp the tie rod housings in the vise and tighten the jamnuts to 100 ft./lbs. using a crow-foot adapter on the end of a torque wrench.

31. Slide the shock dampener rings over the jamnuts.

32. Install the mounting clamp and grommet on the tube end of the steering gear. Align the notch in the clamp with the oil line fitting boss in the tube.

33. Install the breather tube. Make sure the breather tube is not blocked in any way. The breather tube transfers air from one boot to

the other during steering operation. Dust and air could be drawn into the inner tie rod assemblies if the tube is blocked.

34. Install the boot on the mounting bracket side of the tube and housing. Align the hole in the boot with the hole in the mounting grommet and breather tube. Install the boot clamp, but do not tighten it.

35. Install the opposite boot with the hole in the boot aligned with the breather tube. The boot lip must fit over the collar of the housing to seat the tube.

36. Slide the small ends of the boots over the inner tie rod undercuts and secure the boot with the small boot clamps. Tighten the clamps.

37. Secure the large end of the boot to the housing end of the steering gear. Fit the clamp over the groove in the boot with the clamp ear 3/4 in. from the tube. Compress the clamp.

38. Install the adjuster tubes and tie rod ends on the inner tie rods using the alignment marks you made during disassembly. At least 3 threads should be visible at both ends of the adjuster tubes. The difference in the number of threads on each side should be no more than 3.

39. Install the flexible coupling on the stub shaft, flat to flat. Install the pinch bolt and tighten it to 30 ft./lbs.

40. Install the steering gear and compress the boot clamps.

# CHRYSLER FULL-TIME CONSTANT CONTROL TYPE POWER STEERING

### Reconditioning

1. Drain gear by turning worm shaft from limit to limit with oil connections held downward. Thoroughly clean outside.

2. Remove valve body attaching screws, body and three O-rings.

3. Remove pivot lever and spring. Pry under spherical head with a screwdriver.

---
**— CAUTION —**

*Use care not to collapse slotted end of valve lever as this will destroy bearing tolerances of the spherical head.*

---

4. Remove steering gear arm from sector shaft.

5. Remove snap-ring and seal backup washer.

6. Remove seal, using proper tool to prevent damage to relative parts.

7. Loosen gear shaft adjusting screw locknut and remove gear shaft cover nut.

8. Rotate wormshaft to position sector teeth at center of piston travel.

9. Loosen power train retaining nut.

10. Turn worm shaft either to full left or full right (depending on car application) to compress power train parts. Then remove power train retaining nut.

11. Remove housing head tang washer.

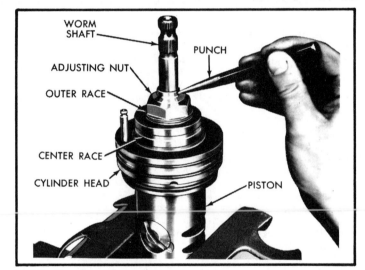

**Staking the wormshaft bearing nut**

12. While holding power train completely compressed, pry on piston teeth with screwdriver, using shaft as a fulcrum, and remove complete power train.

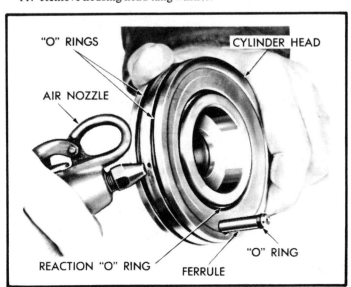

**Removing the reaction seal with air pressure**

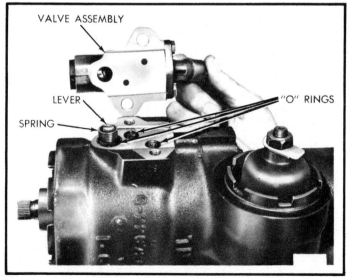

**Removing the valve body**

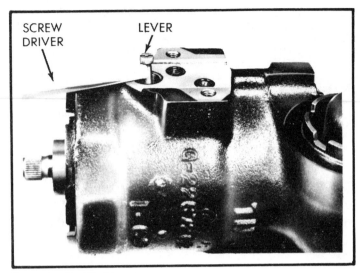

Removing the pivot lever

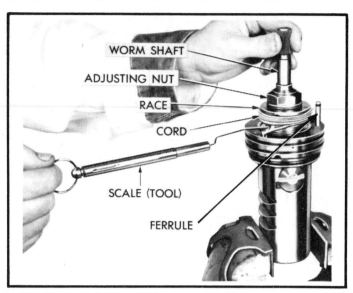

Testing the center bearing preload

─────── CAUTION ───────

*Maintain close contact between cylinder head, center race and spacer assembly and the housing head. This will eliminate the possibility of reactor rings becoming disengaged from their grooves in cylinder and housing head. It will prohibit center spacer from separating from center race and cocking in the housing. This could make it impossible to remove the power train without damaging involved parts.*

13. Place power train is soft-jawed vise in vertical position. The worm bearing rollers will fall out. Use of arbor tool will hold roller when the housing is removed.

14. Raise housing head until wormshaft oil shaft just clears the top of wormshaft and position arbor tool on top of shaft and into seal. With arbor in position, pull up on housing head until arbor is positioned in bearing. Remove when the housing is removed.

15. Remove large O-ring from housing head groove.

16. Remove reaction seal from groove in face of head with air pressure directed into ferrule chamber.

17. Remove reactor spring, reactor ring, worm balancing ring and spacer.

18. While holding wormshaft from turning, turn nut with enough force to release staked portions from knurled section and remove nut.

19. Remove upper thrust bearing race (thin) and upper thrust bearing.

20. Remove center bearing race.

21. Remove lower thrust bearing and lower thrust bearing race (thick).

22. Remove lower reaction ring and reaction spring.

23. Remove cylinder head assembly.

24. Remove O-rings from outer grooves in head.

25. Remove reaction O-ring from groove in face of cylinder head. Use air pressure in oil hole located between O-ring grooves.

26. Remove snap-ring, sleeve and rectangular oil seal from cylinder head counterbore.

27. Test wormshaft operation. Not more than 2 in./lbs. should be required to turn it through its entire travel, and with a 15 ft./lb. side load.

**NOTE: The worm and piston is serviced as a complete assembly and should not be disassembled.**

28. Shaft side play should not exceed 0.008 in. under light pull applied $2\frac{5}{16}$ in. from piston flange.

29. Assemble in reverse of above, noting proper adjustments and preload requirements following.

30. When cover nut is installed, tighten to 20 ft./lbs. torque.

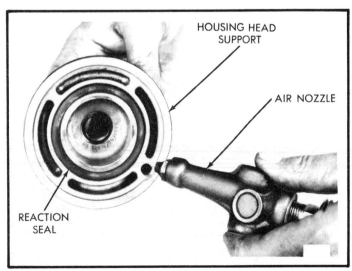

Removing the lower reaction seal with air pressure

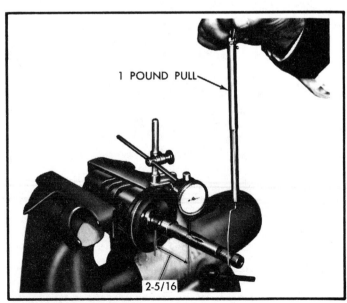

Checking the worm shaft side play

# TROUBLESHOOTING CHRYSLER POWER STEERING

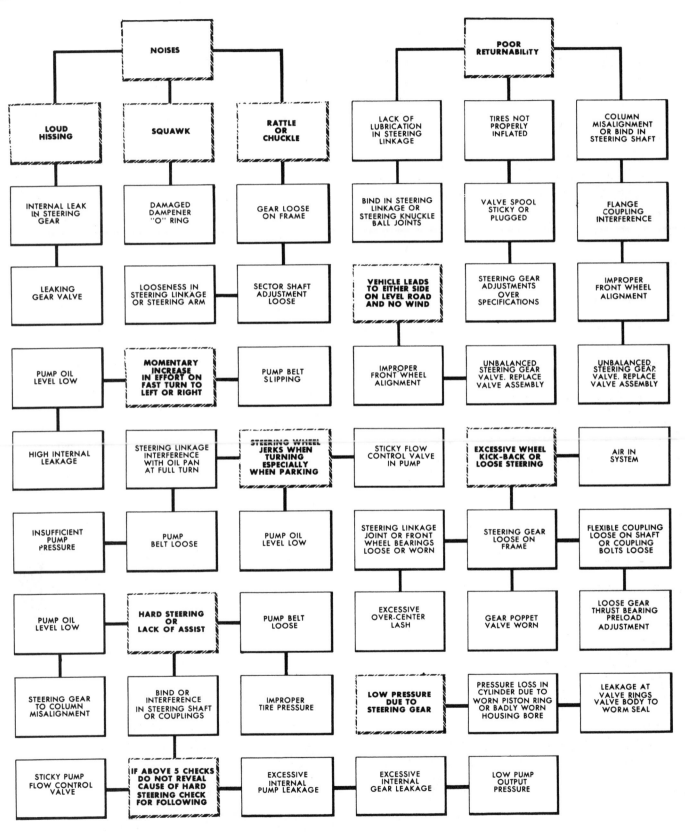

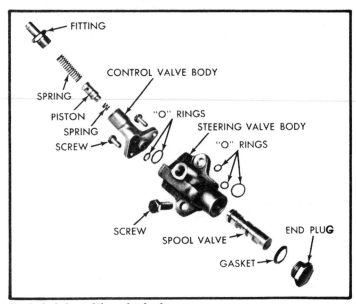

**Exploded view of the valve body**

31. Valve mounting screws should be tightened to 200 in./lbs. torque.

32. With hoses connected, system bled, and engine idling roughly, center valve unit until not self-steering. Tap on head of valve body attaching screws to move valve body up, and tap on end plug to move valve body down.

33. With steering gear on center, tighten gear shaft adjusting screw until lash just disappears.

34. Continue to tighten 3/8 to 1/2 turn and tighten locknut to 50 ft./lbs.

# CHRYSLER CORP. POWER STEERING RACK/PINION ASSEMBLY

## Reliant, Aries, LeBaron, 400, Horizon, Omni

## Power Steering Rack and Pinion Assembly

### OUTER TIE ROD

**Removal**

1. Loosen the rod jam nut.
2. Remove the tie rod from the steering knuckle.
3. Remove the outer tie rod by unscrewing it from the inner tie rod. Count the number of turns to unscrew.

**Installation**

1. Screw the outer tie rod onto the inner tie rod the same number of turns necessary to remove.
2. Expand the outer boot clamp and leave loose on the tie rod.
3. Do not tighten the jam nut until the toe adjustment is made. Do not twist the boot.
4. Torque the jam nut to 50 ft./lbs. and install the outer boot clamp.
5. Be sure the boot is not twisted when done.

### BOOT SEAL

**Removal**

1. With the outer tie rod off, remove the jam nut from the inner tie rod.
2. Expand the outer boot clamp and cut the inner boot clamp and discard.
3. Mark the location of the breather tube on the rubber boot. Remove the boot.

**Installation**

1. Install the boot and inner boot clamp. Align the boot mark and breather tube.
2. Install the boot seal over the housing lip with the hole in the boot aligned with the breather tube.
3. Install the inner boot clamp. Lubricate the tie rod boot groove with a silicone type lubricant before installing the outer clamp.

### INNER TIE ROD

**Removal**

1. With the steering unit out of the vehicle, remove the shock dampener ring from the inner tie rod housing and slide it back on the rack (Horizon and Omni).
2. (Reliant, Aries, LeBaron, 400) Remove the roll pin from the inner tie rod housing.
3. Put a wrench on the tie rod pivot housing flats and turn the housing counterclockwise until the inner tie rod assembly separates from the rack.

**Installation**

1. Install the inner tie rod onto the rack and bottom the threads.
2. Torque the housing while holding the rack with a wrench. Torque to 70 ft./lbs. for the Horizon and Omni vehicles, while torquing the inner tie rod to 60 ft./lbs. for the Reliant, Aries, LeBaron and 400 vehicles. Install roll pin as required.
3. Support the rack and housing and stake the housing in two places for the Horizon and Omni vehicles.
4. Inspect the stake, a 0.010 inch feeler gauge must not pass between the rack and the housing stake on each side.
5. On the Horizon and Omni vehicles, slide the shock dampener over the inner tie rod housing until it engaged.

### RACK BEARING

**Removal**

1. Loosen the lock nut for the adjuster plug.

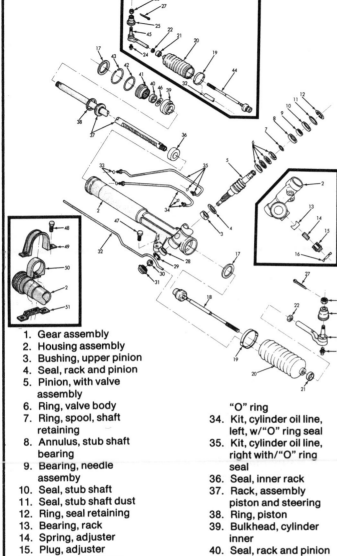

1. Gear assembly
2. Housing assembly
3. Bushing, upper pinion
4. Seal, rack and pinion
5. Pinion, with valve assembly
6. Ring, valve body
7. Ring, spool, shaft retaining
8. Annulus, stub shaft bearing
9. Bearing, needle assemby
10. Seal, stub shaft
11. Seal, stub shaft dust
12. Ring, seal retaining
13. Bearing, rack
14. Spring, adjuster
15. Plug, adjuster
16. Nut, adjuster plug lock
17. Ring, shock dampener
18. Rod assembly, inner tie, left
19. Clamp, boot
20. Boot, rack and pinion
21. Clamp, boot (tie rod end)
22. Nut, hex jam
23. Tie rod, outer, left
24. Fitting, lubrication
25. Seal, tie rod
26. Nut, outer tie rod
27. Pin, cotter
28. Bearing, ball, assembly
29. Ring, pinion bearing retaining
30. Nut, hex lock
31. Cover, dust
32. Tube, breather
33. Seal, cylinder oil line

"O" ring
34. Kit, cylinder oil line, left, w/"O" ring seal
35. Kit, cylinder oil line, right with/"O" ring seal
36. Seal, inner rack
37. Rack, assembly piston and steering
38. Ring, piston
39. Bulkhead, cylinder inner
40. Seal, rack and pinion (bulkhead)
41. Bulkhead, cylinder outer
42. Seal, "O" ring
43. Ring, bulkhead retaining
44. Rod, assembly, inner tie, right
45. Tie rod, outer, right
46. Spring, wave washer
47. Bolt, rack and pinion, steering gear mounting, left
48. Bolt, rack and pinion, steering gear mounting, right
49. Bracket, rack and pinion steering, gear mounting, outer
50. Bushing, rack and pinion steering gear
51. Bracket, rack and pinion steering gear mounting, inner

**Exploded view of power steering rack and pinion - Omni and Horizon**

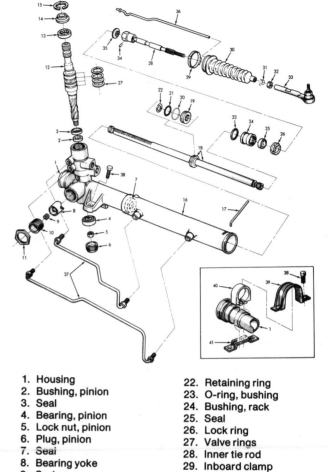

1. Housing
2. Bushing, pinion
3. Seal
4. Bearing, pinion
5. Lock nut, pinion
6. Plug, pinion
7. Seal
8. Bearing yoke
9. Spring
10. plug
11. Lock nut
12. Valve assembly
13. Bearing
14. Seal, shaft
15. Retaining ring
16. Tube assembly
17. Retaining wire
18. Rack assembly
19. Piston
20. Piston ring
21. O-ring
22. Retaining ring
23. O-ring, bushing
24. Bushing, rack
25. Seal
26. Lock ring
27. Valve rings
28. Inner tie rod
29. Inboard clamp
30. Boot
31. Outboard clamp
32. Lock nut
33. Tie rod, outer
34. Spring pin
35. Shock damper
36. Breather tube
37. Oil lines, cylinder
38. Bolt, gear mounting
39. Bracket, gear mounting
40. Bushing, gear mounting
41. Bracket, inner

**Exploded view of power steering rack and pinion - Aries, Reliant, LeBaron, 400**

2. Remove the adjuster plug from the housing. Remove the spring and rack bearing.

### Installation

1. Lubricate the metal parts and install the rack bearing, the spring, the adjuster plug and the lock nut.
2. Turn the adjuster plug in until it bottoms and then back off 40 to 60 degrees.
3. Tighten the lock nut while holding the adjuster plug in place. The torque must be 50 ft./lbs.

## STUB SHAFT SEALS

### Removal

1. Remove the retaining ring and the dust cover.

2. Using a special holder or its equivalent, remove the lock nut from the pinion.

——————————CAUTION——————————
*If the stub shaft is not held, damage to the pinion teeth will occur.*

3. Using the special puller or its equivalent, pull the valve and pinion assembly until flush with the ball bearing assembly. The complete assembly does not have to be removed.

4. Remove the stub shaft dust seal, stub shaft seal, needle bearing and stub shaft bearing annulus.

**NOTE: The bearing and annulus are pressed together and disassembly is required only if bearing replacement is necessary.**

### Installation

1. Lubricate the seals and install in the reverse order of removal, using seal protectors on the pinion shaft. Seal installers are aviilable to assist in seating the seal properly.

2. While holding the stub shaft, firmly seat the lock nut and torque to 26 ft./lbs.

3. Install the retainer and the dust cover.

## VALVE AND PINION ASSEMBLY

### Removal

1. Turn the stub shaft until the rack is equal distance on both sides of the housing, with the pinion fully engaged.

2. Mark the location and angle of the stub shaft flat on the steering housing.

3. With the lock nut off the pinion, use a special puller or its equivalent, and pull the valve and pinion assembly from the housing.

4. Remove the valve body rings.

### Installation

1. Install new rings on the valve body.

2. Lubricate the rings and valve. Install the assembly into the housing. Be sure the rack is equal on both sides of the housing.

3. When the valve and pinion assembly is installed, the stub shaft flat should align with the mark made before disassembly.

4. Hold the stub shaft, install the lock nut and torque to 26 ft./lbs.

## BULKHEAD

### Removal

1. The pinion and valve assembly must be in the housing for this operation.

2. Horizon and Omni vehicles, use a punch in the access hole and remove the bulkhead retaining ring. Discard the ring.

3. Reliant, Aries, LeBaron, 400 vehicles, use a punch to rotate the retaining wire clockwise to expose the end. Pull the retaining wire to remove.

4. Loosen and remove both cylinder lines. Plug fittings at the cylinder.

5. Turn the stub shaft so that the rack moves to the right, forcing the bulkhead from the housing. Use a drain pan to catch the power steering fluid.

6. If the inner rack seal or piston rings are to be replaced, use special seal remover tools for each group of car models.

7. The piston and pinion can be removed.

### Installation

1. Install the inner rack seal with special seal installer tools or equivalent.

2. Install the plastic retainer onto the inner rack seal.

3. Install the bulkhead outer seal into the bulkhead.

4. Install the bulkhead onto the rack.

5. Horizon and Omni vehicles, be sure the open end of the new bulkhead retaining ring is approximately 0.50 from the access hole.

6. Horizon and Omni vehicles, turn the rack to full right turn to fully seat the retaining ring.

7. The remaining models, install the retaining wire by rotating the bulkhead assembly counterclockwise.

## PINION BALL BEARING, UPPER PINION BUSHING AND SEAL

### Removal and Installation

1. The pinion and piston must be out of the housing.

2. The bearing is removed by the use of a drift and hammer. To install, a bearing installer is available.

3. To remove the seal and bushing, a drift and hammer is used. To install, the use of special installing tools are necessary.

# FORD INTEGRAL POWER STEERING GEAR

### Disassembly

1. Hold the steering gear upside down over a drain pan and cycle the input shaft several times to drain the fluid from the gear.

2. Secure the gear in a soft-jawed vise.

3. Remove the nut from the sector shaft adjusting screw.

4. Turn the input shaft to either stop then, turn it back two turns to center the gear.

**NOTE: The indexing flat on the input shaft spline should be facing downward.**

5. Remove the sector shaft cover attaching bolts.

6. Rap the lower end of the sector shaft with a soft-hammer to loosen it, and lift the cover and shaft from the housing **as an assembly**. Discard the O-ring.

7. Turn the sector shaft cover counterclockwise and remove it from the sector shaft adjuster screw.

8. Remove the valve housing attaching bolts and identification tag. Hold the piston to keep it from spinning off the shaft, and lift the valve housing off the steering gear housing. Remove the valve housing and control valve gasket. Discard the gasket.

**NOTE: If valve housing seals are to be replaced, proceed to Step 12. If sector shaft seals are to be replaced go to steering gear housing section. Balls need only to be removed if valve sleeve rings are to be replaced.**

9. With the piston held so that the ball guide faces up, remove the ball guide clamp screws and ball guide clamp. With a finger over the opening in the ball guide, turn the piston so that the ball guide faces down over a clean container. Let the guide tubes drop into the container.

10. Rotate the input shaft from stop to stop until all balls fall from

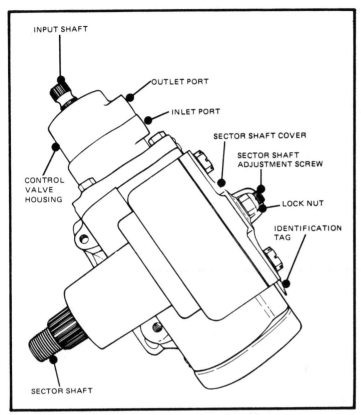

**Ford integral power steering gear assembly - exterior view**

the piston into the container. The valve assembly can then be removed from the piston. Inspect the piston bore to insure all balls have been removed.

11. Install the valve body assembly in the bench mounted holding fixture, and loosen the Allen head race nut screw from the valve housing. Remove the worm bearing race nut.

12. Carefully slide the input shaft, worm and valve assembly out of the valve housing. **Do not cock the spool or it may jam in the housing.**

### Assembly

1. Mount the valve housing in the bench mounted holding fixture with the flanged end up.

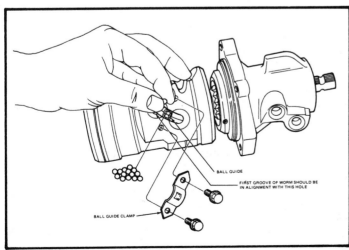

**Assembling ball in piston and piston on worm shaft - Ford integral power steering**

2. Apply a light coat of gear lubricant to the Teflon rings on the valve sleeve.

3. Carefully install the worm shaft and valve in the housing.

4. Install the worm bearing race nut in the housing and torque to specification.

5. Install the Allen head race nut screw through the valve housing and tighten to specification 1.7-2.8 N·m (15-25 in.-lbs.).

6. Place the power cylinder piston on the bench with the ball guide holes facing up. Insert the worm shaft into the piston so that the first groove is in line with the hole nearest the center of the piston.

7. Place the ball guide in the piston. Turning the worm shaft counterclockwise as viewed from the input end of the shaft, place the same balls as removed in Step 9 of Steering Gear Disassembly in the ball guide. A minimum of 27 balls is required. If all the balls have not been inserted upon reaching the left stop, rotate the input shaft in one direction then the other while inserting the remaining balls. DO NOT rotate the input shaft or piston more than three (3) turns from the left stop, or the balls will fall out of the circuit.

8. Secure the guides in the ball nut with the clamp. Tighten screws to specification.

9. Apply petroleum jelly or equivalent to the Teflon seal on the piston.

10. Place a new control valve O-ring on the valve housing.

11. Slide the piston and valve into the gear housing being careful not to damage the piston ring.

12. Align the oil passage in the valve housing with the passage in the gear housing. Place a new O-ring onto the oil passage hole of the gear housing. Install identification tag onto the housing. Install **but do not tighten,** the attaching bolts. Identification tag is to be installed under upper right valve housing bolt.

13. Rotate the ball nut so that the teeth are in the same plane as the sector teeth. Tighten the valve housing attaching bolts to specification.

14. Position the sector shaft cover O-ring in the steering gear housing. Turn the input shaft to center the piston.

15. Apply petroleum jelly or equivalent to the sector shaft journal, and position the sector shaft and cover assembly in the gear housing. Install the sector shaft cover attaching bolts. Tighten the bolts to specification 75-94 M·m (55-70 ft.-lbs.).

16. Attach an in.-lb. torque wrench to the input shaft. Adjust mesh load to specification.

## STEERING GEAR HOUSING

### Disassembly and Assembly

1. Remove the snap ring from the lower end of the housing.

2. Remove dust seal using tools puller attachment and slide hammer.

3. Remove pressure seal in the same manner. Discard the seal.

4. Lubricate the new pressure seal dust seal with clean Ford Polyethylene Grease.

5. Apply Ford Polyethylene Grease to the sector shaft seal bore.

6. Place the dust seal on Sector Shaft Replacement tool so the raised lip of the seal is towards the tool. Place the pressure seal on the tool with lip away from the tool. The flat back side of the pressure seal should be against the flat side of the dust seal.

7. Insert the seal driver tool into the sector shaft bore and drive the tool until the seals clear the snap ring groove. Do not bottom seals against bearing. The seal will not function properly when bottomed against the bearing.

8. Install snap ring in the groove in the housing.

## VALVE HOUSING

### Disassembly and Assembly

1. Remove the dust seal from the rear of the valve housing using puller attachment and slide hammer. Discard the seal.

2. Remove the snap ring from the valve housing.

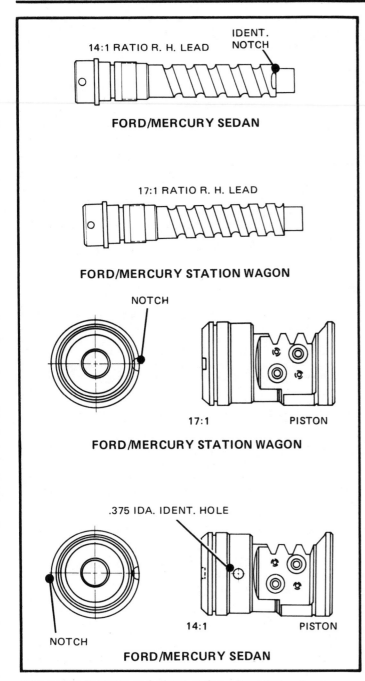

Differences in worm and piston ratios - Ford integral power steering

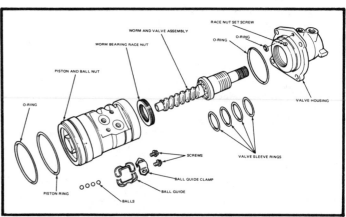

Exploded view of ball nut and housing - Ford integral power steering

3. Turn the bench mounted holding fixture to invert valve housing.

4. Insert tools from the input shaft bearing seal tool in the valve body assembly opposite the oil seal end and gently tap the bearing and seal out of the housing. Discard the seal. **Do not damage the housing when inserting and removing the tools.**

5. Remove the fluid inlet and outlet tube seats with tube seat remover tool if they are damaged.

6. Coat the fluid inlet and outlet tube seats with petroleum jelly or equivalent and install them in the housing with a tube seat installer.

7. Coat the bearing and seal surface of the housing with petroleum jelly or equivalent.

8. Install the bearing with the metal side covering the rollers facing outward. Seat the bearing in the valve housing. Be sure the bearing rotates freely.

9. Dip a new oil seal in gear lubricant, and place it in the housing with the metal side facing outward. Drive the seal into the housing until the outer edge does not quite clear the snap ring groove.

10. Place the snap ring in the housing and drive on the ring until the snap ring seats in its groove.

11. Place the dust seal in the housing with the dished side (rubber side) facing out. Drive the dust seal into place. When properly installed, the seal will be located behind the undercut in the input shaft.

## WORM AND VALVE SLEEVE

### Disassembly and Assembly

1. Remove valve sleeve rings from sleeve by inserting the blade of a small pocket knife under them and cutting them off.

2. Mount the worm end of the worm and valve sleeve assembly into a soft-jawed vise.

3. Install mandrel tool over the sleeve; slide one valve sleeve ring over the tool.

4. Slide the pusher tool over the mandrel; rapidly push down on the pusher tool, forcing the ring down the ramp and into the fourth groove of the valve sleeve. Repeat this step three more times, and each time add one of the spacers under the mandrel tool. By adding the spacer each time, the mandrel tool will line up with the next groove of the valve sleeve.

5. After installing the four valve sleeve rings, apply a light coat of gear lubricant to the sleeve and rings.

6. Install one spacer over the input shaft as a pilot for installing the sizing tube. Slowly install the sizing tube over the sleeve valve end of the worm shaft onto the valve sleeve rings. Make sure that the rings are not being bent over as the tube is slid over them.

7. Remove the sizing tube and check the condition of the rings. Make sure that the rings turn freely in the grooves.

**NOTE: No further service or disassembly of the worm valve assembly is possible.**

## PISTON AND BALL NUT

### Disassembly and Assembly

1. Remove the Teflon piston ring and O-ring from the piston and ball nut. Discard both rings.

2. Dip a **new** O-ring in gear lubricant and install it on the piston and ball nut.

3. Install a **new** Teflon piston ring on the piston and ball nut being careful not to stretch it any more than necessary.

# FORD AND TRW INTEGRAL RACK AND PINION TYPE POWER STEERING

This system was developed to provide a power steering system for Ford Motor Company compact cars equipped with rack and pinion steering.

## RACK YOKE BEARING PRELOAD

### Adjustment

The steering gear must be removed from the car to make this adjustment.

1. Remove the fluid lines from the gear and drain the fluid.
2. Fasten the unit down on the bench.
3. Attach an inch pound torque wrench to the input shaft.
4. Loosen the yoke plug locknut.
5. Attach an inch pound torque wrench to the yoke plug. Tighten the plug to 45-50 in./lbs. with the rack at the center of travel.
6. Back off the yoke plug no more than 45 degrees until the torque required to turn the input shaft is 7-15 in./lbs.
7. Tighten the yoke plug locknut to 44-66 ft./lbs., while holding the plug.
8. Recheck the adjustment after tightening the locknut.

## TIE-ROD ENDS, BELLOWS AND TIE-ROD BALL JOINT SOCKET

### Disassembly and Assembly

1. Loosen the jam nuts adjacent to the tie-rod sockets. Remove the sockets and jam nuts.

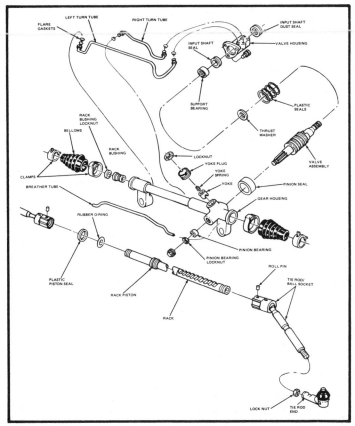

**Exploded view of integral power steering rack and pinion - TRW type**

2. Remove the 4 clamps and remove the bellows and breather tube, after draining the fluid.
3. To remove the tie-rod and ball sockets, drill out the retaining pin in the ball socket.

--- **CAUTION** ---

*This hole must not be drilled deeper than 3/8 in. Remove the tie-rod and ball socket with a spanner wrench.*

4. Remove the locknut, inner thrust bearing, and rack spring from the recess in the end of the rack.

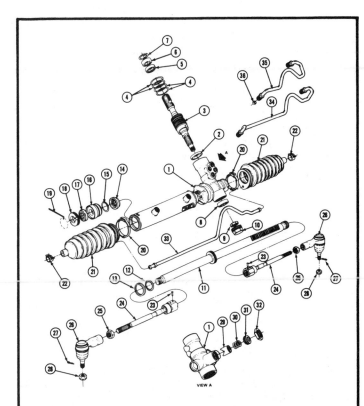

1. Gear housing assembly
2. Pinion seal
3. Valve assembly
4. Plastic rings
5. Input shaft bearing
6. Input shaft seal
7. Snap ring — seal retainer
8. Pinion bearing
9. Pinion bearing locknut
10. Housing cap
11. Rack assembly
12. Back up O-ring (rubber)
13. Piston seal (plastic)
14. Inner rack seal (stepped O.D.)
15. Rack bushing O-ring
16. Rack bushing
17. Outer rack seal
18. Lock-ring
19. Lock-wire
20. Inner bellows clamp
21. Bellows
22. Outer bellows clamp
23. Spiral pin
24. Tie rod assembly
25. Jam nut
26. Tie rod end assembly
27. Cotter pin
28. Castellated nut
29. Rack yoke
30. Yoke spring
31. Yoke plug
32. Yoke plug lock nut
33. Breather tube
34. Right turn transfer tube
35. Left turn transfer tube
36. Copper seal (4 req'd)

**Exploded view of integral power steering rack and pinion - Ford type**

# POWER STEERING
## FORD INTEGRAL RACK & PINION TYPE

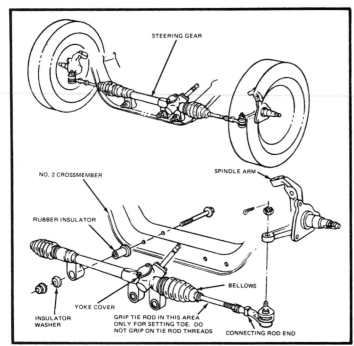

**Attachment location of Ford integral power steering rack and pinion**

To assemble:

5. Install a new rack spring in the recess in the end of the rack. Assemble the ball socket to the tie-rod.

6. Install a new inner thrust bearing in the ball socket.

7. Lubricate the spring, thrust bearing, tie-rodd ball, and the ball socket.

8. Thread a new ball joint locknut onto the end of the rack.

9. Thread the ball socket onto the rack until the socket is tight.

10. Rotate the rod a few times. Adjust the ball socket until the effort required to move the tie-rod end with a spring scale is 4-6 lbs.

11. Hold the ball socket and tighten the locknut to 25-35 ft./lbs. Repeat step 10.

12. Drill a new hole, similar to that made in step 3. It must not enter the notches of the locknut.

13. Install the retaining pin and stake it in place.

14. Install the bellows and the breather tube. Install new clamps. Put 2½ oz. of lubricant into each bellows.

15. Install the jam nuts and tie-rod sockets on the outer ends of the tie-rods.

## INPUT SHAFT AND VALVE ASSEMBLY

### Disassembly and Assembly

1. Remove the fluid lines. Remove the flare gaskets from the ports.

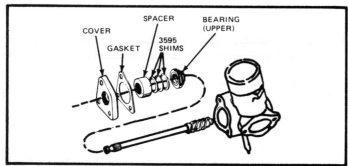

**Pinion bearing cover and shim arrangement**

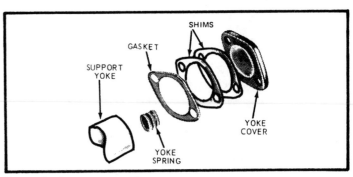

**Support yoke arrangement**

2. Loosen the yoke plug locknut and yoke plug to relieve the rack preload.

3. Remove the pinion bearing plug.

4. Hold the input shaft and remove the pinion bearing locknut.

5. Remove the 3 bolts holding the valve housing to the gear unit. Move the rack to the left stop. Mark the relationship of the input shaft to the valve housing for reinstallation. Carefully work the input shaft and valve assembly out of the gear housing.

6. Remove the pinion bearing from the gear housing with a slide hammer. Remove the bearing-to-gear housing and valve housing-to-gear housing O-rings.

7. Remove the input shaft oil seal from the gear housing.

8. Slide the valve housing over the splined end of the input shaft to remove.

9. Remove the input shaft O-rings, being extremely careful not to damage the lands and grooves.

10. Remove the input shaft needle thrust bearing and the two thrust washers from the inside of the valve housing.

11. Use a slide hammer to remove the input shaft support bearing from the valve housing. Remove the input shaft oil seal also.

12. Pry the input shaft dust seal from the valve housing with a small sharp chisel.

To assemble:

13. Press the input shaft oil seal into the valve housing with a finger. Make sure that the lip faces in and that the seal bottoms in the bore.

14. Fill the input shaft dust seal bore with lubricant and install the dust seal.

15. Lubricate the input shaft support bearing with power steering fluid and install it in the valve housing. Avoid contact with the oil seal.

16. Install the 4 O-rings in the grooves on the valve.

17. Lubricate the two thrust washers and the needle thrust bearing with power steering fluid. Place a thrust washer on each side of the bearing and install over the input shaft.

18. Lubricate the O-rings and valve with power steering fluid. Insert the unit over the valve bore of the housing and push it through until it bottoms and the full spline passes through the dust seal.

19. Install the valve housing to gear housing O-ring on the flange protruding from the gear housing.

20. Use your fingers to install an input shaft oil seal in the gear housing. Make sure that the lip faces the input shaft and valve assembly.

21. Install the pinion bearing in the lower gear housing. Install an O-ring around the bearing adjacent to the gear housing.

22. Move the rack to the left stop. Install the input shaft and valve assembly in the gear housing bore. Align the marks made in step 5.

23. Install the bolts holding the valve housing to the gear housing. Torque to 12-15 ft./lbs.

24. Install the pinion bearing locknut on the pinion shaft. Drill a hole in a spare pinion bearing plus large enough to insert a 9/26 in socket. Thread the plug into the housing bore and tighten to hold the bearing firmly. Hold the input shaft and torque the pinion bearing locknut to 44-66 ft./lbs. Make sure that the rack is is away from the stop while doing this. Remove the drilled bearing plug.

25. Install the pinion bearing plug. Torque it to 60-100 ft./lbs. Stake the plug in place.

26. Install the 4 small flare nuts in the pressure line fittings. Install the pressure lines.

27. Install the tube bracket with the tabs in the slots on the gear housing.

28. Install the yoke plug locknut. Make the rack yoke bearing preload adjustment described earlier, then torque the locknut to 44-66 ft./lbs.

---

**— CAUTION —**

*When the front wheels of the vehicle are suspended completely off the ground, do not turn the wheels quickly or forcefully from lock to lock. This could cause a build-up of hydraulic pressure within the steering gear which could damage or blow out the bellows.*

---

# BENDIX LINKAGE TYPE (FORD) POWER STEERING

The Bendix linkage-type power steering system is a hydraulically controlled linkage-type, composed of an integral pump and fluid reservoir, a control valve, a power cylinder, connecting fluid lines, and the steering linkage. The hydraulic pump, which is driven by a belt turned by the engine, draws fluid from the reservoir and provides fluid pressure through hoses to the control valve and the power cylinder. There is a pressure relief valve to limit the pressures within the steering system to a safe level. After the fluid has passed from the pump to the control valve and the power cylinder, it returns to the reservoir.

The Bendix linkage-type steering system when used in Ford-built cars is called the Ford Non-Integral Power System.

## CONTROL VALVE

### Disassembly and Assembly

1. Clean the outside of the control valve of dirt and fluid.

2. Remove the centering spring cap from the valve housing. The control valve should be put in a soft-faced bench vise during disassembly. Clamp the control valve around the sleeve flange only, to avoid damaging the housing, spool, or sleeve.

3. Remove the nut from the end of the valve spool nut. Remove the washers, spacer, centering spring, adapter, and the bushing from the bolt and valve housing.

4. Remove the two bolts holding the valve housing and the sleeve together. Separate the valve housing and the sleeve.

5. Remove the plug from the sleeve. Push the valve spool out of the centering spring end of the valve housing, and remove the seal from the spool.

6. Remove the spacer, bushing, and seal from the sleeve end.

7. Drive the stop-pin out of the travel regulator stop with a punch and hammer. *Pull the head of the valve spool bolt tightly against the travel regulator stop before driving out the pin.*

8. Turn the travel regulator stop counterclockwise in the valve sleeve to remove the stop from the sleeve.

9. Remove the valve spool bolt, spacer, and rubber washer from the stop.

10. Remove the rubber boot and clamp from the valve sleeve. Slide the bumper, spring, and ball stud seat out of the valve sleeve, and remove the ball stud socket from the sleeve.

11. Remove the return port hose seat and the return port relief valve.

12. Remove the spring plug and O-ring. Then, remove the reaction limiting valve.

13. Replace all worn or damaged hose seats by using an Easy-Out screw extractor or a bolt of proper size as a puller. Tap the existing hole in the hose seat, using a starting tap of the correct size. *Remove all metal chips from the hose seat after tapping.* Place a nut and washer on a bolt of the same size as the tapped hole. The washer must be large enough to cover the hose seat port. Insert the bolt in the tapped hole and remove the hose seat by turning the nut clockwise and drawing the

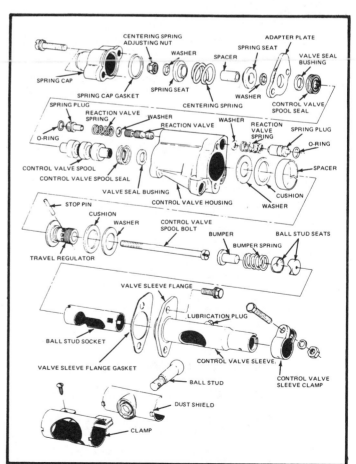

**Control valve disassembled - Maverick, Comet**

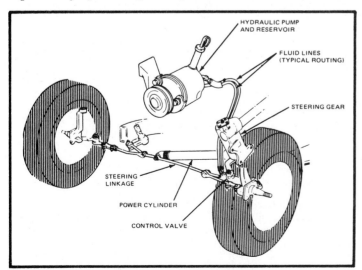

**Non-integral power steering system - Linkage type**

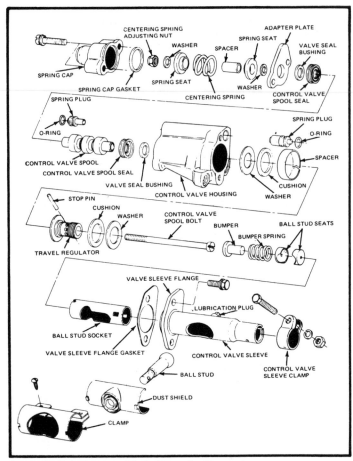

**Control valve disassembled - Granada, Monarch**

sleeve. Be sure the lubrication fitting is turned on tightly and does not bind on the ball stud socket.

21. Insert the valve spool in the valve housing, rotating it while installing.

22. Move the spool toward the centering spring end of the housing, and place the small seat, bushing, and spacer in the sleeve end of the housing.

23. Press the valve spool against the inner lip of the seal and, at the same time, guide the lip of the seal over the spool with a small screwdriver. *Do not nick or scratch the seal or spool during installtion.*

24. Place the sleeve end of the housing on a flat surface so that the seal, bushing, and spacer are at the bottom end, and push down the valve spool until it stops.

25. Carefully install the spool seal and bushing in the centering spring end of the housing. Press the seal against the end of the spool, guiding the seal over the spool with a small screwdriver. *Do not nick or scratch the seal or the spool during installation.*

26. Pick up the housing, and slide the spool back and forth to check for free movement.

27. Place the body gasket and valve sleeve on the housing so that the ball stud is on the same side of the housing as the ports for the two power cylinder lines. Install the two bolts in the sleeve, and torque them to the proper specification.

28. Place the adapter on the centering spring end of the housing, and install the bushing, washers, spacers, and centering spring on the valve spool bolt.

29. Compress the centering spring, and install the nut on the bolt. Tighten the nut snug (90-100 in./lbs.); then, loosen it not more than 1/4 turn. *Do not overtighten, to avoid breaking the stop-pin at the travel regulator stop.*

30. Move the ball stud back and forth to check for free movement.

31. Lubricate the two cap attaching bolts. Install the centering spring cap on the valve housing, and tighten the two cap bolts to the proper torque.

32. Install the nut on the ball stud so that the valve can be put in a vise. Then, push forward on the cap end of the valve to check the valve spool for free movement.

33. Turn the valve around in the vise, and push forward on the sleeve end to check for free movement.

## POWER CYLINDER SEAL

### Removal and Installation

1. Clamp the power cylinder in a vise, and remove the snap-ring from the end of the cylinder. *Do not distort or crack the cylinder in the vise.*

2. Pull the piston rod out all the way to remove the scraper, bushing, and seals. If the seals cannot be removed in this manner, remove them by carefully prying them out of the cylinder with a sharp pick. *Do not damage the shaft or seal seat.*

3. Coat the new seals with power steering fluid and place the parts on the piston rod, which should be lubricated.

4. Push the rod in all the way, and install the parts in the cylinder with a deep socket slightly smaller than the cylinder opening.

bolt out. Install a new hose seal in the port, and thread a bolt of the correct size in the port. Tighten the bolt enough to bottom the seal in the port.

14. Coat all parts of the control valve assembly, except the seals, with power steering fluid. Use grease on the seals.

15. Install the reaction limiting valve, spring, and plug. Install the return port relief valve and hose seat.

16. Insert one of the ball stud seats (flat end first) into the ball stud socket, and insert the threaded end of the ball stud into the socket.

17. Place the socket in the control valve sleeve so that the threaded end of the ball stud can be pulled out through the slot in the sleeve.

18. Place the other ball stud seat, spring, and bumper in the socket. Install and securely tighten the travel regulator stop.

19. Loosen the stop just enough to align the nearest hole in the stop with the slot in the ball stud socket, and install the stop pin in the ball stud socket, travel regulator stop, and valve spool bolt.

20. Install the rubber boot, clamp, and the plug on the control valve

# GM/AMC SAGINAW ROTARY TYPE POWER STEERING

## All A.M.C. Except Pacer

## All G.M.C. Corp. Cars (Except Chevrolet Corvette)

The G.M. Saginaw rotary type power steering gear is designed with all components in one housing.

The power cylinder is an integral part of the gear housing. A

double-acting piston allows oil pressure to be applied to either side of the piston. The one-piece piston and power rack is meshed to the sector shaft.

The hydraulic control valve is composed of a sleeve and valve spool. The spool is held in the neutral position by the torsion bar and spool actuator. Twisting of the torsion bar moves the valve spool, allowing oil pressure to be directed to either side of the power piston, depending on the directional rotation of the steering wheel, to give power assist.

On many General Motors cars a modified version of the system provides variable ratio steering for easier and safer control. The steering gear ratio will vary from a high ratio of about 16:1 while steering straight ahead to a lower gear ratio of about 12.4:1 while making a full turn to either side.

## Checking Steering Effort

Run the engine to attain normal operating temperatures. With the wheels on a dry floor, hook a pull scale to the spoke of the steering wheel at the outer edge. The effort required to turn the steering wheel should be 3½-5 lbs. If the pull is not within these limits, check the hydraulic pressure.

## Pressure Test

To check the hydraulic pressure, disconnect the pressure hose from the gear. Now connect the pressure gauge between the pressure hose from the pump and the steering gear housing. Run the engine to attain normal operating temperatures, then turn the wheel to a full right and a full left turn to the wheel stops.

Hold the wheel in this position only long enough to obtain an accurate reading.

The pressure gauge reading should be within the limits specified. If the pressure reading is less than the minimum pressure needed for proper operation, close the valve at the gauge and see if the reading increases. If the pressure is still low, the pump is defective and needs repair. If the pressure reading is at or near the minimum reading, the pump is normal and needs only an adjustment of the power steering gear or power assist control valve.

## Worm Bearing Preload and Sector Mesh Adjustments

Disconnect the Pitman arm from the sector shaft, then back off on the sector shaft adjusting screw on the sector shaft cover.

Center the steering on the high point, then attach a pull scale to the spoke of the steering wheel at the outer edge. The pull required to keep the wheel moving for one complete turn should be 1/2-2/3 lbs.

If the pull is not within these limits, loosen the thrust bearing locknut and tighten or back off on the valve sleeve adjuster locknut to bring the preload within limits. Tighten the thrust bearing locknut and recheck the preload.

Slowly rotate the steering wheel several times, then center the steering on the high point. Now, turn the sector shaft adjusting screw until a steering wheel pull of 1-1½ lbs. is required to move the worm through the center point. Tighten the sector shaft adjusting screw locknut and recheck the sector mesh adjustment.

Install the pitman arm and draw the arm into position with the nut.

## ADJUSTER PLUG AND ROTARY VALVE

### Removal

1. Thoroughly clean exterior of gear assembly. Drain by holding valve ports down and rotating worm back and forth through entire travel.
2. Place gear in vise.
3. Loosen adjuster plug locknut with punch. Remove adjuster plug.
4. Remove rotary valve assembly by grasping stub shaft and pulling it out.

## ADJUSTER PLUG

### Disassembly

1. Remove upper thrust bearing retainer with screwdriver. Be careful not to damage bearing bore. Discard retainer. Remove spacer, upper bearing and races.
2. Remove and discard adjuster plug O-ring.
3. Remove stub shaft seal retaining ring (Truarc pliers will help) and remove and discard dust seal.

4. Remove stub shaft seal by prying out with screwdriver and discard.
5. Examine needle bearing and, if required, remove same by pressing from thrust bearing end.
6. Inspect thrust bearing spacer, bearing rollers and races.
7. Reassemble in reverse of above.

## ROTARY VALVE

### Disassembly

Repairs are seldom needed. Do not disassemble unless absolutely necessary. If the O-ring seal on valve spool dampener needs replacement, perform this portion of operation only.
1. Remove cap-to-worm O-ring seal and discard.
2. Remove valve spool spring by prying on small coil with small screwdriver to work spring onto bearing surface of stub shaft. Slide spring off shaft. Be careful not to damage shaft surface.
3. Remove valve spool by holding the valve assembly in one hand with the stub shaft pointing down. Insert the end of pencil or wood rod through opening in valve body cap and push spool until it is out far enough to be removed. In this procedure, rotate to prevent jamming. If spool becomes jammed it may be necessary to remove stub shaft, torsion bar and cap assembly.

## ROTARY VALVE

### Reassembly

—————— CAUTION ——————
*All parts must be free of dirt, chips, etc., before assembly and must be protected after assembly.*

1. Lubricate three new back-up O-ring seals with automatic transmission oil and reassemble in the ring grooves of valve body. Assemble three new valve body rings in the grooves over the O-ring seals by carefully slipping over the valve body.

**NOTE: If the valve body rings seem loose or twisted in the grooves, the heat of the oil during operation will cause them to straighten.**

2. Lubricate a new dampener O-ring with automatic transmission oil and install in valve spool groove.
3. Assemble stub shaft torsion bar and cap assembly in the valve body, aligning the groove in the valve cap with the pin in the valve body. Tap lightly with soft hammer until cap is against valve body shoulder. Valve body pin must be in the cap groove. Hold parts together during the remainder of assembly.
4. Lubricate spool. With notch in spool toward valve body, slide the spool over the stub shaft. Align the notch on the spool with the spool drive pin on stub shaft and carefully engage spool in valve body bore. Push spool evenly and with slight rotating motion until spool reaches drive pin. Rotate spool slowly, with some pressure, until notch engages pin. Be sure dampener O-ring seal is evenly distributed in the spool groove.

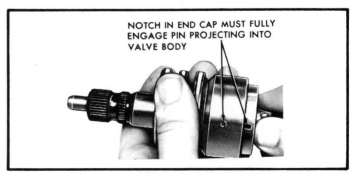

**Installing the stub shaft assembly**

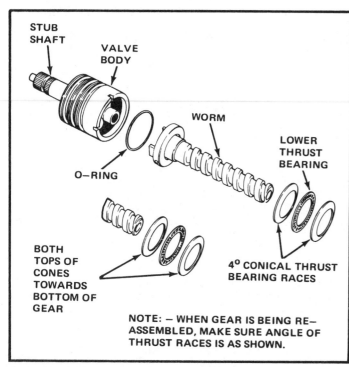

STUB SHAFT

VALVE BODY

WORM

LOWER THRUST BEARING

O–RING

BOTH TOPS OF CONES TOWARDS BOTTOM OF GEAR

4° CONICAL THRUST BEARING RACES

NOTE: – WHEN GEAR IS BEING RE–ASSEMBLED, MAKE SURE ANGLE OF THRUST RACES IS AS SHOWN.

**Correct installation of thrust bearing races**

───── CAUTION ─────
*Use extreme care because spool-to-valve body clearance is very small. Damage is easily caused.*

5. With seal protector tool over stub shaft, slide valve spool spring over stub shaft, with small diameter of spring going over shaft last. Work spring onto shaft until small coil is located in studshaft groove.

6. Lubricate a new cap-to-O-ring seal and install in valve body.

## RACK-PISTON NUT AND WORM ASSEMBLY

### Removal

1. Completely drain the gear assembly and thoroughly clean the outside.

2. Remove pitman shaft assembly, previously described.

3. Rotate housing end plug retaining ring so that one end of ring is over hole in gear housing. Spring one end of ring so screwdriver can be inserted to lift out ring.

4. Rotate stub shaft to full left turn position to force end plug out of housing.

5. Remove and discard housing end plug O-ring seal.

6. Remove rack-piston nut end plug with 1/2 in. square drive.

7. Insert tool in end of worm. Turn stub shaft so that rack-piston nut will go into tool and remove rack-piston nut from gear housing.

8. Remove adjuster plug and rotary valve assemblies as previously described.

9. Remove worm and lower thrust bearing and races.

10. Remove cap-to-O-ring seal and discard.

## RACK-PISTON NUT AND WORM

### Disassembly and Reassembly

1. Remove and discard piston ring and back-up O-ring on rack-piston nut.

2. Remove ball guide clamp and return guide.

3. Place nut on clean cloth and remove ball retaining tool. Make sure all balls are removed.

4. Inspect all parts for wear, nicks, scoring or burrs. If worm or rack-piston nut need replacing, both must be replaced as a matched pair.

5. In assembling, reverse the above.

**NOTE: When assembling, alternate black and white balls, and install guide and clamp. Packing with grease helps in holding during assembly. When new balls are used, various sizes are available and a selection must be made to secure proper torque when making the high point adjustment.**

# G.M. SAGINAW ROTARY TYPE POWER STEERING

The power cylinder is an integral part of the gear housing. A double-acting piston allows oil pressure to be applied to either side of the piston. The one-piece piston and power rack is meshed to the sector shaft.

The hydraulic control valve is composed of a sleeve and valve spool. The spool is held in the neutral position by the torsion bar and spool actuator. Twisting of the torsion bar moves the valve spool, allowing oil pressure to be directed to either side of the power piston, depending on the directional rotation of the steering wheel, to give power assist.

On many American Motors cars a modified version of the G.M. system provides variable ratio steering for easier and safer control. The steering gear ratio will vary from a high ratio of about 16:1 while steering straight ahead to a lower gear ratio of about 12.4:1 while making a full turn to either side.

## ADJUSTER PLUG AND ROTARY VALVE

### Removal

1. Thoroughly clean exterior of gear assembly. Drain by holding valve ports down and rotating worm back and forth through entire travel.

2. Place gear in vise.

3. Loosen adjuster plug locknut with punch. Remove adjuster plug.

4. Remove rotary valve assembly by grasping stub shaft and pulling it out.

## ADJUSTER PLUG

### Disassembly

1. Remove upper thrust bearing retainer with screwdriver. Be careful not to damage bearing bore. Discard retainer. Remove spacer, upper bearing and races.

2. Remove and discard adjuster plug O-ring.

3. Remove stub shaft seal retaining ring (Truarc pliers will help) and remove and discard dust seal.

4. Remove stub shaft seal by prying out with screwdriver and discard.

5. Examine needle bearing and, if required, remove same by pressing from thrust bearing end.

6. Inspect thrust bearing spacer, bearing rollers and races.

7. Reassemble in reverse of above.

## ROTARY VALVE

### Disassembly

Repairs are seldom needed. Do not disassemble unless absolutely

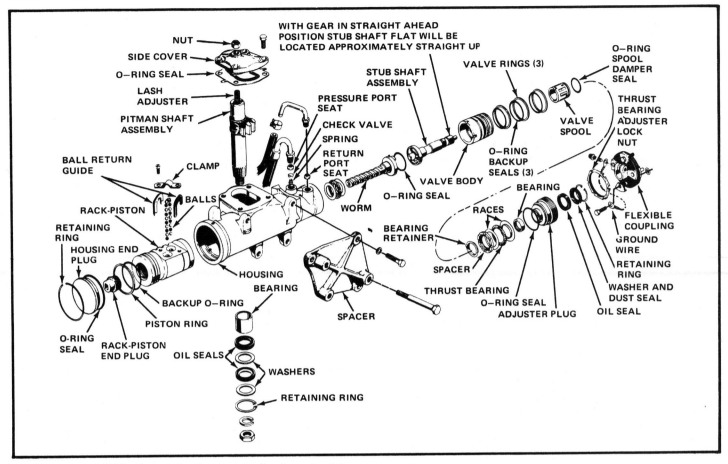

**Exploded view of GM Saginaw rotary type power steering assembly**

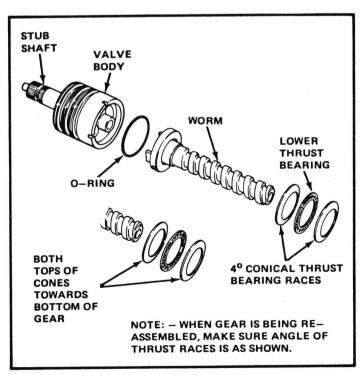

**Exploded view of stub shaft and wormshaft assembly**

necessary. If the O-ring seal on valve spool dampener needs replacement, perform this portion of operation only.

1. Remove cap-to-worm O-ring seal and discard.

2. Remove valve spool spring by prying on small coil with small screw driver to work spring onto bearing surface of stub shaft. Slide spring off shaft. Be careful not to damage shaft surface.

3. Remove valve spool by holding the valve assembly in one hand with the stub shaft pointing down. Insert the end of pencil or wood rod through opening in valve body cap and push spool until it is out far enough to be removed. In this procedure, rotate to prevent jamming. If spool becomes jammed it may be necessary to remove stub shaft, torsion bar and cap assembly.

## ROTARY VALVE

### Reassembly

— CAUTION —

*All parts must be free of dirt, chips, etc., before assembly and must be protected after assembly.*

1. Lubricate three new back-up O-ring seals with automatic transmission oil and reassemble in the ring grooves of valve body. Assemble three new valve body rings in the grooves over the O-ring seals by carefully slipping over the valve body.

**NOTE: If the valve body rings seem loose or twisted in the grooves, the heat of the oil during operation will cause them to straighten.**

2. Lubricate a new dampener O-ring with automatic transmission oil and install in valve spool groove.

# POWER STEERING
## GM SAGINAW ROTARY TYPE

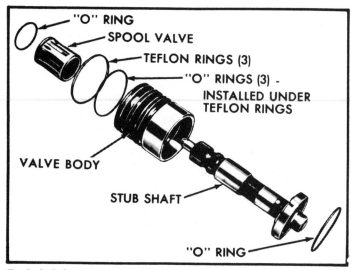

**Exploded view of the valve body and the shaft assembly**

"O" RING
SPOOL VALVE
TEFLON RINGS (3)
"O" RINGS (3) - INSTALLED UNDER TEFLON RINGS
VALVE BODY
STUB SHAFT
"O" RING

3. Assemble stub shaft torsion bar and cap assembly in the valve body, aligning the groove in the valve cap with the pin in the valve body. Tap lightly with soft hammer until cap is against valve body shoulder. Valve body pin must be in the cap groove. Hold parts together during the remainder of assembly.

4. Lubricate spool. With notch in spool toward valve body, slide the spool over the stub shaft. Align the notch on the spool with the spool drive pin on stub shaft and carefully engage spool in valve body bore. Push spool evenly and with slight rotating motion until spool reaches drive pin. Rotate spool slowly, with some pressure, until notch engages pin. Be sure dampener O-ring seal is evenly distributed in the spool groove.

---CAUTION---
*Use extreme care because spool-to-valve body clearance is very small. Damage is easily caused.*

5. With seal protector tool over stub shaft, slide valve spool spring over stub shaft, with small diameter of spring going over shaft last. Work spring onto shaft until small coil is located in stubshaft groove.

6. Lubricate a new cap-to-O-ring seal and install in valve body.

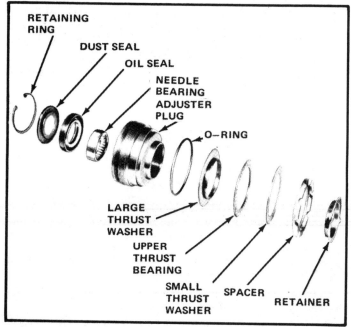

**Adjuster plug assembly sequence**

RETAINING RING
DUST SEAL
OIL SEAL
NEEDLE BEARING ADJUSTER PLUG
O-RING
LARGE THRUST WASHER
UPPER THRUST BEARING
SMALL THRUST WASHER
SPACER
RETAINER

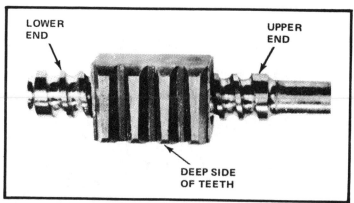

**Installation of ball nut or shaft**

LOWER END
UPPER END
DEEP SIDE OF TEETH

## ADJUSTER PLUG AND ROTARY VALVE

### Installation

1. Align narrow pin slot on valve body with valve body drive pin on the worm. Insert the valve assembly onto gear housing by pressing against valve body with finger tips. Do not press on stub shaft or torsion bar. The return hole in the gear housing should be fully visible when properly assembled.

---CAUTION---
*Do not press on stub shaft as this may cause shaft and cap to pull out of valve body, allowing the spool dampener O-ring seal to slip into valve body oil grooves.*

2. With protector over end of stub shaft, install adjuster plug assembly snugly into gear housing then back plug off approximately one-eighth turn. Install plug locknut but do not tighten. Adjust preload as described in the adjustment section.

3. After adjustment, tighten lock-nut.

## PITMAN SHAFT

### Removal and Installation

1. Completely drain the gear assembly and thoroughly clean the outside.

2. Place gear in vise.

3. Rotate stub shaft until pitman shaft gear is in center position. Remove side cover retaining bolts.

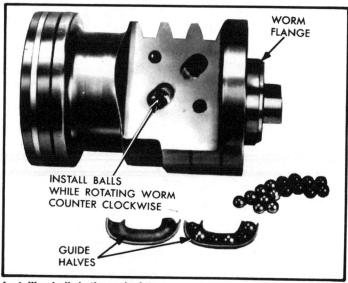

**Installing balls in the rack piston**

WORM FLANGE
INSTALL BALLS WHILE ROTATING WORM COUNTER CLOCKWISE
GUIDE HALVES

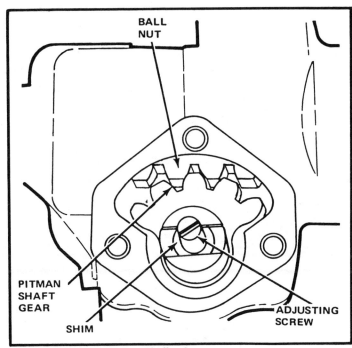

**Position of pitman shaft and ball nut**

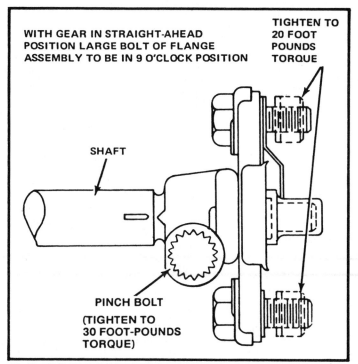

**Shaft and flange steering alignment**

4. Tap end of pitman shaft with soft hammer and slide shaft out of housing.

5. Remove and discard side cover O-ring seal.

6. The seals, washers, retainers and bearings may now be removed and examined.

7. Examine all parts for wear or damage and replace as required.

8. Install in reverse of above. Make proper adjustment as described in adjustment section.

## RACK-PISTON NUT AND WORM ASSEMBLY

### Removal

1. Completely drain the gear assembly and thoroughly clean the outside.

2. Remove pitman shaft assembly, previously described.

3. Rotate housing end plug retaining ring so that one end of ring is over hole in gear housing. Spring one end of ring so screw-driver can be inserted to lift out ring.

4. Rotate stub shaft to full left turn position to force end plug out of housing.

5. Remove and discard housing end plug O-ring seal.

6. Remove rack-piston nut end plug with 1/2 in. square drive.

7. Insert tool in end of worm. Turn stub shaft so that rack-piston nut will go into tool and remove rack-piston nut from gear housing.

8. Remove adjuster plug and rotary valve assemblies as previously described.

9. Remove worm and lower thrust bearing and races.

10. Remove cap-to-O-ring seal and discard.

## RACK-PISTON NUT AND WORM

### Disassembly and Reassembly

1. Remove and discard piston ring and back-up O-ring on rack-piston nut.

2. Remove ball guide clamp and return guide.

3. Place nut on clean cloth and remove ball retaining tool. Make sure all balls are removed.

4. Inspect all parts for wear, nicks, scoring or burrs. If worm or rack-piston nut need replacing, both must be replaced as a matched pair.

5. In assembling, reverse the above.

**NOTE: When assembling, alternate black and white balls, and install guide clamp. Packing with grease helpd in holding during assembly. When new balls are used, various sizes are available and a selection must be made to secure proper torque when making the high point adjustment.**

## RACK-PISTON NUT AND WORM ASSEMBLY

### Installation

1. Install in reverse of removal procedure.
2. In all cases use new O-ring seals.
3. Make adjustments as described in that section.

# GM SAGINAW LINKAGE TYPE POWER STEERING

## Chevrolet Corvette

### CONTROL VALVE

#### Disassembly

1. Place valve assembly in vise with dust cap end up and remove dust cap.

2. Remove adjusting nut.

3. Remove valve-to-adapter bolts and remove valve housing and spool from adapter.

4. Remove spool from housing.

5. Remove spring, reaction spool, washer, reaction spring, and seal. O-ring may now be removed from reaction spool.

6. Remove annulus spacer, valve shaft washer, and plug-to-sleeve key. Remove the ball stud seal and ball stud seal clamp.

# POWER STEERING
## GM SAGINAW LINKAGE TYPE

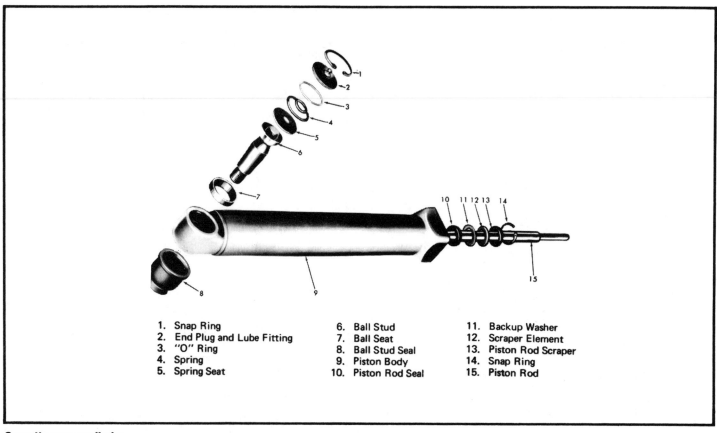

1. Snap Ring
2. End Plug and Lube Fitting
3. "O" Ring
4. Spring
5. Spring Seat
6. Ball Stud
7. Ball Seat
8. Ball Stud Seal
9. Piston Body
10. Piston Rod Seal
11. Backup Washer
12. Scraper Element
13. Piston Rod Scraper
14. Snap Ring
15. Piston Rod

**Corvette power cylinder**

7. Carefully turn adjuster plug out of sleeve. Use care not to nick the top surface.

8. If necessary to replace a connector seat, tap threads in center hole using a 5/16-18 tap. Thread a bolt with a nut and a flat washer into the tapped hole so the washer is against the face of the port boss and the nut is against the washer. Hold the bolt from turning while backing the nut off the bolt. This will force the washer against the port boss face and back out the bolt, drawing the connector seat from the top cover housing. Discard the old connector seat and clean the housing out thoroughly to remove any metal chips. Drive a new connector seat against the housing seat, being careful not to damage either the connector seat or the housing seat.

9. Remove adapter from vise and turn over to allow spring and one of the two ball seats to drop out.

10. Remove ball stud with other ball seat and allow sleeve to fall free.

### Inspection

1. Wash all parts in clean, non-toxic solvent and blow dry with air.

2. Inspect all parts for scratches, burrs, distortion, or excessive wear and replace worn or damaged parts.

3. Replace all seals and gaskets.

**NOTE: Corvette valves incorporate a 55 lb. centering spring which might be inadvertently interchanged with Chevrolet, and Chevelle springs. *They should not be interchanged* as the other springs are only 30 lbs. Corvette valves are stamped with an X on the dust cover.**

### Assembly

1. Replace sleeve and ball seat in adapter, then the ball stud and then the other ball seat and spring. (small end down)

2. Place adapter in vise. Put the shaft through the seat in the adjuster plug and screw adjuster plug into sleeve.

3. Turn plug in until tight, then back off until slot lines up with notches in sleeve.

4. Insert key. Be sure small tangs on end of key fit into notches in sleeve.

5. Install valve shaft washer, annulus spacer, and reaction seal (lip up), spring retainer, reaction spring and spool, then washer and adjustment spring. Install O-ring seal on reaction spool before installing spool on shaft. Install washer with chamfer up.

6. Install seal on valve spool with lip down. Then install spool, being careful not to jam spool in housing.

7. Install housing with spool onto adapter. The side ports should be on the same outside as the ball stud. Bolt the housing to the adapter.

8. Depress the valve spool and turn the locknut into the shaft about four turns. Use a clean wrench or socket.

## POWER CYLINDER

### Disassembly and Reassembly

1. To remove piston rod seal, remove snap-ring and pull out on rod. Remove back-up washer, piston rod scraper, and piston rod seal from rod.

2. To remove the ball stud, depress the end plug and remove the snap ring. Push on the end of the ball stud and the end plug, spring, spring seat, and ball stud and seal may be removed. If the ball seat is to be replaced, it must be pressed out.

3. Reverse disassembly procedure. Be sure snap-ring is properly seated.

# G.M. POWER STEERING RACK AND PINION ASSEMBLY

## 1982 and Later A Body Cars

## Celebrity, 6000, Ciera, Century

## 1980-81 X Body Cars

## 1982 and Later X Body Cars

## Citation, Phoenix, Skylark, Omega

## 1981 and Later Chevette and T1000

A difference exists between the power steering rack and pinion assemblies used on the 1981 and later Chevette and T100, 1980-81 X body cars and the power steering rack and pinions assemblies used on the 1982 and later X body cars. The major difference is in the manner

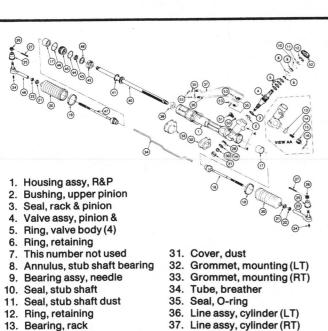

1. Housing assy, R&P
2. Bushing, upper pinion
3. Seal, rack & pinion
4. Valve assy, pinion &
5. Ring, valve body (4)
6. Ring, retaining
7. This number not used
8. Annulus, stub shaft bearing
9. Bearing assy, needle
10. Seal, stub shaft
11. Seal, stub shaft dust
12. Ring, retaining
13. Bearing, rack
14. Spring, adjuster
15. Plug, adjuster
16. Nut, adjuster plug lock
17. Ring, shock dampener
18. Rod assy, inner tie (LT)
19. Clamp, boot
20. Boot, rack & pinion
21. Clamp, boot
22. Nut, hex jam
23. Rod assy, outer tie (LT)
24. Fitting, lubrication
25. Seal, tie rod
26. Nut, hexagon slotted
27. Pin, cotter
28. Bearing assy, ball
29. Ring, retaining
30. Nut, hex lock
31. Cover, dust
32. Grommet, mounting (LT)
33. Grommet, mounting (RT)
34. Tube, breather
35. Seal, O-ring
36. Line assy, cylinder (LT)
37. Line assy, cylinder (RT)
38. Cap, dust
39. Seal, inner rack
40. Rack assy, piston & steering
41. Ring, piston
42. Bulkhead, cylinder inner
43. Seal, rack & pinion (bulkhead)
44. Seal, O-ring
45. Bulkhead, cylinder outer
46. Ring, bulkhead retaining
47. Rod assy, inner tie (RT)
48. Rod assy, outer tie (RT)
49. Spring (wave washer)
50. Seal, O-ring
51. Adapter, O-ring
52. Adapter, seal
53. Seal, O-ring

**Exploded view of power steering rack and pinion - 1980-81 GM X body cars**

1. Housing assy, R&P
2. Bushing, upper pinion
3. Seal, rack & pinion
4. Valve assy, pinion &
5. Ring, valve body (4)
6. Ring, retaining
7. Annulus assembly, stub shaft bearing &
8. Seal, stub shaft
9. Seal, stub shaft dust
10. Ring, retaining
11. Adapter, shield
12. Bearing, rack
13. Spring, adjuster
14. Plug, adjuster
15. Nut, adjuster plug lock
16. Ring, shock dampener
17. Rod assy, inner tie
18. Clamp, boot
19. Boot, rack & pinion
20. Nut, hex jam
21. Rod assy, outer tie
22. Fitting, lubrication
23. Seal, tie rod
24. Nut, hexagon slotted
25. Pin, cotter
26. Bearing assy, ball
27. Ring, retaining
28. Nut, hex lock
29. Cover, dust
30. Seal, O-ring
31. Seal, O-ring
32. Line assy, cylinder (RT)
33. Line assy, cylinder (LT)
34. Grommet, mounting (LH)
35. Tube, breather
36. Ring, retaining
37. Grommet assembly, mounting (RH)
38. Seal, inner rack
39. Rack assy, piston & steering
40. Ring, piston
41. Bulkhead, cylinder inner
42. Seal, rack & pinion (bulkhead)
43. Seal, O-ring
44. Bulkhead, cylinder outer
45. Ring, bulkhead retaining

**Exploded view of power steering rack and pinion - 1981 and later Chevette, T1000**

of attachment of the steering unit to the vehicle body. The early model is attached by wrap around type brackets and rubber grommets, while the later model is attached by bolts through eyelets and grommets, in the steering housing.

Minor variations exist between the three units, both externally and internally, but the basic disassembly and assembly remains the same.

## OUTER TIE ROD

### Removal

1. Loosen the jam nut and remove the tie rod from the steering knuckle.

2. Remove the outer tie rod by turning it off the inner tie rod. Count the number of turns needed to unscrew the outer tie rod.

### Installation

1. Screw the outer tie rod onto the inner tie rod the same number of turns as was needed to remove it.

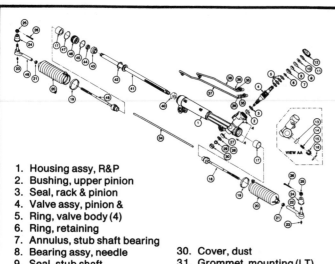

1. Housing assy, R&P
2. Bushing, upper pinion
3. Seal, rack & pinion
4. Valve assy, pinion &
5. Ring, valve body (4)
6. Ring, retaining
7. Annulus, stub shaft bearing
8. Bearing assy, needle
9. Seal, stub shaft
10. Seal, stub shaft dust
11. Ring, retaining
12. Adapter, seal
13. Bearing, rack
14. Spring, adjuster
15. Plug, adjuster
16. Nut, adjuster plug lock
17. Ring, shock dampener
18. Rod assy, inner tie (LT)
19. Clamp, boot
20. Boot, rack & pinion
21. Nut, hex jam
22. Rod assy, outer tie (LT)
23. Fitting, lubrication
24. Seal, tie rod
25. Nut, hexagon slotted
26. Pin, cotter
27. Bearing assy, ball
28. Ring, retaining
29. Nut, hex lock
30. Cover, dust
31. Grommet, mounting (LT)
32. Ring, retaining
33. Grommet assembly, mounting (RT)
34. Tube, breather
35. Seal, O-ring
36. Seal, O-ring
37. Line assy, cylinder (LT)
38. Line assy, cylinder (RT)
39. Cap, dust
40. Seal, inner rack
41. Rack assy, piston & steering
42. Ring, piston
43. Bulkhead, cylinder inner
44. Seal, rack & pinion (bulkhead)
45. Seal, O-ring
46. Bulkhead, cylinder outer
47. Ring, bulkhead retaining
48. Rod assy, inner tie (RT)
49. Rod assy, outer tie (RT)

Exploded view of power steering rack and pinion - 1982 and later GM A body cars

2. Do not tighten the jam nut until the toe-in/out adjustment has been made. Torque to 50 ft./lbs.
3. Be sure the boot is not twisted when done.

## BOOT SEAL AND BREATHER TUBE

### Removal

1. Remove the outer tie rod and the jam nut from the inner tie rod shaft.
2. Cut the boot clamp and discard. Mark the breather tube location on the steering housing before removing the tube.

### Installation

1. Install the breather tube in the same location before removal.
2. Install a new clamp on the boot before installing the boot.
3. Push the boot elbow on the breather tube and engage the boot onto the housing.
4. Secure the boot clamp.
5. Install the jam nut and the outer tie rod. Adjust the toe-in/out as required and tighten the jam nut to 50 ft./lbs.

## INNER TIE ROD

### Removal

1. The steering assembly must be out of the vehicle.

2. Remove the shock damper ring from the inner tie rod housing and slide it back on the rack.
3. Position a wrench on the rack flat to prevent rack damage when removing the tie rod.
4. Position a wrench on the tie rod pivot housing flats.
5. Turn the inner tie rod housing counterclockwise until the tie rod assembly separates.

### Installation

1. Screw the inner tie rod into the steering rack. Be sure the shock damper is positioned on the rack.
2. Torque the tie rod housing to the rack by holding the rack and tie rod with two wrenches. Tighten to 70 ft./lbs. torque. The tie rod must rock freely in the housing before staking.
3. Support the rack and housing and stake the tie rod housing to the rack flat. To inspect the stake, a 0.010 inch feeler gauge should not pass between the rack and the housing stakes on both sides.
4. Slide the shock dampener over the inner tie rod housing until it engages.

## RACK BEARING

### Removal

1. Loosen the adjuster plug lock nut.
2. Remove the adjuster plug, spring and the rack bearing.

### Installation

1. Lubricate the metal parts and install the rack bearing, spring, adjuster plug and the lock nut.
2. Turn the adjuster plug in until it bottoms and then back off 50 to 70 degrees.
3. Check the turning torque of the pinion. The correct turning torque is 8 to 10 in./lbs.
4. Torque the lock nut to 50 ft./lbs.

## STUB SHAFT SEALS

### Removal

1. Remove the retaining ring and the dust cover.
2. While holding the stub shaft, remove the lock nut from the pinion.

——————————— CAUTION ———————————
*If the stub shaft is not held, damage to the pinion teeth will occur.*
————————————————————————————————

3. Using a press, press on the threaded end of the pinion until flush with the ball bearing assembly. Complete removal of the valve and pinion assembly is not necessary.

### Installation

1. Install the shaft protector over the stub shaft and install the stub shaft bearing annulus, needle bearing, stub shaft seal, stub shaft dust seal and the retaining ring.
2. While holding the stub shaft securely, firmly seat the lock nut and torque to 26 ft./lbs.

## VALVE AND PINION ASSEMBLY

### Removal

1. Turn the stub shaft until the rack has equal distance on both sides of the housing, with the pinion fully engaged. The valve and pinion lock nut must be removed.
2. Mark the location of the stub shaft flat surface on the housing.
3. Using a press, press on the threaded end of the pinion until it is possible to remove the valve and pinion assembly.
4. Remove the valve body rings, if the replacement is required.

## Installation

1. Install new valve body rings, if required.
2. Be sure that both ends of the rack are at equal distance from the housing.
3. Install the pinion and valve assembly, being care not to damage the rings during the installation.
4. The valve and pinion assembly must be fully seated and the flat section of the pinion must line up with the previously marked location indicator on the housing.

## BULKHEAD

### Removal

1. Use punch in access hole to remove bulkhead retaining ring.
2. If only the bulkhead, bulkhead O-ring seal or rack seal (bulkhead) are to be replaced. Loosen (LT) fitting and remove cylinder line.
3. Plug (LT) cylinder line hole at cylinder using a finger or plastic cap with 7/16 × 20 internal threads over hole to prevent oil leaking from cylinder.
4. Using a 11/16 inch-12 point socket turn stub shaft. Move rack to the right forcing the bulkhead out of the housing. Use drain pan to catch hydraulic oil from assy.
5. If inner rack seal or piston ring are to be replaced, use rack to remove bulkhead instead of compressed oil method.

### Installation

1. Install the cylinder inner bulkhead, the "O" ring seal, the cylinder outer bulkhead, the bulkhead retaining ring and the shock dampener.

—————————— CAUTION ——————————
*A seal protector should be used on the rack end.*
———————————————————————————————

2. Make sure that the open end of the retaining ring is approximately 0.50 inch from the access hole.
3. Fully seat the retaining ring.

## RACK INNER SEAL, RACK AND PISTON RING

### Removal

1. Remove the rack from the housing.
2. Remove the piston rings and discard.
3. Fit seal remover tool in place and using a long rod, tap the seal from its seat.

### Installation

1. Install new piston ring on rack.

2. Care should be taken not to cut ring at installation.
3. Wrap card stock around end of rack and rack teeth.
4. Coat seal lip with power steering fluid, slide seal with seal lip facing piston on to card stock, slide card and seal over rack teeth.
5. Remove card stock and bottom seal on rack piston.
6. Coat lip of seal insert with power steering fluid and slide on rack with lip facing seal. Be sure insert is fully engaged with seal before installing rack in housing.
7. Coat seal completely with power steering fluid, slide rack and seal in housing, tap on rack with rubber mallet to seat seal.

NOTE: **Seal must be fully seated in housing.**

## BALL BEARING ASSEMBLY

### Removal

1. WIth the piston and pinion assembly out of the housing, remove the bearing retaining ring.
2. Use a drift and gently tap on the bearing and remove it from the housing.

### Installation

1. Using a suitable block, install the bearing in the housing and press it to its seat.
2. Install the retaining ring.

## UPPER PINION BEARING AND SEAL

### Removal

1. With the piston and pinion assembly out of the housing, remove the upper bushing and seal with a drift.

### Installation

1. Install the new bushing to its seat.
2. Install the new seal, using an installer tool. Seat the seal in the housing with the lip of the seal facing up.

## CYLINDER LINES

### Removal and Installation

1. The lines are removed and replaced in a conventional manner, using tubing flare wrenches and using the normal precautions when working with fittings and piping. The fittings should be torqued to 15 ft./lbs.

# G.M. POWER STEERING RACK AND PINION—J BODY CARS

## Cavailier, Cimarron, J-2000, Skyhawk 82-83, Firenza

### OUTER TIE ROD

#### Removal

1. With the steering assembly from the vehicle, loosen the outer rod pinch bolt and turn the tie rod from the adjuster stud, counting the number of turns until the tie rod separates from the adjuster stud.

#### Installation

1. Turn the tie rod onto the adjuster stud the same number of turns as was needed to remove.
2. Tighten the pinch bolt until the toe-out can be verified. Re-loosen and adjust as required.

### INNER TIE ROD AND INNER PIVOT BUSHING

#### Removal

1. Bend back the lock plate tabs and loosen the inner tie rod bolt and remove.

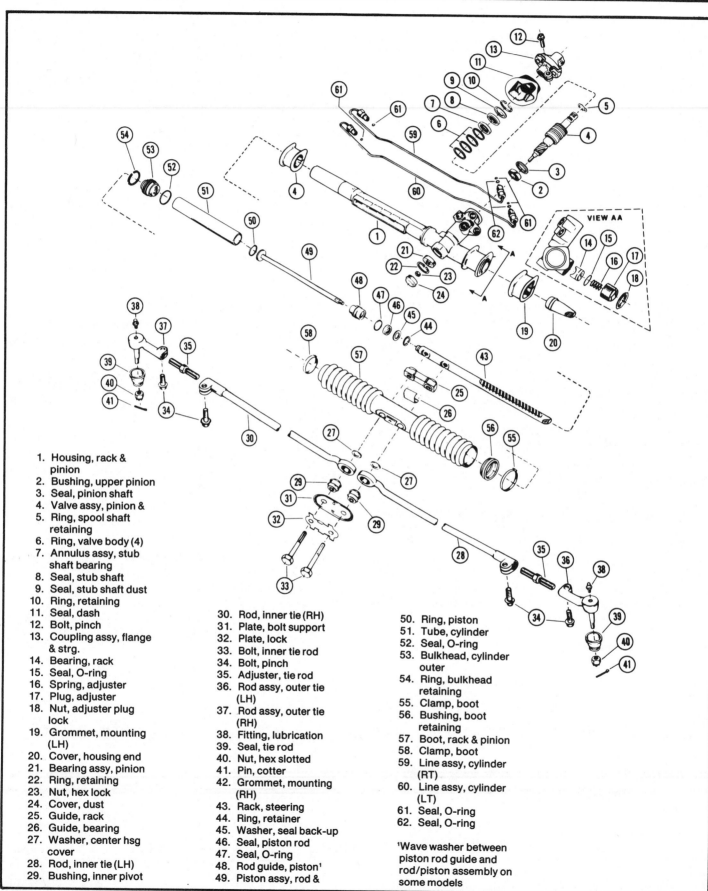

1. Housing, rack & pinion
2. Bushing, upper pinion
3. Seal, pinion shaft
4. Valve assy, pinion &
5. Ring, spool shaft retaining
6. Ring, valve body (4)
7. Annulus assy, stub shaft bearing
8. Seal, stub shaft
9. Seal, stub shaft dust
10. Ring, retaining
11. Seal, dash
12. Bolt, pinch
13. Coupling assy, flange & strg.
14. Bearing, rack
15. Seal, O-ring
16. Spring, adjuster
17. Plug, adjuster
18. Nut, adjuster plug lock
19. Grommet, mounting (LH)
20. Cover, housing end
21. Bearing assy, pinion
22. Ring, retaining
23. Nut, hex lock
24. Cover, dust
25. Guide, rack
26. Guide, bearing
27. Washer, center hsg cover
28. Rod, inner tie (LH)
29. Bushing, inner pivot

30. Rod, inner tie (RH)
31. Plate, bolt support
32. Plate, lock
33. Bolt, inner tie rod
34. Bolt, pinch
35. Adjuster, tie rod
36. Rod assy, outer tie (LH)
37. Rod assy, outer tie (RH)
38. Fitting, lubrication
39. Seal, tie rod
40. Nut, hex slotted
41. Pin, cotter
42. Grommet, mounting (RH)
43. Rack, steering
44. Ring, retainer
45. Washer, seal back-up
46. Seal, piston rod
47. Seal, O-ring
48. Rod guide, piston[1]
49. Piston assy, rod &

50. Ring, piston
51. Tube, cylinder
52. Seal, O-ring
53. Bulkhead, cylinder outer
54. Ring, bulkhead retaining
55. Clamp, boot
56. Bushing, boot retaining
57. Boot, rack & pinion
58. Clamp, boot
59. Line assy, cylinder (RT)
60. Line assy, cylinder (LT)
61. Seal, O-ring
62. Seal, O-ring

[1]Wave washer between piston rod guide and rod/piston assembly on some models

**Exploded view of power steering rack and pinion - GM J body cars**

2. Remove the inner tie rod by sliding it out between the bolt support plate and rack/pinion boot.

**NOTE: If both inner tie rods are to be removed, re-install the inner tie rod bolt in the first tie rod retaining bolt hole to keep the rack and pinion boot and other parts aligned, while the tie rods are out.**

3. With the tie rod disconnected, the pivot bushings can be pressed out and new ones pressed in.

### Installation

1. Be sure the center housing cover washers are fitted into the rack and pinion boot, before rod installation.
2. Remove the locating bolt from the rack and position one inner tie rod assembly in place over the rack. Place the bolt through the lock plate and the tie rod. Place the second inner tie rod in place and install the bolt through the lock and the tie rod.
3. Tighten the inner tie rod bolts to 65 ft. lbs. and bend the lock tabs against the flats of the inner tie rod bolts after torquing.

## FLANGE AND STEERING COUPLING ASSEMBLY

### Removal

1. Loosen and remove the pinch bolt.
2. Remove the coupling.

### Installation

1. Install the flange and steering coupling assembly on the stub shaft.
2. Install the pinch bolt and torque to 37 ft. lbs.

**NOTE: With the flange and steering coupling assembly off the stub shaft, the dash seal can be replaced.**

## HYDRAULIC LINES

### Removal and Installation

For ease of line removal and installation, remove the lines from the valve end first and install them on the cylinder end first. Torque to 13 ft. lbs.

## RACK AND PINION BOOT, RACK GUIDE, BEARING GUIDE, MOUNTING GROMMET, OR HOUSING END COVER

### Removal

1. Separate RH mounting grommet and remove, LH mounting grommet need not be removed unless replacement is required.
2. Cut both boot clamps and discard.
3. For ease of rack and pinion boot removal slide cylinder end of boot toward center of gear enough to expose boot groove in cylinder. Place a rubber band in groove. This fills the groove and allows easy removal of rack and pinion boot from gear.
4. Rack bearing or rack guide can now be removed or replaced if necessary.
5. Remove housing end cover only if damaged.

### Installation

1. Remove boot retaining bushing from rack and pinion boot.
2. Slide new boot clamp on rack and pinion boot. Install boot retaining bushing into rack and pinion boot.
3. Install new bearing guide on rack guide if necessary.
4. Install rack guide on rack.
5. Coat inner lip of boot retaining bushing lightly with grease for ease of assembly.

6. Install boot on housing.
7. Be sure center housing cover washers are in place on boot.
8. For ease of assembly, install inner tie rod bolts through cover washers and rack and pinion boot. Screw into rack lightly. This will keep rack, rack guide, and boot in proper alignment.
9. Slide boot and boot retaining bushing until seated in bushing groove in housing. Crimp new boot clamp.
10. SLide other end of boot into boot groove on cylinder end of housing. Crimp new boot clamp.

## RACK BEARING

### Removal

1. Remove the adjuster plug lock nut.
2. Remove the adjuster plug from the housing.
3. Remove the spring, O-ring seal and the rack bearing.

### Installation

1. Lubricate the metal parts before installation and install in the reverse order of the removal procedure.
2. With the rack centered, tighten the adjuster plug to a torque of 6 to 11 ft. lbs. Back off the adjuster plug 50 to 70 degrees. Check the pinion torque.
3. The pinion torque should be 8 to 16 inch pounds, turning.
4. Assemble the lock nut and while holding the adjusting plug stationary, tighten the adjusting plug lock nut to 50 ft. lbs.

## STUB SHAFT SEALS AND UPPER BEARING

### Removal

1. Remove retaining ring.
2. Remove dust cover.
3. While holding the stub shaft remove lock nut from pinion.

**NOTE: If stub shaft is not held, damage to the pinion teeth will occur.**

4. **Using a press,** press on threaded end of pinion until flush with ball bearing assembly.
5. Complete removal of valve and pinion assembly is not necessary.

**NOTE: Bearing and annulus are pressed together. Disassemble only if bearing replacement is required.**

### Installation

1. Install the seals and bearing in the reverse order of the removal.
2. While holding the stub shaft, firmly seat the lock at 26 ft. lbs.

## VALVE AND PINION ASSEMBLY

### Removal

1. Turn stub shaft until rack guide is equal distance from both sides of housing opening.
2. Mark location of stub shaft flat on housing.
3. **Using a press,** press on threaded end of pinion until it is possible to remove valve and pinion assembly.
4. Remove valve body rings if replacement is necessary.

### Installation

1. Install new valve body rings if required.
2. Care should be taken not to cut rings at installation.
3. Measure rack guide so that it is equal from both sides of housing.
4. Install valve and pinion assembly so that when full seated stub shaft flat and mark on housing line up and the rack guide is centered in housing.

## CYLINDER ASSEMBLY

### Removal

1. The cylinder outer bulkhead and housing must be marked before removal to insure proper location in housing at assembly so cylinder lines will fit correctly.
2. Use a small punch in access hole to unseat retaining ring. Then use a screwdriver to bring retaining ring out enough to be able to remove retaining ring with suitable pliers, discard retaining ring.
3. By threading an inner tie rod bolt into rack, rack can be used in a slide hammer fashion to remove rack and cylinder assembly from housing.

### Installation

1. Replace both O-ring seals before assembly.
2. Using crocus cloth remove burrs or shart edges from retaining ring groove in housing. This must be done to insure that the new O-ring seals are not damaged at assembly.
3. Coat O-ring seals with hydraulic fluid and install rack and cylinder assembly in housing.
4. Line up marks on housing and cylinder outer bulkhead. Gently tap on cylinder outer bulkhead until it is seated far enough in housing to install retaining ring. It may be necessary to use a press to hold bulkhead far enough in housing to install retaining ring.
5. Open end of retaining ring should be approx. 13 millimetres (.50 inch) from access hole. Be sure retaining ring is fully seated in housing.

## PISTON RING

### Removal

1. Hold on to cylinder, using rack and piston rod assembly as a slide hammer. Piston rod guide assembly will disengage from cylinder assembly.

### Installation

1. Install new piston ring coat lightly with hydraulic steering fluid.
2. Slide piston into cylinder assembly.
3. Lightly tap piston rod guide assembly until fully seated in cylinder assembly.

## RACK, PISTON ROD, PISTON ROD GUIDE ASSEMBLY

### Removal

1. Put steering rack in soft-jawed vise.

2. Using a Tool J-29811 or equivalent unscrew rod and piston assembly from rack.
3. Because of close tolerances, it may be necessary to use piston rod guide assembly as a slide hammer to separate rod and piston assembly from rack.
4. Do not remove piston rod guide assembly from rod and piston assembly unless piston rod guide, piston rod seal or rod and piston assembly require replacement, because the piston rod seal will be damaged.

### Installation

1. Install new piston rod seal in piston rod guide, if required.
2. Using crocus cloth remove any burrs or sharp edges from rod and piston assembly. Put seal protector J-29812 or equivalent on rod and piston assembly.
3. Coat piston rod seal with hydraulic fluid and slide piston rod guide assembly on piston and rod assembly.
4. Gently tap rod and piston assembly into rack until threads engage.
5. Tighten to specifications.
6. Stake rack against rod piston flats.

## PINION BEARING ASSEMBLY

### Removal

1. Remove bearing retaining ring.
2. Use drift or punch and gently tap on bearing until bearing is removed.

### Installation

1. Install new pinion bearing assembly. Using a suitable socket, press on outer race. Be careful not to cock bearing in housing.
2. Install retaining ring. Note position of large lug to be sure beveled side of ring is properly located.

## UPPER PINION BEARING AND PINION SHAFT SEAL

### Removal

1. Remove upper pinion bushing and seal with a punch.
2. Remove and discard the bushing and seal.

### Installation

1. Install the new bushing.
2. Install the new seal with an installer tool. Seat the seal with the seal lip facing up.

# 4 Tune-Up

## THE TUNE-UP

The dictionary defines a tune-up as a procedure used to bring a group of things into a harmonious working order, as in tuning an orchestra. An automotive tune-up is an orderly process of inspection, diagnosis, testing, and adjustment that is periodically necessary to maintain peak engine performance or restore the engine to original operating efficiency.

Tests by the Champion Spark Plug Company showed that an average 11.36% improvement in gas economy could be expected after a tune-up. A change to new spark plugs alone provided a 3.44% decrease in fuel use. As for emissions, significantly lower emissions were recorded at idle after a complete tune-up on a car needing service. An average 45.37% reduction of CO (carbon monoxide) emissions was recorded at idle after a complete tune-up; HC (hydrocarbon) emissions were cut 55.5%

The tune-up is also a good opportunity to perform a general preventive maintenance checkout on everything in the engine compartment. Look for failed or faulty components such as loose or damaged wiring, leaking fuel lines, cracked coolant hoses, and frayed fan belts.

## Necessary Tools

In order to perform a proper tune-up, several specific tools are needed; a dwell-tach, a timing light, a spark plug socket, feeler gauges (both the flat type and the round wire type for gapping plugs), and a compression tester. If you have a late-model car with electronic ignition, you won't need a dwell meter since dwell is nonadjustable on these cars. Also keep in mind that some tachometers will not operate on cars equipped with electronic ignition, and neither will some timing lights. So before you buy anything, check to make sure it will work on your particular car.

## Tune-Up Procedures

### COMPRESSION

Along with vacuum gauge readings and spark plug condition, cylinder compression test results are extremely valuable indicators of internal engine condition. Most mechanics automatically check an engine's compression as the first step in a comprehensive tune-up. Obviously, it is useless to try and tune an engine with extremely low or erratic compression readings, since a simple tune-up will not cure the problem. A compression test will uncover many mechanical problems that can cause rough running or poor performance.

### Gasoline Engines

A. Prepare the engine for the test as follows:
1. Run the engine until it reaches operating temperature.
2. Remove the primary lead from the positive terminal on the coil. Remove all high-tension wires from the spark plugs.
3. Clean all dirt and foreign material from around the spark plugs and remove all spark plugs.
4. If a remote starter switch is available, hook it up according to its manufacturer's instructions.
5. Block the throttle wide open.

**Vacuum gauge is one of the best diagnostic tools to determine internal engine condition**

B. Zero the gauge, place it firmly in one of the spark plug holes, and crank the engine for about five compression strokes. Record the reading and the number or position of the cylinder tested. *Release pressure from the gauge.*

C. Repeat the test for all the other cylinders.

D. Evaluate the results. Consult a Chilton service manual for the compression pressure rating of the engine or a guide to acceptable compression. Engines with compression ratios of 8:1–8.5:1 usually produce 140–150 lbs pressure. Higher compression ratios produce up to 175 lbs. The readings should be within 25 percent of each other.

If the test had to be performed on a cold engine because it could not be started, the readings will be considerably lower than normal, even if the engine is in perfect mechanical condition. A substantial pressure should still be produced, and variations in the readings are still indicative of the condition of the engine. If all readings are acceptable, see F.

E. Perform a "wet" compression test if any or all of the cylinders read low. Pour about one teaspoon of engine oil in each of the cylinders with low compression and repeat the test for each cylinder in turn.

F. Further evaluate the results. One or more of the conditions below should apply:

1. All cylinders fall within the specified range of pressures. The engine internal parts are in generally good condition.

2. One or more cylinders produced a low reading in D which was substantially improved by the "wet" compression test. Those cylinders have worn pistons, piston rings, and/or cylinder bores.

3. Two adjacent cylinders (or several pairs whose cylinders are adjacent) have nearly identical low readings, and did not respond to the "wet" compression test. These cylinders share leaks in the head gasket. This may be cross-checked by performing cooling system pressure tests, and by looking at the oil on the dipstick to see if coolant bubbles are present.

4. Compression build-up in one or more cylinders is erratic—it climbs less on some strokes than on others. Normally the pressure rises steadily and then levels off. Erratic pressure build-up indicates sticking valves. This problem may be cross-checked with a timing light. Remove the valve covers. Since this test is run with the engine operating and the valve covers removed, it would be wise to purchase and install special clips that are designed to deflect oil flow to the valve train. Connect a timing light to the spark plug lead of the cylinder suspected of having sticky valves. Aim the timing light at the valves of the cylinder in question. Loosen the distributor and then start the engine and watch the valves. Vary the timing slightly, smoothly, and gradually in order to observe the position of the valve at slightly different points in the rotation of the engine. If there is an erratic motion of either valve, that valve is sticking. Remember to retime the ignition system and remove the oil clips.

## Diesel Engines

The procedure for testing compression on a diesel engine is essentially the same as that for a gasoline engine. However, compression must be tested with a gauge that registers at least 500 to 600 psi. A normal automotive gauge will not do the job because it will not register high enough. The diesel gauge is inserted into the glow plug hole after the glow plug is removed. Some glow plugs may require a special tool to remove them.

## Rotary Engines

Because of the unique design of the rotary engine, special equipment is necessary to measure the rotary engine's compression.

## VACUUM GAUGE READINGS

Strictly speaking, vacuum gauge readings are not a necessary part of the everyday tune-up. Properly used, however, a vacuum gauge is an extremely useful diagnostic tool. Gauge readings and their meanings are given in the accompanying chart.

## SPARK PLUGS

Spark plug life and efficiency depend upon the condition of the engine and the temperatures to which the plug is exposed. Combustion chamber temperatures are affected by many factors such as compression ratio of the engine, air/fuel mixtures, exhaust emission equipment, and the type of driving. Spark plugs are designed and classified by number according to the heat range at which they will operate most efficiently.

### Spark Plug Heat Range

The amount of heat the plug absorbs is determined by the length of the lower insulator. The longer the insulator (or the farther it extends into the engine), the hotter the plug will operate; the shorter the insulator, the cooler it will operate. A plug that absorbs little heat and remains too cool will quickly accumulate deposits of oil and carbon since it is not hot enough to burn them off. This leads to plug fouling and consequently to misfiring. A plug that absorbs too much heat will have no deposits, but, due to the excessive heat, the electrodes will burn away quickly and in some instances, preignition may result. Preignition takes place when plug tips get so hot that they glow sufficiently to ignite the fuel/air mixture before the actual spark occurs. This early ignition will usually cause a pinging during low speeds and heavy loads. In severe cases, the heat may become high enough to start the fuel/air mixture burning throughout the combustion chamber rather than just to the front of the plug as in normal operation. At this time, the piston is rising in the cylinder making its compression stroke. The burning mass is compressed and an explosion results, forcing the piston back down in the cylinder while it is still trying to go up. Obviously, something must go, and it does—pistons are often damaged.

The general rule of thumb for choosing the correct heat range when picking a spark plug is: if most driving is long distance, high speed travel, use a colder plug; if most driving is stop and go, use a hotter plug. Factory-installed plugs are, of course, compromise plugs, since the factory has no way of predicting the type of driving. It should be noted that rarely is there a need to change plugs from the factory-recommended heat range.

### Reading Spark Plugs

Spark plugs are the single most valuable indicator of the engine's internal condition. Study the spark plugs carefully every time they are removed. Compare them to the chart which illustrates the most common plug conditions.

### Replacing Spark Plugs

A set of spark plugs usually requires replacement after about 10,000 miles on cars with conventional ignition systems and after about 20,000 to 30,000 miles on cars with electronic ignition. These figures are dependent on the style of driving, however. The electrode on a new spark plug has a sharp edge, but with use this edge becomes rounded by erosion, causing the plug gap to increase. In normal operation, plug gap increases about 0.001 in. for every 1,000–2,500 miles. As the gap increases, the plug's voltage requirement also increases. It requires a greater voltage to jump the wider gap and about two to three times as much voltage to fire a plug at high speeds than at idle.

Before removing the spark plugs, clean the area around the plugs and seats with compressed air.

Use a wire gauge to check the spark plug gap. Flat gauges are not accurate when used on spark plugs. The correct size feeler gauge should pass through the gap with a slight drag.

If the gap is incorrect, bend the side electrode to adjust the gap. Never bend or try to adjust the center electrode.

Squirt a drop of penetrating oil on the plug before installing it. Don't oil the threads too heavily, and tighten the plug by hand until snug.

**NOTE: When installing plugs that are hard to reach, slip a piece of vacuum hose over the plug and start it by turning the hose.**

# VACUUM DIAGNOSIS CHART

**WHITE POINTER INDICATES STEADY HAND. BLACK POINTER INDICATES FLUCTUATING HAND.**

Normal engine

Late ignition timing

Stuck throttle valve, leaking intake manifold or carburetor gaskets

Leaking head gasket

Worn valve guides

Burnt or leaking valves

Sticking valves

Weak valve springs

Carburetor needs adjustment

Late valve timing

Choked muffler

Normal engine—(opened and closed throttle, rings and valves OK)

## READING SPARK PLUGS

A close examination of spark plugs will provide many clues to the condition of an engine. Keeping the plugs in order according to cylinder location will make the diagnosis even more effective and accurate. The following diagrams illustrate some of the conditions that spark plugs will reveal.

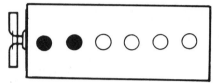

Two adjacent plugs are fouled in a 6-cylinder engine, 4-cylinder engine or either bank of a V-8. This is probably due to a blown head gasket between the two cylinders.

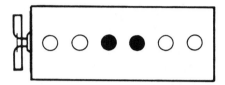

The two center plugs in a 6-cylinder engine are fouled. Raw fuel may be "boiled" out of the carburetor into the intake manifold after the engine is shut-off. Stop-start driving can also foul the center plugs, due to overly rich mixture. Proper float level, a good needle and seat or use of an insulating spacer may help this problem.

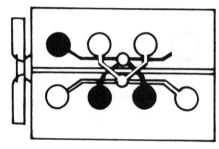

An unbalanced carburetor is indicated. Following the fuel flow on this particular design shows that the cylinders fed by the right-hand barrel are fouled from overly rich mixture, while the cylinders fed by the left-hand barrel are normal.

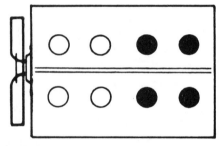

If the four rear plugs are overheated, a cooling system problem is suggested. A thorough cleaning of the cooling system may restore coolant circulation and cure the problem.

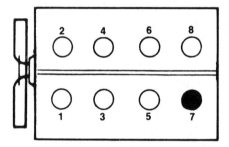

Finding one plug overheated may indicate an intake manifold leak near the affected cylinder. If the overheated plug is the second of two adjacent, consecutively firing plugs, it could be the result of ignition cross-firing. Separating the leads to these 2 plugs will eliminate cross-fire.

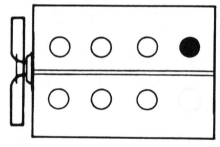

Occasionally, the 2 rear plugs in large, lightly used V-8's will become oil fouled. High oil consumption and smoky exhaust may also be noticed. It is probably due to plugged oil drain holes in the rear of the cylinder head, causing oil to be sucked in around the valve stems. This usually occurs in the rear cylinders first, because the engine slants that way.

### GASKET TYPE PLUGS

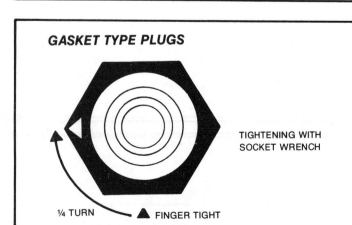

TIGHTENING WITH SOCKET WRENCH

¼ TURN ◢ FINGER TIGHT

### TAPERED SEAT PLUGS (NO GASKET)

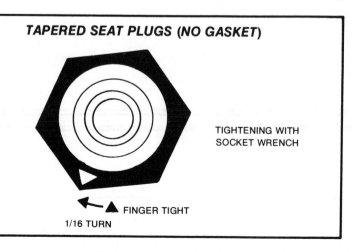

TIGHTENING WITH SOCKET WRENCH

◢ FINGER TIGHT
1/16 TURN

In the absence of specific torque, tighten plugs as shown

## Spark Plug Diagnosis

### NORMAL

Brown to grayish tan color and slight electrode wear. Correct heat range for engine and operating conditions.

**RECOMMENDATION.** Service and reinstall. Replace if over 10,000 miles of service. Also check car maker recommendations.

### MODIFIER DEPOSITS

Powdery white or yellow deposits that build up on shell, insulator and electrodes. This is a normal appearance with certain branded fuels. These materials are used to modify the chemical nature of the deposits to lessen misfire tendences.

**RECOMMENDATION.** Plugs can be cleaned. If replaced, use same heat range.

### OIL DEPOSITS

Oily coating.

**RECOMMENDATION.** Caused by poor oil control. Oil is leaking past worn valve guides or piston rings into the combustion chamber. Hotter spark plug may temporarily relieve problem, but positive cure is to correct the condition with necessary engine repairs.

### CARBON DEPOSITS

Dry soot.

**RECOMMENDATION.** Dry deposits indicate rich mixture or weak ignition. Check for clogged air cleaner, high float level, sticky choke or worn breaker contacts. Hotter plugs will temporarily provide additional fouling protection.

### PREIGNITION

Melted electrodes. Center electrode generally melts first and ground electrode follows. Normally, insulators are white, but may be dirty due to misfiring or flying debris in combustion chamber.

**RECOMMENDATION.** Check for correct plug heat range, overadvanced ignition timing, lean fuel mixtures, clogged cooling system, leaking intake manifold, and lack of lubrication.

### TOO HOT

Blistered, white insulator, eroded electrodes and absence of deposits.

**RECOMMENDATION.** Check for correct plug heat range, overadvanced ignition timing, cooling system level and/or stoppages, lean fuel/air mixtures, leaking intake manifold, sticking valves, and if car is driven at high speeds most of the time.

### HIGH SPEED GLAZING

Insulator has yellowish, varnish-like color. Indicates combustion chamber temperatures have risen suddenly during hard, fast acceleration. Normal deposits do not get a chance to blow off, instead they melt to form a conductive coating.

**RECOMMENDATION.** If condition recurs, use plug type one step colder.

### SPLASHED DEPOSITS

Spotted deposits. Occurs shortly after long delayed tune-up. After a long period of misfiring, deposits may be loosened when normal combustion temperatures are restored by tune-up. During a high-speed run, these materials shed off the piston and head and are thrown against the hot insulator.

**RECOMMENDATION.** Clean and service the plugs properly and reinstall.

Photos courtesy of Champion Spark Plug Co

## Checking and Replacing Spark Plug Cables

Visually inspect the spark plug cables for burns, cuts, or breaks in the insulation. Check the spark plug boots and the nipples on the distributor cap and coil. Replace any damaged wiring. If no physical damage is obvious, the wires can be checked with an ohmmeter for excessive resistance.

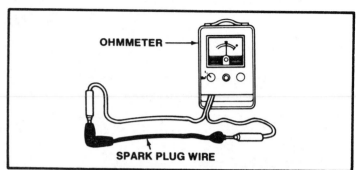

On cars with point-type ignition systems, check the spark plug wires as shown. On electronic ignitions, do not remove the wire. Remove the distributor cap and test the wire through the distributor cap terminal

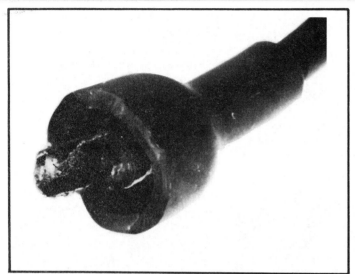

Check both ends of the spark plug wires for corrosion or cracking. The end of this cable is badly corroded and should be replaced

For cars equipped with conventional ignition, the best way to check the wires is to remove the plug wire from the plug *and* the distributor cap and test the wire alone. Simply insert the ends of the ohmmeter in the terminals of the spark plug wire. As a general rule, resistance should not exceed 3,000 to 7,000 ohms per foot. Replace any wire which shows readings well outside these limits.

The procedure for checking plug wires on cars equipped with electronic ignition is slightly different. For one thing *do not,* under any circumstances, pierce the plug wires. Test the wires at their terminals only. When checking the wire, do not remove it from the distributor cap. Test the wire through the distributor cap. If resistance is marginal, remove the wire from the cap carefully and retest it. If resistance is outside the values given, replace the wire.

Resistance values for electronic ignition plug wires vary from manufacturer to manufacturer. However, as a general rule, replace any wire which shows a resistance of over 50,000 ohms total.

When replacing wires on cars with electronic ignition, use plug wires rated for use with electronic ignition *only.* Ordinary plug wires will quickly fail due to the high heat conditions. It should be remembered that wire resistance is a function of length, and that the longer the cable, the greater the resistance. Thus, if the cables on your car are longer than the factory originals, resistance will be higher and quite possibly outside of these limits.

When installing a new set of spark plug cables, replace the cables one at a time. Start by replacing the longest cable first. Install the boot firmly over the spark plug. Route the wire exactly the same as the original. Insert the nipple firmly into the tower on the distributor cap. Repeat the process for each cable.

**NOTE: Engine misfire is sometimes the result of spark plug wires grouped together and running parallel for a long distance. The high voltage tends to jump from wire to wire, and will most likely occur in consecutive firing cylinders that are located close together.**

**Make sure that adjacent cables of consecutively firing cylinders are far apart or crossed at right angles.**

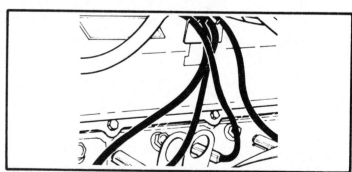

Misfiring can be the result of spark plug leads to adjacent, consecutively firing cylinders running parallel and too close together

## Silicone Lubricants

Modern electronic ignition systems generate extremely high voltages and high heats. The spark plug boots can soften and actually fuse to the ceramic insulator of the spark plugs after long exposures to high temperature and voltage. If this happens, the boot (and possibly the wire) must be replaced.

To help alleviate this condition, many manufacturers are using new silicone compounds (called greases in the trade, although they're not really greases) to slow the deterioration caused by heat and high voltage. The compounds are generally non-conductive, protective lubricants that will not dry out, harden, or melt away.

Test the plug wires inside the cap at the terminals. Test the coil wire at the center terminal

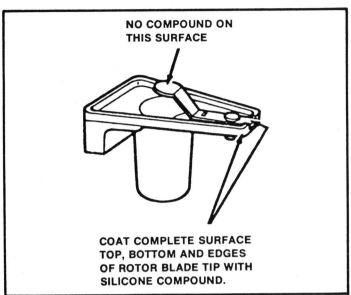

Before installing a new Ford rotor, coat the surfaces with a silicone compound

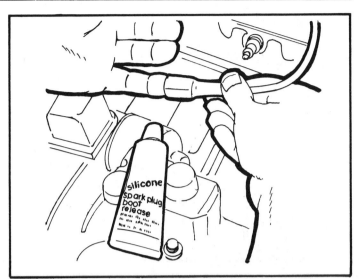

Whenever a spark plug boot is removed, it is wise (and often necessary) to apply a small amount of silicone boot release or similar material to the inside of the boot. This also applies to the distributor end of the spark plug wire

They form a weather-tight seal between rubber or plastic and metal and are found in several typical locations:
- Inside the insulating boots of spark plug wires to improve insulation, prevent aging, and ease removal and installation from the spark plug.
- Inside primary ignition circuit cable connectors to improve insulation.
- On distributor and rotor cap electrodes to improve RFI (Radio Frequency Interference) suppression.
- Under the GM High Energy Ignition (HEI) control module to improve heat transfer.

Most domestic manufacturers supply the silicone compounds through their own parts departments in one-application packages or supply a small quantity of the compounds with the new rotor, cap, or module. Equivalent compounds are also available under General Electric (GE) or Dow Corning (DC) brand names. The most common are the following:
- GE-623, 627, and 628 are interchangeable in automotive applications. They insulate, form a watertight seal, prevent heat aging, and provide RFI suppression when applied to the rotor electrode.
- GE Silicone Spark Plug Boot Release is specially formulated to resist high temperature and ease spark plug boot removal. It must be applied each time the boot is removed or the seal is broken.

- DC-111 is thermally stable and provides a good seal between rubber or plastic and metal. It is similar to GE-623, 627, and 628.
- GE-642 is a heat-transfer compound.
- DC-340 is a heat-transfer compound, similar to GE-642.

The accompanying chart lists specific manufacturers' application points and recommended greases. One nonsilicone application, for the Chrysler spark control computer, is included.

## DISTRIBUTOR SERVICE

Essentially, a distributor performs two functions: It switches primary current on and off at the coil, and it distributes secondary current to the spark plugs through the distributor cap. Switching in a conventional ignition system is accomplished by the breaker points and a condenser. In an electronic ignition system, the primary current is switched on and off by a magnetic pickup unit and a rotating pole piece, called an inductor or reluctor.

The amount of tune-up work involved for the distributor depends on the type of ignition system used. Conventional systems usually require new points and condenser, point gap and dwell adjustments, inspection (and replacement, if necessary) of the primary and secondary wiring, and ignition timing. Electronic igni-

| Application Point | Silicone Compound |
|---|---|
| General Motors: under HEI module | Supplied with new module, or use GE-642 or DC-340 |
| Ford Motor Company: inside spark plug boots, on end of cable when installing new boot, and on rotor and cap electrodes | Ford part number D7AZ-19A331-A or use GE-627 or DC-111 |
| Chrysler Corporation: 1/4" deep within spark control computer connector cavity coating rotor electrode | Use Mopar part number 2932524 or NLGI Grade 2 EP (not a silicone) supplied with new rotor, or use GE-628 or DC-111 |
| American Motors (Prestolite system): distributor primary connector—coat male terminal, fill female 1/4 full | AMC part number 8127445 or GE-623 |
| International Harvester (Prestolite system): coat all secondary terminals, including spark plug boots | IHC part number 472141-C1 only |

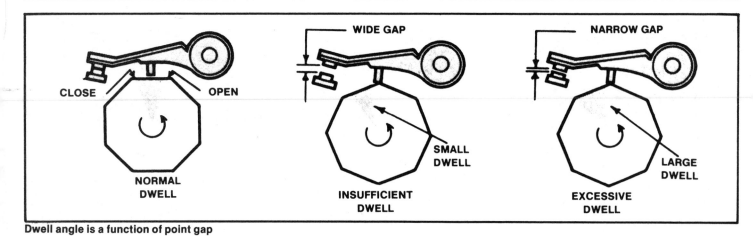

**Dwell angle is a function of point gap**

tion systems eliminate much of this. On these, you need only check the ignition rotor, cap, and wiring. There is no dwell to set, and ignition timing rarely requires adjustment.

## Breaker Points and Condenser

The points and condenser function as a circuit breaker for the primary circuit of the ignition system. The ignition coil must boost the 12 volts of electrical pressure supplied to it by the battery to about 20,000 volts in order to fire the spark plugs. To do this, the coil depends on the points and condenser for assistance.

The coil has a primary and a secondary circuit. When the ignition key is turned to the "on" position, the battery supplies voltage to the primary side of the coil which passes the voltage on to the points. The points are connected to ground to complete the primary circuit. As the cam in the distributor turns, the points open and the primary circuit collapses. The magnetic force in the primary circuit of the coil cuts through the secondary circuit and increases the voltage in the secondary circuit to a level that is sufficient to fire the spark plugs.

When the points open, the electrical charge contained in the primary circuit jumps the gap that is created between the two open contacts of the points. If this electrical charge were not transferred elsewhere, the material on the contacts of the points would melt and that all-important gap between the contacts would start to change. If this gap is not maintained, the points will not have enough voltage to fire the spark plugs. The function of the condenser is to absorb the excessive voltage from the points when they open and thus prevent the points from becoming pitted or burned.

There are two ways to check breaker point gap: with a feeler gauge or with a dwell meter. Either way you set the points, you are adjusting the amount of time (in degrees of distributor rotation) that the points will remain open. If you adjust the points with a feeler gauge, you are setting the maximum distance the points will open when the rubbing block on the points is on a high point of the distributor cam. When you adjust the points with a dwell meter, you are measuring the number of degrees (of distributor cam rotation) that the points will remain closed before they start to open as a high point of the distributor cam approaches the rubbing block of the points.

There are two rules that should always be followed when adjusting or replacing points.

1. The points and condenser are a matched set; never replace one without replacing the other.

2. If you change the point gap or dwell of the engine, you also change the ignition timing. Therefore, if you adjust the points, you must also adjust the timing.

## Points Inspection and Replacement

Remove the distributor cap and the rotor. Insert a screwdriver between the stationary and breaker arms of the points and examine the condition of the contacts. Replace the points if the contacts are

blackened, pitted, or if the metal transfer exceeds that of the specified point gap. Also replace the points if the breaker arm has lost its tension or if the rubbing block is excessively worn. Contact points that have become slightly burned (light gray) may be cleaned with a point file. In order for the points to function properly, the contact faces must be aligned. The alignment must be checked with the points closed. If the contact faces are not centered, bend the stationary arm to suit. Never bend the breaker arm. Discard the points if they cannot be centered correctly.

## Inspect the Secondary Ignition Circuit

Inspect the inside surface of the distributor cap for cracks, carbon tracks, or badly burned contacts. To remove carbon tracks, wash the cap in soap and water and dry thoroughly. Replace the cap if it is cracked or if the contacts are badly eroded.

Inspect the rotor for cracks, excessive burning of the contacts, and mechanical damage, and replace as necessary. Slightly burned contacts should be sanded smooth.

Inspect the spark plug leads and distributor-to-coil high-tension lead for cracks, brittleness, or damaged rubber boots. Replace any deteriorated parts.

While primary wiring is less perishable than the secondary circuit, it should be checked for cracked insulation or loose connections. Tighten connections or replace wires as necessary.

## Adjusting Point Gap

This is a good method of making a preliminary setting of the dwell angle even if a dwell meter is available. Install the points and condenser, making sure all connections are pushed on or screwed together securely. If the mounting screw on the ignition points assembly is also used to make the gap adjustment, tighten it just enough to hold the contacts apart. Rotate the engine until the tip of one of the distributor cams sits squarely under the cam follower on the movable contact arm. Using a leaf type feeler gauge (thickness as specified in the manual), move the base plate of the point assembly back and forth until the gauge just slips between the two contacts when it is forced straight through. If the mounting screw serves as the adjusting lock, the contact assembly may usually be moved by wedging a screwdriver blade between a slot in the contact base plate and a protrusion on the surface of the distributor plate. In assemblies with an adjusting screw which is accessible from outside the distributor, an Allen wrench is inserted into the head of the adjusting screw and rotated to make the adjustment. When the points are pitted, make sure the gauge does not come in contact with the built-up portion of one of the contact surfaces. A wire feeler gauge may help to make the most accurate adjustment when the points are pitted.

**NOTE: Make sure all gauges are clean in gapping the points. An oily gauge will cause rapid point burning.**

## Setting Dwell Angle

The dwell angle is the number of degrees of distributor cam rotation through which the breaker points remain fully closed (conducting electricity). Increasing the point gap decreases dwell, while decreasing the point gap increases dwell.

Using a dwell meter of known accuracy, connect the red lead (positive) wire of the meter to the distributor primary wire connection on the positive (+) side of the coil, and the black ground (negative) wire of the meter to a good ground on the engine (e.g. thermostat housing nut).

The dwell angle may be checked either with the distributor cap and rotor installed and the engine running, or with the cap and rotor removed and the engine cranking at starter speed. The meter gives a constant reading with the engine running. With the engine cranking, the reading will fluctuate between zero degrees dwell and the maximum figure for that angle. While cranking, the maximum figure is the correct one for that setting. Never attempt to change dwell angle while the ignition is on. Touching the point contacts or primary wire connection with a metal screwdriver may result in a 12 volt shock.

To change the dwell angle, loosen the point retaining screw slightly and make the approximate correction. Tighten the retaining screw and test the dwell with the engine cranking. On General Motors V8 engines, dwell angle is set with an Allen wrench through the window in the distributor. If the dwell appears to be correct, install the breaker point protective cover, if so equipped, the rotor and distributor cap, and test the dwell with the engine running. Take the engine through its entire rpm range and observe the dwell meter. The dwell should remain within specifications at all times. Great fluctuation of dwell at different engine speeds indicates worn distributor parts.

Following the dwell angle adjustment, the ignition timing must be checked. A 1° increase in dwell results in the ignition timing being retarded 2° and vice versa.

## IGNITION TIMING

Ignition timing is the measurement in degrees of crankshaft rotation of the instant the spark plugs in the cylinders fire, in relation to the location of the piston, while the piston is on its compression stroke.

Ignition timing is adjusted by loosening the distributor locking device and turning the distributor in the engine.

Ideally, the air/fuel mixture in the cylinder will be ignited (by the spark plug) and just beginning its rapid expansion as the piston passes top dead center (TDC) of the compression stroke. If this happens, the piston will be beginning the power stroke just as the compressed (by the movement of the piston) and ignited (by the spark plug) air/fuel mixture starts to expand. The expansion of the air/fuel mixture will then force the piston down on the power stroke and turn the crankshaft.

It takes a fraction of a second for the spark from the plug to completely ignite the mixture in the cylinder. Because of this, the spark plug must fire before the piston reaches TDC, if the mixture is to be completely ignited as the piston passes TDC. This measurement is given in degrees (of crankshaft rotation) *before* the piston reaches *top dead center* (BTDC). If the ignition timing setting for the engine is six degrees (6°) BTDC, this means that the spark plug must fire at a time when the piston for that cylinder is 6° before top dead center of its compression stroke. However, this only holds true while the engine is at idle speed.

As the engine accelerates from idle, the speed of the engine (rpm) increases. The increase in rpm means that the pistons are now traveling up and down much faster. Because of this, the spark plugs will have to fire even sooner if the mixture is to be completely ignited as the piston passes TDC. To accomplish this, the distributor incorporates means to advance the timing of the spark as engine speed increases.

Conventional distributors have two systems to advance the ignition timing. One is called centrifugal advance and is actuated by spring-controlled weights in the distributor. The advance weights are connected to the movable cam or armature assembly of the distributor shaft. As the engine speed increases, centrifugal force moves the weights outward against the spring tension, causing the movable cam or armature assembly to move ahead in relation to the distributor shaft, and thus causing the ignition point set to open earlier.

The other is called vacuum advance; the vacuum advance unit is contained within the large circular housing on the side of the distributor. The housing surrounds a vacuum-controlled diaphragm which is linked to the advance plate in the distributor. When engine vacuum is applied to the diaphragm, it moves toward the vacuum side (outward), pulling the linkage with it and thus rotating the distributor advance plate. The points riding on the plate are moved ahead in relation to the distributor shaft and consequently timing is advanced.

In addition, some distributors have a vacuum-retard mechanism which is contained in the same housing on the side of the distributor as the vacuum advance. The function of the mechanism is to retard the timing of the ignition spark under certain engine conditions. This causes more complete burning of the air/fuel mixture in the cylinder and consequently lowers exhaust emissions.

Because these mechanisms change ignition timing, it is necessary to disconnect and plug the one or two vacuum lines from the distributor when setting the basic ignition timing. The hoses must be plugged to prevent vacuum leaks that would spoil the accuracy of the timing adjustment.

Most electronic ignition systems also have vacuum advance mechanisms (and, occasionally, vacuum retard units), but most do not have centrifugal advance mechanisms. Instead, they have electronic spark control systems. These systems vary the timing in

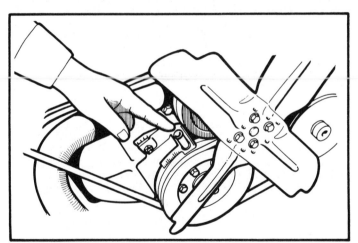

**A magnetic probe provides for much more precise ignition timing than can be achieved with conventional timing lights**

response to engine conditions. Various sensors are used to monitor the engine, including sensors for timing, engine speed, vacuum, barometric pressure, oil and water temperature, exhaust gas oxygen, and throttle position. Information from these sensors is relayed to the spark control unit, which sets the timing according to preprogrammed instructions. Inasmuch as the spark timing is not based on a constant curve, but is instead an infinitely variable system that relates to the engine speed and load conditions, ignition timing and timing advance are not adjustable in any conventional manner, and are not considered as part of a tune-up.

If ignition timing is set too far advanced (BTDC), the ignition and expansion of the air/fuel mixture in the cylinder will try to force the piston down the cylinder while it is still traveling upward. This causes engine "ping," a sound which resembles marbles being dropped into an empty tin can. If the ignition timing is too far retarded (after, or ATDC), the piston will have already started down on the power stroke when the air/fuel mixture ignites and expands. This will cause the piston to be forced down only a portion of its travel, resulting in poor engine performance and lack of power.

Ignition timing adjustment is checked with a timing light. This light is normally connected to the number one (No. 1) spark plug of the engine. The timing light flashes every time an electrical current is sent from the distributor, through the No. 1 spark plug wire, to the spark plug. On most cars, the crankshaft pulley and the front cover of the engine are marked with a timing pointer and a timing scale. (On some German cars, the timing marks are found at the flywheel.) When the timing pointer is aligned with the "0" mark on the timing scale, the piston in No. 1 cylinder is at TDC of its compression stroke. With the engine running, and the timing light aimed at the timing pointer and timing scale, the stroboscopic flashes from the timing light will allow you to check the ignition timing setting of the engine. The timing light flashes every time the spark plug in the No. 1 cylinder of the engine fires. Since the flash from the timing light makes the crankshaft pulley seem stationary for a moment, you will be able to read the exact position of the piston in the No. 1 cylinder.

### Timing the Engine with a Timing Light

1. If the timing light operates from the battery, connect the red lead to the battery positive terminal, and the black lead to a ground. With all lights, connect the trigger lead in series with No. 1 spark plug wire.

2. Disconnect and plug the required vacuum hoses, as in the manufacturer's specifications. Connect the red lead of a tachometer to the distributor side of the coil and the black lead to ground. Start the engine, put the automatic transmission in gear (if required), and read the tachometer. Adjust the carburetor idle screw to the proper speed for setting the timing. Aim the timing light at the crankshaft pulley to determine where the timing point is. If the point is hard to see, it may help to stop the engine and mark it with chalk.

3. Loosen the distributor holding clamp and rotate the distributor slowly in either direction until the timing is correct. Tighten the clamp and observe the timing mark again to determine that the timing is still correct. Readjust the position of the distributor, if necessary.

4. Accelerate the engine in Neutral, while watching the timing point. If the distributor advance mechanisms are working, the timing point should advance as the engine is accelerated. If the engine's vacuum advance is engaged with the transmission in Neutral, check the vacuum advance operation by running the engine at about 1500 rpm and connecting and disconnecting the vacuum advance hose.

### Static Timing

1. Make sure the engine is at the correct temperature for timing adjustment (either fully warmed or cold, as specified in the factory manual or a Chilton repair manual).

2. Locate No. 1 cylinder and trace its wire back to the distributor cap. Then remove the cap.

3. Rotate the engine until the proper timing mark on the crankshaft pulley is lined up with the timing mark on the block. Observe the direction of distributor shaft rotation when the engine is turned in its normal direction of rotation.

**NOTE: On engines with belt-driven camshafts, do not rotate the engine by means of the camshaft pulley bolt. The 2:1 ratio of the camshaft to crankshaft pulleys will cause the belt to slip, changing engine timing.**

4. Connect a test lamp from the coil terminal (the distributor side) to ground. Make sure the tip of the rotor lines up with No. 1 cylinder. If it does not, turn the engine one full revolution and line up the timing marks again.

5. Loosen the clamp that holds the distributor in position and turn the distributor body in the direction of normal shaft rotation until the points close and the test lamp goes out. Now turn the distributor in the opposite direction very slowly, just until the test lamp comes on. Tighten the distributor clamp.

6. To test the adjustment, turn the engine backward until the light again goes out, and then forward just until the light comes back on.

**NOTE: Engines with a belt-driven camshaft must not be rotated backward.**

If the timing marks are lined up, the engine is accurately timed. If the timing is too far advanced, loosen the distributor and turn it just slightly in the direction of shaft rotation, and retighten the clamp. If the timing is retarded, turn the distributor in the opposite direction and then repeat the test. Repeat this procedure until the light comes on just as the two timing marks are aligned.

## CARBURETOR ADJUSTMENTS

Carburetors are fairly complex but since they have relatively few moving parts, they are not normally as vulnerable to the ravages of time as are distributor components. Any recurring carburetor problems indicate incorrect set-up or faulty repair work, since carburetor wear is very gradual.

Essentially, there are only two carburetor adjustments which may be necessary during the course of a normal tune-up: idle speed and idle mixture.

When the engine in your car is running, air/fuel mixture from the carburetor is being drawn into the engine by a partial vacuum that is created by the downward movement of the piston on the intake stroke of the four-stroke cycle of the engine. The amount of air/fuel mixture that enters the engine is controlled by a throttle plate or plates in the bottom of the carburetor. When the engine is not running, the throttle plates are closed, completely blocking off the bottom of the carburetor from the intake manifold and thus the combustion chambers. The throttle plates are connected, through the throttle linkage, to the gas pedal. After you start the engine and put the transmission in gear, you depress the gas pedal to start the car moving. What you actually are doing when you depress the gas pedal is opening the throttle plate in the carburetor to admit more of the fuel/air mixture to the engine. The further you open the throttle plates in the carburetor, the higher the engine speed becomes.

As previously stated, when the engine is not running, the throttle plates in the carburetor are closed. When the engine is idling, it is necessary to open the throttle plates slightly. To prevent having to keep your foot on the gas pedal when the engine is idling, an idle speed adjusting screw was added to the carburetor. This screw has the same effect as keeping your foot slightly depressed on the gas pedal. The idle speed adjusting screw contacts a lever (the throttle lever) on the outside of the carburetor. When the screw is turned in, it opens the throttle plate on the carburetor, raising the idle speed of the engine. This screw is called the curb idle adjusting screw and the procedures in this section will tell you how to adjust it.

Since the early nineteen-seventies, most engines have been equipped with throttle solenoids. Due to the power-robbing effects of emission control systems, car manufacturers have found it necessary to raise the idle speed on almost all engines in order to obtain a smooth idle. Ordinarily, when the key is turned to "off," the current to the spark plugs is cut off, and the engine normally stops running. However, if an engine has a high operating temperature and a high idle speed (conditions common to emission-controlled engines), it is possible for the temperature of the cylinder instead of the spark plug to ignite the fuel/air mixture. When this happens, the engine continues to run after the key is turned off. To solve this problem, a throttle solenoid was added to the carburetor. The solenoid is a cylinder with an adjustable plunger and an electrical lead. When the ignition key is turned to "on," the solenoid plunger extends to contact the carburetor throttle lever and raise the idle speed of the engine. When the ignition key is turned to "off," the solenoid is de-energized and the solenoid plunger falls back from the throttle lever. This allows the throttle lever to fall back and rest on the curb idle adjusting screw. This drops the engine idle speed back far enough so that the engine will not "run-on."

Since it is difficult for the engine to draw the fuel/air mixture from the carburetor with the small amount of throttle plate opening that is present when the engine is idling, an idle mixture passage is provided in the carburetor. This passage delivers fuel/air mixture to the engine from a hole which is located in the bottom of the carburetor below the throttle plates. This idle mixture passage contains an adjusting screw which restricts the amount of fuel/air mixture that enters the engine at idle. The idle mixture screws are capped on late-model cars due to emission-control regulations.

## Idle Speed Adjustment

Generally, the idle speed is adjusted before the idle mixture. Idle speed specifications are found on a decal under the hood (on late model vehicles) or in a service manual.

Connect the tachometer red lead to the negative terminal of the coil and connect the black lead to ground. This procedure is for conventional ignition systems only. Electronic ignition systems generally have specific tach hook-up procedures, and in addition, will not necessarily work with all tachometers. Locate the idle speed screw or the idle solenoid. With the engine at operating temperature, adjust the screw or the solenoid until the correct idle speed is reached. Ordinarily, on cars equipped with idle solenoids, there are two idle speeds listed. The higher of the two speeds is with the solenoid connected, while the lower is obviously with the solenoid disconnected. Set both speeds and then go on to the mixture adjustment.

## Idle Mixture Adjustment

Locate the idle mixture screw or screws. All vehicles manufactured after 1972 have their idle mixture screws capped in accordance with Federal emission control regulations. As a result of this, there is only a very limited range of adjustment possible on these carburetors. To comply with emission regulations, the caps should not be removed or the mixture adjusted without them. After you have found the mixture screws, adjust them according to the manufacturer's instructions. These instructions vary, but in general, the procedure is this:

1. On early vehicles without capped mixture screws, adjust the mixture screw or screws for the highest idle speed you can obtain on the tachometer. An alternative method is to use a vacuum gauge and adjust the screws until the highest possible vacuum reading is obtained.

2. On vehicles with capped mixture screws, adjust the mixture screws (within the limits imposed by the caps) until the highest possible idle is obtained. Then adjust the screws inward from highest idle until the specified rpm drop is obtained.

After the mixture has been adjusted, it is quite often necessary to reset the idle speed with the idle speed screw or solenoid. As a general rule, idle mixture adjustments will raise the idle above that which is called for, necessitating a readjustment with the idle speed screw or solenoid.

## Propane Assisted Idle Adjustment

Stringent emission requirements brought about by the Federal Clean Air Act of 1970 have resulted in lean carburetion and the use of the catalytic converter. The efficiency of the converter is dramatic, but tends to conceal any maladjustment of the carburetor. The emission levels of carbon monoxide (CO) are very low; in fact, too low for precise measurement and readouts by conventional infrared exhaust analyzers. Thus, correct carburetor adjustments have become increasingly difficult.

Every engine will exhibit a characteristic response to the increased richness of fuel/air mixtures. Making the mixture richer will cause the engine rpm to rise to a maximum level, then decrease as the mixture becomes too rich. This procedure is known as the "speed drop" effect. On late model vehicles there may be some difficulty experienced when adjusting the idle mixtures with the speed drop method because of the limiter caps mounted on the idle mixture screws. This is where the propane enrichment procedure simplifies idle mixture adjustments.

Adding propane gas to the fuel being supplied to the engine causes an effect similar to the speed drop method. As the amount of propane being metered in is gradually increased, the mixture becomes richer and the rpm will rise. As it becomes too rich, rpm will drop.

If engine rpm rises to the specified level when propane gas is metered into the engine, the idle mixtures are correct and no further adjustment is necessary. If the highest rpm level is below the specification, idle mixtures are too rich and adjustment is required. If the level is above specification, the mixture is too lean and, once again, an adjustment is required. An rpm level in excess of that specified could also point to the possibility of a vacuum leak somewhere in the system. If a vacuum leak exists, it must be corrected before proceeding with the test, or it will continue to give a false indication of obtainable rpm levels and lean mixtures.

## Propane Tools Required

Only a few tools are needed to perform the propane assisted idle adjustment. Included are:

1. A small standard size propane bottle. The bottle must be fitted with a metering valve which allows a fine adjustment of the propane gas flow.

2. A suitable length of hose to connect the propane source to a designated point on the engine.

3. An accurate tachometer to measure exact idle speeds that are required in the adjustment procedure.

## Propane Supply

If the supply level of propane in the bottle being used to perform the enrichment procedure is not adequate, gas will be fed intermittently causing the engine to "hunt," rather than smoothly increase in speed. Check the supply periodically by tilting the bottle back and forth.

To maintain a uniform gas flow during the procedure, stand the bottle upright. It may also be wise to secure it in some fashion to prevent it from falling into the fan or other dangerous areas.

—CAUTION—

*Propane can be extremely hazardous if not handled properly. Observe all fire and explosion precautions whenever propane is used.*

**NOTE: If equipped with idle limiter caps: During the adjustment procedure, if it is determined that idle mixtures are excessively rich or lean, the idle limiter caps may have to be removed in order to obtain correct settings. If it becomes necessary to remove the limiter caps, they must be replaced with the tang in the full rich position, touching the stop.**

The propane enrichment procedure does not replace the exhaust analyzer, but serves as a time saver. It should be remembered that not all vehicle manufacturers recommend the use of propane to adjust engine speed. Consult the proper manufacturer's specifications manual and follow their instructions.

# FUEL INJECTION ADJUSTMENTS

The introduction of fuel injection systems on many late model cars is due in part to the stringent emission control standards U.S. automakers face. Fuel injection allows a more precise control of the air/fuel mixture than carburetion does. Fuel injectors can be controlled electrically or through fuel pressure to inject exact amounts of fuel, whereas carburetors in most cases supply fuel in a constant stream. Under certain operating conditions (deceleration, etc.), a percentage of carbureted fuel is blown, unburned, out the exhaust pipe, lowering engine efficiency while raising poisonous hydrocarbon emissions into the atmosphere—one reason why carbureted engines usually require more emission control equipment than fuel injected engines.

There are three basic fuel injection systems: direct injection, indirect injection, and throttle body injection.

Direct injection means fuel is injected directly into the combustion chambers by individual fuel injectors, usually screwed into the head much like spark plugs.

On indirect injection systems, the injectors are located behind the intake valves or in special "pre-combustion" chambers instead of in the main combustion chambers.

On both of these systems, the fuel mixes with the air sucked into the combustion chamber by the drawing action of the piston moving down in its cylinder. The air and fuel are mixed in the cylinder head rather than in the intake manifold.

On throttle body injection, the injectors are mounted in a common throttle body (which resembles a carburetor in appearance) and spray fuel into the intake manifold. On this system, each cylinder does not require its own fuel injector, as direct and indirect injection cylinders do. All cylinders are fed by a few (usually two) centrally mounted injectors. The fuel mixes with the incoming air in the throttle body and the mixture is sucked into the cylinders.

By comparison, direct and indirect fuel injection systems allow quicker and more controlled air/fuel mixture adjustments than do throttle body injection systems. However, direct and indirect fuel injection systems are much more costly than the throttle body system.

Fuel injection systems, whether mechanical or electronic, are highly complex units, but, like carburetors, are fairly reliable. Most mechanical fuel injection systems provide for both idle speed and idle mixture adjustments, the only adjustments normally considered to be part of a tune-up. Many electronic systems, however, do not allow mixture adjustments, and newer models may also place idle speed under electronic control.

## Idle Speed Adjustment

Idle speed is adjusted before idle mixture. The engine must be at normal operating temperature, with all electrical accessories shut off, except as noted on the emission control sticker in the engine compartment. Note that on many front drive cars, the electric radiator fan must be off during adjustments. Also, on some cars with electronic ignition, the digital idle stabilizer must be bypassed. See the underhood sticker, or consult the appropriate Chilton manual for details. Attach a tachometer to the engine. Locate the idle speed adjusting screw. Adjust the screw until the correct idle speed is reached. Accelerate the engine briefly to 2500 rpm, then allow it to return to idle. Recheck the idle speed and adjust as necessary. Repeat this sequence until a repeatable idle speed is achieved.

## Idle Mixture Adjustment

Mixture adjustments require the use of a carbon monoxide (CO) meter. Due to the complexity of the adjustments, specific procedures cannot be given here. However, note the following guidelines:

1. The engine must be at normal operating temperature. Oil temperature is more important than coolant temperature. If the engine oil is hot (above normal) the mixture adjustment will be incorrect (too lean).

2. On Bosch units, idle speed and mixture adjustments are interdependent; some juggling of each will probably be required to achieve the specified settings.

3. Do not press down on the idle mixture screw when making adjustments.

4. On Bosch units, remove the adjusting tool between adjustments; the CO reading will be incorrect with the tool in place.

5. Accelerate the engine briefly to 2500 rpm between CO readings. Be sure to remove the adjusting tool before accelerating the engine, or the sensor plate may be damaged.

## VALVE ADJUSTMENT

Periodic valve adjustments are not required on most modern engines with hydraulic valve lifters. In fact, many engines no longer have any provision whatsoever for valve adjustment. Most import car engines, however, have adjustable valves, since a tightly controlled valve lash is the key to wringing horsepower out of these smaller motors. Exact valve adjustment procedures for all engines cannot be given here, but here are some general guidelines:

A. Bring the engine to the condition specified for valve adjustment (cold, hot, running, etc.). Some manufacturers give procedures for both hot and cold adjustment. A note of caution about hot valve adjustment: oil temperature is far more critical to valve adjustment than water temperature; if the adjustment procedure calls for the engine to be at operating temperature, make sure the engine runs for at least fifteen minutes to allow the oil temperature to stabilize and parts to reach their full expansion.

B. Remove the valve or cam cover. If the valves are to be adjusted while the engine is running, oil deflector clips are available which install on the rocker arms and prevent oil spray.

C. If the valves must be adjusted with the engine stopped, follow the manufacturer's instructions for positioning the engine properly. For example, on air-cooled Volkswagens, number one cylinder is adjusted while it is on the compression stroke at TDC, number two with the crankshaft turned 180°, etc.

D. Generally on overhead camshaft engines, valve adjustment is fairly simple. Turn the engine over (either manually or with a remote starter switch) until the highest part of the cam lobe is pointing directly *away* from the rocker arm or lifter. In the event you are unsure of how to do it, consult a service manual.

E. Adjust solid lifters by pushing a leaf type feeler gauge of the specified thickness (consult the manual) between the valve stem and rocker arm. Loosen the locking nut and tighten the screw until a light resistance to the movement of the feeler blade is encountered. Hold the adjusting screw while tightening the locking nut. Some adjusting screws are fitted snugly into the rocker arm so no locknut is required. If the feeler is too snug, the adjustment should be loosened to permit the passage of the blade. Remember that a slightly loose adjustment is easier on the valves than an overly tight one, so adjust the valves accordingly. Always recheck the adjustment after the locknut has been tightened.

Refer to a service manual for details on hydraulic valve lifter adjustment.

F. Replace the valve cover, cleaning all traces of old gasket material from both surfaces and installing a new gasket. Tighten valve cover nuts alternately, in several stages, to ensure proper seating.